Handbook of Statistical Genetics

Third Edition

Volume 2

Handbook of Statistical Genetics

Third Edition

Volume 2

Editors:

D. J. Balding

Imperial College of Science, Technology and Medicine, London, UK

M. Bishop

CNR-ITB, Milan, Italy

C. Cannings

University of Sheffield, UK

Other Wiley Editorial Offices

John Wiley & Sons, Inc. 111 River Street,
Hoboken, NJ 07030, USA

Jossey-Bass, 989 Market Street,
San Francisco, CA 94103-1741, USA

Wiley-VCH Verlag GmbH, Boschstr. 12,
D-69469 Weinheim, Germany

John Wiley & Sons Australia, Ltd, 42 McDougall Street,
Milton, Queensland, 4064, Australia

John Wiley & Sons (Asia) Pte Ltd, 2 Clementi Loop #02-01,
Jin Xing Distripark, Singapore 129809

John Wiley & Sons Canada Ltd, 22 Worcester Road,
Etobicoke, Ontario, Canada, M9W 1LI

Wiley also publishes its books in a variety of electronic formats. Some content that appears in print may not be available in electronic books.

Anniversary logo design: Richard J. Pacifico

Library of Congress Cataloging-in-Publication Data

Handbook of Statistical Genetics / editors, D.J. Balding, M. Bishop, C. Cannings. – 3rd ed.
        p. ; cm.
    Includes bibliographical references and index.
    ISBN 978-0-470-05830-5 (cloth : alk. paper)
    1. Genetics–Statistical methods–Handbooks, manuals, etc. I. Balding, D. J. II. Bishop, M. J. (Martin J.) III. Cannings, C. (Christopher), 1942-
    [DNLM: 1. Genetics. 2. Chromosome Mapping–methods. 3. Genetic Techniques. 4. Genetics, Population. 5. Linkage (Genetics) 6. Statistics–methods.   QH 438.4.S73 H236 2007]
    QH 438.4.S73H36 2007
    576.507′27–dc22

                                                                        2007010263

British Library Cataloguing in Publication Data

A catalogue record for this book is available from the British Library

ISBN-13 978-0-470-05830-5 (HB)

Typeset in 10/12pt Times by Laserwords Private Limited, Chennai, India
Printed and bound in Great Britain by Antony Rowe, Chippenham, Wiltshire, UK.
This book is printed on acid-free paper responsibly manufactured from sustainable forestry in which at least two trees are planted for each one used for paper production.

Contents

VOLUME 2

Contributors

R.E. Ashcroft
Queen Mary's School of Medicine
and Dentistry
University of London
London, UK

G. Barbujani
Dipartimento di Biologia
ed Evoluzione
Università di Ferrara
Ferrara, Italy

M.A. Beaumont
School of Biological Sciences
University of Reading
Reading, UK

J.M. Bowden
Departments of Health Sciences
and Genetics
University of Leicester
Leicester, UK

J.P. Bollback
Department of Biology
University of Rochester
Rochester, NY
USA

J.F.Y. Brookfield
Institute of Genetics
School of Biology
University of Nottingham
Nottingham, UK

P.R. Burton
Departments of Health Sciences
and Genetics
University of Leicester
Leicester, UK

C. Cannings
Division of Genomic Medicine
University of Sheffield
Sheffield, UK

L.R. Cardon
Wellcome Trust Centre for Human
Genetics
University of Oxford
Oxford, UK

C. Cheng
Department of Biostatistics
St. Jude Children's Research Hospital
Memphis, TN
USA

L. Chikhi
Laboratoire Evolution et Diversité
Biologique
Université Paul Sabatier
Toulouse, France

D. Clayton
Cambridge Institute for Medical
Research
University of Cambridge
Cambridge, UK

M. De Iorio
Division of Epidemiology,
Public Health and Primary Care
Imperial College
London, UK

J. Dicks
Department of Computational
and Systems Biology
John Innes Centre
Norwich, UK

F. Dudbridge
MRC Biostatistics Unit
Institute for Public Health
Cambridge, UK

T.M.D. Ebbels
Division of Surgery, Oncology
Reproductive Biology and Anaesthetics
Imperial College
London, UK

L. Excoffier
Zoological Institute
Department of Biology
University of Berne
Berne, Switzerland

D. Gianola
Department of Animal Sciences
Department of Biostatistics
and Medical Informatics
Department of Dairy Science
University of Wisconsin
Madison, WI
USA

N. Goldman
EMBL-European Bioinformatics Institute
Hinxton, UK

M.D. Hendy
Allan Wilson Center for
Molecular Ecology and Evolution
Massey University
Palmerston North, New Zealand

B.R. Holland
Allan Wilson Center for
Molecular Ecology and Evolution
Massey University
Palmerston North, New Zealand

P. Holmans
Department of Psychological
Medicine
Cardiff University
Cardiff, UK

I. Höschele
Virginia Bioinformatics Institute and
Department of Statistics
Virginia Polytechnic Institute and
State University
Blacksburg, VA, USA

F. Hospital
INRA, Université Paris Sud
Orsay, France

W. Huber
Department of Molecular Genome
Analysis
German Cancer Research Center
Heidelberg, Germany

J.P. Huelsenbeck
Department of Biology
University of Rochester
Rochester, NY
USA

R.C. Jansen
Groningen Bioinformatics Centre
University of Groningen
Groningen, The Netherlands

D.P. Klose
Division of Mathematical Biology
National Institute of Medical
Research
London, UK

S.L. Lauritzen
Department of Statistics
University of Oxford
Oxford, UK

A. Lewin
Division of Epidemiology,
Public Health and Primary Care
Imperial College
London, UK

S. Lin
Department of Statistics
Ohio State University
Columbus, OH
USA

Jun S. Liu
Department of Statistics
Harvard University
Cambridge, MA
USA

T. Logvinenko
Department of Statistics
Harvard University
Cambridge, MA
USA

A.S. Macdonald
Department of Actuarial Mathematics
and Statistics
Heriot-Watt University
Edinburgh, UK

P.M. McKeigue
Conway Institute
University College Dublin
Dublin, Ireland

G. McVean
Department of Statistics
Oxford University
Oxford, UK

L. Moreau
INRA, UMR de Génétique Végétale
Ferme du Moulon
France

A.P. Morris
Wellcome Trust Centre for Human Genetics
University of Oxford
Oxford, UK

C. Neuhauser
Department of Ecology,
Evolution and Behavior
University of Minnesota
Saint Paul, MN
USA

M. Nordborg
Molecular and Computational Biology
University of Southern California
Los Angeles, CA
USA

A. Onar
Department of Biostatistics
St. Jude Children's Research Hospital
Memphis, TN
USA

W.R. Pearson
Department of Biochemistry
University of Virginia
Charlottesville, VA
USA

D. Penny
Allan Wilson Center for
Molecular Ecology and Evolution
Massey University
Palmerston North, New Zealand

S.B. Pounds
Department of Biostatistics
St. Jude Children's Research Hospital
Memphis, TN
USA

S. Richardson
Division of Epidemiology,
Public Health and Primary Care
Imperial College
London, UK

F. Rousset
Laboratoire Génétique
et Environnement
Institut des Sciences de l'Évolution
Montpellier, France

G. Savva
Centre for Environmental
and Preventive Medicine
Wolfson Institute
of Preventive Medicine
London, UK

E.E. Schadt
Rosetta Inpharmatics, LLC
Seattle, WA
USA

N.A. Sheehan
Department of Health Sciences
and Genetics
University of Leicester
Leicester, UK

S.K. Sieberts
Rosetta Inpharmatics, LLC
Seattle, WA
USA

K.D. Siegmund
Department of Preventive Medicine
Keck School of Medicine
University of Southern California
Los Angeles, CA
USA

V. Solovyev
Department of Computer Science
University of London
Surrey, UK

T.P. Speed
Department of Statistics
University of California at Berkeley
Berkeley, CA
USA

and

Genetics and Bioinformatics Group
The Walter & Eliza Hall Institute
of Medical Research
Royal Melbourne Hospital
Melbourne, Australia

D.A. Stephens
Department of Mathematics and
Statistics
McGill University
Montreal, Canada

M. Stephens
Departments of Statistics
and Human Genetics
University of Chicago
Chicago, IL
USA

W.R. Taylor
Division of Mathematical Biology
National Institute of Medical
Research
London, UK

M.D. Teare
Mathematical Modelling and Genetic
Epidemiology
University of Sheffield
Medical School
Sheffield, UK

A. Thomas
Department of Biomedical
Informatics
University of Utah
Salt Lake City, UT
USA

E.A. Thompson
Department of Statistics
University of Washington
Seattle, WA
USA

J.L. Thorne
Departments of Genetics
and Statistics
North Carolina State University
Raleigh, NC
USA

M.D. Tobin
Departments of Health Sciences
and Genetics
University of Leicester
Leicester, UK

M. Vingron
Department of Computational Molecular
Biology
Max-Planck-Institute
for Molecular Genetics
Berlin, Germany

A. von Heydebreck
Department of Computational Molecular
Biology
Max-Planck-Institute
for Molecular Genetics
Berlin, Germany

B. Walsh
Department of Ecology and
Evolutionary Biology
Department of Plant Sciences
Department of Molecular
and Cellular Biology
University of Arizona
Tuscon, AZ
USA

B.S. Weir
Department of Biostatistics
University of Washington
Seattle, WA
USA

J. Whittaker
Department of Epidemiology and
Population Health
London School of Hygiene &
Tropical Medicine
London, UK

T.C. Wood
Department of Biochemistry
University of Virginia
Charlottesville, VA
USA

Z. Yang
Department of Biology
University College London
London, UK

H. Zhao
Department of Epidemiology and
Public Health
Yale University School of Medicine
New Haven, CT
USA

Editor's Preface to the Third Edition

In the four years that have elapsed since the highly successful second edition of the *Handbook of Statistical Genetics*, the field has moved on, in some areas dramatically. This is reflected in the present thorough revision and comprehensive updating: 17 chapters are entirely new, 6 providing a fresh approach to topics that had been covered in the second edition, while 11 new chapters cover areas of recent growth, or important topics not previously addressed. These new topics include microarray data analysis (two new chapters to complement the existing one), eQTL analyses and metabonomics. There are also new chapters on graphical models and on pedigrees and genealogies, admixture mapping and genome-wide association studies, cancer genetics, epigenetics, and genetical aspects of insurance. Of the 26 chapters carried over from the second edition, 21 have been revised, among which 5 very substantial revisions are close to being new chapters.

It will be clear from the topics listed above that we continue to define statistical genetics very broadly. Statistics for us goes beyond mathematical models and techniques, and includes the management and presentation of data, as well as its analysis and interpretation. Genetics includes the search for and study of genes implicated in human health and the economic value of plants and animals, the evolution of genes within natural populations, the evolution of genomes and of species, and the analysis of DNA, RNA, gene expression, protein sequence and structure, and now metabonomics. The latter topics probably fall outside even a liberal definition of 'genetics', but we believe they will be of interest to our readers because of their relevance to studies of gene function and because of the statistical methods being used.

We regard more recent terms, such as 'genomics' and 'transcriptomics' as designating new avenues within genetics, rather than as entirely new fields, and we include their statistical aspects as part of statistical genetics. Similarly, we include much of 'bioinformatics', but we do not systematically survey the available genetic databases or computer software, nor methods and protocols for archiving and annotating genetic data. Some pointers to computer software and other Internet resources are given at the end of relevant chapters.

The 43 chapters are intended to be largely independent, so that to benefit from the handbook it is not necessary to read every chapter, nor read chapters sequentially. This structure necessitates some duplication of material, which we have tried to minimise but could not always eliminate. Alternative approaches to the same topic by different authors can convey benefits. The extensive subject and author indexes allow easy reference to topics covered in different chapters.

For those with minimal genetics background, the glossary of genetic terms (newly updated) should be of assistance, while Wiley's *Biostatistical Genetics and Genetic Epidemiology*, edited by Elston, Olson, and Palmer (2002) provides a more substantial resource of definitions and explanations of key terms from both genetics and statistics. For those seeking a more substantial introduction to the foundations of modern statistical methods applied in genetics, we suggest *Likelihood, Bayesian, and MCMC methods in Quantitative Genetics* by Sorenson and Gianola (2003).

We thank the many commentators of the first two editions who were generous in their praise. We have tried to take on board many of the constructive criticisms and suggestions. No doubt many more improvements will be possible for future editions and we welcome comments e-mailed to any of the editors. We are grateful to all of our outstanding set of authors for taking the time to write and update their chapters with care, and for meeting their deadlines (and sometimes ours as well). Finally, we would like to express our appreciation to the staff of John Wiley & Sons for initially proposing the project to us, and for their friendly professionalism in the preparation of both editions. In particular, we thank Martine Bernardes-Silva, Layla Harden, and Kathryn Sharples.

DAVID BALDING
MARTIN BISHOP
CHRIS CANNINGS
August 2007

Glossary of Terms

GLOSSARY OF GENETIC TERMS:
(prepared by Gurdeep Sagoo, University of Sheffield, UK)

N.B. Some of the definitions below assume that the organism of interest is diploid.

Adenine (A): purine base that forms a pair with thymine in DNA and uracil in RNA.

Admixture: arises when two previously isolated populations begin interbreeding.

Allele: one of the possible forms of a gene at a given locus. Depending on the technology used to type the gene, it may be that not all DNA sequence variants are recognised as distinct alleles.

Allele frequency: often used to mean the relative frequency (i.e. proportion) of an allele in a sample or population.

Allelic association: the non-independence, within a given population, of a gamete's alleles at different loci. Also commonly (and misleadingly) referred to as *linkage disequilibrium*.

Alpha helix: a helical (usually right-handed) arrangement that can be adopted by a polypeptide chain; a common type of protein secondary structure.

Amino acid: the basic building block of proteins. There are 20 naturally occurring amino acids in animals which when linked by peptide bonds form polypeptide chains.

Aneuploid cells: do not have the normal number of chromosomes.

Antisense strand: the DNA strand complementary to the coding strand, determined by the covalent bonding of A with T and C with G.

Ascertainment: the strategy by which individuals are identified, selected, and recruited for participation in a study.

Autosome: A chromosome other than the sex chromosomes. Humans have 22 pairs of autosomes plus 2 sex chromosomes.

Backcross: A linkage study design in which the progeny (F1s) of a cross between two inbred lines are crossed back to one of the inbred parental strains.

Bacterial Artificial Chromosome (BAC): a vector used to clone a large segment of DNA (100–200 Kb) in bacteria resulting in many copies.

Base: (abbreviated term for a purine or pyrimidine in the context of nucleic acids), a cyclic chemical compound containing nitrogen that is linked to either a deoxyribose (DNA) or a ribose (RNA).

Base pair (bp): a pair of bases that occur opposite each other (one in each strand) in double stranded DNA/RNA. In DNA adenine base pairs with thymine and cytosine with guanine. RNA is the same except that uracil takes the place of thymine.

Bayesian: A statistical school of thought that, in contrast with the frequentist school, holds that inferences about any unknown parameter or hypothesis should be encapsulated in a probability distribution, given the observed data. Bayes Theorem allows one to compute the posterior distribution for an unknown from the observed data and its assumed prior distribution.

Beta-sheet: is a (hydrogen-bonded) sheet arrangement which can be adopted by a polypeptide chain; a common type of protein secondary structure.

centiMorgan (cM): measure of genetic distance. Two loci separated by 1 cM have an average of 1 recombination between them every 100 meioses. Because of the variability in recombination rates, genetic distance differs from physical distance, measured in base pairs. Genetic distance differs between male and female meioses; an average over the sexes is usually used.

Centromere: the region where the two sister chromatids join, separating the short (p) arm of the chromosome from the long (q) arm.

Chiasma: the visible structure formed between paired homologous chromosomes (non-sister chromatids) in meiosis.

Chromatid: a single strand of the (duplicated) chromosome, containing a double-stranded DNA molecule.

Chromatin: the material composed of DNA and chromosomal proteins that makes up chromosomes. Comes in two types, euchromatin and heterochromatin.

Chromosome: the self-replicating threadlike structure found in cells. Chromosomes, which at certain stages of meiosis and mitosis consist of two identical sister chromatids, joined at the centromere, and carry the genetic information encoded in the DNA sequence.

cis-Acting: regulatory elements and eQTL whose DNA sequence directly influences transcription. The physical location for cis-acting elements will be in or near the gene or genes they regulate.

Clones: genetically engineered identical cells/sequences.

Co-dominance: both alleles contribute to the phenotype, in contrast with recessive or dominant alleles.

Codon: a nucleotide triplet that encodes an amino acid or a termination signal.

Common disease common variant (CDCV) hypothesis: The hypothesis that many genetic variants underlying complex diseases are common, and hence susceptible to detection using current population association study designs. An alternative possibility is that genetic contributions to the causation of complex diseases arise from many variants, all of which are rare.

complementary DNA (cDNA): DNA that is synthesised from a messenger RNA template using the reverse transcriptase enzyme.

Contig: a group of contiguous overlapping cloned DNA sequences.

Cytosine (C): pyrimidine base that forms a pair with guanosine in DNA.

Degrees of freedom (df): This term is used in different senses both within statistics and in other fields. It can often be interpreted as the number of values that can be defined arbitrarily in the specification of a system; for example, the number of coefficients in a regression model. Frequently it suffices to regard df as a parameter used to define certain probability distributions.

Deoxyribonucleic acid (DNA): polymer made up of deoxyribonucleotides linked together by phosphodiester bonds.

Deoxyribose: the sugar compound found in DNA.

Diploid: has two versions of each autosome, one inherited from the father and one from the mother. Compare with haploid.

Dizygotic twins: twins derived from two separate eggs and sperm. These individuals are genetically equivalent to full sibs.

DNA methylation: the addition of a methyl group to DNA. In mammals this occurs at the C-5 position of cytosine, almost exclusively at CpG dinucleotides.

DNA microarray: small slide or 'chip' used to simultaneously measure the quantity of large numbers of different mRNA gene transcripts present in cell or tissue samples.

Depending on the technology used, measurements may either be absolute or relative to the quantities in a second sample.

Dominant allele: results in the same phenotype irrespective of the other allele at the locus.

Effective population size: The size of a theoretical population that best approximates a given natural population under an assumed model. The criterion for assessing the 'best' approximation can vary, but is often some measure of total genetic variation.

Enzyme: a protein that controls the rate of a biochemical reaction.

Epigenetics: the transmission of information on gene expression to daughter cells at cell division.

Epistasis: the physiological interaction between different genes such that one gene alters the effects of other genes.

Epitope: the part of an antigen that the antibody interacts with.

Eukaryote: organism whose cells include a membrane-bound nucleus. Compare with prokaryote.

Exons: parts of a gene that are transcribed into RNA and remain in the mature RNA product after splicing. An exon may code for a specific part of the final protein.

Expression Quantitative Trait Locus (eQTL): a locus influencing the expression of one or more genes.

Fixation: occurs when a locus which was previously polymorphic in a population becomes monomorphic because all but one allele has been lost through genetic drift.

Frequentist: the name for the school of statistical thought in which support for a hypothesis or parameter value is assessed using the probability of the observed data (or more 'extreme' datasets) given the hypothesis or value. Usually contrasted with Bayesian.

Gamete: a sex cell, sperm in males, egg in females. Two haploid gametes fuse to form a diploid zygote.

Gene: a segment (not necessarily contiguous) of DNA that codes for a protein or functional RNA.

Gene expression: the process by which coding DNA sequences are converted into functional elements in a cell.

Genealogy: the ancestral relationships among a sample of homologous genes drawn from different individuals, which can be represented by a tree. Also sometimes used in place of pedigree, the ancestral relationships among a set of individuals, which can be represented by a graph.

Genetic drift: the changes in allele frequencies that occur over time due to the randomness inherent in reproductive success.

Genome: all the genetic material of an organism.

Genotype: the (unordered) allele pair(s) carried by an individual at one or more loci. A multilocus genotype is equivalent to the individual's two haplotypes without the phase information.

Guanine (G): purine base that forms a pair with cytosine in DNA.

Haemoglobin: is the red oxygen-carrying pigment of the blood, made up of two pairs of polypeptide chains called globins (2α and 2β subunits).

Haploid: has a single version of each chromosome.

Haplotype: the alleles at different loci on a chromosome. An individual's two haplotypes imply the genotype; the converse is not true, but in the presence of strong linkage disequilibrium haplotypes may be inferred from genotype with few errors.

Hardy-Weinberg disequilibrium: the non-independence within a population of an individual's two alleles at a locus; can arise due to inbreeding or selection for example. Compare with linkage disequilibrium.

Heritability: the proportion of the phenotypic variation in the population that can be attributed to underlying genetic variation.

Heterozygosity: the proportion of individuals in a population that are heterozygotes at a locus. Also sometimes used as short for expected heterozygosity under random mating, which equals the probability that two homologous genes drawn at random from a population are not the same allele.

Heterozygote: a single-locus genotype consisting of two different alleles.

HIV (Human Immunodeficiency Virus): a virus that causes acquired immune deficiency syndrome (AIDS) which destroys the body's ability to fight infection.

Homology: similarities between sequences that arise because of shared evolutionary history (descent from a common ancestral sequence). Homology of different genes within a genome is called paralogous while that between the genomes of different species is called orthologous.

Homozygote: a single-locus genotype consisting of two versions of the same allele.

Hybrid: the offspring of a cross between parents of different genetic types or different species.

Hybridization: the base pairing of a single stranded DNA or RNA sequence, usually labelled, to its complementary sequence.

ibd: identity by descent; two genes are ibd if they have descended without mutation from an ancestral gene.

Inbred lines: derived and maintained by repeated selfing or brother–sister mating, these individuals are homozygous at essentially every locus.

Inbreeding: either the mating of related individuals (e.g. cousins) or a system of mate selection in which mates from the same geographic area or social group for example are preferred. Inbreeding results in an increase in homozygosity and hence an increase in the prevalence of recessive traits.

Intercross: A linkage study design in which the progeny (F1s) of a cross between two inbred lines are crossed or selfed. This design is also sometimes referred to as an *F2 design* because the resulting individuals are known as F2s.

Intron: non-coding DNA sequence separating the exons of a gene. Introns are initially transcribed into messenger RNA but are subsequently spliced out.

Karyotype: the number and structure of an individual's chromosomes.

Kilobase (Kb): 1000 base pairs.

Linkage: two genes are said to be linked if they are located close together on the same chromosome. The alleles at linked genes tend to be co-inherited more often than those at unlinked genes because of the reduced opportunity for an intervening recombination.

Linkage disequilibrium (LD): the non-independence within a population of a gamete's alleles at different loci; can arise due to linkage, population stratification, or selection. The term is misleading and 'gametic phase disequilibrium' is sometimes preferred. Various measures of linkage disequilibrium exist.

Locus (pl. Loci): the position of a gene on a chromosome.

LOD score: a likelihood ratio statistic used to infer whether two loci reside close to one another on a chromosome and are therefore inherited together. A LOD score of 3 or more is generally thought to indicate that the two loci are close together and therefore linked.

Marker gene: a polymorphic gene of known location which can be readily typed; used for example in genetic mapping.

Megabase (Mb): 1000 kilobases = 1,000,000 base pairs.

Meiosis: the process by which (haploid) gametes are formed from (diploid) somatic cells.

messenger RNA (mRNA): the RNA sequence that acts as the template for protein synthesis.

Microarray: see DNA microarray above.

Microsatellite DNA: small stretches of DNA (usually 1–4 bp) tandemly repeated. Microsatellite loci are often highly polymorphic, and alleles can be distinguished by length, making them useful as marker loci.

Mitosis: the process by which a somatic cell is replaced by two daughter somatic cells.

Monomorphic: a locus at which only one allele arises in the sample or population.

Monozygotic twins: genetically identical individuals derived from a single fertilized egg.

Morgan: 100 centiMorgans.

mtDNA: the genetic material of the mitochondria which consists of a circular DNA duplex inherited maternally.

Mutation: a process that changes an allele.

Negative selection: removal of deleterious mutations by natural selection. Also known as *purifying selection*.

Neutral: not subject to selection.

Neutral evolution: evolution of alleles with nearly zero selective coefficient. When $|Ns| << 1$, where N is the population size and s is the selective coefficient, the fate of the allele is mainly determined by random genetic drift rather than natural selection.

Non-Coding RNA (ncRNA): RNA segments that are coded for in the genome, but not translated into protein product. Composed of many classes, the complete range of functions of these molecules has yet to be characterised, but they have been shown to affect the rate of transcription and transcript degradation.

Nonsynonymous substitution: Nucleotide substitution in a protein-coding gene that alters the encoded amino acid.

Nucleoside: a base attached to a sugar, either ribose or deoxyribose.

Nucleotide: the structural units with which DNA and RNA are formed. Nucleotides consist of a base attached to a five-carbon sugar and mono-, di-, or tri-phosphate.

Nucleotide substitution: the replacement of one nucleotide by another during evolution. Substitution is generally considered to be the product of both mutation and selection.

Oligonucleotide: a short sequence of single-stranded DNA or RNA, often used as a probe for detecting the complementary DNA or RNA.

Open Reading Frame (ORF): a long sequence of DNA with an initiation codon at the 5′-end and no termination codon except for one at the 3′-end.

PCR (polymerase chain reaction): a laboratory process by which a specific, short, DNA sequence is amplified many times.

Pedigree: a diagram showing the relationship of each family member and the heredity of a particular trait through several generations of a family.

Penetrance: the probability that a particular phenotype is observed in individuals with a given genotype. Penetrance can vary with environment and the alleles at other loci for example.

Peptide bond: linkages between amino acids occur through a covalent peptide bond joining the C terminal of one amino acid to the N terminal of the next (with loss of a water molecule).

Phase (of linked markers): the relationship (either coupling or repulsion) between alleles at two linked loci. The two alleles at the linked loci are said to be in coupling if they are present on the same physical chromosome or in repulsion if they are present on different parental homologs.

Phenotype: the observed characteristic under study, may be quantitative (i.e. continuous) such as height, or binary (e.g. disease/no disease), or ordered categorical (e.g. mild/moderate/severe).

Pleiotropy: is the effect of a gene on several different traits.

Polygenic: influenced by more than one gene.

Polymorphic: a locus that is not monomorphic. Usually a stricter criterion is imposed: a locus is polymorphic only if no allele has frequency over 99 %.

Polynucleotide: a polymer of either DNA or RNA nucleotides.

Polypeptide: is a long chain of amino acids joined together by peptide bonds.

Polypeptide chain: A series of amino acids linked by peptide bonds. Short chains are sometimes referred to as oligopeptides or simply peptides.

Polytene: refers to the giant chromosomes that are generated by the successive replication of chromosome pairs without the nuclear division, thus several chromosome sets are joined together.

Population stratification (or population structure): Refers to a situation in which the population of interest can be divided into strata such that an individual tends to be more closely related to others within the same stratum than to other individuals.

Positive selection: fixation, by natural selection, of an advantageous allele with a positive selective coefficient. Also known as *Darwinian selection*.

Proband: an individual through whom a family is ascertained, typically by their phenotype.

Prokaryote: organism whose cells have no nucleus.

Promoter: located upstream of the gene, the promoter allows the binding of RNA polymerase which initiates transcription of the gene.

Protein: a large, complex, molecule made up of one or more chains of amino acids.

Pseudogene: a DNA sequence that is either an imperfect, non-functioning, copy of a gene, or a former gene which no longer functions due to mutations.

Purine and Pyrimidine: are particular kinds of nitrogen containing heterocyclic rings.

Purine: adenine or guanine.

Pyrimidine: cytosine, thymine, or uracil.

QTL (Quantitative Trait Locus): a locus influencing a continuously varying phenotype.

Radiation hybrid: a cell line, usually rodent, that has incorporated fragments of foreign chromosomes that have been broken by irradiation. They are used in physical mapping.

Recessive allele: has no effect on phenotype except when present in homozygote form.

Recombination: the formation of new haplotypes by physical exchange between two homologous chromosomes during meiosis.

Restriction enzyme: recognises specific nucleotide sequences in double-stranded DNA and cuts at a specified position with respect to the sequence.

Restriction fragment: a DNA fragment produced by a restriction enzyme.

Restriction site: a 4–8 bp DNA sequence (usually palindromic) that is recognised by a restriction enzyme.

Retrovirus: an RNA virus whose replication depends on a reverse transcriptase function, allowing the formation of a cDNA copy that can be stably inserted into the host chromosome.

Ribonucleic acid (RNA): polymer made up of ribonucleotides that are linked together by phosphodiester bonds.

Ribosome: a cytoplasmic organelle, consisting of RNA and protein, that is involved in the translation of messenger RNA into proteins.

Ribosomal RNA (rRNA): the RNA molecules contained in ribosomes.

Selection: a process such that expected allele frequencies do not remain constant, in contrast with genetic drift. Alleles that convey an advantage to the organism in its current environment tend to become more frequent in the population (positive, or adaptive, selection), while deleterious alleles become less frequent. Under stabilising (or balancing) selection, allele frequencies tend towards a stable, intermediate value.

Sense strand: the DNA strand in the direction of coding.

Sex-linked: a trait influenced by a gene located on a sex (X or Y) chromosome.

Single nucleotide polymorphism (SNP): a polymorphism consisting of a single nucleotide.

Sister chromatids: two chromatids that are copies of the same chromosome. Non-sister chromatids are different but homologous.

Somatic cell: a non-sex cell.

Synonymous substitution: Nucleotide substitution in a protein-coding gene that does not alter the encoded amino acid.

TATA box: a conserved sequence (TATAAAA) found about 25–30 bp upstream from the start of transcription site in most but not all genes.

Thymine (T): pyrimidine base that forms a pair with adenine in DNA.

trans-Acting: eQTL whose DNA sequence influences gene expression through its gene product. These regulatory elements are often coded for at loci far from or unlinked to the genes they regulate.

Transcription: the synthesis of a single-stranded RNA version of a DNA sequence.

Transition: a mutation that changes either one purine base to the other, or one pyrimidine base to the other.

Translation: the process whereby messenger RNA is 'read' by transfer RNA and its corresponding polypeptide chain synthesized.

Transposon: a genetic element that can move over generations from one genomic location to another.

Transversion: a mutation that changes a purine base to a pyrimidine, or vice-versa.

Uracil (U): pyrimidine base in RNA that takes the place of thymine in DNA, also forming a pair with adenine.

Wild-type: the common, or standard, allele/genotype/phenotype in a population.

Yeast artificial chromosome (YAC): a cloning vector able to carry large (e.g. one megabase) inserts of DNA and replicate in yeast cells.

Zygote: an egg cell that has been fertilized by a sperm cell.

Abbreviations and Acronyms

ABC	Approximate Bayesian Calculation
AD	Alzheimer's Disease
AFLP	Amplified Fragment Length Polymorphism
AGT	Angiotensionogen
AIC	Akaike's Information Criterion
AMOVA	An Analysis of Molecular Variance
ANN	Artificial Neural Network
ANOVA	Analysis of Variance
APM	Affected-Pedigree-Member
ARG	Ancestral Recombination Graph
BAC	Bacterial Artificial Chromosome
BBSRC	Biotechnology and Biological Sciences Research Council
BC	Backcross
BIC	Bayesian Information Criterion
BKYF	Beerli–Kuhner–Yamato–Felsenstein
BLAST	Basic Local Alignment Search Tool
BLUE	Best Linear Unbiased Estimator
BLUP	Best Linear Unbiased Predictor
BMI	Body Mass Index
bp	Base Pairs
CART	Classification and Regression Tree
CASP	Critical Analysis of Structure Prediction
cDNA	Complementary DNA
CEPH	Centre pour l'Etude des Polymorphismes Humains
CGH	Comparative Genomic Hybridization
CHD	Coronary Heart Disease
ChIP	Chromatin Immunoprecipitation
CI	Confidence Interval
CIM	Composite Interval Mapping
CL	Composite Log-Likelihood
CMV	Cytomegalovirus
COGs	Clusters of Orthologous Groups
CTLs	Cytotoxic T Lymphocytes
DAG	Directed Acyclic Graph
df	Degrees of Freedom
DH	Doubled Haploids
DNA	Deoxyribonucleic Acid

EBV	Epstein–Barr Virus
ECHR	European Convention on Human Rights
EC	Extreme Concordant
ECJ	European Court of Justice
ED	Extreme Discordant
EM	Expectation Maximisation
EPD	Eukaryotic Promoter Database
eQTL	Expression Quantitative Trait Loci
EST	Expressed Sequence Tag
FDR	False Discovery Rate
FISH	Fluorescent In Situ Hybridization
FLMs	Finite Locus Models
FM	Fitch–Margoliash Methods
FPM	Finite Polygenic Model
FWER	Family-Wise Error Rate
GA	Genetic Algorithm
GC	Gas Chromatography
GC	Guanine and Cytosine
GEEs	Generalised Estimating Equations
GLM	Generalised Linear Model
GLMM	Generalised Linear Mixed Model
GNG	Gamma-Normal-Gamma
GO	Gene Ontology
GUI	Graphical User Interface
HA	Haemagglutinin
HBV	Hepatitis B Virus
HMM	Hidden Markov Model
HPD	Highest Probability Density
HSV	Herpes Simplex Virus
HTLV	Human T-Cell Lymphotropic Virus Type I
HVRI	Hypervariable Region
HVRII	Hypervariable Region II
HWE	Hardy–Weinberg Equilibrium
IAM	Infinite-Allele Model
ibd	Identical by Descent
ibs	Identical by State
ICRP	International Commission of Radiological Protection
IID	Independent and Identically Distributed
iis	Identity in State
IS	Importance Sampling
Kb	kilobases
KDEs	Kernel Density Estimators
kNN	k-Nearest Neighbour
LC	Liquid Chromatography
LD	Linkage Disequilibrium
LINEs	Long Interspersed Nuclear Elements
LLR	Log-Likelihood Ratio

LogDet	Logarithm of the Determinant
LOH	Loss of Heterozygosity
LR	Likelihood Ratio
LS	Least-Squares
LTR	Long Terminal Repeat
MAI	Marker-Assisted Introgression
MAS	Marker-Assisted Selection
MC	Monte Carlo
MCMC	Markov Chain Monte Carlo
MH	Metropolis–Hastings
ML	Mapping Methods – Maximum Likelihood
MLE	Maximum Likelihood Estimate
MLR	Maximum Likelihood Ratio
MLR	Multiple Linear Regression
MM	Mismatch
MME	Mixed Model Equations
MP	Maximum Parsimony
MRCA	Most Recent Common Ancestor
mRNA	Messenger Ribonucleic Acid
MS	Mass Spectrometry
MSA	Multiple Sequence Alignment
mtDNA	Mitochondrial DNA
MVN	Multivariate Normal
MY	Million Years
MZ	Monozygous
NcRNA	Noncoding Ribonucleic Acid
NJ	Neighbor-Joining
NMR	Nuclear Magnetic Resonance
NP	Non-Deterministic Polynomial
OLS	Ordinary Least Squares
OR	Odds Ratio
ORF	Open Reading Frame
PAC	Product of Approximate Conditionals
PAM	Partitioning Around Medoids
PCs	Principal Components
PCA	Principal Components Analysis
PCR	Polymerase Chain Reaction
PDB	Protein Data Bank
PDF	Probability Density Function
PI	Paternity Index
PIC	Polymorphism Information Content
PKU	Phenylketonuria
PLS	Partial Least-Squares
PM	Perfect Match
PMLE	Pseudo Maximum Likelihood Estimator
PNNs	Probabilistic Neural Networks
PSA	Population-Specific Alleles

QQ   Quantile–Quantile
QTLs   Quantitative Trait Loci
RAPD   Randomly Amplified Polymorphic DNA
RCTs   Randomized Controlled Trials
REML   Residual Maximum Likelihood
RFLP   Restriction Fragment Length Polymorphism
RH   Radiation Hybrid
RIL   Recombinant Inbred Line
SINE   Small Interspersed Nuclear Elements
SIVagm   SIV from African Green Monkeys
SMM   Stepwise Mutation Model
SNP   Single Nucleotide Polymorphism
SPRT   Sequential Probability Ratio Test
STR   Short Tandem Repeat
STRs   Simple Tandem Repeats
STS   Sequence-Tagged Site
SVM   Support Vector Machine
TDT   Transmission/Disequilibrium Test
TF   Transcription Factor
TPM   Two-Phase Model
TRRD   Transcription Regulatory Regions Database
TSG   Tumour Suppressor Gene
TSS   Transcription Start Site
UA   Ultimate Ancestor
UPGMA   Unweighted Pair-Group Method with Arithmetic Mean
WB   Wilson and Balding
WGA   Whole Genome Association
WPC   Weighted Pairwise Correlation
YAC   Yeast Artificial Chromosome

Part 5

Population Genetics

Mathematical Models in Population Genetics

C. Neuhauser

Department of Ecology, Evolution and Behavior, University of Minnesota, Saint Paul, MN, USA

Throughout the history of population genetics, mathematical models have played an important role in elucidating the effects of mutation and selection on the genetic diversity of organisms. Mathematical models provided the theoretical foundation of neo-Darwinism; sophisticated mathematical tools aided Kimura in establishing the neutral molecular theory. Mathematical models in population genetics today are crucial in the development of statistical tools for analyzing molecular data.

This chapter emphasizes models for selection, but also includes the discussion of neutral models. After a brief history of the role of selection in evolution, basic mathematical models are introduced together with the diffusion approximation. A discussion of coalescent theory follows, with primary focus on selection. A short discussion on how to detect selection concludes the chapter.

22.1 A BRIEF HISTORY OF THE ROLE OF SELECTION

> This preservation of favourable variations and the rejection of injurious variations, I call Natural Selection. Variations neither useful nor injurious would not be affected by natural selection, and would be left a fluctuating element, as perhaps we see in the species called polymorphic. (Darwin, 1985)

Charles Darwin was the first to formulate the concept of natural selection and to apply it to evolution and adaptation. Darwin developed his theory of evolution without a knowledge of the source of variation. Today we know that the hereditary information of most organisms is encoded in deoxyribonucleic acid (DNA) and that variation is caused by mutations; the definition of natural selection remains the same, namely the differential reproductive success of different genotypes.

When Darwin first proposed his theory of natural selection, he believed that genetic diversity was primarily driven by natural selection and that variations accumulated

Handbook of Statistical Genetics, Third Edition. Edited by D.J. Balding, M. Bishop and C. Cannings.
© 2007 John Wiley & Sons, Ltd.   ISBN: 978-0-470-05830-5.

gradually over time, as expressed in the quote *Natura non facit saltum* ('Nature makes no leaps'). The concept of gradual evolution met immediately with criticism since the fossil record had not yielded transitional forms at that time that would indicate a gradual change from one species to another, as would be expected under gradual evolution.

Shortly after Darwin's proposal of the nature of evolution, an Austrian monk, Gregor Mendel, carried out experiments on peas in 1865 and discovered the basic rules of inheritance. The importance of Mendel's discoveries was only realized in 1900 when the plant breeders de Vries, Correns, and Tschermak independently obtained plant breeding data that could be interpreted by Mendel's rules of inheritance. The cause of variation, namely mutations, was first described by Hugo de Vries. Based on plant breeding experiments, de Vries concluded that mutations caused drastic, nongradual changes.

The relative importance of mutations versus selection as the driving forces of evolution was a matter of dispute. Proponents of the Darwinian Theory asserted that evolution proceeds by small steps, namely selection operating on small variations; whereas proponents of the Mendelian theory believed that evolution proceeds by large leaps caused by mutations.

This controversy continued until the early 1930s when Fisher (1930), Haldane (1932), and Wright (1931) synthesized the Mendelian theory of inheritance and the Darwinian theory of evolution; this synthesis, called the *neo-Darwinism* or *synthetic theory of evolution*, formed the foundation of the modern theory of population genetics. It emphasizes the importance of natural selection acting on variations caused by mutations in the course of evolution.

When protein sequences became available in the 1960s, it soon became clear that genetic diversity was far greater than had been expected. This prompted Kimura (1968a), and King and Jukes (1969) to question the importance of natural selection as the driving force of evolution. Instead, they proposed that most variation was selectively neutral. This started a heated debate between the proponents of the neutral theory and the selectionists. In subsequent work (Kimura, 1968b; 1977; 1979; 1983; Kimura and Ohta, 1973) developed this idea much further; the controversy has not been resolved. A principal conclusion of the neutral theory is that genetic diversity is largely caused by random genetic drift, implying that the genetic diversity seen in populations is a transient phenomenon: Mutations are introduced at random and they either go to fixation or are lost solely due to stochastic forces.

22.2 MUTATION, RANDOM GENETIC DRIFT, AND SELECTION

Mutant alleles that have little effect on the phenotype of the organism may remain in the population until they either become fixed or lost due to stochastic forces. Other alleles are maintained in or quickly eliminated from a population by selective forces. Mathematical models that are based on the laws of inheritance can illuminate the role and relative importance of stochastic and selective forces. An excellent reference for classical mathematical population genetics is Ewens (2004); many of the models discussed here (and much more) can be found in his book.

The simplest mathematical models track allele frequencies in a randomly mating, monoecious population. Changes in allele frequencies are caused by mutation, random genetic drift, and selection. To understand their effects, we discuss each factor separately at first.

22.2.1 Mutation

Consider a locus with two alleles A_1 and A_2, and let $p(n)$ and $q(n) = 1 - p(n)$ be the frequencies of A_1 and A_2, respectively, at generation n. We assume that generations are nonoverlapping, i.e. in each generation the entire population undergoes random mating; furthermore, A_1 mutates to A_2 at rate u and A_2 to A_1 at rate v, and an allele can mutate at most once per generation. In an infinitely large population, the dynamics can be described by a deterministic equation. In the absence of selective forces and under random mating, the gene frequency of A_1 in the next generation is obtained in the following way: An A_1 allele in generation $n + 1$ could have been either an A_1 allele in generation n that did not mutate (with probability $1 - u$), or an A_2 allele that mutated from A_2 to A_1 (with probability v). Therefore, the gene frequency of A_1 in generation $n + 1$ is given by

$$p(n + 1) = (1 - u)p(n) + v(1 - p(n)). \qquad (22.1)$$

This can be solved in terms of the initial gene frequency of A_1, $p(0)$, and one finds

$$p(n) = \frac{v}{u + v} + \left(p(0) - \frac{v}{u + v} \right) (1 - u - v)^n. \qquad (22.2)$$

The mutation probabilities u and v are typically quite small (of order 10^{-6} or 10^{-5}). In the long run (i.e., when n is very large), the term $(1 - u - v)^n$ approaches zero. For instance, assume $u = v = 10^{-6}$; then $(1 - u - v)^n$ is equal to 0.1353 when $n = 10^6$ and equal to 2.061×10^{-9} when $n = 10^7$. This indicates that changes caused by mutations alone might be quite slow, a realization that prompted population geneticists in the 1930s (Wright, 1931; Haldane, 1932) to suggest that, though mutations are the source of variation, their role in evolution might be limited – see Nei (1987) for a discussion.

Eventually – i.e. in the limit as n tends to infinity – the right-hand side of (22.2) approaches the value $v/(u + v)$. This value is called an *equilibrium*. The equilibrium is also characterized by $p(n + 1) - p(n) = 0$, which means that the gene frequencies do not change over time. If we denote the equilibrium frequency of A_1 by $\hat{p}$, set $p(n) = p(n + 1) = \hat{p}$ in (22.1), and solve for $\hat{p}$, we find (as before)

$$\hat{p} = \frac{v}{u + v}.$$

If $\hat{q} = 1 - \hat{p}$ denotes the equilibrium frequency of A_2, then $\hat{q} = u/(u + v)$. Provided u and v are both positive, we conclude that mutation allows for the maintenance of the two alleles in the population.

22.2.2 Random Genetic Drift

The assumption of an infinite population allowed us to use a deterministic formulation when we investigated the effects of mutation. In a finite population, the random sampling of gametes alone causes changes in gene frequencies. This process is known as *random genetic drift*. To investigate the consequences of random genetic drift, we again look at a single locus with two alleles, A_1 and A_2.

Assume a randomly mating diploid population of size N (or, which is the same in this case, a haploid population of size $2N$) with nonoverlapping generations. Each generation, $2N$ gametes are sampled at random from the parent generation. If $Y(n)$ denotes the number of gametes of type A_1 at generation n, then, in the absence of mutation and selection, the number of A_1 alleles at time $n+1$ is given by the binomial distribution. Namely, the probability that there are j gametes of type A_1 at generation $n+1$, given that there were i gametes of type A_1 at generation n, is

$$P(Y(n+1) = j \mid Y(n) = i) = \binom{2N}{j} p^j (1-p)^{2N-j}, \qquad (22.3)$$

where $p = i/2N$. This model is known as the *Wright–Fisher model*. If we follow a population that evolves according to the model defined in (22.3), we will observe that its behavior is quite unpredictable due to the stochastic nature of the model. In particular, this means that if we follow different populations, all of equal size and each following the dynamics described in (22.3), they will follow different trajectories over time (see Figure 22.1). This discrete-time stochastic process is an example of a Markov chain; we will discuss Markov chains in more detail below. An elementary reference on Markov processes is, for instance, Karlin and Taylor (1975).

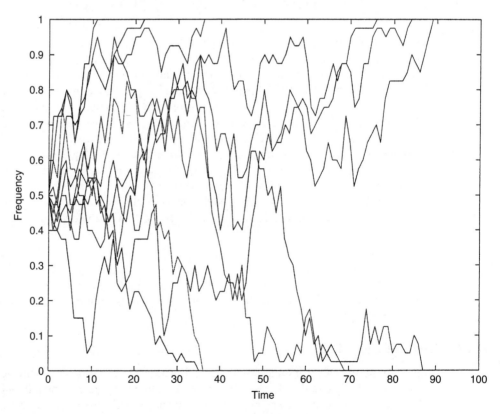

Figure 22.1 Ten trajectories for the neutral Wright–Fisher model when the population size $N = 20$ and the initial gene frequency of A_1 is 0.5.

Since there are no mutations in the model, eventually one of the two alleles will be lost (and the other one fixed), as can be seen in Figure 22.1. The larger the population size, the longer this process of fixation takes. Regardless of the population size, if initially the frequency of A_1 alleles is π_1, then one can show that the probability of fixation of allele A_1 is π_1 (and consequently, the probability of fixation of A_2 is $1 - \pi_1$). This can be understood intuitively, following an argument by Ewens (1972). Namely, after a long enough time, all individuals in the population must have descended from just one of the individuals present at generation 0. The probability that this common ancestor was of type A_1 is equal to the relative frequency of A_1 alleles at generation 0, which is π_1.

22.2.3 Selection

We now turn to the third component, selection. Selection can act on different parts of the life history of an organism; differential fecundity and viability are just two examples. The simplest model of viability selection assumes that selection affects survival between the zygote and adult stage of a diploid organism in a randomly mating population of infinite size, in which generations do not overlap. It is assumed that each genotype has a fixed, specified fitness. Since the population size is infinite, changes in allele frequencies can be described by deterministic equations. In the case of one gene with two alleles, A_1 and A_2, there are three genotypes: $A_1 A_1$, $A_1 A_2$, and $A_2 A_2$. We denote their respective fitnesses by w_{11}, w_{12}, and w_{22}. In the case of viability selection, the fitness w_{ij} reflects the relative survival chances of zygotes of genotype $A_i A_j$. If the population is in Hardy–Weinberg equilibrium and the frequencies of A_1 and A_2 in generation n are $p(n)$ and $q(n) = 1 - p(n)$, respectively, then, ignoring mutations, the gene frequencies at the next generation are

$$p(n+1) = \frac{p(n)[p(n)w_{11} + q(n)w_{12}]}{\bar{w}}$$

and

$$q(n+1) = \frac{q(n)[p(n)w_{12} + q(n)w_{22}]}{\bar{w}},$$

where $\bar{w}$, the average fitness, is chosen so that $p(n+1) + q(n+1) = 1$, i.e.

$$\bar{w} = p^2(n)w_{11} + 2p(n)q(t)w_{12} + q^2(n)w_{22}.$$

The predictions of this model are straightforward. If $w_{11} > w_{12} > w_{22}$ (or $w_{11} < w_{12} < w_{22}$), the case of directional selection, $A_1(A_2)$ becomes fixed in the population. If $w_{11}, w_{22} < w_{12}$, the case of overdominance, a stable polymorphism results. If $w_{12} < w_{11}, w_{22}$, the case of underdominance, the polymorphism is unstable and depending on the initial gene frequencies, either A_1 or A_2 becomes fixed.

22.2.4 The Wright–Fisher Model

The Wright–Fisher model is the basic model for reproduction in a finite population that can utilize several mutation models and selection schemes and is at the heart of many models that describe how gene frequencies evolve in the presence of random drift, mutation, and selection (the neutral version of this model was introduced in (22.3)).

The model for a diploid population is defined as follows. Generations are nonoverlapping and the population size N is held constant. As before, we consider a single locus with two alleles, A_1 and A_2, with genotypes A_1A_1, A_1A_2, and A_2A_2 and respective fitnesses w_{11}, w_{12}, and w_{22}. Furthermore, we assume that the population is randomly mating and in Hardy–Weinberg equilibrium. Suppose there are i genes of type A_1 and $2N - i$ genes of type A_2. Then, assuming selection affects survival between the zygote and adult stage as before, and denoting by $p(n) = i/2N$ the gene frequency of A_1 at generation n, the gene frequency of A_1 after selection becomes

$$\phi_1(n) = \frac{p(n)[p(n)w_{11} + (1 - p(n))w_{12}]}{\bar{w}(n)},$$

where $\bar{w}(n) = p(n)^2 w_{11} + 2p(n)(1 - p(n))w_{12} + (1 - p(n))^2 w_{22}$ is the average fitness. If mutation follows selection, then, assuming symmetric mutation with probability u, the gene frequency of A_1 after mutation becomes

$$\psi_1(n) = \phi_1(n)(1 - u) + (1 - \phi_1(n))u.$$

The N individuals of the next generation are formed by sampling $2N$ independent gametes from the pool according to a binomial resampling scheme. That is, if ψ_1 is the frequency of A_1 after selection and mutation, then the probability that there are j genes of type A_1 in the following generation is

$$\binom{2N}{j} \psi_1^j (1 - \psi_1)^{2N-j}. \tag{22.4}$$

The gene frequency at generation $n + 1$ is then $j/2N$.

This is another example of a discrete-time Markov chain; expression (22.4) is called the *transition probability* of this chain since it describes how the chain evolves from generation to generation. An important feature of this process is that its future depends only on the present state; this property makes this stochastic process a Markov process. The transition probabilities do not depend explicitly on time, and the chain is therefore called *time-homogeneous*. Even for Markov chains with such simple-looking transition probabilities as in (22.4), it is difficult to compute quantities of biological interest exactly, such as the expected time until fixation. For large populations, however, it is often possible to approximate such a chain by a diffusion process. This method was first used by Fisher (1922) and Wright (1931), and later greatly extended by Kimura (1964). (The mathematically rigorous treatment of diffusion processes is due to Kolmogorov, 1931.)

22.3   THE DIFFUSION APPROXIMATION

A diffusion process is a continuous-time stochastic process that tracks a quantity that changes continuously in time and whose future depends only on the present state (the precise definition is somewhat more technical but not needed in the following). An excellent introduction to diffusion processes at an elementary level is Chapter 15 in Karlin and Taylor (1981).

The idea behind using a diffusion process as an approximation of genetic models for large but finite populations is that for many finite-size models, when viewed on a suitable time scale, in the limit as the population size tends to infinity, the change in relative gene frequencies is continuous and results in a well-defined process. The limiting process is typically easier to study than the original process.

We denote the diffusion process by $\{X(t) : t \geq 0\}$. We think of t as representing time and refer to $X(t)$ as the state at time t; for instance, $X(t)$ could be the relative gene frequency of a particular allele at time t. A diffusion process is characterized by two quantities, the mean and the variance of the infinitesimal displacement, called *drift* and *diffusion*, respectively. The displacement during the time interval $(t, t + h)$ is denoted by $\Delta_h X(t) = X(t + h) - X(t)$. Then the drift parameter is defined as

$$a(x, t) = \lim_{h \to 0} \frac{1}{h} E[\Delta_h X(t) \mid X(t) = x].$$

The diffusion parameter is defined as

$$b(x, t) = \lim_{h \to 0} \frac{1}{h} E[(\Delta_h X(t))^2 \mid X(t) = x].$$

The meaning of these quantities is as follows: For small h, $a(x, t)h$ is approximately the mean of the displacement $\Delta_h X(t)$ during the time interval $(t, t + h)$ since

$$E[\Delta_h X(t) \mid X(t) = x] = a(x, t)h + o(h).$$

The quantity $b(x, t)h$ is approximately the variance of the displacement $\Delta_h X(t)$ during the interval $(t, t + h)$ for small h since

$$\text{var}[\Delta_h X(t) \mid X(t) = x] = E[(\Delta_h X(t))^2 \mid X(t) = x] - \left(E[\Delta_h X(t) \mid X(t) = x]\right)^2$$

$$= b(x, t)h - (a(x, t)h)^2 + o(h) = b(x, t)h + o(h).$$

In all of our examples, the infinitesimal drift and diffusion parameters will not depend on t; this is so since the underlying Markov chains will be time homogeneous. In such cases, we can simply write $a(x)$ and $b(x)$ instead of $a(x, t)$ and $b(x, t)$.

As a first example, we consider the neutral Wright–Fisher model with symmetric mutation for a randomly mating diploid population of size N (i.e. $2N$ gametes). We assume a one-locus model with two alleles, A_1 and A_2, with mutation probability u, and denote by $Y(n)$ the number of A_1 gametes at generation n. The transition probabilities are given by

$$P(Y(n + 1) = j \mid Y(n) = i) = \binom{2N}{j} \psi_1^j (1 - \psi_1)^{2N-j},$$

where

$$\psi_1 = \frac{i}{2N}(1 - u) + \frac{2N - i}{2N} u.$$

To compute the drift and diffusion parameter, we define the scaled process

$$X_N(t) = \frac{Y(\lfloor 2Nt \rfloor)}{2N}, \qquad t \geq 0,$$

where $\lfloor 2Nt \rfloor$ is the largest integer less than or equal to $2Nt$. To find the infinitesimal drift parameter, we compute

$$2N E\left[X_N\left(t + \frac{1}{2N}\right) - X_N(t) \,\middle|\, X_N(t) = \frac{i}{2N} \right]$$

$$= E\left[Y(\lfloor 2Nt \rfloor + 1) - i \mid Y(\lfloor 2Nt \rfloor) = i\right]$$

$$= 2N\psi_1 - i = i(1 - u) + (2N - i)u - i = 2Nu\left(1 - \frac{i}{N}\right).$$

We set $h = 1/2N$, implying that we measure time in units of $2N$ generations, and let N tend to infinity (or h tends to 0). To do this, we also need to scale the mutation parameter, namely, we assume that $\lim_{N \to \infty} 4Nu = \theta$. We then find, with $x = i/2N$,

$$a(x) = \lim_{h \to 0} \frac{1}{h} E\left[X_N(t + h) - X_N(t) \mid X_N(t) = x\right] = \frac{\theta}{2}(1 - 2x).$$

To find the infinitesimal diffusion parameter, we compute

$$2NE\left[\left(X_N\left(t + \frac{1}{2N}\right) - X_N(t)\right)^2 \,\middle|\, X_N(t) = \frac{i}{2N}\right]$$

$$= \frac{1}{2N} E\left[(Y(\lfloor 2Nt \rfloor) - i)^2 \mid Y(\lfloor 2Nt \rfloor) = i\right]$$

$$= \frac{1}{2N} 2N\psi_1(1 - \psi_1),$$

since $2N\psi_1(1 - \psi_1)$ is the variance of a binomial distribution with parameters ψ_1 and $2N$. With $x = i/2N$ and $\lim_{N \to \infty} 4Nu = \theta$, we see that $\psi_1 \to x$ as $N \to \infty$. Hence the infinitesimal diffusion parameter is

$$b(x) = x(1 - x).$$

To obtain nontrivial limits of the drift and diffusion parameters, we needed to assume that $\lim_{N \to \infty} 4Nu$ exists (we denoted the limit by θ). Of course, the mutation probability per gene per generation, u, does not depend on the population size but rather is a fixed number. We therefore cannot expect that this limit exists in reality (though we can stipulate that it exists in a mathematical model). How, then, should we interpret this limit? First, the diffusion limit is an *approximation* to the real model when the population size is fixed but finite. Second, we mentioned earlier that the mutation probability is typically quite small, namely of the order of 10^{-5} or 10^{-6}. We should therefore interpret the existence of the limit $\lim_{N \to \infty} 4Nu$ as a guide to when the approximation might be good; namely, we expect the approximation to be good when the population size is of the order of the reciprocal of the mutation probability.

In the literature, one typically finds $4Nu = \theta$ instead of $\lim_{N \to \infty} 4Nu = \theta$. Both are to be interpreted in the same way. In the following, we will adopt the convention of writing $4Nu = \theta$ instead of $\lim_{N \to \infty} 4Nu = \theta$.

Below, we will need the diffusion limit of a haploid Wright–Fisher model with mutations and selection. It is a one-locus model with two alleles A_1 and A_2 for a haploid population of size N. Mutations occur at birth with probability u; i.e. the offspring of an individual is of the same type with probability $1 - u$ and of the other type with probability u. Furthermore, allele A_2 has a selective advantage with selection parameter s. Let $Y_1(n)$ denote the number of individuals of type A_1 at generation n. Then the transition probabilities are given by

$$P[Y_1(n+1) = j \mid Y_1(n) = i] = \binom{N}{j} \psi_1^j (1 - \psi_1)^{N-j}, \qquad (22.5)$$

where

$$\psi_1 = \frac{p(1-u) + (1-p)(1+s)u}{p + (1-p)(1+s)}, \qquad p = \frac{i}{N}.$$

If we set $\theta = 2Nu$ and $\sigma = 2Ns$, measure time in units of N generations, and let N tend to infinity, then

$$a(x) = -\frac{\sigma}{2} x(1-x) + \frac{\theta}{2}(1 - 2x), \qquad (22.6)$$

$$b(x) = x(1-x). \qquad (22.7)$$

(Recall that for a haploid population of size N, there are only N gametes; so instead of the factor $2N$, we only have a factor of N in the above scaling.)

The selection parameter s is scaled in the same way as the mutation parameter u. This has an important implication: The diffusion limit for a model with selection can only be carried out provided the selection intensity is of the order of the reciprocal of the population size. Since the diffusion limit is only a good approximation for large populations, the diffusion limit is only useful for weak selection.

What is the advantage of using the diffusion limit? As mentioned earlier, it is rarely possible to obtain exact results of biological interest for the original Markov chain. Using the diffusion limit allows one to take advantage of a number of analytical tools that facilitate the computation of various quantities. We provide two such applications, namely the computation of quantities associated with fixation, and the Kolmogorov forward equation, which can be used to find the stationary distribution.

22.3.1 Fixation

The diffusion approximation can be used to compute certain functionals associated with the process, such as the probability of fixation or the mean time to fixation.

Suppose $X(t)$ denotes the gene frequency at time t. We define $T(y)$ as the (random) time until $X(t)$ reaches y. For $0 \leq x_1 < x < x_2 \leq 1$, we define the probability that the process reaches x_2 before x_1 when starting at x,

$$u(x) = P[T(x_2) < T(x_1) \mid X(0) = x],$$

and the average time until the process reaches either x_1 or x_2,

$$v(x) = E[\min(T(x_1), T(x_2)) \mid X(0) = x].$$

One can show that, for $0 \leq x_1 < x < x_2 \leq 1$, $u(x)$ satisfies the differential equation

$$0 = a(x)\frac{du}{dx} + \frac{1}{2}b(x)\frac{d^2u}{dx^2}, \qquad \text{with } u(x_1) = 0 \text{ and } u(x_2) = 1,$$

and $v(x)$ satisfies

$$-1 = a(x)\frac{dv}{dx} + \frac{1}{2}b(x)\frac{d^2v}{dx^2}, \qquad \text{with } v(x_1) = v(x_2) = 0;$$

see Karlin and Taylor (1981) for more detail.

We will apply this to the diffusion approximation of the neutral Wright–Fisher model of genetic drift considered in (22.3), which has no mutation and which is called the *random drift model*; in this case, $a(x) = 0$ since u (and hence θ) is equal to 0, and $b(x) = x(1 - x)$. In the random drift model, eventually, one of the two alleles will be lost. We assume that the initial frequency of one of the alleles is x. By solving the respective differential equations, one can show that the probability of fixation of a particular allele is equal to its initial frequency, and that the mean time until fixation in the diffusion limit is $v(x) = -2[x \ln x + (1 - x) \ln(1 - x)]$, $0 < x < 1$. For instance, when $x = 0.5$, $v(x)$ is equal to 1.39; since time is measured in units of $2N$ generations, this means that it takes approximately $2.78N$ generations until fixation when the population size N is large.

22.3.2 The Kolmogorov Forward Equation

One can show that in the diffusion limit, the conditional probability density that the gene frequency is x at time t given that it was p at time 0, denoted by $\chi(p, x; t)$, satisfies the Kolmogorov forward equation or Fokker–Planck equation

$$\frac{\partial \chi(p, x; t)}{\partial t} = \frac{1}{2} \frac{\partial^2}{\partial x^2}[b(x, t)\chi(p, x; t)] - \frac{\partial}{\partial x}[a(x, t)\chi(p, x; t)].$$

This equation is analytically much more tractable and is the starting point for many investigations of quantities of biological interest. In the time-homogeneous case, we simply replace $a(x, t)$ and $b(x, t)$ by $a(x)$ and $b(x)$.

Solving the Fokker–Planck equation for a given initial gene frequency allows one to study how gene frequencies change over time. A particular important case is when the distribution of gene frequencies reaches an equilibrium. In this case, we expect that as t tends to infinity the limit of the conditional probability density $\chi(p, x; t)$ converges, and that the limit does not depend on the initial state p. We denote this limit by $\rho(x)$ (if it exists), which is then called a *stationary density*. In the time-homogeneous case, $\rho(x)$ would then satisfy

$$0 = \frac{1}{2} \frac{d^2}{dx^2}[b(x)\rho(x)] - \frac{d}{dx}[a(x)\rho(x)]. \tag{22.8}$$

The haploid Wright–Fisher model with mutation and selection introduced earlier has an equilibrium. Using $a(x)$ and $b(x)$ in (22.6) and (22.7), we can compute the stationary density by integrating (22.8). This yields

$$\rho(x) = Kx^{\theta-1}(1 - x)^{\theta-1}e^{-\sigma x}, \qquad 0 \le x \le 1, \tag{22.9}$$

where K is a normalizing constant so that $\int_0^1 \rho(x)\,dx = 1$. The density in (22.9) is a special case of Wright's formula (Wright, 1949) and was derived by Kimura (1956) using the diffusion approximation method.

22.3.3 Random Genetic Drift Versus Mutation and Selection

As emphasized above, when a population is finite, the random sampling of gametes introduces a stochastic component into the model. However, the importance of this stochastic component, which we called *genetic drift*, depends on the strength of mutation

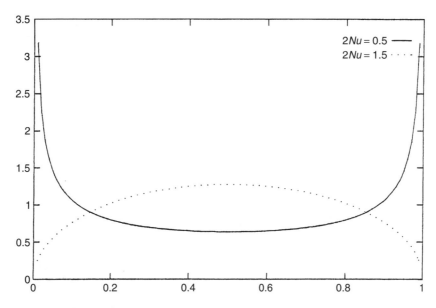

Figure 22.2   The stationary density when $2Nu = 0.5$ and 1.5.

and selection relative to the population size. In general, genetic drift is the more important component in smaller populations.

Let us compare mutation and genetic drift in a population of size N, which evolves according to the one-locus, two-allele haploid Wright–Fisher model with symmetric mutation probability u per gene per generation defined in (22.5). We assume no selection. The stationary density is given by (22.9) when we set $2Nu = \theta$ and $\sigma = 0$. The graph of this function is U-shaped when $2Nu \leq 1$ and unimodal with a peak at $\frac{1}{2}$ when $2Nu \geq 1$ (see Figure 22.2). We see that if $Nu \ll 1$, the population is typically fixed for one or the other allele, meaning that genetic drift is the dominant force; whereas, if $Nu \gg 1$, both alleles will be simultaneously present, meaning that mutational forces are dominant.

To compare random genetic drift versus selection, we again assume the one-locus two-allele Wright–Fisher model with symmetric mutation defined in (22.5), as above, but now we allow allele A_2 to have a selective advantage $1 + s$ over allele A_1. The stationary density is given by (22.9) with $\theta = 2Nu$ and $\sigma = 2Ns$. One can show that if $Ns \ll 1$, selection does not have much of an effect, and the population behaves almost as neutral; whereas if $Ns \gg 1$, selection strongly biases the distribution of alleles towards the favored type, implying that selective forces are dominant (see Figure 22.3).

22.4   THE INFINITE ALLELE MODEL

The models considered so far have a simple genetic structure: One locus with two alleles and mutation changing one allele into the other type and back. This is not a realistic assumption since a gene consists of a large number of nucleotides and thus a mutation occurring at one nucleotide site will likely not result in a type already present in the population, but rather in a novel allele. This prompted Kimura and Crow (1964) to

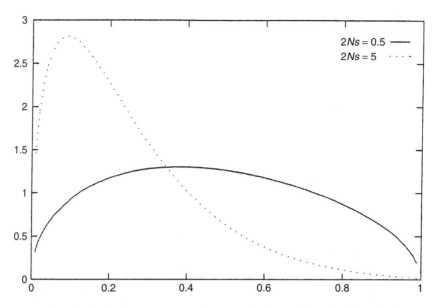

Figure 22.3 The stationary density when $2Nu = 1.5$, $2Ns = 0.5$ and 5.

introduce a new model, the model of infinitely many alleles, or (for short) the infinite allele model.

22.4.1 The Infinite Allele Model with Mutation

The population consists of N diploid individuals and evolves according to a one-locus, neutral Wright–Fisher model with nonoverlapping generations. Mutations occur with probability u per gene per generation, but now every mutation results in a new allele not previously seen in the population.

It is interesting to note that this model was introduced before Kimura proposed his neutral theory; in fact, Kimura and Crow point out that '[i]t is not the purpose of this article to discuss the plausibility of such a system'. Their goal was rather to determine the number of alleles that can be maintained in a population under the extreme case that each mutation would generate a novel allele. This would then provide an upper bound for situations in which mutations could also result in alleles that are already present in the population.

A frequently used measure of genetic diversity is the probability of sampling two alleles of the same type. In the infinite allele model, each allele arises only once, and, therefore, genes in a homozygous individual are identical by descent. If we denote by $F(m)$ the probability that two gametes are identical by descent at generation m, then

$$F(m+1) = \left[\frac{1}{2N} + \left(1 - \frac{1}{2N}\right) F(m)\right] (1 - u)^2.$$

In equilibrium, $F(m)$ does not depend on m. Denoting the equilibrium by $\hat{F}$, we find

$$\hat{F} = \frac{(1-u)^2}{2N - (1-u)^2(2N-1)}.$$

The infinite allele model is best studied in the diffusion limit when time is measured in units of $2N$ generations. We set $\theta = 4Nu$ and let N tend to infinity. We find

$$\hat{F} = \frac{1}{\theta + 1}.$$

We conclude from this that in equilibrium the frequency of homozygotes is a decreasing function of mutation pressure.

Since new alleles are constantly introduced into the population and other alleles get lost due to random drift, the actual alleles present are changing over time – even in equilibrium. That is, in equilibrium, the distribution of a particular gene is not stable, but rather the number of alleles attains a steady state. In addition, if we look at the most common allele (whose type will change over time), the second most common allele, and so on, we would find that their frequencies are stable in equilibrium.

22.4.2 Ewens's Sampling Formula

Ewens (1972) investigated the equilibrium properties of samples taken from a population that evolves according to the infinite allele model. He defined the allelic partition of a sample: Denote by a_i the number of alleles present exactly i times in a sample. The vector $(a_1, a_2, \ldots)$ then denotes the allelic partition. If the sample size is n, then $a_{n+1} = a_{n+2} = \cdots = 0$. The number of different alleles in a sample of size n, K_n, is then

$$K_n = \sum_{i=1}^{n} a_i,$$

and the sample size $n = \sum_{i=1}^{n} i a_i$. Ewens obtained the distribution of the allelic partition of a sample in equilibrium in the diffusion limit,

$$P_\theta(a_1, a_2, \ldots, a_n) = \frac{n!}{\theta_{(n)}} \prod_{j=1}^{n} \left(\frac{\theta}{j}\right)^{a_j} \frac{1}{a_j!},$$

where $\theta = 4Nu$ and $\theta_{(n)} = \theta(\theta + 1) \cdots (\theta + n - 1)$. (Karlin and McGregor, 1972, gave a formal derivation of this formula.)

An interesting conclusion of this formula is that under neutrality, alleles are not equally likely in equilibrium, but rather the partition is quite lopsided, with a few common alleles and all others relatively uncommon.

Ewens's sampling formula allows one to find the probability distribution of the number of alleles in a sample of size n. Let K_n denote the random variable that counts the number of alleles in a sample of size n in equilibrium; then

$$E[K]_n = \sum_{j=1}^{n} \frac{\theta}{\theta + j - 1}.$$

For large n, this is asymptotically $\theta \ln n$. Furthermore, the variance of K_n for n large is asymptotically $\theta \ln n$ as well.

22.4.3 The Infinite Allele Model with Selection and Mutation

Ethier and Kurtz (1987; 1994) gave a general version of the infinite allele model with selection. Again, the population is diploid and of size N ($2N$ gametes). To distinguish alleles from each other, we assign each allele a number chosen at random from the interval $(0, 1)$. (This also ensures that each mutation results in a novel allele.) Denote the type of the ith gene by x_i. In each generation, $2N$ gametes are chosen in pairs so that the probability that the ith and the jth gene are selected is

$$P(i, j) = \frac{w_N(x_i, x_j)}{\sum_{1 \leq l, m \leq 2N} w_N(x_l, x_m)},$$

where

$$w_N(x, y) = 1 + \frac{1}{2N} \sigma(x, y),$$

and $\sigma(x, y)$ is a symmetric function of x and y. In the next step, one of the two genes is chosen at random and subjected to mutation; it remains the same type with probability $1 - u$ and mutates to a novel type with probability u, chosen at random from $(0,1)$. This gene is then one of the $2N$ gametes in the next generation.

Ethier and Kurtz (1994) showed that the stationary density of the infinite allele model with selection described above is absolutely continuous with respect to the stationary density of the neutral infinite allele model. Joyce (1995) used this result to show that, for large sample sizes, the allele counts and the total number of alleles under this selection scheme are nearly the same as under neutrality. This convergence, however, is rather slow, and, for small sample sizes, selection has a substantial effect (Li, 1977).

22.5 OTHER MODELS OF MUTATION AND SELECTION

We include three more frequently used models which include mutation and/or selection. The last model mentioned here is a model with overlapping generations.

22.5.1 The Infinitely Many Sites Model

This model was introduced by Watterson (1975) as a model to approximate DNA sequences (see also Ethier and Griffiths, 1987; Griffiths, 1989). Each individual is described by a string of infinitely many, completely linked sites. The mutation probability per site is assumed to be very small so that the total number of mutations per individual per generation is finite and one can assume that no backmutations occur. It is enough to keep track of the segregating sites since none of the other sites carries any information. Each individual is thus represented by just a finite string of sites. The assumption of complete linkage means that there is no recombination between different strings. This assumption is typically made for mitochondrial DNA, which is haploid and maternally inherited.

22.5.2 Frequency-dependent Selection

The fitness of a genotype depends on its frequency (and possibly on the frequencies of other genotypes). An example of this type of selection is gametophytic self-incompatibility, found in many flowering plants. A simple model for a haploid population with minority-advantage frequency-dependent selection is as follows (Wright and Dobzhansky, 1946; Clarke, 1976; Takahata and Nei, 1990): We assume one locus with infinitely many alleles. The fitness of allele A_k is $1 - ax_k$, where x_k is the frequency of A_k in the population and a is a positive constant. If $F = \sum_{k=1}^{\infty} x_k^2$, then the allelic type A_k contributes a fraction $x_k/(1 - aF)$ to the next generation. Offspring are of the same type with probability $1 - u$ and of a novel type with probability u. If $\theta = 2Nu$ and $\alpha = 2Na$, then in the diffusion scaling, the drift term is $a(x) = -[\alpha x(x - F) + \theta x]/2$ and the diffusion term is $b(x) = x(1 - x)$. Frequency-dependent selection of this form is a powerful mechanism for the maintenance of a polymorphism.

22.5.3 Overlapping Generations

The Wright–Fisher model assumes that generations are discrete (or nonoverlapping). To model overlapping generations, we need to allow individuals to reproduce asynchronously. Moran (1958; 1962) introduced a haploid model in which reproduction occurs continuously in time: Each individual produces an offspring at an exponential rate depending on its fitness. The offspring then undergoes mutation according to a specified mutation process. To keep the population size constant, the offspring then chooses one of the individuals in the population at random and replaces it. In this way, one can define continuous-time analogs for each discrete-time model. It turns out that the corresponding models have the same diffusion limits under suitable scaling of the parameters in the respective processes.

There are many other models that address various aspects of population dynamics. For instance, there is a large amount of literature on population genetic models that take geographic substructure into account (see **Chapter 28**).

22.6 COALESCENT THEORY

22.6.1 The Neutral Coalescent

Looking at the genealogical relationships of a sample of genes is a powerful way to study population dynamics and to infer population parameters, such as the mutation parameter. This method was introduced by Kingman (1982a; 1982b) for neutral models (see **Chapter 25**). A typical genealogy is shown in Figure 22.4. When going back in time, the ancestral lines coalesce until only one line is left; this is the line of the most recent common ancestor.

An important feature of this process is that the family tree and the mutation process can be treated separately under neutrality. We discuss this for a haploid population of size N. Under neutrality, each individual in generation n chooses one parent at random from the previous generation. The ancestral lines of individuals who choose the same parent coalesce.

The dynamics of this genealogical process is not difficult to derive. We will again take advantage of the diffusion approximation, which will allow us to obtain exact results in the limit as the population size goes to infinity. Under neutrality, to simulate the family tree of a sample of size n, we follow the ancestral line of each individual back in time. Suppose now that there are $j \leq n$ ancestral lines left at some time in the past. Each line will choose one ancestor at random from the population. If the population size is

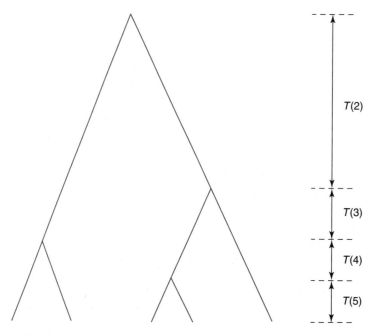

Figure 22.4   The genealogical relationship of a sample of genes.

N, then the probability that the j genes will have no common ancestors in the previous generation is

$$\prod_{i=1}^{j-1}\left(1-\frac{i}{N}\right)=1-\frac{\binom{j}{2}}{N}+O\left(\frac{1}{N^2}\right).$$

To prepare for the diffusion limit, we will measure time in units of N generations. If $T(j)$ denotes the time between the coalescing events where the size of the genealogy goes from j to $j-1$ measured in units of N generations. Then for any $t>0$,

$$P(T(j)>t)=\left(\prod_{i=1}^{j-1}\left(1-\frac{i}{N}\right)\right)^{Nt}\longrightarrow \exp\left[-\binom{j}{2}t\right],$$

as N tends to infinity. We find that in the diffusion limit only pairwise coalescence events occur, and that $T(j)$ is exponentially distributed with parameter $\binom{j}{2}$. The resulting process is called *the coalescent* (Figure 22.4).

An important quantity is the time it takes until a sample of size n is traced back to its most recent common ancestor (MRCA). If we denote this (random) time by $T_{\text{MRCA}}(n)$, then

$$T_{\text{MRCA}}(n)=T(n)+T(n-1)+\cdots+T(2).$$

Since $T(j)$ is exponentially distributed with parameter $\binom{j}{2}$, the expected value of $T(j)$ is $1/\binom{j}{2}$, and hence,

$$E[T_{\mathrm{MRCA}}](n) = \sum_{j=2}^{n} E[T(j)] = \sum_{j=2}^{n} \frac{2}{j(j-1)}$$

$$= 2 \sum_{j=2}^{n} \left(\frac{1}{j-1} - \frac{1}{j} \right) = 2 \left(1 - \frac{1}{n} \right).$$

From

$$E[T(2)] = 1 \quad \text{and} \quad \lim_{n \to \infty} E[T_{\mathrm{MRCA}}(n)] = 2,$$

we conclude that, on average, the amount of time it takes to reach the MRCA for a very large sample is only about twice that for a sample of size 2.

Using independence, we can compute the variance of $T_{\mathrm{MRCA}}(n)$. We find

$$\mathrm{var}(T_{\mathrm{MRCA}}(n)) = \sum_{j=2}^{n} \mathrm{var}(T(j)) = \sum_{j=2}^{n} \left(\frac{2}{j(j-1)} \right)^2$$

$$= 8 \sum_{j=1}^{n-1} \frac{1}{j^2} - 4 \left(1 - \frac{1}{n} \right) \left(3 + \frac{1}{n} \right)$$

$$\longrightarrow \frac{8\pi^2}{6} - 12 \approx 1.16 \qquad \text{as } n \to \infty.$$

Since $\mathrm{var}(T(2)) = 1$, we see that the coalescence time $T(2)$ has by far the biggest contribution to the variance.

The mutation process is superimposed on the coalescent. Using the same scaling as in the previous section, namely $\theta = 2Nu$, and measuring time in units of N generations, the mutation process can be described by a Poisson process that puts down mutation events independently on all branches at rate $\theta/2$. It is straightforward to include other mutation schemes in the model. For instance, the coalescent for the neutral infinite allele model has the same structure but we stipulate that at mutation events a novel allele is created.

To simulate a sample of size n, we thus first simulate the genealogical tree, then put mutations on the tree, and finally assign a type to the MRCA and run the process down the tree to obtain a sample of size n. For further details on the neutral coalescent theory, see **Chapter 25**.

It is important to realize that simulating samples of a given size is different from making inferences of population parameters based on a given sample. The coalescent is used for statistical inference; this is described in **Chapter 26**.

22.6.2   The Ancestral Selection Graph

The argument above relied crucially on the assumption that individuals choose their offspring *at random* in the previous generation. This no longer holds under selection. Initial attempts to include selection in the genealogical approach assumed that the gene frequencies were known at all times in the past (Hudson and Kaplan, 1988; Hudson *et al.*, 1988); this is covered in detail in **Chapter 25**. This approach is successful if the population is in equilibrium and the stationary distribution is known. It turns out, however, that this assumption is not needed when selection is weak. The ancestral selection graph (Neuhauser and Krone, 1997; Krone and Neuhauser, 1997) provides a framework that allows one to study genealogies of samples under selection without knowing the gene frequencies at all times in the past.

We will show how to obtain genealogies for samples under selection using the haploid model for a population of size N, which was defined in (22.5). We follow one locus with two alleles, A_1 and A_2, and assume symmetric mutation with mutation probability u per gene per generation. One of the alleles (A_2) has a selective advantage s.

The genealogy under selection is embedded in a graph, called the *ancestral selection graph*, which has both coalescing and branching events (see Figure 22.5). The coalescing events have the same interpretation as in the neutral case, namely, two ancestral lines join together at the time of their common ancestor. The branching events result in additional ancestral lines. Since the number of actual ancestral lines cannot increase, they constitute potential ancestral lines, reflecting the fact that individuals with a higher fitness have more offspring.

Figure 22.6 illustrates ancestral lines and explains why branching events occur in the presence of selection. If 3 has a selective advantage, then 2' and 3' have a common ancestor, namely 3; whereas if 3 does not have a selective advantage, then 2 is the ancestor of 2' and 3 is the ancestor of 3'. Since, when constructing genealogies, the gene frequencies in the past are not known, both possibilities need to be carried back until all ancestral lines coalesce and one ancestor, called the *ultimate ancestor* (UA), results. Knowing the type of the UA then allows one to extract the genealogy from the ancestral selection graph. This is illustrated in Figure 22.7, where type 2 has a selective advantage over type 1. We stipulate that a selectively advantageous type on an incoming branch displaces the type on the continuing branch. We see from Figure 22.7 that the MRCA is not necessarily the UA.

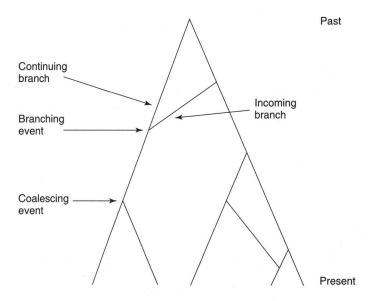

Figure 22.5 The ancestral selection graph with its coalescing and branching structure.

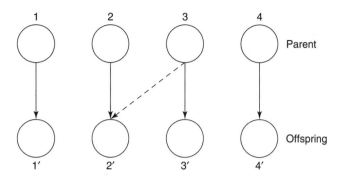

Figure 22.6 Branching events come from differential reproductive success.

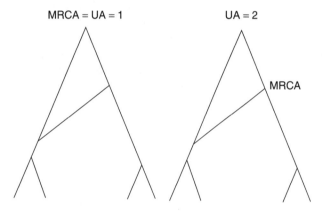

Figure 22.7 Extracting the embedded genealogy from the ancestral selection graph.

The ancestral selection graph can be defined for a large class of models (Neuhauser and Krone, 1997; Neuhauser, 1999), but only in the diffusion limit do we obtain a graph that has a simple structure. The diffusion limit requires that selection is weak, i.e. s is of the order of the reciprocal of the population size (more precisely, $\lim_{N \to \infty} 2Ns = \sigma$). In this limit, the ancestral selection graph for the model defined in (22.5) has the following dynamics. If there are j branches, then

$$\text{coalescing: } j \to j - 1 \quad \text{at rate } \binom{j}{2};$$

$$\text{branching: } j \to j + 1 \quad \text{at rate } \frac{\sigma}{2} j.$$

It follows that coalescing events occur at the same rate as in the neutral coalescent, but, in addition, branching events occur.

The ancestral selection graph allows one to separate the mutation process from the genealogical process as in the neutral case provided the mutation parameter scales with the reciprocal of the population size (i.e, $\lim_{N \to \infty} 2Nu = \theta$). The mutation process is superimposed on the graph: Mutation events occur according to a Poisson process with rate $\theta/2$ independently along each branch of the ancestral selection graph.

The ancestral selection graph described here is for the simplest case of a one-locus two-allele model with symmetric mutation. It can be extended to many other models of selection, including diploid models (Neuhauser and Krone, 1997) and frequency dependent selection (Neuhauser, 1999). The graphs are somewhat more complicated in their branching structure, but have straightforward and intuitive interpretations.

Straightforward simulations of this process are computationally intensive. Recent advances (Slade, 2000a; 2000b; Stephens and Donnelly, 2003; Barton *et al.*, 2004) have led to algorithms that are much less computationally intensive. These advances rely on modeling genealogies conditioned on allele frequencies in the sample.

22.6.3   Varying Population Size

So far, we have always assumed that the population size is constant. There is ample evidence, for instance, that the human population underwent a large expansion in the past. The literature on changing populations sizes, and in particular on bottlenecks, is extensive (Nei *et al.*, 1977; Watterson, 1984; 1989; Slatkin and Hudson, 1991; Rogers and Harpending, 1992; Marjoram and Donnelly, 1994; random environment models are discussed in Donnelly, 1986). It is not difficult to incorporate a changing population size into the coalescent. This was done for the neutral coalescent in Griffiths and Tavaré (1994) and extended to the selection case in Neuhauser and Krone (1997).

To derive the coalescent with varying population size, we begin with the neutral Wright–Fisher model for a haploid population whose size at time 0 (the present) is equal to $N = N(0)$. Denote by $T_N(2)$ the coalescence time of two genes. Then

$$P(T_N(2) > \tau) = \prod_{j=1}^{\tau} \left(1 - \frac{1}{N(j)}\right),$$

where $N(j)$ is the population size j generations in the past. Measuring time in units of N generations, we find, in the limit as N tends to infinity,

$$\lim_{N\to\infty} P(T_N(2) > \lfloor Nt \rfloor) = \lim_{N\to\infty} \prod_{j=1}^{\lfloor Nt \rfloor} \left(1 - \frac{1}{N(j)}\right)$$

$$= \exp\left[-\int_0^t \lambda(u)\,du\right],$$

where $\lambda(u)$, the coalescent intensity function, is defined as follows: For $N(j)$ large,

$$\sum_{j=1}^{\lfloor Nt \rfloor} \log\left(1 - \frac{1}{N(j)}\right) \approx -\sum_{j=1}^{\lfloor Nt \rfloor} \frac{1}{N(j)}.$$

If we define $f_N(x) = N(j)/N$ for $x = j/N$, then

$$\lim_{N\to\infty} \sum_{j=1}^{\lfloor Nt \rfloor} \frac{1}{N(j)} = \lim_{N\to\infty} \sum_{x=1/N}^{t} \frac{1}{f_N(x)} \frac{1}{N} = \int_0^t \lambda(u)\,du.$$

That is, $\lambda(u) = 1/f(u)$ with $f(u) = \lim_{N\to\infty} f_N(u)$. If we define

$$\Lambda(t) = \int_0^t \lambda(u) \, du,$$

then, as $N \to \infty$, $T_N(2)/N$ converges in distribution to a random variable $T(2)$ with

$$P(T(2) > t) = \exp[-\Lambda(t)].$$

We can now define the coalescent with varying population size. We conclude from the above calculation that at the rescaled time t in the past, each pair of branches coalesces at rate $\lambda(t)$, where $\lambda(t)$ is the coalescent intensity function defined above. If there are j branches present at time t in the past, then a coalescing event occurs at rate $\binom{j}{2}\lambda(t)$. Note that if the population size is constant, i.e. $N(j) = N$ for all $j \geq 0$, then $\lambda(t) = 1$ for all $t \geq 0$, and we obtain the neutral coalescent for fixed population size. A population that expanded rapidly leads to a star-shaped genealogy.

Mutation and selection are not affected by varying population sizes. As in the case of fixed population size, mutation events occur at rate $\theta/2$ along each branch, and branching events occur at rate $j\sigma/2$ if j branches are present.

22.7   DETECTING SELECTION

Explaining molecular differences between individuals and determining the causes of molecular evolution remain a challenging task. Darwin emphasized the role of natural selection as the driving force of evolution, whereas Kimura championed the neutral theory. The availability of molecular data makes it possible to study genetic diversity directly, which should help to resolve the neutrality–selectionist controversy.

A number of statistical tests have been devised to detect selection. This section focuses on tests that are based on coalescent theory, or can at least be understood using coalescent theory. **Chapter 12** discusses statistical methods for phylogenetic analysis of protein coding DNA sequences; this method relies on the ratio of synonymous versus nonsynonymous substitutions.

The null hypothesis for statistical tests of selection typically assumes that the population is neutral. If the observed data deviate too much from what is expected under neutrality, the hypothesis of neutrality is rejected. However, this does not tell one what type of selection acted on the population or if the deviation from the neutral expectation resulted from selection or other forces – for instance, changes in population size, temporally varying environments, or other factors.

One of the first tests was Watterson's (1978) homozygosity test of neutrality. This is based on Ewens's sampling formula and tests whether the observed homozygosity in the sample agrees with that predicted by Ewens's sampling formula. Neither population size nor the mutation parameter needs to be known to apply this test since sample size and the number of distinct alleles in the sample are sufficient to compute the probability of a particular allelic composition given the number of alleles observed in the sample. This test, however, is not very powerful (Gillespie, 1991).

Watterson's test works for allozyme data, which were the prevalent data available then. As DNA sequence data became widespread, there was a need for statistical tests based on this type of data. Hudson *et al.* (1988) developed a test, the HKA test that compares regions of the genome of two species. This test was developed to detect polymorphism that is maintained by balancing selection.

Tajima (1989) developed a test for DNA sequence data that compares the number of segregating sites and the average number of pairwise nucleotide differences. Under neutrality, the expected number of segregating sites S in a sample of size n is given by (Watterson, 1975)

$$E[S] = \theta \sum_{k=1}^{n-1} \frac{1}{k}, \tag{22.10}$$

where $\theta = 4Nu$, N is the (diploid) population size and u is the mutation probability per gene per generation. The expected number of pairwise nucleotide differences K is given by (Tajima, 1983)

$$E[K] = \theta, \tag{22.11}$$

where $\theta = 4Nu$ as above.

Selection affects these two quantities differently. The number of segregating sites ignores allele frequencies but the presence of deleterious alleles, which occur in low frequencies, affects this number strongly. The number of pairwise nucleotide differences, on the other hand, is primarily affected by allele frequencies and this is less sensitive to deleterious alleles, which occur at low frequencies.

Both (22.10) and (22.11) suggest a method for estimating θ under neutrality. If the neutrality assumption is violated, however, the two methods should yield different estimates of θ, as explained above. The difference is measured by

$$d = \hat{K} - \frac{\hat{S}}{c_n},$$

where $\hat{K}$ and $\hat{S}$ are the observed pairwise nucleotide differences and the observed number of segregating sites, respectively, and $c_n = \sum_{k=1}^{n-1} k^{-1}$. Tajima (1989) computed the variance of d and derived an estimate of var(d), $\hat{V}(d)$, based on the observed number of segregating sites. He then proposed the test statistic

$$D = \frac{d}{\sqrt{\hat{V}(d)}}.$$

Though Tajima did not base his test on the coalescent, properties of both S and K can be computed using the coalescent. Namely, under the infinite sites model, the number of segregating sites, S, is equal to the total number of mutations in the genealogy of the sampled genes. The total branch length of the genealogical tree is $T_{\text{tot}} = \sum_{j=2}^{n} jT(j)$, where $T(j)$ is exponentially distributed with mean $1/\binom{j}{2}$. Mutations occur independently along branches according to a Poisson process with rate $\theta/2$. Hence,

$$E[S] = \frac{\theta}{2} \sum_{j=2}^{n} j \frac{1}{\binom{j}{2}} = \theta \sum_{j=2}^{n} \frac{1}{j-1} = \theta \sum_{k=1}^{n-1} \frac{1}{k},$$

as in (22.10). The pairwise nucleotide differences can also be derived using the coalescent. Since a site is affected by a mutation at most once under the infinite site model, a pairwise difference occurs if a mutation occurs on either lineage to their common ancestor. Hence,

$$E[K] = \frac{\theta}{2} 2E[T(2)] = \theta,$$

as in (22.11).

A test that is directly based on the coalescent is the test by Fu and Li (1993). They divide the coalescent into external and internal branches (an external branch is a segment that ends in a tip on the genealogy). External branches correspond to younger parts of the tree, internal branches to older parts of the tree. In the case of purifying selection, they expect an excess number of mutations in the external branches since deleterious alleles occur at low frequencies, whereas in the case of balancing selection they expect the opposite. In essence, their test is based on the assumption that branch lengths in the genealogy are affected by selection.

All tests introduced above are based on the assumption that selection will have an effect on the proposed test statistics. Since it is typically not possible to compute the statistical properties of the proposed test statistics under alternatives, a frequent approach is to use simulations to assess the power of proposed tests. This was done, for instance, in Simonsen *et al.* (1995) for Tajima's test statistic and the test proposed by Fu and Li (1993). They simulated alternative hypotheses, such as selective sweeps, population bottlenecks, and population subdivision. In general, they found that Tajima's test was more powerful.

In Golding (1997) and Neuhauser and Krone (1997) the effects of selection on branch lengths were investigated. They found that the shape of the genealogy (i.e. both its topology and the coalescence times) under purifying selection differs only little from that expected under neutrality. Przeworski *et al.* (1999) confirmed their results. Slade (2000a) investigated the effects of selection on branch lengths using the enhanced algorithm of constructing graphs conditioned on the sample. He found as well that the effect of selection on branch lengths is weak. Barton and Etheridge (2004) investigated the effects of both purifying and balancing selection and found that branch lengths are only significantly altered under selection if selection is very strong and deleterious mutations are common in the case of purifying selection or mutation is weak in the case of balancing selection. Overdominant (balancing) selection changes the genealogy (Kaplan *et al.*, 1988). Thus, when using the shape of the genealogy to construct a statistical test for selection, the power of the test depends rather crucially on the alternative.

Acknowledgments

The author was partially supported by National Science Foundation grant DMS-97-03694 and DMS-00-72262.

REFERENCES

Barton, N.H. and Etheridge, A.M. (2004). The effect of selection on genealogies. *Genetics* **166**, 1115–1131.

Barton, N.H., Etheridge, A.M. and Sturm, A.K. (2004). Coalescence in a random background. *Annals of Applied Probability* **14**, 754–785.

Clarke, B. (1976). *Genetic Aspects of Host – Parasite Relationships*, A.E.R. Taylor and R.M. Muller, eds. Blackwell, Oxford, pp. 87–103.

Darwin, C. (1985). *The Origin of Species*. Penguin, Harmondsworth, First published in 1859.

Donnelly, P. (1986). A genealogical approach to variable-population size models in population genetics. *Journal of Applied Probability* **23**, 283–296.

Ethier, S.N. and Griffiths, R.C. (1987). The infinitely-many sites model as a measure valued diffusion. *Annals of Probability* **15**, 515–545.

Ethier, S.N. and Kurtz, T.G. (1987). The infinitely-many alleles model with selection as a measure-valued diffusion. In *Stochastic Methods in Biology, Lecture Notes in Biomathematics, Vol. 70*, M. Kimura, G. Kallianpur and T. Hida, eds. Springer-Verlag, Berlin, pp. 72–86.

Ethier, S.N. and Kurtz, T.G. (1994). Convergence to Fleming-Viot processes in the weak atomic topology. *Stochastic Processes and their Applications* **54**, 1–27.

Ewens, W.J. (1972). The sampling theory of selectively neutral alleles. *Theoretical Population Biology* **3**, 87–112.

Ewens, W.J. (2004). *Mathematical Population Genetics: Theoretical Introduction*, Vol. 1, 2nd Revised edition, Springer-Verlag.

Fisher, R.A. (1922). On the dominance ratio. *Proceedings of the Royal Society of Edinburgh* **42**, 321–341.

Fisher, R.A. (1930). *The Genetical Theory of Natural Selection*, 1st edition. Clarendon, Oxford.

Fu, Y.-X. and Li, W.-H. (1993). Statistical tests of neutrality of mutations. *Genetics* **133**, 693–709.

Gillespie, J.H. (1991). *The Causes of Molecular Evolution*. Oxford University Press, Oxford.

Golding, G.B. (1997). The effect of purifying selection on genealogies. In *Progress in Population Genetics and Human Evolution, IMA Volumes in Mathematics and Its Applications, Vol. 87*, P. Donnelly and S. Tavaré, eds. Springer-Verlag, New York, pp. 271–285.

Griffiths, R.C. (1989). Genealogical-tree probabilities in the infinitely-many sites model. *Journal of Mathematical Biology* **27**, 667–680.

Griffiths, R.C. and Tavaré, S. (1994). Sampling theory for neutral alleles in a varying environment. *Philosophical Transactions of the Royal Society of London Series B* **334**, 403–410.

Haldane, J.B.S. (1932). *The Causes of Evolution*. Longmans, Green, London.

Hudson, R.R. and Kaplan, N.L. (1988). The coalescent process in models with selection and recombination. *Genetics* **120**, 831–840.

Hudson, R.R., Kreitman, M. and Aguadé, M. (1988). A test of neutral molecular evolution based on nucleotide data. *Genetics* **116**, 153–159.

Joyce, P. (1995). Robustness of the Ewens sampling formula. *Journal of Applied Probability* **32**, 609–622.

Kaplan, N.L., Darden, T. and Hudson, R.R. (1988). The coalescent process in models with selection. *Genetics* **120**, 819–829.

Karlin, S. and McGregor, J. (1972). Addendum to a paper by W. Ewens. *Theoretical Population Biology* **3**, 113–116.

Karlin, S. and Taylor, H.M. (1975). *A First Course in Stochastic Processes*. Academic Press, San Diego, CA.

Karlin, S. and Taylor, H.M. (1981). *A Second Course in Stochastic Processes*. Academic Press, San Diego, CA.

Kimura, M. (1956). Stochasting processes in population genetics. Ph.D. thesis, University of Wis-consin, Madison.

Kimura, M. (1964). Diffusion models in population genetics. *Journal of Applied Probability* **1**, 177–232.

Kimura, M. (1968a). Evolutionary rate at the molecular level. *Nature* **217**, 624–626.

Kimura, M. (1968b). Genetic variability maintained in a finite population due to mutational production of neutral and nearly neutral isoalleles. *Genetical Research* **11**, 247–269.

Kimura, M. (1977). Preponderance of synonymous changes as evidence for the neutral theory of molecular evolution. *Nature* **267**, 275–276.

Kimura, M. (1979). Model of effectively neutral mutations in which selective constraint is incorporated. *Proceedings of the National Academy of Sciences of the United States of America* **76**, 3440–3444.

Kimura, M. (1983). *The Neutral Theory of Molecular Evolution*. Cambridge University Press, Cambridge.

Kimura, M. and Crow, J.F. (1964). The number of alleles that can be maintained in a finite population. *Genetics* **49**, 725–738.

Kimura, M. and Ohta, T. (1973). Mutation and evolution at the molecular level. *Genetics (Supplement)* **73**, 19–35.

King, J.L. and Jukes, T.H. (1969). Non-Darwinian evolution: random fixation of selectively neutral mutations. *Science* **164**, 788–798.

Kingman, J.F.C. (1982a). The coalescent. *Stochastic Processes and their Applications* **13**, 235–248.

Kingman, J.F.C. (1982b). On the genealogy of large populations. *Journal of Applied Probability* **19A**, 27–43.

Kolmogorov, A. (1931). Über die analytische methoden inder Wahrscheinlichkeit-srechnung. *Mathematische Annalen* **104**, 415–458.

Krone, S.M. and Neuhauser, C. (1997). Ancestral processes with selection. *Theoretical Population Biology* **51**, 210–237.

Li, W.-H. (1977). Maintenance of genetic variability under mutation and selection pressures in a finite population. *Proceedings of the National Academic of Sciences of the United States of America* **74**, 2509–2513.

Marjoram, P. and Donnelly, P. (1994). Pairwise comparison of mitochondrial DNA sequences in subdivided populations and implications for early human evolution. *Genetics* **136**, 673–683.

Moran, P.A.P. (1958). Random processes in genetics. *Proceedings of the Cambridge Philosophical Society* **54**, 60–71.

Moran, P.A.P. (1962). *The Statistical Processes of Evolutionary Theory*. Clarendon Press, Oxford.

Nei, M. (1987). *Molecular Evolutionary Genetics*. Columbia University Press, New York.

Nei, M., Maruyama, T. and Chakraborty, R. (1977). The bottleneck effect and genetic variability in populations. *Evolution* **29**, 1–10.

Neuhauser, C. (1999). The ancestral selection graph and gene genealogy under frequency-dependent selection. *Theoretical Population Biology* **56**, 203–214.

Neuhauser, C. and Krone, S.M. (1997). The genealogy of samples in models with selection. *Genetics* **145**, 519–534.

Przeworski, M., Charlesworth, B. and Wall, J.D. (1999). Genealogies and weak purifying selection. *Molecular Biology and Evolution* **16**, 246–252.

Rogers, A. and Harpending, H. (1992). Population growth makes waves in the distribution of pairwise genetic differences. *Molecular Biology and Evolution* **9**, 552–569.

Simonsen, K.L., Churchill, G.A. and Aquadro, C.F. (1995). Properties of statistical tests of neutrality for DNA polymorphism data. *Genetics* **141**, 413–429.

Slade, P.F. (2000a). Simulation of selected genealogies. *Theoretical Population Biology* **57**, 35–49.

Slade, P.F. (2000b). Most recent common ancestor probability distributions in gene genealogies under selection. *Theoretical Population Biology* **58**, 291–305.

Stephens, M. and Donnelly, P. (2003). Ancestral inference in population genetics models with selection. *Australian and New Zealand Journal of Statistics* **45**, 395–430.

Slatkin, M. and Hudson, R.R. (1991). Pairwise comparisons of mitochondrial DNA sequences in stable and exponentially growing populations. *Genetics* **129**, 555–562.

Tajima, F. (1983). Evolutionary relationship of DNA sequences in finite populations. *Genetics* **105**, 437–460.

Tajima, F. (1989). Statistical method for testing the neutral mutation hypothesis by DNA polymorphism. *Genetics* **123**, 585–595.

Takahata, N. and Nei, M. (1990). Allelic genealogy under overdominant and frequency-dependent selection and polymorphism of major histocompatibility complex loci. *Genetics* **124**, 967–978.

Watterson, G.A. (1975). On the number of segregating sites in genetical models without recombination. *Theoretical Population Biology* **7**, 256–276.

Watterson, G.A. (1978). The homozygosity test of neutrality. *Genetics* **88**, 405–417.

Watterson, G.A. (1984). Allele frequencies after a bottleneck. *Theoretical Population Biology* **26**, 387–407.

Watterson, G.A. (1989). The neutral allele model with bottlenecks. In *Mathematical Evolutionary Theory*, M.W. Feldman, ed. Princeton University Press, Princeton, NJ, pp. 26–40.

Wright, S. (1931). Evolution in Mendelian populations. *Genetics* **16**, 97–159.

Wright, S. (1949). In *Genetics, Paleontology and Evolution*, G.L. Jepson, G.G. Simpson and E. Mayr, eds. Princeton University Press, Princeton, NJ, pp. 365–389.

Wright, S. and Dobzhansky, T. (1946). Genetics of natural populations. XII. Experimental reproduction of some of the changes caused by natural selection in certain populations of *Drosophila pseudoobscura*. *Genetics* **31**, 125–145.

23

Inference, Simulation and Enumeration of Genealogies

C. Cannings

Division of Genomic Medicine, University of Sheffield, Sheffield, UK

and

A. Thomas

Department of Biomedical Informatics, University of Utah, Salt Lake City, UT, USA

If we confine our attention to species, such as humans, with two sexes and no selfing, a genealogy is simply a set of individuals and the specification of two binary relations, *mother of* and *father of*. Of course, each individual has at most one mother and one father specified and the set must observe the restrictions imposed by the temporal aspects. This can be most conveniently specified as a list of individual – father – mother triplets with either or both of the latter two fields allowed to be null. Genealogies lend themselves very naturally to graphical representation and anyone involved in genealogical work or plant and animal genetics will likely have drawn a pedigree at some stage to keep track of relationships. These relationships follow a logic that allows us to represent and quantify correlations between pairs or sets of individuals, or their genes, and thus lets us develop algebras of relationship that we discuss here. We also consider two representations of genealogies as graphs that are of primary importance in genetics, discuss their properties and significance and consider the correspondence between them.

23.1 GENEALOGIES AS GRAPHS

Perhaps the neatest way of representing a genealogy is as a *marriage node graph* (Figure 23.1). This is a directed graph $\mathbf{G} = \mathbf{G}(\mathbf{V}, \mathbf{E})$, where the set of nodes $\mathbf{V}$ is partitioned into three subsets: $\mathbf{M}$ the marriages, $\mathbf{f}$ the females and $\mathbf{m}$ the males. The union of $\mathbf{f}$ and $\mathbf{m}$ constitutes $\mathbf{I}$, the set of individuals. The edges, $\mathbf{E}$, which are directed, are similarly partitioned into three subsets, $\mathbf{fM}$ with each element from a female to a marriage, $\mathbf{mM}$ with each element from a male to a marriage and $\mathbf{R}$ with each element

Handbook of Statistical Genetics, Third Edition. Edited by D.J. Balding, M. Bishop and C. Cannings.
© 2007 John Wiley & Sons, Ltd.   ISBN: 978-0-470-05830-5.

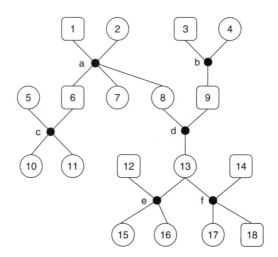

Figure 23.1 A marriage node graph with 18 individuals linked by six marriage nodes. Males are conventionally represented by rectangles, and females by circles. Edges are directed from top to bottom.

from a marriage to an individual, either a male or a female. The restrictions required to make this directed graph a valid genealogy (see Figure 23.1) is simply that (1) there are no cycles, (2) a marriage node may have at most one input edge in **fM** and at most one in **mM**, and (3) an individual node has at most one input in **R**. This representation is well suited for use when drawing a genealogy, particularly when it is complex with many loops. When edges are drawn as angled brackets, instead of straight lines, we see that the result is a picture in the more familiar traditional format, as shown in Figure 23.2.

The second representation is the *moral graph*. This is an undirected graph with a vertex for each individual, edges connecting individuals to their parents, and their parents to each other, so that it is the union of the triangles connecting offspring – father – mother triplets. Although not usually as clear as the marriage node graph when drawn, the structure of the moral graph is important in determining the computational complexity of probability calculations made on genealogies. The name comes from the graphical modelling field where, initially, a directed graph is defined with connections to a dependent, or *daughter*, variable from the variables on which it depends: the *parent* variables. This graph is then *moralised* by connecting, or *marrying*, the parents and discarding the information on direction (Lauritzen and Spiegelhalter, 1988). Although for probability calculations on pedigrees the vertices represent variables for properties of individuals rather than the individuals themselves, the parent – offspring analogy used generally in graphical modelling, in effect, becomes literal for pedigree analysis. An example is given in Figure 23.3.

23.2 RELATIONSHIPS

23.2.1 The Algebra of Pairwise Relationships

It was pointed out earlier that for a genealogy we need only two binary relationships, *mother of* and *father of*, and everything else should follow from that. In practice of

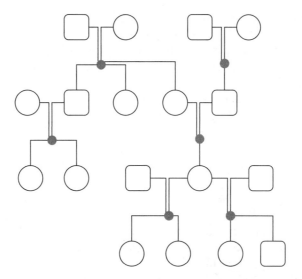

Figure 23.2   The marriage node graph in Figure 23.1 rendered in the more traditional format. As pedigrees get more looped and complex, this format quickly becomes unreadable, and the format in Figure 23.1 is clearer.

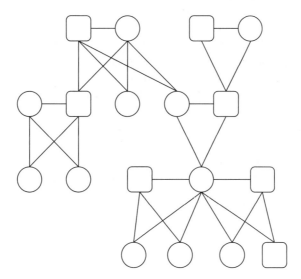

Figure 23.3   The moral graph for the pedigree shown in Figure 23.1.

course there are many other relationships which are defined, though these vary greatly from society to society. The study of relationships, or kinship, is of major import in anthropology, since many of the obligations in societies of all types are based on kinship, though the match between terminology and kinship as represented by the genealogy is not always precise. However, the focus there is often on providing an explanation of why certain individuals are considered to have a particular kinship descriptor.

Here, we briefly comment on the logic of relationships as defined by relative positions in the directed graph of the genealogy. Following Atkins (1974) and Cannings and Thompson (1981), suppose that we denote an edge from a marriage A to an individual B by ARB (R for reproduction) and an edge from an individual B to a marriage A by BMA (M for marriage) where the direction is important. The relationship between any two individuals can then be represented by a string of Rs and Ms, together with their reverses $\overline{R}$ and $\overline{M}$, if the directed graph is a tree, or by a set of such strings if there are loops within the genealogy. We abstract the string of Rs and Ms to define a type of relationship.

Ambiguity can arise by virtue of the possibility of retracing edges in the graph. As a simple example, following Cannings and Thompson (1981), consider a small genealogy, where A, B, C and D are individuals, and a and b are marriages as in Figure 23.4. Now consider the relationship defined with respect to individual A by $\overline{RM}MR$. Following the edges as defined we can go from A to a, and then from a to D. At this point there is a choice: from D to a or from D to b. If the choice is D to a, then a further choice will take us to A or B. The possible paths are therefore AaDaA, AaDaB and AaDbC, so that the relationship defined is self, sib or half-sib.

It is possible to use this simple system to solve issues of relationship. For example, as discussed by Kendall (1971), there is a well-known riddle regarding relationships: *brothers and sisters have I none but this man's father is my father's son*. It is then necessary to differentiate between males and females. Thus, we split R into S (reproduction producing a son) and D (reproduction producing a daughter), and M into H, the husband of a marriage, and W, the wife of a marriage. Thus, this man's father is $\overline{SH}$, while my father's son is $\overline{RH}HS$, so the relationship between me and this man is summarised by equating these two expressions. We have $\overline{SH} = \overline{RH}HS$. The relationship between the man and me is thus $\overline{SHSH}HR$, so we have two possible relationships (as derived by Kendall) (1) the man is the grandson of my father by a different woman to my mother (I may be male or female), (2) the relationship reduces to $\overline{SHSR}$ so that, since I have no siblings, I must be the man's father.

There are other systems of representation for genealogies used within the anthropological literature, but the focus is usually different from the biological. The interested reader is referred to White and Harary (2001), and references therein, for details of the *P-Systems* approach and other graph systems used in anthropology.

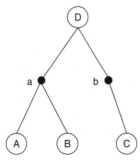

Figure 23.4 A simple pedigree connecting two sibs and a cousin to their common ancestor.

23.2.2 Measures of Genetic Relationship

Here, we discuss measuring and representing the genetic relationship between pairs or sets of individuals, or the genes within an individual. We shall assume throughout that each individual is diploid, i.e. has two genes at each locus (we discuss only the autosomal case, sex-linked genes are treated similarly).

Each individual then has two copies of the gene at any locus of interest. We shall consider sets of copies of the gene in question taken in some specified way (e.g. one taken randomly from each of the individuals A, B and C). However, in accordance with common practice, we shall refer to a set of genes, rather than a set of copies of a gene.

23.2.2.1 Identity by Descent

A set of genes at some locus are said to be *identical by descent* (IBD) if they are all copies of some single ancestral gene, having been passed down by reproduction from that ancestral gene. We shall abbreviate identical by descent by IBD, when appropriate.

Note that identity by descent is an equivalence relation (reflexive, symmetric and transitive), so the set of genes are partitioned into subsets such that genes in the same subset are IBD.

Our measures of relationship are all based on this notion and defined in terms of the probability that the set of genes (at some locus) in question have some specific pattern of identity by descent. This notion was developed by the early work of Cotterman (1940) and Malécot (1948).

23.2.2.2 Pairs of Individuals

In the preceding section, we have referred to the genes *at some locus*, since the definition of identity by descent is referring to a specific set of genes at a specific locus. We now move on to defining and calculating probabilities of various quantities relating to IBD. Here, we can drop the *at some locus* requirement since the probabilities are identical for each locus with similar inheritance (i.e. autosomal or sex-linked). The probabilities are essentially properties of the genealogies.

Definition. The probability that the two distinct genes of an individual A (i.e. that derived from the individual's mother and that from the father) are identical by descent is called the *inbreeding coefficient*, and is denoted by $\mathbf{f}(A)$.

Definition. The probability that a gene randomly selected from individual A and a gene randomly selected from individual B are identical by descent is called the *kinship coefficient* of A and B, and is denoted by $\mathbf{K}(AB)$.

It follows immediately that, for A = B, we have

$$\mathbf{K}(AA) = \frac{1}{2} + \mathbf{f}(A). \tag{23.1}$$

Definition. If individuals A and B are related in some specific way, REL say, then we refer to their kinship coefficient as the *kinship coefficient* of the relationship REL and write this as $\mathbf{K}(REL)$.

Note that if we say two individuals are related in some specific way then it is assumed that the original ancestors of the individuals in the defining genealogy are unrelated.

For example, for a pair of first cousins, which we denote by FC, who share two grandparents, it is assumed that these two individuals are unrelated. We shall consider the pair-wise relationships U = unrelated; S = siblings; FC = first cousins; UN = uncle – niece; DFC = double first cousins. Additionally, we have the pair-wise relationship $\mathbf{I}$ of an individual with himself.

A fundamental identity allows one to develop a calculus of kinship coefficients deriving that for one relationship in terms of others. Suppose we have a pair of individuals A and B, where B is not an ancestor of A, whose parents are P(A) (father of A), M(A) (mother of A), and similarly P(B) and M(B). Then just using the rules of Mendel one can deduce that

$$\mathbf{K}(AB) = \frac{\mathbf{K}(AP(B)) + \mathbf{K}(AM(B))}{2}. \tag{23.2}$$

The restriction that B is not an ancestor of A is required in order to ensure that there is only one route from the parents of A to individual B. A simple counterexample (see e.g. Lange, 2002) is provided by the case of a three-generation sib mating scheme, where A is in the third generation and B in the second. In order to use recursion in this case one should use the parents of A (which of course include B) rather than the parents of B.

In a similar manner, provided B is not an ancestor of A, nor A of B,

$$\mathbf{K}(AB) = \frac{\mathbf{K}(P(A)P(B)) + \mathbf{K}(P(A)M(B)) + \mathbf{K}(M(A)P(B)) + \mathbf{K}(M(A)M(B))}{4}. \tag{23.3}$$

Hence, for example

$$\mathbf{K}(S) = \frac{\mathbf{K}(I) + \mathbf{K}(U)}{2}$$

$$\mathbf{K}(FC) = \frac{3\mathbf{K}(U) + \mathbf{K}(S)}{4}$$

$$\mathbf{K}(DFC) = \frac{\mathbf{K}(U) + \mathbf{K}(S)}{2}$$

$$\mathbf{K}(UN) = \frac{\mathbf{K}(U) + \mathbf{K}(S)}{2}. \tag{23.4}$$

We may wish to consider coefficients of kinship for a set $\mathbf{W}$ of individuals $I_i; i = 1, k$, who may not all be different. Then, provided I_1 is not an ancestor of any of the other I_is, we can write

$$\mathbf{K}(\mathbf{W}) = \frac{\mathbf{K}(\mathbf{WM1}) + \mathbf{K}(\mathbf{WF1})}{2}, \tag{23.5}$$

where $\mathbf{WM1}$ is formed from $\mathbf{W}$ by removing the first individual and replacing with his/her mother, and $\mathbf{WF1}$ similarly replacing with the father.

23.2.3 Identity States for Two Individuals

As pointed out earlier, the reason that the upward recursion of coefficients of kinship breaks down is because it (potentially) ignores a route back to the earlier ancestors. This problem, and many others, can be dealt with by using more complete specification of the states of sets of individuals.

The simplest non-trivial example of the set of identity states occurs for two individuals. We label the genes of the first individual as 1 and 2 and those of the second individual as 3 and 4. We then need to identify those genes that are identical by descent, which is done by considering the distinct partitions of the four genes into equivalence (by identity) classes. There are 15 partitions $(1234), (123, 4)*(124, 3), (134, 2)*(234,1), (12, 34), (13, 24)* (14, 23), (12, 3, 4), (13, 2, 4)*(14, 2, 3)*(23,1, 4)*(24,1, 3), (34,1, 2), (1, 2, 3, 4)$, where the $*$ indicates that the partitions are equivalent if we do not require to differentiate between the genes within an individuals. Corresponding to each partition, there is what is termed an *identity state*. Thus, the partition $(12, 34)$ is written as state $(1, 1, 2, 2)$ in the obvious way. For ease of reference, we number the nine states obtained in the latter case as per Table 23.1. Figure 23.5 shows the usual diagrammatic representation of those states.

Example. Suppose that two individuals (one male and one female) whose genes at a particular locus have no identity by descent, i.e their identity state is $(1, 2, 3, 4)$, have two offspring. Now each offspring receives a gene from each parent so that in terms of the identity of the genes by descent each will be one of $(1, 3), (1, 4), (2, 3), (2, 4)$ with probabilities $1/4$, independently of the other. Collecting up and reducing to their equivalent identity states, we have that with probabilities $1/4$, $1/2$ and $1/4$ the states will be $(1, 2, 1, 2)$, $(1, 2, 1, 3)$ or $(1, 2, 3, 4)$ respectively. Note that the values in the final states do not relate directly to those values in the parents' states.

23.2.4 More Than Two Individuals

For multiple individuals, Cockerham (1971) generalised to higher-order identities on up to four individuals, while Thompson (1974) used the methods of group theory to discuss the case for arbitrary numbers of individuals, giving examples with up to five.

Table 23.1 Canonical expressions for identity states.

Number	State
1	(1,1,1,1)
2	(1,1,1,2)
3	(1,2,1,1)
4	(1,1,2,2)
5	(1,2,1,2)
6	(1,1,2,3)
7	(1,2,3,3)
8	(1,2,1,3)
9	(1,2,3,4)

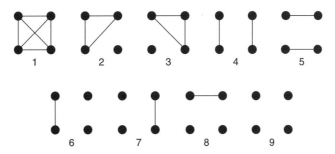

Figure 23.5 Diagram of the identity states of two individuals. The two left-most circles correspond to the genes of the first individual.

Perhaps the neatest treatment is that of Karigl (1981), who generalised the notion of kinship to apply to arbitrary sets of individuals. Thus, suppose that we have sets of individuals $S_1, S_2 \ldots S_k$ possibly overlapping. Then, define $K(S_1, S_2, \ldots, S_k)$ as the probability that a set of genes taken independently one from each of the individuals of set S_i are identical by descent for every i. Note that

$$K(S_1, S_2, \ldots, S_k) = K(P_1(S_1), P_2(S_2), \ldots, P_k(S_k)), \tag{23.6}$$

where $P_i(S_i)$ is a permutation of S_i, and also

$$K(S_1, S_2, \ldots, S_k) = K\left(S_{\rho(1)}, S_{\rho(2)}, \ldots, S_{\rho(k)}\right), \tag{23.7}$$

where $(\rho(1), \rho(2), \ldots, \rho(k))$ is a permutation of $\{1, 2, \ldots, k\}$.

Karigl gave the expressions for small numbers of individuals, up to four, which we repeat here in our notation. We shall write $K(AB, CD)$ for the case where $k = 2$, $S_1 = \{A, B\}$ and $S_2 = \{C, D\}$.

We have already defined $K(AA)$ and $K(AB)$, and given expressions in terms of the parents of A, given A is not an ancestor of B.

We have, where A is not an ancestor of B, C or D,

$$K(ABC) = \frac{K(P(A)BC) + K(M(A)BC)}{2}$$

$$K(AAB) = \frac{K(AB) + K(M(A)P(A)B)}{2}$$

$$K(AAA) = \frac{1 + 3K(M(A)P(A))}{4}$$

$$K(ABCD) = \frac{K(P(A)BCD) + K(M(A)BCD)}{2}$$

$$K(AABC) = \frac{K(ABC) + K(P(A)M(A)BC)}{2}$$

$$K(AAAB) = \frac{K(AB) + 3K(P(A)M(A)B)}{4}$$

$$K(AAAA) = \frac{1 + 7K(P(A)M(A))}{8}$$

$$K(AB, CD) = \frac{K(P(A)B, CD) + K(M(A)B, CD)}{2}$$

$$K(AA, BC) = \frac{K(P(A)M(A), BC) + BC}{2}$$

$$K(AB, AC) = \frac{2K(ABC) + K(P(A)B, M(A)C) + K(M(A)B, F(A)C)}{4}$$

$$K(AA, AB) = \frac{K(AB) + K(P(A)M(A)B)}{2}$$

$$K(AA, AA) = \frac{1 + 3K(P(A)M(A))}{4}. \tag{23.8}$$

In fact, it is relatively easy to express a general recurrence relationship for any $\mathbf{K}(\mathbf{S}_1, \mathbf{S}_2, \ldots, \mathbf{S}_k)$. Supposing that A is not an ancestor of any other individual under consideration, and $\mathbf{S}_i = \mathbf{A}^{r_i} \cup \mathbf{T}_i$, where $\mathbf{T}_i$ contains no As. Then

$$\mathbf{K}(\mathbf{S}_1, \mathbf{S}_2, \ldots, \mathbf{S}_k) = \frac{1}{2^R} \sum_j \prod_{l=1}^{k} \mathbf{K}(\mathbf{m}^{j_l} \mathbf{f}^{r_l - j_l} \mathbf{T}_l), \qquad (23.9)$$

R being the total number of As, and $\mathbf{j} = (j_1, j_2, \ldots j_k), 1 \leq j_i \leq r_i$ For neatness of expression, $\mathbf{M}(\mathbf{A})$ has been replaced by m, and $\mathbf{P}(\mathbf{A})$ by f. We then replace each expression $m^{j_i} f^{r_i - j_i}$ by m if $j_i = r_i = 1$, by mm if $j_i = r_i > 1$, by f if $j_i = 0, r_i = 1$, by ff if $j_i = 0, r_i > 1$ and every other expression by mf. We then collect up within each expression for $\mathbf{K}()$ all the terms that have a single m and merge the corresponding $\mathbf{T}_i$s. Similar procedure is adopted for terms with mm, f, ff and fm. Finally, we replace all the mm and ff terms by A, and collect up; note that this last step does not imply that a term with mm is equal to one with A, but rather that the two terms with mm and ff together equal 2 times one with A.

As an example, consider

$$16 \times \mathbf{K}(\mathbf{AAB}, \mathbf{AAC}) = \mathbf{K}(\mathbf{mmB}, \mathbf{mmC}) + 2\mathbf{K}(\mathbf{mmB}, \mathbf{mfC}) + \mathbf{K}(\mathbf{mmB}, \mathbf{ffC})$$

$$+ 2\mathbf{K}(\mathbf{mfB}, \mathbf{mmC}) + 4\mathbf{K}(\mathbf{mfB}, \mathbf{mfC}) + 2\mathbf{K}(\mathbf{mfB}, \mathbf{ffC})$$

$$+ \mathbf{K}(\mathbf{ffB}, \mathbf{mmC}) + 2\mathbf{K}(\mathbf{ffB}, \mathbf{mfC}) + \mathbf{K}(\mathbf{ffB}, \mathbf{ffC}), \qquad (23.10)$$

leading to

$$\mathbf{K}(\mathbf{AAB}, \mathbf{AAC}) = \frac{\mathbf{K}(\mathbf{AB}, \mathbf{AC}) + \mathbf{K}(\mathbf{AB}, \mathbf{mfC}) + \mathbf{K}(\mathbf{mfB}, \mathbf{AC}) + \mathbf{K}(\mathbf{mfB}, \mathbf{mfC})}{4}.$$
$$(23.11)$$

Karigl (1981) gives expressions for the nine coefficients of identity in terms of the above. The main benefit of the extended coefficients is that they readily allow the recursive derivation of the coefficients of kinship for individuals on a genealogy. Note that each K-coefficient is expressed in terms of K-coefficients with at most the same number of entries. This therefore allows the use of these to calculate the coefficients given above recursively each time taking A as the most recent individual, thus ensuring that the condition on ancestry is met. Working up through the genealogy finally gives us expressions involving the founders, which will have known values. Karigl (1981) gives a more substantial example for a pedigree of 19 individuals with several loops, and made available programmes to carry out the calculations.

23.2.5   Example: Two Siblings Given Parental States

One often needs to calculate the probabilities of the identity states of a pair of siblings, particularly as part of genetic counselling when the first sibling has some particular genetic disorder and the second is a future offspring about whom one wishes to make a probability statement regarding his/her genetic state. All the information is captured in the probabilities of the gene identity states. We could use Karigl's expressions or proceed directly. Suppose that the parental pair have probabilities $\Pi = (\pi_1, \pi_2, \pi_3, \pi_4, \pi_5, \pi_6, \pi_7, \pi_8, \pi_9)$, where π_i is the probability of the ith state. Then, if a pair of offspring have probabilities Π^*, we can write $\Pi_* = A\Pi$ where

$$
A = \begin{pmatrix}
1 & 1/4 & 1/4 & 0 & 1/8 & 0 & 0 & 1/16 & 0 \\
0 & 1/4 & 1/4 & 0 & 1/4 & 0 & 0 & 1/8 & 0 \\
0 & 1/4 & 1/4 & 0 & 1/4 & 0 & 0 & 1/8 & 0 \\
0 & 0 & 0 & 0 & 1/8 & 0 & 0 & 0 & 0 \\
0 & 1/4 & 1/4 & 1 & 1/4 & 1/2 & 1/2 & 3/16 & 1/4 \\
0 & 0 & 0 & 0 & 0 & 0 & 0 & 1/16 & 0 \\
0 & 0 & 0 & 0 & 0 & 0 & 0 & 1/16 & 0 \\
0 & 0 & 0 & 0 & 0 & 1/2 & 1/2 & 3/8 & 1/2 \\
0 & 0 & 0 & 0 & 0 & 0 & 0 & 0 & 1/4
\end{pmatrix},
\tag{23.12}
$$

the columns of A adding to unity, the ninth column corresponding to the example above.

In passing, observe that the matrix above is of upper block diagonal form with block sizes 1, 4, 3 and 1, which correspond to the number of distinct identity states with 1, 2, 3 and 4 distinct genes (by descent). Clearly the process of reproduction can only maintain or reduce the number of distinct genes.

The above matrix can be used as the transition matrix of the Markov chain corresponding to the process of repeated sib mating, see Figure 23.9 (though with columns adding to unity rather than the more usual rows), which is a scheme sometimes used in plant and animal breeding. It is possible in this case, and in more complex examples, to calculate π through time via the eigenvectors and eigenvalues. Fisher (1949) discusses this case in considerable detail, taking the seven identity states, which are relevant when there is no need to differentiate between states 2 and 3, or between 6 and 7. It is then straightforward to calculate the seven eigenvalues $1, 1/4, (1 \pm \sqrt{5})/4, 1/2, -1/8, 1/4)$ corresponding to the blocks and to express exactly the value of π. The dominant eigenvalue is $(1 + \sqrt{5})/4$ and indicates the rate of approach to homozygosity.

Consideration of two loci has been addressed by Cockerham and Weir (1968), Weir and Cockerham (1969) and Dennison (1975) among others, and for two and three loci by Thompson (1988). These in the main do not add much to our understanding of the graphical aspects of the genealogies, adding to the complexity but not the conceptual framework, and so are not discussed further here.

23.3 THE IDENTITY PROCESS ALONG A CHROMOSOME

23.3.1 Theory of Junctions

In reality, it is chromosomes that are passed from parent to offspring through the genealogy, so ideally we wish to examine the nature of the identity by descent process along the chromosomes of individuals. This was treated by Fisher (1949) through his theory of junctions, which we discuss in more detail below. This has become more relevant as density of information on the human, and other species, genomes has increased dramatically. Given a set of n individuals, there are $2n$ copies of each chromosome, so the identity state for some particular locus is specified in the obvious way described above.

The description of the identity state of the chromosome as a whole is then given by specifying the identity state at each point along the chromosome; i.e. by specifying a set of intervals within each of which there is some specific identity state. The description of the probabilities associated with the possible realisations is through a random process along the chromosome. This is a continuous time Markov chain (time here denoting distance along the chromosome from one end), with the states corresponding to the identity states.

In Fisher's theory of junctions, a *junction* is a point on a chromosome at which, due to a recombination event, the genetic material on the two sides of the junction has descended by different routes from the ancestors. If the ancestral origin of these two tracts is different, then the junction is termed *external*, and if it is the same, then the junction is termed *internal*. The occurrence of junctions is assumed to be driven by a Poisson process at each segregation, and thus overall by a Poisson process. Whenever a junction occurs, it can then be treated like a mutation and its persistence or loss is governed by the appropriate process for that single locus.

Fisher (1949; 1954; 1959), Bennett (1953) and Gale (1964) investigated the expected number of distinct chromosomal regions (i.e. each separated by a recombination) for a variety of systems of inbreeding: repeated selfing, repeated sib mating, repeated parent – offspring mating and others. Various authors, such as Franklin (1977), Guo (1995), have considered the proportion of the genome which is homozygous by descent in inbred individuals.

23.3.2   Random Walks

In practice, it is usually necessary to concentrate on genealogies of some specific, and relatively simple, structure in order to make much progress. For example, Donnelly (1983) demonstrated that for a pair of first cousins the identity state process along a chromosome was isomorphic to a random walk on the vertices of a three-cube. The transition rates being twice the recombination rate (assuming these to be identical in males and females), six of the vertices corresponding to identity states with zero identity, and two, neighbouring, vertices corresponding to having identity state $(1, 2, 1, 3)$.

Cannings (2003) discusses a general method of deriving the transitions matrix for the continuous time (position along the chromosome) process over the set of identity states. Given the transition matrix for some set of chromosomes at generation n, one can derive that for some new set derived in some specified (Mendelian) manner from these incorporating both the segregation and the recombination process. One can then use the transition matrix to deduce, or estimate numerically, any required variable associated with that particular set of chromosomes.

23.3.3   Other Methods

Bickeboller and Thompson (1996a; 1996b) provided approximations for an arbitrary number of half-sibs, using the Poisson clumping heuristic. Stefanov (2000; 2002; 2004) has provided exact methods for a variety of relationships (grandparent, full- and half-sibs and great grandparental) and Stefanov and Ball (2005) have refined the method in the context of half-sibs.

23.4   STATE SPACE ENUMERATION

23.4.1   Applying the *Peeling* Method

We now proceed to an enumeration problem on genealogies. We suppose that each individual in the genealogy is to be labelled with a genotype and we then enumerate the number of possible such labellings subject to Mendel's first law.

Camp *et al.* (1994) demonstrated how one can use the peeling algorithm to count the number of assignments of genotypes to an arbitrary genealogy that are consistent with Mendel's first law. The rates of growth can be illustrated most clearly in regular mating patterns. At the stage of the peeling algorithm that incorporates the information at a particular nuclear family, the other sections of the pedigree are said to be *below* if connected to the marriage through an offspring and *above* if connected by a parent. For zero-loop pedigrees, these sections are disjoint, although they are not in general. We introduce two functions on these parts, essentially the R-functions of Cannings *et al.* (1978),

$$N_X^U(i) = \text{number of possible states in genealogy and above } X \text{ where } X \text{ has genotype } i$$
(23.13)

and

$$N_X^L(i) = \text{number of possible states in genealogy below } X \text{ given } X \text{ has genotype } i.$$
(23.14)

Consider the nuclear family shown in Figure 23.6, where it is assumed that there are further individuals joined at A, B, C and D who are not shown. Then, supposing that there

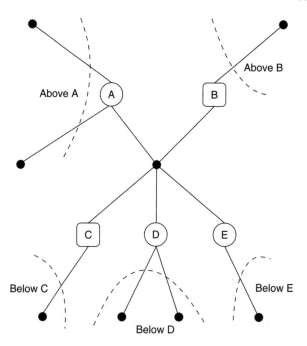

Figure 23.6   The sections of a pedigree above and below a marriage node and attached individuals.

are only two alleles a and b, and indexing the genotypes $aa = 1$, $ab = 2$ and $bb = 2$, then we can write

$$
\begin{aligned}
N_A^L(1) = {} & N_B^U(1)N_C^L(1)N_D^L(1)N_E^L(1) \\
& + N_B^U(2)N_C^L(1,2)N_D^L(1,2)N_E^L(1,2) \\
& + N_B^U(3)N_C^L(2)N_D^L(2)N_E^L(2),
\end{aligned}
\tag{23.15}
$$

where

$$
N_I^V(\mathbf{S}) = \Sigma_{k \in \mathbf{S}} N_I^V(k).
\tag{23.16}
$$

This is easily seen! For example, A has genotype 1 and B has genotype 2, i.e. the mating is $aa \times ab$ each of the offspring is either aa or ab. Similarly

$$
\begin{aligned}
N_A^L(2) = {} & N_B^U(1)N_C^L(1,2)N_D^L(1,2)N_E^L(1,2) \\
& + N_B^U(2)N_C^L(1,2,3)N_D^L(1,2,3)N_E^L(1,2,3) \\
& + N_B^U(3)N_C^L(2,3)N_D^L(2,3)N_E^L(2,3),
\end{aligned}
\tag{23.17}
$$

and $N_A^L(3)$ is obtained from $N_A^L(1)$ by replacing 1 by 3 everywhere.

We also have

$$
\begin{aligned}
N_C^U(1) = {} & N_A^U(1)N_B^U(1)N_D^L(1)N_E^L(1) + N_A^U(1)N_B^U(2)N_D^L(1,2)N_E^L(1,2) \\
& + N_A^U(2)N_B^U(1)N_D^L(1,2)N_E^L(1,2) + N_A^U(2)N_B^U(2)N_D^L(1,2,3)N_E^L(1,2,3)
\end{aligned}
\tag{23.18}
$$

and

$$
\begin{aligned}
N_C^U(2) = {} & N_A^U(1)N_B^U(2)N_D^L(1,2)N_E^L(1,2) + N_A^U(2)N_B^U(1)N_D^L(1,2)N_E^L(1,2) \\
& + N_A^U(2)N_B^U(2)N_D^L(1,2,3)N_E^L(1,2,3) + N_A^U(1)N_B^U(3)N_D^L(2)N_E^L(2) \\
& + N_A^U(3)N_B^U(1)N_D^L(2)N_E^L(2).
\end{aligned}
\tag{23.19}
$$

Using expressions of the above form, we can easily peel through any genealogy until one has all the information, on the number of possibilities, concentrated on one of the individuals. This then yields the total required.

23.4.2   Recursions

In order to understand the relationship between the *shape* of a genealogy and the number of possible genotype assignments, we examine some regular genealogies. These are made by adding successive nuclear families in a prescribed manner.

Our first example is illustrated in Figure 23.7, where nuclear families with two offspring are joined as shown (it is assumed that there is nothing attached to the Bs and Cs). Now, we write $\mathbf{N}_A^t = (N_{A_t}^L(1), N_{A_t}^L(2))$ (there is no need to keep track of $N_{A_t}^L(3)$ separately since it equals $N_{A_t}^L(1)$). Collecting up the necessary terms, we easily obtain

$$
\mathbf{N}_A^{t+1} = \mathbf{M}\mathbf{N}_A^t,
\tag{23.20}
$$

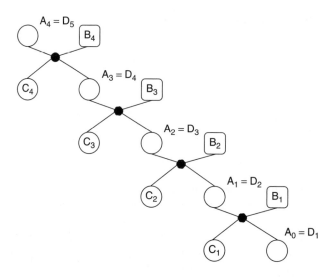

Figure 23.7   A descending line of regular nuclear families.

where

$$\mathbf{M} = \begin{bmatrix} 3 & 3 \\ 10 & 7 \end{bmatrix}. \tag{23.21}$$

Thus, the dominant eigenvalue is $\lambda = 5 + \sqrt{34} \approx 10.83$, so the number of assignments grows approximately at rate 2.21 per individual.

Our second example is illustrated in Figure 23.8 and leads to

$$\mathbf{N}_A^{t+1} = \mathbf{M}\mathbf{N}_A^t, \tag{23.22}$$

where

$$\mathbf{M} = \begin{bmatrix} 2 & 4 \\ 8 & 9 \end{bmatrix} \tag{23.23}$$

with a dominant eigenvalue of $\lambda = (11 + \sqrt{177})/2 \approx 12.15$ giving a growth rate of 2.30 per individual.

Camp *et al.* (1999) prove that the rate 2.30 per individual is the maximum possible among genealogies constructed by adding a new nuclear family with two offspring at each stage, no matter how these are joined on. This information could be of importance in the context of making calculations on a genealogy by, for example MCMC, where knowledge of the size of the sample space could be useful.

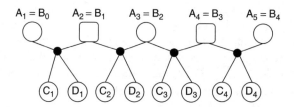

Figure 23.8   A marriage chain.

23.4.3   More Complex Linear Systems

Figure 23.9 shows a repeated sib mating system sometimes used in animal genetics. Here, we link our basic nuclear unit to the existing genealogy at two individuals, as shown, and because the system is growing linearly the recursion is also linear. Here, we switch to the case where there is an arbitrary number, l, of alleles. We work with the seven permutationally distinct states for a pair of unordered sibs

$$(11, 11), (11, 12), (11, 22), (12, 12), (12, 13), (11, 23), (12, 34) \qquad (23.24)$$

(the identity states), noting that these represent respectively

$$l, 4_lC_2, 2_lC_2, {}_lC_2, 6_lC_3, 6_lC_3, 6_lC_4, \qquad (23.25)$$

underlying states.

Now, we can again write

$$\mathbf{N}_S^{t+1} = \mathbf{MN}_S^t, \qquad (23.26)$$

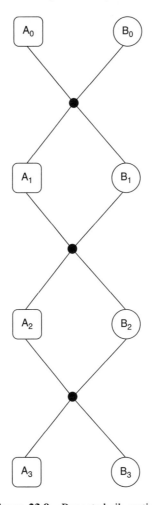

Figure 23.9   Repeated sib mating.

where

$$\mathbf{M} = \begin{bmatrix} 1 & 2(l-1) & 0 & (l-1) & 2_{(l-1)}C_2 & 0 & 0 \\ 0 & 2 & 0 & 1 & 2(l-2) & 0 & 0 \\ 0 & 0 & 0 & 1 & 0 & 0 & 0 \\ 0 & 4 & 2 & 1 & 6(l-1) & 4(l-2) & 4_{(l-2)}C_2 \\ 0 & 0 & 0 & 0 & 6 & 2 & 2(l-3) \\ 0 & 0 & 0 & 0 & 2 & 0 & 0 \\ 0 & 0 & 0 & 0 & 0 & 0 & 4 \end{bmatrix}. \tag{23.27}$$

Note that the matrix is block diagonal, essentially since one can only move from a state with the same or fewer distinct elements. Further, the eigenvalues of $\mathbf{M}$ are independent of l. For $l = 2$, the dominant eigenvalue $\lambda \approx 3.7785$ and for $l \geq 3$ $\lambda \approx 6.60$. Thus, since two individuals are added at each stage the respective rates are ≈ 1.944 and 2.569 respectively.

23.4.4   A Non-linear System

As discussed briefly earlier the addition of units in such a way that the number of individuals increases linearly leads to a linear equation for the number of configurations. Figure 23.10 shows a simple non-linear case. The basic unit is a mating with a single offspring A. At each stage, each of the founder members (those who have no parents) are replaced by one of the basic units. The recursive equations here have a bilinear form, rather than linear, so the methods as above are not applicable. Application of the theory of sub-multiplicative sequences, however, gives a rate of growth (for a system with two alleles) of 1.89.

23.5   MARRIAGE NODE GRAPHS

23.5.1   Drawing Marriage Node Graphs

Since, as mentioned earlier, marriage node graphs are a good format for producing drawings of pedigrees, it is worth considering methods for positioning the vertices on a page so as to produce a clear visualisation. Several approaches have been used, for

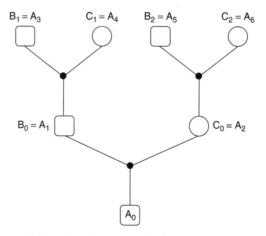

Figure 23.10   A pedigree growing in a non-linear fashion.

instance, Thomas (1994) described using simulated annealing to position vertices on a lattice so as to minimise the total line length, or the amount of ink on a page. A class of methods that works well for general graphs is that of force-directed methods. In these schemes, a quantity representing forces acting between vertices is calculated and minimised, usually using the Newton–Raphson iteration. For example, Fruchterman and Reingold (1991) minimise the quantity

$$\sum_{i=1}^{n}\sum_{j\sim i}d_{i,j}^{2}+\alpha\sum_{i=1}^{n}\sum_{j\neq i}\frac{1}{d_{i,j}},\tag{23.28}$$

where $d_{i,j}$ is the absolute value of the distance in the plane between the ith and jth vertices. The first sum is over connected pairs and represents an attractive force between adjacent vertices. The second sum is a repulsive force pushing apart all pairs.

An attractive feature of this approach is that it lends itself very well to animated graphics with which we can see the vertices move to their optimal positions, and with a few mouse clicks and drags even intervene in the process. Seeing how vertices move from an arbitrary starting point to their final positions is often as informative as the static final picture itself. The first term of 23.28 requires $O(e)$ time to compute, where e is the number of edges, and since pedigrees are relatively sparse graphs this is a quick calculation. However, the second term requires $O(n^2)$ time. This can be greatly improved by ignoring the small repulsions between distant vertices so that the function to minimise becomes

$$\sum_{i=1}^{n}\sum_{j\sim i}d_{i,j}^{2}+\alpha\sum_{i=1}^{n}\sum_{j\neq i:|d_{i,j}<\gamma|}\frac{1}{d_{i,j}},\tag{23.29}$$

since we can sort vertices into bins depending on their current position and only look for near neighbours in adjacent bins. Since the vertices near any particular vertex tend to be those connected to it, the whole of 23.29 can be computed in time approaching $O(e)$, greatly speeding up calculations. Unfortunately now, the Newton–Raphson scheme does not converge, but oscillates between solutions as the distances between some vertices fluctuate around γ. While this is not a problem when trying to produce a static picture, it leads to an unpleasant flickering effect in animated graphics. We can overcome this by replacing the repulsion term with a function that goes to zero smoothly, i.e. with zero derivative, at a finite value γ. The following modified target function achieves this

$$\sum_{i=1}^{n}\sum_{j\sim i}d_{i,j}^{2}+\alpha\sum_{i=1}^{n}\sum_{j\neq i:|d_{i,j}<\gamma|}\frac{(\gamma-d_{i,j})^{2}}{d_{i,j}}.\tag{23.30}$$

While the above works well for general graphs, for pedigrees and other directed acyclic graphs, it is nice to have ancestors shown above their descendants. Adding extra terms penalizing the difference between the vertical distance connecting two adjacent vertices and an ideal value θ achieves this, giving the final target function:

$$\sum_{i=1}^{n}\sum_{j\sim i}d_{i,j}^{2}+\alpha\sum_{i=1}^{n}\sum_{j\neq i:|d_{i,j}<\gamma|}\frac{(\gamma-d_{i,j})^{2}}{d_{i,j}}$$

$$+\beta\left\{\sum_{i}\sum_{j:j\rightsquigarrow i}(y_{i}-y_{j}-\theta)^{2}+\sum_{i}\sum_{j:i\rightsquigarrow j}(y_{i}-y_{j}+\theta)^{2}\right\},\tag{23.31}$$

where $i \rightsquigarrow j$ indicates a directed edge from i to j. Something similar can also be achieved by giving each vertex a generation number and fixing the vertical coordinate accordingly. However, for looped pedigrees, this is not a well-defined value as multiple paths connecting an ancestor to an inbred descendant may be of different lengths. Moreover, the repulsive force does not allow vertices fixed on the same horizontal line to cross as it becomes infinite if two vertices coincide exactly.

An animated programme for drawing pedigrees, and also one for general graphs, are available at Alun Thomas's Internet site. These allow the user to interactively change the γ, β and θ parameters and to move and fix specific vertices using mouse controls. A screen shot of the programme is given in Figure 23.11. The pedigree shown is that of the population of Tristan da Cunha as collected in 1962 (Roberts, 1971). Although this contains only 273 people, the multiple, complex relationships between them make it challenging for any pedigree drawing method.

23.5.2 Zero-loop Pedigrees

A *zero-loop* pedigree is one in which there are no loops caused by inbreeding, multiple mating or relative exchange in marriage and corresponds to the case when the marriage

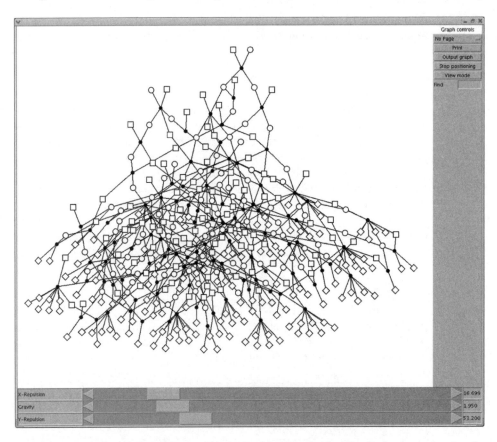

Figure 23.11 A screen shot of the graphical user interface for the ViewPed programme showing a marriage node graph of the pedigree of Tristan da Cunha collected in the 1960s.

node graph is a tree or a collection of unconnected trees: a *forest*. In genetic studies that focus on relatively small pedigrees involving perhaps three or four generations of individuals ascertained from a large, out-bred population, zero-loop pedigrees are the norm. In many cases, when looped pedigrees are found they are either discarded or the loops are avoided because the additional computational complexity that they introduce makes it difficult or impossible to analyse the pedigree. Zero-loop pedigrees are, therefore, of particular interest in genetics.

As discussed below and elsewhere in this book (see **Chapter 24**), probability calculations on pedigrees were the first applications of what became graphical modelling methods. An interesting and non-standard application of graphical modelling can answer the question: *How many ways can the same tree structure be interpreted as a marriage node graph?*. While this may seem an odd question, enumeration is intimately tied up with simulation and the answer will give us an efficient way to simulate zero-loop pedigrees.

Given a specific tree, we can assign vertices roles as either individuals or marriages in only two ways: starting with any vertex designate it an individual, its neighbours as marriages, their neighbours as individuals and so on. To make the other assignment, simply change each vertex's role. Thus, by dealing with each possibility in turn we can assume that we have the structure of the tree and know which vertices are individuals and which are marriages and all that remains is to count in how many ways we can assign individuals as parents or offspring of the marriages. In effect, we have to count the number of ways in which we can assign direction to the edges while following the rules that allow us to interpret the graph as a marriage node graph, which are as follows:

- A marriage node is included only if both parents are specified and there is at least one offspring from the marriage. This is equivalent to insisting that the first element of a triplet is non-null, and that either both or neither of the second and third elements are null.
- An individual must be listed as descended from no more than one marriage node. Individuals whose parents' marriage is not listed are the founders of the pedigree.

For any edge (i, j), let $y_{i,j}$ be 1 if it connects an individual down to its own marriage and 0 if it connects it to its parents' marriage, and for each vertex i let $C_i = \{y_{k,j} : k = i \text{ or } j = i\}$, i.e. the set of direction indicators for the edges that are incident to vertex i. Letting $I[\ldots]$ be an indicator function taking the value 1 when the condition inside the braces is true and 0 when it is false, an allocation of directions $y = \{y_{i,j}\}$ makes a proper marriage node graph if and only if

$$\prod_{\text{individual } i} I\left[\sum_{y \in C_i}(1 - y) \leq 1\right] \prod_{\text{marriage } j} I\left[\sum_{y \in C_j} y = 2\right] I\left[\sum_{y \in C_j}(1 - y) \geq 1\right] = 1.$$

(23.32)

This product defines a Markov graph on the indicator variables and so it can be manipulated using the usual forward – backward methods of graphical modelling. For example, the forward *collect evidence* step computes the total number of allocations of y that result in a valid marriage node graph. If we follow this with a backward simulation

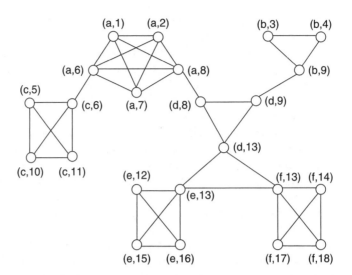

Figure 23.12   The Markov graph for the direction indicators for the edges of the tree structure of the marriage node graph shown in Figure 23.1. The numbers and letters correspond to the individual and marriage vertices of the marriage node graph. Each edge is represented by a letter – number pair reflecting the bipartite structure of the original tree.

step (Dawid, 1992), we sample one of these allocations from a distribution that gives equal probability to each possible allocation.

Figure 23.12 shows the Markov graph for the edge direction indicators corresponding to the tree structure of the marriage node graph in Figure 23.1. You might observe that this graph is *triangulated*, i.e. it contains no cycles of length 4 or more that are un-chorded. Triangulated graphs are amenable to graphical modelling methods, and non-triangulated graphs first have to be made triangulated by inserting additional edges. This triangulation process consists of adding sufficient new edges – *fill-ins* – across loops to break them up into three-cycles. In order to minimise the computations needed to deal with the graphical model, we would like to find the triangulation that makes the largest clique in the graph as small as possible. This is, however, in the general case a significant problem which Markov graphs that are already triangulated avoid.

That the graph in Figure 23.12 is triangulated is not an accident, but is due to the way in which it was constructed. It is a defining property of triangulated graphs that they can be represented as the *intersection graph* of a family of subtrees of a tree (Golumbic, 1980). That is, each vertex of the triangulated graph corresponds to a contiguous subtree of some tree and vertices are connected if and only if the corresponding subtrees intersect. In this case, the subtrees simply comprise the edges of the original tree including the vertices at each end, so that they intersect only when edges have a common end point. The consequence of this is that we can read off the computational requirements for performing the graphical modelling operations described above without having to first triangulate the graph. In particular, we know that the largest clique in the graph will be equal in size to the highest degree of any vertex in the original tree, and so the computational time and storage needed for enumeration and simulation grows as $2^{\max_i |C_i|}$.

Prüfer's constructive bijection (Prüfer, 1918) from the set of trees of size n onto a set of n^{n-2} vectors $(x_1, \ldots x_{n-2})$, where $x_i \in \{1, 2, \ldots n\}$ can be used to generate trees uniformly at random. Thomas and Cannings (2004) modified this slightly to generate random trees with n individuals and m marriages such that each marriage node has a degree of at least 3, as required for the above conditions. This also allows controlling the numbers of offspring per marriage and marriage per individual according to specific distributions.

23.6 MORAL GRAPHS

23.6.1 Significance for Computation

While a moral graph for a genealogy can be defined without reference to genetic variables, it is also the Markov graph corresponding to the joint distribution of genotypes at a genetic locus for all the individuals in a pedigree. These probabilities are functions of the way that genes arrive at the top of the pedigree, the way they are passed from generation to generation and the way they are expressed as phenotypes of observed individuals. If $g = \{g_1, \ldots g_n\}$ are the genotypes of pedigree members, and $x = \{x_1, \ldots x_n\}$ is the set of observed phenotypes

$$P(x|g) = \prod_{i \in F} \pi(g_i) \prod_{i \in \bar{F}} \tau(g_i | g_{f_i}, g_{m_i}) \prod_{i=1}^{n} \rho(x_i | g_i), \qquad (23.33)$$

where F are the founders of the pedigree and $\pi(g_i)$ is the frequency of genotype g_i in the population from which the founder is drawn, $\tau(g_i | g_{f_i}, g_{m_i})$ encodes the probabilities for transmission of genes from parents f_i and m_i to offspring i and $\rho(x_i | g_i)$ is the probability of the individual's phenotype given the individual's genotype, usually called the *penetrance* function, Each of the three products in this equation assumes conditional independences. The first assumes that founders are independently sampled from some population. The factorisation of the overall transmission probability into the second product is a consequence of the *offspring conditionally independent given genotypes of parents, or OCIGGOP* property (Thompson, 1986) for Mendelian inheritance. The third product assumes that the phenotype is expressed independent of any shared non-genetic factors. The only terms involving multiple genotypes are in the second product and the factors involve the triplets specifying the pedigree. Hence, the moralised graph for the Bayesian network that 23.33 defines is the same as that we have already defined for the pedigree. Probabilities can be calculated on genealogies using standard graphical modelling techniques, although it should be noted that much of this methodology was commonplace in genetics due to the work of Elston and Stewart (1971) and Cannings *et al.* (1978) well in advance of general applications of graphical modelling (Lauritzen and Spiegelhalter, 1988).

The structure of the moral graph is vital in organizing computations efficiently. As Figures 23.13 and 23.14 shows, moral graphs are not necessarily triangulated graphs and, in general, it is a significant problem to find a triangulation of the moral graph with a sufficiently small maximum clique size to make computations feasible, although simulated annealing and heuristic searches such as the *greedy algorithm* have proved to be effective for this (Thomas, 1986). However, the moral graph is not a general graph and has substantial structure.

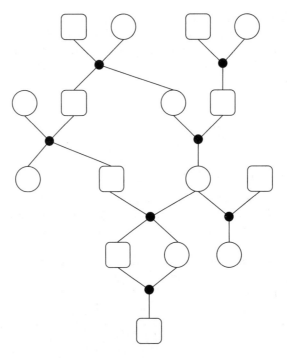

Figure 23.13   A marriage node graph with loops.

23.6.2   Derivation from Marriage Node Graphs

Clearly, starting with a marriage node graph, we can read off the offspring – parent triplets and hence construct the corresponding moral graph. The converse of this is not true since the moral graph is undirected so that different pedigrees can have the same moral graph. However, it is also possible to define the moral graph directly from the marriage node graph as an intersection graph of subtrees as follows. For each individual, define a subtree consisting of his/her vertex, the edge up to but not including their parents' marriage vertex, the edges down to and including each of their own marriages and finally a portion, say a half, of the edge down from each of their own marriages to each of their offspring. This is illustrated in Figure 23.15.

The subtree for any individual intersects with the subtree for a spouse at their marriage node and the upper portion of the edges down from it. Parent and offspring subtrees intersect on the upper portion of the edge down from the connecting marriage node. There are no other ways for subtrees to intersect; hence, the intersection graph is the moral graph.

An immediate consequence of this is that when a pedigree is zero-loop the marriage node graph is a tree and so, from the characterisation of triangulated graphs used above, the moral graph must be triangulated. Thus, for the zero-loop pedigrees used in most genetic studies, no search for an optimal triangulation is necessary. For all but trivial pedigrees, we know that the graph contains a cliques of three vertices corresponding to the offspring – parent triplet cliques. If we can show that there are no larger cliques, then from standard graphical modelling methods we know that the resources required to compute probabilities on a zero-loop pedigree will be, at worst, proportional to tk^3, where

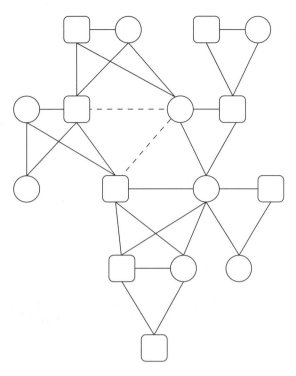

Figure 23.14 The moral graph corresponding to the looped pedigree shown in Figure 23.13. The dotted lines are fill-ins needed to make the graph triangulated.

t is the number of triplets and k is the number of possible genotypes. We can do this using a colouring argument since, for any graph, we know that the size of the largest clique must be no larger than the smallest number of colours that are needed to colour each vertex so that no two adjacent vertices have the same colour (Golumbic, 1980). In fact, it is a property of triangulated graphs that their largest clique size and the smallest possible colouring number must always be equal.

This can be achieved directly from the intersection graph. Starting with an arbitrary marriage node, colour the offspring subtrees all colour 1 and the parent subtrees colours 2 and 3. Working away from the starting marriage node to adjacent ones in a recursive fashion, because the marriage node graph is a tree, at each stage only one of the subtrees adjacent to the marriage node is coloured. If this is an offspring subtree, colour the other offspring subtrees the same colour and allocate the parents the two remaining colours in some order. If it is a parent subtree, allocate the other parent one of the two remaining colours and all offspring subtree the final colour. In this way, each subtree is coloured so that any intersecting subtrees are of different colours. Hence, by giving the vertices of the moral graph the same colour at its corresponding subtree, we can find a three colouring of the graph. Therefore, we can conclude that probability computations on genetic loci in zero-loop pedigrees require at worst tk^3 resources. If fact, we can usually do better than this and make computations using resources of order $t4k^2$ since for any pair of parental genotypes under simple Mendelian inheritance at most four genotypes are possible for any offspring. This model precludes any probability of germ line mutation.

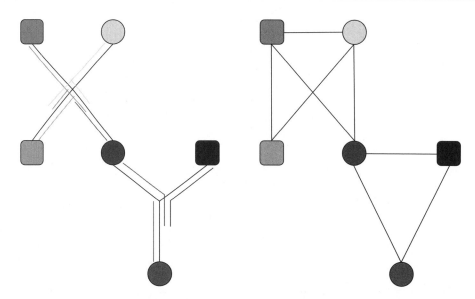

Figure 23.15 A small marriage node graph showing the intersecting subtrees from which a moral graph can be constructed. The colours of the subtrees in the marriage node graph correspond to the individual vertices in the moral graph. Subtrees that intersect in the marriage node graph correspond to connected vertices in the moral graph.

23.6.3   Four Colourability and Triangulation

While it seems obvious that moral graphs of zero-loop pedigrees should be three colourable, it is perhaps surprising at first glance that the moral graph for any pedigree, however large, complex or inbred, can be four coloured. However, this is also straight-forward to show. To do this, first notice that if we make the sub graph of the moral graph consisting of males and the edges between them, then this must be a forest, i.e. a tree or collection of trees. Similarly, the sub graph of females and connecting edges must also be a forest. In effect, we have constructed the ancestral graphs of Y-chromosomes and mitochondria in females respectively. These forests can each be coloured in two colours by taking each founder and giving it one colour, its offspring the second colour, their offspring the first colour again and so on. Since the moral graph is the union of these two forests and a set of edges connecting males to females, these two colourings of the sub graphs can be combined to give a four colouring of the moral graph. That four colours are also necessary in general is illustrated in Figure 23.16, which includes a sib mating that results in a four-clique and which therefore requires four colours. It should also be noted that genealogies are not necessarily planar so that this result is quite separate from the classical four-colour theorem. As a counter example, consider three males each mated with the same three females. The moral graph for this will contain the complete bipartite graph on two sets of three vertices, which cannot be contained in a planar graph.

The consequences of this general result are not as straightforward as in the case of a zero-loop pedigree because a general moral graph needs to be triangulated by adding fill-ins before the computational steps can be read from it. The addition of these fill-ins in general requires connecting vertices of the same colour, thus destroying the colourability

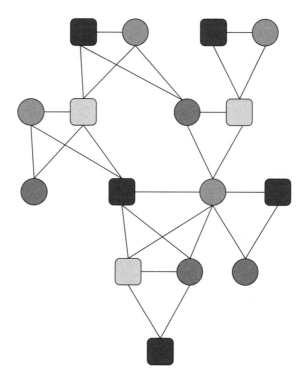

Figure 23.16 A four colouring of the moral graph shown in Figure 23.14. Any genealogy can be four coloured using the same scheme.

and the constraint we have on the largest clique size. However, the colouring can help with heuristic rules for preferring some fill-ins to others. For example, fill-ins connecting males to females never destroy the colourability. Since any loop in the graph must contains at least one individual of each sex, we can always fill it in so that the largest cycles in the graph are four cycles. Unfortunately, except for some simple cases, this leaves an optimisation problem on the same scale as the original.

REFERENCES

Atkins, J.R. (1974). *Grafik: A Multipurpose Kinship Metalanguage*. Mouton and Co.

Bennett, J.H. (1953). Junctions in inbreeding. *Genetica* **26**, 392–406.

Bickeboller, H. and Thompson, E.A. (1996a). Distribution of genome shared IBD by half-sibs: approximation by the Poisson clumping heuristic. *Theoretical Population Biology* **50**, 66–90.

Bickeboller, H. and Thompson, E.A. (1996b). The probability distribution of the amount of an individual's genome surviving to the following generation. *Genetics* **143**, 1043–1049.

Camp, N., Cannings, C. and Sheehan, N. (1994). The number of genotypic assignments on a genealogy. 1: the method and simple examples. *IMA Journal of Mathematics Applied in Medicine and Biology* **11**, 95–106.

Camp, N., Cannings, C. and Sheehan, N. (1999). The number of genotypic assignments on a genealogy 2. general linear systems. *IMA Journal of Mathematics Applied in Medicine and Biology* **16**, 213–236.

Cannings, C. (2003). The identity by descent process along the chromosome. *Human Heredity* **56**, 126–130.

Cannings, C. and Thompson, E.A. (1981). *Genealogical and Genetic Structure*. Cambridge University Press.

Cannings, C., Thompson, E.A. and Skolnick, M.H. (1978). Probability functions on complex pedigrees. *Annals of Applied Probability* **10**, 26–61.

Cockerham, C.C. (1971). Higher order probability functions of identity of alleles by descent. *Genetics* **69**, 235–246.

Cockerham, C.C. and Weir, B.S. (1968). Sib-mating with two linked loci. *Genetics* **60**, 629–640.

Cotterman, C.W. (1940). A calculus for statistico genetics. Ph.D. Thesis, Ohio State University.

Dawid, A.P. (1992). Applications of a general propogation algorithm for probabilistic expert systems. *Statistics and Computing* **2**, 25–36.

Dennison, C. (1975). Probability and genetic relationship: two loci. *Annals of Human Genetics* **39**, 89–104.

Donnelly, K.P. (1983). The probability that related individuals share some section of the genome identical by descent. *Theoretical Population Biology* **23**, 34–63.

Elston, R.C. and Stewart, J. (1971). A general model for the genetic analysis of pedigree data. *Human Heredity* **21**, 523–542.

Fisher, R.A. (1949). *The Theory of Inbreeding*. Oliver and Boyd, Edinburgh.

Fisher, R.A. (1954). A fuller theory of 'junctions' in inbreeding. *Heredity* **8**, 187–197.

Fisher, R.A. (1959). An algebraically exact examination of junction formation and transmission in parent-offspring inbreeding. *Heredity* **13**, 179–186.

Franklin, I.R. (1977). The distribution of the proportion of the genome which is homozygous by descent in inbred individuals. *Theoretical Population Biology* **11**, 60–80.

Fruchterman, T. and Reingold, E. (1991). Graph drawing by force-directed placement. *Software Practice and Experience* **21**, 1129–1164.

Gale, J.C. (1964). Some applications of the theory of junctions. *Biometrics* **20**, 85–117.

Golumbic, M.C. (1980). *Algorithmic Graph Theory and Perfect Graphs*. Academic Press.

Guo, S.-W. (1995). Proportion of genome shared identical by descent by relatives: concept, computation and applications. *American Journal of Human Genetics* **56**, 1468–1476.

Karigl, G. (1981). A recursive algorithm for the calculation of identity coefficients. *Annals of Human Genetics* **45**, 299–305.

Kendall, D.G. (1971). The algebra of genealogy. *Mathematical Spectrum* **4**, 7–8.

Lange, K. (2002). *Mathematical and Statistical Methods for Genetic Analysis*. Springer.

Lauritzen, S.L. and Spiegelhalter, D.J. (1988). Local computations with probabilities on graphical structures and their applications to expert systems. *Journal of the Royal Statistical Society, Series B* **50**, 157–224.

Malécot, G. (1948). *Les Mathematiques de l'Heredite*. Masson et Cie.

Prüfer, H. (1918). Neuer beweis eines satzes uber permutationen. *Archiv fur Mathematik und Physik* **27**, 142–144.

Roberts, D.F. (1971). The demography of Tristan da Cunha. *Population Studies* **25**, 469–475.

Stefanov, V.T. (2000). Distribution of genome shared IBD by two individuals in grandparent relationship. *Genetics* **156**, 1403–1410.

Stefanov, V.T. (2002). Statistics on continuous IBD data: exact distribution evaluation for a pair of full(half)-sibs and a pair of a (great-)grandchild with a (great-)grandparent. *BMC Genetics* **3**, 7.

Stefanov, V.T. (2004). Distribution of the amount of genetic material from a chromosome surviving to the following generation. *Journal of Applied Probability* **41**, 345–354.

Stefanov, V.T. and Ball, F. (2005). Evaluation of identity-by-descent probabilities for half-sibs on continuous genome. *Mathematical Biosciences* **196**, 215–225.

Thomas, A. (1986). Optimal computation of probability functions for pedigree analysis. *IMA Journal of Mathematics Applied in Medicine and Biology* **3**, 167–178.

Thomas, A. (1994). Linkage analysis on complex pedigrees by simulation. *IMA Journal of Mathematics Applied in Medicine and Biology* **11**, 79–93.

Thomas, A. and Cannings, C. (2004). Simulating realistic zero loop pedigrees using a bipartite Prüfer code and graphical modelling. *Mathematical Medicine and Biology* **21**, 335–345.

Thompson, E.A. (1974). Gene identities and multiple relationships. *Biometrics* **30**, 667–680.

Thompson, E.A. (1986). *Pedigree Analysis in Human Genetics*. The Johns Hopkins University Press, Baltimore, MD.

Thompson, E.A. (1988). Two-locus and three-locus gene identity by descent in pedigrees. *IMA Journal of Mathematics Applied in Medicine and Biology* **5**, 261–281.

Weir, B.S. and Cockerham, C.C. (1969). Pedigree mating with two linked loci. *Genetics* **61**, 923–940.

White, D.R. and Harary F. (2001). P-systems: a structural model for kinship studies. *Connections* **24**, 22–33.

24

Graphical Models in Genetics

S.L. Lauritzen

Department of Statistics, University of Oxford, Oxford, UK

and

N.A. Sheehan

Department of Health Sciences and Genetics, University of Leicester, Leicester, UK

In this chapter, graphical models are introduced and used as a natural way to formulate and address problems in genetics and related areas. Local computational algorithms on graphical models are presented and their relationship with the traditional peeling algorithms discussed. The potential of graphical model representations is explored and illustrated using examples in linkage and association analysis, pedigree uncertainty, forensic identification, and causal inference from observational data.

24.1 INTRODUCTION

Graphs appear in a number of different contexts in genetics to convey information, e.g. about population development and evolution, and relationships between genes and individuals. This chapter focuses on the role of probabilistic graphical models (Lauritzen, 1996) within genetics, and in particular, on aspects of genetics which involve *pedigree analysis*, i.e. the analysis of genetic information among related individuals.

Probabilistic graphical models have their origin in genetics, in path analysis (Wright, 1921; 1923; 1934), which explicitly studies the propagation of hereditary properties through a family tree. They form a natural general framework to express and manipulate a number of important aspects of statistical genetics, e.g. computational algorithms such as 'peeling' (Elston and Stewart, 1971; Cannings *et al.*, 1978; Lander and Green, 1987), but have applications beyond that; e.g. in forensic genetics where complex issues of identification can be naturally expressed in terms of graphical models, in genetic epidemiology where the notion of Mendelian instruments helps to identify causal effects of genes, and in the study of regulatory networks, where graphs are naturally suited

Handbook of Statistical Genetics, Third Edition. Edited by D.J. Balding, M. Bishop and C. Cannings.
© 2007 John Wiley & Sons, Ltd.   ISBN: 978-0-470-05830-5.

to represent information about the interaction of genes. The latter area is developing particularly rapidly at the time of writing.

In this chapter, we describe basic elements of graphical models, especially Bayesian networks and their use for representing genetic information in pedigrees. We give a relatively detailed description of local computation algorithms in graphs and their use in a number of contexts, in particular, forensic applications, quantitative trait locus (QTL) and linkage analysis, and the handling of pedigree uncertainty. We outline basic elements of causal inference in graphical models and Mendelian randomization, and finally touch upon recent developments in using graphical models for genome-wide association studies and for identifying regulatory networks and patterns of associations in gene expression data.

24.2 BAYESIAN NETWORKS AND OTHER GRAPHICAL MODELS

24.2.1 Graph Terminology

We shall consider a *graph* $\mathcal{G} = (V, E)$ to consist of a finite set V of *vertices* or *nodes* and a set E of *edges* or *links* representing relationships between the nodes. Edges can be either *directed* with arrows indicating the direction of the link, or *undirected*. If there is an undirected edge between a node a and a node b, we say that a and b are *neighbours*, and we say that a is a *parent* of b and b is a *child* of a, if there is a directed edge from a to b. In contrast with the biological interpretation of these terms, a node in a graph can have more than two parents (e.g. node 7 in Figure 24.1).

A *trail* in a graph is defined to be a sequence of edges, each having a node in common with both preceding and succeeding edges. A *path* is a trail with no edges violating the direction of the trail. If all edges of the trail are undirected, the path is *undirected*, and the path is *directed* if all edges are directed. If there is a path from node a to node b (i.e. we can arrive at b by following arrows from a), we say that a is a (graph) *ancestor* of b and b is a (graph) *descendant* of a. A trail beginning and ending with the same node is a *cycle* or *loop*. If all the edges of a graph are directed, it is a *directed graph*, and if it has no directed cycles, it is a *directed acyclic graph* or DAG. An example of a DAG is displayed in Figure 24.1.

A graph is *connected* if there is a trail between any pair of nodes. A connected graph with no cycles is a *tree*. In this paper, unless otherwise stated, it will be assumed that all graphs are connected.

24.2.2 Conditional Independence

Graphical models (Lauritzen, 1996) exploit graphs to express assumptions of conditional independence to simplify specification, modelling, and analysis of high dimensional

Figure 24.1 A directed acyclic graph. [Reproduced with permission from Lauritzen, S. L. (2000). Causal inference from graphical models. In Barndorff-Nielsen, O. E., Cox, D. R. and Kluppelberg, C., editors, Complex Stochastic Systems, chapter 2, pages 63–107. Chapman & Hall.]

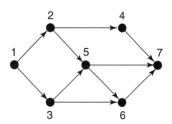

problems (Pearl, 1988; Lauritzen and Spiegelhalter, 1988; Cowell *et al.*, 1999). The conditional independence assumptions essentially split the problem into smaller manageable components.

Random variables X and Y are *conditionally independent* given the random variable Z if the conditional distribution of X given Y and Z is the same as the conditional distribution of X given Z alone, i.e.

$$\mathcal{L}(X \mid Y, Z) = \mathcal{L}(X \mid Z), \tag{24.1}$$

and we then write $X \perp\!\!\!\perp Y \mid Z$. In other words, if $X \perp\!\!\!\perp Y \mid Z$, the value of Y cannot be used to improve the prediction of X once Z is known.

Conditional independence can equivalently be expressed in terms of factorization of the corresponding probability density or probability mass function as

$$X \perp\!\!\!\perp Y \mid Z \quad \Longleftrightarrow \quad f(x, y, z)f(z) = f(x, z)f(y, z) \tag{24.2}$$

$$\Longleftrightarrow \quad \exists a, b\colon f(x, y, z) = a(x, z)b(y, z). \tag{24.3}$$

24.2.3 Elements of Bayesian Networks

A *Bayesian network* is a DAG with node set V, where the nodes represent random variables, $X = (X_v)_{v \in V}$, having some joint probability distribution function of the form:

$$f(x) = \prod_{v \in V} f(x_v \mid x_{\mathrm{pa}(v)}), \tag{24.4}$$

with $\mathrm{pa}(v)$ denoting the set of parent nodes of the node v and $x_A = (x_v)_{v \in A}$ for any subset $A \subseteq V$. It then holds that any node, given the values at its parents, is conditionally independent of all nodes which are not descendants. This is known as the *directed local Markov property*. The local Markov property for the DAG in Figure 24.1 yields, e.g. $4 \perp\!\!\!\perp \{1, 3, 5, 6\} \mid 2$, $5 \perp\!\!\!\perp \{1, 4\} \mid \{2, 3\}$, and $3 \perp\!\!\!\perp \{2, 4\} \mid 1$.

Further independencies can be deduced from the *global directed Markov property* which gives a complete description of independence relationships associated with a Bayesian network. In fact, the factorization (24.4) is equivalent to either of the local or global directed Markov properties; see (Section 3.2.2) Lauritzen (1996) for details. Note that through (24.4), the joint distribution of a Bayesian network is completely specified from the associated DAG and the conditional distributions of each node, given its parents.

24.2.4 Object-oriented Specification of Bayesian Networks

Bayesian networks involving pedigrees are composed of repeated structures each of identical composition, making these amenable to object-oriented specification. An *object-oriented* Bayesian network (OOBN) (Koller and Pfeffer, 1997) is based on a DAG as above, but each node in the DAG can itself be an OOBN. Simple nodes inside an OOBN can be *internal*, *input*, or *output nodes*. Directed links from an OOBN A to another OOBN B identify output nodes of A with input nodes of B. Each OOBN is typically an *instance* of a *class* of identical OOBNs, thus representing repeated patterns in an efficient manner. In the next section, we shall show examples of such representations of pedigree information. We refer to Dawid *et al.* (2007) for a detailed description of the use of OOBNs in networks representing problems in forensic genetics.

24.3   REPRESENTATION OF PEDIGREE INFORMATION

A pedigree is a group of individuals together with a full specification of all the relationships between them (Thompson, 1986). Individuals without parents in the pedigree are *founders*, are unrelated by definition, and can either be recent or belong to some baseline ancestral generation of interest. Pedigree members with mutual offspring in the pedigree are *spouses* and every spouse pairing is a *marriage*.

24.3.1   Graphs for Pedigrees

A standard diagrammatic representation of a pedigree is shown in Figure 24.2.

A pedigree can also be expressed as a directed graph (Lange and Elston, 1975), the simplest of which is depicted in Figure 24.3(a), where the nodes denote pedigree members, and the arcs connect individuals to their offsprings. A natural extension leads to the *marriage node* graph of Figure 24.3(b) (Thomas, 1985) which has two kinds of node, individual and marriage nodes, and two kinds of arc, connecting an individual to his marriages and connecting a marriage to the offspring of that marriage, respectively (Lange and Elston, 1975; Cannings *et al.*, 1978). Directions on the arcs can be omitted since direction is always *down* from parents to offspring via the relevant marriage node. Consequently, since an individual cannot be his own biological ancestor or descendant, pedigree graphs are always DAGs as there are no directed cycles. Undirected cycles or *loops* can arise, however. These include inbreeding loops, marriage rings, exchange loops, multiple marriage loops, and all kinds of interconnecting combinations of the above (Cannings *et al.*, 1978), where the presence of loops can depend on the particular graphical representation. For example, the loop $14 - 10 - 13 - 9 - 14$ formed by siblings 13 and 14 in the graph of Figure 24.3(a) does not feature in the marriage node graph representation of Figure 24.3(b), but the marriage loop connecting individuals 6, 10, 14, 15, 12, 7, 11 remains.

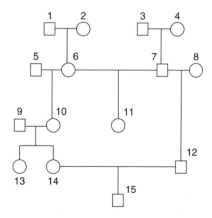

Figure 24.2   A standard representation of a simple pedigree of 14 individuals. As is consistent with common usage, females are represented by circles and males by squares. Individuals 1, 2, 3, and 4 are the baseline founders, while 5, 8, and 9 are recent founders who have married in. Individuals 11, 13, and 15 are *finals* in that they have no marriages.

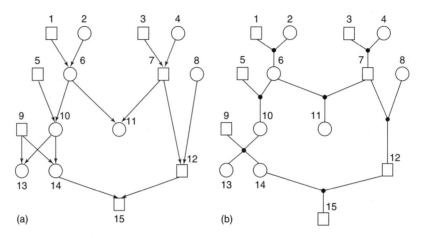

Figure 24.3 The pedigree of Figure 24.2 drawn (a) with nodes representing individuals and directed edges (arcs) connecting individuals to their offsprings and (b) as a marriage node graph with edge directions omitted.

24.3.2 Pedigrees and Bayesian Networks

The heredity of a trait between individuals in a pedigree has a natural expression as a Bayesian network with the graph nodes now representing random variables for which a joint probability distribution satisfying the factorisation in (24.4) can be defined. There are several ways of designing a Bayesian network for a pedigree and these various representations have different properties.

24.3.2.1 Segregation Network

Using the pedigree of Figure 24.2 and a single-locus, discrete genetic trait as an example, we begin with the most direct and complete representation–the *segregation network* (Lauritzen and Sheehan, 2003).

For each individual $i = 1, \ldots, m$, there are two nodes representing the trait genes inherited from his father and mother, respectively. The underlying random variables can assume any of the a allelic types in the relevant genetic system. Following common usage (Thompson, 2001), we will use 0 to label maternal inheritance and 1 for paternal inheritance. Thus the node labelled i^1 is identified with the random variable L_{i^1} assigning the allelic type of the gene inherited by the individual i from his father. For each nonfounder, arcs are directed from the two genes in the father to the paternal gene in the individual and the individual's maternal gene is likewise a (graph) child of the two genes in his mother. An additional node representing the *meiosis* or *segregation* indicator (Thompson, 1994; Sobel and Lange, 1996) is added as a parent to each nonfounder gene node. This is a binary node assuming the values 1 and 0 according to whether the inherited gene is a copy of the paternal or maternal gene in the corresponding parent. In this way, the allelic type of each nonfounder gene is a deterministic function of its (graph) parents. Specifically,

$$L_{i^1} = f(l_{p_i^1}, l_{p_i^0}, S_{p_i,i}) = \begin{cases} l_{p_i^1} & \text{if } S_{p_i,i} = 1 \\ l_{p_i^0} & \text{if } S_{p_i,i} = 0, \end{cases} \tag{24.5}$$

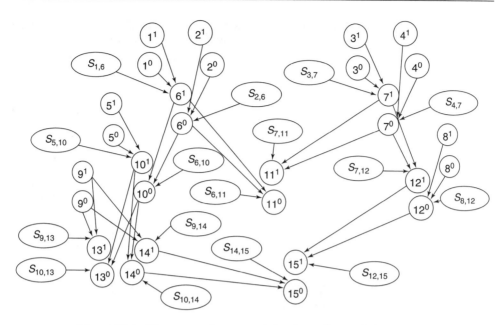

Figure 24.4 The segregation network for the pedigree of Figure 24.2.

for the paternally inherited gene and

$$L_{i^0} = f(l_{m_i^1}, l_{m_i^0}, S_{m_i,i}) = \begin{cases} l_{m_i^1} & \text{if } S_{m_i,i} = 1 \\ l_{m_i^0} & \text{if } S_{m_i,i} = 0, \end{cases} \qquad (24.6)$$

for the maternally inherited gene, where m_i and p_i are the labels of the mother and father of the individual i, and $S_{m_i,i}$ and $S_{p_i,i}$ are binary random variables assigning indicators for the segregations to i from the mother and father, respectively. The resulting graph is shown in Figure 24.4.

The laws of inheritance are encoded by letting the segregation indicators be independent with probabilities

$$P(S_{p_i,i} = 1) = \sigma_1, \text{ and } P(S_{m_i,i} = 1) = \sigma_0, \qquad (24.7)$$

where $\sigma_1 = \sigma_0 = 1/2$ if inheritance is Mendelian. The assumption of random union of gametes, and hence Hardy–Weinberg proportions for founder genotypes, is implied as the graph clearly indicates that founder genes are independent of each other and of the segregation indicators.

The segregation network can be given a simple object-oriented specification using a master network representing each individual in the pedigree as an OOBN, the class *founder* representing founders of the pedigree, and *schild* representing children as in Figure 24.5. Each instance of the *founder* class is itself an OOBN having nodes that represent the allelic types of the founder, chosen at random from the population and the associated genotype, since this may possibly be observed, see (a) in Figure 24.6. The OOBNs of the class *schild* in Figure 24.6(b) represent the transmission of genetic information from parent to child through yet another OOBN of the class *meiosis* displayed in Figure 24.6(c), which directly describes the segregation of alleles from

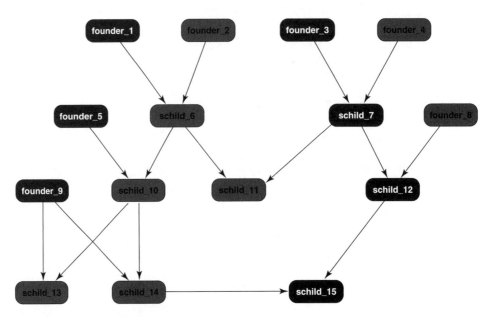

Figure 24.5 The master OOBN for the pedigree of Figure 24.2. There are two OOBN classes: *founder* and *schild*.

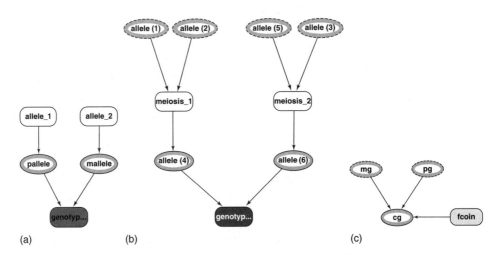

Figure 24.6 (a) The *founder* OOBN has two output nodes determined from the allele class, the latter representing a random allele from the relevant population; (b) the class *schild* has parental alleles as input nodes and represents the transmission of alleles to children via the class *meiosis*; (c) the OOBN of class *fcoin* in the *meiosis* class represents the segregation indicator.

parent to child using the OOBN of the class *fcoin* as the segregation indicator. The latter contains only a single output node with two equally probable values, but defining it as a class object makes it simple to express that this is repeated throughout the pedigree.

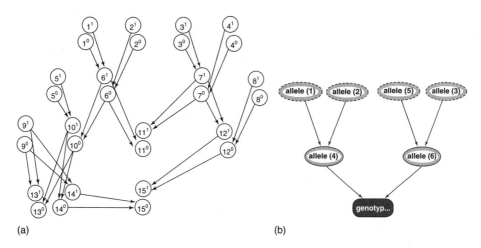

Figure 24.7 (a) Allele network representation of the pedigree; (b) the corresponding modification of the *schild* object to *achild*.

24.3.2.2 Allele Network

In some contexts, the full details of the segregation network may not be required. A convenient reduction is provided by removing the segregation indicators and associated arcs. Figure 24.7(a) shows the corresponding single-locus network for the pedigree of Figure 24.2. The conditional probability distribution of the allelic type of the paternally inherited gene given the (graph) parents, for example, are easily derived from (24.5), (24.6), and (24.7) to be

$$P(L_{i^1} = l \mid l_{p_{i1}}, l_{p_{i0}}) = \begin{cases} \sigma_1 & \text{if } l = l_{p_{i1}} \\ 1 - \sigma_1 & \text{if } l = l_{p_{i0}} \end{cases}.$$

The maternally inherited gene is handled analogously. We call this the *allele network* (Lauritzen and Sheehan, 2003), but note that it also features in Jensen (1997) and Thompson and Heath (2000). In the object-oriented representation, the corresponding reduction is made by replacing the OOBN *schild* with a simpler object, *achild*, which represents the transmission of alleles probabilistically rather than using the explicit process of meiosis, see Figure 24.7(b).

24.3.2.3 Genotype Network

The visually most parsimonious standard representation, although not necessarily the most useful, is the *genotype network* and uses the graph of Figure 24.3(a) as the underlying DAG with the nodes now representing the genotypes of the individuals rather than the individuals themselves. This representation features in Heath (2003) and Spiegelhalter (1990), is the 'genotype representation' of Jensen (1997), and is the 'genotype network' of Lauritzen and Sheehan (2003). We emphasize that the assumption of Mendelian inheritance is necessary for this representation to be valid, see Lauritzen and Sheehan (2003) for further details.

24.3.2.4 Adding Phenotypic Information

Each of the above networks specifies the inheritance relationships without referring to the observational situation in any given context. Although the genotypes may be identifiable from the phenotypes in many cases, they are often not identifiable and only partial information is available in some situations. To accommodate such data, an extra node can be added for each individual for whom phenotypic information is available and, depending on the purpose of the analysis (e.g. genetic counselling where a future child might be of interest), possibly for some unobserved individuals as well. We let Y_i denote the variable associated with the phenotype of the individual i. In the allele and segregation networks, the node carrying the phenotype Y_i has the two alleles of the individual i as parents, whereas in the genotype network, Y_i has the genotype G_i as its only parent. The conditional distribution of the phenotype Y_i, given its (graph) parents, is the *penetrance distribution* and may take the form of a deterministic relationship (e.g. $Y_i = G_i$) or a more complicated function. If the genotype is itself observable, i.e. $Y_i = G_i$, then we can omit this extra node in the genotype network. For the most parsimonious representation (the genotype network), the local Markov property of the network augmented with phenotypic information is ensured by the phenotype Y_i of any individual being conditionally independent of other variables in the network, given the genotype G_i of that individual. There are some traits for which this conditional independence assumption is clearly not reasonable. A woman's risk of pre-eclampsia is higher for a first pregnancy than a subsequent pregnancy, for example. Genetic imprinting, by which paternal and maternal alleles have differential influence on the phenotype also violates this assumption. In the latter case, one could simply define a network in terms of *ordered* genotypes for which the Markov property would then hold. Alternatively, the augmented allele and segregation networks both contain sufficiently detailed information for the Markov property to hold for more complex genotype–phenotype relationships.

24.4 PEELING AND RELATED ALGORITHMS

Almost every problem associated with pedigree analysis or other complex genetic applications involves a difficult computation. This could be the computation of a likelihood, the probability of an individual having a specific allele, genotype or haplotype, or some other characteristic of the system under investigation. Superficially, such computations seem too complex to be feasible at all and indeed many are not. However, there are a number of related computational algorithms that exploit the local structure of the system. These algorithms yield drastic reductions in the computational complexity. In genetic applications, such computation is typically referred to as *peeling* (Elston and Stewart, 1971; Cannings *et al.*, 1978; Lander and Green, 1987). See also Thompson (2001; 2000), Heath (2003), and Lauritzen and Sheehan (2003) for further discussion.

The peeling algorithms are special cases, or variants, of general algorithms for so-called local computation on graphs (Cowell *et al.*, 1999). In this section, we describe and explain a general algorithm and how it can be applied in this context. The algorithm can be seen as having two phases. During the first phase, a suitable computational structure is established. In the second phase, the computations themselves are executed. The first phase is sometimes referred to as *compilation*, the latter as *propagation*.

24.4.1 Compilation

The compilation process involves the collection of groups of variables into *cliques* so that computations can be performed locally, i.e. only involving functions of sets of variables belonging to the same clique. At the next stage, these cliques are organized in a tree structure, the *junction tree*, which is used to coordinate the local computations in a consistent way to yield the desired correct global result. Finally the numbers to be used in the calculations are associated with the relevant location in the junction tree. The various steps of the compilation process are described in further detail below.

24.4.1.1 From Bayesian Network to Undirected Graph

The local computation algorithms are based on undirected graphs. The first step, therefore, is to transform the Bayesian network structure into an undirected graph. This is done by removing the directions from the existing edges and adding further undirected edges between all pairs of graph parents with a common (graph) child node. The latter process is referred to as *moralising* the graph, i.e. by 'marrying' the (graph) parents. In the resulting *moral graph*, all sets of the form $\{v\} \cup \mathrm{pa}(v)$ are *complete* in the graph, meaning that all pairs of elements are connected with edges. The factorization (24.4) can therefore be written as

$$f(x) = \prod_{v \in V} f(x_v \mid x_{\mathrm{pa}(v)}) = \prod_{C \in \mathcal{C}} \phi_C(x_C), \qquad (24.8)$$

where $\mathcal{C}$ denotes the set of *cliques* of the moral graph, i.e. the maximal complete subsets of nodes, and the functions ϕ are the *potentials*. To obtain this factorization, we just collect factors $f(x_v \mid x_{\mathrm{pa}(v)})$ with $\{v\} \cup \mathrm{pa}(v)$ in the clique C, so that the potential ϕ_C is a product of these factors. Since $\{v\} \cup \mathrm{pa}(v)$ is complete in the moral graph, this can always be done. Heath (2003) uses the term *dependency graph* for the moral graph.

Figure 24.8(a) shows a Bayesian allele network corresponding to a modification of the pedigree in Figure 24.2, where the phenotypes of individuals 11 and 15 have been explicitly represented. The corresponding moral graph is displayed in Figure 24.8(b).

24.4.1.2 Triangulation

Computational difficulties associated with pedigree analysis are related to the cycles of the moral graph rather than to the loops of the pedigree graph. The next step of the compilation process is addressing this problem through *triangulation* of the graph by adding *fill-in* edges to the moral graph until all cycles involving more than three nodes have *chords*. When this has been done, and only then, the cliques of the resulting graph can be arranged in a junction tree, see details below.

A triangulation of the graph in Figure 24.8(b) is displayed in Figure 24.9, where six fill-in edges have been added. Such a triangulation is most often found by using an ordering for node *elimination*; when a node is eliminated, fill-in edges are added between any pairs of the node's neighbours, which are not already connected by an edge. The node is then removed together with all its neighbours. The notion of an elimination ordering is identical to what is known as a *peeling sequence*, and the term *peeling* refers to the

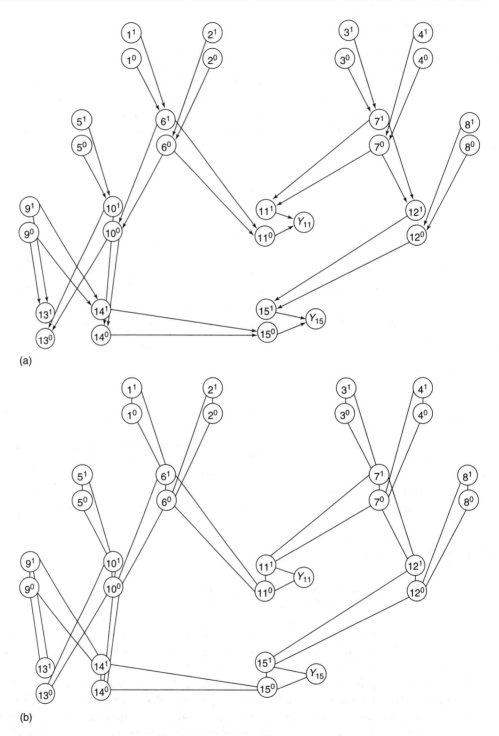

Figure 24.8   (a) Bayesian allele network with phenotypic information on two individuals and (b) its associated moral graph.

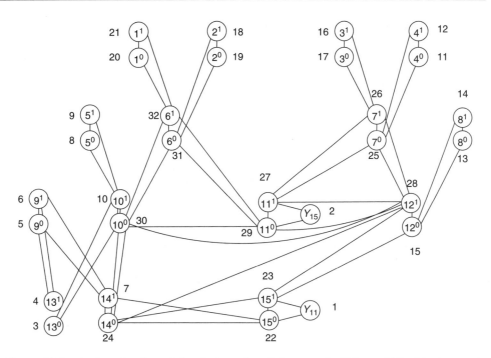

Figure 24.9 A triangulated graph for the Bayesian allele network of Figure 24.8 with phenotypic information represented for two individuals. The numbers $1, \ldots, 32$ indicate the corresponding node-elimination ordering.

elimination process, where one node is 'peeled off' at a time. The factorization (24.8) clearly implies a similar factorization

$$f(x) = \prod_{C \in \mathcal{C}} \phi_C(x_C), \qquad (24.9)$$

where $\mathcal{C}$ now denotes the set of cliques in the triangulated graph, since cliques in the moral graph are complete in any graph with more edges.

Figure 24.9 also displays the elimination order used to produce the given triangulation. A triangulation is not unique and the goal is to generate *cliques* (maximal sets of pairwise connected nodes) which are as small as possible. Optimizing this step is known to be NP (nondeterministic polynomial) complete (Yannakakis, 1981), but Jensen (2002) has implemented an algorithm which, in most cases, runs at reasonable computational speed and is guaranteed to return an optimal triangulation. This is based on the work of (Shoiket and Geiger, 1997; Berry *et al.*, 2000; Bouchitté and Todinca, 2001).

The triangulation step is crucial, as this determines the computational complexity, and thus whether exact computations are at all feasible or whether approximate methods such as Markov chain Monte Carlo (MCMC) will be required.

24.4.1.3 Constructing the Junction Tree

Once the graph has been triangulated, the cliques can easily be identified and connected in what is known as a *junction tree*. This is a tree having the set $\mathcal{C}$ of cliques of a triangulated graph as nodes, and satisfying the further property that

$$C \cap D \subseteq E \text{ for all } C, D, E \in \mathcal{C} \text{ with } E \text{ between } C \text{ and } D, \qquad (24.10)$$

where E is between C and D if it lies on the unique path from C to D. A junction tree for the triangulated graph in Figure 24.9 is displayed in Figure 24.10.

24.4.1.4 Loading the Junction Tree

The next step is to identify the *potentials* ϕ_C in the factorization (24.9). This is done by collecting factors of the form $f(x_v \mid x_{pa(v)})$ in (24.8) into cliques which contain both v and $pa(v)$. For each node v, at least one such clique exists and we choose one of them, say C, and *assign* the node v to C. If $V(C)$ denotes the set of nodes which are assigned to C, we let $\phi_C(x_C) \equiv 1$ for $V(C) = \emptyset$, otherwise,

$$\phi_C(x_C) = \prod_{v \in V(C)} f(x_v \mid x_{pa(v)}),$$

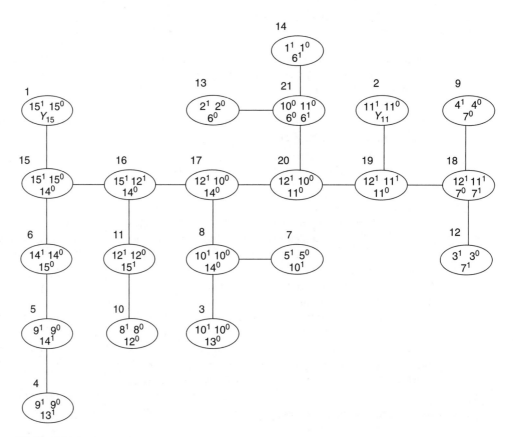

Figure 24.10 Junction tree for the triangulation of the Bayesian allele network shown in Figure 24.9.

whereby (24.9) is clearly satisfied with the joint distribution expressible as a product of potentials over the cliques. This concludes the general part of the compilation process.

24.4.1.5 Incorporating Observations

The compilation process described above has not yet taken account of the data available for the analysis in question. The representation (24.9) gives the joint probability of an arbitrary configuration of variables in the network. However, we want the probability of configurations which are consistent with the observations. This can be obtained if, for all v where $X_v = x_v^*$ is observed, we find a clique C with $v \in C$ and modify the potential there to ϕ_C^* by changing appropriate values to zero. More precisely, we let

$$\phi_C^*(x_C) = \begin{cases} \phi_C(x_C) & \text{if } x_v = x_v^* \\ 0 & \text{otherwise.} \end{cases} \tag{24.11}$$

This then implies that $\prod_C \phi_C^*(x_C)$ is equal to the joint probability of an arbitrary configuration x which is consistent with the observations. The process of forming ϕ^* from ϕ is often referred to as *entering evidence*. If we denote the set of observed nodes with E, we have

$$f(x \mid x_E^*) = \frac{\prod_{C \in \mathcal{C}} \phi_C^*(x_C)}{Z(x_E^*)}, \tag{24.12}$$

where the normalizing constant $Z(x_E^*)$ is the probability of the observations, obtained by summing over all configurations which are consistent with the observations:

$$f(x_E^*) = Z(x_E^*) = \sum_{x: x_E = x_E^*} \prod_{C \in \mathcal{C}} \phi_C(x_C) = \sum_x \prod_{C \in \mathcal{C}} \phi_C^*(x_C). \tag{24.13}$$

This also yields the *likelihood* when comparing different models.

24.4.2 Propagation

In the second part of the algorithm, often referred to as *propagation of evidence*, the actual computations with numbers are made, and the probabilities of interest are calculated. In particular, the sum in (24.13) must be calculated with more sophisticated techniques than brute force, since the number of terms in the sum grows exponentially with the number of nodes in the network.

There are several variants of the general algorithm of which we describe the HUGIN procedure (Jensen *et al.*, 1990), which represents a refinement of the algorithm of Lauritzen and Spiegelhalter (1988). Another variant, known as the *Shafer–Shenoy procedure* (Shenoy and Shafer, 1990), is closer to what is known as *peeling*, but it includes the more general variant used in Thompson (1981) to derive gene probabilities for all individuals in the pedigree.

24.4.2.1 The HUGIN Procedure

With every branch of the junction tree between neighbours C and D, we associate a *separator* $S = C \cap D$. The algorithm used in the software HUGIN (Andersen *et al.*, 1989) makes specific use of the separators by storing a single potential ψ_S along every branch of the junction tree. Initially, all these separator potentials are identically set to be equal to unity, so the factorization (24.12) implies that

$$f(x \mid x_E^*) \propto \frac{\prod_{C \in \mathcal{C}} \psi_C(x_C)}{\prod_{S \in \mathcal{S}} \psi_S(x_S)}, \tag{24.14}$$

where $\mathcal{S}$ is the set of separators and, initially, $\psi_C = \phi_C^*$ after evidence has been entered.

When a message is sent from C to D via the separator $S = C \cap D$, the following operations are performed:

$$\psi_C^{\downarrow S}(x_C) = \sum_{y_{C \setminus S}} \psi_C(x_S, y_{C \setminus D})$$

$$\tilde{\psi}_D(x_D) = \psi_D(x_D) \frac{\psi_C^{\downarrow S}(x_S)}{\psi_S(x_S)}$$

$$\tilde{\psi}_S(x_S) = \psi_C^{\downarrow S}(x_S),$$

i.e. first the *S-marginal* $\psi_C^{\downarrow S}$ of ψ_C is calculated by summing out over all variables not in S, then the clique potential ψ_D is modified by multiplication with the 'likelihood ratio' $\psi_C^{\downarrow S}/\psi_S$, and finally the separator potential ψ_S is replaced with $\psi_C^{\downarrow S}$. The potential from the clique which sends the message is unmodified, i.e. $\tilde{\psi}_C = \psi_C$. Since we have

$$\frac{\tilde{\psi}_C(x_C)\tilde{\psi}_D(x_D)}{\tilde{\psi}_S(x_S)} = \frac{\psi_C(x_C)\left(\psi_D(x_D)\dfrac{\psi_C^{\downarrow S}(x_S)}{\psi_S(x_S)}\right)}{\psi_C^{\downarrow S}(x_C)} = \frac{\psi_C(x_C)\psi_D(x_D)}{\psi_S(x_S)},$$

the factorization (24.14) remains valid at all times during the computational procedure.

Messages are now sent between neighbours in the tree according to a specific *schedule*. An efficient message passing schedule allows a clique to send exactly one message to each of its neighbours only after it has already received messages from all its other neighbours. Such a message passing schedule can be implemented via a local control. Alternatively, one can use a global control by first choosing a root R, then making an inward pass through the junction tree, known as COLLECTEVIDENCE, by which messages are sent from the leaves inwards towards R, and subsequently making an outward pass, DISTRIBUTEEVIDENCE, which sends messages in the reverse direction from the root towards the leaves. The first of these phases is illustrated in Figure 24.11.

When exactly two messages have been sent along every branch of the junction tree in an efficient schedule, it holds that

$$f(x_A \mid x_E^*) = \psi_A(x_A)/Z(x_E^*), \quad \text{for all } A \in \mathcal{C} \cup \mathcal{S}. \tag{24.15}$$

The marginal probability and normaliser Z can therefore be found as

$$f(x_E^*) = Z(x_E^*) = \sum_{x_S} \psi_S(x_S),$$

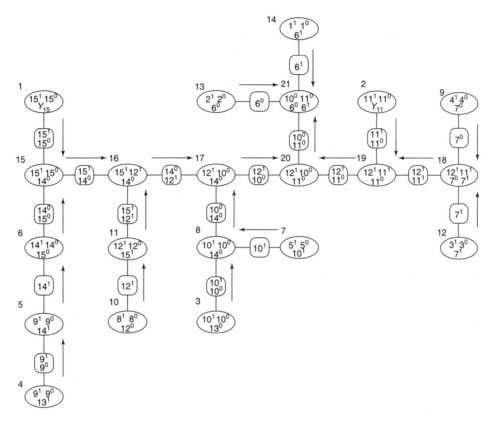

Figure 24.11 The first of the two computational phases in the HUGIN procedure. During COLLECTEVIDENCE, messages are sent towards the root C_{21}.

from any of the separator potentials ψ_S. In particular, the separator with the smallest associated state space can be chosen to calculate this sum.

24.4.3 Random and Other Propagation Schemes

Some generalizations of the message passing schemes described above use different definitions of the marginalisation operation $\downarrow$ and the multiplication in the basic factorization and message computation, but otherwise work in essentially the same fashion (Shenoy and Shafer, 1990; Lauritzen and Jensen, 1997). For example, replacing summation with maximization in (24.13) still yields a valid propagation scheme, known as *max-propagation*. Then, after CollectEvidence, the (max) normalisation constant Z is the probability of the most probable configuration of all variables in the network, and this configuration will be identified after DISTRIBUTEEVIDENCE (Dawid, 1992). Since the relation (24.14) remains invariant in the HUGIN procedure (also under max-propagation), one can easily switch between propagation modes.

Another important generalisation is the *random propagate* algorithm described by Dawid (1992). This begins with CollectEvidence to a root R using sum-marginalisation, but in the reverse step, a Monte Carlo sample is drawn as follows. After COLLECTEVIDENCE, the potential ψ_R is proportional to the conditional probability distribution of the variables in the root clique, given the evidence, cf (24.15). Hence, a random configuration $\check{x}_R$ can readily be sampled according to this distribution. The root clique now passes this configuration on to each of its neighbours C as $\check{x}_{R \cap C} = \check{x}_S$, where $S = C \cap R$ is the separator between C and R. After this has been done, each of the neighbouring cliques C chooses a random configuration $\check{x}_{C \setminus S}$ of the remaining variables according to a probability distribution which is proportional to $\psi_C(x_{C \setminus S}, \check{x}_S)$. When the neighbouring cliques have sampled their configurations in this way, they in turn pass on the chosen configuration to their neighbours, and so on. When the sampling stops at the leaves of the junction tree, a configuration $\check{x}$ has been correctly generated from the conditional distribution $f(x \mid x_E^*)$, given the evidence. This procedure is the general version of what (Thompson, 2000, page 95) describes as a variation of the Baum (1972) algorithm, and forms an essential step in many Monte Carlo based computational schemes which are relevant for genetic analyses. In particular, any sampling scheme which carries out a block update on several variables jointly and conditionally on the values of the remaining variables in the network makes use of the *random propagate* algorithm.

24.4.4 Computational Shortcuts

Computational issues have been considered by geneticists for a long time. As a result, a number of shortcuts have been developed which speed up computations beyond the efficiency intrinsic to the local computation algorithms themselves. These shortcuts are all associated with pre-processing before the compilation and propagation steps, and have the purpose of eventually leading to a reduction of the total size of the state spaces associated with the cliques of the final junction tree. Several of these pre-processing steps are, for example, described in Sheehan (2000), Fishelson and Geiger (2002), and Heath (2003). We refer to Lauritzen and Sheehan (2003) for details of these procedures known as, e.g. *trimming, forcing, excluding, allele recoding,* and *delayed triangulation.*

24.5 PEDIGREE ANALYSIS AND BEYOND

We will now use Bayesian network representations for some specific problems involving pedigree analysis.

24.5.1 Single-point Linkage Analysis

Consider a diallelic dichotomous disease segregating through a population with alleles D and d, and affected and normal phenotypes. Typically, we will have some observed phenotypes for the disease and some individuals will be typed at the marker locus. As in Kong (1991), we assume a recessive model for disease with complete penetrance whereby dd homozygotes are always affected and are never normal while both other genotypes are always normal and never affected. Assume that allele frequencies for both loci are known, segregation is Mendelian, founder genotype frequencies are in Hardy–Weinberg Proportions, and the founder population is in linkage equilibrium. The only unknown quantities are the unobserved genotypes and phenotypes, and r, the recombination fraction between the two loci.

To construct a graphical model for the single-point linkage problem, it suffices to focus on a nuclear family comprising a father 1, mother 2, and their offspring 3.

This construction is then replicated for all parent–child triplets in the pedigree, e.g. using an object-oriented method of specification. We will use the segregation network representation as we need to explicitly refer to phase information for linkage. Beginning with the marker locus–the 'α locus'–for each parent $i = 1, 2$, we create two nodes, $i^{1\alpha}$ and $i^{0\alpha}$, for the paternal and maternal genes of the individual with values drawn from the marker allele frequency distribution (e.g. multinomial). Note that this random assignment immediately deals with the fact that phase is unknown in the parents and we have to integrate it out by summing over all possibilities. We can assume that segregation from both parents is Mendelian at this 'first' locus, i.e.

$$P(S_{1,3}^\alpha = 1) = P(S_{2,3}^\alpha = 0) = 1/2, \tag{24.16}$$

although this is not necessary. As in Section 24.3, the paternally inherited allele of the offspring, $3^{1\alpha}$, is a (graph) child of both alleles in the father, $1^{1\alpha}$ and $1^{0\alpha}$, and of $S_{1,3}^\alpha$. Likewise, the maternally inherited allele of 3 is a (graph) child of both genes of 2 and $S_{2,3}^\alpha$. The genotype node is a (graph) child of both genes of individual nodes.

Labelling the disease locus as δ, the graph is extended by adding two nodes $i^{1\delta}$ and $i^{0\delta}$ for each parent, 1 and 2, exactly as above. The unobserved disease genotype, G_i^δ, is a child node of the corresponding gene nodes and a (graph) parent of the observable phenotype, Y_i^δ with link specified by the penetrance function. In this case, the penetrance probabilities are either 0 or 1. For the offspring, 3, we have gene nodes and a segregation indicator exactly as for the marker locus with the difference now being that we must take account of the linkage between the loci. In particular, the value of the segregation indicators $S_{1,3}^\delta$ and $S_{1,3}^\alpha$ are dependent via the recombination fraction, r. This dependence can be modelled with an undirected link between the corresponding nodes. Formally this would lead to a chain graph representation (Lauritzen, 1996) rather than a DAG. However, for the sake of exposition, we use here the equivalent non-symmetric description through the conditional distribution of $S_{1,3}^\delta$, given $S_{1,3}^\alpha$, specifically:

$$S_{1,3}^\delta \sim \begin{cases} Ber(1-r) & \text{if } S_{1,3}^\alpha = 1 \\ Ber(r) & \text{if } S_{1,3}^\alpha = 0, \end{cases} \tag{24.17}$$

and similarly for $S_{2,3}^\delta$. To complete the graph in Figure 24.12, we now add nodes G_3^δ and Y_3^δ for the offspring's unobserved genotype and phenotype with links defined exactly as above.

Note that this is a full specification of the model similar to that described elsewhere (Kong, 1991; Jensen and Kong, 1999; Thomas *et al.*, 2000). It is important to note that further derivation of the relevant joint and marginal distributions is not necessary as these are a direct result of the induced factorisation (24.4). The graph in Figure 24.12 provides a clear visual representation of the model.

24.5.2 QTL Mapping

Sheehan *et al.* (2002) extend the linkage scenario described above to the problem of detecting a QTL from possibly incomplete marker data for a simple example involving two flanking loci. Two markers are considered with known map positions and it is hypothesised that there is a diallelic QTL somewhere between the two. The trait of interest is any trait measured on a continuum with an associated polygenic effect unlinked to the QTL.

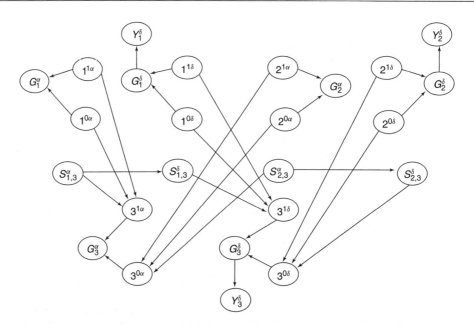

Figure 24.12   The segregation network for two linked loci on individuals 1 (father), 2 (mother), and 3 (offspring). Note that the information on linkage is contained in the directed edge between the segregation indicators $S_{1,3}^{\alpha}$ and $S_{1,3}^{\delta}$ and similarly between $S_{2,3}^{\alpha}$ and $S_{2,3}^{\delta}$.

Marker data are available on a *half-sib* design, common in animal breeding applications, comprising several families each with a single sire and up to 100 offsprings. Trait data are available only on the offspring. In contrast with the two-locus linkage example above, no information is given on the mothers (dams) of these offspring and hence the maternal segregations are all ignored.

The phenotype record on offspring j of sire i is a realisation of the random variable Y_{ij}. The effect of the unobserved genotype at the QTL is q_{ij}, where q_{ij} can have three possible values μ_1, μ_2, μ_3 corresponding to each of the three genotypes. A normal linear mixed model for the data is

$$Y_{ij} = Z_i + q_{ij} + E_{ij},$$

where Z_i represents the average additive genetic effect of the ith sire on the phenotypes of his offspring, which cannot be explained by the QTL. Let σ_a^2 be the total additive genetic variance unexplained by the QTL and σ_e^2 be the environmental variance. We have that $Z_i \sim N(0, \sigma_z^2) \forall i$ where $\sigma_z^2 = \frac{1}{4}\sigma_a^2$, the sire variance component, since half the genes of an offspring are shared with its sire (Falconer and Mackay, 1996). The remaining unexplained variation is picked up by the residual term, $E_{ij} \sim N(0, \frac{3}{4}\sigma_a^2 + \sigma_e^2)$. Assuming no genetic interference, only one of the unknown marker-QTL recombination fractions, or equivalently the QTL map location λ_Q, is required to parameterise the problem.

Figure 24.13 shows the graphical model for this trivial QTL-mapping problem for one sire and two offspring. The marker loci are labelled α and β while the QTL locus is now δ. The model is essentially an extension of the single-point linkage problem in Figure 24.12 to a two-point problem. Gene nodes are added for the third locus in an analogous fashion with the segregation indicator for inheritance from the sire linked to the previous value

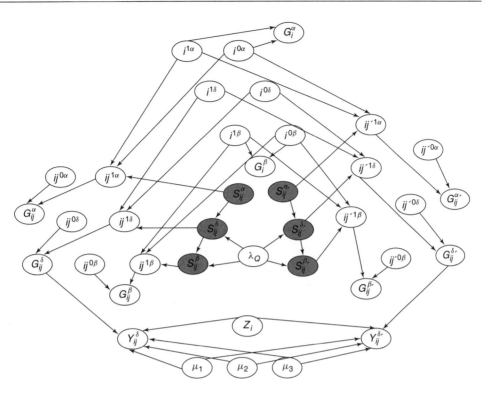

Figure 24.13 The graphical model for the QTL-mapping problem depicting a sire, i with two daughters, ij and ij'. [Reproduced from Sheehan *et al.* (2002). International Statistical Review, 68:83–110 with permission from Blackwell Publishing.]

via the recombination fraction between the second and third loci, as described in (24.17) above. This assumes that there is no genetic interference and that recombinations in adjacent intervals are independent. Maternal genes in the offspring are randomly drawn from the population as there is no information on the dams. Sire and offspring have marker genotype nodes while only offspring have trait genotype and phenotype nodes. Covariance between the offspring is reflected in the genes they share with their sire, and they are duly connected by the sire effect node. Note that this creates a cycle in the graph which becomes increasingly complex computationally when more offspring are added and, for a typical half-sib design with a sire having up to 100 offspring, the relevant Bayesian network for this problem features many long cycles despite the simplicity of the pedigree structure and the mapping problem under consideration (Sheehan *et al.*, 2002). The half-sib design is a zero-loop pedigree (or *tree*) and genetic mapping problems do not get any simpler than this one. The graphical model highlights the computational complexity implicit in the mixed-model approach to this analysis and clarifies why these analyses are challenging on more complex problems.

24.5.3 Pedigree Uncertainty

The flexibility of a graphical modelling approach to applications in genetics is powerfully demonstrated when the pedigree is not fixed and known, or when other circumstances should be integrated into the analysis. Hansen and Pedersen (1994) elegantly handle incomplete paternity information in a two-locus inheritance model for fur colour in foxes from pedigrees supplied by Scandinavian fur farms where there is uncertainty with some of the litter paternities. It is common breeding practice to mate a female with two males in order to increase the chances of fertilisation and hence, it is not always possible to determine which male actually fathered the resulting pups. Indeed, two males could father pups in the same litter. The pedigree declares the most likely candidate as the father (to all pups) and registers the second sire as an alternative whenever there is doubt.

The phenotypic record on each fox is a subjective classification of fur colour on a scale from $1-8$. From analogous models for mice and sheep (Adalsteinsson *et al.*, 1987), a model for genetic inheritance of fur colour was proposed involving two diallelic loci, α (with alleles A and a) and ε (E, e), possibly on the same chromosome with unknown recombination fraction.

Although there are some loops, the fox pedigrees are generally small enough for exact likelihood calculation with simple models (Skjøth *et al.*, 1994). The usual method for handling paternity uncertainty is to compare the likelihoods for all possible pedigrees (Thompson, 1986). There is only one alternative father for each of a small number of litters, but as each pup in the litter could have been fathered by either of the two candidates, this problem would require the consideration of 2^{21} pedigrees. Skjøth *et al.* (1994) circumvent this problem by estimating paternal genotypes from the phenotypic information and choosing the most likely individual but this is not very satisfactory. Although standard statistical genetics programs will not accept a pedigree where an individual can have more than one biological father, this presents no difficulty for a general graphical model. Hansen and Pedersen (1994) exploit this to incorporate all the paternity information by defining a binary node, W_i, for each queried pup:

$$W_i = \begin{cases} 1 & \text{if stated father is the true father} \\ 0 & \text{if alternative father is the true father,} \end{cases}$$

and $W_i \sim Ber(p_w)$, where p_w is to be estimated.

Again, the graphical model for this problem is essentially the two-locus linkage model described above (Section 24.5.1) *except* for those cases where paternity is uncertain. We thus focus on the quadruplet comprising the mother, 1, declared father, 2, offspring, 3, and alternative father, 4. For illustration, we have simplified the model by assuming a known penetrance matrix and known allele frequencies at both loci. As before, paternal and maternal alleles, $i^{1\alpha}, i^{0\alpha}, i^{1\varepsilon}, i^{0\varepsilon}$, are assigned to the founders $i = 1, 2$, and 4 at both loci. Mother–offspring segregation indicators are assigned for each locus, $S_{1,3}^{\alpha}, S_{1,3}^{\varepsilon}$, and indicators for paternal inheritance are $S_{f_3,3}^{\alpha}, S_{f_3,3}^{\varepsilon}$, where $f_3 = 2$ if $W_3 = 1$ and $f_3 = 4$, otherwise. In contrast with our representation of Section 24.5.1, Hansen and Pedersen (1994) consider segregation at both loci jointly, so we have phase indicators for inheritance, $S_{1,3}$ and $S_{f_3,3}$, where

$$S_{1,3}, S_{f_3,3} = \begin{cases} (0, 0) & \text{with probability} & (1\text{-}r)/2 \\ (0, 1) & \text{with probability} & r/2 \\ (1, 0) & \text{with probability} & r/2 \\ (1, 1) & \text{with probability} & (1\text{-}r)/2 \end{cases}.$$

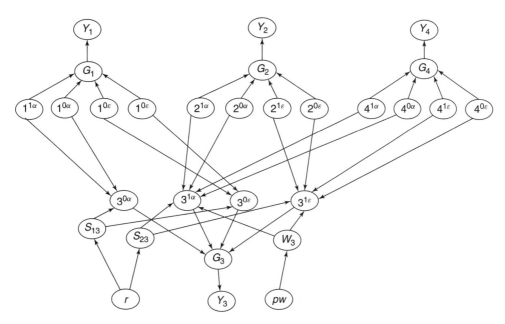

Figure 24.14 The graphical model for the fox data depicting a mother, 1, father 2, offspring 3, and alternative father 4. [Adapted from Hansen and Pedersen (1994).]

Note that this is the same model as described in (24.16) and (24.17), since $S_{1,3} = (S_{1,3}^\alpha, S_{1,3}^\varepsilon)$. The earlier parameterisation, involving more nodes with fewer states, is more flexible when considering more than two loci and is generally better for computational purposes. All four alleles of individual i are graph parents of the node G_i representing the two-locus genotype and this, in turn, is a graph parent of Y_i, the fur colour phenotype. Figure 24.14 shows the corresponding graphical model for this problem.

Despite the simplifications, the graph in Figure 24.14 is more complicated than those shown earlier in this section in that it has many more cycles. The advantage is that questions about paternity, genetic inheritance, and linkage can all be addressed from this one graph, whereas these would typically require separate considerations using standard pedigree software.

An interesting class of questions focuses on the determination of pedigrees from observable genetic information, see, e.g. Egeland *et al.* (2000), Cowell and Mostad (2003), Steel and Hein (2006), and Sheehan and Egeland (2007).

24.5.4 Forensic Applications

Graphical models, or *probabilistic expert systems*, have been shown to be particularly useful in forensic applications (Dawid *et al.*, 2002; Cowell, 2003; Taroni *et al.*, 2006) and have been adapted to handle a wide range of routine problems in forensic inference. The central problem here is to infer the identity of an individual based on the given evidence which possibly includes DNA profile information. This can be done by calculating relative likelihoods for the various competing hypotheses (Dawid and Mortera, 1996), but these become computationally intensive when information is imperfect or missing (Dawid and

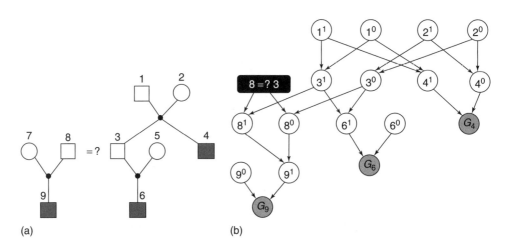

Figure 24.15 The simple paternity problem of Dawid *et al.* (2002) represented here (a) by two marriage node graphs, where individuals shaded in grey denote those for whom DNA evidence is available, and (b) as a graphical model. The three grey nodes in (b) represent the observed genotypes and the black node is the 'query' node.

Mortera, 1998) and especially when the possibility of observing a mutation from one generation to the next is considered (Dawid *et al.*, 2001).

Consider the inheritance claim case in Figure 24.15(a) from Dawid *et al.* (2002), where a man, 9, claims to be the son of the diseased individual 3 and hence entitlement to part of his estate. It is known that 3 had a (undisputed) child, 6, and that 6 and 9 had different mothers. There is no DNA information on 3 since he is dead and buried, nor on either of the two mothers, but we have DNA profile samples from both offspring 6 and 9, and from the brother of the diseased, 4. This problem can be formally expressed as a case of disputed paternity with 3 as the putative father and 9 as the disputed child. As is common for such applications, attention is on just two competing hypotheses: either the true father, 8, of the disputed child, 9, really is one and the same as the putative father, 3, *or* the true father can be considered as randomly drawn from the general population.

Forensic markers are usually selected to be unlinked so we only need a model for a single marker and the overall likelihood is the product over all markers. Figure 24.15(b) shows the single marker representation used by Dawid *et al.* (2002). This uses the allele network rather than the segregation network but, as indicated earlier, the latter is superfluous in the absence of linkage. Our notation is as before with the marker labels omitted, so i^0 and i^1 represent the random variables assigning maternal and paternal genes of the individual i, and G_i assigns the genotype of i at the marker. Untyped individuals who are not directly of interest (i.e. 1, 2, 5, and 7) are only represented by the genes they contribute to the next generation, which, in the absence of any information, are assumed on the contrary to be randomly drawn from the population.

The black node in Figure 24.15(b) which is a (graph) parent of both genes in 8, is the 'query' or 'target' node (Dawid *et al.*, 2002). This is a binary node and is 1 if the true father of the disputed child is the putative father, i.e. if individual 8 is the same as 3 in Figure 24.15(a) and the two genes in 8 are hence copies of the corresponding two in 3. Otherwise, the men are different individuals and the genes of 8 are drawn randomly from the population. The advantage of using the query node for this application is that the quantity of interest to the court–the likelihood ratio in favour of paternity–can be read off directly. However, we note that despite the different emphasis of the analysis, this node is essentially the same as the paternity indicator, W_i, in Figure 24.14 determining which of the two possible alternatives fathered the individual i. For the forensic example, although a specific alternative is often not available, determination of paternity is crucial: for the fox example, a genetic model for inheritance of fur colour was the focus and the uncertain paternity was a nuisance factor that had to be taken into account. The same graphical modelling approach can be used to address both questions.

The area of forensic genetics yields a variety of problems which can clearly benefit from the flexibility and modularity of a graphical modelling environment. For example, in criminal cases such as assault and rape cases, it is not uncommon to observe crime trace evidence which represents a *mixture* of DNA from an unknown number of individuals. Mortera *et al.* (2003) consider an example where there is a victim 1, a suspect 2, and exactly two contributors, 6 and 7, to the trace evidence (i.e. there were three alleles present for at least one marker in the mix). To complicate matters, the suspect has left the country and is not available for typing. However, his brother 3 has been found and has given a sample. Figure 24.16 shows the allele network used by Mortera *et al.* (2003) for this identification problem, where individuals 4 and 5 are the parents of the two brothers. Four standard competing hypotheses concerning the makeup of the set contributing to the mixture (Weir *et al.*, 1997) are as follows:

1. the suspect and victim have both contributed;

2. the suspect and an unknown contributor are represented;

3. the victim and an unknown contributor are represented;

4. both contributors to the mixture are unknown.

Figure 24.16 shows that we have genotype data on 1 and 3. Unobserved genotypes of the two contributors, 6 and 7, are graph parents of the node representing the observed

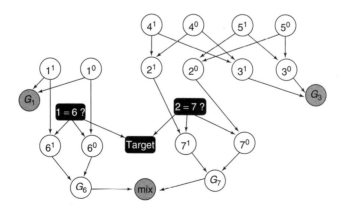

Figure 24.16 The mixture network of Mortera *et al.* (2003). The three grey nodes represent the observed genotypes and the black nodes are the 'query' and 'target' nodes respectively. [Reprinted from Theoretical Population Biology, 63(3), Mortera J. *et al.*, Probabilistic expert systems, 191–205, copyright (2003) with permission from Elsevier.]

mixed trace. We now have two binary query nodes defined exactly as before which address the two questions: is the victim the first contributor ($1 = 6$?) and is the suspect the second ($2 = 7$?). If the answer is 'yes', the genes of the two individuals are identical. Otherwise, the genes of the contributor are drawn randomly from the population gene pool. The 'target' node is the logical conjunction of these two, and has four states corresponding to the four hypotheses above. Thus, competing standard hypotheses can all be considered from the same network and separate pairwise comparisons, as are routinely performed, are not necessary. In principle, this approach can be extended to deal with multiple contributors, possible silent alleles in the mixture, and multiple missing individuals (Mortera *et al.*, 2003), and information on the amount of DNA contributed by each individual can be exploited for separation of DNA profiles (Cowell *et al.*, 2007).

24.5.5 Bayesian Approaches

These models lend themselves readily to Bayesian analysis and interpretation (Spiegelhalter, 1998), where all unknown quantities such as data, latent variables, and model parameters can be regarded as random variables and thus represented as extra nodes in the graph with their associated distributions. Exact local computation algorithms fail as more complex distributional types need to be accommodated, so alternative methods for calculating the quantities of interest, such as MCMC (Hastings, 1970; Metropolis *et al.*, 1953) must be entertained. It is well known that the popular single-site Gibbs sampler Geman and Geman (1984) can mix very slowly in complex models involving both discrete and continuous nodes, even when the sampler is theoretically irreducible (Janss *et al.*, 1995; Heath, 1997; Jensen and Kong, 1999; Lund and Jensen, 1999). Some kind of blocking or joint updating of variables is hence required in order to sample more efficiently. Lund and Jensen (1999), e.g. use graphical models for a Bayesian formulation of a mixed inheritance model, Sheehan *et al.* (2002) extend the model in Figure 24.13 to a full Bayesian analysis for the QTL mapping problem, and Hansen and Pedersen (1994) take a Bayesian approach to the fox problem of Figure 24.14. They all invoke *random propagation* as described in Section 24.4.3 as an essential part of the associated block update for the discrete part of the model.

24.6 CAUSAL INFERENCE

Graphical models, especially DAGs, have natural causal interpretations and hence provide a formal framework to discuss causal concepts. We will illustrate this using *Mendelian randomisation*, an approach to understanding aetiological relationships in observational studies, by way of an example. Inferring causation from observed associations is often a problem with epidemiological data as it is not always clear which of two variables is the cause, which is the effect, or whether both are common effects of a third unobserved variable or confounder. In the case of experimental data, causal inference is facilitated either by using randomisation or experimental control. In many biological settings, it is not possible to randomly assign values of a hypothesised 'cause' to experimental units for ethical, financial, or practical reasons. In epidemiological applications, for example, randomised controlled trials (RCTs) to evaluate the effects of exposures such as smoking, alcohol consumption, physical activity, and complex nutritional regimes are unlikely to be carried out.

24.6.1 Causal Concepts

As in Pearl (1995), Lauritzen (2000), and Dawid (2002; 2003), we will regard causal inference to be about predicting the effect of *interventions* in a given system. If X is the cause under investigation and Y the response, the question of interest is whether intervening on X has an effect on Y. By intervening on X, we mean that we can set X (or more generally its distribution) to any value we choose without affecting the distributions of the other variables in the system, except through the resulting changes in X. This is clearly idealistic and may not always be justifiable. The *causal effect* of X on Y is a function of the distributions of Y under different interventions in X. It is well known that this is not necessarily equal to the usual conditional distribution $P(Y|X = x)$ which just describes a statistical dependence (Pearl, 2000; Lauritzen, 2000). We will follow Pearl (2000) and use the notation $P(Y|\mathrm{do}(X = x))$ to clarify that conditioning is on intervention in X.

The average causal effect (ACE) is defined as the difference in expectations under different settings of X:

$$ACE(x_1, x_2) = E(Y|\mathrm{do}(X = x_1)) - E(Y|\mathrm{do}(X = x_2)). \qquad (24.18)$$

X is regarded as causal for Y if the ACE is non-zero for some values x_1, x_2. If X is binary, the unique ACE is given by $E(Y|\mathrm{do}(X = 1)) - E(Y|\mathrm{do}(X = 0))$. If Y is continuous, a popular assumption is that the causal dependence of Y on X is linear (possibly after suitable transformations), i.e. $E(Y|\mathrm{do}(X = x)) = \alpha + \beta x$. In this case, the ACE is $\beta(x_1 - x_2)$ and can simply be summarised by β, which is now interpreted as the average effect of increasing X by one unit through some intervention. In the more general cases of more than two categories and/or nonlinear dependency, the ACE is not necessarily summarised by a single parameter, and one may want to choose a different causal parameter altogether (Didelez and Sheehan, 2007).

A causal parameter is *identifiable* if it can be estimated consistently from obtainable information on the joint distribution of the observed variables. Mathematically, this amounts to being able to express the parameter in terms that do not involve the intervention (i.e. the 'do' operation) by only using 'observational' terms that can be estimated from data. In the presence of unknown confounders, e.g. parameters of $P(Y|\mathrm{do}(X = x))$ cannot be estimated directly from observations that represent $P(Y|X = x)$. In the rare case of known confounders, it can be shown that the intervention distribution can be re-expressed in observational terms and can thus be estimated from the observed data by adjusting for these confounders (Pearl, 1995; 2000; Lauritzen, 2000; Dawid, 2002).

24.6.2 Mendelian Randomisation

Mendelian randomisation has been proposed as a method to test for, or estimate, the causal effect of an exposure or phenotype on a disease when confounding is believed to be likely and not fully understood (Davey Smith and Ebrahim, 2003; Katan, 2004). It exploits the idea that a well-understood genotype, known to affect levels of the exposure, affects the disease status only indirectly and is assigned randomly (given the parents' genes) at meiosis, independently of the possible confounding factors. It is well known in the econometrics and causal literature (Bowden and Turkington, 1984; Angrist *et al.*, 1996; Pearl, 2000; Greenland, 2000) that these properties define an *instrumental* variable (IV), but they are *minimal* conditions in the sense that unique identification of the causal effect of the phenotype on the disease status is possible only in the presence of additional, fairly strong assumptions.

The core conditions that characterise an IV have been given in many different forms, using counterfactual variables (Angrist *et al.*, 1996; Robins, 1997), linear structural equations (Goldberger, 1972; Pearl, 2000, Chapter 7), or conditional independence statements, as we will use here. Our notation and terminology closely follow Greenland *et al.* (1999) and Dawid (2002). We now present these conditions together with a graphical way of depicting and checking the relevant conditional independencies.

Let X and Y be defined as above with the causal effect of X on Y being of primary interest. Furthermore, let G be the variable that we want to use as the instrument (the genotype in our case) and let U be an unobservable variable that will represent the confounding between X and Y. The 'core conditions' that G has to satisfy are the following

1. $G \perp\!\!\!\perp U$, i.e. G must be (marginally) independent of the confounding between X and Y;

2. $G \not\perp\!\!\!\perp X$, i.e. G must not be (marginally) independent of X; and

3. $Y \perp\!\!\!\perp G \mid (X, U)$, i.e. conditional on X and the confounder U, the instrument and the response are independent.

Because U is not observable, these assumptions cannot be formally tested and have to be justified on the basis of subject- matter or background knowledge. Moreover, the above assumptions do not imply any testable conditional independencies regarding the instrument G. Figure 24.17 shows the unique DAG involving G, X, Y, and U that satisfies the core conditions 1–3 with corresponding factorisation

$$p(y, x, u, g) = p(y|u, x)p(x|u, g)p(u)p(g). \qquad (24.19)$$

Note that DAGs only represent conditional dependencies and independencies: they are not causal in themselves despite the arrow suggesting a 'direction' of dependence. We say that the DAG has a *causal* interpretation with respect to the relationship between X and Y, or, more specifically, the DAG is causal with respect to intervention in X, if we believe that an intervention in X does not change any of the other factors in the joint

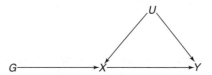

Figure 24.17   The directed acyclic graph (DAG) representing the core conditions for G to be an instrument. [Reproduced from Didelez, V. and Sheehan, N. A. (2007) Mendelian randomisation as an instrumental variable approach to causal inference. Statistical Methods in Medical Research, by permission of Sage Publication Ltd.]

distribution (24.19) (see Pearl, 2000, page 23). This means that

$$p(y, u, g | \mathrm{do}(X = x_0)) = p(y | u, x_0) p(u) p(g), \qquad (24.20)$$

assuming that $p(y | u, x_0) = p(y | u, \mathrm{do}(X = x_0))$.

The limitations of Mendelian randomisation, from the perspective of complicating features leading to poor estimation of the required genotype–phenotype and genotype–disease associations, have been discussed in detail in several places in the literature (Davey Smith and Ebrahim, 2003; 2004; Thomas and Conti, 2004; Davey Smith *et al.*, 2005). More crucially, biological complications can sometimes violate one or more of the core conditions 1–3 so that Figure 24.17 no longer applies. In order to understand what any added complexity implies with regard to meeting these conditions, the relevant conditional independencies can be easily checked using DAGs that are ideally dictated by the biology. For instance, when our chosen gene G_1 is in linkage disequilibrium with another gene G_2 which has a direct or indirect influence on the disease Y, condition 3, $Y \perp\!\!\!\perp G_1 | (X, U)$, might be violated as shown in Figure 24.18(a), or else condition 1, $G \perp\!\!\!\perp U$, might be violated as shown in Figure 24.18(b).

It is possible that the core conditions may still hold for the chosen instrument G_1 in the presence of genetic heterogeneity, as illustrated in Figure 24.19, if none of the other genes influence Y in any way other than via their effect on X. If instead, the situation is similar to Figure 24.18(b), however, the core conditions may be violated as already explained. Note that in Figure 24.19, genetic heterogeneity could weaken the $G_1 - X$ association and thus G_1 would be a poor instrument. If the chosen instrument G has pleiotropic effects and, in particular, if it is associated with another intermediate phenotype which also affects the disease Y (Figure 24.20(a)), condition 3, $Y \perp\!\!\!\perp G | (X_1, U)$, is again violated if we do not also condition on X_2. Moreover, a genetic polymorphism under study might have pleiotropic effects that influence confounding factors like consumption of tobacco or alcohol, for example, (Davey Smith and Ebrahim, 2003). This is represented in Figure 24.20(b) and violates condition 1. In the presence of population stratification, we see in Figure 24.21(a) that condition 3, $Y \perp\!\!\!\perp G_1 | (X, U)$, is again violated: we need to condition on the population subgroup as well. However, if the effect of population stratification is to cause an association between allele frequencies and phenotype levels, as in Figure 24.21(b), all conditions for G to be an instrument are still satisfied, and, in this situation, the $G - X$ association may in fact be strengthened, as a result.

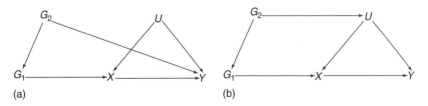

(a)   (b)

Figure 24.18 Linkage disequilibrium where the chosen instrument G_1 is associated with another genotype G_2 which directly influences the outcome Y, as in (a), or influences Y indirectly via the confounder U, as in (b). [Reproduced from Didelez, V. and Sheehan, N. A. (2007) Mendelian randomisation as an instrumental variable approach to causal inference. Statistical Methods in Medical Research, by permission of Sage Publication Ltd.]

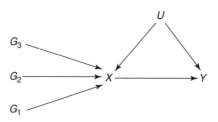

Figure 24.19 Genetic heterogeneity showing three genes which are all associated with the intermediate phenotype X, but none of which has an effect on the disease Y except through X. [Reproduced from Didelez, V. and Sheehan, N. A. (2007) Mendelian randomisation as an instrumental variable approach to causal inference. Statistical Methods in Medical Research, by permission of Sage Publication Ltd.]

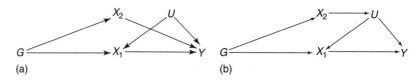

Figure 24.20 An example of pleiotropy where the instrument G is associated with both X_1 and X_2 and (a) both have a direct effect on the outcome Y of interest, or (b) where X_1 has a direct effect but X_2 has an indirect effect via the confounder U. [Reproduced from Didelez, V. and Sheehan, N. A. (2007) Mendelian randomisation as an instrumental variable approach to causal inference. Statistical Methods in Medical Research, by permission of Sage Publication Ltd.]

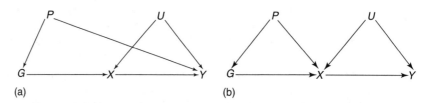

Figure 24.21 Two examples of population stratification where (a) one of the conditions for G to be an instrument is violated and (b) all conditions are satisfied. [Reproduced from Didelez, V. and Sheehan, N. A. (2007) Mendelian randomisation as an instrumental variable approach to causal inference. Statistical Methods in Medical Research, by permission of Sage Publication Ltd.]

24.7 OTHER APPLICATIONS

24.7.1 Graph Learning for Genome-wide Associations

The above examples all assume that the graphical structure of the problem is known. It is also possible to learn from the data and estimate the graphical model from observations of a joint distribution (Cowell *et al.*, 1999). Modelling linkage disequilibrium, or the tendency for alleles observed at one genetic locus to be correlated with those observed at another locus, is an important issue in statistical genetics with the increasing availability of single nucleotide polymorphism (SNP) data over dense marker maps for large numbers of individuals. In particular, the search for genetic determinants of complex diseases in genome-wide association studies is based on the idea that the ancestral disease-bearing mutations in a population will be flanked by segments of chromosome that will show less variability amongst those individuals with the disease than amongst those unaffected. Linkage disequilibrium is a function of physical distance between loci on a chromosome, but much heterogeneity has been observed in the correlations between adjacent loci, long range correlations are quite common, and completely uncorrelated loci can be interspersed between regions of tightly linked loci. Existing methods for multilocus haplotype analysis exploiting excess sharing amongst affected individuals around a disease locus do not scale up to the datasets that we can now reasonably anticipate (Verzilli *et al.*, 2006).

Thomas and Camp (2004) suggest the use of undirected graphs to model dependencies between genetic loci allowing for higher order interactions. Their method assumes proximal loci (e.g. SNPs within a single gene) and requires haploid data. The space of decomposable, or triangulated, graphs is searched to find the best-fitting model within that class using a simulated annealing algorithm, where the objective function is the maximised log-likelihood penalised according to the number of degrees of freedom. A simple perturbation rule for searching this space is provided by which two vertices of the existing graph are selected, and are connected if they are not already connected, disconnected otherwise. However, the authors acknowledge that mixing properties are greatly enhanced when other perturbation rules are included. The method is extended to deal with unphased diploid data (i.e. genotype data) by Thomas (2005), where the haplotype estimation step is incorporated iteratively with the estimation of the graphical model. Verzilli *et al.* (2006) apply these ideas to identify patterns of multilocus genotypes around a disease locus and thereby avoid the problems inherent in the estimation of haplotypes for large numbers of loci. They consider case-control data where the unphased genotypes, together with a binary disease status indicator, form the vertices of the graph. Edges between genotypes thus reflect the linkage disequilibrium structure, and edges between genotypes and the disease status indicator would suggest the presence of a disease susceptibility locus somewhere near these loci. Computational efficiency is achieved by the restriction of the search space to the set of triangulated graphs and by putting a limit on the size of the cliques. The set of possible graphs is restricted even further by imposing a prior that restricts the physical distance between clique members. Proposed moves around this space act via changes to the set of cliques and separators in the current graph thus avoiding the need to check that the new proposal is decomposable.

Although some restrictions on the set of graphs connecting over 500 000 genotypes and a disease status indicator are obviously required to make the graph learning exercise feasible, the above restrictions, as conceded by all the authors, are mainly driven by the desire to facilitate the computations and provide a solution in reasonable time. Consequently, all kinds of configurations are ruled out by these restrictions which may not be biologically implausible. For example, there does not seem to be any strong reason to believe that the graph in Figure 24.22(b) is an unreasonable model for four linked loci, where the locus at each end is associated with the disease. The complexity of the biology underlying most complex diseases is still not well understood, however, and it is thus difficult to suggest a more sensible restriction on the set of models to be explored.

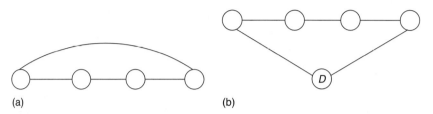

(a)   (b)

Figure 24.22 Two graphical models reflecting (a) associations between genotypes and (b) associations between genotypes and a disease indicator. Both are disallowed because they are not decomposable.

24.7.2 Gene Networks

Other recent and important applications of graphical models in genetics are concerned with inferring regulatory mechanisms involving genes, typically based on expression data measured using microarray technology. The graphical relationships involved in this type of model describe regulatory mechanisms, in principle, based on a causal interpretation of the relevant graphical models , such as those described in Section 24.6. The applications of graphical models in this context are generally exploratory and serve to exploit the massive data available to conjecture potential relationships which must subsequently be investigated further using other forms of biological subject-matter knowledge and experiments. One direction of research uses relatively sophisticated model specifications with some simplifying structures (Friedman, 2004; Segal *et al.*, 2003a; 2003b), exploiting the notion of probabilistic relational models (Friedman *et al.*, 1999) which add repeated relational structure to the conditional independence structure of graphical models.

Another approach uses less sophisticated models, essentially undirected Gaussian graphical models, exploiting the simplicity and detailed statistical understanding of these to make efficient and well-founded search procedures, either using Bayesian ideas (Jones *et al.*, 2005) or other graphical model selection algorithms (Spirtes *et al.*, 1993) to conjecture interesting relationships between genes under study. For an early example of this type of research, we refer to West *et al.* (2001). The methodology is currently developing rapidly, see, e.g. Schäfer and Strimmer (2005) and the corresponding software implementation as described in Schäfer *et al.* (2006). Markovetz (2006) provides an up-to-date online bibliography of the area.

REFERENCES

Adalsteinsson, S., Hersteinsson, P. and Gunnarsson, E. (1987). Fox colors in relation to colors in mice and sheep. *The Journal of Heredity* **78**, 235–237.

Andersen, S.K., Olesen, K.G., Jensen, F.V. and Jensen, F. (1989). HUGIN – a shell for building Bayesian belief universes for expert systems. *Proceedings of the 11th International Joint Conference on Artificial Intelligence*. Morgan Kaufmann Publishers, San Mateo, CA, pp. 1080–1085.

Angrist, J., Imbens, G. and Rubin, D. (1996). Identification of causal effects using instrumental variables. *Journal of the American Statistical Association* **91**(434), 444–455.

Baum, L.E. (1972). An equality and associated maximization technique in statistical estimation for probabilistic functions of Markov processes. *Inequalities* **3**, 1–8.

Berry, A., Bordat, J.-P. and Cogis, O. (2000). Generating all the minimal separators of a graph. *International Journal of Foundations of Computer Science* **11**, 397–403.

Bouchitté, V. and Todinca, I. (2001). Treewidth and minimum fill-in: grouping the minimal separators. *SIAM Journal on Computing* **31**, 212–232.

Bowden, R. and Turkington, D. (1984). *Instrumental Variables*. Cambridge University Press, Cambridge.

Cannings, C., Thompson, E.A. and Skolnick, M.H. (1978). Probability functions on complex pedigrees. *Advances in Applied Probability* **10**, 26–61.

Cowell, R.G. (2003). A probabilistic expert system for forensic identification. *Forensic Science International* **134**, 196–206.

Cowell, R.G., Dawid, A.P., Lauritzen, S.L. and Spiegelhalter, D.J. (1999). *Probabilistic Networks and Expert Systems*. Springer-Verlag, New York.

Cowell, R.G., Lauritzen, S.L. and Mortera, J. (2007). Identification and separation of DNA mixtures using peak area information. *Forensic Science International* **166**, 28–34.

Cowell, R.G. and Mostad, P. (2003). A clustering algorithm using DNA marker information for subpedigree reconstruction. *Journal of Forensic Sciences* **48**, 1239–1248.

Davey Smith, G. and Ebrahim, S. (2003). Mendelian randomization: can genetic epidemiology contribute to understanding environmental determinants of disease? *International Journal of Epidemiology* **32**, 1–22.

Davey Smith, G. and Ebrahim, S. (2004). Mendelian randomization: prospects, potentials, and limitations. *International Journal of Epidemiology* **33**, 30–42.

Davey Smith, G., Ebrahim, S., Lewis, S., Hansell, A., Palmer, L. and Burton, P. (2005). Genetic epidemiology and public health: hope, hype, and future prospects. *Lancet* **366**, 1484–1498.

Dawid, A.P. (1992). Applications of a general propagation algorithm for probabilistic expert systems. *Statistics and Computing* **2**, 25–36.

Dawid, A.P. (2002). Influence diagrams for causal modelling and inference. *International Statistical Review* **70**, 161–189.

Dawid, A.P. (2003). Causal inference using influence diagrams: the problem of partial compliance. In *Highly Structured Stochastic Systems*, P.J. Green, N.L. Hjort and S. Richardson, eds. Oxford University Press, Oxford, pp. 45–81.

Dawid, A.P. and Mortera, J. (1996). Coherent analysis of forensic identification evidence. *Journal of the Royal Statistical Society, Series B* **58**, 425–443.

Dawid, A.P. and Mortera, J. (1998). Forensic identification with imperfect evidence. *Biometrika* **85**, 835–849.

Dawid, A.P., Mortera, J. and Pascali, V.L. (2001). Non-fatherhood or mutation? A probabilistic approach to parental exclusion in paternity testing. *Forensic Science International* **124**, 55–61.

Dawid, A.P., Mortera, J., Pascali, V.L. and van Boxel, D. (2002). Probabilistic expert systems for forensic inference from genetic markers. *Scandinavian Journal of Statistics* **29**, 577–595.

Dawid, A.P., Mortera, J. and Vicard, P. (2007). Object-oriented Bayesian networks for complex forensic DNA profiling problems. *Forensic Science International* **169**, 195–205.

Didelez, V. and Sheehan, N.A. (2007). Mendelian randomisation as an instrumental variable approach to causal inference. *Statistical methods in Medical Research* (in press).

Egeland, T., Mostad, P.F., Mervåg, B. and Stenersen, M. (2000). Beyond traditional paternity and identification cases. Selecting the most probable pedigree. *Forensic Science International* **110**, 47–59.

Elston, R.C. and Stewart, J. (1971). A general model for the genetic analysis of pedigree data. *Human Heredity* **21**, 523–542.

Falconer, D.S. and Mackay, T.F.C. (1996). *Introduction to Quantitative Genetics*, 4th edition. Longman Group Ltd.

Fishelson, M. and Geiger, D. (2002). Exact genetic linkage computations for general pedigrees. *Bioinformatics* **18**, S189–S198.

Friedman, N. (2004). Inferring cellular networks using probabilistic graphical models. *Science* **303**(5659), 799–805.

Friedman, N., Getoor, L., Koller, D. and Pfeffer, A. (1999). Learning probabilistic relational models. *Proceedings of the Sixteenth International Joint Conference on Artificial Intelligence (IJCAI-99), Stockholm, Sweden*. Morgan Kaufman, San Francisco, CA, pp. 1300–1307.

Geman, S. and Geman, D. (1984). Stochastic relaxation, Gibbs distributions, and the Bayesian restoration of images. *IEEE Transactions on Pattern Analysis and Machine Intelligence* **6**, 721–741.

Goldberger, A. (1972). Structural equation methods in the social sciences. *Econometrica* **40**, 979–1001.

Greenland, S. (2000). An introduction to instrumental variables for epidemiologists. *International Journal of Epidemiology* **29**, 722–729.

Greenland, S., Robins, J.M. and Pearl, J. (1999). Confounding and collapsibility in causal inference. *Statistical Science* **14**, 29–46.

Hansen, B. and Pedersen, C.B. (1994). Analysing complex pedigrees using Gibbs sampling. A theoretical and empirical investigation. Technical Report R-94-2032, Institute for Electronic Systems, Aalborg University.

Hastings, W.K. (1970). Monte Carlo sampling methods using Markov chains and their applications. *Biometrika* **57**, 97–109.

Heath, S.C. (1997). Markov chain Monte Carlo segregation and linkage analysis for oliogenic models. *American Journal of Human Genetics* **61**, 748–760.

Heath, S.C. (2003). Genetic linkage analysis using Markov chain Monte Carlo techniques. In *Highly Structured Stochastic Systems*, P.J. Green, N.L. Hjort and S. Richardson, eds. Oxford University Press, Oxford.

Janss, L.L.G., Thompson, R. and Van Arendonk, J.A.M. (1995). Application of Gibbs sampling for inference in a mixed major gene-polygenic inheritance model in animal populations. *Theoretical and Applied Genetics* **91**, 1137–1147.

Jensen, C.S. (1997). Blocking Gibbs sampling for inference in large and complex Bayesian networks with applications in genetics. Ph.D. thesis, Aalborg University, Aalborg.

Jensen, F. (2002). *HUGIN API Reference Manual Version 5.4*. HUGIN Expert Ltd., Aalborg.

Jensen, C.S. and Kong, A. (1999). Blocking Gibbs sampling for linkage analysis in large pedigrees with many loops. *American Journal of Human Genetics* **65**, 885–901.

Jensen, F.V., Lauritzen, S.L. and Olesen, K.G. (1990). Bayesian updating in causal probabilistic networks by local computation. *Computational Statistics Quarterly* **4**, 269–282.

Jones, B., Carvalho, C., Dobra, A., Hans, C., Carter, C. and West, M. (2005). Experiments in stochastic computation for high-dimensional graphical models. *Statistical Science* **20**, 388–400.

Katan, M.B. (2004). Commentary: Mendelian randomization, 18 years on. *International Journal of Epidemiology* **33**, 10–11.

Koller, D. and Pfeffer, A. (1997). Object-oriented Bayesian networks. In *Proceedings of the 13th Annual Conference on Uncertainty in Artificial Intelligence*, D. Geiger and P. Shenoy, eds. Morgan Kaufmann Publishers, San Francisco, CA, pp. 302–313.

Kong, A. (1991). Efficient methods for computing linkage likelihoods of recessive diseases in inbred pedigrees. *Genetic Epidemiology* **8**, 81–103.

Lander, E.S. and Green, P. (1987). Construction of multilocus genetic linkage maps in humans. *Proceedings of the National Academy of Sciences of the United States of America* **84**, 2363–2367.

Lange, K. and Elston, R.C. (1975). Extensions to pedigree analysis. I. Likelihood calculations for simple and complex pedigrees. *Human Heredity* **25**, 95–105.

Lauritzen, S.L. (1996). *Graphical Models*. Clarendon Press, Oxford.

Lauritzen, S.L. (2000). Causal inference from graphical models. In *Complex Stochastic Systems*, Chapter 2, O.E. Barndorff-Nielsen, D.R. Cox and C. Kluppelberg, eds. Chapman Hall, pp. 63–107.

Lauritzen, S.L. and Jensen, F.V. (1997). Local computation with valuations from a commutative semigroup. *Annals of Mathematics and Artificial Intelligence* **21**, 51–69.

Lauritzen, S.L. and Sheehan, N.A. (2003). Graphical models for genetic analysis. *Statistical Science* **18**, 489–514.

Lauritzen, S.L. and Spiegelhalter, D.J. (1988). Local computations with probabilities on graphical structures and their application to expert systems (with discussion). *Journal of the Royal Statistical Society, Series B* **50**, 157–224.

Lund, M.S. and Jensen, C.S. (1999). Blocking Gibbs sampling in the mixed inheritance model using graph theory. *Genetics, Selection, Evolution* **31**, 3–24.

Markovetz, F. (2006). A bibliography on learning causal networks of gene interactions. Manuscript: http://www.molgen.mpg.de/~markowet/docs/network-bib.pdf.

Metropolis, N., Rosenbluth, A.W., Rosenbluth, M.N., Teller, A.H. and Teller, E. (1953). Equations of state calculations by fast computing machines. *Journal of Chemical Physics* **21**, 1087–1092.

Mortera, J., Dawid, A.P. and Lauritzen, S.L. (2003). Probabilistic expert systems for DNA mixture profiling. *Theoretical Population Biology* **63**, 191–205.

Pearl, J. (1988). *Probabilistic Inference in Intelligent Systems*. Morgan Kaufmann Publishers, San Mateo, CA.

Pearl, J. (1995). Causal diagrams for empirical research. *Biometrika* **82**, 669–710.

Pearl, J. (2000). *Causality*. Cambridge University Press.

Robins, J. (1997). Causal inference from complex longitudinal data. In *Latent Variable Modeling with Applications to Causality*, M. Berkane, ed. Springer-Verlag, New York, pp. 69–117.

Schäfer, J., Opgen-Rhein, R. and Strimmer, K. (2006). Reverse engineering genetic networks using the GeneNet package. *R News* **6**, 50–53.

Schäfer, J. and Strimmer, K. (2005). An empirical Bayes approach to inferring large-scale gene association networks. *Bioinformatics* **21**(6), 754–764.

Segal, E., Shapira, M., Regev, A., Pe'er, D., Botstein, D., Koller, D. and Friedman, N. (2003a). Module networks: identifying regulatory modules and their condition-specific regulators from gene expression data. *Nature Genetics* **34**(2), 166–176.

Segal, E., Wang, H. and Koller, D. (2003b). Discovering molecular pathways from protein interaction and gene expression data. *Bioinformatics* **19**(Suppl. 1), i264–i272.

Sheehan, N.A. (2000). On the application of Markov chain Monte Carlo methods to genetic analyses on complex pedigrees. *International Statistical Review* **68**, 83–110.

Sheehan, N. and Egeland, T. (2007). Structured incorporation of prior information in relationship estimation problems. *Annals of Human Genetics* (in press).

Sheehan, N.A., Guldbrandtsen, B., Lund, M.S. and Sorensen, D.A. (2002). Bayesian MCMC mapping of quantitative trait loci in a half-sib design: a graphical model perspective. *International Statistical Review* **70**, 241–267.

Shenoy, P.P. and Shafer, G. (1990). Axioms for probability and belief-function propagation. In *Uncertainty in Artificial Intelligence 4*, R.D. Shachter, T.S. Levitt, L.N. Kanal and J.F. Lemmer, eds. North-Holland, Amsterdam, pp. 169–198.

Shoiket, K. and Geiger, D. (1997). A practical algorithm for finding optimal triangulations. *Proceedings of the Fourteenth National Conference on Artificial Intelligence*. AAAI Press, Menlo Park, CA, pp. 185–190.

Skjøth, F., Lohi, O. and Thomas, A.W. (1994). Genetic models for the inheritance of the silver mutation of foxes. *Genetical Research* **64**, 11–18.

Sobel, E. and Lange, K. (1996). Descent graphs in pedigree analysis: applications to haplotyping, location scores, and marker-sharing statistics. *American Journal of Human Genetics* **58**, 1323–1337.

Spiegelhalter, D.J. (1990). Fast algorithms for probabilistic reasoning in influence diagrams, with applications in genetics and expert systems. In *Influence Diagrams, Belief Nets and Decision Analysis*, R.M. Oliver and J.Q. Smith, eds. John Wiley & Sons, Chichester, pp. 361–384.

Spiegelhalter, D.J. (1998). Bayesian graphical modelling: a case-study in monitoring health outcomes. *Applied Statistics* **47**, 115–133.

Spirtes, P., Glymour, C. and Scheines, R. (1993). *Causation, Prediction and Search*. Springer-Verlag, New York. Reprinted by MIT Press.

Steel, M. and Hein, J. (2006). Reconstructing pedigrees: a combinatorial perspective. *Journal of Theoretical Biology* **240**, 360–367.

Taroni, F., Aitken, C., Garbolino, P. and Biedermann, A. (2006). *Bayesian Networks and Probabilistic Inference in Forensic Science*. John Wiley & Sons, Chichester.

Thomas, A. (1985). Data structures, methods of approximation and optimal computation for pedigree analysis. Ph.D. thesis, Cambridge University.

Thomas, A. (2005). Characterizing alleleic associations from unphased diploid data by graphical modeling. *Genetic Epidemiology* **29**, 23–35.

Thomas, A. and Camp, N.J. (2004). Graphical modeling of the joint distribution of alleles at associated loci. *American Journal of Human Genetics* **74**, 1088–1101.

Thomas, D. and Conti, D. (2004). Commentary: the concept of "Mendelian randomization". *International Journal of Epidemiology* **33**, 21–25.

Thomas, A., Gutin, A., Abkevich, V. and Bansal, A. (2000). Multilocus linkage analysis by blocked Gibbs sampling. *Statistics and Computing* **10**, 259–269.

Thompson, E.A. (1981). Pedigree analysis of Hodgkin's disease in a newfoundland genealogy. *Annals of Human Genetics* **45**, 279–292.

Thompson, E.A. (1986). *Pedigree Analysis in Human Genetics*. The Johns Hopkins University Press, Baltimore, MD.

Thompson, E.A. (1994). Monte Carlo likelihood in genetic mapping. *Statistical Science* **9**(3), 355–366.

Thompson, E.A. (2000). *Statistical Inference from Genetic Data on Pedigrees, Volume 6 of NSF-CBMS regional Conference Series in Probability and Statistics*. Institute of Mathematical Statistics and the American Statistical Association.

Thompson, E.A. (2001). Monte Carlo methods on genetic structures. In *Complex Stochastic Systems*, O.E. Barndorff-Nielsen, D.R. Cox and C. Klüppelberg, eds. Chapman Hall/CRC Press, London/Boca Raton, FL, pp. 176–218.

Thompson, E.A. and Heath, S.C. (2000). Estimation of conditional multilocus gene identity among relatives. In *Statistics in Molecular Biology and Genetics*, F., Seiller-Moiseiwitsch, ed. *IMS Lecture Notes*. Institute of Mathematical Statistics, American Mathematical Society, pp. 95–113.

Verzilli, C.J., Stallord, N. and Whittaker, J.C. (2006). Bayesian graphical models for genomewide association studies. *American Journal of Human Genetics* **79**, 100–112.

Weir, B.S., Triggs, C.M., Starling, L., Stowell, L.I., Walsh, K.A.J. and Buckleton, J.S. (1997). Interpreting DNA mixtures. *Journal of Forensic Sciences* **42**, 213–222.

West, M., Blanchette, C., Dressman, H., Huang, E., Ishida, S., Spang, R., Zuzan, H., Olson, J.A., Marks, J.R. and Nevins, J.R. (2001). Predicting the clinical status of human breast cancer by using gene expression profiles. *Proceedings of the National Academy of Sciences of the United States of America* **98**, 11462–11467.

Wright, S. (1921). Correlation and causation. *Journal of Agricultural Research* **20**, 557–585.

Wright, S. (1923). The theory of path coefficients: a reply to Niles' criticism. *Genetics* **8**, 239–255.

Wright, S. (1934). The method of path coefficients. *Annals of Mathematical Statistics* **5**, 161–215.

Yannakakis, M. (1981). Computing the minimum fill-in is NP-complete. *SIAM Journal on Algebraic and Discrete Methods* **2**, 77–79.

25

Coalescent Theory

M. Nordborg

Molecular and Computational Biology, University of Southern California, Los Angeles, CA, USA

The coalescent process is a powerful modeling tool for population genetics. The allelic states of all homologous gene copies in a population are determined by the genealogical and mutational history of these copies. The coalescent approach is based on the realization that the genealogy is usually easier to model backward in time, and that selectively neutral mutations can then be superimposed afterwards. A wide range of biological phenomena can be modeled using this approach.

Whereas almost all of classical population genetics considers the future of a population given a starting point, the coalescent considers the present, while taking the past into account. This allows the calculation of probabilities of sample configurations under the stationary distribution of various population genetic models, and makes full likelihood analysis of polymorphism data possible. It also leads to extremely efficient computer algorithms for generating simulated data from such distributions, data which can then be compared with observations as a form of exploratory data analysis.

25.1 INTRODUCTION

The stochastic process known as 'the coalescent' has played a central role in population genetics since the early 1980s, and results based on it are now used routinely to analyze DNA sequence polymorphism data. In spite of this, there is no comprehensive textbook treatment of coalescent theory. For biologists, the most widely used source of information is probably Hudson's seminal review (Hudson, 1990), which, along with a few other book chapters (Donnelly and Tavaré, 1995; Hudson, 1993; Li, 1997) and various unpublished lecture notes, is all that is available beyond the primary literature. Furthermore, since the field is very active, many relevant results are not generally available because they have not yet been published. They may be due to appear sometime in the indefinite future in a mathematical journal or obscure conference volume, or they may simply never have been written down. As a result of all this, there is a considerable gap between the theory that is available, and the theory that is being used to analyze data.

Handbook of Statistical Genetics, Third Edition. Edited by D.J. Balding, M. Bishop and C. Cannings.
© 2007 John Wiley & Sons, Ltd.   ISBN: 978-0-470-05830-5.

The present chapter is intended as an up-to-date introduction suitable for a wider audience. The focus is on the stochastic process itself, and especially on how it can be used to model a wide variety of biological phenomena. I consider a basic understanding of coalescent theory to be extremely valuable – even essential – for anyone analyzing genetic polymorphism data from populations, and will try to defend this view throughout. First of all, such an understanding can in many cases provide an intuitive feeling for how informative polymorphism data are likely to be (the answer is typically 'Not very'). When intuition is not enough, the coalescent provides a simple and powerful tool for exploratory data analysis through the generation of simulated data. The efficacy of the coalescent as a simulation tool is also the basis for promising statistical methods that use rejection algorithms to compute likelihoods (e.g., Weiss and von Haeseler, 1998). Various more sophisticated inference methods are described in **Chapter 26**.

25.2 THE COALESCENT

The word 'coalescent' is used in several ways in the literature, and it will also be used in several ways here. Hopefully, the meaning will be clear from the context. The coalescent – or, perhaps more appropriately, the coalescent approach – is based on two fundamental insights, which are the topic of Section 25.2.1. Section 25.2.2 then describes the stochastic process known as the coalescent, or sometimes Kingman's coalescent in honor of its discoverer (Kingman, 1982a; 1982b; 1982c). This process results from combining the two fundamental insights with a convenient limit approximation.

The coalescent will be introduced in the setting of the Wright–Fisher model of neutral evolution, but it applies more generally. This is one of the main topics for the remainder of the chapter. First of all, many different neutral models can be shown to converge to Kingman's coalescent. Second, more complex neutral models often converge to coalescent processes analogous to Kingman's coalescent.

The coalescent was described by Kingman (1982a; 1982b; 1982c), but it was also discovered independently by Hudson (1983) and by Tajima (1983). Indeed, arguments anticipating it had been used several times in population genetics (reviewed by Tavaré, 1984).

25.2.1 The Fundamental Insights

The first insight is that since selectively neutral variants by definition do not affect reproductive success, it is possible to separate the neutral mutation process from the genealogical process. In classical terms, 'state' can be separated from 'descent'.

To see how this works, consider a population of N clonal organisms that reproduce according to the neutral Wright–Fisher model, that is to say, generations are discrete, and each new generation is formed by randomly sampling N parents with replacement from the current generation. The number of offspring contributed by a particular individual is thus binomially distributed with parameters N (the number of trials) and $1/N$ (the probability of being chosen), and the joint distribution of the numbers of offspring produced by all N individuals is symmetrically multinomial. Now consider the random genealogical relationships (i.e. 'who begat whom') that result from reproduction in this setting. These can be represented graphically, as shown in Figure 25.1. Going forward in

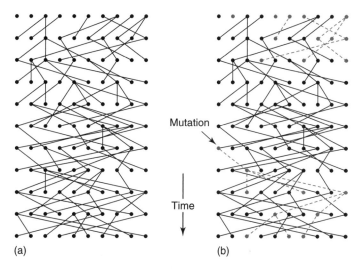

(a)   (b)

Figure 25.1 The neutral mutation process can be separated from the genealogical process. The genealogical relationships in a particular 10-generation realization of the neutral Wright–Fisher model (with population size $N = 10$) are shown on the left. On the right, allelic states of have been superimposed (so-called 'gene dropping').

time, lineages branch whenever an individual produces two or more offspring, and end when there is no offspring. Going backward in time, lineages coalesce whenever two or more individuals were produced by the same parent. They never end. If we trace the ancestry of a group of individuals back through time, the number of distinct lineages will decrease and eventually reach one, when the most recent common ancestor (MRCA) of the individuals in question is encountered. None of this is affected by neutral genetic differences between the individuals.

As a consequence, the evolutionary dynamics of neutral allelic variants can be modeled through so-called 'gene dropping' ('mutation dropping' would be more accurate): given a realization of the genealogical process, allelic states are assigned to the original generation in a suitable manner, and the lines of descent then simply followed forward in time, using the rule that offspring inherit the allelic state of their parent unless there is a mutation (which occurs with some probability each generation). In particular, the allelic states of any group of individuals (for instance, all the members of a given generation) can be generated by assigning an allelic state to their MRCA and then 'dropping' mutations along the branches of the genealogical tree that leads to them. Most of the genealogical history of the population is then irrelevant (cf. Figures 25.1 and 25.2).

The second insight is that it is possible to model the genealogy of a group of individuals backward in time without worrying about the rest of the population. It is a general consequence of the assumption of selective neutrality that each individual in a generation can be viewed as 'picking' its parent at random from the previous generation. It follows that the genealogy of a group of individuals may be generated by simply tracing the lineages back in time, generation by generation, keeping track of coalescences between lineages, until eventually the MRCA is found. It is particularly easy to see how this is done for the Wright–Fisher model, where individuals pick their parents independently of each other.

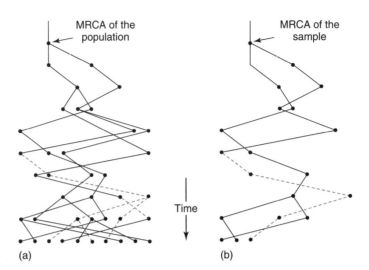

Figure 25.2 The genetic composition of a group of individuals is completely determined by the group's genealogy and the mutations that occur on it. The genealogy of the final generation in Figure 25.1 is shown on the left, and the genealogy of a sample from this generation is shown on the right. These trees could have been generated backward in time without generating the rest of Figure 25.1.

In summary, the joint effects of random reproduction (which causes 'genetic drift') and random neutral mutations in determining the genetic composition of a group of clonal individuals (such as a generation or a sample thereof) may be modeled by first generating the random genealogy of the individuals backward in time, and then superimposing mutations forward in time. This approach leads directly to extremely efficient computer algorithms (cf. the 'classical' approach which is to simulate the *entire*, usually very large population forward in time for a long period of time, and then to look at the final generation). It is also mathematically elegant, as Section 25.2.2 will show. However, its greatest value may be heuristic: the realization that the pattern of neutral variation observed in a population can be viewed as the result of random mutations on a random tree is a powerful one, which profoundly affects the way we think about data.

In particular, we are almost always interested in biological phenomena that affect the genealogical process, but do not affect the mutation process (e.g. population subdivision). From the point of view of inference about such phenomena, the observed polymorphisms are only of interest because they contain information about the unobserved underlying genealogy. Furthermore, the underlying genealogy is only of interest because it contains information about the evolutionary process that gave rise to it. In statistical terms, almost all inference problems that arise from polymorphism data can be seen as 'missing data' problems.

It is crucial to understand this, because no matter how many individuals we sample, there is still only a *single* underlying genealogy to estimate. It could of course be that this single genealogy contains a lot of information about the interesting aspect of the evolutionary process, but if it does not, then our inferences will be as good as one would normally expect from a sample of size one!

Another consequence of the above is that it is usually possible to understand how model parameters affect polymorphism data by understanding how they affect genealogies. For this reason, I will focus on the genealogical process and only discuss the neutral mutation process briefly toward the end of the chapter.

25.2.2   The Coalescent Approximation

The previous subsection described the conceptual insights behind the coalescent approach. The sample genealogies central to this approach can be conveniently modeled using a continuous-time Markov process known as the coalescent (or Kingman's coalescent, or sometimes 'the n-coalescent' to emphasize the dependence on the sample size). We will now describe the coalescent and show how it arises naturally as a large-population approximation to the Wright–Fisher model. Its relationship to other models will be discussed later.

Figure 25.2 is needlessly complicated because the identity (i.e. the horizontal position) of all ancestors is maintained. In order to superimpose mutations, all we need to know is which lineage coalesces with which, and when. In other words, we need to know the topology, and the branch lengths. The topology is easy to model: because of neutrality, individuals are equally likely to reproduce; therefore all lineages must be equally likely to coalesce. It is convenient to represent the topology as a sequence of coalescing equivalence classes: two members of the original sample are equivalent at a certain point in time if and only if they have a common ancestor at that time (see Figure 25.3). But what about the branch lengths, that is, the coalescence times?

Follow two lineages back in time. We have seen that offspring pick their parents randomly from the previous generation, and that, under the Wright–Fisher model, they do so independently of each other. Thus, the probability that the two lineages pick the same parent and coalesce is $1/N$, and the probability that they pick different parents and remain distinct is $1 - 1/N$. Since generations are independent, the probability that they remain distinct more than t generations into the past is $(1 - 1/N)^t$. The expected coalescence time is N generations. This suggests a standard continuous-time diffusion

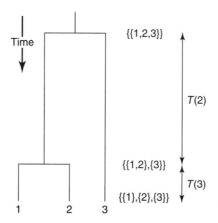

Figure 25.3   The genealogy of a sample can be described in terms of its topology and branch lengths. The topology can be represented using equivalence classes for ancestors. The branch lengths are given by the waiting times between successive coalescence events.

approximation, which is good as long as N is reasonably large (see **Chapter 22**). Rescale time so that one unit of scaled time corresponds to N generations. Then the probability that the two lineages remain distinct for more than τ units of scaled time is

$$\left(1 - \frac{1}{N}\right)^{[N\tau]} \longrightarrow e^{-\tau}, \tag{25.1}$$

as N goes to infinity ($[N\tau]$ is the largest integer less than or equal to $N\tau$). Thus, in the limit, the coalescence time for a pair of lineages is exponentially distributed with mean 1.

Now consider k lineages. The probability that none of them coalesce in the previous generation is

$$\prod_{i=0}^{k-1} \frac{N-i}{N} = \prod_{i=1}^{k-1}\left(1 - \frac{i}{N}\right) = 1 - \frac{\binom{k}{2}}{N} + O\left(\frac{1}{N^2}\right), \tag{25.2}$$

and the probability that more than two do so is $O(1/N^2)$. Let $T(k)$ be the (scaled) time till the first coalescence event, given that there are currently k lineages. By the same argument as above, $T(k)$ is in the limit exponentially distributed with mean $2/[(k(k-1))]$. Furthermore, the probability that more than two lineages coalesce in the same generation can be neglected. Thus, under the coalescent approximation, the number of distinct lineages in the ancestry of a sample of (finite) size n decreases in steps of one back in time, so $T(k)$ is the time from k to $k-1$ lineages (see Figure 25.3).

In summary, the coalescent models the genealogy of a sample of n haploid individuals as a random bifurcating tree, where the $n-1$ coalescence times $T(n), T(n-1), \ldots, T(2)$ are mutually independent, exponentially distributed random variables. Each pair of lineages coalesces independently at rate 1, so the total rate of coalescence when there are k lineages is 'k choose 2'. A concise (and rather abstract) way of describing the coalescent is as a continuous-time Markov process with state space $\mathcal{E}_n$ given by the set of all equivalence relations on $\{1, \ldots, n\}$, and infinitesimal generator $Q = (q_{\xi\eta})_{\xi,\eta\in\mathcal{E}_n}$ given by

$$q_{\xi\eta} := \begin{cases} -k(k-1)/2, & \text{if } \xi = \eta, \\ 1, & \text{if } \xi \prec \eta, \\ 0, & \text{otherwise,} \end{cases} \tag{25.3}$$

where $k := |\xi|$ is the number of equivalence classes in ξ, and $\xi \prec \eta$ if and only if η is obtained from ξ by coalescing two equivalence classes of ξ.

It is worth emphasizing just how efficient the coalescent is as a simulation tool. In order to generate a sample genealogy under the Wright–Fisher model as described in Section 25.2.1, we would have to go back in time some N generations, checking for coalescences in each of them. Under the coalescent approximation, we simply generate $n-1$ independent exponential random numbers and, independently of these, a random bifurcating topology.

What do typical coalescence trees look like? Figure 25.4 shows four examples. It is clear that the trees are extremely variable, both with respect to topology and branch lengths. This should come as no surprise considering the description of the coalescent just given: the topology is independent of the branch lengths; the branch lengths are independent, exponential random variables; and the topology is generated by randomly picking lineages to coalesce (in this sense all topologies are equally likely).

Note that the trees tend to be dominated by the deep branches, when there are few ancestors left. Because lineages coalesce at rate 'k choose 2', coalescence events occur much more rapidly when there are many lineages (intuitively speaking, it is easier for lineages to find each other then). Indeed, the expected time to the MRCA (the height of the tree) is

$$ \mathrm{E}\left[\sum_{k=2}^{n} T(k)\right] = \sum_{k=2}^{n} \mathrm{E}[T(k)] = \sum_{k=2}^{n} \frac{2}{k(k-1)} = 2\left(1 - \frac{1}{n}\right), \tag{25.4} $$

while $\mathrm{E}[T(2)] = 1$, so the expected time during which there are only two branches is greater than half the expected total tree height. Furthermore, the variability in $T(2)$ accounts for most of the variability in tree height. The dependence on the deep branches becomes increasingly apparent as n increases, as can be seen by comparing Figures 25.4 and 25.5.

The importance of realizing that there is only a single underlying genealogy was emphasized earlier. As a consequence of the single genealogy, sampled gene copies from a population must almost always be treated as dependent, and increasing the sample size is therefore often surprisingly ineffective (the point is well made by Donnelly, 1996). Important examples of this follow directly from the basic properties of the coalescent. Consider first the MRCA of the population. One might think that a large sample is needed to ensure that the deepest split is included, but it can be shown (this and related results can be found in Saunders *et al.*, 1984) that the probability that a sample of size n contains the MRCA of the whole population is $(n - 1)/(n + 1)$. Thus even a small sample is likely to

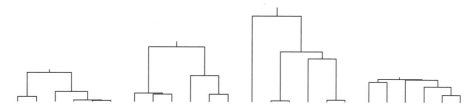

Figure 25.4 Four realizations of the coalescent for $n = 6$, drawn on the same scale (the labels 1–6 should be assigned randomly to the tips).

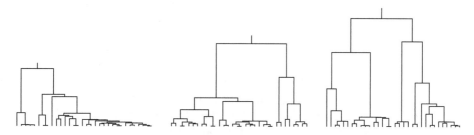

Figure 25.5 Three realizations of the coalescent for $n = 32$, drawn on the same scale (the labels 1–32 should be assigned randomly to the tips).

contain it and the total tree height will quickly stop growing as n increases. Second, the number of distinct lineages decreases rapidly as we go back in time. This severely limits inferences about ancient demography (e.g. Nordborg, 1998). Third, since increasing the sample size only adds short twigs to the tree (cf. Figure 25.5), the expected total branch length of the tree, $T_{tot}(n)$ grows very slowly with n. We have

$$E[T_{tot}(n)] = E\left[\sum_{k=2}^{n} kT(k)\right] = \sum_{k=1}^{n-1} \frac{2}{k} \sim 2(\gamma + \log n), \tag{25.5}$$

as $n \to \infty$ ($\gamma \approx 0.577\,216$ is Euler's constant). Since the number of mutations that are expected to occur in a tree is proportional to $E[T_{tot}(n)]$, this has important consequences for estimating the mutation rate, as well as for inferences that depend on estimates of the mutation rate. Loosely speaking, it turns out that a sample of n copies of a gene often has the statistical properties one would expect of a random sample of size $\log n$, or even of size 1 (which is not much worse than $\log n$ in practice).

25.3   GENERALIZING THE COALESCENT

This section will present ideas and concepts that are important for generalizing the coalescent. The following sections will then illustrate how these can be used to incorporate greater biological realism.

25.3.1   Robustness and Scaling

We have seen that the coalescent arises naturally as an approximation to the Wright–Fisher model, and that it has convenient mathematical properties. However, the real importance of the coalescent stems from the fact that it arises as a limiting process for a wide range of neutral models, *provided time is scaled appropriately* (Kingman, 1982b; 1982c; Möhle, 1998b; 1999). It is thus robust in this sense.

This is best explained through an example. Recall that the number of offspring contributed by each individual in the Wright–Fisher model is binomially distributed with parameters N and $1/N$. The mean is thus 1, and the variance is $1 - 1/N \to 1$, as $N \to \infty$. Now consider a generalized version of this model in which the mean number of offspring is still 1 (as it must be for the population size to remain constant), but the limiting variance is σ^2, $0 < \sigma^2 < \infty$ (perhaps giants step on 90 % of the individuals before they reach reproductive age). It can be shown that this process also converges to the coalescent, provided time is measured in units of N/σ^2 generations. We could also measure time in units of N generations as before, but then $E[T(2)] = 1/\sigma^2$ instead of $E[T(2)] = 1$, and so on. Either way, the expected coalescence time for a pair of lineages is N/σ^2 generations. The intuition behind this is clear: increased variance in reproductive success causes coalescence to occur faster (at a higher rate). In classical terms, 'genetic drift' operates faster. By changing the way we measure time, this can be taken into account, and the standard coalescent process obtained.

The remarkable fact is that a very wide range of biological phenomena (overlapping generations, separate sexes, mating systems – several examples will be given below) can likewise be treated as a simple linear change in the time scale of the coalescent. This has important implications for data analysis. The good news is that we may often be able to justify using the coalescent process even though 'our' species almost certainly does not reproduce according to a Wright–Fisher model (few species do). The bad news is that biological phenomena that can be modeled this way will never be amenable to inference based on polymorphism data alone. For example, σ^2 in the model above could never be estimated from polymorphism data unless we had independent information about N (and vice versa).

Of course, we could not even estimate N/σ^2 without external data. It is important to realize that all parameters in coalescent models are scaled, and that only scaled parameters can be directly estimated from the data. In order to make any kind of statement about unscaled quantities, such as population numbers, or ages in years or generations, external information is needed. This adds considerable uncertainty to the analysis. For example, an often used source of external information is an estimate of the neutral mutation probability per generation. Roughly speaking, this estimate is obtained by measuring sequence divergence between species, and dividing by the estimated species divergence time (Li, 1997). The latter is in turn obtained from the fossil record and a rough guess of the generation length. It should be clear that it is not appropriate to treat such an estimate as a known parameter when analyzing polymorphism data. However, it should also be noted that interesting conclusions can often be drawn directly from scaled parameters (e.g. by looking at relative values). Such analyses are likely to be more robust, given the robustness of the coalescent.

Because the generalized model above converges with the same scaling as a Wright–Fisher model with a population size of N/σ^2, it is sometimes said that it has an 'effective population size', $N_e = N/\sigma^2$. Models that scale differently would then have other effective population sizes. Although convenient, this terminology is unfortunate for at least two reasons. First, the classical population genetics literature is full of variously defined 'effective population sizes', only some of which are effective population sizes in the sense used here. For example, populations that are subdivided or vary in size cannot in general be modeled as a linear change in the time scale of the coalescent. Second, the term is inevitably associated with real population sizes, even though it is simply a scaling factor. To be sure, N_e is always a function of the real demographic parameters, but there is no direct relationship with the total population size (which may be smaller as well as much, much larger). Indeed, as we shall see in Section 25.7, it is now clear that N_e must vary between chromosomal regions in the same organism!

25.3.2 Variable Population Size

Real populations vary in size over time. Although the coalescent is not robust to variation in the population size in the sense described above (i.e. there is no 'effective population size'), it is nonetheless easy to incorporate changes in the population size, at least if we are willing to assume that we know what they were – that is, if we assume that the variation can be treated deterministically. Since a rigorous treatment of these results can be found in the review by Donnelly and Tavaré (1995), also in **Chapter 22** this volume, I will try to give an intuitive explanation.

Imagine a population that evolves according to the Wright–Fisher model, but with a different population size in each generation. If we know how the size has changed over time, we can trace the genealogy of a sample precisely as before. Let $N(t)$ be the population size t generations ago. Going back in time, lineages are more likely to coalesce in generations when the population is small than in generations when the population is large. In order to describe the genealogy by a continuous-time process analogous to the coalescent, we must therefore allow the rate of coalescence to change over time. However, since the time scale used in the coalescent directly reflects the rate of coalescence, we may instead let this scaling change over time. In the standard coalescent, t generations ago corresponds to t/N units of coalescence time, and τ units of coalescence time ago corresponds to $[N\tau]$ generations. When the population size is changing, we find instead that t generations ago corresponds to

$$g(t) := \sum_{i=1}^{t} \frac{1}{N(i)} \tag{25.6}$$

units of coalescence time, and τ units of coalescence time ago corresponds to $[g^{-1}(\tau)]$ generations (g^{-1} denotes the inverse function of g). It is clear from equation (25.6) that many generations go by without much coalescence time passing when the population size is large, and, conversely, that much coalescence time passes each generation when the population is small. Let $N(0)$ go to infinity, and assume that $N(t)/N(0)$ converges to a finite number for each t, to ensure that the population size becomes large in every generation. It can be shown that the variable population size model converges to a coalescent process with a *nonlinear* time scale in this limit (Griffiths and Tavaré, 1994). The scaling is given by (25.6). Thus, a sample genealogy from the coalescent with variable population size can be generated by simply applying g^{-1} to the coalescence times of a genealogy generated under the standard coalescent.

An example will make this clearer. Consider a population that has grown exponentially, so that, backwards in time, it shrinks according to $N(t) = N(0)e^{-\beta t}$ (note that this violates the assumption that the population size be large in every generation – this turns out not to matter greatly). Then

$$g(t) \approx \int_0^t \frac{1}{N(s)} \, ds = \frac{e^{\beta t} - 1}{N(0)\beta}, \tag{25.7}$$

and

$$g^{-1}(\tau) \approx \frac{\log(1 + N(0)\beta\tau)}{\beta}. \tag{25.8}$$

The difference between this model and one with a constant population size is shown in Figure 25.6. When the population size is constant, there is a linear relationship between real and scaled time. The genealogical trees will tend to look like those in Figures 25.4 and 25.5. When the population size is changing, the relationship between real and scaled time is nonlinear, because coalescences occur very slowly when the population was large, and more rapidly when the population was small. Genealogies in an exponentially growing population will tend to have most coalescences early in the history. Since all branches will then be of roughly equal length, the genealogy is said to be 'starlike'.

Models of exponential population growth have often been used in the context of human evolution (e.g. Rogers and Harpending, 1992; Slatkin and Hudson 1991). Marjoram and Donnelly (1997) have pointed out that some of the predictions from such models (e.g. the starlike genealogies) depend crucially on exponential growth from a *very* small

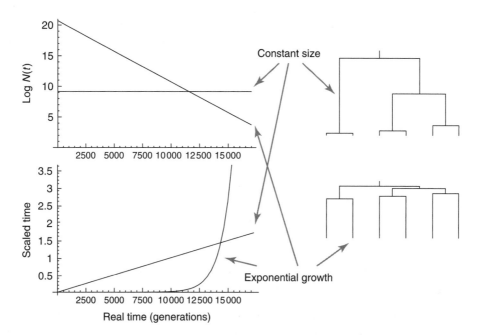

Figure 25.6 Variable population size can be modeled as a standard coalescent with a nonlinear time scale. Here, a constant population is compared to one that has grown exponentially. As the latter population shrinks backward in time, the scaled time begins to run faster, reflecting the fact that coalescences are more likely to have taken place when the population was small. Note that the trees are topologically equivalent and differ only in the branch lengths.

size – unrealistically small for humans. However, other predictions are more robust. For example, the argument in the previous paragraph explains why it may be reasonable to ignore growth altogether when modeling human evolution, even though growth has clearly taken place: if the growth was rapid and recent enough, no scaled time would pass, and no coalescence occur. In classical terms, exponential growth stops genetic drift.

Finally, it should be pointed that it is not entirely clear how general the nonlinear scaling approach to variable population sizes is. It relies, of course, on knowing the historical population sizes, but it also requires assumptions about the type of density regulation (Marjoram and Donnelly, 1997).

25.3.3 Population Structure on Different Time Scales

Real populations are also often spatially structured, and it is obviously important to be able to incorporate this in our models. However, structured models turn out to be even more important than one might have expected from this, because many biological phenomena can be thought of as analogous to population structure (Nordborg, 1997; Rousset, 1999a). Examples range from the obvious, like age structure, to the more abstract, like diploidy and allelic classes.

The following model, which may be called the 'structured Wright–Fisher model', turns out to be very useful in this context. Consider a clonal population of size N, as before, but let it be subdivided into patches of fixed sizes N_i, $i \in \{1, \ldots, M\}$, so that $\sum_i N_i = N$. In every generation, each individual produces an effectively infinite number of propagules. These propagules then migrate among the patches independently of each other, so that with probability m_{ij}, $i, j \in \{1, \ldots, M\}$, a propagule produced in patch i ends up in patch j. We also define the 'backward migration' probability, b_{ij}, $i, j \in \{1, \ldots, M\}$, that a randomly chosen propagule in patch i after dispersal was produced in patch j; it is easy to show that

$$b_{ij} = \frac{N_j m_{ji}}{\sum\limits_k N_k m_{ki}}. \tag{25.9}$$

The next generation of adults in each patch is then formed by random sampling from the available propagules.

Thus the number of offspring a particular individual in patch i contributes to the next generation in patch j is binomially distributed with parameters N_j and $b_{ji} N_i^{-1}$. The joint distribution of the numbers of offspring contributed to the next generation in patch j by *all* individuals in the current generation is multinomial (but no longer symmetric).

Just like the unstructured Wright–Fisher model, the genealogy of a finite sample in this model can be described by a discrete-time Markov process. Lineages coalesce in the previous generation if and only if they pick the same parental patch, and the same parental individual within that patch. A lineage currently in i and a lineage currently in j 'migrate' (backward in time) to k and coalesce there with probability $b_{ik} b_{jk} N_k^{-1}$.

It is also possible to approximate the model by a continuous-time Markov process. The general idea is to let the total population size, N, go to infinity with time scaled appropriately, precisely as before. However, we now also need to decide how M, N_i, and b_{ij} scale with N. Different biological scenarios lead to very different choices in this respect, and it is often possible to utilize convergence results based on separation of time scales (Möhle, 1998a; Kaj, *et al.*, 1991; Nordborg, 1997; 1999; Nordborg and Donnelly, 1997; Nordborg and Krone, 2002; Wakeley, 1999). This technique will be exemplified in what follows.

25.4 GEOGRAPHICAL STRUCTURE

Genealogical models of population structure have a long history. The classical work on identity coefficients (see **Chapter 28**) concerns genealogies when $n = 2$, and the coalescent was also quickly used for this purpose (for early work see Slatkin, 1987; Strobeck, 1987; Tajima, 1989a; Takahata, 1988).

Since geographical structure is reviewed in **Chapter 28**, we will mainly use it to introduce some of the scaling ideas that are central to the coalescent. The discussion will be limited to the structured Wright–Fisher model (which is a matrix migration model when viewed as a model of geographic subdivision). Most coalescent modeling has been done in this setting (reviewed in Wilkinson-Herbots, 1998 and Hudson, 1998). For time-scale approximations different from the ones discussed below, see Takahata (1991) and Wakeley (1999). An important variant of the model considers isolation: gene flow which stopped completely at some point in the past, for example due to speciation (e.g. Wakeley, 1996). For models of continuous environments and isolation-by-distance, see Barton and Wilson (1995) and Wilkins and Wakeley (2002).

25.4.1 The Structured Coalescent

Assume that M, $c_i := N_i/N$, and $B_{ij} := 2Nb_{ij}$, $i \neq j$, all remain constant as N goes to infinity. Then, with time measured in units of N generations, the process converges to the so-called 'structured coalescent', in which each pair of lineages in patch i coalesces independently at rate $1/c_i$, and each lineage in i 'migrates' (backward in time) independently to j at rate $B_{ij}/2$ (Herbots, 1994; Notohara, 1990; Wilkinson-Herbots, 1998). The intuition behind this is as follows (an excellent discussion of how the scaled parameters should be interpreted can be found in **Chapter 22**). By assuming that B_{ij} remains constant, we assure that the backward per-generation probabilities of *leaving* a patch (b_{ij}, $i \neq j$), are $O(1/N)$. Similarly, by assuming that c_i remains constant, we assure that all per-generation coalescence probabilities are $O(1/N)$. Thus, in any given generation, the probability that all lineages remain in their patch, without coalescing, is $1 - O(1/N)$. Furthermore, the probabilities that more than two lineages coalesce, that more than one lineage migrates, and that lineages both migrate and coalesce, are all $O(1/N^2)$ or smaller. In the limit $N \to \infty$, the only possible events are pairwise coalescences within patches, and single migrations between patches.

These events occur according to independent Poisson processes, which means the following. Let k_i denote the number of lineages currently in patch i. Then the waiting time till the first event is exponentially distributed with rate given by the sum of the rates of all possible events, that is,

$$h(k_1, \ldots, k_M) = \sum_i \left(\frac{\binom{k_i}{2}}{c_i} + \sum_{j \neq i} k_i \frac{B_{ij}}{2} \right). \tag{25.10}$$

When an event occurs, it is a coalescence in patch i with probability

$$\frac{\binom{k_i}{2}/c_i}{h(k_1, \ldots, k_M)}, \tag{25.11}$$

and a migration from i to j with probability

$$\frac{k_i B_{ij}/2}{h(k_1, \ldots, k_M)}. \tag{25.12}$$

In the former case, a random pair of lineages in i coalesces, and k_i decreases by one. In the latter case, a random lineage moves from i to j, k_i decreases by one, and k_j increases by one. A simulation algorithm would stop when the MRCA is found, but note that this single remaining lineage would continue migrating between patches if followed further back in time.

Structured coalescent trees generally look different from standard coalescent trees. Whereas variable population size only altered the branch lengths of the trees, population structure also affects the topology. If migration rates are low, lineages sampled from the same patch will tend to coalesce with each other, and a substantial amount of time can then pass before migration allows the ancestral lineages to coalesce (see Figure 25.7). Structure will often increase the mean and, equally importantly, the variance in time to the MRCA considerably (discussed in the context of human evolution by Marjoram and Donnelly, 1997).

25.4.2 The Strong-migration Limit

It is intuitive that weak migration, which corresponds to strong population subdivision, can have a large effect on genealogies. Conversely, we would expect genealogies in models with strong migration to look much like standard coalescent trees. This intuition turns out to be correct, except for one important difference: the scaling changes. Strong migration is thus one of the phenomena that can be modeled as a simple linear change in the time scale of the coalescent. It is important to understand why this happens.

Formally, the strong-migration limit means that $\lim_{N \to \infty} N b_{ij} = \infty$ because the per-generation migration probabilities, b_{ij}, are not $O(1/N)$. Since the coalescence probabilities *are* $O(1/N)$, this means that, for large N, migration will be much more likely than coalescence. As $N \to \infty$, there will in effect be infinitely many migration events between coalescence events. This is known as separation of time scales: migration occurs on a faster time scale than does coalescence. However, coalescences can of course still only occur when two lineages pick a parent in the same patch. How often does this happen? Because lineages jump between patches infinitely fast on the coalescence time scale, this is determined by the stationary distribution of the migration process (strictly speaking, this assumes that the migration matrix is ergodic). Let π_i be the stationary probability that a lineage is in patch i. A given pair of lineages then co-occur in i a fraction π_i^2 of the time. Coalescence in this patch occurs at rate $1/c_i$. Thus the total rate at which pairs of lineages coalesce is $\alpha := \sum_i \pi_i^2/c_i$. Pairs coalesce independently of each other just as in the standard model, so the total rate when there are k lineages is $\binom{k}{2}\alpha$. If time is measured in units of $N_e = N/\alpha$ generations, the standard coalescent is retrieved (Nagylaki, 1980; Notohara, 1993).

It can be shown that $\alpha \geq 1$, with equality if and only if $\sum_{j \neq i} N_i b_{ij} = \sum_{j \neq i} N_j b_{ji}$ for all i. This condition means that, going forward in time, the number of emigrants equals the number of immigrants in all populations, a condition known as 'conservative

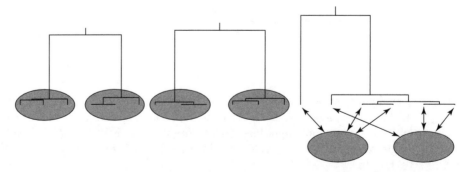

Figure 25.7 Three realizations of the structured coalescent in a symmetric model with two patches, and $n = 3$ in each patch (labels should be assigned randomly within patches). Lineages tend to coalesce within patches – but not always, as shown by the rightmost tree.

migration' (Nagylaki, 1980). Thus we see that, unless migration is conservative, the effective population size with strong migration is smaller than the total population size. The intuitive reason for this is that when migration is nonconservative, some individuals occupy 'better' patches than others, and this increases the variance in reproductive success among individuals. The environment has 'sources' and 'sinks' (Pulliam, 1988; Rousset, 1999b). Conservative migration models (like Wright's island model) have many simple properties that do not hold generally (Nagylaki, 1982; 1998; Nordborg, 1997).

25.5 SEGREGATION

Because everything so far has been done in an asexual setting, it has not been necessary to distinguish between the genealogy of an organism and that of its genome. This becomes necessary in sexual organisms. Most obviously, a diploid organism that was produced sexually has two parents, and each chromosome came from one of them. The genealogy of the genes is thus different from the genealogy (the pedigree) of the individuals: the latter describes the *possible* routes the genes could have taken (and is largely irrelevant – cf. Figure 25.9). This is simply Mendelian segregation viewed backwards in time, and it is the topic of this section. It is usually said that diploidy can be taken into account by simply changing the scaling from N to $2N$; it will become clear from what follows why, and in what sense, this is true.

The other facet of sexual reproduction, genetic recombination, turns out to have much more important effects. Genetic recombination causes ancestral lineages to branch, so that the genealogy of a sample can no longer be represented by a single tree: instead it becomes a collection of trees, or a single, more general type of graph. We will continue to ignore recombination until Section 25.6 (it makes sense to discuss diploidy first).

Sex takes many forms: I will first consider organisms that are hermaphroditic and therefore potentially capable of fertilizing themselves (this includes most higher plants and many mollusks), and thereafter discuss organisms with separate sexes (which includes most animals and many plants). Further examples can be found in Nordborg and Krone (2002).

25.5.1 Hermaphrodites

The key to modeling diploid populations is the realization that a diploid population of size N can be thought of as a haploid population of size $2N$, divided into N patches of size 2. In the notation of the structured Wright–Fisher model above, $M = N$, $N_i = 2$, and $c_i = 2/N$. Thus, in contrast to the assumptions for the structured coalescent, both M and c_i depend on N. This leads to a convenient convergence result based on separation of time scales (Nordborg and Donnelly, 1997; for a formal proof, see Möhle, 1998a), that can be described as follows (cf. Figure 25.8).

If time is scaled in units of $2N$ generations, then each pair of lineages 'coalesces' into the same individual at rate 2. Whenever this happens, there are two possibilities: either the two lineages pick the same of the two available (haploid) parents, or they pick different ones. The former event, which occurs with probability $\frac{1}{2}$, results in a real coalescence, whereas the latter event, which also occurs with probability $\frac{1}{2}$, simply results in the two distinct lineages temporarily occupying the same individual. Let S be the probability

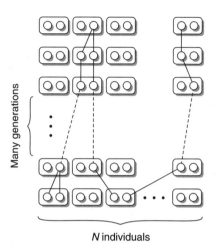

Many generations

N individuals

Figure 25.8 The coalescent with selfing. On the coalescent time scale, lineages within individuals instantaneously coalesce (probability F), or end up in different individuals (probability $1 - F$).

that a fertilization occurs through selfing, and $1 - S$ the probability that it occurs through outcrossing. If the individual harboring two distinct lineages was produced through selfing (probability S), then the two lineages must have come from the same individual in the previous generation, and again pick different parents with probability $\frac{1}{2}$ or coalesce with probability $\frac{1}{2}$. If the individual was produced through outcrossing, the two lineages revert to occupying distinct individuals. Thus the two lineages will rapidly either coalesce or end up in different individuals. The probability of the former outcome is

$$\frac{S/2}{S/2 + 1 - S} = \frac{S}{2 - S} =: F, \tag{25.13}$$

and that of the latter $1 - F$. Thus each time a pair of lineages coalesces into the same individual, the total probability that this results in a coalescence event is $\frac{1}{2} \times 1 + \frac{1}{2} \times F = (1 + F)/2$, and since pairs of lineages coalesce into the same individual at rate 2, the rate of coalescence is $1 + F$. On the chosen time scale, all states that involve two or more pairs occupying the same individual are instantaneous.

Thus, the genealogy of a random sample of gene copies from a population of hermaphrodites can be described by the standard coalescent if time is scaled in units of

$$2N_e = \frac{2N}{1 + F} \tag{25.14}$$

generations (cf. Pollak, 1987). If individuals are obligate outcrossers, $F = 0$, and the correct scaling is $2N$.

It should be noted that a sample from a diploid population is not a random sample of gene copies, because both copies in each individual are sampled. This is easily taken into account. It follows from the above that the two copies sampled from the same individual will instantaneously coalesce with probability F, and end up in different individuals with probability $1 - F$. The number of distinct lineages in a sample of $2n$ gene copies from n

individuals is thus $2n - X$, where X is as a binomially distributed random variable with parameters n and F. This corresponds to the well-known increase in the frequency of homozygous individuals predicted by classical population genetics. Note that this initial 'instantaneous' process has much nicer statistical properties than the coalescent, and that most of the information about the degree of selfing comes from the distribution of variability within and between individuals (Nordborg and Donnelly, 1997).

25.5.2 Males and Females

Next consider a diploid population that consists of N_m breeding males and N_f breeding females so that $N = N_m + N_f$. The discussion will be limited to *autosomal* genes, that is, genes that are not sex-linked. With respect to the genealogy of such genes, the total population can be thought of as a haploid population of size $2N$, divided into two patches of size $2N_m$ and $2N_f$, respectively, each of which is further divided into patches of size 2, as in the previous section. Clearly, a lineage currently in a male has probability $\frac{1}{2}$ of coming from a male in the previous generation, and probability $\frac{1}{2}$ of coming from a female. Within a sex, all individuals are equally likely to be chosen. The model looks a lot like a structured Wright–Fisher model with $M = 2$, $c_m = N_m/N$, $c_f = N_f/N$, and $b_{mf} = b_{fm} = \frac{1}{2}$, the only difference being that two distinct lineages in the same individual must have come from individuals of different sexes in the previous generation, and thus do not migrate independently of each other. However, because states involving two distinct lineages in the same individual are instantaneous, this difference can be shown to be irrelevant. Pairs of lineages in different individuals (regardless of sex) coalesce in the previous generation if and only if both members of the pair came from: (a) the same sex; (b) the same diploid individual within that sex; and (c) the same haploid parent within that individual. This occurs with probability

$$\frac{1}{4} \times \frac{1}{N_m} \times \frac{1}{2} + \frac{1}{4} \times \frac{1}{N_f} \times \frac{1}{2} = \frac{N_m + N_f}{8N_m N_f}, \tag{25.15}$$

or, in the limit $N \to \infty$, with time measured in units of $2N$ generations, and c_m and c_f held constant, at rate $\alpha = (4c_m c_f)^{-1}$ (in accordance with the strong-migration limit result above). Alternatively, if time is measured in units of

$$2N_e = \frac{2N}{\alpha} = \frac{8N_m N_f}{N_m + N_f}, \tag{25.16}$$

generations, the standard coalescent is obtained (cf. Wright, 1931). Note that if $N_m = N_f = N/2$, the correct scaling is again the standard one of $2N$.

25.6 RECOMBINATION

In the era of genomic polymorphism data, the importance of modeling recombination can hardly be overemphasized. When viewed backward in time, recombination (in the broad sense that includes phenomena such as gene conversion and bacterial conjugation in

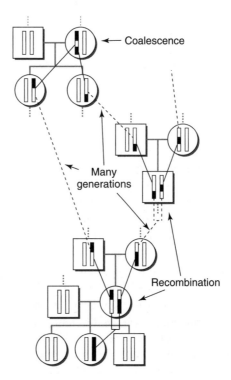

Figure 25.9 The genealogy of a DNA segment (colored black) subject to recombination both branches and coalesces. Note also that the genealogy of the sexually produced *individuals* (the pedigree) is very different from the genealogy of their *genes*.

addition to crossing over) causes the ancestry of a chromosome to spread out over many chromosomes in many individuals. The lineages branch, as illustrated in Figure 25.9. The genealogy of a sample of recombining DNA sequences can thus no longer be represented by a single tree: it becomes a graph instead. Alternatively, since the genealogy of each point in the genome (each base pair, say) *can* be represented by a tree, the genealogy of a sample of sequences may be envisioned as a 'walk through tree space'.

25.6.1 The Ancestral Recombination Graph

As was first shown by Hudson (1983), incorporating recombination into the coalescent framework is in principle straightforward. The following description is based on the elegant 'ancestral recombination graph' of Griffiths and Marjoram (1996; 1997), which is closely related to Hudson's original formulation (for other approaches, see Simonsen and Churchill, 1997; Wiuf and Hein, 1999b).

Consider first the ancestry of a single ($n = 1$) chromosomal segment from a diploid species with two sexes and an even sex ratio. As shown in Figure 25.9, each recombination event (depicted here as crossing over at a point – we will return to whether this is reasonable below) in its ancestry means that a lineage splits into two, when going backward in time. Recombination spreads the ancestry of the segment over many chromosomes, or rather over many 'chromosomal lineages'. However, as also shown in

Figure 25.9, these lineages will coalesce in the normal fashion, and this will tend to bring the ancestral material back together on the same chromosome (Wiuf and Hein, 1997).

To model this, let the per-generation probability of recombination in the segment be r, define $\rho := \lim_{N \to \infty} 4Nr$, and measure time in units of $2N$ generations. Then the (scaled) time till the first recombination event is exponentially distributed with rate $\rho/2$ in the limit as N goes to infinity. Furthermore, once recombination has created two or more lineages, we find that these lineages undergo recombination independently of one another, and that simultaneous events can be neglected. This follows from standard coalescent arguments analogous to those presented for migration above. The only thing that may be slightly nonintuitive about recombination is that *the lineages we follow never recombine with each other* (the probability of such an event is vanishingly small): they always recombine with the (infinitely many) nonancestral chromosomes.

Each recombination event increases the number of lineages by one, and because lineages recombine independently, the total rate of recombination when there are k lineages is $k\rho/2$. Each coalescence event decreases the number of lineages by one, and the total rate of coalescence when there are k lineages is $k(k-1)/2$, as we have seen previously. Since lineages are 'born' at a linear rate, and 'die' at a quadratic rate, the number of lineages is guaranteed to stay finite and will even hit one occasionally – there will then temporarily be a single ancestral chromosome again (Wiuf and Hein, 1997).

A sample of n lineages behaves in the same way. Each lineage recombines independently at rate $\rho/2$, and each pair of lineages coalesces independently at rate 1. The number of lineages *will* hit one occasionally. The segment in which this first occurs is known as the 'Ultimate' MRCA, because, as we shall see, each point in the sample may well have a younger MRCA.

The genealogy of a sample of n lineages back to the Ultimate MRCA can thus be described by a branching and coalescing graph (an 'ancestral recombination graph') that is analogous to the standard coalescent. A realization for $n = 6$ is shown in Figure 25.10.

What does a lineage in the graph look like? For each point in the segment under study, it must contain information about *which* (if any) sample members it is ancestral to. It is convenient to represent the segment as a (0,1) interval (this is just a coordinate system that can be translated into base pairs or whatever is appropriate). An ancestral lineage can then be represented as a set of elements of the form {interval, labels}, where the intervals are those resulting from all recombinational breakpoints in the history of the sample (Fisher's 'junctions' (Fisher, 1965) for aficionados of classical population genetics) and the labels denote the descendants of that segment (using the 'equivalence class' notation introduced previously). An example of this notation is given in Figure 25.10. Note that pieces of a given chromosomal lineage will often be ancestral to no one in the sample. Indeed, recombination in a nonancestral piece may result in an entirely nonancestral lineage!

So far nothing has been said about where or how recombination breakpoints occur. This has been intentional, to emphasize that the ancestral recombination graph does not depend on (most) details of recombination. It is possible to model almost any kind of recombination (including, for example, various forms of gene conversion) in this framework. But of course the graph has no meaning unless we interpret the recombination events somehow. To proceed, we will assume that each recombination event results in crossing over at a point, x, somewhere in (0,1). How x is chosen is again up to the modeler: it could be a fixed point; it could be a uniform random variable; or it could be drawn from some other distribution (perhaps centered around a 'hotspot'). In any case, a

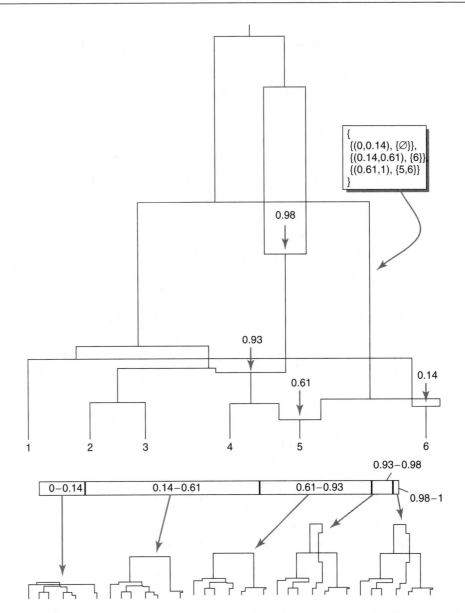

Figure 25.10 A realization of an ancestral recombination graph for $n = 6$. There were four recombination events, which implies $6 + 4 - 1 = 9$ coalescence events. Each recombination was assumed to lead to crossing over at a point, which was chosen randomly in $(0, 1)$. Four breakpoints (or 'junctions') implies five embedded trees, which are shown underneath. The tree for a particular chromosomal point is extracted from the graph by choosing the appropriate path at each recombination event. I have followed the convention that one should 'go left' if the point is located 'to the left' of (is less than) the breakpoint. Note that the two rightmost trees are identical. The box illustrates notation that may be used to represent ancestral lineages in the graph. The lineage pointed to is ancestral to: no (sampled) segment for the interval $(0, 0.14)$; segment 6 for the interval $(0.14, 0.61)$; and segments 5 and 6 for the interval $(0.61, 1)$.

breakpoint needs to be generated for each recombination event in the graph. We also need to know which branch in the graph carries which recombination 'product' (remember that we are going backward in time). With breaks affecting a point, a suitable rule is that the left branch carries the material to the 'left' of the breakpoint (i.e. in $(0, x)$), and the right branch carries the material to the 'right' (i.e. in $(x, 1)$).

Once recombination breakpoints have been added to the graph, it becomes possible to extract the genealogy for any given point by simply following the appropriate branches. Figure 25.10 illustrates how this is done. An ancestral recombination graph contains a number of embedded genealogical trees, each of which can be described by the standard coalescent, but which are obviously not independent of each other. An alternative way of viewing this process is thus as a 'walk through tree space' along the chromosome (Wiuf and Hein, 1999). The strength of the correlation between the genealogies for linked points depends on the scaled genetic distance between them, and goes to zero as this distance goes to infinity. The number of embedded trees equals the number of breakpoints plus one, but many of these trees may (usually will) be identical (cf. the two rightmost trees in Figure 25.10). Note also that the embedded trees vary greatly in height. This means that some pieces will have found their MRCA long before others. Indeed, it is quite possible for every piece to have found its MRCA long before the Ultimate MRCA. A number of interesting results concerning the number of recombination events and the properties of the embedded trees are available (see Griffiths and Marjoram, 1996; 1997; Hudson, 1983; 1987; Hudson and Kaplan, 1985; Kaplan and Hudson, 1985; Pluzhnikov and Donnelly, 1996; Simonsen and Churchill, 1997; Wiuf and Hein, 1999a,b).

25.6.2 Properties and Effects of Recombination

It probably does not need to be pointed out that the stochastic process just described is extremely complicated. At least I have found that whereas it is possible to develop a fairly good intuitive understanding of the random trees generated by the standard coalescent, the behavior of the random recombination graphs continues to surprise. It may therefore be worth questioning first of all whether it is necessary to incorporate recombination. It would seem reasonable that recombination could be ignored if it is sufficiently rare in the segment studied (e.g. if the segment is very short). But what is 'sufficiently rare'? Consider a pair of segments. The probability that they coalesce before either recombines is

$$\frac{1}{1 + 2(\rho/2)} = \frac{1}{1 + \rho} \qquad (25.17)$$

(cf. equation (25.11)). In order for recombination not to matter, we would need to have $\rho \approx 0$. It is thus the *scaled* recombination rate that matters, not the per-generation recombination probability. Estimates based on comparing genetic and physical maps indicate that the average per-generation per-nucleotide probability of recombination is of roughly the same order of magnitude as the average per-generation per-nucleotide probability of mutation (which can be estimated in various ways). This means that the scaled mutation and recombination rates will also be of the same order of magnitude, and, thus, that recombination can be ignored when mutation can be ignored. In other words, as long we restrict our attention to segments short enough not to be polymorphic, we do not need to worry about recombination!

Of course, both recombination and mutation rates vary widely over the genome, so regions where recombination can be ignored almost certainly exist. Unfortunately, whereas direct estimates of recombination probabilities (genetic distances) are restricted to large scales, estimates of the recombination rate from polymorphism data are extremely unreliable (Griffiths and Marjoram, 1996; Hudson, 1987; Hudson and Kaplan, 1985; Wakeley, 1997; Wall, 2000; Fearnhead and Donnelly, 2001; McVean *et al.*, 2002). The latter problem is unavoidable. The main reason is the usual one that there is only a single realization of the underlying genealogy. Thus, for example, numerous recombination events in a particular region of a gene do not necessarily mean that it is a recombinational hotspot: it could just be that that region has a deep enough genealogy for multiple recombination events to have had time to occur. This is the same problem that affects estimates of the mutation rate.

However, there are also problems peculiar to recombination. It is important to realize that most recombination events are undetectable (Hudson and Kaplan, 1985). Recombination in sequence data has often been inferred by identifying 'tracts' that have obviously moved from one sequence to another. The presence of such tracts is actually indicative of *low* rather than of high recombination rates (Maynard Smith, 1999). Even a moderate amount of recombination will wipe out the tracts. Recombination can then only be 'detected' through the 'four-gamete test' (Hudson and Kaplan, 1985): the four linkage configurations AB, Ab, aB, and ab for two linked loci can only arise through recombination or repeated mutation (which is more likely is debatable (Eyre-Walker *et al.*, 1999; Templeton *et al.*, 2000)). Furthermore, recombination events can clearly only be detected if there is sufficient polymorphism. However, many recombination events can *never* be detected even with infinite amounts of polymorphism (Griffiths and Marjoram, 1997; Hudson and Kaplan, 1985; Nordborg, 2000). Consider, for example, the two rightmost trees in Figure 25.10. These trees are identical. This means that the recombination event that gave rise to them cannot possibly leave any trace.

The phenomenon of undetectable breakpoints turns out to have special relevance for models with inbreeding. The 'forward' intuition that corresponds to undetectable recombination events is that these events took place in homozygous individuals. Inbreeding increases the frequency of homozygous individuals, and can therefore have a considerable effect on the recombination graph. It turns out that this effect can also be modeled as a scaling change, but this time of the recombination rate. Thus, for example, the recombination graph in a partially selfing hermaphrodite reduces to the standard recombination graph if we introduce an 'effective recombination rate', $\rho_e :=$ $\rho(1 - F)$ (Nordborg, 2000). Recombination breaks up haplotypes much less efficiently in inbreeders.

So far, we have only discussed the problems associated with recombination. It must be remembered that recombination is the only thing that allows us to get around the 'single underlying genealogy'. Unlinked loci will, with respect to most questions, provide independent samples. Of course this also applies within a segment: if ρ were infinite, then each base pair would be an independent locus (Pluzhnikov and Donnelly, 1996). High rates of recombination are thus an enormous advantage for many purposes.

Finally, it should be noted that since crossing over is mechanistically tied to gene conversion, there is reason to question the applicability of the simple model used above at the intragenic scale (Andolfatto and Nordborg, 1998; Nordborg, 2000). However, the ancestral recombination graph is quite general, and more realistic recombination models have been developed (Wiuf, 2000; Wiuf and Hein, 2000). Models of other kinds of recombination, such as bacterial transformation (Hudson, 1994) and intergenic gene conversion (Bahlo, 1998), also exist.

25.7   SELECTION

The coalescent depends crucially on the assumption of selective neutrality, because if the allelic state of a lineage influences its reproductive success, it is not possible to separate 'descent' from 'state'. Nonetheless, it turns out that it is possible to circumvent this problem, and incorporate selection into the coalescent framework. Two distinct approaches have been used. The first is an elegant extension of the coalescent process, known as the 'ancestral selection graph' (Krone and Neuhauser, 1997; Neuhauser and Krone, 1997). The genealogy is generated backward in time, as in the standard coalescent, but it contains branching as well as coalescence events. The result is a genealogical graph that is superficially similar to the one generated by recombination. Mutations are then superimposed forward in time, and, with knowledge of the state of each branch, the graph is 'pruned' to a tree by preferentially removing bad branches (i.e. those carrying selectively inferior alleles). In a sense, the ancestral selection graph allows the separation of descent from state by including 'potential' descent: lineages that might have lived, had their state allowed it.

The second approach is based on two insights. First, a polymorphic population may be thought of as subdivided into *allelic classes* within which there is no selection. Second, if we know the historical sizes of these classes, then they may be modeled as analogous to patches, using the machinery described above. Lineages then 'mutate between classes' rather than 'migrate between patches'. This approach was pioneered in the context of the coalescent by Kaplan *et al.* (1988). Knowing the past class sizes is the same as knowing the past allele frequencies, so it is obviously not possible to study the dynamics of the selectively different alleles themselves using this approach. However, it is possible to study the effects of selection on the underlying genealogical structure, which is relevant if we wish to understand how linked neutral variants are affected.

It is not entirely clear how the two approaches relate to each other. Since the second approach requires knowledge of the past allele frequencies, it may be viewed as some kind of limiting (strong selection) or, alternatively, conditional version of the selection graph (Nordborg, 1999). However, whereas the second approach would be most appropriate for very strong, deterministic selection, the selection graph requires all selection coefficients to be $O(1/N)$. This is an area of active research.

The ancestral selection graph is described in **Chapter 22**, and will not be discussed here. The second approach, which might be called the 'conditional structured coalescent', will be illustrated through three simple but very different examples.

25.7.1   Balancing Selection

By 'balancing selection' is meant any kind of selection that tends to maintain two or more alleles in the population. The effect of such selection on genealogies has been studied by a number of authors (Barton and Navarro, 2002; Hey, 1991; Hudson and Kaplan, 1988; Kaplan *et al.*, 1988; Kaplan *et al.*, 1991; Kelly and Wade, 2000; Navarro and

Barton, 2002; Nordborg, 1997; 1999; Schierup *et al.*, 2001; Takahata, 1990; Vekemans and Slatkin, 1994). We will limit ourselves to the case of two alleles, A_1 and A_2, maintained at constant frequencies p_1 and $p_2 = 1 - p_1$ by strong selection. Alleles mutate to the other type with some small probability v per generation, and we define the scaled rate $v := 4Nv$. Reproduction occurs according to a diploid Wright–Fisher model, as for the recombination graph above.

Consider a segment of length ρ that contains the selected locus. Depending on the allelic state at the locus, the segment belongs to either the A_1 or the A_2 allelic class. Say that it belongs to the A_1 allelic class. Trace the ancestry of the segment a single generation back in time. It is easy to see that its creation involved an $A_2 \to A_1$ mutation with probability

$$\frac{vp_2}{vp_2 + (1-v)p_1} = \frac{v}{4N} \times \frac{p_2}{p_1} + O\left(\frac{1}{N^2}\right), \qquad (25.18)$$

(cf. equation (25.9)), and involved recombination with probability $r = \rho/(4N)$. Thus the probability that neither happens is $1 - O(1/N)$, and the probability of two events, for example both mutation and recombination, is $O(1/N^2)$ and can be neglected. If nothing happens, then the lineage remains in the A_1 class. If there was a mutation, the lineage 'mutates' to the A_2 allelic class. If there was a recombination event, we have to know the genotype of the individual in which the event took place.

Because the lineage we are following is A_1, we know that the individual must have been either an A_1A_1 homozygote or an A_1A_2 heterozygote. What fraction of the A_1 alleles was produced by each genotype? In general, this will depend on their relative fitness as well as their frequencies. Let x_{ij} be the frequency of A_iA_j individuals, and w_{ij} their relative fitness. Then the probability that an A_1 lineage was produced in a heterozygote is

$$\frac{w_{12}x_{12}/2}{w_{12}x_{12}/2 + w_{11}x_{11}}. \qquad (25.19)$$

If we can ignore the differences in fitness, and assume Hardy–Weinberg equilibrium (see Nordborg, 1999, for more on this), (25.19) simplifies to

$$\frac{p_1p_2}{p_1p_2 + p_1^2} = p_2. \qquad (25.20)$$

Thus the probability that an A_1 lineage 'meets' and recombines with an A_2 segment is equal to the frequency of A_2 segments, which is intuitive. The analogous reasoning applies to A_2 lineages, which recombine with A_1 segments with probability p_1, and with members of their own class with probability p_2. It should be noted that the above can be made rigorous using a model that treats genotypes as well as individuals as population structure (Nordborg, 1999).

What happens when the lineage undergoes recombination? If it recombines in a homozygote, then both branches remain in the A_1 allelic class. However, if it recombines in a heterozygote, then one of the branches (the one *not* carrying the ancestry of the selected locus) will 'jump' to the A_2 allelic class. The other branch remains in the A_1 allelic class.

When more than two lineages exist, coalescences may occur, but only within allelic classes (remember that since mutation is $O(1/N)$ it is impossible for lineages to mutate and coalesce in the same generation).

If time is measured in units of $2N$ generations, and we let N go to infinity, the model converges to a coalescent process with the following types of events:

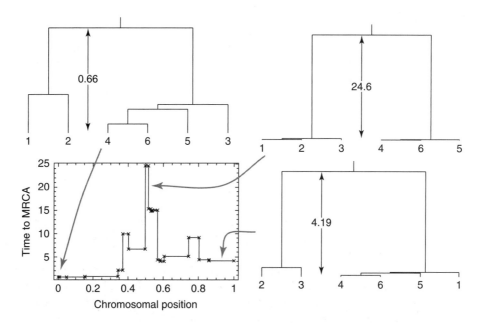

Figure 25.11 Selection will have a local effect on genealogies. A realization of the coalescent with recombination and strong balancing selection is shown. Lineages 1–3 belong to one allelic class, and lineages 4–6 to the other. The selected locus is located in the middle of the region. The plot shows how the time to the MRCA varies along the chromosome (the crosses denote cross-over points). The three extracted trees exemplify how the topology and branch lengths are affected by linkage to the selected locus. Note that the trees are not drawn to scale (the numbers on the arrows give the heights).

- each pair of lineages in the A_i allelic class coalesces independently at rate $1/p_i$;

- each lineage in A_i recombines with a segment in class j at rate ρp_j;

- each lineage in A_i mutates to A_j, $j \neq i$, at rate $\nu p_j / p_i$.

The process may be stopped either when the Ultimate MRCA is reached, or when all points have found their MRCA.

  This model has some very interesting properties. Consider a sample that contains both types of alleles. Since coalescence is only possible within allelic classes, the selected locus (in the strict sense of the word, i.e. the 'point' in the segment where the selectively important difference lies) cannot coalesce without at least one mutation event. If mutations are rare, then this may have occurred a very long time ago. In other words, the polymorphism may be ancient. All coalescences will occur within allelic classes before mutation allows the final two lineages to coalesce. The situation is similar to strong population subdivision (see Figure 25.7). However, this is only true for the locus itself: linked pieces may 'recombine away' and coalesce much earlier. This will usually result in a local increase in the time to MRCA centered around the selected locus, as illustrated in Figure 25.11. Because the expected number of mutations is proportional to the height of the tree, this may lead to a 'peak of polymorphism' (Hudson and Kaplan, 1988).

25.7.2 Selective Sweeps

Next consider a population in which favorable alleles arise infrequently at a locus, and are rapidly driven to fixation by strong selection. Each such fixation is known as a 'selective sweep' for reasons that will become apparent. This process can be modeled using the framework developed above, if we know how the allele frequencies have changed over time. Of course we do not know this, but if the selection is strong enough, it may be reasonable to model the increase in frequency of a favorable allele deterministically (Kaplan *et al.*, 1989).

Consider a population that is currently not polymorphic, but in which a selective sweep recently took place. During the sweep, there were two allelic classes just as in the balancing selection model above. The difference is that these classes changed in size over time. In particular, the class corresponding to the allele that is currently fixed in the population will *shrink* rapidly back in time. The genealogy of the selected locus itself (in the 'point' sense used above) will therefore behave as if it were part of a population that has expanded from a very small size (cf. Figure 25.6). Indeed, unlike 'real' populations, the allelic class *will* have grown from a size of 1. A linked point must have grown in the same way, unless recombination in a heterozygote took place between the point and the selected locus. Whether this happens or not will depend on how quickly the new allele increased. Typically, it depends on the ratio r/s, where s is the selective advantage of the new allele, and r is the relevant recombination probability.

The result of such a fixation event is thus to cause a local 'genealogical distortion', just like balancing selection. However, whereas the distortion in the case of balancing selection looks like population subdivision, the distortion caused by a fixation event looks like population growth. Close to the selected site, coalescence times will have a tendency to be short, and the genealogy will have a tendency to be starlike (cf. Figure 25.6). Note that a single recombination event in the history of the sample can change this, and that the variance will consequently be enormous (note the variance in time to MRCA in Figure 25.11). Shorter coalescence times mean less time for mutations to occur, so a local reduction in variability is expected. This is obvious: when the new allele sweeps through the population and fixes, it causes linked neutral alleles to 'hitchhike' along and also fix (Maynard Smith and Haigh, 1974). Repeated selective sweeps can thus decrease the variability in a genomic region (Kaplan *et al.*, 1989; Simonsen *et al.*, 1995; Kim and Stephan, 2002; Przeworski, 2002). Because each sweep is expected to affect a bigger region the lower the rate of recombination is, this has been proposed as an explanation for the correlation between polymorphism and local rate of recombination that is observed in many organisms (Begun and Aquadro, 1992; Nachman, 1997; Nachman *et al.*, 1998).

25.7.3 Background Selection

We have seen that selection can affect genealogies in ways reminiscent of strong population subdivision and of population growth. It is often difficult to distinguish statistically between selection and demography for precisely this reason (Fu and Li, 1993; Tajima, 1989b). It is also possible for selection to affect genealogies in a way that is completely undetectable, that is, as a linear change in time scale. This appears to be the case for selection against deleterious mutations, at least under some circumstances (Charlesworth *et al.*, 1995; Hudson and Kaplan, 1994; 1995; Nordborg, 1997; Nordborg *et al.*, 1996).

The basic reason for this is the following. Strongly deleterious mutations are rapidly removed by selection. Looking backward in time, this means that each lineage that carries a deleterious mutation must have a nonmutant ancestor in the near past. On the coalescent time scale, lineages in the deleterious allelic class will 'mutate' (backward in time) to the 'wild-type' allelic class instantaneously. The process looks like a strong-migration model, with the wild-type class as the source environment, and the deleterious class as the sink environment: the presence of deleterious mutations increases the variance in reproductive success. The resulting reduction in the effective population size is known as 'background selection' (Charlesworth *et al.*, 1993).

More realistic models with multiple loci subject to deleterious mutations, recombination, and several mutational classes turn out to behave similarly. The strength of the background selection effect at a given genomic position will depend strongly on the local rate of recombination, which determines how many mutable loci influence a given point. Thus, deleterious mutations have also been proposed as an explanation for the correlation between polymorphism and local rate of recombination referred to above (Charlesworth *et al.*, 1993). The 'effective population size' would thus depend on the mutation, selection, and recombination parameters in each genomic region.

It should be pointed out that, unlike the many limit approximations presented in this chapter, the idea that background selection can be modeled as a simple scaling is not mathematically rigorous. However, we would rather hope that selection against deleterious mutations can be taken care of this way, because given that amino acid sequences are conserved over evolutionary time, practically all of population genetics theory would be in trouble otherwise!

25.8   NEUTRAL MUTATIONS

Not much has been said about the neutral mutation process because it is trivial from a mathematical point of view. Once we know how to generate the genealogy, mutations can be added afterwards according to a Poisson process with rate $\theta/2$, where θ is the scaled per-generation mutation probability. Thus, if a particular branch has length τ units of scaled time, the number of mutations that occur on it will be Poisson-distributed with mean $\tau\theta/2$ (and they occur with uniform probability along the branch). It is also possible to add mutations while the genealogy is being created, instead of afterwards. This can in some circumstances lead to much more efficient algorithms (see, for example, the 'urn scheme' described by Donnelly and Tavaré, 1995), although from the point of view of simulating samples, all coalescent algorithms are so efficient that such fine-tuning does not matter. However, it can matter greatly for the kinds of inference methods described in **Chapter 26**.

It should be noted that the mutation process is just as general as the recombination process. Almost any neutral mutation model can be used. A useful trick is so-called 'Poissonization': let mutation events occur according to a simple Poisson process with rate $\theta/2$, but once an event occurs, determine the *type* of event through some kind of transition matrix which includes mutation back to self (i.e. there was no mutation). This allows models where the mutation probability depends on the current allelic state.

The only restriction is that in order to interpret samples generated by the coalescent as samples from the relevant stationary distribution (which incorporates demography, migration, selection at linked sites, etc.), we need to be able to choose the type of the MRCA from the stationary distribution of the mutation process (alone, since demography, for example, does not affect samples of size $n = 1$). In many cases, such as the infinite-alleles model (each mutation gives rise to a new allele) or the infinite-sites model (each mutation affects a new site), the state of the MRCA does not matter, since all we are interested in is the number of mutational changes.

25.9 CONCLUSION

25.9.1 The Coalescent and 'Classical' Population Genetics

The differences between coalescent theory and 'classical' population genetics have frequently been exaggerated or misunderstood. First, the basic models do not differ. The coalescent is essentially a diffusion model of lines of descent. This can be done forward in time, for the whole population (e.g. Griffiths, 1980), but it was realized in the early 1980s that it is easier to do it backward in time. Second, the coalescent is not limited to finite samples. Everything above has been limited to finite samples because it is mathematically much easier, but it is likely that all of it could be extended to the whole population. Of course, it is essential for the independence of events that the number of lineages be finite, but in the whole-population coalescent the number of lineages becomes finite infinitely fast (it is an 'entrance boundary', e.g. Griffiths, 1984). Third, classical population genetics is not limited to the whole population. A sample of size $n = 6$ from a K-allele model, say, could be obtained either through the coalescent, or by first drawing a population from the stationary distribution found by Wright (1949), and then drawing six alleles conditional on this population. Note, however, that it would be rather more difficult (read 'impossible') to use the second approach for most models. Fourth, the coalescent is in no sense tied to sequence data: any mutation model can be used. The impression that it is came about doubtless because models for sequence evolution such as the infinite-sites model are indeed impossibly hard to analyze using classical methods (Ethier and Griffiths, 1987).

I would argue that the real difference is more philosophical. As has been pointed out by Ewens (1979; 1990), essentially all of classical population genetics is 'prospective', looking forward in time. Another way of saying this is that it is conditional: given the state in a particular generation, what will happen? This approach is fine when modeling is done to determine 'how evolution might work' (which is what most classical population genetics was about). It is usually not suitable for statistical analysis of data, however. Wright considered how 'heterozygosity' would decay from the same starting point in infinitely many identical populations, that is to say, he took the expectation over evolutionary realizations. Data, alas, come from a single time-slice of one such realization. The coalescent forces us to acknowledge this, and allows the utilization of modern statistical methods, such as the calculation of likelihoods for samples.

25.9.2 The Coalescent and Phylogenetics

If the differences between coalescent theory and classical population genetics have sometimes been exaggerated, the differences between coalescent theory and phylogenetics have not always been fully appreciated. The central role played by trees in both turns out to be very misleading.

To be able to compare them, we need to model speciation. This has usually been done using an 'isolation' model in which randomly mating populations split into two completely isolated ones at fixed times in the past. The result is a 'species tree', within which we find 'gene trees' (see Figure 25.12). The model is quite simple: lineages will tend to coalesce within their species, and can only coalesce with lineages from other species back in the ancestral species.

Molecular phylogenetics attempts to estimate the species tree by estimating the genealogy of homologous sequences from the different species, that is, by estimating the gene tree. The species tree is assumed to exist and is treated as a model parameter.

In addition, the standard methods rely on all branches in the species tree being very long compared to within-species coalescence times. This means that the coalescent can be ignored: regardless of how we sample, all (neutral) gene genealogies will rapidly coalesce within their species, and thereafter have the same topology as the species tree. Furthermore, the variation in the branch lengths caused by different coalescence times in the ancestral species will be negligible compared to the lengths of the interspecific branches. There is no need to sample more than one individual per species, and recombination is completely irrelevant. Gene trees perfectly reflect the species tree.

It is of course widely acknowledged that gene trees and species trees may differ (see Avise, 1994; Avise and Ball, 1990; Hey, 1994; Hudson, 1992; Li, 1997; Nei, 1987; Takahata, 1989; Wu, 1991; phylogenetic methods are discussed in **Chapters 15** and **16**). Nonetheless, phylogenetic methods would not work unless the interspecific branches usually were long enough for the gene trees to reflect the species tree closely. Indeed, in many situations, the problem is the opposite: the branches are so long that repeated mutations have erased much of the phylogenetic information.

Figure 25.12   A gene tree within a species tree.

Phylogenetic inference can thus be viewed as a 'missing data' problem just like population genetic inference: polymorphisms contain information about an unobserved genealogy, which in turn provides information about an evolutionary model. However, note that in phylogenetics, there is relatively little doubt about what the right model is (it is typically an isolation model that gives rise to a species tree, as in Figure 25.12). Furthermore, because of the long branch lengths, the gene genealogies, although random variables with a coalescent distribution under the model, can be treated as parameters (which, among other things, means that we do not need to know the sizes of ancestral populations to estimate divergence times). None of this is true when analyzing population genetic data (which, strictly speaking, means any data for which the 'long branch' assumption above is not fulfilled). Unfortunately, the considerable success and popularity of phylogenetics (coupled with the ready availability of user-friendly software) has sometimes led to the inappropriate application of phylogenetic methods. It is important to remember that a genealogical tree from a population (or several populations that have not been isolated for a very long time) does not have an obvious interpretation: it certainly contains information about the process that gave rise to it, but usually less than we would hope (see Rosenberg and Nordborg, 2002).

25.9.3 Prospects

A theme of this review has been the versatility and generality of the coalescent model. Considerable theoretical progress is being made, especially in the areas of statistical inference and modeling selection. At the same time, we are entering the era of genomic polymorphism data. This wealth of information will make it increasingly possible to evaluate whether the models constructed by population genetics over the years actually fit the data. It seems likely that we will find that the data is in many ways much less informative than imagined (as happened before in population genetics; see Lewontin, 1991); it also seems likely that we will discover new phenomena that require new models. Either way, the importance of a rigorous statistical approach to analyzing genetic polymorphism data can hardly be overstated.

Acknowledgments

I wish to thank David Balding, Bengt Olle Bengtsson, Malia Fullerton, Jenny Hagenblad, Maarit Jaarola, Martin Lascoux, Claudia Neuhauser, François Rousset, Matthew Stephens, and Torbjörn Säll for comments on the original version of this chapter. The second edition benefited from a large number of readers: thanks to all.

REFERENCES

Andolfatto, P. and Nordborg, M. (1998). The effect of gene conversion on intralocus associations. *Genetics* **148**, 1397–1399.

Avise, J.C. (1994). *Molecular Markers, Natural History and Evolution*. Chapman & Hall, New York.

Avise, J.C. and Ball, R.M. (1990). In *Oxford Surveys in Evolutionary Biology*, Vol. 7, D. Futuyama, and J. Antonovics, eds. Oxford University Press, Oxford, pp. 45–67.

Bahlo, M. (1998). Segregating sites in a gene conversion model with mutation. *Theoretical Population Biology* **54**, 243–256.

Barton, N.H. and Navarro, A. (2002). Extending the coalescent to multilocus systems: the case of balancing selection. *Genetical Research* **79**, 129–139.

Barton, N.H. and Wilson, I. (1995). Genealogies and geography. *Proceedings of the Royal Society London Series B* **349**, 49–59.

Begun, D.J. and Aquadro, C.F. (1992). Levels of naturally occurring DNA polymorphism correlate with recombination rates in *Drosophila melanogaster*. *Nature* **356**, 519–520.

Charlesworth, B., Morgan, M.T. and Charlesworth, D. (1993). The effect of deleterious mutations on neutral molecular variation. *Genetics* **134**, 1289–1303.

Charlesworth, D., Charlesworth, B. and Morgan, M.T. (1995). The pattern of neutral molecular variation under the background selection model. *Genetics* **141**, 1619–1632.

Donnelly, P. (1996). In *Variation in the Human Genome*, Ciba Foundation Symposium No. 197. John Wiley & Sons, Chichester, pp. 25–50.

Donnelly, P. and Tavaré, S. (1995). Coalescents and genealogical structure under neutrality. *Annual Review of Genetics* **29**, 401–421.

Ethier, S.N. and Griffiths, R.C. (1987). The infinitely many sites model as a measure valued diffusion. *Annals of Probability* **5**, 515–545.

Ewens, W.J. (1979). *Mathematical Population Genetics*. Springer-Verlag, Berlin.

Ewens, W.J. (1990). *Mathematical and Statistical Development of Evolutionary Theory*, S. Lessard, ed. Kluwer Academic, Dordrecht, pp. 177–227.

Eyre-Walker, A., Smith, N.H. and Maynard Smith, J. (1999). How clonal are human mitochondria? *Proceedings of the Royal Society London Series B* **266**, 477–483.

Fearnhead, P. and Donnelly, P. (2001). Estimating recombination rates from population genetic data. *Genetics* **159**, 1299–1318.

Fisher, R.A. (1965). *Theory of Inbreeding*, 2nd edition. Oliver and Boyd, Edinburgh.

Fu, Y.-X. and Li, W.-H. (1993). Statistical tests of neutrality of mutations. *Genetics* **133**, 693–709.

Griffiths, R.C. (1980). Lines of descent in the diffusion approximation of neutral Wright-Fisher models. *Theoretical Population Biology* **17**, 37–50.

Griffiths, R.C. (1984). Asymptotic line-of-descent distributions. *Journal of Mathematical Biology* **21**, 67–75.

Griffiths, R.C. and Marjoram, P. (1996). Ancestral inference from samples of DNA sequences with recombination. *Journal of Computational Biology* **3**, 479–502.

Griffiths, R.C. and Marjoram, P. (1997). In *Progress in Population Genetics and Human Evolution*, P. Donnelly and S. Tavaré, eds. Springer-Verlag, New York, pp. 257–270.

Griffiths, R.C. and Tavaré, S. (1994). Sampling theory for neutral alleles in a varying environment. *Philosophical Transactions of the Royal Society of London Series B* **344**, 403–410.

Griffiths, R.C. and Tavaré, S. (1998). The age of a mutant in a general coalescent tree. *Stochastic Models* **14**, 273–295.

Herbots, H.M. (1994). Stochastic models in population genetics: genealogy and genetic differentiation in structured populations. PhD thesis, University of London.

Hey, J. (1991). A multi-dimensional coalescent process applied to multi-allelic selection models and migration models. *Theoretical Population Biology* **39**, 30–48.

Hey, J. (1994). *Molecular Ecology and Evolution: Approaches and Applications*, B. Schierwater, B. Streit, G.P. Wagner, R.Desalleeds. Birkhäuser, Basel, pp. 435–449.

Hudson, R.R. (1983). Properties of a neutral allele model with intragenic recombination. *Theoretical Population Biology* **23**, 183–201.

Hudson, R.R. (1987). Estimating the recombination parameter of a finite population model without selection. *Genetical Research, (Cambridge)* **50**, 245–250.

Hudson, R.R. (1990). In *Oxford Surveys in Evolutionary Biology*, Vol. 7, D. Futuyma and J. Antonovics, eds. Oxford University Press, Oxford, pp. 1–43.

Hudson, R.R. (1992). Gene trees, species trees and the segregation of ancestral alleles. *Genetics* **131**, 509–512.

Hudson, R.R. (1993). *Mechanisms of Molecular Evolution*, A.G. Takahata and N. Clark, eds. Japan Scientific Societies Press, Tokyo, pp. 23–36.

Hudson, R.R. (1994). Analytical results concerning linkage disequilibrium in models with genetic transformation and conjugation. *Journal of Evolutionary Biology* **7**, 535–548.

Hudson, R.R. (1998). Island models and the coalescent process. *Molecular Ecology* **7**, 413–418.

Hudson, R.R. and Kaplan, N.L. (1985). Statistical properties of the number of recombination events in the history of a sample of DNA sequences. *Genetics* **111**, 147–164.

Hudson, R.R. and Kaplan, N.L. (1988). The coalescent process in models with selection and recombination. *Genetics* **120**, 831–840.

Hudson, R.R. and Kaplan, N.L. (1994). *Non-neutral Evolution: Theories and Molecular Data*, G.B. Golding, ed. Chapman & Hall, New York, pp. 140–153.

Hudson, R.R. and Kaplan, N.L. (1995). Deleterious background selection with recombination. *Genetics* **141**, 1605–1617.

Kaj, I., Krone, S.M. and Lascoux, M. (1991). Coalescent theory for seed bank models. *Journal of Applied Probability* **38**, 285–301.

Kaplan, N.L. and Hudson, R.R. (1985). The use of sample genealogies for studying a selectively neutral *m*-loci model with recombination. *Theoretical Population Biology* **28**, 382–396.

Kaplan, N.L., Darden, T. and Hudson, R.R. (1988). The coalescent process in models with selection. *Genetics* **120**, 819–829.

Kaplan, N.L., Hudson, R.R. and Iizuka, M. (1991). The coalescent process in models with selection, recombination and geographic subdivision. *Genetical Research, Cambridge* **57**, 83–91.

Kaplan, N.L., Hudson, R.R. and Langley, C.H. (1989). The 'hitch-hiking' effect revisited. *Genetics* **123**, 887–899.

Kelly, J.K. and Wade, M.J. (2000). Molecular evolution near a two-locus balanced polymorphism. *Journal of Theoretical Biology* **204**, 83–101.

Kim, Y. and Stephan, W. (2002). Detecting a local signature of genetic hitchhiking along a recombining chromosome. *Genetics* **160**, 765–777.

Kingman, J.F.C. (1982a). The coalescent. *Stochastic Processes and their Applications* **13**, 235–248.

Kingman, J.F.C. (1982b). In *Exchangeability in Probability and Statistics*, G. Koch and F. Spizzichino, eds. North-Holland, Amsterdam, pp. 97–112.

Kingman, J.F.C. (1982c). In*Essays in Statistical Science: Papers in Honour of P.A.P. Moran*, J. Gani and E.J. Hannan, eds. (Journal of Applied Probability, special, Vol. 19A)Applied Probability Trust, Sheffield, pp. 27–43.

Krone, S.M. and Neuhauser, C. (1997). Ancestral processes with selection. *Theoretical Population Biology* **51**, 210–237.

Lewontin, R.C. (1991). Twenty-five years ago in genetics: electrophoresis in the development of evolutionary genetics: milestone or millstone? *Genetics* **128**, 657–662.

Li, W.-H. (1997). *Molecular Evolution*. Sinauer, Sunderland, MA.

Marjoram, P. and Donnelly, P. (1997). *Progress in Population Genetics and Human Evolution*, S. Donnelly, P. Tavaré, eds. Springer-Verlag, New York, pp. 107–131.

Maynard Smith, J. (1999). The detection and measurement of recombination from sequence data. *Genetics* **153**, 1021–1027.

Maynard Smith, J. and Haigh, J. (1974). The hitchhiking effect of a favourable gene. *Genetical Research, Cambridge* **23**, 23–35.

McVean, G., Awadalla, P. and Fearnhead, P. (2002). A coalescent-based method for detecting and estimating recombination from gene sequences. *Genetics* **160**, 1231–1241.

Möhle, M. (1998a). A convergence theorem for Markov chains arising in population genetics and the coalescent with selfing. *Advances in Applied Probability* **30**, 493–512.

Möhle, M. (1998b). Robustness results for the coalescent. *Journal of Applied Probability* **35**, 438–447.

Möhle, M. (1999). Weak convergence to the coalescent in neutral population models. *Journal of Applied Probability* **36**, 446–460.

Nachman, M.W. (1997). Patterns of DNA variability at X-linked loci in Mus domesticus. *Genetics* **147**, 1303–1316.

Nachman, M.W., Bauer, V.L., Crowell, S.L. and Aquadro, C.F. (1998). DNA variability and recombination rates at X-linked loci in humans. *Genetics* **150**, 1133–1141.

Nagylaki, T. (1980). The strong-migration limit in geographically structured populations. *Journal of Mathematical Biology* **9**, 101–114.

Nagylaki, T. (1982). Geographical invariance in population genetics. *Journal of Theoretical Biology* **99**, 159–172.

Nagylaki, T. (1998). The expected number of heterozygous sites in a subdivided population. *Genetics* **149**, 1599–1604.

Navarro, A. and Barton, N.H. (2002). The effects of multilocus balancing selection on neural variability. *Genetics* **161**, 849–863.

Nei, M. (1987). *Molecular Evolutionary Genetics*. Columbia University Press, New York.

Neuhauser, C. and Krone, S.M. (1997). The genealogy of samples in models with selection. *Genetics* **145**, 519–534.

Nordborg, M. (1997). Structured coalescent processes on different time scales. *Genetics* **146**, 1501–1514.

Nordborg, M. (1998). On the probability of Neanderthal ancestry. *American Journal of Human Genetics* **63**, 1237–1240.

Nordborg, M. (1999). In *Statistics in Molecular Biology and Genetics, IMS Lecture Notes Monograph Series*, Vol. 33, F. Seillier-Moiseiwitsch, ed. Institute of Mathematical Statistics, Hayward, CA, pp. 56–76.

Nordborg, M. (2000). Linkage disequilibrium, gene trees, and selfing: an ancestral recombination graph with partial self-fertilization. *Genetics* **154**, 923–929.

Nordborg, M. and Donnelly, P. (1997). The coalescent process with selfing. *Genetics* **146**, 1185–1195.

Nordborg, M. and Krone, S.M. (2002). *Modern Developments in Theoretical Population Genetics*, M. Slatkin, M. Veuille, eds. Oxford University Press, Oxford, pp. 194–232.

Nordborg, M., Charlesworth, B. and Charlesworth, D. (1996). The effect of recombination on background selection. *Genetical Research, Cambridge* **67**, 159–174.

Notohara, M. (1990). The coalescent and the genealogical process in geographically structured populations. *Journal of Mathematical Biology* **29**, 59–75.

Notohara, M. (1993). The strong-migration limit for the genealogical process in geographically structured populations. *Journal of Mathematical Biology* **31**, 115–122.

Pluzhnikov, A. and Donnelly, P. (1996). Optimal sequencing strategies for surveying molecular genetic diversity. *Genetics* **144**, 1247–1262.

Pollak, E. (1987). On the theory of partially inbreeding finite populations. I. Partial selfing. *Genetics* **117**, 353–360.

Przeworski, M. (2002). The signature of positive selection and randomly chosen loci. *Genetics* **160**, 1179–1189.

Pulliam, H.R. (1988). Sources, sinks, and population regulation. *American Naturalist* **132**, 652–661.

Rogers, A.R. and Harpending, H. (1992). Population growth makes waves in the distribution of pairwise genetic differences. *Molecular Biology Evolution* **9**, 552–569.

Rosenberg, N.A. and Nordborg, M. (2002). Genealogical trees, coalescent theory, and the analysis of genetic polymorphisms. *Nature Reviews Genetics* **3**, 380–390.

Rousset, F. (1999a). Genetic differentiation in populations with different classes of individuals. *Theoretical Population Biology* **55**, 297–308.

Rousset, F. (1999b). Genetic differentiation within and between two habitats. *Genetics* **151**, 397–407.

Saunders, I.W., Tavaré, S. and Watterson, G.A. (1984). On the genealogy of nested subsamples from a haploid population. *Advances in Applied Probability* **16**, 471–491.

Schierup, M.H., Mikkelsen, A.M. and Hein, J. (2001). Recombination, balancing selection and phylogenies in MHC and self-incompatibility genes. *Genetics* **159**, 1833–1844.

Simonsen, K.L. and Churchill, G.A. (1997). A Markov chain model of coalescence with recombination. *Theoretical Population Biology* **52**, 43–59.

Simonsen, K.L., Churchill, G.A. and Aquadro, C.F. (1995). Properties of statistical tests of neutrality for DNA polymorphism data. *Genetics* **141**, 413–429.

Slatkin, M. (1987). The average number of sites separating DNA sequences drawn from a subdivided population. *Theoretical Population Biology* **32**, 42–49.

Slatkin, M. (1996). Gene genealogies within mutant allelic classes. *Genetics* **143**, 579–587.

Slatkin, M. and Hudson, R.R. (1991). Pairwise comparisons of mitochondrial DNA sequences in stable and exponentially growing populations. *Genetics* **129**, 555–562.

Slatkin, M. and Rannala, B. (1997a). Estimating the age of alleles by use of intraallelic variability. *American Journal of Human Genetics* **60**, 447–458.

Slatkin, M. and Rannala, B. (1997b). The sampling distribution of disease-associated alleles. *Genetics* **147**, 1855–1861.

Strobeck, C. (1987). Average number of nucleotide differences in a sample from a single subpopulation: a test for population subdivision. *Genetics* **117**, 149–153.

Tajima, F. (1983). Evolutionary relationship of DNA sequences in finite populations. *Genetics* **105**, 437–460.

Tajima, F. (1989a). DNA polymorphism in a subdivided population: the expected number of segregating sites in the two-subpopulation model. *Genetics* **123**, 229–240.

Tajima, F. (1989b). Statistical method for testing the neutral mutation hypothesis by DNA polymorphism. *Genetics* **123**, 585–595.

Takahata, N. (1988). The coalescent in two partially isolated diffusion populations. *Genetical Research, Cambridge* **52**, 213–222.

Takahata, N. (1989). Gene genealogy in three related populations: consistency probability between gene and population trees. *Genetics* **122**, 957–966.

Takahata, N. (1990). A simple genealogical structure of strongly balanced allelic lines and trans-species polymorphism. *Proceedings of the National Academy of Sciences (USA)* **87**, 2419–2423.

Takahata, N. (1991). Genealogy of neutral genes and spreading of selected mutations in a geographically structured population. *Genetics* **129**, 585–595.

Tavaré, S. (1984). Line-of-descent and genealogical processes, and their applications in population genetic models. *Theoretical Population Biology* **26**, 119–164.

Templeton, A.R., Clark, A.G., Weiss, K.M., Nickerson, D.A., Boerwinkle, E. and Sing, C.F. (2000). Recombinational and mutational hotspots within the human lipoprotein lipase gene. *American Journal of Human Genetics* **66**, 69–83.

Vekemans, X. and Slatkin, M. (1994). Gene and allelic genealogies at a gametophytic self-incompatibility locus. *Genetics* **137**, 1157–1165.

Wakeley, J. (1996). Distinguishing migration from isolation using the variance of pairwise differences. *Theoretical Population Biology* **49**, 369–386.

Wakeley, J. (1997). Using the variance of pairwise differences to estimate the recombination rate. *Genetical Research, (Cambridge)* **69**, 45–48.

Wakeley, J. (1999). Nonequilibrium migration in human history. *Genetics* **153**, 1863–1871.

Wall, J.D. (2000). A comparison of estimators of the population recombination rate. *Molecular Biology Evolution* **17**, 156–163.

Weiss, G. and von Haeseler, A. (1998). Inference of population history using a likelihood approach. *Genetics* **149**, 1539–1546.

Wilkins, J.F. and Wakeley, J. (2002). The coalescent in a continuous, finite, linear population. *Genetics* **161**, 873–888.

Wilkinson-Herbots, H.M. (1998). Genealogy and subpopulation differentiation under various models of population structure. *Journal of Mathematical Biology* **37**, 535–585.

Wiuf, C. (2000). A coalescence approach to gene conversion. *Theoretical Population Biology* **57**, 357–367.

Wiuf, C. and Donnelly, P. (1999). Conditional genealogies and the age of a neutral mutant. *Theoretical Population Biology* **56**, 183–201.

Wiuf, C. and Hein, J. (1997). On the number of ancestors to a DNA sequence. *Genetics* **147**, 1459–1468.

Wiuf, C. and Hein, J. (1999a). The ancestry of a sample of sequences subject to recombination. *Genetics* **151**, 1217–1228.

Wiuf, C. and Hein, J. (1999b). Recombination as a point process along sequences. *Theoretical Population Biology* **55**, 248–259.

Wiuf, C. and Hein, J. (2000). The coalescent with gene conversion. *Genetics* **155**, 451–462.

Wright, S. (1931). Evolution in Mendelian populations. *Genetics* **16**, 97–159.

Wright, S. (1949). *Genetics, Palaeontology, and Evolution*, G.L. Jepson, G.G. Simpson and E. Mayr, eds. Princeton University Press, Princeton, NJ, pp. 365–389.

Wu, C.-I. (1991). Inferences of species phylogeny in relation to segregation of ancient polymorphisms. *Genetics* **127**, 429–435.

26

Inference Under the Coalescent

M. Stephens

Departments of Statistics and Human Genetics, University of Chicago, Chicago, IL, USA

This chapter introduces some modern statistical methods for inference from molecular population genetic data. The methods are based on the use of the coalescent to model the genealogy relating a random sample of chromosomes from a population, and help us to say something about the demographic and genetic factors that shaped the evolution of populations from which the samples were taken. The chapter focuses primarily on 'full-data' methods, which aim to make full use of the high resolution of modern genetic data, rather than relying on simple summaries such as pair-wise differences, or sample heterozygosity. These methods are often somewhat complex, and the aim of this chapter is to provide practical guidelines for those who wish to apply these methods using existing software, and sufficient theoretical background to understand how the methods work, at least in outline. We also include a brief review of other popular approaches that provide computationally attractive alternatives to full-data methods. In describing these methods, the chapter provides an introduction to importance sampling and Markov chain Monte Carlo – statistical methods that are likely to play a major role in the analyses of future genetic data.

26.1 INTRODUCTION

Genetic data collected from populations are potentially useful for answering a variety of interesting questions. These questions could relate to, for example, the following:

- The genetic forces, such as mutation and recombination, which have affected the evolution of a particular chromosomal segment or locus.

- The historical relationships amongst different subpopulations. In humans this includes the relationships amongst different continental groups, and the times and routes of major migrations.

- The ancestry of the sampled chromosomes, including, for example, the ages of particular mutations, which are carried by some of the chromosomes.

This chapter describes some modern statistical methods for answering these kinds of questions.

Handbook of Statistical Genetics, Third Edition. Edited by D.J. Balding, M. Bishop and C. Cannings.
© 2007 John Wiley & Sons, Ltd.   ISBN: 978-0-470-05830-5.

Historically, inference in these settings has been based on summaries of the data, such as pair-wise differences, or sample heterozygosity. In contrast, here we focus primarily on methods that aim to make use of the full genetic data in order to provide more accurate inference than is possible using simple summary statistics. These modern methods make use of a theoretical innovation known as the *coalescent* (see **Chapter 25** for an introduction). Both methodological and computational advances have made these 'full-data' methods tractable for some problems, although for other applications, particularly those involving more than a small amount of recombination, full-data approaches remain computationally intractable. In these cases, inference is generally performed using summary statistics, or through other recently developed approximate approaches, some of which we briefly review towards the end of the chapter. Some of these approximate approaches are based on applying full-data approaches to smaller subsets of the data, and then combining information across subsets. Thus, even where the full-data methods themselves are intractable, they can nevertheless help provide a path to an effective practical solution.

The chapter is intended for those who wish to use and understand full-data coalescent-based inference methods. We give practical guidelines, with theoretical background and examples. For ease of exposition, we focus on the simplest models, but most of the principles we discuss will continue to hold in more complex settings. While the focus of this chapter is on inference under the coalescent, many of the concepts and ideas discussed are of more general interest. For example, the next section gives a brief introduction to the ideas underlying likelihood inference and a discussion of the relative merits of maximum-likelihood and Bayesian approaches. Later sections include descriptions of computationally intensive statistical methods, such as importance sampling (IS) and Markov chain Monte Carlo (MCMC), which have proved useful in a wide range of statistical applications, including some related to genetics. Most of the comments we make on these methods apply quite generally, and not only to problems involving the coalescent.

26.1.1 Likelihood-based Inference

Broadly speaking, likelihood-based methods of drawing inference from data proceed by treating the observed data as having arisen from some random process, or *model*, certain aspects of which are unknown. We refer to these unknown quantities as *parameters*. Typically, the aim of a statistical analysis is to use the data to estimate the parameters of the model, and (importantly) to assess the degree of uncertainty associated with these estimates.

More explicitly, likelihood-based inference requires calculation of the probability, $P(\mathcal{D} \mid \psi)$, of observing data $\mathcal{D}$ if the parameters of the model take the value ψ. The likelihood $L(\psi)$ is defined to be this quantity considered as a function of ψ:

$$L(\psi) = P(\mathcal{D} \mid \psi). \qquad (26.1)$$

In the population genetics context, the data $\mathcal{D}$ are typically the genetic types of a random sample of chromosomes.[1] from a population Their distribution depends in a complex way on many unknown parameters, relating to both the demographic history of the population (e.g. population size, migration rates) and the underlying genetic mechanisms

---

[1] In this chapter, when we refer to chromosomes we typically mean 'chromosomal segments'.

(e.g. mutation and recombination rates). It is not usually possible to write down an explicit expression for the likelihood of these parameters.

For example, consider an idealised population, evolving according to the *Wright–Fisher model* (see also **Chapter 22**). In other words, the population evolves in non-overlapping generations, with constant size N chromosomes, with the number of offspring of the chromosomes in each generation being symmetric multinomial, independently of their genetic types (i.e. evolution is neutral). Assume a simple mutation model with K possible genetic types, or *alleles*, with the distribution of the type of the offspring of a parent of type i being given by

$$P(\text{offspring of type } j \mid \text{parent of type } i) = (1 - \mu)\delta_{ij} + \mu p_{ij}, \qquad (26.2)$$

where $\delta_{ij} = 1$ if $i = j$ and 0 otherwise, and $P = (p_{ij})$ is a known transition matrix (so p_{ij} is the probability that an allele of type i mutates to type j, given that a mutation occurs). Thus with probability $1 - \mu$ the offspring is an identical copy of the parent, but with probability μ a mutation occurs according to the transition matrix P. Taken together, these demographic and mutation processes specify a model for the evolution of the population, with two unknown parameters, N and μ. The models are amongst the simplest imaginable, avoiding the complications of fluctuating population size, selection or recombination, for example. Nevertheless, if we examine the genetic types $\mathcal{D}$ of a random sample of N chromosomes taken from the population after it has been evolving for a long period of time, and ask, 'how should we calculate the likelihood $L(N, \mu) = P(\mathcal{D} \mid N, \mu)$?' the answer is far from obvious.

In this case it turns out that the probability $P(\mathcal{D} \mid N, \mu)$ actually depends on N and μ through only their product[2] $N\mu$, and not the individual values of N and μ. As a result, it is common to define the *scaled mutation parameter*[3] $\theta = 2N\mu$, and write the probability $P(\mathcal{D} \mid N, \mu)$ as $P(\mathcal{D} \mid \theta)$. We can then concentrate on calculating the likelihood $L(\theta)$ for the single parameter θ. For concreteness, the remainder of the chapter concentrates on this problem, though most of the methods described can be extended to more complex problems with more parameters (see Section 26.4.7 of this chapter).

Although reduction of the model to a single parameter simplifies things slightly, it is still not possible to write down an explicit expression for the likelihood. However, using methods described in later sections we can accurately approximate it. Figure 26.1 shows an approximation of the (log) likelihood surface for data on the β-globin gene from Harding *et al.* (1997). We now consider how such a likelihood surface can be used to estimate the parameter θ and to assess the uncertainty associated with this estimate. There are two common approaches: the maximum-likelihood approach, and the Bayesian approach.

26.1.1.1   *Maximum-likelihood Inference*

Perhaps the most common likelihood-based approach to parameter estimation is to estimate θ by the maximum-likelihood estimate, $\widehat{\theta}$, which is that value of θ that maximises

---

[2] Actually this is only an approximation that is derived by assuming N is large, and μ is small, but it is an extremely accurate approximation in practice.

[3] Our definition is based on our earlier definition of N as the number of *chromosomes* in the population. An alternative is to define N to be the number of *individuals* in the population, in which case there are $2N$ chromosomes if the individuals are diploid, and $\theta = 4N\mu$.

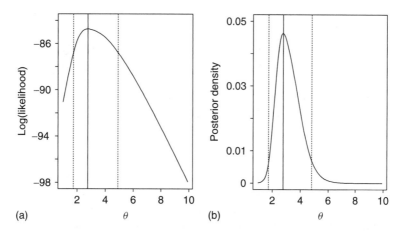

Figure 26.1   (a) Estimated log likelihood for the β-globin data from Harding *et al.* (1997). The vertical solid line indicates the maximum-likelihood estimate, while the vertical dotted lines indicate the approximate 95 % confidence interval found by taking values of θ for which the likelihood is within 2 log likelihood units of the maximum. (b) Posterior density of θ for the same data, with uniform prior on θ. The vertical solid line indicates the posterior mode, while the vertical dotted lines indicate a 95 % credible region. In this case, the 95 % confidence interval and the 95 % credible region are almost identical.

the likelihood $L(\theta)$. The uncertainty associated with this estimate is typically expressed by giving a 95 % *confidence region* for θ, which is a set of values of θ that has the rather tricky and non-intuitive interpretation that if you sampled data from the model repeatedly, and created confidence regions in the same way, for 95 % of datasets, the region would contain the true value of θ. The trickiness of this interpretation must be the source of more confusion than almost any other basic statistical concept. Without dwelling on the issue, we note that it is *not* correct to say that the probability that θ lies in the confidence region is 0.95, which is the intuitive interpretation most people would like to place on the interval: in order to make such statements, one must take a Bayesian approach, as outlined below.

There is a rather general and elegant statistical theory underlying the maximum-likelihood approach to inference. The theory states that *under suitable conditions* $\widehat{\theta}$ will be asymptotically[4] normally distributed with mean being the 'true' value of θ, and that the log likelihood ratio statistic

$$\Lambda = -2\log \frac{L(\theta_0)}{L(\widehat{\theta})}, \tag{26.3}$$

has asymptotically a χ-squared distribution, under repeated sampling from the model, if θ_0 is the 'true' value of θ. A useful consequence of this is that an approximate 95 % confidence interval for a one-dimensional parameter θ may be obtained by considering those θ for which the log likelihood is within two units of the maximum log likelihood.

---

[4] The term *asymptotically* here means 'as the amount of data tends to infinity'. The implicit hope is that asymptotic results will be good approximations for reasonably large amounts of data. This is, of course, not always the case.

The maximum likelihood estimate of θ for the β-globin data (Figure 26.1) is $\theta = 2.75$, and an approximate 95 % confidence interval found in this way is [1.73, 4.91].

Note that we have emphasised the phrase 'under suitable conditions' in the previous paragraph. This is because in the types of applications considered here, these suitable conditions often fail to hold, destroying the validity of the asymptotic theory, and making the construction of valid confidence intervals much more difficult. The reason the conditions often fail to hold in genetics applications is that, unlike most statistical problems where observed data are independent, the types of randomly sampled chromosomes are *not* independent, owing to the fact that they share ancestry. In the same way that you and your parents are related, and your genetic types are not independent, randomly sampled chromosomes are also (more distantly) related, and their types are not independent. These relationships between sampled chromosomes are of course exactly those that lead to the coalescent. While this lack of independence does not in itself necessarily preclude the standard theory from applying, in many settings it is known that the standard theory does not hold, and in other settings it is unclear. The standard method for obtaining confidence intervals from the likelihood, and indeed any procedure that relies on the asymptotic χ-squared distribution of the likelihood ratio statistic, must therefore be used with caution in population genetics applications.

Proponents of the maximum-likelihood approach sometimes claim that it has the advantage (over the Bayesian approach described below) of 'objectivity', in that the opinion of the researcher should not affect the results obtained. However, this is only partially true since specification of the likelihood function itself typically requires subjective judgements to be made about what to include in the model. For example, is it plausible to assume that all sites mutate at an equal rate? Should we assume that the population has grown linearly or exponentially? Can we ignore selection and/or recombination? The answers to such questions are often unclear, and in practice it is necessary to make some subjective modelling assumptions in order to proceed. Furthermore, the maximum-likelihood approach provides no coherent way for making use of information we have about parameters in the model from sources other than the data: from archaeological or anthropological sources, or previous genetics studies for example. The Bayesian approach to inference allows such information to be included in the analysis.

26.1.1.2 *Bayesian Inference*

Bayesian methods also make use of the likelihood in performing inference for parameters in the model, but they allow (indeed require) the incorporation of 'prior information' about the model parameters. Formally, information about the parameters θ must be expressed by specifying a *prior* or *pre-data* distribution $P(\theta)$ for θ. The distribution $P(\theta)$ is weighted towards those θ that are considered most likely according to our prior information. This prior information is then combined with the likelihood by multiplying them together to give the *posterior* or *post-data* distribution $P(\theta \mid \mathcal{D})$:

$$P(\theta \mid \mathcal{D}) = L(\theta)P(\theta)/P(\mathcal{D}). \qquad (26.4)$$

From this equation we see that $P(\theta \mid \mathcal{D})$ will be large for values of θ that are both well-supported by the data (i.e. have high likelihood) *and* are consistent with our prior information (high $P(\theta)$).

The post-data distribution represents our beliefs about the parameters, taking into account both our prior information and the observed data. It is often convenient to summarise these beliefs in some way. For example, it is common practice to report a point estimate for θ (usually the mode of the post-data distribution) and to specify a 95 % credible region for θ, which is a set of values Θ satisfying $P(\theta \in \Theta \mid \mathcal{D}) = 0.95$. Unlike a confidence region, a 95 % credible region has the natural interpretation that the probability that θ is in the region is 0.95. Another nice feature is that the Bayesian approach does not rely on asymptotic arguments, and so is valid in settings where the standard likelihood theory fails.

For the β-globin data, taking our prior distribution on θ to be uniform in the range 0–10, thus favouring no particular value of θ in that range above any others,[5] the posterior distribution of θ is as shown in Figure 26.1. The mode of the distribution is at $\theta = 2.75$, and the 95 % credible region with the highest posterior probability is [1.73, 4.82]. Thus, in this case, the Bayesian and maximum-likelihood approaches provide similar results. In later examples, results from the two approaches differ rather more than they do here.

Despite the apparent advantages of the Bayesian approach over the Maximum-likelihood approach, there is still considerable resistance to it from some quarters. Concerns are often centred on the 'subjective' nature of the Bayesian method. Since the post-data distribution depends on the specified prior distribution, different researchers with differing prior beliefs may come to differing conclusions. In some cases, given enough data, the post-data distribution is dominated by the likelihood, and the conclusions drawn become relatively insensitive to the prior distribution used. Nevertheless, in practice it is often the case that conclusions *do* depend quite sensitively on the prior distribution, and it is important to recognise this fact. Such a result may seem unsatisfactory, but is simply a reflection of the fact that there is sometimes insufficient information in the data to make robust inferential statements.

It may be clear from the above discussion that the author's preferences lean towards the Bayesian approach to inference. However, the methods we examine may be used to compute likelihood surfaces, and thus allow either a Bayesian or maximum-likelihood approach to be taken.

26.2 THE LIKELIHOOD AND THE COALESCENT

In the following sections, we consider methods for accurately approximating the likelihood $L(\theta)$. These methods make use of ideas from coalescent theory, and those unfamiliar with these ideas will find it helpful to study the companion **Chapter 25**, before proceeding.

The coalescent, and related processes, describe the distribution of the unknown tree $\mathcal{T}$ relating a random sample of chromosomes. Here we wish to be deliberately vague about what we mean by the 'tree': it might be simply the genealogy relating the sampled chromosomes (see, e.g. Figure 26.3 and accompanying text in **Chapter 25**), or it might also include the mutations on the branches of the genealogy. What matters

---

[5] This prior is convenient for the purposes of illustration. As we discuss later, rather than placing a prior distribution on $\theta = 2N\mu$, it makes more sense to consider priors for N and μ separately.

is that if we knew the tree then we could calculate[6] the probability of the data $P(\mathcal{D} \mid T, \theta)$.

We emphasise that although the tree relating the sampled individuals plays a major role in the computational methods we are considering here, it is a role that is different to that of the tree in phylogenetic analyses (see for example Huelsenbeck, 2001 this volume). In population genetics applications, interest typically focuses on the genetic and evolutionary forces that have affected the evolution of the genetic region under study, and the tree relating randomly sampled individuals from those populations is of interest only in so far as it informs us about these parameters. Nevertheless, many population genetics studies have focused on the problem of estimating quantities relating to the tree (i.e. ancestral inference). In particular, estimating the time since the most recent common ancestor, T_{MRCA}, of the sampled chromosomes, and the ages of mutations in the tree, have become common objectives.[7] Formally, these questions involve the post-data distribution[8] of the tree, $P(T \mid \mathcal{D})$, which is often easy to find using methods we describe later. We anticipate that in the future focus will shift away from estimating the tree, and towards estimating parameters in a model for the demographic history of populations under study (e.g. times of major migration events in human history).

In attempting to approximate the likelihood $L(\theta)$, it turns out to be helpful to write it as an average over all possible trees:

$$L(\theta) = P(\mathcal{D} \mid \theta) = \sum_{T} P(\mathcal{D} \mid T, \theta) P(T \mid \theta). \qquad (26.5)$$

This is helpful because every term in this sum is easy to calculate: we noted earlier that we can calculate $P(\mathcal{D} \mid T, \theta)$, and the pre-data distribution of the tree relating the sampled individuals, $P(T \mid \theta)$, is given by basic coalescent theory (**Chapter 25**). However, in most cases (26.5) cannot be directly used to evaluate the likelihood, because the number of terms in the sum is so huge that it simply cannot be calculated in a reasonable amount of time. For example, the number of tree topologies relating 10 chromosomes is 2571912000.

In fact, since the branch lengths of the tree T are generally continuous quantities, the sum above ought really to be written as an integral:

$$L(\theta) = P(\mathcal{D} \mid \theta) = \int P(\mathcal{D} \mid T, \theta) P(T \mid \theta) \, dT. \qquad (26.6)$$

This makes exact computation of the likelihood even harder, and most of the remainder of this chapter is spent examining efficient methods of approximating this integral, and others like it. Note that the integration takes place over the space of all possible T, which is a big space! Numerical integration methods, such as Gaussian quadrature, which typically work well for integrals in one or two dimensions, are difficult to apply to problems such as

---

[6] In the case where T is the genealogy of the sampled chromosomes this calculation can be done using the *peeling* algorithm (Felsenstein, 1981).

[7] As noted by Wilson and Balding (1998), while the T_{MRCA} of a sample of chromosomes may not be the most important time in human history, it has become central to the interpretation of genetic samples through its widespread use.

[8] In the coalescent framework, the tree T is the result of a random process (the evolution of the population), and not a parameter in the model. This is true even when taking the maximum-likelihood approach to inference. Thus, it is technically correct to talk of the distribution of T, rather than the likelihood of T.

these, and the approximation of such high-dimensional integrals is now most commonly performed using Monte Carlo (i.e. simulation-based) methods.

A naive Monte Carlo method of approximating (26.6) is based on the idea that, if a random variable X has density $f_X(x)$, then the mean of any function of X, $g(X)$ say, may be approximated by simulating many values $X^{(1)}, X^{(2)}, \ldots, X^{(M)}$ from the distribution with density f_X, and forming the average:

$$E(g(X)) = \int g(x) f_X(x) \, \mathrm{d}x \approx \frac{1}{M} \sum_{i=1}^{M} g(X^{(i)}). \qquad (26.7)$$

With suitable conditions on the function g, this approximation will be good provided M is sufficiently large (formally the error tends to 0 as M tends to infinity). Applying this idea to (26.6) gives

$$L(\theta) = \int P(\mathcal{D} \mid \mathcal{T}, \theta) P(\mathcal{T} \mid \theta) \, \mathrm{d}\mathcal{T} \approx \frac{1}{M} \sum_{i=1}^{M} P(\mathcal{D} \mid \mathcal{T}^{(i)}, \theta), \qquad (26.8)$$

where[9] $\mathcal{T}^{(1)}, \mathcal{T}^{(2)}, \ldots, \mathcal{T}^{(M)} \sim P(\mathcal{T} \mid \theta)$. The approximation (26.8) is easy to implement, since $P(\mathcal{T} \mid \theta)$ can be simulated from using coalescent methods, and $P(\mathcal{D} \mid \mathcal{T}^{(i)}, \theta)$ is easily calculated. Unfortunately, the approximation is also almost useless for problems involving samples of more than a few chromosomes. The reason for this is that $P(\mathcal{D} \mid \mathcal{T}^{(i)}, \theta)$ will be very small for all but a few of the $\mathcal{T}^{(i)}$ we simulate. Only the few larger values will contribute significantly to the sum (26.8), but almost all of our effort will be spent calculating the very small terms that contribute negligible amounts to the sum. Typically, the proportion of trees that actually contribute significantly to the sum is less than one in a million, and in such cases the method becomes hopelessly inefficient.

26.3   IMPORTANCE SAMPLING

Importance sampling (IS) (see Ripley, 1987, for background) is a standard statistical method for improving the efficiency of Monte Carlo integration. The idea is simple, and based on rectifying the inefficiency of the approximation (26.8) by concentrating simulation and computational effort on the more 'important' trees, which are those trees for which $P(\mathcal{D} \mid \mathcal{T}, \theta)$ is relatively large, or in other words those trees that are more consistent with the observed data. IS was first used in this context by Griffiths and Tavaré (1994a; 1994b; 1994c), though they derived their method in a rather different way, as a method of solving recursive equations. Similar methods have since been applied, by themselves and others, to a variety of genetic systems and demographic models (see, for example, Griffiths and Marjoram, 1996; Griffiths and Tavaré, 1997; 1999; Nielsen, 1997; Bahlo and Griffiths, 2000). The connection between the Griffiths–Tavaré method and IS was pointed out by Felsenstein et al. (1999), and Stephens and Donnelly (2000) show how this observation can be exploited to develop substantially more efficient algorithms.

---

[9] The notation $\sim$ in what follows means 'are distributed as'.

The IS method is based on rewriting the integral (26.6) as

$$
L(\theta) = \int P(\mathcal{D} \mid \mathcal{T}, \theta) P(\mathcal{T} \mid \theta) \, d\mathcal{T}
$$

$$
= \int P(\mathcal{D} \mid \mathcal{T}, \theta) \frac{P(\mathcal{T} \mid \theta)}{Q(\mathcal{T})} Q(\mathcal{T}) \, d\mathcal{T}, \tag{26.9}
$$

where $Q(\cdot)$ is any distribution satisfying the condition

$$
Q(\mathcal{T}) > 0 \text{ whenever } P(\mathcal{D} \mid \mathcal{T}, \theta) P(\mathcal{T} \mid \theta) > 0. \tag{26.10}
$$

(This condition is required when multiplying by $Q(\mathcal{T})/Q(\mathcal{T})$ to obtain (26.9) above, to avoid multiplying by 0/0, which is undefined.) The distribution $Q(\cdot)$ is referred to as the IS distribution, or proposal distribution. Straightforward Monte Carlo approximation of (26.9) gives

$$
L(\theta) \approx \frac{1}{M} \sum_{i=1}^{M} P(\mathcal{D} \mid \mathcal{T}^{(i)}, \theta) \frac{P(\mathcal{T}^{(i)} \mid \theta)}{Q(\mathcal{T}^{(i)})}, \tag{26.11}
$$

where $\mathcal{T}^{(1)}, \mathcal{T}^{(2)}, \ldots, \mathcal{T}^{(M)} \sim Q(\mathcal{T})$.

By choosing the distribution Q carefully, this method of approximation is much more efficient than (26.8). For example, when estimating $L(\theta)$ the optimal choice Q_θ^* for Q depends on θ and is the post-data distribution of the tree $\mathcal{T}$ given the genetic data $\mathcal{D}$ and the parameter θ:

$$
Q_\theta^*(\mathcal{T}) = P(\mathcal{T} \mid \mathcal{D}, \theta) = \frac{P(\mathcal{T} \mid \theta) P(\mathcal{D} \mid \mathcal{T}, \theta)}{P(\mathcal{D} \mid \theta)} = \frac{P(\mathcal{T}, \mathcal{D} \mid \theta)}{P(\mathcal{D} \mid \theta)}. \tag{26.12}
$$

Intuitively this choice of Q directs more sampling and computational effort towards trees that have larger values of $P(\mathcal{D} \mid \mathcal{T}, \theta)$, and thus avoids the inefficiency associated with the naive approximation (26.8) discussed earlier. In fact, for $Q = Q_\theta^*$, every term in the sum (26.11) is the same:

$$
\frac{P(\mathcal{D} \mid \mathcal{T}^{(i)}, \theta) P(\mathcal{T}^{(i)} \mid \theta)}{Q_\theta^*(\mathcal{T}^{(i)})} = \frac{P(\mathcal{D}, \mathcal{T}^{(i)} \mid \theta)}{P(\mathcal{D}, \mathcal{T}^{(i)} \mid \theta)/P(\mathcal{D} \mid \theta)} = L(\theta), \tag{26.13}
$$

and the variance of the estimator (26.11) is 0, or in other words the approximation becomes exact. This seems almost too good to be true, and in most situations this is indeed the case, the problem being that *we do not know the distribution required in* (26.12). In fact, to implement the estimate (26.11) we must be able to do two things:

1. Simulate (directly) from $Q(\cdot)$.

2. Calculate $Q(\mathcal{T}^{(i)})$ for each $\mathcal{T}^{(i)}$.

For $Q = Q_\theta^*$ we are not able to do either of these things.

Nevertheless, the observation that $Q_\theta^*(\mathcal{T}) = P(\mathcal{T} \mid \mathcal{D}, \theta)$ is helpful, as it gives us an insight into which choices of Q may be sensible. In particular, it seems that when attempting to estimate $L(\theta)$ at a particular value of θ, a good strategy would be to choose Q to closely approximate Q_θ^*. This is the strategy pursued in Stephens and Donnelly

(2000), who found it to be a very successful way of developing good IS methods in many cases.

26.3.1 Likelihood Surfaces

The approximation (26.11) allows us to approximate the likelihood surface $L(\theta)$ at all values of θ, using samples from a single $Q(\mathcal{T})$. However, since the optimal IS function (26.12) depends on θ, it is clear that no one choice of Q can be optimal for all θ simultaneously, and that the efficiency of the resulting estimator for $L(\theta)$ may vary with θ. Suppose that we have constructed an IS function $Q_{\theta_0}(\mathcal{T})$ to approximate $Q^*_{\theta_0}(\mathcal{T})$, for some fixed value θ_0, and consider using this IS function to estimate the likelihood curve $L(\theta)$ at all values of θ, using (26.11). A strategy along these lines was adopted in Griffiths and Tavaré (1994c), which refers to θ_0 as the 'driving value' for θ, and has since been frequently employed in implementations of IS schemes. Intuitively we might expect this IS function to be most efficient in estimating $L(\theta)$ for values of θ near θ_0, and less efficient for θ far away from θ_0. For reasons discussed in Stephens and Donnelly (2000), this can tend to cause such methods to underestimate the likelihood for values of θ away from the driving value θ_0. This can have two rather undesirable consequences. First, it can cause the estimate of the likelihood surface to have its maximum near θ_0, even when the true likelihood surface has its maximum elsewhere. Second, if the driving value θ_0 is near the maximum of the true likelihood surface then the estimate of the likelihood may tend to be more highly peaked about θ_0 than the true likelihood.[10] The severity of these effects is rather difficult to predict: in some cases they cause few problems, while in others the effects can be extreme. It is important to bear this in mind when using these methods.

In order to avoid such problems we might try using a different IS function to estimate the likelihood at each value of θ. In other words, we might use an IS function Q_{θ_i} to estimate the likelihood $L(\theta_i)$ on a grid of values, $\theta = \theta_1, \theta_2, \ldots, \theta_R$, say. In order to obtain efficient estimates at all values of θ we need Q_{θ_i} to be a good approximation to the optimal IS function $Q^*_{\theta_i}$. This 'point-wise' approach to estimating the likelihood surface is somewhat wasteful in that it requires a set of samples from Q_{θ_i} for each i, but at each point in the grid it uses only one of these samples to estimate the likelihood. A more efficient approach, which is in some sense a combination of the 'point-wise' and 'driving value' approaches, is to use a sample from a single IS function

$$Q(\mathcal{T}) = \frac{1}{R} \sum_{i=1}^{R} Q_{\theta_i}(\mathcal{T}), \tag{26.14}$$

to estimate the likelihood surface at a grid of values of θ. A stratified sample from this IS function could be obtained by pooling samples of equal size from each of the $Q_{\theta_i}(\mathcal{T})$. It can be shown that using the IS function (26.14) is guaranteed to be more efficient (by some reasonable criterion) than using the point-wise approach. However, while this more efficient approach is straightforward in principle, it needs to be coded into the software being used. Where this is not the case, it is advisable to at least compare estimates of the likelihood surfaces using different driving values in order to check that this is not causing

---

[10] This is important because it would cause estimated confidence intervals, or credible regions to be too small.

difficulties. If the surfaces found using different driving values are very different, then a 'point-wise' approach may be safer, if rather time consuming.

26.3.2 Ancestral Inference

IS schemes also provide a straightforward way of approximating the post-data distribution of the tree $P(\mathcal{T} \mid \mathcal{D}, \theta)$. Define the *weight* $w^{(i)}$ associated with tree $\mathcal{T}^{(i)}$ by

$$
w^{(i)} = \frac{W^{(i)}}{\sum_{j=1}^{M} W^{(j)}},
\tag{26.15}
$$

where

$$
W^{(i)} = P(\mathcal{D} \mid \mathcal{T}^{(i)}, \theta) \frac{P(\mathcal{T}^{(i)} \mid \theta)}{Q(\mathcal{T}^{(i)})}.
\tag{26.16}
$$

It is easy to show that the distribution with weight $w^{(i)}$ on tree $\mathcal{T}^{(i)}$ is an approximation to the post-data distribution $P(\mathcal{T} \mid \mathcal{D}, \theta)$. As a result it is straightforward to approximate quantities of interest relating to the tree by forming a weighted average of these quantities over the sampled trees. For example, the expected value of T_{MRCA} can be approximated by

$$
E(T_{\text{MRCA}}) \approx \sum_{i} w^{(i)} T_{\text{MRCA}}(\mathcal{T}^{(i)}).
\tag{26.17}
$$

26.3.3 Application and Assessing Reliability

One of the most important considerations when applying computationally intensive methods such as these is whether the results obtained are reliable. When applying IS methods, this comes down to whether the number of iterations M is sufficiently large for the approximations being made to be reasonably accurate. Unfortunately the value of M required varies drastically from problem to problem, and so no simple answer exists to the question, 'How many iterations must I use?'. However, a simple and generally effective procedure is to run the algorithm several times with different seeds for the pseudo-random number generator, and to check that the same results are obtained each time. If the results differ greatly with each run then the runs are too short. Here we confine ourselves to a single illustrative example: a more detailed discussion and more examples can be found in Stephens and Donnelly (2000), and in Fearnhead and Donnelly (2001), both of which give illuminating examples of how inaccurate results can be, even with large amounts of computing time, and how these inaccuracies are often evident in multiple runs of an algorithm.

Our illustration comes from applying the Griffiths–Tavaré method, implemented in the program genetree, to the β-globin data from Harding *et al.* (1997), where a more comprehensive analysis and discussion of the biological significance of the results can be found. Five different approximations to the likelihood surface, based on runs of length $M = 10\,000$, using a driving value[11] of $\theta = 5.0$, are shown in Figure 26.2(a). The

---

[11] Earlier we warned that care must be taken when using a 'driving value' to approximate the likelihood surface. In this case we checked that using point-wise estimates, or different driving values, gives very similar answers. In general using a driving value approach with genetree tends to give reasonable results in models without migration or growth. However, once growth and migration are included in a model more care may be required.

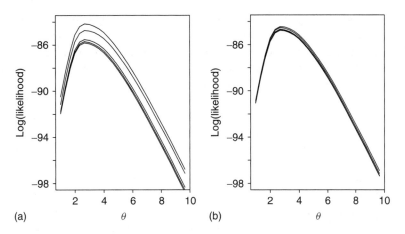

Figure 26.2 Estimated log likelihoods for the β-globin data from Harding *et al.* (1997). (a) Five different curves, each obtained using a different seed for the random number generator and $M = 10\,000$ iterations. (b) Five different curves, each obtained using a different seed for the random number generator and $M = 1\,000\,000$ iterations.

reasonably large differences between the runs indicate that the runs are perhaps rather too short to give reliable results. Five approximations based on runs of length $1\,000\,000$ shown in part (b) of the figure show much less variation. We conclude that for these data genetree requires perhaps millions of iterations to deliver reliable results. This sort of requirement is typical. Indeed for problems involving migration between subpopulations and/or growth, many more iterations are typically required, which can result in the method becoming prohibitively time consuming (perhaps months or even years of computer time). In order to tackle such problems, improved IS methods along the lines of Stephens and Donnelly (2000) may be required.

A notable feature of the curves in Figure 26.2 is that although many of the curves are at different *heights*, they are all a similar 'shape', and peak at the same place. It is not really well understood why exactly this occurs, but it often appears to be the case, at least in settings not involving population growth, migration or recombination. It is important to stress that when sufficient iterations have been performed, all curves should in theory be both the same height *and* shape. If they are not all the same height, it is a sign that the algorithm has not been run for long enough. In some cases, as here, the results obtained when the algorithm *has* been run long enough do not differ in most practical terms from the results obtained from a short run. However, in other cases they may differ, and you can only find out by doing the longer runs!

26.4 MARKOV CHAIN MONTE CARLO

26.4.1 Introduction

MCMC has recently become a very popular technique in computational statistics. Essentially it is a method for producing samples from a distribution that is not easy to simulate directly. For example, we noted that to answer questions about the tree relating the sampled chromosomes, it would be helpful to simulate from the conditional distribution of the tree given the data, $P(T \mid D)$. Although this is difficult to achieve directly, it is relatively straightforward, at least in principle, using MCMC methods. Further, as we will also see later, these methods can also be used to approximate the likelihood surface for θ. However, we begin by explaining the basics of MCMC techniques.

Suppose we wish to generate trees from some distribution $\pi(T)$ that is difficult (or impossible) to simulate from directly (e.g. $\pi(T) = P(T \mid D)$). Informally, MCMC methods achieve this by starting at some initial tree $T^{(0)}$, and moving randomly from tree to tree in such a way that, in the long run, the frequency with which trees are visited is proportional to $\pi(T)$. The question that MCMC methods answer is how can we move randomly from genealogy to genealogy so that this is achieved? Here we consider perhaps the most important and widely used MCMC method: the Metropolis–Hastings algorithm. The idea is that rather arbitrary methods of moving randomly from genealogy to genealogy can be modified to create an algorithm with the required properties. Let Q denote some method of moving randomly from tree to tree, with $Q(T \to T')$ being the probability that, if we start at T then we move to T'. The following algorithm ensures that, in the long run, the frequency with which trees are visited is proportional to $\pi(T)$.

Starting at some initial point, iterate the following steps:

1. Given the ith genealogy $T^{(i)}$, draw a genealogy T' from $Q(T^{(i)} \to T')$.

2. With probability A, given by

$$A = \min\left(1, \frac{\pi(T')Q(T' \to T^{(i)})}{\pi(T^{(i)})Q(T^{(i)} \to T')}\right), \tag{26.18}$$

accept the proposed genealogy, and set $T^{(i+1)} = T'$. Otherwise reject the proposed genealogy and set $T^{(i+1)} = T^{(i)}$.

Note that when a proposed genealogy is 'rejected' in the above algorithm, the new tree is the same as the current tree. It is important that these duplicate trees are counted, and not simply discarded. The distribution Q is often referred to as the *proposal distribution*, and the probability A is referred to as the *acceptance probability*. Calculating A requires being able to calculate the transition probabilites Q (which is usually easy) and the ratio $\pi(T')/\pi(T)$ which is also often easy, even if $\pi(T)$ itself cannot be calculated. For example, in the case $\pi(T) = P(T \mid D, \theta)$ we have

$$\frac{\pi(T')}{\pi(T)} = \frac{P(T' \mid D, \theta)}{P(T \mid D, \theta)} = \frac{P(T', D \mid \theta)}{P(T, D \mid \theta)} = \frac{P(T' \mid \theta)P(D \mid T', \theta)}{P(T \mid \theta)P(D \mid T, \theta)}, \tag{26.19}$$

which we can calculate just as we could calculate the terms of the sum (26.5).

Under relatively weak conditions on Q (see Gilks *et al.*, 1996, for example), which are usually easily satisfied in practice, it is straightforward to show that $T^{(1)}, T^{(2)}, T^{(3)}, \ldots$ form a Markov Chain with stationary distribution $\pi(\cdot)$, and that $T^{(b)}$, for sufficiently large b, is approximately a sample from $\pi(\cdot)$, the distribution we wished to sample from. Furthermore, for sufficiently large k, $T^{(b)}, T^{(b+k)}, T^{(b+2k)}, \cdots$ may be treated as approximately independent samples from $\pi(\cdot)$. The value b is known as the *burn-in*, while k is referred to as the *thinning interval*.

Why does this algorithm work? Some helpful intuition can be obtained by considering the special case where the proposal distribution $\mathcal{Q}$ is symmetric, in that $\mathcal{Q}(T \to T') = \mathcal{Q}(T' \to T)$ for all T and T'. In this case, the acceptance probability for moving from T to T' becomes $A = \min(1, \pi(T')/\pi(T))$. Thus, moves that are accepted tend to be *to* trees for which $\pi(T')$ is relatively large, and moves that are rejected tend to be *from* trees for which $\pi(T)$ is relatively large. As a Result, the algorithm tends to spend more time exploring those trees with large values of π than those with small values, which is at least a necessary condition for it to work.

26.4.2 Choosing a Good Proposal Distribution

Although in theory the above algorithm works for *any* choice of proposal distribution $\mathcal{Q}$ that satisfies the required conditions, in practice choice of $\mathcal{Q}$ can have a very strong effect on the efficiency of the method, or in other words on what values of b and k are 'sufficiently large'. Even for good choices of $\mathcal{Q}$, the values of b and k required can be large for problems of moderate size: tens of thousands, or millions, for example. For bad choices of $\mathcal{Q}$, the values of b and k required are so large as to make the approach infeasible. How to choose a good proposal $\mathcal{Q}$ is therefore an important issue.

Unfortunately, it is often difficult to predict in advance whether one particular $\mathcal{Q}$ will perform better than another, although there are some general guidelines. For example, it is desirable that $\mathcal{Q}$ at least occasionally proposes moves to 'good' trees, which are those with a high value for π. Otherwise almost every proposed tree will be rejected, and the algorithm will remain stuck where it is. Similarly, algorithms that always propose only small changes to the current tree will tend to explore the set of all possible trees rather slowly. Algorithms that exhibit this kind of behaviour are sometimes called *sticky*. Conversely, algorithms that move freely between very different trees are said to 'mix well'. Ideally then we want our proposal distribution to propose moves that (1) are to trees that are very different from the current tree, and (2) have a high probability of acceptance. It is typically difficult to find a proposal that achieves both these aims: proposals that propose big moves tend to have low acceptance rates. A common strategy is to simply experiment with different schemes, trying various *ad hoc* modifications until a scheme that performs well (in that it moves quickly between different plausible trees) is found. This strategy has practical merit, and is acceptable since in principle these *ad hoc* modifications do not affect the validity of the algorithm. Nevertheless it often remains tricky to assess how well a scheme is performing, even after it has been implemented and the results can be examined.

26.4.3 Likelihood Surfaces

There are many ways of using MCMC methods to calculate likelihood surfaces for θ, and we now examine some of these in detail. Although it is possible to use MCMC methods to estimate the likelihood surface itself, in practice it is much easier to estimate a *relative* likelihood surface, that is, a function $\widetilde{L}(\theta)$ that satisfies $\widetilde{L}(\theta) = \alpha L(\theta)$ for some unknown constant α. Knowledge of a relative likelihood surface is sufficient for most applications: in particular it allows the parameters to be estimated using either the maximum-likelihood or Bayesian approach. We distinguish between two types of methods here: those that keep θ fixed and those that allow θ to vary.

26.4.3.1 θ Fixed

Suppose $T^{(1)}, T^{(2)}, \ldots, T^{(M)}$ is an MCMC sample from $P(T \mid \mathcal{D}, \theta_0)$ for some fixed θ_0. Then we might attempt to estimate the likelihood surface using $P(T \mid \mathcal{D}, \theta_0)$ as an IS distribution:

$$
\begin{aligned}
L(\theta) &= \int P(\mathcal{D} \mid \theta, T) P(T \mid \theta)\, \mathrm{d}T \\
&= \int \frac{P(\mathcal{D} \mid \theta, T) P(T \mid \theta)}{P(T \mid \mathcal{D}, \theta_0)} P(T \mid \mathcal{D}, \theta_0)\, \mathrm{d}T \\
&\approx \frac{1}{M} \sum_{i=1}^{M} \frac{P(\mathcal{D} \mid \theta, T^{(i)}) P(T^{(i)} \mid \theta)}{P(T^{(i)} \mid \mathcal{D}, \theta_0)}.
\end{aligned}
\tag{26.20}
$$

However, this estimator is of little use as we cannot evaluate $P(T^{(i)} \mid \mathcal{D}, \theta_0)$. In fact we noted earlier that we could not use $P(T \mid \mathcal{D}, \theta_0)$ as an IS function since we could neither simulate from it directly, nor calculate it explicitly. MCMC methods allow us to simulate from it indirectly, but do not solve the problem of calculating it explicitly. However, by writing

$$
P(T^{(i)} \mid \mathcal{D}, \theta_0) = \frac{P(T^{(i)}, \mathcal{D} \mid \theta_0)}{P(\mathcal{D} \mid \theta_0)} = \frac{P(T^{(i)}, \mathcal{D} \mid \theta_0)}{L(\theta_0)},
\tag{26.21}
$$

and substituting into (26.20) we obtain

$$
\frac{L(\theta)}{L(\theta_0)} \approx \frac{1}{M} \sum_{i=1}^{M} \frac{P(\mathcal{D}, T^{(i)} \mid \theta)}{P(\mathcal{D}, T^{(i)} \mid \theta_0)},
\tag{26.22}
$$

which we *can* evaluate, and which gives us an estimate of a relative likelihood surface, $\tilde{L}(\theta) = \alpha L(\theta)$, where the unknown constant $\alpha = [L(\theta_0)]^{-1}$.

What can be said of the efficiency of the estimate (26.22)? Well, recall that $P(T \mid \mathcal{D}, \theta_0)$ was identified as the optimal IS function for estimating $L(\theta)$ at $\theta = \theta_0$, but that this efficiency will be reduced for other values of θ. Similarly it is the optimal IS function for estimating $\tilde{L}(\theta)$ at $\theta = \theta_0$, and efficiency will be reduced for other values of θ. However, the value of $\tilde{L}(\theta)$ at $\theta = \theta_0$ is known by definition to be $L(\theta_0)/L(\theta_0) = 1.0$, so 'optimality' at this point is very easy to achieve. What is of more concern is the potential lack of efficiency for other values of θ, which can cause severe problems. We repeat the warning that lack of efficiency in these methods needs to be taken seriously, as it can easily lead to estimates of the (relative) likelihood that are wrong by orders of magnitude. In particular, likelihood surfaces estimated in this way may have a peak near θ_0 even when the true likelihood surface has its peak away from θ_0, or, if the true likelihood surface has a peak near θ_0, may be more sharply peaked about θ_0 than the true surface. A short theoretical exploration of this problem in a simple context is given in Stephens (1999).

One approach to addressing this problem is based on the fact that the estimator (26.22) for $\tilde{L}(\theta)$ will be most efficient, and hence most reliable, for values of θ that are 'close' to θ_0. Suppose we wish to estimate the relative likelihood on a grid of values of $\theta = (\theta_1, \theta_2, \ldots, \theta_R)$, where $\theta_1 < \theta_2 < \cdots < \theta_R$. If θ_2 is close enough to θ_1

then we can accurately estimate $L(\theta_2)/L(\theta_1)$ using (26.22) and a sample of trees from $P(\mathcal{T} \mid \mathcal{D}, \theta_1)$. Similarly, if θ_3 is close enough to θ_2 then we can accurately estimate $L(\theta_3)/L(\theta_2)$ using (26.22) and a sample of trees from $P(\mathcal{T} \mid \mathcal{D}, \theta_2)$. We can then estimate

$$\frac{L(\theta_3)}{L(\theta_1)} = \frac{(L(\theta_3)}{L(\theta_2))} \times \frac{(L(\theta_2)}{L(\theta_1))},$$

and continuing in this way can estimate $L(\theta_i)/L(\theta_1)$ for $i = 1, 2, \ldots, R$, and thus construct an estimate of the relative likelihood surface.

Note that the above procedure requires MCMC samples from $P(\mathcal{T} \mid \mathcal{D}, \theta_i)$ for $i = 1, 2, \ldots, (R-1)$, and thus requires us to run $R-1$ independent Markov chain simulations. In fact, given that we have to generate these samples anyway, there is a more efficient way of combining them than the naive method we have just described. It is an iterative method due to Geyer (1991), who named it 'reverse logistic regression'. Its implementation in this context is described by Kuhner *et al.* (1995).

26.4.3.2 θ Varying

MCMC schemes can also be used to produce an MCMC sample

$$(\theta^{(1)}, \mathcal{T}^{(1)}), (\theta^{(2)}, \mathcal{T}^{(2)}), \ldots, (\theta^{(M)}, \mathcal{T}^{(M)}),$$

from $P(\theta, \mathcal{T} \mid \mathcal{D})$ for any given prior distribution on θ. Instead of simply wandering around different $\mathcal{T}$, now the algorithm must also explore different possible values for θ. One way to achieve this is to define a proposal distribution $\mathcal{Q}$ that proposes a move from the current pair of values $(\theta, \mathcal{T})$ to a new pair of values $(\theta', \mathcal{T}')$, and then accepts or rejects this proposal in the same way as before (replacing $\pi(\mathcal{T})$ with $\pi(\mathcal{T}, \theta)$). More commonly, methods that alternately propose changing $\mathcal{T}$ and changing θ can be developed, as in Wilson and Balding (1998).

This approach arises most naturally from the Bayesian approach to inference, when we would be interested in the post-data distribution for θ. However, since the post-data distribution for θ is proportional to the prior $\pi(\theta)$ times the likelihood $L(\theta)$, it is straightforward to obtain an estimate of a relative likelihood surface from an estimate of the post-data density, simply by dividing by the prior density (which is usually easy to calculate). The simplest (though not the best) method of estimating the post-data density of θ is to plot a histogram of the θ values in an MCMC sample $(\theta^{(1)}, \mathcal{T}^{(1)}), (\theta^{(2)}, \mathcal{T}^{(2)}), \ldots, (\theta^{(M)}, \mathcal{T}^{(M)})$. Thus these methods can provide a useful computational tool for likelihood-based inference in general, and need not be confined to Bayesian analyses. (However, it is worth noting that this approach to obtaining a likelihood surface is probably practical only if the likelihood involves at most two parameters–for larger numbers of parameters it will typically be difficult to obtain reliable estimates of the post-data density from an MCMC sample.)

26.4.4 Ancestral Inference

The use of MCMC methods to perform ancestral inference is relatively straightforward. For example, if $\mathcal{T}^{(1)}, \mathcal{T}^{(2)}, \ldots, \mathcal{T}^{(M)}$ is an MCMC sample from $P(\mathcal{T} \mid \mathcal{D}, \theta)$ for some fixed value of θ, then the expectation of T_{MRCA} given this value of θ can be approximated by the straightforward Monte Carlo method:

$$E(T_{\mathrm{MRCA}} \mid \mathcal{D}, \theta) = \int T_{\mathrm{MRCA}}(\mathcal{T}) P(\mathcal{T} \mid \mathcal{D}, \theta) \, \mathrm{d}\mathcal{T} \tag{26.23}$$

$$\approx \frac{1}{M} \sum_{i=1}^{M} T_{\mathrm{MRCA}}(\mathcal{T}^{(i)}). \tag{26.24}$$

Provided the MCMC scheme mixes reasonably well, this approximation will be accurate for moderate values of M. Other quantities of interest relating to the ancestry of the sample, such as the expected age of a particular mutation, can be approximated in a similar way. There is though a slight complication: what value of θ should be used to make these estimates? In a maximum-likelihood framework, it is natural to use the maximum-likelihood estimate for θ. However, this approach ignores the fact that in practice we do not know the true value of θ, and there may be considerable uncertainty associated with any estimate of θ. There does not seem to be an obvious way around this problem without moving to the Bayesian framework, which provides a coherent way of taking into account the uncertainty in our estimate of θ. For example, if $(\mathcal{T}^{(1)}, \theta^{(1)}), (\mathcal{T}^{(2)}, \theta^{(2)}), \ldots, (\mathcal{T}^{(M)}, \theta^{(M)})$ is an MCMC sample from $P(\mathcal{T}, \theta \mid \mathcal{D})$, then

$$E(T_{\mathrm{MRCA}} \mid \mathcal{D}) = \int T_{\mathrm{MRCA}}(\mathcal{T}) P(\mathcal{T} \mid \mathcal{D}) \, \mathrm{d}\mathcal{T} \tag{26.25}$$

$$\approx \frac{1}{M} \sum_{i=1}^{M} T_{\mathrm{MRCA}}(\mathcal{T}^{(i)}), \tag{26.26}$$

is an estimate of the expected value of T_{MRCA} that takes into account the uncertainty associated with θ.

26.4.5 Example Proposal Distributions

As we noted earlier, in principle there is a huge amount of flexibility in the choice of proposal distribution $\mathcal{Q}$. Not surprisingly then, several different suggestions for the choice of suitable $\mathcal{Q}$ have been made. We examine two of the earliest suggestions here. Examples of other MCMC schemes include Beaumont (1999), and Nielsen (2000).

The examples we discuss below are illustrated in Figures 26.3 and 26.4. These figures show trees relating a sample of five individuals, each typed at a single micro-satellite locus, for which each individual's genetic type may be represented by an integer (shown at the bottom of the tree). In what follows, it is worth remembering that alleles that are closer numerically, are likely to be more closely related (so a '7' is likely to be more closely related to a '6' than to a '3').

26.4.5.1 *The Beerli–Kuhner–Yamato–Felsenstein (BKYF) 'Conditional Coalescent' Proposal*

This is the proposal described in Beerli and Felsenstein (1999), and Felsenstein *et al.* (1999). They refer to it as a 'conditional coalescent' proposal, because a new tree is formed from the current tree, by removing one of the branches, and resimulating it from

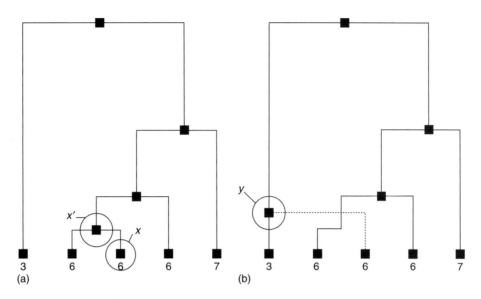

Figure 26.3 Illustration of the BKYF 'conditional coalescent' proposal, which is a proposal distribution for moving from one tree (a) to another (b), as explained in the text.

the appropriate coalescent process, *conditional* on all the other branches staying fixed. For the case of samples from a single panmictic population, it proceeds (in outline) as follows:

1. Choose a node, x, uniformly at random from all nodes other than the root on the current tree.

2. Remove the parent node of x (marked x' in Figure 26.3) from the tree, together with the lineage joining x to x'.

3. Starting from x, simulate a lineage towards the root of the tree. At any given time this lineage coalesces at constant rate with each lineage existing in the tree at that time. (If necessary, the lineage above the current root is extended backwards in time.)

4. Eventually the branch from x will coalesce with another branch, to create a new node (marked y in Figure 26.3).

A more detailed description of this proposal, including a natural extension for samples from several different populations, is given in Beerli and Felsenstein (1999).

26.4.5.2 *The Wilson and Balding (WB) 'Branch-swapping' Proposal*

In order to design their proposal distribution, Wilson and Balding (1998) augmented their tree to include details of the genetic types at every node. In outline, the proposal distribution proceeds as follows:

1. Choose a node, x, uniformly at random from all nodes other than the root on the current tree.

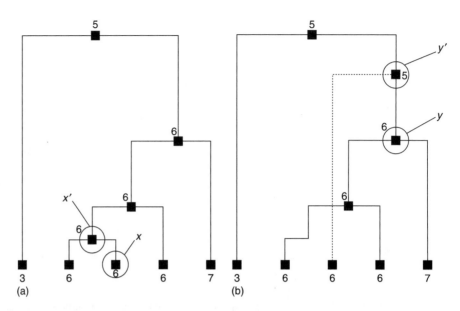

Figure 26.4 Illustration of the Wilson and Balding 'branch-swapping' algorithm, which is a proposal distribution for moving from one tree (a) to another (b), as explained in the text.

2. Remove the parent node of x (marked x' in Figure 26.3) from the tree, together with the lineage joining x to x'.

3. Choose a node, y, above which to attach the branch that formerly attached x to the tree, weighting the choice of y towards nodes that have a type that is similar to the type at x.

4. Attach the branch leading from x somewhere above y, creating a new node (labelled y' in Figure 26.4) where the join occurs.

5. Choose a type label for the newly created node, weighting this choice towards those types that are close to the types at both x and y (i.e. favouring types between the types at x and y).

26.4.5.3 Comparing the BKYF and WB Proposals

The Beerli–Kuhner–Yamato–Felsenstein (BKYF) and Wilson and Balding (WB) proposal distributions have an important qualitative difference. By augmenting the genealogy to include the genetic types at the internal nodes, Wilson and Balding were able to design an 'intelligent' proposal distribution that tends to make sensible choices for destinations to which branches may be moved. In particular, step 3, which tends to reattach the branch leading to x at a sensible point in the tree, would be impossible without information on the types at the internal nodes. In contrast, the behaviour of the BKYF proposal does not even depend on the observed genetic types at the tips of the tree, and as a consequence moves to very 'unlikely' trees may often be proposed (and will usually be rejected). For example, the move illustrated in Figure 26.3 is to a tree that is rather inconsistent with the observed data, and so would most likely be rejected. Such a move has reasonable probability under the BKYF proposal, but small probability under the WB proposal. One might wonder whether there is ever a good reason for choosing a less 'intelligent' scheme. The answer is that the intelligence of the WB scheme comes at a cost: the expanded sample space, which results from including the types at the internal nodes. In order for the MCMC scheme to explore the space properly it is now necessary for it to explore not only all plausible trees, but also all plausible configurations for the internal node types. Another potential drawback of more 'intelligent' schemes is that more computation may be required to propose each move, and so fewer moves are proposed in a fixed time.

The trick then is to get a good balance between the sizes of the space to be explored, the computational effort required for each proposal, and the 'intelligence' of the proposal distribution. In the author's experience the WB proposal seems to strike this balance nicely for micro-satellite data. In order to apply a similar approach to sequence data, it might be worth including the types at the internal nodes at the segregating sites only, as this should allow intelligent proposals to be designed while keeping the sample space down to a manageable size.

26.4.6   Application and Assessing Reliability

The application of an MCMC scheme is illustrated using the `micsat` program of Wilson and Balding (1998), and the example files (`cooper.inp` and `cooper.dat`), which are supplied with `micsat`. The data is from Cooper *et al.* (1996), and pertains to 60 males from Nigeria, Sardinia and East Anglia, typed at five micro-satellites on the Y chromosome. Wilson and Balding (1998) refers to these data as the NSE dataset.

Wilson and Balding (1998) use `micsat` to provide a Bayesian analysis of these data. In particular, they estimate the post-data distributions of θ and T_{MRCA} given certain priors on N and μ. Histograms of these distributions, based on the output obtained from the example files supplied with `micsat`, are shown in Figure 26.5. (We show the distribution of T_{MRCA} measured in units of generations. This can be converted to years by multiplying by a generation time, as we see later.) As might be expected, the results are very similar

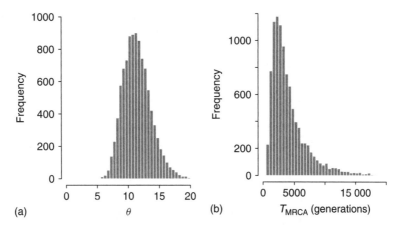

Figure 26.5   Results of a Bayesian analysis of the NSE data, based on output of `micsat` using example files `cooper.inp` and `cooper.dat`. (a) Histogram of post-data distribution of θ. (b) Histogram of post-data distribution of T_{MRCA}.

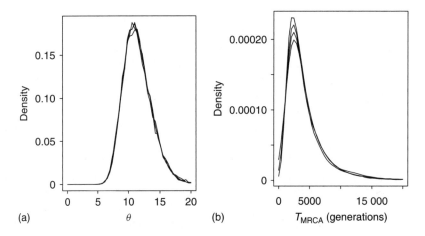

Figure 26.6 Estimates of post-data distribution of (a) θ and (b) T_{MRCA} based on output of `micsat` with example files `cooper.inp` and `cooper.dat`. Each of the four estimates shown is based on a run using a different seed for the pseudo-random number generator (which is set in the file `cooper.inp`). The similarity of the estimates to one another gives grounds for optimism that the MCMC runs are sufficiently long.

to those in Wilson and Balding (1998), which the reader should consult for a discussion of their biological significance. Here we concentrate on some of the statistical issues, the first and most important being *how can we be sure that the MCMC scheme has been run for long enough to give reliable results?* This question should always be asked when results from an MCMC method are being considered. Perhaps the most straightforward and practical method of investigating this is to re-run the algorithm several times, using different seeds for the pseudo-random number generator, and different initial trees, and compare the results obtained. Although more sophisticated diagnostic methods exist (see Brooks and Roberts, 1998, for example), none are foolproof, and this is the approach the author would recommend for the novice user. It has the practical advantage that it can be applied even when the program implementing the MCMC scheme gives the user only a summary of the results, which is sometimes the case (though `micsat` allows you to inspect the raw MCMC sample). Figure 26.6 shows estimated post-data distributions for θ and T_{MRCA} from four further runs of `micsat`, each using a different seed. Very similar results are obtained in each case, giving grounds for optimism that the MCMC runs are sufficiently long.

Another issue is to what extent the results depend on the use of the Bayesian method, and choice of prior. Wilson and Balding (1998) show how their results vary with different choices of prior for the parameters N and μ. It is natural to specify independent priors for N and μ, rather than a prior for θ directly, in view of the following:

1. There is often information about likely values for μ from sources other than the data being considered. For example, Wilson and Balding (1998) based their prior for μ on pedigree studies, in which 3 mutations were observed in 1491 meioses.

2. There is often information about likely values for N from studies at other loci: under the assumption of neutrality all loci should share the same value for N (except of

course that the X chromosome and Y chromosome should have values for N that are respectively 3/4 and 1/4 of the value of N for autosomal loci).

Note also that the priors for N and μ affect the distribution of T_{MRCA}, and that T_{MRCA} cannot be estimated unless there is some information from another source on either N or μ.

Here we compare the estimated distribution of T_{MRCA} obtained with the 'high-variance' priors from Wilson and Balding (1998), with that obtained by a maximum-likelihood approach. An estimate of the (relative) likelihood for θ can be obtained by taking an estimate of the post-data density for θ from the histogram in Figure 26.5, and dividing it by the prior density for θ. Maximising this gives an estimate of the maximum-likelihood estimate for θ, which for these data is around $\theta = 11.0$. In this case, the maximum likelihood estimate is very close to the post-data mode (the peak of the post-data density) for θ, but this is not always the case. On the basis of the observation of 3 mutations in 1491 meioses in a pedigree, the maximum-likelihood estimate of μ is $3/1491 \approx 0.002$, and substituting these point estimates into $\theta = 2N\mu$ gives a point estimate of 2734 for N. The natural maximum likelihood approach to estimating the distribution of T_{MRCA} is to take these estimates as given, which can be achieved using `micsat` by placing priors on N and μ, which are very peaked about these point estimates.

Finally, in order to convert T_{MRCA} from generations to years, it is necessary to multiply by what we believe to be the generation time. Opinions differ on what the appropriate value for generation time should be. Although 20 years has been commonly used, there is evidence suggesting that the appropriate value may be nearer 27 Weiss (1973) or even higher Tremblay and Vézina (2000). The uncertainty in this value ought really to be taken into account in a proper analysis, and in a Bayesian framework this can be achieved straightforwardly by placing on the generation time. As an illustration, Figure 26.7 shows the distribution of T_{MRCA} if we assume a uniform prior distribution on generation time between 25 and 35 years. This distribution is rather less peaked than is obtained using a point estimate of 27.5 years for the generation time in the maximum-likelihood framework (same figure). This is a consequence of the maximum likelihood method ignoring the inherent uncertainty in estimates of N and μ and in the estimated generation time. For these reasons, the author endorses a Bayesian approach for such questions.

26.4.7 Extensions to More Complex Demographic and Genetic Models

The examples given above have all been restricted to the assumption of a constant-sized panmictic population. Fortunately, almost everything that we have said about the simple case also applies to these more complex cases. Typically, the number of parameters in these more complex models is larger, which makes plotting likelihood surfaces (for example) rather harder, but the basic principles underlying IS and MCMC are unchanged. When dealing with these more complex models, some important questions to bear in mind are as follows:

- Have I run the method long enough to obtain reliable results? As we have stressed, multiple runs using different seeds and starting points can be a big help here.

- Is my model appropriate, and will it allow me to answer the questions I am interested in? For example, if you are interested in historical rates of migration between subpopulations, it makes sense to try to fit a model with migration, rather than a model in which populations split some time in the past, with no subsequent migration.

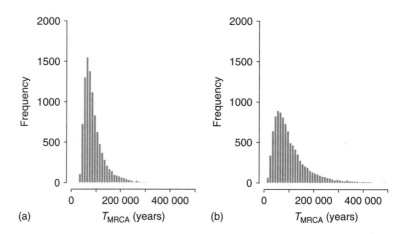

Figure 26.7 Comparison of the post-data distribution of T_{MRCA} for the NSE dataset using maximum-likelihood and Bayesian approaches. (a) Post-data distribution of T_{MRCA} using the maximum-likelihood point estimates, $\theta = 11.0$ and $\mu = 0.02$, and a generation time of 27.5 years. (b) Post-data distribution of T_{MRCA} using a fully Bayesian approach, with the priors on N and μ used by Wilson and Balding (1998), and a uniform distribution on generation time between 20 and 35 years. The Bayesian approach allows for uncertainty in the estimates of μ, N and the generation time, and as a result gives a more diffuse distribution for T_{MRCA}.

26.5 OTHER APPROACHES

As noted in the introduction, the full-data methods that are the main focus of this chapter remain computationally demanding. In many settings they remain computationally intractable, and in other settings they may be tractable, but no software is available implementing them. As a result, there remains considerable interest in alternative approaches that are computationally simpler and (often) easier to implement, without sacrificing too much in the way of accuracy. Here we review some of these methods (see also Marjoram and Tavaré, 2003).

26.5.1 Rejection Sampling and Approximate Bayesian Computation

Rejection sampling provides a convenient way to perform inference based on summaries of the observed data – see Tavaré *et al.* (1997), Li and Fu (1999), and Pritchard *et al.* (1999) for examples of this approach within the context of coalescent-based inference. Rejection sampling has the advantage of being relatively easy to implement, provided one has the ability to simulate data from models of interest. Several software packages are now available for simulating population data under a wide range of different modelling assumptions, including models involving factors such as selection, which can be difficult to incorporate into full-data inference methods. See the section on software for a selection of simulation packages.

In outline, the rejection sampling approach is as follows:

1. Simulate a value for the parameter θ from its prior distribution.

2. Simulate data given the value of θ obtained in step 1, from $P(\mathcal{D} \mid \theta)$.

3. Compute a summary, S, of the simulated data, and compare this summary with the same summary computed for the observed data (S_{obs} say). If $S = S_{obs}$ 'accept' (i.e. save) the simulated value of θ; otherwise reject it.

The values of θ that are saved (rather than rejected) in step 3 provide a sample of draws from the posterior distribution of θ given S_{obs}. (The intuition here is that the values are simulated from the prior distribution, and accepted with probability given by the likelihood $P(S_{obs}|\theta)$, resulting in samples from the posterior, which is proportional to the prior times the likelihood.) The above steps are repeated until sufficient samples of θ have been obtained to perform desired inference.

Note that the *statistical* efficiency of this approach depends on whether the summary used contains most of the information in the data regarding θ. In most cases, there is little theory to guide appropriate choice of summary, and we are left with intuition, and experimentation, in order to find 'good' summaries. Typically the choice of good summary will depend heavily on which parameters are of primary interest. For example, the number of polymorphic sites in a region seems likely to contain substantial information about the mutation rate in the region, but less information about the recombination rate. In principle, one can use as complex a summary of the data as one would like – it could involve multiple summary statistics, up to the whole data itself. However, the *computational* efficiency of the approach becomes very poor as the summary S becomes more complex. This is because if the summary is of very high dimension then one will never obtain $S = S_{obs}$ in step 3, and so the algorithm will never produce any accepted values of θ. Computational efficiency also depends on the choice of prior distribution, and will be greatest when the prior distribution is concentrated on areas of the parameter space that are consistent with the observed data. Of course, one should not change one's prior distribution in order to make it consistent with the observed data: the point of the prior is that it encapsulates information available *before* one has observed the data. However, this consideration does mean that one might want to avoid using simple but unrealistic prior distributions that include many values that the user would consider implausible.

Depending on the choice of summary statistic, the rejection sampling approach can be a very simple and effective approach to inference. However, in many cases one will run into the problem, as noted above, of very rarely accepting any simulated values of θ, which makes the approach computationally impractical. One way to alleviate this problem is to relax the condition in step 3 that $S = S_{obs}$, and to accept values of θ where S is 'close to' S_{obs}. In other words, to accept θ when $d(S, S_{obs}) < \delta$ where $d(\cdot, \cdot)$ is some measure of distance between two values of the summary statistic, and δ is some threshold that controls how similar the values need to be in order to accept θ. Typically, one would choose δ as small as possible while keeping the acceptance rate high enough to make the algorithm computationally tractable. This modified rejection sampling scheme is probably more widely used in practice than the standard rejection scheme. Formally, it provides samples from the distribution $P(\theta|d(S, S_{obs}) < \delta)$, rather than the distribution $P(\theta|S = S_{obs})$ sampled from by the standard scheme.

Two other approaches to inference, related to the above rejection sampling scheme, are worth noting. The first, due to Beaumont *et al.* (2002) (see also **Chapter 30**), is based on the ideas of adjusting, and weighting sampled values of θ. The adjustment and weights are designed both to correct the sample so that it more closely follows the distribution $P(\theta|S = S_{obs})$ rather than $P(\theta|d(S, S_{obs}) < \delta)$, and to improve computational efficiency (by weighting the samples, rather than simply accepting or rejecting them). The second, from Marjoram *et al.* (2003), incorporates the accept/reject steps into an MCMC scheme. One motivation for doing this is to try to avoid problems suffered by rejection sampling in situations where the prior distribution is very different from the posterior distribution for θ. In such cases, as noted above, the standard rejection scheme becomes very inefficient, spending most of its time simulating values of θ from the prior that are completely inconsistent with the observed data, and then rejecting them. The MCMC version can avoid this problem because once it has found values of θ that are consistent with the data it concentrates on proposing (and then accepting or rejecting) near-by values of θ, which will also tend to be somewhat consistent with the data. However, this MCMC approach also introduces potentially serious problems of its own: the acceptance rate for proposed moves can become very small when exploring the tails of the distribution of θ (see Sisson *et al.* (2007), who also suggest a way to circumvent this problem).

26.5.2 Composite Likelihood Methods

Composite likelihood methods are being widely used in recent years, particularly for estimating recombination rates from population data under coalescent-based models. The use of composite likelihoods in this context was pioneered by Hudson (2001), who suggested estimating recombination rates in a region by computing likelihoods for all pairs of available markers, and then multiplying these likelihoods together to form what is known as a '*composite*' *likelihood*. In other words, the composite likelihood for the population recombination rate, ρ, given observed genotype data G, is

$$L_{\text{comp}}(\rho; G) = \prod_{i,j} P(G_i, G_j \mid \rho), \tag{26.27}$$

where G_i denotes the genotype data at marker i. The great advantage of this approach is that it requires only the ability to compute the probability of data at two markers, which is considerably easier than computing the probability of data at several markers. Indeed, with current computing power, one can accurately estimate the probability of any given sample configuration at two markers in n individuals (at least if these markers are bi-allelic) by simply simulating two-marker genotypes in n individuals huge numbers of times, and measuring how often each configuration is observed (see Hudson, 2001). This kind of approach is impractical for several markers, because the number of configurations grows to be too large; but for two markers the number of possible configurations is manageable.

The composite likelihood (26.27) is sometimes referred to as the 'pair-wise' composite likelihood, as it involves a product over all pairs of sites. Alternative composite likelihood approaches are also possible: for example, Fearnhead and Donnelly (2002) form a composite likelihood for a region by multiplying together likelihoods obtained from non-overlapping subregions. Each of these non-overlapping subregions itself contains multiple markers, and so they use the kind of IS methods described above to approximate the likelihood for each subregion. The advantage of this approach over simply using IS to approximate the likelihood for the whole region is that the likelihood for each subregion is easier to accurately approximate (because it contains fewer markers and less recombination). Compared to the pair-wise composite likelihood method, this approach typically requires considerably more computation and is much more complicated to implement; however, it has the potential advantage that it takes into account information at multiple markers simultaneously, which could lead to gains in efficiency.

Both the pair-wise and region-wise composite likelihoods share the property that they are obtained by multiplying together likelihoods obtained from subsets of the data that are not actually independent. One consequence of ignoring dependence in this way is that composite likelihoods are typically more peaked about their maximum than they 'should' be (i.e. than they would be if the dependence was taken into account). As a result, it can be difficult to obtain confidence intervals or other measures of uncertainty for parameter estimates; however, the parameter estimates themselves appear to perform well in practice (see Fearnhead, 2003 for a theoretical treatment), and are widely used. See, for example, McVean *et al.* (2004), Myers *et al.* (2005), and **Chapter 27**.

26.5.3   Product of Approximate Conditionals (PAC) Models

Another approach to inference from population data was introduced by Li and Stephens (2003), who suggested approximating the likelihood of the observed data $\mathcal{D}$ by first writing this likelihood as a product of conditional distributions, and then developing approximations to these conditional distributions that are easy to compute. Specifically, in the context of estimating the population-scaled recombination rate ρ, from observed haplotypes $H_1, \ldots, H_n$, they write

$$L(\rho) = P(H_1, \ldots, H_n \mid \rho) = P(H_1|\rho)P(H_2|H_1, \rho) \ldots P(H_n|H_1, H_2, \ldots, H_{n-1}, \rho),$$
(26.28)

and then propose approximations to the conditional distributions on the right-hand side of this equation, which can be used to create an approximate likelihood, which they call aproduct of approximate conditionals (PAC) likelihood. This can be viewed as an attempt to approximate inference under a coalescent-like model, in that the approximate conditional distributions used are motivated by attempting to approximate the true conditional distributions under a coalescent model. One advantage of this kind of approach over the composite likelihood methods is that (26.28) is actually based on a probability distribution, and so, unlike composite likelihoods, the PAC likelihood has about the right amount of peakedness about its maximum (see Li and Stephens, 2003). This makes it possible to obtain (approximate) interval estimates for parameters (e.g. confidence or credible intervals). One disadvantage of the PAC compared with the pair-wise composite likelihood is that it requires haplotypes to be known, or estimated, and cannot be applied directly to unphased genotype data. However, actually the PAC model by itself provides a way to estimate haplotypes from genotype data (Stephens and Scheet, 2005). Further, and perhaps unexpectedly, the pair-wise composite likelihood appears to be more accurate when applied to haplotypes estimated using the PAC model than when applied directly to unphased genotypes (Smith and Fearnhead, 2005).

In addition to estimating recombination rates, the PAC approach has also been used to estimate the population-scaled mutation rate for micro-satellites (Cornuet and Beaumont, 2007; Chaudhuri and Stephens, 2006), and it seems likely that it could also be helpful in other contexts involving, for example, migration or selection.

26.6   SOFTWARE AND WEB RESOURCES

Among the methods discussed above, the rejection sampling methods based on summary statistics are perhaps the only ones that are algorithmically simple enough to be implemented, and adapted to specific contexts, by researchers with limited computing experience. Most of the other methods are more complex, and so choice of method in a particular context will depend largely on the availability of software that can handle the types of data and models of interest. Where more than one program is available, a conservative recommendation would be to use all available methods to check that they give the same answers. This can give a useful check that the software is behaving properly, and that you are using it correctly.

Here we give a brief selection of the software currently available on the World Wide Web, implementing some of the methods described in this chapter, and their extensions to more complex genetic and demographic settings. Some of these packages are also included in the review by Excoffier and Heckel (2006).

26.6.1 Population Genetic Simulations

Simulating data from coalescent models is a key step in implementing the rejection sampling approaches outlined above, and can also be useful for testing other more complex approaches to inference. A number of programs are available.

- ms (Hudson, 2002), available from http://home.uchicago.edu/~rhudson1/source.html, can perform coalescent-based simulation of sequence data under a range of neutral models, including models with migration, population expansion and bottlenecks, with recombination and/or gene conversion.

- FREGENE, available from, http://www.ebi.ac.uk/projects/BARGEN, uses forwards simulation to simulate data from coalescent-like models.

- cosi, (Schaffner *et al.*, 2005), available from http://www.broad.mit.edu/~sfs/cosi/, performs coalescent-based simulations. A novel feature is the availability of parameter values that aim to generate data that match, in certain ways, data from modern human populations.

- Selsim, (Spencer and Coop, 2004), available from http://www.stats.ox.ac.uk/~spencer/SelSim/Controlfile.html, simulates data under a coalescent model with recombination, allowing for the presence of a single bi-allelic site that has experienced natural selection.

26.6.2 Inference Methods

26.6.2.1 General

- The program micsat (Wilson and Balding, 1998) is available from http://www.mas.ncl.ac.uk/~nijw/. It is an MCMC scheme for a Bayesian analysis of micro-satellite data from a constant-sized panmictic population, based on the proposal distribution outlined earlier. A more powerful and flexible MCMC program, batwing, is also now available from the same address, and can deal with single nucleotide polymorphisms as well as micro-satellites, and more complicated demographic models involving populations splitting from each other some time in the past, with no subsequent migration.

- The programs Lamarc, available from http://evolution.genetics.washington.edu/lamarc.html, implements MCMC approaches for estimating population growth rates and recombination and migration rates from both sequence and micro-satellite data. Earlier versions of the software were based on keeping parameters of the model fixed, and use methods described in Section 26.4.3.1 to calculate likelihood surfaces. Current versions are also able to perform Bayesian analyses; parameter estimates obtained using the Bayesian approach may be more reliable in some contexts (Stephens, 1999; Beerli, 2006).

- The program `genetree` (see for example Griffiths and Tavaré, 1994a) is available from `http://www.stats.ox.ac.uk/mathgen/software.html`. The software implements an IS algorithm, and includes options allowing estimation of migration and growth rates in structured populations.

- The program `IM`, available from `http://lifesci.rutgers.edu/~heylab/`, implements an MCMC approach to the estimation of divergence times and migration rates between populations.

26.6.2.2 *Estimating Recombination Rates*

Although some of the above packages (e.g. Lamarc) incorporate recombination, most of the above packages are most appropriate either for genomic regions that are assumed to contain little or no recombination (eg mitochondrial DNA, Y chromosome, and possibly small autosomal regions), or for analysing multiple unlinked loci. In recent years, there has been considerable interest in estimating fine-scale recombination rates across larger genomic regions using population data (see **Chapter 27**). The following list is a selection of 'coalescent-based' software available.

- The programs `maxdip` and `maxhap`, available from `http://home.uchicago.edu/~rhudson1/source.html`, implement the pair-wise composite likelihood approach to estimating recombination rates (both crossover and gene conversion) from unphased and phased single nucleotide polymorphism (SNP) data. These programs assume that the recombination rates across the region to be analysed are constant.

- The program `LDhat`, implementing the composite likelihood approach for estimating (varying) recombination rates from McVean *et al.* (2004), is available from `http://www.stats.ox.ac.uk/~mcvean/LDhat/`.

- The program `PHASE`, available from `http://www.stat.washington.edu/stephens/software.html`, implements a PAC likelihood approach to estimating (varying) recombination rates and identifying recombination hotspots.

- There are several programs implementing IS and composite likelihood methods for estimating recombination rates available from Paul Fearnhead at `http://www.maths.lancs.ac.uk/~fearnhea/software/`.

Acknowledgments

I thank Karen Ayres, David Balding, Malia Fullerton, Rosalind Harding, and Magnus Nordborg for their helpful comments on an earlier draft. This work was supported by a grant from the Wellcome Trust (ref 057416).

REFERENCES

Bahlo, M. and Griffiths, R.C. (2000). Inference from gene trees in a subdivided population. *Theoretical Population Biology* **57**, 79–95.

Beaumont, M. (1999). Detecting population expansion and decline using microsatellites. *Genetics* **153**, 2013–2029.

Beaumont, M.A., Zhang, W. and Balding, D.J. (2002). Approximate Bayesian computation in population genetics. *Genetics* **162**(4), 2025–2035.

Beerli, P. (2006). Comparison of Bayesian and maximum-likelihood inference of population genetic parameters. *Bioinformatics* **22**(3), 341–345.

Beerli, P. and Felsenstein, J. (1999). Maximum likelihood estimation of migration rates and effective population numbers in two populations using a coalescent approach. *Genetics* **152**(2), 763–773.

Brooks, S.P. and Roberts, G.O. (1998). Convergence assessment techniques for Markov chain Monte Carlo. *Statistics and Computing* **8**, 319–335.

Chaudhuri, A.R. and Stephens, M. (2006). Fast and accurate estimation of the population scaled mutation parameters, θ, from microsatellite data. *Genetics*, in press.

Cooper, G., Amos, W., Hoffman, D. and Rubinsztein, D.C. (1996). Network analysis of human Y microsatellite haplotypes. *Human Molecular Genetics* **5**, 1759–1766.

Cornuet, J.M. and Beaumont, M.A. (2007). A note on the accuracy of pac-likelihood inference with microsatellite data. *Theoretical Population Biology* **71**(1), 12–19.

Excoffier, L. and Heckel, G. (2006). Computer programs for population genetics data analysis: a survival guide. *Nature Reviews Genetics* **7**(10), 745–758.

Fearnhead, P. (2003). Consistency of estimators of the population-scaled recombination rate. *Theoretical Population Biology* **64**(1), 67–79.

Fearnhead, P.N. and Donnelly, P. (2001). Estimating recombination rates from population genetic data. *Genetics* **159**, 1299–1318.

Fearnhead, P.N. and Donnelly, P. (2002). Approximate likelihood methods for estimating local recombination rates. *Journal of the Royal Statistical Society Series B* **64**, 657–680.

Felsenstein, J. (1981). Evolutionary trees from DNA sequences: a maximum likelihood approach. *Journal of Molecular Evolution* **17**, 368–376.

Felsenstein, J., Kuhner, M.K., Yamato, J. and Beerli, P. (1999). Likelihoods on coalescents: a Monte Carlo sampling approach to inferring parameters from population samples of molecular data. In *Statistics in Molecular Biology and Genetics*, Volume 33 of *IMS Lecture Notes–Monograph Series*, F. Seillier-Moiseiwitsch, ed. Institute of Mathematical Statistics and American Mathematical Society, Hayward, CA, pp. 163–185.

Geyer, C. (1991). *Reweighting Monte Carlo mixtures*. Technical Report No. 568, School of Statistics, University of Minnesota. Available from http://stat.umn.edu/PAPERS/tr568r.html.

Gilks, W.R., Richardson, S. and Spiegelhalter, D.J. (1996). Introducing Markov chain Monte Carlo. In *Markov Chain Monte Carlo in Practice*, W.R. Gilks, S. Richardson and D.J. Spiegelhalter, eds. Chapman & Hall, London.

Griffiths, R.C. and Marjoram, P. (1996). Ancestral inference from samples of DNA sequences with recombination. *Journal of Computational Biology* **3**, 479–502.

Griffiths, R.C. and Tavaré, S. (1994a). Ancestral inference in population genetics. *Statistical Science* **9**, 307–319.

Griffiths, R.C. and Tavaré, S. (1994b). Sampling theory for neutral alleles in a varying environment. *Philosophical Transactions of the Royal Society London Series B* **344**, 403–410.

Griffiths, R.C. and Tavaré, S. (1994c). Simulating probability distributions in the coalescent. *Theoretical Population Biology* **46**, 131–159.

Griffiths, R.C. and Tavaré, S. (1997). Computational methods for the coalescent. In *Progress in Population Genetics and Human Evolution*, Chapter 10, Volume 87 of *IMA Volumes in*

Mathematics and its Applications, P. Donnelly and S. Tavaré, eds. Springer Verlag, Berlin, pp. 165–182.

Griffiths, R.C. and Tavaré, S. (1999). The ages of mutations in gene trees. *Annals of Applied Probability* **9**, 567–590.

Harding, R.M., Fullerton, S.M., Griffiths, R.C., Bond, J., Cox, M.J., Schneider, J.A., Moulin, D.S. and Clegg, J.B. (1997). Archaic African and Asian lineages in the genetic ancestry of modern humans. *American Journal of Human Genetics* **60**, 772–789.

Hudson, R.R. (2001). Two-locus sampling distribution and their application. *Genetics* **159**, 1805–1817.

Hudson, R.R. (2002). Generating samples under a Wright–Fisher neutral model of genetic variation. *Bioinformatics* **18**, 337–338.

Huelsenbeck, J. (2001). Likelihood analysis of phylogenetic trees. In *Application of the Likelihood Function in Phylogenetic Analysis*. John Wiley & Sons.

Kuhner, M.K., Yamato, J. and Felsenstein, J. (1995). Estimating effective population size and mutation rate from sequence data using Metropolis–Hastings sampling. *Genetics* **140**, 1421–1430.

Li, N. and Stephens, M. (2003). Modelling linkage disequilibrium, and identifying recombination hotspots using snp data. *Genetics* **165**, 2213–2233.

Li, W.-H. and Fu, Y.-X. (1999). Coalescent theory and its application in population genetics. In *Statistics in Genetics*, M.E. Halloran and S. Geisser, eds. Springer, pp. 45–80.

Marjoram, P., Molitor, J., Plagnol, V. and Tavare, S. (2003). Markov chain Monte Carlo without likelihoods. *Proceedings of the National Academy of Sciences of the United States of America* **100**(26), 15324–15328.

Marjoram, P. and Tavaré, S. (2006). Modern computational approaches for analysing molecular genetic variation data. *Nature Reviews Genetics* **7**(10), 759–770.

McVean, G.A.T., Myers, S.R., Hunt, S., Deloukas, P., Bentley, D.R. and Donnelly, P. (2004). The fine-scale structure of recombination rate variation in the human genome. *Science* **304**, 581–584.

Myers, S., Bottolo, L., Freeman, C., McVean, G. and Donnelly, P. (2005). A fine-scale map of recombination rates and hotspots across the human genome. *Science* **310**(5746), 321–324.

Nielsen, R. (1997). A likelihood approach to population samples of microsatellite alleles. *Genetics* **146**, 711–716.

Nielsen, R. (2000). Estimation of population parameters and recombination rates from single nucleotide polymorphisms. *Genetics* **154**, 931–942.

Pritchard, J.K., Seielstad, M.T., Perez-Lezaun, A. and Feldman, M.W. (1999). Population growth of human Y chromosomes: a study of Y chromosome microsatellites. *Molecular Biology and Evolution* **16**(12), 1791–1798.

Ripley, B.D. (1987). *Stochastic Simulation*. John Wiley & Sons, New York.

Schaffner, S.F., Foo, C., Gabriel, S., Reich, D., Daly, M.J. and Altshuler, D. (2005). Calibrating a coalescent simulation of human genome sequence variation. *Genome Research* **15**(11), 1576–1583.

Sisson, S.A., Fan, Y. and Tanaka, M.M. (2007). Sequential Monte Carlo without likelihoods. *Proceedings of the National Academy of Sciences* **104**, 1760–1765.

Smith, N. and Fearnhead, P. (2005). A comparison of three estimators of the population-scaled recombination rate: accuracy and robustness. *Genetics* **171**(4), 2051–2062.

Spencer, C.C. and Coop, G. Selsim: a program to simulate population genetic data with natural selection and recombination. *Bioinformatics* **20**(18), 3673–3675.

Stephens, M. (1999). Problems with computational methods in population genetics, 273–276. Available from http://www.stats.ox.ac.uk/~stephens.

Stephens, M. and Donnelly, P. (2000). Inference in molecular population genetics. *Journal of the Royal Statistical Society Series B* **62**, 605–655.

Stephens, M. and Scheet, P. (2005). Accounting for decay of linkage disequilibrium in haplotype inference and missing data imputation. *American Journal of Human Genetics* **76**, 449–462.

Tavaré, S., Balding, D.J., Griffiths, R.C. and Donnelly, P. (1997). Inferring coalescence times from DNA sequence data. *Genetics* **145**, 505–518.

Tremblay, M. and Vézina, H. (2000). New estimates of intergenerational time intervals for the calculation of age and origins of mutations. *American Journal of Human Genetics* **66**, 651–658.

Weiss, K. (1973). Demographic models for anthropology. *American Antiquity* **38**, 1–186.

Wilson, I.J. and Balding, D.J. (1998). Genealogical inference from microsatellite data. *Genetics* **150**, 499–510.

27

Linkage Disequilibrium, Recombination and Selection

G. McVean

Department of Statistics, Oxford University, Oxford, UK

Every chromosome carries a unique sequence of DNA; however, certain combinations of variants are shared between individuals. The extent of this sharing is referred to as *linkage disequilibrium* and its distribution in natural populations can be informative about diverse processes, from the distribution of recombination to the influence of adaptive evolution. This chapter aims to provide a foundation for the empirical analysis of linkage disequilibrium discussing how it can be measured, how it relates to underlying genealogical processes and how to perform inference about underlying molecular, historical and evolutionary processes. A central idea is that linkage disequilibrium is most naturally understood in terms of the genealogical structure underlying a sample of chromosomes.

27.1 WHAT IS LINKAGE DISEQUILIBRIUM?

Every human genome has a unique DNA sequence. This is, in part, because each individual inherits a few hundred novel mutations that occurred in the germ lines of their parents (Nachman and Crowell, 2000). But it is also because the meiotic processes of chromosomal segregation and recombination shuffle existing variation: the result of mutations in our ancestors' germ lines. Consequently, while every genome may be unique, certain combinations of variants are shared: sometimes by just a few individuals, sometimes by a large fraction of the population. The term *linkage disequilibrium* (LD) is broadly used to refer to the non-random sharing (or lack thereof) of combinations of variants. It would, perhaps, be better to talk about 'haplotype structure' or 'allelic association' (both terms also used in this chapter). Nevertheless, although LD neither requires linkage (physical association on a chromosome) nor is particularly a disequilibrium (e.g. one can discuss the equilibrium level of LD), the term *LD* has stuck.

To be clear about what LD means here, consider the three example data sets in Figure 27.1. In each case the marginal allele frequencies at each polymorphic site are approximately equal; what is different between the data sets is the degree of structuring

Handbook of Statistical Genetics, Third Edition. Edited by D.J. Balding, M. Bishop and C. Cannings.
© 2007 John Wiley & Sons, Ltd. ISBN: 978-0-470-05830-5.

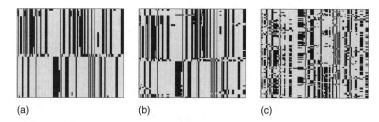

(a)   (b)   (c)

Figure 27.1   Haplotype patterns with different levels of linkage disequilibrium. Each panel shows a sample of 100 chromosomes drawn from a population with (a) no, (b) low and (c) high crossover. Each row is a chromosome and the rarer allele at each site has been shaded black. Data sets all have approximately the same set of marginal allele frequencies, so the only difference is in terms of the degree of structuring, or linkage disequilibrium.

of the variation or LD. Specifically, Figure 27.1(a) shows a very high level of structuring, Figure 27.1(c) shows a very low level of structuring and Figure 27.1(b) shows something in between. Such differences in how genetic variation is structured point to differences in the underlying biological processes experienced by the populations from which the samples are drawn. Here, the primary difference is in the crossing-over rate; the data have been simulated with zero, some and lots of crossing over respectively. However, other molecular and historical processes, including mutation, natural selection, geographical isolation and changes in population size, will also influence the structuring of genetic variation in populations. It is the goal of population genetics to make inference about such processes from observations of genetic variation in contemporary populations. Naturally, we wish to use the relevant information contained in patterns of LD. The aim of this chapter is to explore how we can understand patterns of LD observed in empirical data.

It is also worthwhile to give a more formal definition of LD. Consider a sample of chromosomes where polymorphism has been observed at a series of three loci, x, y and z. For simplicity each locus is assumed to have only two alleles (A/a, B/b and C/c respectively). The obvious description of the sample is in terms of the number of times we observe each haplotype, n_{ABC}, n_{abc}, etc., or alternatively their sample proportions, f_{ABC}, f_{abc}, etc. However, we can equivalently describe the data in terms of the marginal allele frequencies at each locus, f_A, f_a, etc. and a series of terms that reflect the extent to which combinations of alleles are found more or less frequently than expected assuming independence. For example:

$$f_{ABC} = f_A f_B f_C + f_A D_{BC} + f_B D_{AC} + f_C D_{AB} + D_{ABC}, \quad (27.1)$$

where

$$D_{AB} = f_{AB} - f_A f_B$$
$$D_{AC} = f_{AC} - f_A f_C$$
$$D_{BC} = f_{BC} - f_B f_C$$
$$D_{ABC} = f_{ABC} - f_A D_{BC} - f_B D_{AC} - f_C D_{AB} - f_A f_B f_C. \quad (27.2)$$

The D terms, which are referred to as LD coefficients, therefore measure the difference between the observed frequency of pairs or triples of alleles and that expected from the marginal allele frequencies and other D terms of lower order (e.g. the D terms for triples contain the D terms for pairs, etc.). Similar expressions apply to any data set of any complexity (in terms of number of loci and number of alleles at each locus). However, the number of terms clearly explodes as the number of loci increases. Nevertheless, the point should be clear: patterns of genetic variation can be described in terms of the marginal allele frequencies and a series of terms relating to the degree of association between pairs, triples, etc. of alleles. These terms are broadly referred to as *LD*.

Before progressing, there is a rather subtle (but ultimately profound) point to make. The previous paragraph deals with how to describe genetic variation within a sample of chromosomes. Historically, population genetics has focused more on describing genetic variation within *populations*: idealised entities consisting of (effectively) infinite numbers of individuals whose genetic composition can be described in terms of allele frequencies and coefficients of LD, just as in the sample. While the notion of a population is very helpful (and is used extensively in this chapter) a focus on the sample has three benefits. First, the sample is all that we have; although we can of course use the sample to make inferences about populations. Second, in reality there is no such thing as a population, just a set of individuals related through a complex and unknown pedigree. Third, thinking about the history of the sample, specifically the genealogical history, provides a coherent way of linking what we observe in patterns of genetic variation to underlying biological and historical processes (Hudson, 1990). For these reasons, this chapter focusses heavily on the interpretation of LD within a sample.

The rest of this chapter is divided into three parts. In the first, the method of summarising LD in empirical data is explored. In the second, how simple probabilistic models of genealogical history can be used to explore the effects of various molecular and historical processes on patterns of LD is dealt with. Finally, in the third part, the problem of inference is considered: how we can learn about such processes from empirical data.

27.2 MEASURING LINKAGE DISEQUILIBRIUM

As stated above, our ultimate aim is to make inference about underlying biological and historical processes from patterns of genetic variation, including the structuring of variants, or LD. To motivate the problem, consider the three data sets shown in Figure 27.2. Each panel shows the inferred haplotypes for single nucleotide polymorphism (SNP) data from the same 100 kb surrounding the *Lactase* gene on human chromosome 2, but from three different samples of individuals: individuals of European ancestry living in Utah (referred to as *CEU*), individuals of Yoruba origin from Nigeria (referred to as *YRI*) and a combination of Han Chinese from Beijing and Japanese people living in Tokyo (referred to collectively as *CHB/JPT*); the data is from the International HapMap Project (The International HapMap Consortium, 2005). This gene is important because a particular mutation in the promoter, found at high frequencies in Europe, is associated (in Europe) with the ability to digest lactose in milk persisting until adulthood, whereas in most non-pastoralist populations from Africa, Asia and the Americas this ability ceases between 5 and 10 years. The high frequency of this mutation is thought to have been the result of strong selection for lactose persistence associated with the innovation and subsequent importance of dairy farming. This hypothesis is strongly supported by the genetic variation data, which shows European populations dominated by a single haplotype in marked comparison to the other HapMap populations (Bersaglieri *et al.*, 2004; Hollox *et al.*, 2001). Such patterns are suggestive of a recent selective sweep (Maynard Smith

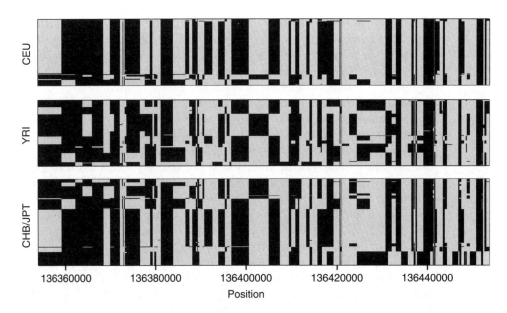

Figure 27.2 Haplotype structure in a 100-kb region surrounding the *Lactase* gene on human chromosome 2 for the four HapMap populations (The International HapMap Consortium, 2005). CEU = Individuals of European origin from Utah; YRI = Yoruba from Nigeria; CHB/JPT = Han Chinese from Beijing and Japanese from the Tokyo region (60 unrelated individuals in each of CEU and YRI, 90 in CHB/JPT). The CEU panel is dominated by a single haplotype that extends over the entire region. Much higher haplotype diversity is found in the other populations (see also Table 27.1).

and Haigh, 1974) in which the haplotype on which the beneficial mutation arose was also dragged to high frequency. However, it is worth noting that other pastoralist populations, such as the Fulani and bedouin, where lactose persistence also occurs at high frequency, do not show the same patterns and most likely carry other mutations within the *Lactase* region (Ingram *et al.*, 2007).

However, suppose we knew nothing about the gene and its function. Faced with such haplotype data, how might we begin to assemble a coherent picture of the underlying processes? By way of an aside it should be noted that, throughout, the haplotype 'phase' is typically assumed to be known (i.e. for diploid species the sequence of each chromosome in an individual is known separately), through a combination of experimental techniques, genotyping in pedigrees and/or statistical analysis (e.g., Marchini *et al.*, 2006). Our first approach might be to present the data graphically, as in Figure 27.2. This shows the marked differences between populations and is clearly the most complete representation of the genetic variation data. Nevertheless, it would also be useful to have some low-dimensional summaries of the data that could be used to compare populations, or perhaps this region to some other region in the genome.

The aim of this section is to introduce a range of low-dimensional summaries of LD that could be applied to such data. In reality, we would also use low-dimensional summaries of the data that are functions of the SNP allele frequencies rather than their structuring. For example, statistics like Tajima's D (Tajima, 1989), Fay and Wu's H (Fay and Wu, 2000), Fu and Li's D (Fu and Li, 1993), F_{ST}, etc. Indeed, it doesn't make much sense to separate out the analysis of LD from the analysis of allele frequencies. However, the diversity of measures of LD is sufficient to justify separate consideration. One thing that must be stressed, however, is that no single summary of LD is 'best' in the sense that it captures all information about underlying processes (e.g. no single-number summary is *sufficient* in the statistical sense that it captures all information about some parameter of interest). Different summaries are more or less useful for identifying the effects of different underlying processes. In the analysis of empirical data it is therefore important to stress the use of multiple summaries, each of which may give some more insight.

27.2.1   Single-number Summaries of LD

A natural starting point in the analysis of LD would be to ask whether there are any single-number statistics of the data that are informative about LD in the same way that, for example, the average pairwise diversity or Tajima's D are used in the analysis of allele frequency data. If the region is relatively short, one useful summary is simply the number of unique haplotypes observed (the more LD the fewer haplotypes). Similarly, haplotype homozygosity (the probability that two haplotypes picked without replacement from the sample are identical) is an indication of how skewed the haplotype frequencies are. There is, however, a problem with such summaries. As the length of the region surveyed increases (or more SNPs are typed), we would eventually expect to reach a point where every haplotype is unique (and haplotype homozygosity is 0). We might therefore want a summary whose value is not so arbitrarily determined by the length of the region analysed.

There are many possibilities for such statistics. One approach is to attempt to break the data up into a series of blocks, each of which represents a region with low numbers of haplotypes and high haplotype homozygosity (Anderson and Novembre, 2003; Gabriel *et al.*, 2002; Wang *et al.*, 2002; Zhang *et al.*, 2002). The number of such blocks is therefore a measure of the degree of structure in the data. However, any choice about how to break the data up is arbitrary and the concept of such 'haplotype blocks' has been little used in recent years. A related idea, motivated by the design of association studies to map the genetic basis of phenotypic variation, is to identify 'tag' SNPs that capture (in ways that are discussed below) variation within a region (Carlson *et al.*, 2004; de Bakker *et al.*, 2005; Johnson *et al.*, 2001). The number of tag SNPs required for a region is therefore a measure of how structured the variation is (e.g. if only two distinct haplotypes were observed, only one tag SNP would be required). However, there is no single most useful measure of how well variation is 'captured' and differences in experimental design mean that estimated numbers of tag SNPs are hard to compare between studies.

Another approach to summarising LD within a region is to estimate the influence of recombination. For example, non-parametric techniques (Hudson and Kaplan, 1985; Myers and Griffiths, 2003; Song *et al.*, 2005) can be used to estimate the minimum number of historical recombination events consistent with the data under the infinite-sites assumption. Similarly, parametric, coalescent-based techniques (discussed later) can be used to estimate the 'population recombination rate', or more accurately the 'population crossover rate', $4N_ec$, under the standard neutral model, where c is the genetic distance across the region and N_e is the effective population size (Stumpf and McVean, 2003). Broadly speaking, a high crossover rate will tend to result in little genetic structuring (like Figure 27.1c), while low rates will tend to result in data sets like that in Figure 27.1(a). However, the association is far from perfect, particularly if the region of interest has experienced adaptive evolution, the demographic history is not well approximated by a randomly mating population of constant size (e.g. there have been dramatic changes in population size or geographical subdivision) or there is considerable gene conversion.

By way of example, single-number summaries of LD for the *Lactase* gene region of Figure 27.2 are presented in Table 27.1. The strong structuring of the CEU sample is shown clearly in the reduced number of haplotypes, the increased homozygosity and the smaller number of detectable recombination events relative to the other populations. The similarity in haplotype numbers and detectable recombination events between YRI and CHB/JPT is complicated by the larger sample size of the latter; sub-samples of 120 chromosomes from CHB/JPT typically show values intermediate between YRI and CEU. Interestingly, the parametric estimates of the population crossover rates show the rate in CEU to be slightly higher than CHB/JPT, a pattern generally seen across the genome (Myers *et al.*, 2006; The International HapMap Consortium, 2005) and perhaps not one expected for a gene where natural selection has acted so strongly. It is important to note that the model-based estimates of the population crossover rate typically assume neutrality (and a very simple demographic history). The action of natural selection is just one of many historical forces that can result in different summaries of LD giving apparently conflicting indications as to the relative amount of LD.

27.2.2   The Spatial Distribution of LD

Any single-number summary of LD will fail to capture heterogeneity in the observed structuring of variation. Some genomic regions may have greater structuring than others or there may be systematic patterns in the relationship between genomic location and LD. For example, crossing over during meiosis will tend to lead to systematically lower levels of LD for variants at distantly separated loci compared to that for those at closely situated ones. Alternatively, gene conversion and/or mutational hotspots might create variants that are much more randomly distributed than their neighbours.

There are two approaches to summarizing the spatial distribution of LD (i.e. how it changes along the chromosome). One possibility is to make inferences about the spatial nature of the underlying biological or evolutionary processes (crossing over, gene conversion, mutation, natural selection, etc.). For example, LD in humans is strongly influenced by the concentration of meiotic crossing-over events into short regions called *recombination hotspots*. Consequently, inferences about the underlying recombination landscape reflect, at least in part, the distribution of LD along a chromosome (Jeffreys *et al.*, 2001; 2005; McVean *et al.*, 2005; The International HapMap Consortium, 2005).

Table 27.1   Statistics of LD for the *Lactase* gene.

	YRI	CEU	CHB/JPT
Number of chromosomes	120	120	180
Number of unique haplotypes	34	18	35
Haplotype homozygosity	0.05	0.53	0.15
Recombination events*	23	10	23
Estimated $4N_e c$/kb[†]	0.12	0.10	0.07

* Lower bound on the minimum number of recombination events estimated by the method of Myers and Griffiths (2003).

[†] Estimated using the method of McVean *et al.* (2002) assuming a constant crossover rate and $\theta = 0.001$ per site.

The alternative is to make summaries of LD for subsets of the data (e.g. pairs of sites) and show, usually graphically, spatial patterns in these summaries. The following discussion focuses on two-locus summaries of LD as these are the most widely used summaries of LD for genetic variation data.

Consider a pair of loci, at which exactly two different alleles are observed in the population; these being A/a at the first locus and B/b at the second. These are most naturally thought of as SNPs, but they might also be insertion–deletion polymorphisms or restriction fragment length polymorphism (RFLPs). For the moment assume that the haplotype phase of the alleles is known. As described above, the standard coefficient of LD between the alleles at the two loci is defined as

$$D_{AB} = f_{AB} - f_A f_B$$
$$= f_{AB} f_{ab} - f_{Ab} f_{aB}, \qquad (27.3)$$

where f_{AB} is the frequency of haplotypes carrying the A and B alleles and f_A is the marginal allele frequency of allele A. D_{AB} therefore measures the difference between the frequency of the AB haplotype and that expected if the haplotype frequencies were simply given by the product of the marginal allele frequencies. Any deviation from this expectation results in a non-zero value for D_{AB}, with a positive value indicating that the AB haplotype is found more often than expected assuming independence and a negative value indicating that it is found less frequently than expected. Although (27.3) focuses on the AB haplotype, the coefficient of LD for any other haplotype is given by the simple relationship $D_{AB} = -D_{aB} = -D_{Ab} = D_{ab}$.

As described above, the coefficient is computed from the sample haplotype frequencies. However, we might also be interested in asking how the sample coefficient relates to that of the population (if we believe that one exists). If we let D_{AB} be the population coefficient, the sample coefficient $\hat{D}_{AB}$ has the properties (Hill, 1974)

$$\hat{D}_{AB} = \hat{f}_{AB} - \hat{f}_A \hat{f}_B$$

$$E\left[\hat{D}_{AB}\right] = \left(\frac{n-1}{n}\right) D_{AB}$$

$$\mathrm{Var}(\hat{D}_{AB}) = \frac{1}{n}[f_A f_a f_B f_b + (f_A - f_a)(f_B - f_b)D_{AB} - D_{AB}^2]. \qquad (27.4)$$

Here $\hat{f}_{AB}$ means the obvious estimate of f_{AB} (the population frequency) from the sample; i.e. n_{AB}/n, where n_{AB} is the number of AB haplotypes in the sample. The most important point about (27.4) is that the variance in the estimate is strongly influenced by the allele frequencies at the two loci. Furthermore, the range of values $\hat{D}_{AB}$ can take is strongly influenced by the allele frequencies. If we arbitrarily define the A and B alleles to be the rarer alleles at each locus and enforce (without loss of generality) $\hat{f}_B \le \hat{f}_A$, it follows that

$$-\hat{f}_A \hat{f}_B \le \hat{D}_{AB} \le \hat{f}_a \hat{f}_B. \qquad (27.5)$$

The strong dependency on allele frequency of the standard coefficient of LD is an undesirable property because it makes comparison between pairs of alleles with different allele frequencies difficult. Consequently, several other measures of LD have been proposed that (at least in some ways) are less sensitive to marginal allele frequencies (Hedrick, 1987).

The most useful of these is the r^2 measure (Hill and Robertson, 1968). Consider assigning an allelic value, X_A, which is 1 if the allele at the first locus is A and 0 if the allele is a. Also assign an allelic value, X_B, with equivalent properties at the second locus. The quantity measured by (27.3) can then be interpreted as the covariance in allelic value between the loci. A standard way to transform the covariance is to measure the Pearson correlation coefficient,

$$r_{AB} = \frac{\text{Cov}(X_A, X_B)}{\sqrt{\text{Var}(X_A)\text{Var}(X_B)}} = \frac{D_{AB}}{\sqrt{f_A f_a f_B f_b}}. \tag{27.6}$$

In fact, for several reasons (not least because (27.6) has an arbitrary sign depending on how the allelic values are assigned), it is actually more useful to consider the square of the correlation coefficient,

$$r^2 = \frac{D^2}{f_A f_a f_B f_b}. \tag{27.7}$$

The r^2 measure has many useful properties. First, as indicated by the lack of subscripts for D in (27.7), it has the same value however the alleles at the two loci are labelled. Second, as described later, there are simple relationships between r^2 and two features of interest, the power of association studies (Chapman *et al.*, 2003; Pritchard and Przeworski, 2001) and properties of the underlying genealogical history (McVean, 2002). Third, there is a direct relationship between the sample estimate of r^2, obtained by replacing population values by the sample values in (27.7), and the power to detect significant association, i.e. to reject the null hypothesis $H_0: D = 0$. An obvious test to consider is the contingency table test where, under the null hypothesis, the test statistic

$$X^2 = \sum_{ij} \frac{(O_{ij} - E_{ij})^2}{E_{ij}}, \tag{27.8}$$

is asymptotically χ^2 distributed with 1 df as the sample size tends to infinity. Here O_{ij} and E_{ij} are, respectively, the observed and expected counts of the ij haplotype where the expectation is calculated assuming independence between the loci. The relationship between (27.8) and r^2 is

$$X^2 = n\hat{r}^2. \tag{27.9}$$

Consequently the null hypothesis of no association can be rejected at a specified level, α, if $n\hat{r}^2$ is greater than the appropriate critical value of the test statistic. Another test that might be considered is the likelihood ratio test, where the test statistic

$$\Lambda = 2\log\left(\frac{L(D = \hat{D})}{L(D = 0)}\right), \tag{27.10}$$

is also asymptotically χ^2 distributed with 1 df under the null hypothesis. Here $L(D)$ indicates the likelihood of the LD coefficient calculated using the multinomial distribution and the specified value of D. Using a Taylor expansion to approximate the logarithm, it can be shown that

$$\Lambda = 2n \sum_{ij} (\hat{f}_i \hat{f}_j + \hat{D}_{ij}) \log\left(1 + \frac{\hat{D}_{ij}}{\hat{f}_i \hat{f}_j}\right)$$

$$= n\hat{r}^2 + o(D^3). \tag{27.11}$$

Consequently the test statistics (and hence the power) of the contingency table and likelihood ratio tests are approximately equal and a function only of the sample size and observed r^2. Note that for small sample sizes the χ^2 approximation is unlikely to hold. In these circumstances it is possible to use standard permutation procedures or exact tests to estimate the significance of the observed correlation.

Although r^2 has many useful properties, it is far from the only statistic in use. For example, the $|D'|$ measure (Lewontin, 1964) is defined as the absolute value of the ratio of the observed D to the most extreme value it could take given the observed allele frequencies

$$|D'| = \begin{cases} \dfrac{-\hat{D}_{AB}}{\min(\hat{f}_A\hat{f}_B, \hat{f}_a\hat{f}_b)} & \hat{D}_{AB} < 0 \\[4mm] \dfrac{\hat{D}_{AB}}{\min(\hat{f}_A\hat{f}_b, \hat{f}_a\hat{f}_B)} & \hat{D}_{AB} > 0 \end{cases}. \tag{27.12}$$

The main use of $|D'|$ is that it measures the evidence for recombination between the loci. A feature of (27.12) is that $|D'|$ can only be less than 1 if all four possible haplotypes are observed in the sample. If the mutation rate is low, such that repeat or back mutation is unlikely, then if all four possible haplotypes are observed, it can be inferred that at least one recombination event must have occurred in the history of the sample (Hudson and Kaplan, 1985). Conversely, if anything less than the four combinations are observed the data are compatible with a history in which no recombination has occurred. Furthermore, the greater the rate of recombination between the loci, the more likely the alleles are to be in linkage equilibrium. So a value of $|D'| = 1$ can be interpreted as evidence for no recombination, while a value near 0 can be interpreted as evidence for considerable recombination. There is, however, a problem with such an interpretation for rare alleles (Figure 27.3). Even if all four combinations are present in the population, it may be unlikely to see all four in a finite sample if at least one haplotype is at low frequency (Devlin and Risch, 1995; Hedrick, 1987; Lewontin, 1988). For this reason, the interpretation of a $|D'|$ of 1 is highly dependent on the sample allele frequencies and the construction of confidence intervals for $|D'|$ is highly recommended (Gabriel *et al.*, 2002). Furthermore, if the primary interest of a study is to learn about recombination, it makes considerably more sense to use non-parametric or parametric approaches to learn about recombination directly.

As discussed above, an informative approach to summarising the spatial structure of LD is to compute two-locus statistics for all pairs of polymorphic loci and to represent these values graphically. Figure 27.4 shows three example data sets and their corresponding LD plots that demonstrate how the spatial distribution of LD can vary. In Figure 27.4(a)

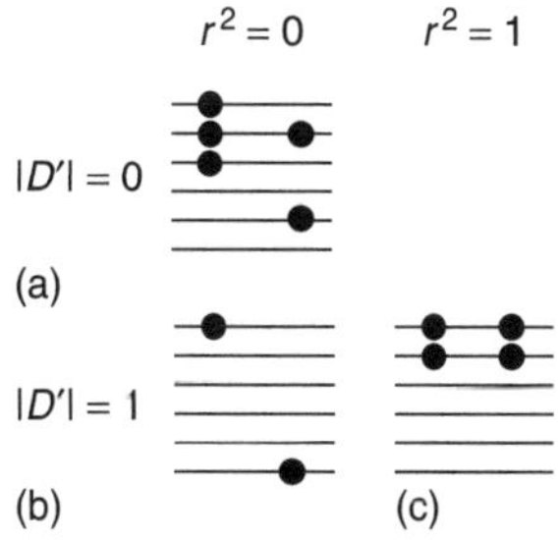

Figure 27.3 The relationship between sample configuration and the r^2 and $|D'|$ measures of two-locus LD. Each panel shows a configuration that corresponds to either high or low values of the two measures. Each bar is a chromosome and each circle represents an allele. For the diagonal plots the measures agree. However, for (b), r^2 is near 0 while $|D'|$ is 1, demonstrating how the two measures focus on different aspects of the sample configuration.

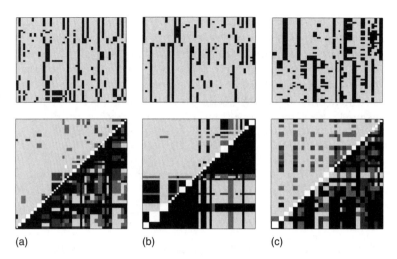

Figure 27.4 The spatial structure of LD. In each panel the upper plot shows the haplotypes and the lower plot shows a matrix of pairwise r^2 (upper left) and $|D'|$ values (bottom right) shaded by magnitude from black (values near one) to light grey (values near zero). (a) In a region of constant crossover rate sites close to each other show strong LD, which gradually decreases the further apart they are ($n = 50, \theta = 10, C = 30$). (b) In a region with a central crossover hotspot, LD appears blocklike, with regions of very high LD separated by points at which the association breaks down ($n = 50, \theta = 10, C = 50$ concentrated on a single central hotspot). (c) Where LD is generated by the mixture of individuals from two differentiated populations there is no spatial structure to LD; sites near or far can be in strong LD ($n = 25$ in each population, 80 loci, data simulated under a β-binomial model of differentiation (Balding and Nichols, 1995) with $F_{ST} = 0.9$).

there is a tendency for closely situated alleles to show strong to moderate LD, while more distant ones show much weaker LD. In Figure 27.4(b) there are two strong blocks of LD separated by a point at which LD breaks down almost completely. In Figure 27.4(c) there is no apparent spatial structure to LD: alleles can be in strong association whether they are near each other or far away. These differences in LD patterns reflect different underlying processes: a region with a moderate and constant crossover rate, a region with a strong crossover hotspot and low background crossover rate and a series of unlinked loci sampled from two highly differentiated populations, respectively. Although the pictures are somewhat noisy, it is clear that an understanding of the spatial distribution of LD can greatly help in the interpretation of underlying processes.

27.2.3 Various Extensions of Two-locus LD Measures

The previous situation considered how to measure LD in the setting where loci are bi-allelic and the allelic phase is known. But in many situations neither may be true. If the sampled individuals are diploid and the haplotype phase of alleles is unknown, it is possible to estimate the haplotype frequencies, e.g. by maximum likelihood (Weir, 1996). For two bi-allelic loci, lack of phase information adds remarkably little uncertainty to estimates of LD (Hill, 1974), because the only two-locus genotype where phase cannot be accurately inferred is when the individual is heterozygous at both loci. For multi-allelic loci (such as microsatellites), haplotype estimation can also be achieved by maximum likelihood, e.g. by the expectation maximisation (EM) algorithm.

The key problem for multi-allelic systems is how to summarise LD. One approach, motivated by the relationship between r^2 and the χ-square test in the bi-allelic case, is to use the statistic (Hill, 1975)

$$Q = \frac{1}{n} \sum_{i=1}^{k} \sum_{j=1}^{l} \frac{(O_{ij} - E_{ij})^2}{E_{ij}} = \sum_{i=1}^{k} \sum_{j=1}^{l} \frac{\hat{D}_{ij}^2}{\hat{f}_i \hat{f}_j}. \tag{27.13}$$

Here, as before, O_{ij} and E_{ij} are the observed and expected haplotype counts when there are k and l alleles respectively at the two loci. Again, under the null hypothesis $H_0 : D_{ij} = 0$, for all i and j, Qn will be approximately χ^2 distributed, though with $(k-1)(l-1)$ degrees of freedom. Of course, as the number of alleles increases, so that the expected haplotype counts tend to decrease, it becomes more important to use permutation methods or exact tests, rather than the χ^2 approximation, to assess significance. There are also multi-allele versions of $|D'|$ (Hedrick, 1987). Again, it is informative to present the result of such analyses graphically.

27.2.4 The Relationship between r^2 and Power in Association Studies

Before trying to understand what influences the distribution of LD (and how we can learn about these processes), it is worth discussing an important, though perhaps rather misunderstood, relationship between measures of LD and the power of association studies (Chapman *et al.*, 2003; Pritchard and Przeworski, 2001). An association study is an experiment that aims to map the genetic basis of phenotypic variation by comparing the genetic variation between individuals with a disease (cases) and without the disease (controls) with the aim of identifying variants that are enriched among sufferers of the disease; similar approaches can be taken for quantitative traits (see **Chapter 37**). Typically, only a subset of all polymorphic loci will have been analysed. Nevertheless, if the causative locus does not happen to be among those analysed, there is still some hope of identifying the locus through the LD that might exist between the causative variant and nearby variants that have been analysed in the study (Hirschhorn and Daly, 2005).

To see what reduction in power this leads to we need to specify a way in which the statistical analysis of the data will be performed. The simplest (though almost certainly not the most powerful) approach is to measure the difference in allele frequencies between cases and controls and assess the significance by means of a contingency table analysis. For simplicity, the diploid nature of most eukaryotes is ignored, though the results shown here generalise to the diploid case for diseases with additive (or multiplicative) risk (Chapman *et al.*, 2003). Table 27.2 shows the frequency of each case/allele combination for a simple disease model where the risk allele, A, increases the chance of getting the disease by a factor e^γ. The population frequencies of the two alleles are f_A and f_a and the proportion of the subjects who are cases and controls is ϕ_D and ϕ_C, respectively. Under this model, the sample counts are drawn from these frequencies (usually with fixed row totals) and the association between the disease and genetic variation at the locus could be summarised by the sample covariance between allelic status and disease status, a statistic called T here

$$T = \text{Cov}(A, D) = \tfrac{1}{n^2}(n_{AD}n_{aC} - n_{AC}n_{aD}). \tag{27.14}$$

Where n_{AD} is the number of individuals with the disease and the A allele, etc., and n is the total number of individuals. Note that this statistic is also equivalent to the standard

Table 27.2 Frequencies of each case/allele status under a simple disease model.

Status	Allele		Totals
	A	a	
Disease	$\dfrac{\phi_D f_A e^\gamma}{f_a + f_A e^\gamma}$	$\dfrac{\phi_D f_a}{f_a + f_A e^\gamma}$	ϕ_D
Control*	$\approx \phi_C f_A$	$\approx \phi_C f_a$	ϕ_C

* It is assumed that the allele risk and disease prevalence are sufficiently small so as not to create a skew in allele frequencies in the control set.

coefficient of LD between the allelic and disease status (27.3). However, we also wish to know whether the observed value of T lies outside the range we would expect by chance if there were no causative association (i.e. whether we can reject the null model at some specified significance level). Under the null hypothesis, $\gamma = 0$ (i.e. when there is no causal association), we know from (27.4) that

$$E[T] = 0$$

$$\mathrm{Var}(T) = \frac{f_A f_a \phi_D \phi_C}{n}. \tag{27.15}$$

So under the null hypothesis the statistic

$$Z = \frac{T - E[T]}{\sqrt{\mathrm{Var}(T)}} = \frac{T}{\sqrt{\hat{f}_A \hat{f}_a \phi_D \phi_C}} n^{\frac{1}{2}} \tag{27.16}$$

is asymptotically normally distributed with mean 0 and variance 1 as the sample size tends to infinity. Note the hats on the frequencies indicate that these are estimates from the sample. Under the alternative, $\gamma \neq 0$, but assuming that γ is small such that $e^\gamma \approx 1 + \gamma$ and the variance of the test statistic does not change appreciably, it follows that

$$E[T] \approx \gamma \times f_A f_a \phi_D \phi_C$$

$$Z = \frac{T}{\sqrt{\hat{f}_A \hat{f}_a \phi_D \phi_C}} n^{\frac{1}{2}} \sim \mathrm{Normal}\left(n^{\frac{1}{2}} [f_A f_a \phi_D \phi_C]^{\frac{1}{2}} \gamma, 1\right). \tag{27.17}$$

Equivalently, the standard test statistic for the 2×2 contingency table, Z^2, is approximated by the χ^2 distribution with 1 df and a non-centrality parameter of $n^{\frac{1}{2}} [f_A f_a \phi_D \phi_C]^{\frac{1}{2}} \gamma$. Now consider a marker locus with alleles B and b. The entries in Table 27.2 can be updated in a simple manner by noting that

$$\Pr(B|D) = \frac{\Pr(D \ \& \ B)}{\Pr(D)}$$

$$= \frac{\Pr(B)[\Pr(A|B)\Pr(D|A) + \Pr(a|B)\Pr(D|a)]}{\Pr(D)}$$

$$= \frac{f_{AB}e^{\gamma} + f_{aB}}{f_A e^{\gamma} + f_a}$$

$$\approx f_B + D_{AB}\gamma. \tag{27.18}$$

Substituting this term into Table 27.2 and an equivalent one for the b allele, $\Pr(b|D) \approx f_b - D_{AB}\gamma$, shows that the test statistic at the marker locus

$$Z = \frac{\hat{f}_{BD}\hat{f}_{bC} - \hat{f}_{BC}\hat{f}_{bD}}{\sqrt{\hat{f}_B \hat{f}_b \phi_D \phi_C}} n^{\frac{1}{2}} \tag{27.19}$$

is approximately normally distributed

$$Z \sim \text{Normal}\left(n^{\frac{1}{2}}[\phi_D \phi_C]^{\frac{1}{2}}\gamma \frac{D_{AB}}{\sqrt{f_B f_b}}, 1\right)$$

$$\sim \text{Normal}\left(n^{\frac{1}{2}}[f_A f_a \phi_D \phi_C]^{\frac{1}{2}}\gamma \frac{D_{AB}}{\sqrt{f_A f_a f_B f_b}}, 1\right) \tag{27.20}$$

The last term in the mean of the distribution should look familiar as (27.6), the correlation coefficient between the A and B alleles at the two loci. Finally, imagine two experiments. In the first, we type the causative locus in a total of n_1 individuals. In the second, we type the marker locus in n_2 individuals. From (27.17) and (27.20) it should be clear that the distributions of the test statistics across replicates of the two experiments will be the same if and only if

$$n_2 = \frac{n_1}{r^2}. \tag{27.21}$$

The subscript on the square of the correlation coefficient has again been dropped to indicate that it is the same for all pairs of alleles. Equation (27.21) implies that to achieve the same power in the experiment at the marker locus the sample size has to be increased by a factor of $1/r^2$.

This is a very elegant result and clearly has implications for the design of association studies. Nevertheless, two critical issues need to be appreciated. The first is that while the r^2 result can be used to define experiments with equal power, it does not follow that in a given experiment typing a marker locus relative to a causative locus results in a loss of power of $100 \times (1 - r)\,\%$. The relationship between the non-centrality parameter and the power is non-linear. Consequently, analysing a marker with $r^2 = 0.5$ to an unanalysed causative allele may result in a drop in power of much less than 50 % if the power to detect the disease allele (if typed) is very high or a drop of much more than 50 % if the power is intermediate or low. The second point is that the r^2 result, at least as stated, only applies to one, probably suboptimal, way of analysing the data for association. Indeed, it only considers an experiment where a single marker has been studied. In reality, more complex models will be fitted to substantially more complex data sets in which the multiple-testing issue becomes important (because many markers will be analysed). It is therefore clear that the magnitude of pairwise r^2 values between alleles is only one factor in determining the power of an association study.

27.3   MODELLING LD AND GENEALOGICAL HISTORY

It should be clear by now that many different forces can influence the distribution of
LD. These include molecular processes such as mutation and recombination, historical
processes such as natural selection and population history and various aspects of experimental design (marker and subject ascertainment). If we are to have any hope of making
useful inferences about the underlying processes from empirical data, we need to have a
coherent framework within which to assess the way in which they can affect the patterns
of variation we observe. One approach is to use simple probabilistic models to explore
the distribution of patterns of LD we might observe under different scenarios. The aim
of this section is to introduce such models and illustrate how they can be used to provide
insights from empirical data. A key feature is the idea that genealogical models, specifically the coalescent with recombination (see **Chapter 25**), provide a flexible and intuitive
approach to modelling genetic variation. The relation between features of the underlying
genealogical history and properties of LD is also shown.

27.3.1   A Historical Perspective

Before introducing the coalescent perspective it is useful to provide a brief historical
sketch of mathematical treatments of LD (see also **Chapter 22**). These approaches have
given considerable insight into the nature of LD and, in contrast to most of the Monte
Carlo based coalescent work, do provide simple analytical results about quantities of
interest. The description given below is not chronological. All of these models assume a
population of constant size and random mating.

27.3.1.1   *The Relationship between LD and Two-locus IBD*

Intuitively, the level of LD observed between two loci must be related to the extent to
which they share a common ancestry (i.e. the extent to which the two loci have been
co-transmitted in genealogical history). Indeed this is really the point of coalescent
modelling. Completely linked loci share exactly the same ancestry and typically show
high levels of LD, while unlinked loci have independent ancestries and typically show
low LD. One way of quantifying the degree of shared ancestry between two loci is
two-locus identity by descent (IBD). Single-locus IBD measures the probability that two
chromosomes sampled at random from a population share a common ancestor before some
defined point in the past (note that all chromosomes ultimately share a common ancestor
so the equilibrium value of IBD is 1). The two-locus version simply extends the notion to
measure the probability that two chromosomes sampled at random share a single common
ancestor at both loci before some defined point in the past (and that there has been no
crossover on either pathway from the sample chromosomes to the common ancestor).
Note that IBD does not refer to identity in state (i.e. whether the two chromosomes
carry the same alleles) rather it refers to relatedness. Over time, IBD within a population
increases through genetic drift and decreases through recombination. An expression for
the change in two-locus IBD, $Q = \Pr(\text{same ancestor})$, over time can be obtained (Sved,
1971) as a function of the diploid population size, N, and the per generation probability
of a crossing-over event occurring between the two loci, c. Note that in the literature c
can refer to either the genetic map distance between two loci (for details of how genetic

maps are constructed see **Chapter 1**) or the probability of crossing over (which cannot take a value greater than 0.5). In all cases considered here, c is sufficiently small that the genetic map distance and probability of crossover are approximately the same.

$$Q_{t+1} = \frac{1}{2N}(1-c)^2 + \left(1 - \frac{1}{2N}\right) Q_t (1-c)^2. \tag{27.22}$$

Solving for the equilibrium state gives

$$\tilde{Q} \approx \frac{1}{1+C}, \tag{27.23}$$

where $C = 4Nc$. Note that the same result can be obtained more naturally by thinking of the process backwards in time. Two-locus IBD is just the probability that the first 'event' that happens to the chromosomal segment flanked by the two loci on two randomly sampled chromosomes is a common ancestor event. That is, they coalesce before any crossover event occurs.

How does two-locus IBD relate to LD? Sved (1971) proposed the following argument. Consider sampling two chromosomes known to be identical in state at one locus proportionally to the allele frequencies at that locus. Both loci are bi-allelic with alleles A/a at the first and B/b at the second. It can be shown that the probability that these chromosomes are also identical at the second locus, P_{H}, is a function only of the homozygosity at the second locus, $F_B = f_B^2 + f_b^2$ where f_B is the frequency of the B allele etc. and the r^2 measure of LD between the alleles at the two loci

$$P_{\mathrm{H}} = r^2 + (1 - r^2)F_B. \tag{27.24}$$

If the crossover rate is large relative to the mutation rate then two chromosomes that show IBD at both loci are likely to also be homozygous at both loci (i.e. under these conditions IBD also implies identity in state). Consequently,

$$P_{\mathrm{H}} \approx \tilde{Q} + (1 - \tilde{Q})F_B. \tag{27.25}$$

Combining expressions gives

$$r^2 \approx \tilde{Q} = \frac{1}{1+C}. \tag{27.26}$$

In short, the argument suggests that the expected value of r^2 is near 1 for very small crossover rates and approaches $1/C$ for $C >> 1$.

Although this approximation is a useful heuristic (Chakravarti *et al.*, 1984) and is widely quoted (Jobling *et al.*, 2004), it is limited in application (Weir and Hill, 1986) because two chromosomes may share a common ancestor yet be different in allelic state due to a more recent mutation. Implicit within (27.25) is the assumption that allele frequencies do not change over time. For these reasons, (27.26) will only be a good approximation when the time scale over which two chromosomes at two loci may share a common ancestor before recombining is very short, which is only true for large values of C.

27.3.1.2 Matrix Methods and Diffusion Approximations

There are two alternative, and rather more rigorous, approaches to obtaining results about the expected value of LD statistics under simple population models. One approach is to use matrix recursions to describe the change in moments of LD statistics over time (Hill, 1975; 1977; Hill and Robertson, 1966; 1968). The other is to use a diffusion approximation (see also **Chapter 22**), replacing the discrete nature of genes in populations by a continuous space of allele frequencies (Ohta and Kimura, 1969a; 1969b; 1971). Although these methods appear somewhat different at first, they are actually closely related and can be used to examine both the dynamics of change in LD over time and to obtain expressions for quantities of interest at equilibrium. For example, although it is not possible to calculate the expected value of r^2 at equilibrium, it is possible to calculate a related quantity

$$\sigma_d^2 = \frac{E[D_{AB}^2]}{E[f_A(1 - f_A)f_B(1 - f_B)]},\qquad(27.27)$$

for a pair (or a set of equivalent pairs) of bi-allelic loci, where f_A and f_B are the allele frequencies at two loci. Under the infinite-sites model (Karlin and McGregor, 1967; Kimura, 1969), the diffusion approximation leads to the solution (Ohta and Kimura, 1971)

$$\sigma_d^2 = \frac{10 + C}{22 + 13C + C^2}.\qquad(27.28)$$

The same result can be obtained as the limit under low mutation rate from models of bi-allelic loci with a symmetric rate of mutation between alleles (Hill, 1975; Ohta and Kimura, 1969a). Like the expression of Sved this result predicts that for large C, the expected value of r^2 is approximately $1/C$. The main difference between the predictions of (27.26) and (27.28) is for small C where (27.28) predicts a value considerably less than 1 (a value of 5/11 for $C = 0$). Figure 27.5 compares estimates from Monte Carlo coalescent simulation. Neither approximation provides a particularly accurate prediction for the expected value of r^2, unless rare variants (loci where the rare variant is less than 10 % in frequency) are excluded. Nevertheless, (27.28) does predict the general shape of the decrease in average r^2 with increasing C.

27.3.2 Coalescent Modelling

The single most striking feature about Figure 27.5 is just how noisy LD is; the mean value of r^2 between loci at a given genetic distance captures very little of the complexity of the full distribution. This has two implications. First, it is hard to obtain an intuitive understanding of LD by thinking about 'expected' values of LD statistics. Second, the analysis of empirical data by comparing observed LD statistics to their 'expected' values is likely to be only weakly informative.

In order to capture the full complexity of LD patterns it is necessary to use stochastic modelling techniques to simulate the types of patterns one might observe under difference scenarios. The advent of coalescent modelling (Hudson, 1983b; Kingman, 1982), and specifically the coalescent with recombination (Hudson, 1983a), has led to a revolution in the way genetic variation data is approached. Coalescent models focus on properties of the sample by considering the genealogical history that relates a set of chromosomes to each other (see **Chapter 25**). For recombining data, the ancestral recombination

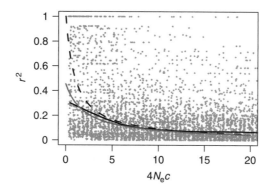

Figure 27.5 Analytical approximations for the expected r^2 between pairs of alleles as a function of the population genetic distance, $4N_e c$, between sites. Sved's approximation, (27.26): dashed black line. Ohta and Kimura's approximation, (27.28): solid grey line. Also shown (grey dots) are the values of r^2 between pairs of alleles at the corresponding genetic distance obtained from a single infinite-sites coalescent simulation with 50 chromosomes, $\theta = 100$, $C = 100$ and the sliding average (solid black line) for all pairs of sites where the minor allele is at frequency of at least 10 %. Although the Ohta and Kimura approximation performs quite well at predicting the mean r^2, its predictive power for any pair of sites is extremely poor due to high variance.

graph (ARG) generalises the idea of a 'tree' that relates individuals to each other to a complex network in which the trees at different positions along a chromosome are embedded (Griffiths, 1991). Figure 27.6 shows an ARG for two loci. Looking back in time there are two types of events: coalescent events, in which two lineages find a common ancestor, and recombination events, in which an ancestral lineage splits. The marginal trees at the two loci are embedded in the larger graph and the trees at the two loci, though different, do share some clades. Mutations that lead to variation within the sample occur along the branches of the ARG. Consequently, the structure of LD reflects the correlation structure of the underlying genealogical history. In other words, from a coalescent perspective the best way of understanding genetic variation is to think about the structure of the underlying genealogy of the sample. Different evolutionary forces have different effects on the shape and correlation structure of these underlying genealogies. Although this view is strongly driven by a neutral perspective (the mutations we observe have not themselves influenced genealogical history), the effects of certain types of natural selection, such as selective sweeps or balancing selection, on patterns of linked neutral variation can also be considered from a coalescent or genealogical viewpoint.

In addition to the various theoretical insights that coalescent theory has made possible, a genealogical approach has greatly enabled the analysis of genetic variation through the ability to simulate data under various evolutionary models. In the next three sections the use of a genealogical framework to understand the distribution of LD is described.

27.3.2.1 LD Patterns in the Absence of Recombination

It may sound strange, but we can actually learn a lot about LD by studying its behaviour in regions where there is no recombination (Slatkin, 1994). Consider the three data

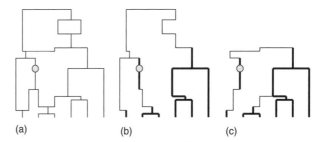

Figure 27.6 An example ancestral recombination graph for two loci and six chromosomes. (a) The history of the chromosomes can be traced back in time to coalescent events (where pairs meet their common ancestors) and crossover events, which split the lineage into two. The resulting trees at the two loci, (b) and (c), share some branches (indicated by the thick lines). Because of recombination, a mutation on a particular shared internal branch, indicated by the circle, will be present in different members of the sample depending on which locus it occurs at.

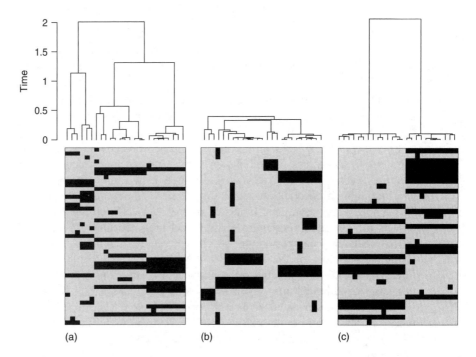

Figure 27.7 Linkage disequilibrium in non-recombining regions. Despite each example having no recombination, there can be low, moderate or high LD depending on the structure of the underlying genealogy. Data simulated with either (a) a constant population size, (b) a strongly growing population or (c) a population that has experienced a severe recent bottleneck.

sets and their corresponding genealogies in Figure 27.7 corresponding respectively to simulations with a constant population size, a growing population and a population that has experienced a strong recent bottleneck (though not one strong enough to wipe out all pre-existing variation). Apart from the differences in the numbers of polymorphic sites, there are also strong differences in their structuring. First note that any two mutations

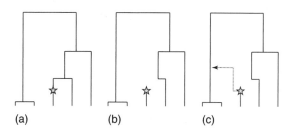

(a)   (b)   (c)

Figure 27.8 Crossover events as a point process along the sequence (Wiuf and Hein, 1999). (a) At any point along the sequence there is an underlying genealogical tree. The distance to the next crossover event is exponentially distributed with rate dependent on the total branch length of the current tree and the crossover rate. Having chosen the physical position of the crossover event, the position on the tree, indicated by the star, is chosen uniformly along the branch lengths. (b) In an approximation to the coalescent process, called the *SMC* (McVean and Cardin, 2005), the branch immediately above the recombination point is erased leaving a floating lineage. (c) The floating lineage coalesces back into the remaining tree and the process is repeated. In the coalescent, no branch can be erased.

at two distinct sites that occur on the same branch in the tree will occur in exactly the same members of the sample, and hence be in 'perfect' LD, i.e. $r^2 = 1$. Next note that mutations on different branches will be only very weakly correlated, $r^2 << 1$, if the branches are in different parts of the tree and particularly if one or both are near the tips. It follows that the extent of LD reflects the extent to which the tree shape is dominated by a few long branches, as in the case of the bottlenecked population (on which lots of mutations occur and which are therefore in perfect association), or is not, as in the case of the growing population (mutations occur on branches in different parts of the tree, particularly towards the tips, and typically show weak association). It follows that (at least for r^2) LD is strongest for the bottlenecked population and weakest for the growing one. The constant-size population shows a mixture of highly correlated and weakly correlated alleles, reflecting the distribution of mutations all across the tree. Of course, because of the inherent stochasticity in the coalescent process it would be unwise to make inferences about population history from a single non-recombining locus that showed the patterns in Figure 27.8(b or c).

Digressing slightly, it is interesting to note how classical population genetics theory concerning the distribution of genetic variation under the infinite-alleles model can be related to study the structure of LD in infinite-sites models without recombination. For example, the number of distinct haplotypes in the sample, H, is equivalent to the number of distinct alleles, k. The classic result then gives an expectation for these quantities for the case of constant-sized neutral populations (Ewens, 1972),

$$E[k] = \sum_{i=0}^{n-1} \frac{\theta}{i + \theta}. \tag{27.29}$$

Here $\theta = 4N_e u L$ is the population mutation rate over the region of interest, where u is the per site, per generation mutation rate and L is the number of sites. Similarly, the Ewens sampling formula (Ewens, 1972) describes the distribution of the numbers of each distinct haplotype conditional on the total number of observed haplotypes

$$\Pr(n_1, n_2, \ldots, n_k | k, n) \propto \frac{1}{n_1 n_2 \ldots n_k}. \tag{27.30}$$

This result inspired the first statistical test for the hypothesis that the region of interest is evolving neutrally: the Ewens–Watterson homozygosity test (Watterson, 1977; 1978), which compares the observed (haplotype) homozygosity to the distribution expected from the above formula. Indeed, thinking about the effect of recombination on this test first led to the development of the coalescent model with recombination (Hudson, 1983a; Strobeck and Morgan, 1978). Recombination tends to increase the number of observed alleles (haplotypes) and reduces the skew in allele (haplotype) frequency resulting in a systematic decrease in (haplotype) homozygosity.

27.3.2.2 LD in Recombining Regions

When recombination occurs within a region, different positions may have different, though correlated, trees (Figure 27.6). This raises a series of questions. How does recombination change tree structure? What is the relationship between correlation in tree structure and LD? How should we measure the correlation in trees? To answer these questions it is first helpful to think about how trees 'evolve' along a sequence through recombination (Wiuf and Hein, 1999). The following argument gives a sense of how genealogical history changes because of recombination, specifically crossover events, by considering how to simulate a sequence of trees along a unit region (i.e. a region where the start is labelled 0, and the end labelled 1) over which there is a constant crossover rate of C. Readers should be aware that what follows is an approximation, referred to as the *spatially Markov coalescent* (SMC) (McVean and Cardin, 2005). Nevertheless, the distribution of data generated by the SMC process is almost indistinguishable from that generated by the true coalescent process (Marjoram and Wall, 2006; McVean and Cardin, 2005).

1. Simulate a coalescent tree (i.e. no recombination) at the far left-hand edge of the region. The total tree length (in units of $2N_e$ generations) is T_L.

2. The distance along the region to the next crossover event is exponentially distributed with rate $T_L C / 2$. If the position of the next crossover event lies within the unit region, continue, otherwise stop.

3. Choose a point within the tree to recombine uniformly along its branches. Erase the remainder of the branch immediately above the chosen point.

4. Allow the recombined lineage to coalesce back to the remaining lineages at a rate proportional to the number of non-recombined lineages present (note this has to be updated if there are coalescent events among these). Note also that the recombined lineage could coalesce beyond the most recent common ancestor (MRCA) of the current tree.

5. Now assign the total length of the new tree to T_L. Return to step 2.

These steps are illustrated in Figure 27.8. Several features should be clear from this approximation. First, trees change over the region by small steps. Of course, lots of steps (resulting from large crossover rates) result in lots of changes, so the tree at the right hand of the sequence looks very different from that at the left hand. Second, because of the structure of coalescent trees, many of the crossover events occur deep in the tree when there are relatively few lineages. These often have remarkably little effect on the distribution of genetic variation and crossover events during the phase when only two lineages are present are essentially undetectable. Third, if a pair or group of sequences shares a very recent common ancestor, this part of the tree will persist over considerable genetic distances. Consequently, it is often possible to identify pairs or small groups of sequences that are identical over extremely long regions. To illustrate the effect consider the statistic T_S, the sum of the lengths of the branches shared by the trees at the start and end of a sequence. The expectation of T_S under the SMC is

$$E[T_S] = 2 \sum_{i=1}^{n-1} \frac{1}{i+C}. \tag{27.31}$$

For $n > C$ the proportion of the total expected tree length that is shared is approximately $1 - \log(C)/\log(n)$ (note that in the coalescent the shared time is likely to be slightly higher). Consequently, for large sample sizes a considerable proportion of the tree can be shared even when the total crossover rate is very high. For example, with 1000 sequences, over 15 % of the total expected time is shared by points separated by $C = 400$, which in humans of European origin corresponds to about 1 cM. Of course, if the tree shape is strongly dominated by large recent clades, as e.g. happens if there has been a recent and strong but partial selective sweep, such parts of the tree can persist over much greater genetic distances. These considerations emphasise the fact that LD is very heterogeneous, not just between genomic regions but also between different individuals within a sample.

27.3.2.3 *LD in Populations with Geographical Subdivision and/or Admixture*

Although the notion of a tree changing along the sequence provides useful insights into the nature of LD, there is one aspect that, at least at first glance, it fails to describe: LD between alleles at unlinked loci (e.g. on different chromosomes) in structured populations. As demonstrated in Figure 27.4, when the sample contains individuals from one or more populations LD can exist even between unlinked loci because of differences in allele frequencies between populations. More generally, any deviation from random mating will lead to systematic spatial structuring of genetic variation and LD even between unlinked loci. How can we understand this phenomenon within a genealogical context?

   The answer is simple. LD is determined by the correlation in genealogical history along a chromosome. Recombination acts to break down such correlations, but if there are biases induced by geographic or cultural factors in terms of which lineages can coalesce, some correlation will persist indefinitely. Figure 27.9 illustrates this idea by considering independent coalescent trees sampled from a pair of populations that diverged some time ago. Despite their independence, both trees show a strong clustering of individuals from the same populations. In short, while directly sharing genealogical trees results in

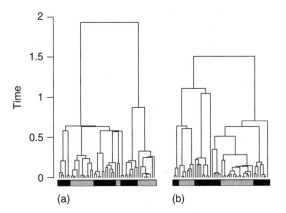

Figure 27.9 Correlations in genealogical history at unlinked loci. Individuals from two populations, indicated by the grey and black bars below, have been sampled at two unlinked loci. Although
the trees are independent, there is nevertheless genealogical correlation because a pair of individuals
sampled from within a population are likely to have a more recent common ancestor at both loci
than a pair of chromosomes where one from each population has been sampled. Data simulated
with 25 chromosomes sampled from each of two populations of equal size that diverged exactly
N_e generations ago from an ancestral population of the same size.

genealogical correlation and hence LD, genealogical correlations can also arise indirectly
by forces shaping the nature of coalescence.

27.3.3 Relating Genealogical History to LD

It should be clear by now that LD is a reflection of correlation in the genealogical history
of samples. Informally, if the coalescence time for a pair of sequences sampled at a
given locus is informative about the coalescence time for the same pair of sequences at
a second locus (relative to the sample as a whole), we expect variation at the two loci
to exhibit significant LD. But exactly what is the relationship? Is it possible to be more
quantitative about which aspects of genealogical correlation relate to which measures
of LD?

A partial answer to this question comes from studying the quantity σ_d^2; see (27.27). It
is well known that D^2 between alleles at a pair of loci, x and y, can be written in terms
of two-locus identity coefficients (Hudson, 1985; Strobeck and Morgan, 1978).

$$D_{xy}^2 = F(C_{ij}^x, C_{ij}^y) - 2F(C_{ij}^x, C_{ik}^y) + F(C_{ij}^x, C_{kl}^y). \qquad (27.32)$$

The three terms relate to the probabilities of identity at the two loci for pairs of
chromosomes sampled in different ways. Consider sampling four chromosomes from
a population and labelling them i, j, k and l. The first identity coefficient compares
chromosomes i and j at both the x and the y loci. The second compares chromosomes
i and j at the x locus, but i and k at the y locus. The third compares chromosomes i and j
at the x locus and k and l at the y locus. The expectation of these identity coefficients
over evolutionary replicates therefore determines the average strength of the LD.

What is the relationship between identity coefficients and genealogical history? The
key point is that each identity coefficient relates to the probability that no mutation has

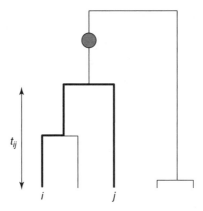

Figure 27.10 Identity in state and genealogical history. Two chromosomes, i and j, will be identical in state if no mutation occurs on those branches of the tree that lead to their common ancestor(s), at time t_{ij}. It follows that two-locus identity in state reflects whether mutations at the two trees fall on the branches to the pair's common ancestor. In the limit of a low mutation rate, but conditioning on segregation at the two loci in a sample of size n, identity in state can be written as a function of the correlations in coalescence time at the two loci, see (27.34).

occurred on the branches of the genealogy that link two chromosomes to their common ancestor (see Figure 27.10). If we condition on a single mutation occurring at both loci within a sample of size n and let the mutation rate tend towards 0, we get the result (McVean, 2002)

$$E[F(C_{ij}^x, C_{ij}^y)] = \lim_{\theta \to 0} \frac{E[(T_x - 2t_{ij}^x)(T_y - 2t_{ij}^y)e^{-\theta(T_x+T_y)/2}]}{E[T_x T_y e^{-\theta(T_x+T_y)/2}]}$$
$$= \frac{E[(T_x - 2t_{ij}^x)(T_y - 2t_{ij}^y)]}{E[T_x T_y]}, \tag{27.33}$$

where T_x is the sum of the branch lengths in the tree at locus x and t_{ij}^x is the coalescent time for chromosomes i and j at locus x. Similar results can be found for the other identity coefficients and for the denominator in (27.27). Combining these equations, we get the following result:

$$\sigma_d^2 = \frac{\rho(t_{ij}^x, t_{ij}^y) - 2\rho(t_{ij}^x, t_{ik}^y) - \rho(t_{ij}^x, t_{kl}^y)}{E[t]^2/\text{Var(t)} + \rho(t_{ij}^x, t_{kl}^y)}, \tag{27.34}$$

where $\rho(t_{ij}^x, t_{kl}^y)$ indicates the Pearson correlation coefficient between the coalescent time for chromosomes i and j at locus x and the coalescent time for chromosomes k and l at locus y and the first term in the denominator is the ratio of the square of the expected coalescence time for a pair of sequences to its variance.

Equation (27.34) has two important features. First, the required correlations in time to the common ancestor can be obtained from coalescent theory (Griffiths, 1981; Pluzhnikov and Donnelly, 1996), replicating the result of (27.28). More importantly, (27.34) gives a way of understanding the behaviour of LD, or perhaps more appropriately r^2, under more complex population genetic models. For example, the increase in LD that accompanies population bottlenecks can largely be understood through its effect on the ratio of the mean coalescence time to its standard deviation. Bottlenecks increase the variance in coalescent time considerably, leading to a reduction in the denominator of (27.34) and an increase in LD (McVean, 2002). It is also possible to describe the behaviour of LD under models of population structure (McVean, 2002; Wakeley and Lessard, 2003) and even selective sweeps (McVean, 2007) using the same approach.

27.4 INFERENCE

Coalescent modelling provides a framework within which to understand patterns of LD (see also **Chapter 26**). But what we are usually interested in is making inferences about underlying processes from the patterns of genetic variation observed in an experiment. The aim of this section is to introduce key concepts in statistical inference and to explore how these relate to understanding patterns of genetic variation.

27.4.1 Formulating the Hypotheses

Let us return to the example of the *Lactase* gene. Various summaries of the data have been presented: graphically in Figure 27.2 and numerically in Table 27.1. While the observed patterns are suggestive of various underlying processes, we need to approach the analysis of the data in a more rigorous fashion. Indeed, we need to start by asking ourselves why we are analysing this region in the first place. Do we want to categorically prove or disprove the hypothesis that it has experienced a selective sweep? Do we want to identify the simplest model that provides an adequate explanation for the data? Do we want to use these data as an aid to design an experiment to understand the genetic basis of lactose intolerance? The motivation determines, and is inseparable from, the approach to statistical inference.

Suppose that our goal is to determine whether a selective sweep has happened in the region. How might we go about doing this? Suppose that we knew exactly the process that shapes genetic variation in the absence of selection, which is referred to as Ω (a complete description of mutation rates, crossover and gene conversion rates, changes in population size, geographic structuring and experimental design). In effect this means being able to accurately characterise the distribution of sampled genetic variation in the absence of selection, from which we can calculate the probability (density) of the collected data (the likelihood of Ω). Suppose, in addition, that we can calculate the likelihood under a model that includes Ω, but also a selective sweep, which is referred to as Π (in what follows it is important that both Ω and Π have specified, though possibly multidimensional, values). Then the Neyman–Pearson lemma (e.g., Casella and Berger, 1990) tells us that the most powerful test compares the likelihood of the data with the sweep to the likelihood of the data without the sweep. Specifically, the test compares the likelihood ratio test statistic against the quantiles of the null distribution such that the null is rejected if

$$\Lambda(x) = \frac{L(\Pi, \Omega; x)}{L(\Omega; x)} > k, \quad \text{where } \Pr(\Lambda(X) > k | H_0) = \alpha. \qquad (27.35)$$

Here x refers to the data, X refers to the random variable of which x is a realisation and α is the size of specified rejection region (e.g. $\alpha = 0.01$).

Although (27.35) looks useful, unfortunately careful consideration reveals many problems. First, it is very unlikely that we can exactly specify the component values of Ω. In fact, the only source of information about many of its details comes from the data we have just collected. Second, a related problem is that we would not want to restrict ourselves to considering one specific value of Π, rather we would like to estimate the time of origin of the mutation, its location in the sequence and the strength and nature of the selection pressure from the data. This raises the problem that estimation of Π may influence estimates of Ω. Third, there is no simple way of calculating the likelihood given the data. Fourth, we can, of course, never really know Ω, all we can hope to do is to capture its key features. Finally, though perhaps least importantly, even if we could estimate all these parameters from the data, (27.35) does not actually tell us which hypothesis test or other method of model choice is the most powerful in practice (e.g. maximum likelihood, etc.).

So how might we proceed? Although it seems daunting, we have to try to learn about the components of Ω from the data and through any other sources of information. For example, we can refer back to the experimenter's notebook to see how the experiment was designed. We can also learn about aspects of Ω from genetic variation in other regions of the genome. For example, the population history, which should affect all loci more or less equally, and the mutation rate, which is largely predictable from the DNA sequence (see **Chapter 31**). To learn about some processes, such as the fine-scale structure of recombination rate variation, we may have to use the data, but in the case of *Lactase* we have three populations, only one of which is hypothesised to have experienced selection. In general, when estimating the components of Ω we are looking for estimators with four key properties:

- Statistical sufficiency: the estimator uses as much of the information in the data about the parameter of interest as possible.
- Relative efficiency: the estimator with the lowest mean-square error is preferred. Accurate estimates of uncertainty are also important.
- Robustness: the methods are not strongly influenced by deviations from the underlying assumptions.
- Computational tractability: estimation can be achieved within a reasonable time frame (perhaps no more than a month!).

It should be stressed that in most population genetics problems fully efficient inference is currently beyond the limits of computational tractability (see **Chapter 26**). For this reason, most current approaches to estimating parameters from genetic variation use methods that are less efficient but computationally tractable.

27.4.2 Parameter Estimation

To illustrate various approaches to parameter estimation in the context of understanding LD the problem of how to estimate the crossover rate, or rather the population crossover rate, over a region is considered. All of the methods described consider only a neutrally evolving population of constant size. The point of covering the various estimators is to give an indication of the range of possible approaches.

27.4.2.1 Moment Methods

The first population genetic method for estimating the population crossover rate (Hudson, 1987; Wakeley, 1997) used a method of moments approach. Using the results mentioned above that relate LD coefficients to two-locus sampling identities, Hudson derived an expression for the expected sample variance of the number of nucleotide differences between pairs of sequences under the infinite-sites model. If the number of pairwise differences between sequences i and j is π_{ij} then under the infinite-sites model

$$\bar{\pi} = \frac{1}{n^2} \sum_{i,j}^{n} \pi_{ij}$$

$$S_{\pi}^2 = \frac{1}{n^2} \sum_{i,j}^{n} \pi_{ij}^2 - \bar{\pi}^2$$

$$E[\bar{\pi}] = (1 - 1/n)\theta$$

$$E[S_{\pi}^2] = f(\theta, C, n), \tag{27.36}$$

where $f(\theta, C, n)$ is a known function of the two unknown parameters and the sample size (Hudson, 1987). To obtain a point estimate of C, θ is replaced with a point estimate from the sample (also obtained by the method of moments) and the equation is solved (if possible) for C. Similar approaches could be constructed for other single-number summaries of the data (e.g. those in Table 27.1), although Monte Carlo methods would have to be used to obtain estimates.

The great strength of this estimator is its simplicity. Unfortunately, the estimator also has very poor properties (bias, high variance, lack of statistical consistency, undefined values). This is partly because of the inherent stochasticity of the coalescent process; so for any given value of C and θ, there is a huge amount of variation in the observed patterns of variation (hence any estimator of C is expected to have considerable variance). But it is partly also because it only uses a small fraction of the total information about crossover present in the data. It is not known how well moment estimators from other sample properties might perform.

27.4.2.2 Likelihood Methods

As suggested above, a useful quantity to compute is the likelihood of the parameter (proportional to the probability, or probability density, of observing the data given the specified parameter value). Ideally, we would like to calculate the probability of observing the data given the coalescent model and specified values of the population parameters, perhaps choosing the values that maximise the likelihood as point estimates. While this problem has received considerable attention (Fearnhead and Donnelly, 2001; Kuhner et al., 2000; Nielsen, 2000), currently such approaches are only computationally feasible for small to moderate sized data sets (Fearnhead et al., 2004). A common feature of all these methods is the use of Monte Carlo techniques (particularly Markov chain Monte Carlo and importance sampling). For example, in importance sampling, a proposal function, Q, is used to generate coalescent histories, H (a series of coalescent, mutation and recombination events), compatible with the data for given values of θ and C. The likelihood of the data can be estimated from

$$L(\theta, C; x) = E_Q \left[\frac{\Pr(H|\theta, C)}{Q(H|\theta, C, x)} \Pr(x|H) \right], \qquad (27.37)$$

where $\Pr(H|\theta, C)$ is the coalescent probability of the history (note that this is not a function of the data), $\Pr(x|H)$ is the probability of the data given the history (which is always exactly one here) and $Q(H|\theta, C, x)$ is the proposal probability for the history (note that this is a function of the data). The main problem is that unless Q is close to the optimal proposal scheme,

$$Q_{\text{OPT}} = \Pr(H|\theta, C, x), \qquad (27.38)$$

the large variance in likelihood estimates across simulations makes obtaining accurate estimates of the likelihood nearly impossible.

For larger data sets a more practical alternative is to calculate the likelihood from some informative summary (or summaries) of the data using Monte Carlo techniques (Beaumont *et al.*, 2002; Marjoram *et al.*, 2003; Padhukasahasram *et al.*, 2006; Wall, 2000). Wall's estimator, based on the number of haplotypes and a non-parametric estimate of the minimum number of detectable recombination events, while conditioning on the number of segregating sites, is good in terms of having low bias and variance comparable to the best alternatives. A great strength of these methods is that they can provide useful summaries of uncertainty in estimates, e.g. through estimating the Bayesian posterior distribution of the parameter. However, because some information is thrown away, the resulting uncertainty will be greater than the true uncertainty. Furthermore, although likelihood surfaces computed from a summary of the data are expected to mirror likelihood surfaces computed from full data, such agreement is not guaranteed and, for some pathological data sets, they may disagree considerably.

27.4.2.3 *Approximating the Likelihood*

Although it may not be practical to calculate the coalescent likelihood of the entire data, it is possible to approximate the likelihood function through multiplying the likelihoods for subsets of the data (Fearnhead *et al.*, 2004; Hey and Wakeley, 1997; Hudson, 2001; McVean *et al.*, 2002; 2004; Wall, 2004), the resulting quantity being referred to as a *composite likelihood*. In effect, the idea is to treat subsets of the data as if they were independent of each other whereas in reality these subsets (pairs, triples or non-overlapping sets) are clearly not. Nevertheless, finding the value of the crossover parameter that maximises this function appears to provide estimates that are at least as accurate as any other approximate approach.

Perhaps the greatest strength of the composite likelihood approaches lies in their flexibility, particularly in the use of estimating variable crossover rates and identifying crossover hotspots (Fearnhead and Donnelly, 2002; Fearnhead and Smith, 2005; McVean *et al.*, 2004; Myers *et al.*, 2005). For example, consider Hudson's composite likelihood approach, which considers pairs of sites. For each pair of sites it is possible to pre-compute the coalescent likelihood for a specified value of θ per site and a range of crossover rates between the sites, e.g. using the importance sampling approach (McVean *et al.*, 2002).

The likelihood of a genetic map $g = \{g_1, g_2, \ldots, g_m\}$, where each g_i is the map position of the ith site in a set of m ordered sites, is approximated by

$$L_C(\theta, g; x) = \prod_{i, j > i} L(\theta, g_j - g_i; x_{ij}). \qquad (27.39)$$

Here L_C indicates the composite likelihood, L indicates the coalescent likelihood for the genetic distance between loci i and j and x_{ij} is the data at those two sites. In practice, the value of θ is estimated previously using a moment method. Because of the pre-computation, searching over the space of g is computationally feasible. McVean *et al.* (2004) used a Monte Carlo technique called *reversible jump Markov chain Monte Carlo* (Green, 1995) to explore the space of possible genetic maps. The greatest weakness of this approach is the difficulty in assessing uncertainty. Specifically, because of non-independence between the pairs of sites, the composite likelihood surface is typically more sharply peaked compared to the true likelihood surface (even though it uses less information), resulting in considerable underestimation of uncertainty.

27.4.2.4 *Approximating the Coalescent*

An alternative approach to approximating the coalescent likelihood function is to devise an alternative model for the data, motivated by an understanding of the coalescent, but under which it is straightforward to calculate the likelihood (Crawford *et al.*, 2004; Li and Stephens, 2003). The product of approximate conditionals (PAC) scheme uses the following decomposition to motivate an approximate model (see **Chapter 26**). Consider the data $x = \{x_1, x_2, \ldots, x_n\}$, where x_i is the ith haplotype in a sample of size n. The likelihood function can be written as the product of a series of conditional distributions (the dependence on θ has been dropped for simplicity)

$$L(C; x) = \Pr(x_1 | C) \Pr(x_2 | x_1, C) \ldots \Pr(x_n | x_1, x_2, \ldots, x_{n-1}, C). \qquad (27.40)$$

Of course, knowing the conditional probabilities is equivalent to knowing the coalescent likelihood. However, the PAC scheme approximates the conditional probabilities by considering the kth haplotype as an imperfect mosaic of the previous $k - 1$. Specifically, the model employed has the structure of a hidden Markov model in which the underlying state at a given nucleotide position refers to which of the $k - 1$ other chromosomes the kth is derived from. The transition probabilities are a function of the crossover rate and the emission probabilities are a function of the mutation rate and the sample size. This model captures many of the key features of genetic variation, such as the relatedness between chromosomes, how this changes through crossover and how, as the sample size increases, additional chromosomes tend to look more and more like existing ones, but it is entirely non-genealogical.

This approach was actually derived from the work mentioned above that uses impor-tance sampling to calculate the full coalescent likelihood of the data (Fearnhead and Donnelly, 2001; Stephens and Donnelly, 2000). Because of the Markovian structure of the coalescent, each history that is compatible with the data can be broken down into a series of events that can be sampled sequentially. The optimal proposal density chooses an event (coalescence, mutation and crossover), e, according to its probability given the data

$$Q_{OPT}(e|\theta, C, x) = \Pr(e|\theta, C)\frac{\Pr(x + e|\theta, C)}{\Pr(x|\theta, C)}, \qquad (27.41)$$

where $\Pr(e|\theta, C)$ is the coalescent probability of the event and $x + e$ is the original data, x, modified by event e. The key insight is to note that $x + e$ and x are very similar. For example, if e is a coalescent event between the ith and jth sequences (which are also identical) it follows that

$$\frac{\Pr(x + e|\theta, C)}{\Pr(x|\theta, C)} = \frac{\Pr(x_1, \ldots, x_i, \ldots, x_k|\theta, C)}{\Pr(x_1, \ldots, x_i, x_j, \ldots, x_k|\theta, C)}$$

$$= \frac{1}{\Pr(x_j|x_1, \ldots, x_i, \ldots, x_k, \theta, C)}. \qquad (27.42)$$

This is exactly the same conditional probability approximated in the PAC scheme. For other types of events, similar expressions involving the conditional distributions can be found (Fearnhead and Donnelly, 2001).

The strengths of the PAC scheme are its computational tractability and the sense that it uses much of the information in the data about recombination. It is also very flexible and can potentially be extended to include features such as geographical structuring and gene conversion. Nevertheless, because the model is not the coalescent, parameter estimates are typically biased in ways that are unpredictable (and the estimated uncertainty in estimates may not necessarily reflect the 'true' uncertainty). The approximation also introduces an order dependency into the likelihood function (the true conditionals would, of course, give rise to the same likelihood no matter how the sequences were considered). This can be partly overcome by averaging inference over multiple orderings.

27.4.3 Hypothesis Testing

Aside from estimating parameters, it is also often of interest to make categorical statements about whether particular hypotheses can be rejected. For example, in the case of *Lactase*, we want to attempt to categorically reject the hypothesis that the European population has not experienced adaptive evolution. More generally, we would like to make some choice about which model provides a better explanation of the data: one with selection or one without it.

Historically, most approaches to detecting the influence of natural selection in population genetic data have used a slightly different approach. Specifically, a battery of neutrality tests including Tajima's D (Tajima, 1989), Fu and Li's D (Fu and Li, 1993), Fay and Wu's H (Fay and Wu, 2000), the Hudson, Kreitman, Aguadé (HKA) test (Hudson *et al.*, 1987), the haplotype partition test (Hudson *et al.*, 1994), the extended haplotype homozygosity test (Sabeti *et al.*, 2002; Voight *et al.*, 2006) and more are used to reject the null hypothesis of a neutral model, without rigorously demonstrating that a model that includes natural selection is a better fit to the data. All of these methods are designed to look for patterns in the data that are inconsistent with neutral evolution (Sabeti *et al.*, 2006). Furthermore, Monte Carlo simulation under different scenarios can be used to assess power and whether the methods are robust to deviations from the assumed model. A second class of methods look for large between-population differences in genetic variation; e.g. striking changes in allele frequency (Akey *et al.*, 2002; 2004; Lewontin and

Krakauer, 1973). Such methods aim to identify geographically restricted selective sweeps, as thought to be the case for *Lactase*. However, their major drawback is that it is very hard to accurately specify the null model, because this requires accurate specification of the demographic histories of the populations under study. For this reason, population differentiation–based tests typically rely on comparing a locus to an empirical distribution derived from the analysis of multiple genomic regions, with the implicit assumption that loci with unusual patterns have been targeted by selection. It is currently unclear how well empirical methods perform at identifying true selected loci (Teshima *et al.*, 2006), though such methods have identified strong candidates for local selective sweeps (McVean *et al.*, 2005; Sabeti *et al.*, 2006).

There is, however, a class of methods that explicitly attempt to fit a model of a selective sweep to patterns of genetic variation (Coop and Griffiths, 2004; Kim and Stephan, 2002; Nielsen *et al.*, 2005; Przeworski, 2003). For example, one method uses a theoretical approximation to predict how allele frequencies are distorted around a selected allele, which is fitted to data by a composite likelihood approach (Nielsen *et al.*, 2005). There are also related methods that attempt to date the age of a potentially selected mutation, e.g. by examining the amount of variation found 'under' a mutation (Slatkin, 2000; Slatkin and Rannala, 2000; Toomajian *et al.*, 2003). In effect, the age of a mutation is all that can ever be estimated from genetic variation. The evidence for adaptive evolution comes from observing recent mutations that exist at much higher frequency in a population than expected from neutral mutations of the same age.

To understand the signal these methods are looking for, consider the genealogy in Figure 27.11, in which a recent mutation has risen to high frequency over a short timescale. The effect is to create drastically different genealogical histories for different subsets of the chromosomes. Those under the selected mutation have a very short time to their MRCA compared to the much deeper tree for the other sequences. The impact of this genealogical distortion on genetic variation will be to dramatically reduce genetic variation among those chromosomes that carry the beneficial mutation. Furthermore, because of the short amount of time in the tree under the mutation, it will be some considerable genetic distance from the selected site before crossover has allowed the genealogy to return to the neutral distribution. These features are particularly striking in the *Lactase* example, where the haplotype structure around the selected mutation persists for several hundred kilobases from the gene. However, it is also a region of very low crossover, which is probably why the signal is so remarkable. In regions of high crossover the power to detect recent adaptive evolution may be considerably lower.

27.5 PROSPECTS

The aim of this chapter has been to introduce a way of thinking about patterns of genetic variation – the coalescent with recombination – and to explore how to analyse such data. The reader should be aware that the field of population genetic inference is extremely dynamic, with new methods appearing at a high rate (Marjoram and Tavare, 2006). The reader should also be aware that all the methods discussed above for estimating crossover rates and detecting natural selection have considerable limitations and none are optimal. It seems likely that the next few years will see intensive research in the development of more sophisticated and perhaps fully genealogical approaches to the analysis of genetic

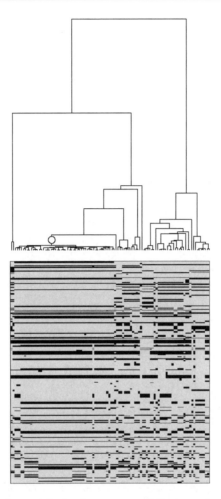

Figure 27.11 The effect of a partial selective sweep on linked genetic variation. A recent selected mutation distorts genealogical history creating a clade within the tree at the selected site with a very recent common ancestor. This clade 'persists' for long genetic distances leading to a marked haplotype structuring amongst those individuals that carry the beneficial mutation. The signature is, however, slowly eroded by crossover. Here the haplotypes are shown vertically, with the genealogy at the selected locus at the top and the selected mutation indicated by the circle. Data were simulated using the program SelSim (Spencer and Coop, 2004) for 100 chromosomes, 50 of which carry the beneficial mutation which has a scaled selective advantage of $4N_e s = 400, \theta = 200, C = 50$.

variation. Indeed, some genealogical methods are already appearing, particularly in the field of association mapping (Larribe *et al.*, 2002; Minichiello and Durbin, 2006; Zollner and Pritchard, 2005).

Acknowledgments

Many thanks to David Balding, Dick Hudson, Molly Przeworski, Raazesh Sainudiin and Jay Taylor for comments on an earlier version of the chapter.

RELATED CHAPTERS

Chapter 1; **Chapter 22**; **Chapter 25**; **Chapter 26**; **Chapter 31**; and **Chapter 37**.

REFERENCES

Akey, J.M., Eberle, M.A., Rieder, M.J., Carlson, C.S., Shriver, M.D., Nickerson, D.A. and Kruglyak, L. (2004). Population history and natural selection shape patterns of genetic variation in 132 genes. *PLoS Biology* **2**, e286.

Akey, J.M., Zhang, G., Zhang, K., Jin, L. and Shriver, M.D. (2002). Interrogating a high-density SNP map for signatures of natural selection. *Genome Research* **12**, 1805–1814.

Anderson, E.C. and Novembre, J. (2003). Finding haplotype block boundaries by using the minimum-description-length principle. *American Journal of Human Genetics* **73**, 336–354.

de Bakker, P.I., Yelensky, R., Pe'er, I., Gabriel, S.B., Daly, M.J. and Altshuler, D. (2005). Efficiency and power in genetic association studies. *Nature Genetics* **37**, 1217–1223.

Balding, D.J. and Nichols, R.A. (1995). A method for quantifying differentiation between populations at multi-allelic loci and its implications for investigating identity and paternity. *Genetica* **96**, 3–12.

Beaumont, M.A., Zhang, W. and Balding, D.J. (2002). Approximate Bayesian computation in population genetics. *Genetics* **162**, 2025–2035.

Bersaglieri, T., Sabeti, P.C., Patterson, N., Vanderploeg, T., Schaffner, S.F., Drake, J.A., Rhodes, M., Reich, D.E. and Hirschhorn, J.N. (2004). Genetic signatures of strong recent positive selection at the lactase gene. *American Journal of Human Genetics* **74**, 1111–1120.

Carlson, C.S., Eberle, M.A., Rieder, M.J., Yi, Q., Kruglyak, L. and Nickerson, D.A. (2004). Selecting a maximally informative set of single-nucleotide polymorphisms for association analyses using linkage disequilibrium. *American Journal of Human Genetics* **74**, 106–120.

Casella, G. and Berger, R.L. (1990). *Statistical Inference*. Duxbury, Belmont, California.

Chakravarti, A., Buetow, K.H., Antonarakis, S.E., Waber, P.G., Boehm, C.D. and Kazazian, H.H. (1984). Nonuniform recombination within the human beta-globin gene cluster. *American Journal of Human Genetics* **36**, 1239–1258.

Chapman, J.M., Cooper, J.D., Todd, J.A. and Clayton, D.G. (2003). Detecting disease associations due to linkage disequilibrium using haplotype tags: a class of tests and the determinants of statistical power. *Human Heredity* **56**, 18–31.

Coop, G. and Griffiths, R.C. (2004). Ancestral inference on gene trees under selection. *Theoretical Population Biology* **66**, 219–232.

Crawford, D.C., Bhangale, T., Li, N., Hellenthal, G., Rieder, M.J., Nickerson, D.A. and Stephens, M. (2004). Evidence for substantial fine-scale variation in recombination rates across the human genome. *Nature Genetics* **36**, 700–706.

Devlin, B. and Risch, N. (1995). A comparison of linkage disequilibrium measures for fine-scale mapping. *Genomics* **29**, 311–322.

Ewens, W.J. (1972). The sampling theory of selectively neutral alleles. *Theoretical Population Biology* **3**, 87–112.

Fay, J.C. and Wu, C.I. (2000). Hitchhiking under positive Darwinian selection. *Genetics* **155**, 1405–1413.

Fearnhead, P. and Donnelly, P. (2001). Estimating recombination rates from population genetic data. *Genetics* **159**, 1299–1318.

Fearnhead, P. and Donnelly, P. (2002). Approximate likelihood methods for estimating local recombination rates. *Journal of the Royal Statistical Society, Series B* **64**, 657–680.

Fearnhead, P., Harding, R.M., Schneider, J.A., Myers, S. and Donnelly, P. (2004). Application of coalescent methods to reveal fine-scale rate variation and recombination hotspots. *Genetics* **167**, 2067–2081.

Fearnhead, P. and Smith, N.G. (2005). A novel method with improved power to detect recombination hotspots from polymorphism data reveals multiple hotspots in human genes. *American Journal of Human Genetics* **77**, 781–794.

Fu, Y.X. and Li, W.H. (1993). Statistical tests of neutrality of mutations. *Genetics* **133**, 693–709.

Gabriel, S.B., Schaffner, S.F., Nguyen, H., Moore, J.M., Roy, J., Blumenstiel, B., Higgins, J., DeFelice, M., Lochner, A., Faggart, M., Liu-Cordero, S.N, Rotimi, M., Adeyemo, A., Cooper, R, Ward, R., Lander, E.S., Daly, M.J., Atshuler, D. (2002). The structure of haplotype blocks in the human genome. *Science* **296**(5576), 2225–2229.

Green, P. (1995). Reversible jump Markov chain Monte Carlo computation and Bayesian model determination. *Biometrika* **82**, 711–732.

Griffiths, R.C. (1981). Neutral two-locus multiple allele models with recombination. *Theoretical Population Biology* **19**, 169–186.

Griffiths, R.C. (1991). The two-locus ancestral graph. In *Selected Proceedings on the Symposium on Applied Probability*, I.V. Basawa and R.L. Taylor, eds. Institute of Mathematical Statistics, Hayward, CA, pp. 100–117.

Hedrick, P.W. (1987). Gametic disequilibrium measures: proceed with caution. *Genetics* **117**, 331–341.

Hey, J. and Wakeley, J. (1997). A coalescent estimator of the population recombination rate. *Genetics* **145**, 833–846.

Hill, W.G. (1974). Estimation of linkage disequilibrium in randomly mating populations. *Heredity* **33**, 229–239.

Hill, W.G. (1975). Linkage disequilibrium among multiple neutral alleles produced by mutation in finite population. *Theoretical Population Biology* **8**, 117–126.

Hill, W.G. (1977). Correlation of gene frequencies between neutral linked genes in finite populations. *Theoretical Population Biology* **11**, 239–248.

Hill, W.G. and Robertson, A. (1966). The effect of linkage on limits to artificial selection. *Genetical Research* **8**, 269–294.

Hill, W.G. and Robertson, A. (1968). Linkage disequilibrium in finite populations. *Theoretical and Applied Genetics* **38**, 226–231.

Hirschhorn, J.N. and Daly, M.J. (2005). Genome-wide association studies for common diseases and complex traits. *Nature Reviews Genetics* **6**, 95–108.

Hollox, E.J., Poulter, M., Zvarik, M., Ferak, V., Krause, A., Jenkins, T., Saha, N., Kozlov, A.I. and Swallow, D.M. (2001). Lactase haplotype diversity in the Old World. *American Journal of Human Genetics* **68**, 160–172.

Hudson, R.R. (1983a). Properties of a neutral allele model with intragenic recombination. *Theoretical Population Biology* **23**, 183–201.

Hudson, R.R. (1983b). Testing the constant-rate neutral model with protein data. *Evolution* **37**, 203–217.

Hudson, R.R. (1985). The sampling distribution of linkage disequilibrium under an infinite allele model without selection. *Genetics* **109**, 611–631.

Hudson, R.R. (1987). Estimating the recombination parameter of a finite population model without selection. *Genetical Research* **50**, 245–250.

Hudson, R.R. (1990). Gene genealogies and the coalescent process. *Oxford Surveys in Evolutionary Biology*. Oxford University Press, Oxford, pp. 1–44.

Hudson, R.R. (2001). Two-locus sampling distributions and their application. *Genetics* **159**, 1805–1817.

Hudson, R.R., Bailey, K., Skarecky, D., Kwiatowski, J. and Ayala, F.J. (1994). Evidence for positive selection in the superoxide dismutase (*Sod*) region of *Drosophila melanogaster*. *Genetics* **136**, 1329–1340.

Hudson, R.R. and Kaplan, N.L. (1985). Statistical properties of the number of recombination events in the history of a sample of DNA sequences. *Genetics* **111**, 147–164.

Hudson, R.R., Kreitman, M. and Aguade, M. (1987). A test of neutral molecular evolution based on nucleotide data. *Genetics* **116**, 153–159.

Ingram, C.J., Elamin, M.F., Mulcare, C.A., Weale, M.E., Tarekegn, A., Raga, T.O., Bekele, E., Elamin, F.M., Thomas, M.G., Bradman, N., Swallow, D.M. (2007). A novel polymorphism associated with lactose tolerance in Africa: multiple causes for lactase persistence? *Human Genetics* **120**(6), 779–788.

The International HapMap Consortium. (2005). A haplotype map of the human genome. *Nature* **437**, 1299–1320.

Jeffreys, A.J., Kauppi, L. and Neumann, R. (2001). Intensely punctate meiotic recombination in the class II region of the major histocompatibility complex. *Nature Genetics* **29**, 217–222.

Jeffreys, A.J., Neumann, R., Panayi, M., Myers, S. and Donnelly, P. (2005). Human recombination hot spots hidden in regions of strong marker association. *Nature Genetics* **37**, 601–606.

Jobling, M.A., Hurles, M.E. and Tyler-Smith, C. (2004). *Human Evolutionary Genetics: Origins, Peoples and Disease*. Garland Science, New York.

Johnson, G.C., Esposito, L., Barratt, B.J., Smith, A.N., Heward, J., Di Genova, G., Ueda, H., Cordell, H.J., Eaves, I.A., Dudbridge, F., Twells, R.C., Payne, F., Hughes, W., Nutland, S., Stevens, H., Carr, P., Tuomilehto-Wolf, E., Tuomilehto, J., Gough, S.C., Clayton, D.G., Todd, J.A. (2001). Haplotype tagging for the identification of common disease genes. *Nature Genetics* **29**, 233–237.

Karlin, S. and McGregor, J.L. (1967). The number of mutant forms maintained in a population. *Proceedings of the Fifth Berkeley Symposium on Mathematical Statistics and Probability* **4**, 415–438.

Kim, Y. and Stephan, W. (2002). Detecting a local signature of genetic hitchhiking along a recombining chromosome. *Genetics* **160**, 765–777.

Kimura, M. (1969). The number of heterozygous nucleotide sites maintained in a finite population due to steady flux of mutations. *Genetics* **61**, 893–903.

Kingman, J.F. (1982). The coalescent. *Stochastic Processes and their Applications* **13**, 235–248.

Kuhner, M.K., Yamato, J. and Felsenstein, J. (2000). Maximum likelihood estimation of recombination rates from population data. *Genetics* **156**, 1393–1401.

Larribe, F., Lessard, S. and Schork, N.J. (2002). Gene mapping via the ancestral recombination graph. *Theoretical Population Biology* **62**, 215–229.

Lewontin, R.C. (1964). The interaction of selection and linkage. I. General considerations; heterotic models. *Genetics* **49**, 49–67.

Lewontin, R.C. (1988). On measures of gametic disequilibrium. *Genetics* **120**, 849–852.

Lewontin, R.C. and Krakauer, J. (1973). Distribution of gene frequency as a test of the theory of the selective neutrality of polymorphisms. *Genetics* **74**, 175–195.

Li, N. and Stephens, M. (2003). Modeling linkage disequilibrium and identifying recombination hotspots using single-nucleotide polymorphism data. *Genetics* **165**, 2213–2233.

Marchini, J., Cutler, D., Patterson, N., Stephens, M., Eskin, E., Halperin, E., Lin, S., Qin, Z.S., Abecassis, H., Munro, H.M. and Donnelly, P. (2006). A comparison of phasing algorithms for trios and unrelated individuals. *American Journal of Human Genetics* **78**, 437–450.

Marjoram, P., Molitor, J., Plagnol, V. and Tavare, S. (2003). Markov chain Monte Carlo without likelihoods. *Proceedings of the National Academy of Sciences of the United States of America* **100**, 15324–15328.

Marjoram, P. and Tavare, S. (2006). Modern computational approaches for analysing molecular genetic variation data. *Nature Reviews Genetics* **7**, 759–770.

Marjoram, P. and Wall, J.D. (2006). Fast "coalescent" simulation. *BMC Genetics* **7**, 16.

Maynard Smith, J. and Haigh, J. (1974). The hitch-hiking effect of a favourable gene. *Genetical Research* **23**, 23–35.

McVean, G. (2007). The structure of linkage disequilibrium around a selective sweep. *Genetics* **175**, 1395–1406.

McVean, G.A. (2002). A genealogical interpretation of linkage disequilibrium. *Genetics* **162**, 987–991.

McVean, G., Awadalla, P. and Fearnhead, P. (2002). A coalescent-based method for detecting and estimating recombination from gene sequences. *Genetics* **160**, 1231–1241.

McVean, G.A. and Cardin, N.J. (2005). Approximating the coalescent with recombination. *Philosophical Transactions of the Royal Society of London, Series B: Biological Sciences* **360**, 1387–1393.

McVean, G.A., Myers, S.R., Hunt, S., Deloukas, P., Bentley, D.R. and Donnelly, P. (2004). The fine-scale structure of recombination rate variation in the human genome. *Science* **304**, 581–584.

McVean, G., Spencer, C.C. and Chaix, R. (2005). Perspectives on human genetic variation from the HapMap project. *PLoS Genetics* **1**, e54.

Minichiello, M.J. and Durbin, R. (2006). Mapping trait Loci by use of inferred ancestral recombination graphs. *American Journal of Human Genetics* **79**, 910–922.

Myers, S., Bottolo, L., Freeman, C., McVean, G. and Donnelly, P. (2005). A fine-scale map of recombination rates and hotspots across the human genome. *Science* **310**, 321–324.

Myers, S.R. and Griffiths, R.C. (2003). Bounds on the minimum number of recombination events in a sample history. *Genetics* **163**, 375–394.

Myers, S., Spencer, C.C., Auton, A., Bottolo, L., Freeman, C., Donnelly, P. and McVean, G. (2006). The distribution and causes of meiotic recombination in the human genome. *Biochemical Society Transactions* **34**, 526–530.

Nachman, M.W. and Crowell, S.L. (2000). Estimate of the mutation rate per nucleotide in humans. *Genetics* **156**, 297–304.

Nielsen, R. (2000). Estimation of population parameters and recombination rates from single nucleotide polymorphisms. *Genetics* **154**, 931–942.

Nielsen, R., Williamson, S., Kim, Y., Hubisz, M.J., Clark, A.G. and Bustamante, C. (2005). Genomic scans for selective sweeps using SNP data. *Genome Research* **15**, 1566–1575.

Ohta, T. and Kimura, M. (1969a). Linkage disequilibrium at steady state determined by random genetic drift and recurrent mutation. *Genetics* **63**, 229–238.

Ohta, T. and Kimura, M. (1969b). Linkage disequilibrium due to random genetic drift. *Genetical Research (Cambridge)* **13**, 47–55.

Ohta, T. and Kimura, M. (1971). Linkage disequilibrium between two segregating nucleotide sites under the steady flux of mutations in a finite population. *Genetics* **68**, 571–580.

Padhukasahasram, B., Wall, J., Marjoram, P. and Nordborg, M. (2006). Estimating recombination rates from SNPs using summary statistics. *Genetics* **174**, 1517–1528.

Pluzhnikov, A. and Donnelly, P. (1996). Optimal sequencing strategies for surveying molecular genetic diversity. *Genetics* **144**, 1247–1262.

Pritchard, J.K. and Przeworski, M. (2001). Linkage disequilibrium in humans: models and data. *American Journal of Human Genetics* **69**, 1–14.

Przeworski, M. (2003). Estimating the time since the fixation of a beneficial allele. *Genetics* **164**, 1667–1676.

Sabeti, P.C., Reich, D.E., Higgins, J.M., Levine, H.Z., Richter, D.J., Schaffner, S.F., Gabriel, S.B., Platko, J.V., Patterson, N.J., McDonlad, G.J., Ackerman, H.C., Campbell, S.J., Atshuler, D., Cooper, R., Kwiatkowski, D., Ward, R., Lander, E.S. (2002). Detecting recent positive selection in the human genome from haplotype structure. *Nature* **419**(6909), 832–837.

Sabeti, P.C., Schaffner, S.F., Fry, B., Lohmueller, J., Varilly, P., Shamovsky, O., Palma, A., Mikkelsen, T.S., Altshuler, D., Lander, E.S. (2006). Positive natural selection in the human lineage. *Science* **312**(5780), 1614–1620.

Slatkin, M. (1994). Linkage disequilibrium in growing and stable populations. *Genetics* **137**, 331–336.

Slatkin, M. (2000). Allele age and a test for selection on rare alleles. *Philosophical Transactions of the Royal Society of London, Series B: Biological Sciences* **355**, 1663–1668.

Slatkin, M. and Rannala, B. (2000). Estimating allele age. *Annual Review of Genomics and Human Genetics* **1**, 225–249.

Song, Y.S., Wu, Y. and Gusfield, D. (2005). Efficient computation of close lower and upper bounds on the minimum number of recombinations in biological sequence evolution. *Bioinformatics* **21**(Suppl. 1), i413–i422.

Spencer, C.C. and Coop, G. (2004). SelSim: a program to simulate population genetic data with natural selection and recombination. *Bioinformatics* **20**, 3673–3675.

Stephens, M. and Donnelly, P. (2000). Inference in molecular population genetics. *Journal of the Royal Statistical Society, Series B* **62**, 605–655.

Strobeck, C. and Morgan, K. (1978). The effect of intragenic recombination on the number of alleles in a finite population. *Genetics* **88**, 829–844.

Stumpf, M.P. and McVean, G.A. (2003). Estimating recombination rates from population-genetic data. *Nature Reviews Genetics* **4**, 959–968.

Sved, J.A. (1971). Linkage disequilibrium and homozygosity of chromosome segments in finite populations. *Theoretical Population Biology* **2**, 125–141.

Tajima, F. (1989). Statistical method for testing the neutral mutation hypothesis by DNA polymorphism. *Genetics* **123**, 585–595.

Teshima, K.M., Coop, G. and Przeworski, M. (2006). How reliable are empirical genomic scans for selective sweeps? *Genome Research* **16**, 702–712.

Toomajian, C., Ajioka, R.S., Jorde, L.B., Kushner, J.P. and Kreitman, M. (2003). A method for detecting recent selection in the human genome from allele age estimates. *Genetics* **165**, 287–297.

Voight, B.F., Kudaravalli, S., Wen, X. and Pritchard, J.K. (2006). A map of recent positive selection in the human genome. *PLoS Biology* **4**, e72.

Wakeley, J. (1997). Using the variance of pairwise differences to estimate the recombination rate. *Genetical Research* **69**, 45–48.

Wakeley, J. and Lessard, S. (2003). Theory of the effects of population structure and sampling on patterns of linkage disequilibrium applied to genomic data from humans. *Genetics* **164**, 1043–1053.

Wall, J.D. (2000). A comparison of estimators of the population recombination rate. *Molecular Biology and Evolution* **17**, 156–163.

Wall, J.D. (2004). Estimating recombination rates using three-site likelihoods. *Genetics* **167**, 1461–1473.

Wang, N., Akey, J.M., Zhang, K., Chakraborty, R. and Jin, L. (2002). Distribution of recombination crossovers and the origin of haplotype blocks: the interplay of population history, recombination, and mutation. *American Journal of Human Genetics* **71**, 1227–1234.

Watterson, G.A. (1977). Heterosis or neutrality? *Genetics* **85**, 789–814.

Watterson, G.A. (1978). The homozygosity test of neutrality. *Genetics* **88**, 405–417.

Weir, B.S. (1996). *Genetic Data Analysis II. Methods for Discrete Population Genetic Data*, 2nd edition. Sinauer Associates, Sunderland, MA.

Weir, B.S. and Hill, W.G. (1986). Nonuniform recombination within the human beta-globin gene cluster. *American Journal of Human Genetics* **38**, 776–781.

Wiuf, C. and Hein, J. (1999). Recombination as a point process along sequences. *Theoretical Population Biology* **55**, 248–259.

Zhang, K., Deng, M., Chen, T., Waterman, M.S. and Sun, F. (2002). A dynamic programming algorithm for haplotype block partitioning. *Proceedings of the National Academy of Sciences of the United States of America* **99**, 7335–7339.

Zollner, S. and Pritchard, J.K. (2005). Coalescent-based association mapping and fine mapping of complex trait loci. *Genetics* **169**, 1071–1092.

28

Inferences from Spatial Population Genetics

F. Rousset

Laboratoire Génétique et Environnement, Institut des Sciences de l'Évolution, Montpellier, France

This chapter reviews theoretical models and statistical methods for inference from genetic data in subdivided populations. With few exceptions, these methods are based on neutral models of genetic differentiation and have been mainly concerned with estimation of dispersal rates. However, simulation-based methods allow to draw inferences under models involving additional demographic processes such as changes in dispersal rates over time. The formulation and main results of migration matrix, island, and isolation-by-distance models, are briefly described. The definition and basic properties of F-statistics are reviewed, and moment methods for their estimation are contrasted with likelihood methods. Then, the application of the different methodologies to simple biological scenarios is reviewed. Their practical performance is discussed in light of comparisons with demographic estimates, as well as of their robustness to different assumptions and of concepts of separation of timescale.

28.1 INTRODUCTION

Since the advent of molecular markers in population genetics, there have been many efforts to define methods of inference from the spatial genetic structure of populations. This chapter can only review a small selection of them including, in particular, some recent developments of simulation-based likelihood methods, and also of less sophisticated methods in so far as they provide analytical insight and proven performance in realistic conditions. With few exceptions, I will focus on allele frequency data; some methods for other types of data are described in **Chapter 29**.

The perspective taken in this review is that studies of spatial population structure are conducted in order to make inferences about parameters considered important for the evolution of natural populations, for example, for the dynamics of adaptation. Thus,

Handbook of Statistical Genetics, Third Edition. Edited by D.J. Balding, M. Bishop and C. Cannings.
© 2007 John Wiley & Sons, Ltd. ISBN: 978-0-470-05830-5.

all such analyses should ultimately be based on models of evolution in subdivided populations. This would lead to the identification of important parameters in such processes and to the formulation of appropriate statistical models to estimate them (assuming it is useful to estimate them in order to test the models). In this perspective, the material reviewed below may seem imperfect not only because the statistical models are approximate but also because the important evolutionary parameters are not always clearly identified.

In all inferences, we will consider a total sample from a population structured by restricted dispersal in a number of demes (a technical term used in the analysis of the models) or subpopulations (a somewhat looser term). The population concept must be carefully distinguished from another concept of 'population' often considered in statistics, which actually refers to the probability distribution of samples under some model. In general the value of a variable in the biological population is not the expected value of this variable in this statistical 'population', in other words, this is not an expected value in a theoretical model. In practice the word parameter is used for both, but here it will be used only for the value in the theoretical model.

A statistical corollary is that by sampling only one locus, one may compute estimates which will approach the value in the biological population, rather than the parameter value, as more individuals are sampled. In other words, it will approach a value that will depend on the realized genealogy in the biological population, and this will be a random variable. The usual solution to this problem is to analyze several loci with different genealogies, assumed independent. For a nonrecombining DNA, it may not be very useful to sequence longer fragments: Since the whole DNA has the same genealogy, any estimate will depend on the single realized random genealogy in the biological population sampled (see **Chapter 25**).

28.2 NEUTRAL MODELS OF GEOGRAPHICAL VARIATION

The major models considered for statistical analysis describe the evolution of neutral genetic polymorphisms; among models of selected markers, statistical analysis will be considered only for clines.

28.2.1 Assumptions and Parameters

We consider a set of subpopulations each with N_i adults, and with dispersal rates m_{ij} from subpopulation j to subpopulation i. These dispersal rates are defined as the probability that an offspring had its parent in some subpopulation: Thus they are defined by looking backward in time (backward dispersal rates), rather than by looking where offspring go (forward dispersal rates). Forward and backward rates will differ, for example, when individuals that disperse at longer distances have a higher probability of dying before reproduction.

These models are known as *migration-matrix models*, with migration matrix (m_{ij}). Limit cases of these models when all deme sizes $\to \infty$, all backward rates $\to 0$, with their products $N_i m_{ij}$ remaining finite, have been described as 'structured coalescents' (see e.g. **Chapter 25**). With many subpopulations, the number of parameters may be large. However, some symmetric structure is usually assumed, as in the island and isolation-by-distance models developed below. Further, the migration matrix, as well as the subpopulation sizes N_i, are supposed to be invariant in time. These assumptions allow for more detailed mathematical analysis. Simulation-based methods have allowed to investigate more complex historical scenarios involving range expansions, interruptions of gene flow, and so on. A relatively well-worked case is the isolation-with-migration model (Nielsen and Wakeley, 2001), according to which an initially panmictic population differentiates at some time T in the past into two subpopulations that will keep on exchanging migrants at rate m until the time of sampling.

The island model (Wright, 1931a) with n_d subpopulations is the simplest form of migration-matrix model: For different subpopulations i, j, the dispersal rate is supposed to be independent of i, j and may be written as $m_{ij} = m/(n_d - 1)$ where m is the total dispersal rate; $m_{ii} = 1 - m$. The subpopulation sizes $N_i = N$ are also supposed to be independent of i. The infinite island model is the limit process as $n_d \rightarrow \infty$. This is the most often considered model, because of its ease of analysis. However, it should be noticed that most of the results of the infinite island model with $N_i = N$ can easily be extended to infinite island model with N_i different for different subpopulations and with total dispersal rate into each subpopulation i being a function of i (see the discussion of equation (28B.1)). Thus the main defining assumption of such island models is that immigrants may come with equal probability from any of the other subpopulations.

Dispersal is often localized in space, so that immigrants preferentially come from close populations. Two kinds of models that take this into account have been considered, one for demes on a discrete lattice, and one for 'continuous' populations (e.g. Malécot, 1951; 1967). In a continuous population, the local density may fluctuate in space and time, but there is no rigorous mathematical analysis of models incorporating such fluctuations. In the lattice models, different demes are arranged on a regularly spaced lattice and the dispersal rates are a function of the distance between demes. There is a fixed number of adults, N, in every generation on each node of the lattice. Thus the position of individuals is rigidly fixed and density does not fluctuate. The island model may be recovered as a specific case.

In models of isolation by distance, the parameter σ^2 often appears (e.g. Malécot, 1967; Nagylaki, 1976; Sawyer, 1977). This is an average squared distance between parent and offspring. In two dimensions this is the average square of the projection of the two-dimensional (vectorial) distance on an axis, also known as the *axial distance* (Figure 28.1). This parameter is a measure of the speed at which two lineages descending from a common ancestor depart from each other in space. The models as formulated above may be generalized to include age- or stage-structure, and it is possible to generalize some of the results for island and isolation by distance models given below in terms of concept of effective dispersal rate and effective deme size or effective population density, albeit through some approximations (Rousset, 1999a; 2004, Chapter 9). Then, effective dispersal is the asymptotic rate of increase of the second moment of distance between two independently dispersing gene lineages per unit time. The definition of population density also needs to be generalized. First, it is actually not simply a density but a rate of coalescence per surface and per unit time (Rousset, 1999a). In the basic models, it can be computed as the expected number of coalescence events per generation among all pairs of genes in the total population, divided by the total surface occupied by the population (or habitat length in linear habitats). With age structure, it can be computed as a weighted

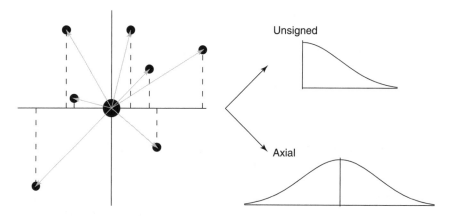

Figure 28.1 σ^2 in two dimensions. One considers the two-dimensional dispersal distances (gray arrows) between one parent (large central dot) and different offspring (or different parent–offspring pairs). The projection of these vectors on two axes yield signed axial distances on each axis. In terms of variance, σ^2 is the variance of the distribution of one such axial dispersal distance (bottom right). This is *not* the variance of the unsigned dispersal distance (top right).

average of such among all pairs, these being weighted by reproductive value weights as in the computation of the effective dispersal parameter.

28.3   METHODS OF INFERENCE

With few exceptions, explicit formulas for the likelihood of samples under the models formulated above are not available. This section therefore focuses on moment methods for which explicit analytical results are available, and on simulation methods for likelihood inference.

28.3.1   *F*-statistics

Moment methods are based on the analysis of moments of order k of allele frequencies. By far the most common of them (analysis of variance) consider only squares of allele frequencies or equivalently frequencies of identical pairs of genes. This is the basis for the theory of F-statistics in population genetics. Autocorrelation methods (e.g. Sokal and Wartenberg, 1983; Epperson and Li, 1997) are constructed from pair-wise comparisons of genes or genotypes, hence there should be essentially the same information in such statistics as in the more standard moment methods. The relationship between autocorrelation methods and some of the methods described below is discussed by Hardy and Vekemans (1999).

28.3.1.1   *Probabilities of Identity and F-statistics*

To define genetic identity, we consider a pair of homologous genes and ask whether they descend without mutation from their most recent common ancestor. If no mutation has occurred since the coalescence of ancestral lineages, there is *identity by descent* (IBD).

By *identity in state* (IIS) of a pair of genes we simply consider whether they have the same sequence (if the alleles are distinguished by their sequence), the same length (if the alleles are distinguished by the number of repeats of a microsatellite motif), the same electrophoretic mobility, etc. In short, we only look at the allelic state of a gene. IBD is a specific case of IIS for the infinite allele model, in which each allele produced by mutation is considered different from preexisting alleles. The generic notation Q will be used to denote expected values of IIS under any model.

If we consider a population structured in any way (age, geography, etc.), one may always define Q_w, the IIS probability within a class of genes (for example among individuals of some age class, in the same subpopulation, etc.), and Q_b, the IIS probability between two different classes of individuals. In a generic way one may then define:

$$F \equiv \frac{Q_w - Q_b}{1 - Q_b}.$$

(28.1)

Such quantities are known as *F-statistics*, but Q_w, Q_b, and F as defined above are parameters. That is, Q_w and Q_b are expectations under independent replicates of the stochastic process considered, and F is a function of these parameters. In other words, they are functions of the parameters that define the model under study, such as subpopulation sizes, mutation rates, migration rates, etc. If (say) deme size is by itself random, then F and the Qs, being expectations in the process considered, are function of the parameters of the distribution of deme size. In models of spatial genetic structure, Q_w and Q_b are generally not 'the value in the (biological) population'. Alternative definitions of F-statistics, as values in biological populations, have been used in the literature (e.g. Nei, 1986; see Nagylaki, 1998 for further discussion), but analytical results below hold only with the present parametric definitions.

Let Q_2 be the IIS probability within subpopulations, and Q_3 be the IIS probability between subpopulations. The well-known F_{ST} parameter, originally considered by Wright, is best defined as

$$F_{ST} \equiv \frac{Q_2 - Q_3}{1 - Q_3}.$$

(28.2)

F-statistics may be described as correlations of genes within classes with respect to genes between classes, that is as intraclass correlations (Cockerham and Weir, 1987).

28.3.1.2 *Generic Methods for Estimation and Testing*

The estimation of F-statistics is described at length in the literature (see e.g. **Chapter 29**, Weir, 1996) so I will confine myself to emphasizing a few easily missed points.

A simple way to estimate parameters such as F_{ST} is to estimate each of the probabilities of identity by the corresponding frequencies $\hat{Q}$ of identical pairs of genes in the sample, computed by simple counting. Thus F_{ST} may be estimated by

$$\hat{F} \equiv \frac{\hat{Q}_2 - \hat{Q}_3}{1 - \hat{Q}_3},$$

(28.3)

where $\hat{Q}_2$ and $\hat{Q}_3$ are by definition the frequencies of identical pair of genes in the sample, within and between deme, respectively. This simple approach to defining estimators may be easily adapted to a number of different settings, given that the parameters to be estimated may be expressed as functions of probabilities of IIS. With balanced samples, this approach directly yields Cockerham's estimator of F_{ST} (Cockerham, 1973; Weir and Cockerham, 1984). This estimator has been developed by analogy with the methods of analysis of variance, and this analogy has proved difficult to understand. Appendix A details the nature of the analogy and its relationship with 28.3.

It is easy to test for differentiation (nonzero F_{ST}) by the usual exact tests for contingency tables either applied to gametic or genotypic data. These are standard statistical techniques and their application to genetic data has been discussed elsewhere (e.g. Weir, 1996; Goudet et al., 1996; Rousset and Raymond, 1997). A general set of techniques to draw confidence intervals from the moment estimators are the bootstrap (e.g. Efron and Tibshirani, 1993) and related techniques based on the resampling of loci. However, the simplest applications of resampling techniques may be misleading. This may be apparent when they lead to symmetric confidence intervals while the variance of the estimator is expected to be sensitive to the parameter value, in which case more involved uses of the bootstrap (DiCiccio and Efron, 1996) may be required.

28.3.1.3   Why F-statistics?

Wright was the first to note that such measures of genetic structure appear in some theoretical models of adaptation, and his ideas remain among the most influential in population genetics. He used them to quantify his 'shifting balance' model Wright (1931a; 1931b), which remains controversial today (Coyne et al., 1997). Nevertheless, F-statistics are useful descriptors of selection in one-locus models (Rousset, 2004). Wright also used them to estimate demographic parameters (Dobzhansky and Wright, 1941) and they have become a standard tool to 'estimate gene flow' or for merely descriptive studies of genetic population structure. Such studies are not always very convincing and may be questioned on statistical grounds. Two major objections are (1) the connection between such measures and the likelihood-based framework of statistics (e.g. Cox and Hinkley, 1974; Lehmann and Casella, 1998) is not obvious; and (2) although F_{ST} bears a simple relationship with the 'number of migrants' Nm in the infinite island model, it is not always clear how this would extend to more general models of population structure. Also, with the definition given above in terms of IIS, F_{ST} might be expected to depend on mutation processes at the loci considered, and how this affects estimation of dispersal parameters is not clear.

One of the main attractions of F-statistics may be their robustness to several factors. In an infinite island model, the ancestral lineages of two genes sampled within the same deme coalesce within this deme in a recent past with probability $\approx F_{ST}$; with probability $\approx 1 - F_{ST}$ the lineages separate (looking backward) in different demes as a result of immigration and will take a long time to coalesce. This implies that F_{ST} will depend mostly on recent events. Before considering the implications in detail, we will see how this argument can be generalized under isolation by distance.

We consider the probability $c_{j,t}$ that two genes coalesce at time t in the past. The j index corresponds to the type of pair of genes considered (e.g. $j = 2$ or 'w' for genes within demes and $j = 3$ or 'b' between demes as above). Identity by descent, here denoted $\dot{Q}$, has been defined as the probability that there has not been any mutation since the common

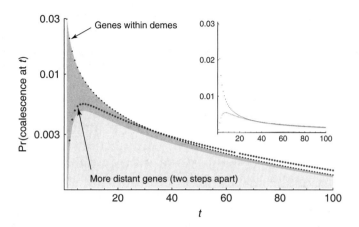

Figure 28.2   This figure compares the distributions of coalescence times in demes two steps apart (thick gray points) and within the same deme (thin black points). The inset shows the distributions on a linear y scale. The distribution for genes within demes is decomposed in two areas, the light gray one whose height is a constant times the height of the other distribution (hence it is shifted on the log scale), and the remainder (dark gray) which is the excess probability of coalescence in recent generations. The distribution were computed for 100 demes of $N = 10$ haploid individuals, with dispersal rate $m = 1/4$.

ancestor. Thus

$$\dot{Q}_j = \sum_{t=1}^{\infty} c_{j,t}(1-u)^{2t}. \tag{28.4}$$

(Malécot, 1975, (28.6); Slatkin, 1991). To understand the properties of F-statistics, we compare the distributions of coalescence times $c_{w,t}$ and $c_{b,t}$ of the pairs of genes that define these parameters. We can view the area covered by the probability distribution of coalescence time of the more related pair of genes as the sum of two 'probability areas', one part which is a smaller copy of the area covered by the probability distribution function of coalescence of less related genes, the other part being the remainder of the area for more related genes (Figure 28.2). This second part decreases faster than probabilities of coalescence (it is approximately $O(c_{.,t}/t)$, Rousset, 2006), and is therefore concentrated on the recent past. As a first approximation, the value of the corresponding F-statistics is this excess probability of recent coalescence. Let us call ω the value of $1 - c_{w,t}/c_{b,t}$ for large t. This will also be the excess probability of recent coalescence (as can be deduced from the fact that both distributions must sum up to 1). Then it can be shown that

$$Q_{\text{w}:k} \approx (1-\omega)Q_{\text{b}:k} + \omega\pi_k \Rightarrow \frac{Q_{\text{w}:k} - Q_{\text{b}:k}}{\pi_k - Q_{\text{b}:k}} \approx \omega \approx F, \tag{28.5}$$

where π_k is the expected frequency of allele k in the model considered. One may obtain this result by considering that with probability $1 - \omega$, the probability of identity of pairs of genes 'within' is the same as the probability of identity of genes 'between' (this corresponds to the proportional parts of the distributions of coalescence times), and with probability ω (the excess recent probability mass) the coalescence event has occurred recently in a common ancestor, of allelic type k with probability π_k.

The above result should not be overinterpreted. Expressions of the form "probability of identity equals $(1 - F)p^2 + Fp$" for F independent of p are not generally valid unless p is the expectation π_k and the probability of identity is the process expectation (Rousset, 2002), although they also correctly describe the conditional probability of identity given p in the infinite island model, even in cases where p is a random variable.

The above logic will be valid as long as mutations can be neglected within the time span covered by the probability mass. This time span is shorter for higher migration rates m in the island model, or for high σ^2 relative to spatial distance in isolation by distance models, so in practice F-statistics are weakly dependent on mutation rates at small spatial scales. The same argument can also be used to show that F-statistics more quickly recover their stationary values than probabilities of identity under the same conditions after a single demographic perturbation, a fact noticed by several authors (e.g. Crow and Aoki, 1984; Slatkin, 1993; Pannell and Charlesworth, 1999). This kind of approximate independence is important for statistical applications since it makes F-statistics analyses at a small spatial scale interpretable despite the fact that past demographic history and mutation processes are generally not known. At a local scale, F_{ST} is also only weakly dependent on the total population size.

28.3.2 Likelihood Computations

With few exceptions, likelihood computation in population genetics are based on 'coalescent' arguments, i.e. they derive the probability of the sample from consideration of the sequence of states that relate the individuals in the sample to their common ancestor (e.g. Kingman, 1982; Hudson, 1990, see **Chapters 25**, and **26**). This sequence of events may be the genealogy, G, of the sample that includes information about the coalescence time of ancestral lineages and about which lineages coalesce. In other cases it may be a 'gene tree' H, which takes into account the relative timing of coalescence and mutation events, as well as the nature of mutation events, i.e. the states before and after mutation, but which does not take into account the time between events, nor which lineage, among several with identical state, was involved in each event (see **Chapter 26**, Griffiths and Tavaré, 1995).

Coalescent arguments are used to estimate the likelihood by simulation using importance sampling algorithms. In one class of algorithms (see **Chapter 26**, Beerli and Felsenstein, 1999), the likelihood of the parameters $\mathcal{P}$ as a function of the data D may be written as

$$L(\mathcal{P}; D) = \sum_G \Pr(D|G; \mathcal{P}) \Pr(G; \mathcal{P}), \tag{28.6}$$

where the sum is over all possible genealogies G,

$$= \sum_G \Pr(D|G; \mathcal{P}) \frac{\Pr(G; \mathcal{P})}{f(G)} f(G), \tag{28.7}$$

for any distribution $f(G)$ such that $f(G) > 0$ when $\Pr(G; \mathcal{P}) > 0$

$$= \mathcal{E}\left[\Pr(D|G; \mathcal{P}) \frac{\Pr(G; \mathcal{P})}{f(G)} \right], \tag{28.8}$$

where $\mathcal{E}$ is an expectation over sample paths of a Markov chain with stationary distribution $f(G)$

$$\approx \frac{1}{s} \sum_{i=1}^{s} \Pr(D|G(i); \mathcal{P}) \frac{\Pr(G(i); \mathcal{P})}{f(G(i))}, \qquad (28.9)$$

where the sum is over the sample path of a Markov chain with stationary distribution $f(G)$.

In neutral models, given the genealogy G, the data only depend on the mutation process with parameters $\mathcal{N}$, while the genealogy itself does not depend on the mutation process but only on demographic parameters $\mathcal{D}$ (with $\mathcal{P} = (\mathcal{N}, \mathcal{D})$). Thus we may choose the importance sampling function

$$g \equiv \Pr(G|D; \mathcal{P}_0) = \frac{\Pr(G; \mathcal{D}_0)\Pr(D|G; \mathcal{N})}{L(\mathcal{P}_0; D)}, \qquad (28.10)$$

for some value $\mathcal{D}_0$ of $\mathcal{D}$ and for $\mathcal{P}_0 = (\mathcal{N}, \mathcal{D}_0)$. Then from 28.8

$$L(\mathcal{P}; D) = \mathcal{E}\left[L(\mathcal{P}_0; D) \frac{\Pr(G; \mathcal{D})\Pr(D|G; \mathcal{N})}{\Pr(G; \mathcal{D}_0)\Pr(D|G; \mathcal{N})} \right] = L(\mathcal{P}_0; D)\mathcal{E}\left[\frac{\Pr(G; \mathcal{D})}{\Pr(G; \mathcal{D}_0)} \right], \quad (28.11)$$

for any $\mathcal{N}$. We try to find the maximum likelihood estimate (MLE) of $\mathcal{P}$ or equivalently the maximum value of

$$\frac{L(\mathcal{P}; D)}{L(\mathcal{P}_0; D)} = \mathcal{E}\left[\frac{\Pr(G; \mathcal{D})}{\Pr(G; \mathcal{D}_0)} \right], \qquad (28.12)$$

which may be estimated by

$$\frac{L(\mathcal{P}; D)}{L(\mathcal{P}_0; D)} \approx \frac{1}{s} \sum_{i=1}^{s} \frac{\Pr(G(i); \mathcal{D})}{\Pr(G(i); \mathcal{D}_0)}, \qquad (28.13)$$

where the $G(i)$s are generated by a Markov chain with stationary distribution g. Thus an algorithm to find the maximum must define such a Markov chain (for parameters $\mathcal{P}_0$), and compute $\Pr(G(i); \mathcal{D})/\Pr(G(i); \mathcal{D}_0)$ for different $\mathcal{P}$ values and for this single Markov chain.

Beerli and Felsenstein (1999) have used the importance sampling function (28.10) to estimate the ratio (28.12). They define a Markov chain on genealogies G, and use the Metropolis-Hastings algorithm (Hastings, 1970) to ensure that the importance sampling function g is the stationary distribution of this chain. Their (28.14) shows that the transition probabilities of this chain are determined by the probabilities $\Pr(D|G; \mathcal{N})$, which for sequence data may be computed as previously described (e.g. Swofford and Olsen, 1990).

Griffiths and Tavaré (1994) have proposed a different class of algorithms. They derive recursions for the stationary probability $\Pr(D'|S')$ of a sample D' given sample size S' (here a vector and subsample sizes in different populations) over a time interval (typically the interval between two genealogical or mutation events). For any state D of ancestors in the previous time, $\Pr(D'|S')$ is the probability that a sample of size S' derives from a sample of size S, times the stationary probability of ancestral states $\Pr(D|S)$, times the forward probability that a sample D' derives from a sample D given the sample sizes:

$$\Pr(D'|S') = \sum_D \Pr(D'|D, S, S') \Pr(S|S') \Pr(D|S). \tag{28.14}$$

It is relatively straightforward to express the backward transition probabilities $\Pr(S|S')$ and the forward transition probabilities $\Pr(D'|D, S, S')$ in terms of model parameters $\mathcal{P}$ (further details are lengthy; see de Iorio and Griffiths, 2004b). An importance sampling algorithm is then derived by writing $\Pr(D'|D, S, S') \Pr(S|S')$ in the form $w(D', D) p(D|D')$ where the $p(D|D')$ define an absorbing Markov chain going backward over possible ancestral histories,

$$\Pr(D') = \sum_D w(D', D) p(D|D') \Pr(D). \tag{28.15}$$

Iterating this recursion until the ancestor of the whole sample shows that

$$\Pr(D) = \mathcal{E}_p \left[\prod_{D_i} w(D_i, D_{i-1}) \right], \tag{28.16}$$

where the product is over successive states D_i of the ancestry of the sample. Then w is an importance sampling weight and p is a proposal distribution (compare 28.8).

Different choices of p, and of the implied w, are possible. Griffiths and Tavaré (1994) describe one such choice and show that the likelihood for different values of $\mathcal{P}$ may be obtained by running the Markov chain for only one value of $\mathcal{P}$. However, more recent works have aimed to optimize the choice of p for each parameter value analyzed so that the variance of $\prod_{D_i} w(D_i, D_{i-1})$ among runs of the Markov chain would be minimal. An optimal choice of the proposal distribution would be such that any realization of the Markov chain would give exactly the likelihood of the sample. This occurs when the proposal distribution has transition probabilities given by the reverse probabilities in the biological process considered, $\Pr(D|D') = \Pr(D'|D) \Pr(D)/\Pr(D')$ (see **Chapter 26**). In cases of interest these reverse probabilities cannot be computed from this formula since the aim is precisely to evaluate the probabilities $\Pr(D)$. Nevertheless, the variance of $\prod_{D_i} w(D_i, D_{i-1})$ should be low if good approximations for $\Pr(D)/\Pr(D')$ are used. de Iorio and Griffiths (2004a; 2004b) write such ratios as simple functions of the probabilities π that an additional gene sampled from a population is of a given allelic type, conditional on the result of previous sampling, and propose approximations $\hat{\pi}$ for them, from which approximations for $\Pr(D)/\Pr(D')$ and for the proposal distribution follow. Their approximation scheme method applies in principle for any stationary migration-matrix model and any Markov mutation model (for allele frequency data). The $\hat{\pi}$ are not given in closed form but as solutions of a system of $n_d \times K$ linear equations for a model of n_d populations and K allelic types. By assuming independence between the mutation and genealogical processes, this can be reduced to a system of n_d equations holding for each of the different eigenvalues of the mutation matrix, a technique used by de Iorio *et al.* (2005) to analyze a two-demes model with stepwise mutation.

Despite substantial improvement over previous proposals, the computation times of the latter algorithm would remain prohibitively long in many practical applications. A method known as product of approximate conditional likelihoods (PAC-likelihood,

Li and Stephens, 2003) has been proposed to derive heuristic approximations of the likelihood estimation. Here a sample of genotypes (g_k) is described as a sequential addition of genotypes, so that the likelihood of an ordered sample is the product of the probabilities π that an additional genotype is g_i given previously added genotypes were $g_1, \ldots, g_{i-1}$. Approximations are then considered for these conditional probabilities. This was originally applied to inference of recombination rate in a panmictic population. Using de Iorio and Griffiths's approximations $\hat{\pi}$ in one-locus models, it turns out to perform well under stepwise mutation, where the expectation of the PAC-likelihood statistic was indistinguishable from the likelihood, but computation was much faster (Cornuet and Beaumont, 2007). In a linear stepping stone model, one can find small differences between the expectation of the PAC-likelihood statistic and the likelihood, but estimation of model parameters based on PAC-likelihood is essentially equivalent to maximum likelihood (ML) estimation while again requiring far less computation than likelihood computation via the importance sampling algorithm (F.R., unpublished data).

28.4 INFERENCE UNDER THE DIFFERENT MODELS

In this section I review the implementation and application of the different methodologies in specific cases. Published genetic and demographic data from Gainj- and Kalam-speaking people of New Guinea (Wood *et al.*, 1985; Long *et al.*, 1986) will conveniently illustrate several conclusions.

28.4.1 Migration-Matrix Models

For any migration matrix at stochastic equilibrium, the distribution of frequencies p_{ki} of allele k in each deme i follows some probability distribution with (parametric) covariances which can be written as

$$\mathrm{E}[(p_{ki} - \pi_k)(p_{ki'} - \pi_k)] = Q_{ii':k} - \pi_k^2, \qquad (28.17)$$

where $Q_{ii':k}$ is the expected frequency of pairs of genes in demes i, i' that are of allelic type k. For mutation models assuming identical mutation rates between K alleles, this is also $(Q_{ii'} - \sum_{k=1}^{K} \pi_k^2)/K$, where $Q_{ii'}$ is the probability of IIS of pairs of genes in demes i, i'. Probabilities of IIS $Q_{ii':k}$ or $Q_{ii'}$ may be derived from the probabilities of IBD for various mutation models (Markov chain models, or stepwise mutation models; see e.g. Tachida, 1985; Rousset, 2004). For any migration matrix model, probabilities of IBD within and among demes can be computed as solutions of a linear system of equations (see e.g. Nagylaki, 1982 or Rousset, 1999a; 2004 for details and examples). In principle, the demographic parameters can be estimated by inverting such relationships. This approach has been taken seriously only in a few cases, in particular the island and isolation-by-distance model, as detailed below, and does not generate likelihood expressions in a straightforward way, although some heuristic likelihood formulas have been proposed (Tufto *et al.*, 1996) by using Gaussian approximations for the distribution of allele frequencies, with the covariances given above.

28.4.2 Island Model

In the island model, one has the well-known approximation $F_{ST} \approx 1/(1 + 4Nm)$ (with $2N$ genes per deme, Wright, 1969). This has led to the usage of computing F_{ST}s and expressing the results in terms of 'estimates of Nm', i.e. in terms of $(1/\hat{F} - 1)/4$. This usage is often problematic. The worst sin is to estimate an F_{ST} between a pair of samples far apart, to translate it into a nonzero 'Nm', and to conclude that the populations must have exchanged migrants in the recent past. In the context of the island model, an F_{ST} between a pair of subpopulations is not a function of the number of migrants exchanged specifically between these two subpopulations. More generally, in many models of population structure, it is expected that subpopulations that never exchange migrants will have nonzero 'Nm' values. It may be seen from the above definition of F_{ST} that its maximum value is the probability of identity within demes Q_2 (when $Q_3 = 0$), which results in a minimum possible value of 'Nm' of $(1/Q_2 - 1)/4$ which may well be > 1 even for demes that never directly exchange migrants. Likewise, the practice of equating $Nm > 1$ to panmixia and $Nm < 1$ to divergence is not useful.

Likelihood functions for allele frequency data may be derived relatively easily by diffusion techniques for the infinite island model (see (28B.5)) and can in principle be recovered by coalescent arguments (Balding and Nichols, 1994). These sampling formulas allow analytical insight, and may be used to define estimators of Nm as well as to discuss efficient estimation of F_{ST} by moment methods (See Appendix B). Kitada *et al.* (2000) have implemented likelihood estimation under this model. Approximation have also been considered such as the 'pseudo maximum likelihood estimator' (PMLE, Rannala and Hartigan, 1996) of the number of migrants in an island model (see (28B.6)). These authors found that this estimator of Nm generally (though not always) had lower mean square error than the moment estimator $(1/\hat{F} - 1)/4$, depending on sampling scheme and Nm values. The MLE is also biased when the number of sampled populations is small and some corrections have been proposed (Kitakado *et al.*, 2006). For the New Guinea population, application of pseudo maximum likelihood (PML) estimation yields an estimate of 10.2 migrants per generation. This is one-fourth of the average value, 41.87, that can be computed from maternal and paternal dispersal rates and total subpopulations sizes (Tables 1–3 in Wood *et al.* (1985)), but it is closer to one third if we take 'effective size' considerations into account following Storz *et al.* (2001).

In comparison with the simulation methods for likelihood computation, it should be noted that no mutation model has to be considered here. Wright's formula (28B.1) is an approximation for low mutation, common to the different 'Markov chain' models of mutation, and so is the likelihood formula (28B.5). In this respect it is analogous to the methods based on F-statistics.

28.4.3 Isolation by Distance

Here analytical insight is available only for the moment methods. The results reviewed here are not tied to a Gaussian model of dispersal. We consider

$$a(\mathbf{r}) \equiv \frac{Q_0 - Q_{\mathbf{r}}}{1 - Q_0}, \tag{28.18}$$

which is $F_{ST}/(1 - F_{ST})$ at (vectorial) distance $\mathbf{r}$. Approximations for F_{ST} immediately follow from those for $a(\mathbf{r})$. I will use a dot on a or Q to emphasize that the results given hold strictly only for IBD, but the differences with IIS do not affect the main practical conclusions drawn below (Rousset, 1997).

Two cases are usually considered, the one-dimensional model for populations in a linear habitat, and the two-dimensional model. In one dimension, at distance r,

$$\dot{a}(r) \approx \frac{A_1}{4N\sigma} + \frac{1 - e^{\frac{-(2u)^{1/2}r}{\sigma}}}{4N\sigma(2u)^{1/2}} \overset{r\,\text{small}}{\approx} \frac{A_1}{4N\sigma} + \frac{r}{4N\sigma^2} \approx \frac{A_1}{4D\sigma} + \frac{r}{4D\sigma^2}, \quad (28.19)$$

where A_1 is a constant determined by the dispersal distribution, but not by N nor u. Its definition is given by Sawyer(1977, eq. 2.4).

In two dimensions, for genes at Euclidian distance r,

$$\dot{a}(\mathbf{r}) \approx \frac{-\ln((2u)^{1/2}) - K_0((2u)^{1/2}r/\sigma) + 2\pi A_2}{4N\pi\sigma^2} \overset{r\,\text{small}}{\approx} \frac{\ln(r/\sigma) - 0.116 + 2\pi A_2}{4N\pi\sigma^2}$$

$$\approx \frac{\ln(r/\sigma) - 0.116 + 2\pi A_2'}{4D\pi\sigma^2}, \quad (28.20)$$

where K_0 is the modified Bessel function of second kind and zero order (e.g. Abramovitz and Stegun, 1972), and A_2 is of the same nature as A_1 above. Its definition is given by Sawyer, (1977, eq. 3.4); see also Rousset (1997 eq. A11).

In the last two equations the first expression is given for σ measured in the length unit of the model (i.e. one interdeme distance on the lattice), the second is the small distance/low mutation limit of the first, and the third is the second for any length unit. They are in terms of population density D per length or surface unit, and the A_2' constant depends on the length unit.

In the same equations, the second approximation shows a linear relationship between $\dot{a}(r)$ and geographical distance in one dimension, and between $\dot{a}(r)$ and the logarithm of geographical distance in two dimensions. In both cases, the slope of this relationship is a function of $D\sigma^2$.

These different expressions emphasize two points. First, differentiation is a function of the A constants, which are not simple functions of σ^2 but also of other features of the dispersal distribution. In fact, when the total migration rate is low, the differentiation between adjacent subpopulations is close to that expected under an island model with the same total number of migrants. This confirms that σ^2 is not the only relevant parameter of the dispersal distribution. Second, the value of A_2' depends on the spatial unit chosen to measure σ and D. A method of inference from A_2' values that would not take into account the discrepancy between the length unit used and the idealized interdeme distance would therefore be internally incoherent.

The above approximations allow a relatively simple description of the expected differentiation in these models as well as relatively simple estimation of $D\sigma^2$ from genetic data. Estimates of $a(r)$ at different distances may be obtained in some cases as estimates of $F_{ST}/(1 - F_{ST})$ for pairs of samples, and simply regressed to spatial distance (Rousset, 1997, as implemented in GENEPOP, Raymond and Rousset, 1995). An estimate of $1/(D\sigma^2)$ may be deduced from the slope of the regression. Two early applications of this method yielded estimates about twice the demographic estimate (Rousset, 1997). For the New Guinea population, the regression equation $F_{ST}/(1 - F_{ST}) \doteq 0.0191 + 0.0047 \ln(\text{distance in km.})$ provides an estimate of $D\sigma^2$ which is about twice the demographic estimate (after application of effective density correction following Storz et al. (2001), and after correction of clerical errors affecting the reported σ^2 of females and males in Rousset (1997), which should be 3.1 and 0.76 km^2 respectively).

When the migration rate is low, an estimate of the number of immigrants per generation may also be computed by the '$(1/\hat{F} - 1)/4$' method, taking the value of the estimated regression equation at the distance between the closest subpopulations as an estimate of $\hat{F}/(1 - \hat{F})$. The estimate of number of migrants in the New Guinea population is then 11.5, close to the pseudo maximum likelihood estimate (10.2, see above). This result illustrates the approximate convergence of estimates by different methods and under different dispersal models to Wright's classic result, even though the dispersal rate in this study is not precisely low (its average value being 0.43 from demographic data).

The regression of $F_{ST}/(1 - F_{ST})$ to distance is not always applicable, particularly when there are no recognizable demes of several individuals, as for 'continuous' populations. A variant based on the comparison of pairs of individuals has been designed to address this problem (Rousset, 2000). Simulations have shown that this method performs reasonably well when σ is small (a few times interindividual distance at most) and when most individuals are sampled within an area of about $20\sigma \times 20\sigma$ (Leblois et al., 2003,2004). For higher dispersal, the variant considered by Vekemans and Hardy (2004) provides more accurate upper confidence bounds for $D\sigma^2$ (Watts et al., 2007). Several comparisons have found agreement within a factor of two with independently derived demographic estimates (Rousset, 2000; Sumner et al., 2001; Winters and Waser, 2003; Fenster et al., 2003; Broquet et al., 2006; Watts et al., 2007). Whether this is considered an important discrepancy or not will depend on the accuracy expected from such analyses, but this is certainly much better than usually reported (see e.g. Slatkin, 1994; Koenig et al., 1996). They actually go against an earlier long stream of reported discrepancies between genetic and demographic estimates, which needs explaining.

Part of the discrepancies hinge on misunderstandings of the models. For example, Wright assumed that the value of F-statistics under isolation by distance (Wright, 1946) was determined by the 'neighborhood size'. The value of this parameter would be $4D\pi\sigma^2$ under the assumption of two-dimensional Gaussian dispersal and its more general definition would be a function of 'the chance that two uniting gametes came from the same individual' (Wright, 1946). A third common 'definition' found in the literature is that the 'neighborhood size' would be the size of a subpopulation that would behave as a panmictic unit. It is not clear in which respect the subpopulation would behave as a panmictic unit nor whether there is a subpopulation that behaves as a panmictic unit in some useful sense. In any case Wright's measures do not correctly predict the value of unambiguously defined parameters in unambiguously defined models. In the analysis of Malécot's model, neither $D\sigma^2$ (because of the important A_2 term), nor the more generally defined neighborhood, determine the differentiation alone (Rousset, 1997). In one dimension, Wright proposed that $D\sigma$ was the important parameter, but the above results show that $D\sigma^2$ is important. One must give up the idea that $D\sigma^2$ equals neighborhood equals a number of individuals. In one dimension, $D\sigma^2$ scales as number of individuals times a length, not as a number of individuals, since density is a number of individuals per unit length.

The neighborhood concept was an attempt to account for different families of dispersal distributions. On the other hand, it has recurrently been assumed that differentiation is essentially a function of σ^2 and not of other features of the dispersal distribution. If so, it would be easy to seemingly improve on the regression method by considering only a family of dispersal distributions with a single parameter, completely determined by σ^2, for example a discretized Gaussian. In this case, F_{ST} or $a(r)$ values, not simply their increase with distance, would contain information about σ^2. But such improvements would not be robust to misspecification of the dispersal distribution.

To explain reported discrepancies, it has often been argued that genetic patterns are highly sensitive to long-distance dispersal, which occurrence is easily missed in demographic studies. While some genetic patterns are indeed affected by long-distance dispersal (e.g. Austerlitz et al., 2000), this is much less so for the patterns considered in the regression analyses, and this contributes to their concordance with demographic estimates. If a fraction m of immigrants come from an infinite distance (so that the 'true' σ^2 is infinite and does not predict any local pattern of differentiation), such migrants will be unrelated to their neighbors, and these migration events are analogous to mutation events. Hence we can deduce the effect of such immigrants from the effect of mutation, e.g.

$$\dot{a}(\mathbf{r}) \approx \frac{-\ln(\sqrt{2m}) - K_0(\sqrt{2m}r/\hat{\sigma})}{4D\pi\hat{\sigma}^2} + \text{constant}, \tag{28.21}$$

where $\hat{\sigma}$ is the parameter of the dispersal distribution for the fraction $1 - m$ of locally dispersing individuals. This result implies that there is an approximately linear increase of differentiation, determined by $D\hat{\sigma}^2$, roughly up to distance $0.56\hat{\sigma}/\sqrt{2m}$ (from Figure 3 in Rousset (1997)). For example if we ignore a 1 % (respectively, 0.1 %) tail of the distribution of dispersal distance in a demographic study which estimates $\hat{\sigma} = 10$ distance units, the prediction of increase of differentiation with distance will reach 20 % error at $0.56\hat{\sigma}/\sqrt{2m} = 39.6$ (respectively, 125) distance units. This is a wide overestimate of the error for any data set spread over such a distance, but more accurate predictors of bias will depend on the distribution of spatial distances in the sample.

Naive application of testing methodology has been another factor contributing to confusion. The absence of a pattern of isolation by distance (null slope of the regression, $D\sigma^2$ infinite) may be tested by the exact permutation procedure known as the *Mantel test* (Mantel, 1967; see Rousset and Raymond, 1997, for a simple description). In practice, nonsignificant test results have often been interpreted as evidence that dispersal is not localized. However, the Mantel test has often been applied in conditions of low power. In many populations with localized dispersal, the value of $D\sigma^2$ will be large, and thus expected patterns of isolation by distance (increase of differentiation with distance) will be weak (particularly in two-dimensional habitats), even though differentiation will be inferred by classical tests for differentiation.

Finally, variation in expected gene diversity due to spatial heterogeneity of demographic parameters may result in larger variation in expected differentiation than that due to isolation by distance. For example, if several demes with very small deme size and restricted dispersal are clustered in space, they will have low expected gene diversity and will show a larger differentiation between them than with more distant demes with higher expected diversity, and the above methods will obviously fail unless this heterogeneity is taken into account (Rousset, 1999b).

28.4.4 Likelihood Inferences

Maximum likelihood methods have not been developed and tested to a comparable level. The migration matrix models have been implemented for allele frequency and sequence data in the software MIGRATE. A more restricted set of models based on the 'isolation-with-migration' scenario has been implemented in the software IM (Nielsen and Wakeley,

2001; Hey, 2005). In the coalescent algorithms as in previous inferences based on per-locus information from samples taken at one time, it is not possible to estimate the deme sizes, mutation rates, and immigration rates separately: only products of deme size with migration probability and mutation probability, and (as a consequence) the ratio of migration and mutation probabilities, can be estimated. In the latest version of IM it is possible to analyze the divergence between two populations in terms of their deme sizes scaled relative to mutation rate ($N_1\mu$ and $N_2\mu$), of immigration rates in each of them m_1 and m_2, either scaled relative to deme size (e.g. $N_1 m_1$) or relative to mutation rate (e.g. m_1/μ), of the scaled size of the ancestral population, and of relative fractions of this ancestral which contributed to the two descendant populations.

In a two-populations setting, one study has compared ML estimates with demographic estimates and with some other genetic estimation methods (Wilson *et al.*, 2004). Both ML and the two-locus method of Vitalis and Couvet (2001) produced estimates roughly in agreement with demographic data, while F_{ST} did not. It appears difficult to estimate all parameters of a four-demes migration-matrix model (Beerli, 2006). Likewise, it has not been possible to choose among different scenarios of colonization of the Americas using the IM software (Hey, 2005). The effect of unsampled demes on estimation of dispersal between two sampled demes has been investigated by Beerli (2004), for sequence data (100 000 bp per individual). As expected, the biases increase with immigration from the unsampled population(s), but it was found that estimates of immigration rate between the two populations were hardly affected by an equal total immigration rate from unsampled population(s) (estimates of scaled populations sizes were more affected). Abdo *et al.* (2004) argued that confidence intervals given by MIGRATE are not accurate, a fact attributed to too short run times of the algorithm (Beerli, 2006).

The above simulation scenarios are rather distinct from those considered in the previous section, and it would be hard to perform ML analyses of the New Guinea data using current software. With MIGRATE for example, estimation of a full migration matrix from the New Guinea data has been attempted (R. Leblois and F.R., unpublished results), yielding larger estimates of dispersal than inferred by the regression method described above and from demographic data. Attempts to estimate fewer parameters could be more successful. In addition, one can question the convergence of the Markov chain algorithm, a problem which remains with no easy solution (e.g. Brooks and Gelman, 1998). In this respect, the importance sampling algorithms of Griffiths and collaborators are more convenient as estimates are derived from independent runs of an absorbing Markov chain (28.15), so traditional techniques based on independent variables apply.

28.5   SEPARATION OF TIMESCALES

The properties of F-statistics illustrate the more general idea of separation of timescales, in which some events occur at a much faster rate than others. For example, in an island model, when the number of demes n_d increases indefinitely, the rate of coalescence of ancestral lineages of genes sampled in different demes decreases as $1/n_d$, while for genes sampled within the same deme, the probability of coalescence of ancestral lineages in some recent generation is nonvanishing in this limit. Thus the events in the genealogy of a sample can be described as a sum of two processes, a fast process by which lineages either coalesce within the demes they are sampled or separate in distinct demes, at rate $O(1)$ as

$n_d \rightarrow \infty$, and a slow process by which genes in different demes coalesce at rate $O(1/n_d)$ as $n_d \rightarrow \infty$ (e.g. Wakeley and Aliacar, 2001). This slow process is a rescaled version of the well-studied coalescence process for an unstructured population (the n-coalescent, Kingman, 1982; see **Chapter 25**).

Under a separation of timescales, the likelihood can be expressed in terms of distinct terms for fast and slow processes (see Nordborg, 1997 for a population with selfing), and in the above example, analytical results or simulation techniques for the n-coalescent can be used as soon as ancestral lineages have separated in different demes in a simulation of the ancestry of a sample. The traditional formula giving probability of identity conditional on allele frequency as $(1 - F)p^2 + Fp$ also results from such a separation of timescales, provided that allele frequency does not change in the total population until the completion of the fast process.

However, convergence to the n-coalescent may hold under sometimes restrictive conditions. No convergence to the n-coalescent has been found in one-dimensional models of isolation by distance (Cox, 1989; Wilkins, 2004). In two-dimensional models on a lattice of size $L \times L$, the genealogy of genes sampled far enough (at distance $O(L)$) from each other converges to that of an unstructured population (the n-coalescent) with coalescence events occurring at rate $O(1/[L^2 \ln(L)])$ as $L \rightarrow \infty$ (Cox, 1989; Zähle *et al.*, 2005). A separation of timescales may hold if the genealogy of genes sampled closer in space compounds a process of 'fast' coalescence (in less than $O[L^2 \ln(L)]$ generations) and the n-coalescent. The closest results are those of Zähle *et al.* according to which genes at distance $O(L^\beta)$ for $0 < \beta \leq 1$ either coalesce in less than $L^2/2$ generations with probability $1 - \beta$, or follow the scaled n-coalescent with probability β. This differs qualitatively from the island model where the fast process becomes negligible in finite time, so defining a finite time span for a fast process from these results seems less than straightforward. Instead, they suggest applying n-coalescent approximations in a backward simulation algorithm when ancestral lineages are distant enough relative to the total size of the lattice, although the minimal distance to consider is itself not obvious since all results stand for $\beta > 0$ only, i.e. for spatial separation of genes increasing with L. As shown in Figure 28.2, a separation of timescales holds for fixed distance and fixed $N\sigma^2$, with the fast process vanishing in finite time, but the slower process is not an n-coalescent.

Some moment methods have attempted to extract more information from the data by taking into account allele size (for microsatellites) or DNA sequence divergence. However, in an island model, most of the information about Nm is in whether genes sampled within a deme have their most recent common ancestor within this deme (in which case they are identical, unless mutation rates are higher than migration rates). Otherwise, the ancestral lineages separate in different demes, and in this case the allelic divergence may contain little useful information about dispersal rates. This may explain why moment methods based on microsatellite allele size often yield estimates of demographic parameters with higher variance and mean square error than estimates derived from allelic identity statistics (Gaggiotti *et al.*, 1999; Balloux and Goudet, 2002; Leblois *et al.*, 2003) despite their lower asymptotic bias (Slatkin, 1995), except when mutation rates are larger than migration rates (Balloux *et al.*, 2000) or when genetic correlations are not determined by a distinctly fast process (Tsitrone *et al.*, 2001). A key issue in evaluating the performance of likelihood methods will be the extent to which they specifically capture the information from fast processes and how much better they are than pair-wise identity methods in this respect.

28.6   OTHER METHODS

As likelihood computations are often difficult, simulation methods based on other summary statistics have been developed. In essence, the likelihood of the data is substituted with the probability of observing in simulations values of a summary statistic S sufficiently close to its value observed in the data. These methods are reviewed by Beaumont (see **Chapter 30**) and have been applied to some subdivided population scenarios (Estoup *et al.*, 2004; Hamilton *et al.*, 2005). There is a huge collection of other methods of analysis of spatial patterns of genetic variation in the literature. Here again a small selection is presented on the basis of their impact in the field or of their perceived practical validity.

28.6.1   Assignment and Clustering

Of particular interest are assignment methods, which aim to assign individuals to their subpopulations of origin. For example, in an ecological perspective it may be useful to know whether immigrants differ from residents in some aspects of their behavior, and therefore to individually identify immigrants; further it might be of interest to identify their habitat of origin. Sometimes there will be independent information about the potential source populations. A more challenging problem (on which this section will focus) is to estimate dispersal rates by inferring the number of subpopulations and then assigning individuals to them (see **Chapter 30** for other aspects of these methods). In this perspective, the traditional problem of clustering becomes part of the assignment task.

Assignment methods stem from the idea that individuals are more likely to originate from populations with higher frequencies of the alleles they possess. Thus, considering an haploid organism for simplicity, for each individual one can compute a statistic such as

$$\prod_{k=1}^{K} \tilde{q}_{ki}^{x_k}, \tag{28.22}$$

where $\tilde{q}_{ki}$ is the observed frequency for allele k in sample i (possibly with some correction such as excluding the focal individual itself), and $x_k = 1$ if the individual bears allele k and $x_k = 0$ otherwise. This is usually viewed as a likelihood statistic (e.g. Paetkau *et al.*, 1995). If each locus is considered independent of the others, multilocus statistics are the product of single locus statistics, and the individual is assigned to the sample i that maximizes the multilocus statistic.

It can be expected that, given some differentiation between different subpopulations, this method will preferentially assign an individual to its original subpopulation. It is also expected that such assignments will be more accurate when differentiation is higher. More generally, if the individuals are correctly assigned to their subpopulations of origin, then we could estimate from the same data and the same likelihood formulas the dispersal rates between each of them in the last round of dispersal, independently of past dispersal rates. Conversely if we cannot consistently estimate the dispersal rates in this way, this implies that there is no way to consistently assign individuals to their populations of origin. So we consider the question whether we can estimate the dispersal rate in order to address the question whether immigrants can be assigned to their population of origin.

Consider the allele frequencies in the subpopulations before the last round of dispersal (p_{ki} for allele k in subpopulation i), and the dispersal rate for this last round of dispersal

($m_{ii'}$ from subpopulation i' to i). For each locus and each individual in sample i the likelihood is approximately $\prod_{k=1}^{K} q_{ki}^{x_{ki}}$ where $q_{ki} = \sum_{i'} m_{ii'} p_{ki'}$ is the frequency of allele k in subpopulation i after the last round of dispersal. For the total sample from n_d demes the likelihood is therefore proportional to

$$\prod_{i=1}^{n_d} \prod_{k=1}^{K} q_{ki}^{n_{ki}}. \tag{28.23}$$

Since one can always explain the data by some model assuming that there was no dispersal in the last round, there is not enough information in the data to separately estimate the dispersal rates and the p_{ki}s, at least when each locus is considered independent of the others as done previously. This implies that it is not possible to consistently assign individuals to their populations of origin without introducing additional assumptions or additional knowledge.

The most obvious assumption is to assume no linkage disequilibrium within populations before dispersal, which is a good approximation for many organisms (though not highly selfing ones). In principle, one- and two-locus measures of genetic association contain information which allows to estimate separately dispersal rates and subpopulation sizes. A numerical method has been developed to use such information (Vitalis and Couvet, 2001). Assignment methods may be viewed as using multilocus associations, in that their current formulation relies on assuming no within-population disequilibria. Additional assumptions have been made, for example specifying a prior probability model for the distribution of allele frequencies, often a Dirichlet distribution as per an island model (see (28B.1); Rannala and Mountain, 1997; Falush *et al.*, 2003; Wilson and Rannala, 2003; Corander *et al.*, 2003). In this respect, although an asserted aim of such methods is to infer migration rates without the many assumptions of other methods, the stationary island model is still lurking in the background.

How do these methods perform in practice? Cornuet *et al.* (1999) reported $>75\%$ correct assignment probabilities by Rannala and Mountain's method for populations diverged since several hundred generations. Évanno *et al.* (2005) found that the 'most likely' number of populations reported by the program STRUCTURE was a biased estimator of the actual number of populations. Waples and Gaggiotti (2006) reported that this program correctly identified the number of populations when the number of immigrants per population was less than 5, mutation rates were high, and for large sample size (20 loci genotype in 50 individuals from each population), but quickly degraded in other conditions. The latter study also reported similar problems with the methods implemented in the programs BAPS (Corander *et al.*, 2003) and IMMANC (Rannala and Mountain, 1997), and found that a method based on traditional contingency tests of spatial structure performed better than these different methods in identifying the number of populations. In a comparison with mark-recapture data, Berry *et al.* (2004) found good performance in estimating dispersal. In this study, the populations were known in advance and at most 4, the number of immigrants was known from mark-recapture data to be small, and the latter information was somehow used as prior information in STRUCTURE.

Most of the recently formulated methods have been presented as 'Bayesian', a label which in practice covers various compromises between subjective Bayesian (e.g. Lindley, 1990) and frequentist (Neyman, 1977) statistics, to the point of being uninformative. However, numerous accounts of supposed differences between Bayesian and frequentist methods have drawn attention away from the real issue, which is the criterion by which to measure the performance of statistical methods. It is always possible to generate 'better than previous' methods by changing the measure of average performance. In the context of estimation of dispersal rates, the results of Beerli (2006) illustrate this, where performance averaged over a prior distribution for model parameters was better than that of maximum likelihood ignoring the prior distribution, a predictable result from textbook statistical theory (Cox and Hinkley, 1974, Chapter 11; Lehmann and Casella, 1998, Chapter 4). Likewise, the performance of an assignment method defined in terms of priors over allele frequencies may be evaluated in terms of its performance for any given allele frequency, or of averaged performance over the distribution of allele frequencies. The context of scientific inference should determine the appropriate measure of performance, but the averaged measure can balance misleading assignment inferences in some species with more efficient inferences in some other species, depending on the spectrum of allele frequencies in different species. One may need to know when this is the case, and comparing results for two choices of the prior distribution is not enough in this respect.

As seen above, a more prosaic problem with assignment methods is that it is difficult from available information to give simple bounds on the frequency of erroneous inferences, even averaged over prior distributions.

28.6.2 Inferences from Clines

An important class of models of spatial structure for selected genes are the cline models. Clines arise in two contexts: selection for two distinct alleles or genotypes in two adjacent habitats exchanging migrants, or selection against hybrid genotypes between two taxa ('tension zones'). Theoretical models predict the shape of clines, notably the steepness in the cline center, as a function of the σ parameter defined above and of one or several selection coefficients (for tension zones: Bazykin, 1969; Barton, 1979; for selection variable in space: Nagylaki, 1975). Additional information on dispersal and selection is given by linkage disequilibria between loci. Such methods, therefore, depend on assumptions specific to each case study (e.g. external information on recombination rates and epistasis between loci).

A typical expression for the shape of a cline, i.e. for allele frequency p at distance $x - x_0$ from the center of the cline, is $p = \left(1 + \exp((x_0 - x)(2s)^{1/2}/\sigma)\right)^{-1}$ (Bazykin, 1969; Barton, 1979). In the center of the cline, it holds approximately for different models of selection (Barton and Gale, 1993) and is relatively insensitive to the shape of the dispersal distribution, as a small rate of unaccounted long-distance immigration has little effect on the shape of the center of the cline (Rousset, 2001). Inferences about s and σ separately are possible by considering multilocus clines and by taking linkage disequilibria into account. Then, the expected shapes of clines must be computed numerically. Drift is neglected relative to selection: Allele frequencies in the different subpopulations are fixed values, functions of the parameters defining the demography and the selection regime, not random variables as in the neutral models.

The likelihood function is then given by multinomial sampling in each subpopulation. Thus, the statistical model is conceptually straightforward and relatively easily tailored to the specific demography and selection regime of different organisms, at least when only a few loci subject to selection need be considered and when the expected frequencies of each genotype are directly computed using recursion equations for specific values of the parameters to be estimated (e.g. Lenormand *et al.*, 1999). Variants of the Metropolis algorithm (Metropolis *et al.*, 1953) including simulated annealing (Kirkpatrick *et al.*, 1983) have been used to find MLEs (a software ANALYSE is distributed by Barton and S. J. Baird, `http://helios.bto.ed.ac.uk/evolgen/Mac/Analyse/`). Sites *et al.* (1995) reported an agreement with demographic estimates of σ similar to that of isolation-by-distance analyses reported above.

28.7   INTEGRATING STATISTICAL TECHNIQUES INTO THE ANALYSIS OF BIOLOGICAL PROCESSES

Several of the methods reviewed in this chapter have been both widely used and widely criticized. Much of the criticisms rest on the difficulty of formulating precise quantitative models of population genetic processes. Theoretical studies of robustness are important, but may themselves overlook factors that turn out to be important in natural populations. For these reasons, this review has emphasized comparisons with demographic estimates. Independent demographic estimates may have their own problems, but it will be hard to detect misapplications of the genetic inferences if no such comparison is made. This last section discusses some of these problems and partial solutions to them.

We have seen that many discrepancies between 'models and data' inferred from empirical studies using F-statistics derive from various misunderstandings of the models. The more successful studies, in terms of comparisons with independent estimates, were conducted at a small spatial scale (between 1 and 20 σ). This is somehow unavoidable because of the need for good demographic data in these comparisons, but one may expect larger discrepancies over larger spatial scales, for many reasons: spatial variation of demographic parameters should be taken into account; the effects on genetic differentiation of some demographic events such as range expansions will be more likely to be observed (Slatkin, 1993); mutation will have measurable consequences (e.g. Estoup *et al.*, 1998); and selection variable in space may also affect the markers.

A frequently raised concern is the possible nonneutrality of the markers used. On the positive side a number of authors have realized that divergent selection would increase levels of differentiation between different subpopulations. Thus potentially selected loci may be detected in a first step by 'weak' statistical approaches such as classifying loci as showing structure or not by conventional significance tests (see Kreitman, 2000 for tests of selection at a molecular level not specifically using geographical information). Formal statistical evidence for selection may be obtained by other experiments in a second step. This approach has proven efficient (e.g. Feder *et al.*, 1997). Lewontin and Krakauer (1973) proposed a quantitative test of selection from the heterogeneity of F_{ST} estimates. This procedure was inadequate in several ways, but there have been more recent attempts to refine the detection of candidate selected loci (e.g. Beaumont and Nichols, 1996; Vitalis *et al.*, 2001; Beaumont and Balding, 2004).

Another often expressed criticism of the models and analyses reviewed above is that they assume equilibrium, while the populations are often not at demographic equilibrium, i.e. population sizes and migration rates fluctuate in time. If so, it is not clear what is estimated by such techniques: the present demography, an average over 'recent' times, a 'long-term' average, or none of them? If the fluctuations can be described as a fast process, they may be described by some effective size correction. In other cases, such as the range expansion of a species, this cannot be so, and either modeling the demographic expansion, or using methods insensitive to it, are the only coherent alternatives. To understand this, it suffices to note that spatial patterns of pairs of genes approach equilibrium faster the smaller the spatial scale considered. Therefore, if the effect of a demographic event was captured by some effective size correction, the effective size would differ at different scales. Hence this effect cannot be described by a single effective size characterizing the total population.

One way to avoid some assumptions of equilibrium is to analyze sequential samples. All the above methods assume that samples have been taken at one point in time, yet sometimes temporal information is available. See e.g. Robledo-Arnuncio *et al.* (2006) for inferences of dispersal distributions from mother–offspring data, Wang and Whitlock (2003) for estimation of immigration rate and deme size from samples over wider time spans, and Ewing and Rodrigo (2006) for implementation of Markov chain algorithms for inference of changes in demographic parameters from sequential samples.

The idea of estimating dispersal parameters such as σ is also open to difficulties. By allowing an arbitrarily small fraction of immigrants to come from far enough, it is easy to design cases where the theoretical σ value would be arbitrarily large, and where long-distance immigrants would have arbitrarily small effect on the likelihood of samples. The question that one must address prior to statistical analysis is how important such long-distance immigrants are for population processes. For example, the speed of range expansions is known to be affected by the most extreme long-distance migrants in a way generally not characterized by the σ parameter (Mollison, 1977; Clark *et al.*, 2001), so if one is interested in characterizing such processes, not only it will be difficult to estimate σ but this may be irrelevant. One the other hand, some processes of local adaptation (as may lead to allele frequency clines, for example) are not very sensitive to long-distance migrants, and then approximations ignoring them are not only adequate but required to formulate good statistical inferences.

Current methods of estimation still have low range of applications, low efficiency, or both. In principle, this can be improved by the development of likelihood methods, yet this leaves room for different methodologies, and it is unclear how far research practices will be improved. One common theme is that genealogical structure is affected by events occurring at different timescales, and that inferences based on models of the faster processes could be relatively independent to uncontrolled historical processes, and therefore perhaps more reliable. It is not yet clear how much complexity we can add in the models, for given data, nor where will be the limit between reliable and unreliable inference; further, the answer will likely differ whether sequence data or allele frequency data are considered.

Acknowledgments

I thank D. Balding, M. Beaumont, M. Lascoux, R. Leblois, R. Vitalis, and M. Nordborg for comments or discussion related to this or previous versions of this chapter.

RELATED CHAPTERS

Chapter 25; **Chapter 26**; **Chapter 29**; and **Chapter 30**.

REFERENCES

Abdo, Z., Crandall, K.A. and Joyce, P. (2004). Evaluating the performance of likelihood methods for detecting population structure and migration. *Molecular Ecology* **13**, 837–851.

Abramovitz, M. and Stegun, I.A. (eds) (1972). *Handbook of Mathematical Functions.* Dover Publications, New York.

Austerlitz, F., Mariette, S., Machon, N., Gouyon, P.-H. and Godelle, B. (2000). Effects of colonization processes on genetic diversity: differences between annual plants and tree species. *Genetics* **154**, 1309–1321.

Balding, D.J. (2003). Likelihood-based inference for genetic correlation coefficients. *Theoretical Population Biology* **63**, 221–230.

Balding, D.J. and Nichols, R.A. (1994). DNA profile match probability calculation: how to allow for population stratification, relatedness, database selection and single bands. *Forensic Science International* **64**, 125–140.

Balloux, F., Brünner, H., Lugon-Moulin, N., Hausser, J. and Goudet, J. (2000). Microsatellites can be misleading: an empirical and simulation study. *Evolution* **54**, 1414–1422.

Balloux, F. and Goudet, J. (2002). Statistical properties of population differentiation estimators under stepwise mutation in a finite island model. *Molecular Ecology* **11**, 771–783.

Barton, N.H. (1979). The dynamics of hybrid zones. *Heredity* **43**, 341–359.

Barton, N.H. and Gale, K.S. (1993). Genetic analysis of hybrid zones. In *Hybrid Zones and the Evolutionary Process*, R.G. Harrison, ed. Oxford University Press, Oxford, pp. 13–45.

Bazykin, A.D. (1969). Hypothetical mechanism of speciation. *Evolution* **23**, 685–687.

Beaumont, M.A. and Balding, D.J. (2004). Identifying adaptive genetic divergence among populations from genome scans. *Molecular Ecology* **13**, 969–980.

Beaumont, M.A. and Nichols, R.A. (1996). Evaluating loci for use in the genetic analysis of population structure. *Proceedings of the Royal Society of London, Series B* **263**, 1619–1626.

Beerli, P. (2004). Effect of unsampled populations on the estimation of population sizes and migration rates between sampled populations. *Molecular Ecology* **13**, 827–836.

Beerli, P. (2006). Comparison of Bayesian and maximum-likelihood inference of population genetic parameters. *Bioinformatics* **22**, 341–345.

Beerli, P. and Felsenstein, J. (1999). Maximum likelihood estimation of migration rates and effective population numbers in two populations using a coalescent approach. *Genetics* **152**, 763–773.

Berry, O., Tocher, M.D. and Sarre, S.D. (2004). Can assignment tests measure dispersal?. *Molecular Ecology* **13**, 551–561.

Brooks, S. and Gelman, A. (1998). General methods for monitoring convergence of iterative simulations. *Journal of Computational and Graphical Statistics* **7**, 434–455.

Broquet, T., Johnson, C.A., Petit, E., Burel, F. and Fryxell, J.M. (2006). Dispersal kurtosis and genetic structure in the American marten, *Martes americana*. *Molecular Ecology* **15**, 1689–1697.

Chakraborty, R. (1992). Multiple alleles and estimation of genetic parameters: computational equations showing involvement of all alleles. *Genetics* **130**, 231–234.

Chuang, C. and Cox, C. (1985). Pseudo maximum likelihood estimation for the Dirichlet-multinomial distribution. *Communications in Statistics Part A: Theory and Methods* **14**, 2293–2311.

Clark, J.S., Lewis, M. and Horváth, L. (2001). Invasion by extremes: population spread with variation in dispersal and reproduction. *American Naturalist* **157**, 537–554.

Cockerham, C.C. (1973). Analyses of gene frequencies. *Genetics* **74**, 679–700.

Cockerham, C.C. and Weir, B.S. (1987). Correlations, descent measures: drift with migration and mutation. *Proceedings of the National Academy of Sciences of the United States of America* **84**, 8512–8514.

Corander, J., Waldmann, P. and Sillanpää, M. (2003). Bayesian analysis of genetic differentiation between populations. *Genetics* **163**, 367–374.

Cornuet, J.M. and Beaumont, M.A. (2007). A note on the accuracy of PAC-likelihood inference with microsatellite data. *Theoretical Population Biology* **71**, 12–19.

Cornuet, J.-M., Piry, S., Luikart, G., Estoup, A. and Solignac, M. (1999). New methods employing multilocus genotypes to select or exclude populations as origins of individuals. *Genetics* **153**, 1989–2000.

Cox, J.T. (1989). Coalescing random walks and voter model consensus times on the torus in $\mathbb{Z}^d$. *Annals of Probability* **17**, 1333–1366.

Cox, D.R. and Hinkley, D.V. (1974). *Theoretical Statistics*. Chapman & Hall, London.

Coyne, J.A., Barton, N.H. and Turelli, M. (1997). A critique of Sewall Wright's shifting balance theory of evolution. *Evolution* **51**, 643–671.

Crow, J.F. and Aoki, K. (1984). Group selection for a polygenic behavioural trait: estimating the degree of population subdivision. *Proceedings of the National Academy of Sciences of the United States of America* **81**, 6073–6077.

DiCiccio, T.J. and Efron, B. (1996). Bootstrap confidence intervals (with discussion). *Statistical Science* **11**, 189–228.

Dobzhansky, T. and Wright, S. (1941). Genetics of natural populations. V. Relations between mutation rate and accumulation of lethals in populations of *Drosophila pseudoobscura*. *Genetics* **26**, 23–51.

Efron, B. and Tibshirani, R.J. (1993). *An Introduction to the Bootstrap*. Chapman & Hall, Boca Raton, FL.

Epperson, B.K. and Li, T.-Q. (1997). Gene dispersal and spatial genetic structure. *Evolution* **51**, 672–681.

Estoup, A., Beaumont, M., Sennedot, F., Moritz, C. and Cornuet, J.-M. (2004). Genetic analysis of complex demographic scenarios: spatially expanding populations of the cane toad, *Bufo marinus*. *Evolution* **58**, 2021–2036.

Estoup, A., Rousset, F., Michalakis, Y., Cornuet, J.-M., Adriamanga, M. and Guyomard, R. (1998). Comparative analysis of microsatellite and allozyme markers: a case study investigating microgeographic differentiation in brown trout (*Salmo trutta*). *Molecular Ecology* **7**, 339–353.

Évanno, G., Regnaut, S. and Goudet, J. (2005). Detecting the number of clusters of individuals using the software STRUCTURE: a simulation study. *Molecular Ecology* **14**, 2611–2620.

Ewing, G. and Rodrigo, A. (2006). Coalescent-based estimation of population parameters when the number of demes changes over time. *Molecular Biology and Evolution* **23**, 988–996.

Falush, D., Stephens, M. and Pritchard, J.K. (2003). Inference of population structure using multilocus genotype data: linked loci and correlated allele frequencies. *Genetics* **164**, 1567–1587.

Feder, J.L., Roethele, J.B., Wlazlo, B. and Berlocher, S.H. (1997). Selective maintenance of allozyme differences between sympatric host races of the apple maggot fly. *Proceedings of the National Academy of Sciences of the United States of America* **94**, 11417–11421.

Fenster, C.B., Vekemans, X. and Hardy, O.J. (2003). Quantifying gene flow from spatial genetic structure data in a metapopulation of *Chamaecrista fasciculata* (Leguminosae). *Evolution* **57**, 995–1007.

Gaggiotti, O.E., Lange, O., Rassmann, K. and Gliddon, C. (1999). A comparison of two indirect methods for estimating average levels of gene flow using microsatellite data. *Molecular Ecology* **8**, 1513–1520.

Goudet, J., Raymond, M., De Meeüs, T. and Rousset, F. (1996). Testing differentiation in diploid populations. *Genetics* **144**, 1931–1938.

Griffiths, R.C. and Tavaré, S. (1994). Simulating probability distributions in the coalescent. *Theoretical Population Biology* **46**, 131–159.

Griffiths, R.C. and Tavaré, S. (1995). Unrooted genealogical tree probabilities in the infinitely-many-sites model. *Mathematical Biosciences* **127**, 77–98.

Hamilton, G., Currat, M., Ray, N., Heckel, G., Beaumont, M. and Excoffier, L. (2005). Bayesian estimation of recent migration rates after a spatial expansion. *Genetics* **170**, 409–417.

Hardy, O.J. and Vekemans, X. (1999). Isolation by distance in a continuous population: reconciliation between spatial autocorrelation analysis and population genetics models. *Heredity* **83**, 145–154.

Hastings, W.K. (1970). Monte Carlo sampling methods using Markov chains and their applications. *Biometrika* **57**, 97–109.

Hey, J. (2005). On the number of New World founders: a population genetic portrait of the peopling of the Americas. *PLoS Biology* **3**, e193.

Hudson, R.R. (1990). Gene genealogies and the coalescent process. *Oxford Surveys in Evolutionary Biology* **7**, 1–44.

de Iorio, M. and Griffiths, R.C. (2004a). Importance sampling on coalescent histories. *Advances in Applied Probability* **36**, 417–433.

de Iorio, M. and Griffiths, R.C. (2004b). Importance sampling on coalescent histories. II. Subdivided population models. *Advances in Applied Probability* **36**, 434–454.

de Iorio, M., Griffiths, R.C., Leblois, R. and Rousset, F. (2005). Stepwise mutation likelihood computation by sequential importance sampling in subdivided population models. *Theoretical Population Biology* **68**, 41–53.

Kingman, J.F.C. (1982). On the genealogy of large populations. *Journal of Applied Probability* **19A**, 27–43.

Kirkpatrick, S., Gelatt, C. and Vecchi, M. (1983). Optimization by simulated annealing. *Science* **220**, 671–680.

Kitada, S., Hayashi, T. and Kishino, H. (2000). Empirical Bayes procedure for estimating genetic distance between populations and effective population size. *Genetics* **156**, 2063–2079.

Kitakado, T., Kitada, S., Kishino, H. and Skaug, H.J. (2006). An integrated-likelihood method for estimating genetic differentiation between populations. *Genetics* **173**, 2073–2082.

Koenig, W.D., Van Vuren, D. and Hooge, P.N. (1996). Detectability, philopatry, and the distribution of dispersal distances in vertebrates. *Trends in Ecology and Evolution* **11**, 514–517.

Kreitman, M. (2000). Methods to detect selection in populations with applications to the human. *Annual Review of Genomics and Human Genetics* **1**, 539–559.

Leblois, R., Estoup, A. and Rousset, F. (2003). Influence of mutational and sampling factors on the estimation of demographic parameters in a "continuous" population under isolation by distance. *Molecular Biology and Evolution* **20**, 491–502.

Leblois, R., Rousset, F. and Estoup, A. (2004). Influence of spatial and temporal heterogeneities on the estimation of demographic parameters in a continuous population using individual microsatellite data. *Genetics* **166**, 1081–1092.

Lehmann, E.L. and Casella, G. (1998). *Theory of Point Estimation*. Springer-Verlag, New York.

Lenormand, T., Bourguet, D., Guillemaud, T. and Raymond, M. (1999). Tracking the evolution of insecticide resistance in the mosquito *Culex pipiens*. *Nature* **400**, 861–864.

Lewontin, R.C. and Krakauer, J. (1973). Distribution of gene frequency as a test of the theory of the selective neutrality of polymorphisms. *Genetics* **74**, 175–195.

Li, N. and Stephens, M. (2003). Modeling linkage disequilibrium and identifying recombination hotspots using single-nucleotide polymorphism data. *Genetics* **165**, 2213–2233. Correction: **167**, 1039.

Lindley, D.V. (1990). The present position in Bayesian statistics. *Statistical Science* **5**, 44–89.

Long, J.C. (1986). The allelic correlation structure of Gainj- and Kalam-speaking people. I. The estimation and interpretation of Wright's *F*-statistics. *Genetics* **112**, 629–647.

Long, J.C., Naidu, J.M., Mohrenweiser, H.W., Gershowitz, H., Johnson, P.L. and Wood, J.W. (1986). Genetic characterization of Gainj- and Kalam-speaking peoples of Papua New Guinea. *American Journal of Physical Anthropology* **70**, 75–96.

Malécot, G. (1951). Un traitement stochastique des problèmes linéaires (mutation, linkage, migration) en génétique de population. *Annales de l'Université de Lyon A* **14**, 79–117.

Malécot, G. (1967). Identical loci and relationship. In *Proceedings of the Fifth Berkeley Symposium on Mathematical Statistics and Probability*, Vol. 4, L.M. Le Cam and J. Neyman, eds. University of California Press, Berkeley, CA, pp. 317–332.

Malécot, G. (1975). Heterozygosity and relationship in regularly subdivided populations. *Theoretical Population Biology* **8**, 212–241.

Mantel, N. (1967). The detection of disease clustering and a generalized regression approach. *Cancer Research* **27**, 209–220.

Metropolis, N., Rosenbluth, A.W., Rosenbluth, M.N., Teller, A.H. and Teller, E. (1953). Equation of state calculations by fast computing machines. *Journal of Chemical Physics* **21**, 1087–1092.

Mollison, D. (1977). Spatial contact models for ecological and epidemic spread. *Journal of the Royal Statistical Society, Series B* **39**, 283–326.

Nagylaki, T. (1975). Conditions for the existence of clines. *Genetics* **80**, 595–615.

Nagylaki, T. (1976). The decay of genetic variability in geographically structured populations. II. *Theoretical Population Biology* **10**, 70–82.

Nagylaki, T. (1982). Geographical invariance in population genetics. *Journal of Theoretical Biology* **99**, 159–172.

Nagylaki, T. (1998). Fixation indices in subdivided populations. *Genetics* **148**, 1325–1332.

Nei, M. (1986). Definition and estimation of fixation indices. *Evolution* **40**, 643–645.

Neyman, J. (1977). Frequentist probability and frequentist statistics. *Synthese* **36**, 97–131.

Nielsen, R. and Wakeley, J. (2001). Distinguishing migration from isolation: a Markov chain Monte Carlo approach. *Genetics* **158**, 885–896.

Nordborg, M. (1997). Structured coalescent processes on different time scales. *Genetics* **146**, 1501–1514.

Paetkau, D., Calvert, W., Stirling, I. and Strobeck, C. (1995). Microsatellite analysis of population structure in Canadian polar bears. *Molecular Ecology* **4**, 347–354.

Pannell, J.R. and Charlesworth, B. (1999). Neutral genetic diversity in a metapopulation with recurrent local extinction and recolonization. *Evolution* **53**, 664–676.

Rannala, B. and Hartigan, J.A. (1996). Estimating gene flow in island populations. *Genetical Research (Cambridge)* **67**, 147–158.

Rannala, B. and Mountain, J.L. (1997). Detecting immigration by using multilocus genotypes. *Proceedings of the National Academy of Sciences of the United States of America* **94**, 9197–9201.

Raufaste, N. and Bonhomme, F. (2000). Properties of bias and variance of two multiallelic estimators of F_{ST}. *Theoretical Population Biology* **57**, 285–296.

Raymond, M. and Rousset, F. (1995). GENEPOP version 1.2: population genetics software for exact tests and ecumenicism. *The Journal of Heredity* **86**, 248–249.

Robledo-Arnuncio, J.J., Austerlitz, F. and Smouse, P.E. (2006). A new method of estimating the pollen dispersal curve independently of effective density. *Genetics* **173**, 1033–1046.

Rousset, F. (1997). Genetic differentiation and estimation of gene flow from F-statistics under isolation by distance. *Genetics* **145**, 1219–1228.

Rousset, F. (1999a). Genetic differentiation in populations with different classes of individuals. *Theoretical Population Biology* **55**, 297–308.

Rousset, F. (1999b). Genetic differentiation within and between two habitats. *Genetics* **151**, 397–407.

Rousset, F. (2000). Genetic differentiation between individuals. *Journal of Evolutionary Biology* **13**, 58–62.

Rousset, F. (2001). Genetic approaches to the estimation of dispersal rates. In *Dispersal*, J. Clobert, E. Danchin, A.A. Dhondt and J.D. Nichols, eds. Oxford University Press, Oxford, pp. 18–28.

Rousset, F. (2002). Inbreeding and relatedness coefficients: what do they measure?. *Heredity* **88**, 371–380.

Rousset, F. (2004). *Genetic Structure and Selection in Subdivided Populations*. Princeton University Press, Princeton, NJ.

Rousset, F. (2006). Separation of time scales, fixation probabilities and convergence to evolutionarily stable states under isolation by distance. *Theoretical Population Biology* **69**, 165–179.

Rousset, F. and Raymond, M. (1997). Statistical analyses of population genetic data: old tools, new concepts. *Trends in Ecology and Evolution* **12**, 313–317.

Sawyer, S. (1977). Asymptotic properties of the equilibrium probability of identity in a geographically structured population. *Advances in Applied Probability* **9**, 268–282.

Searle, S.R. (1971). *Linear Models*. John Wiley & Sons, New York.

Sites, J.W. Jr., Barton, N.H. and Reed, K.M. (1995). The genetic structure of a hybrid zone between two chromosome races of the *Sceloporus grammicus* complex (Sauria, Phrynosomatidae) in central Mexico. *Evolution* **49**, 9–36.

Slatkin, M. (1991). Inbreeding coefficients and coalescence times. *Genetical Research (Cambridge)* **58**, 167–175.

Slatkin, M. (1993). Isolation by distance in equilibrium and non-equilibrium populations. *Evolution* **47**, 264–279.

Slatkin, M. (1994). Gene flow and population structure. In *Ecological Genetics*, L.A. Real, ed. Princeton University Press, Princeton, NJ, pp. 3–17.

Slatkin, M. (1995). A measure of population subdivision based on microsatellite allele frequencies. *Genetics* **139**, 457–462.

Smouse, P. and Williams, R.C. (1982). Multivariate analysis of HLA-disease associations. *Biometrics* **38**, 757–768.

Sokal, R. and Wartenberg, D.E. (1983). A test of spatial autocorrelation analysis using an isolation-by-distance model. *Genetics* **105**, 219–237.

Storz, J.F., Ramakrishnan, U. and Alberts, S.C. (2001). Determinants of effective population size for loci with different modes of inheritance. *The Journal of Heredity* **92**, 497–502.

Sumner, J., Estoup, A., Rousset, F. and Moritz, C. (2001). 'Neighborhood' size, dispersal and density estimates in the prickly forest skink (*Gnypetoscincus queenslandiae*) using individual genetic and demographic methods. *Molecular Ecology* **10**, 1917–1927.

Swofford, D. and Olsen, G. (1990). Phylogeny reconstruction. In *Molecular Systematics*, D. Hillis and C. Moritz, eds. Sinauer Associates, pp. 411–501.

Tachida, H. (1985). Joint frequencies of alleles determined by separate formulations for the mating and mutation systems. *Genetics* **111**, 963–974.

Tsitrone, A., Rousset, F. and David, P. (2001). Heterosis, marker mutational processes, and population inbreeding history. *Genetics* **159**, 1845–1859.

Tufto, J., Engen, S. and Hindar, K. (1996). Inferring patterns of migration from gene frequencies under equilibrium conditions. *Genetics* **144**, 1911–1921.

Vekemans, X. and Hardy, O.J. (2004). New insights from fine-scale spatial genetic structure analyses in plant populations. *Molecular Ecology* **13**, 921–934.

Vitalis, R. and Couvet, D. (2001). Estimation of effective population size and migration rate from one- and two-locus identity measures. *Genetics* **157**, 911–925.

Vitalis, R., Dawson, K. and Boursot, P. (2001). Interpretation of variation across marker loci as evidence of selection. *Genetics* **158**, 1811–1823.

Wakeley, J. and Aliacar, N. (2001). Gene genealogies in a metapopulation. *Genetics* **159**, 893–905. Correction in *Genetics* **160**, 1263.

Wang, J. and Whitlock, M.C. (2003). Estimating effective population size and migration rates from genetic samples over space and time. *Genetics* **163**, 429–446.

Waples, R.S. and Gaggiotti, O. (2006). What is a population? An empirical evaluation of some genetic methods for identifying the number of gene pools and their degree of connectivity. *Molecular Ecology* **15**, 1419–1439.

Watts, P.C., Rousset, F., Saccheri, I.J., Leblois, R., Kemp, S.J. and Thompson, D.J. (2007). Compatible genetic and ecological estimates of dispersal rates in insect (*Coenagrion mercuriale*: Odonata: Zygoptera) populations: analysis of 'neighbourhood size' using an improved estimator. *Molecular Ecology* **16**, 737–751.

Weir, B.S. (1996). *Genetic Data Analysis II*. Sinauer Associates, Sunderland, MA.

Weir, B.S. and Cockerham, C.C. (1984). Estimating F-statistics for the analysis of population structure. *Evolution* **38**, 1358–1370.

Wilkins, J.F. (2004). A separation-of-timescales approach to the coalescent in a continuous population. *Genetics* **168**, 2227–2244.

Wilson, A.J., Hutchings, J.A. and Ferguson, M.M. (2004). Dispersal in a stream dwelling salmonid: inferences from tagging and microsatellite studies. *Conservation Genetics* **5**, 25–37.

Wilson, G.A. and Rannala, B. (2003). Bayesian inference of recent migration rates using multilocus genotypes. *Genetics* **163**, 1177–1191.

Winters, J.B. and Waser, P.M. (2003). Gene dispersal and outbreeding in a philopatric mammal. *Molecular Ecology* **12**, 2251–2259.

Wood, J.W., Smouse, P.E. and Long, J.C. (1985). Sex-specific dispersal patterns in two human populations of highland New Guinea. *American Naturalist* **125**, 747–768.

Wright, S. (1931a). Evolution in Mendelian populations. *Genetics* **16**, 97–159. Reprinted in Wright (1986), pp. 98–160.

Wright, S. (1931b). Statistical theory of evolution. *Journal of the American Statistical Society* **26**(Suppl.), 201–208. Reprinted in Wright (1986), pp. 89–96.

Wright, S. (1946). Isolation by distance under diverse systems of mating. *Genetics* **31**, 39–59. Reprinted in Wright (1986), pp. 444–464.

Wright, S. (1949). Adaptation and selection. In *Genetics, Paleontology and Evolution*, G.L. Jepson, G.G. Simpson and E. Mayr, eds. Princeton University Press, pp. 365–389. Reprinted in Wright (1986), pp. 546–570.

Wright, S. (1969). *Evolution and the Genetics of Populations. II. The Theory of Gene Frequencies*. University of Chicago Press, Chicago, IL.

Wright, S. (1986). *Evolution: Selected Papers*. University of Chicago Press, Chicago, IL.

Zähle, I., Cox, J.T. and Durrett, R. (2005). The stepping stone model, II: genealogies and the infinite sites model. *Annals of Applied Probability* **15**, 671–699.

APPENDIX A: ANALYSIS OF VARIANCE AND PROBABILITIES OF IDENTITY

Here we detail the relationship between classical formulas for estimators of F-statistics and expressions in terms of frequency of identical pairs of genes. In the framework considered here, negative 'components of variance' (which are actually not variances) arise naturally.

We use the following notation: the total sample is made up of samples of n_i ($i = 1, \ldots, n_s$) individuals in n_s subpopulations; $X_{ij:k}$ is an indicator variable for gene j ($j = 1, \ldots, n_i$) in sample i being of allelic type k ($k = 1, \ldots, K$), i.e. $X_{ij:k} = 1$ if the sampled gene is of type k and $X_{ij:k} = 0$ otherwise; standard dot notation is used for sample averages: e.g. for weights w_j, $\bar{X}_. \equiv (\sum_j w_j X_j)/(\sum_j w_j)$ is a weighted average of the Xs. Here the weighting for each individual will be simply 1. A discussion of optimal weighting with respect to allele frequencies or samples sizes will be given in Appendix B. The indicator variables are given a single index (X_j) if no reference is made to a specific sample. π_k is the expected frequency of allele k, the expectation of $X_{j:k}$ over independent replicates of some evolutionary process (typically a mutation–drift stationary equilibrium, but this is in no way required).

For haploid data the statistical model is generally described as

$$X_{ij:k} = \mu + \alpha_i + \varepsilon_{ij}, \qquad (28A.1)$$

$\mu = \pi_k$ here, α_i is a random effect with zero mean and variance σ_a^2, and ε_{ij} is a random effect with zero mean and variance σ_e^2. It is also assumed that $E[\alpha_i \alpha_{i'}] = 0$ for $i \neq i'$ and that $E[\varepsilon_{ij}\varepsilon_{ij'}] = 0$ for $j \neq j'$ (e.g. Searle, 1971, p. 384), but this is precisely what we will not do here, in order to obtain the most general analogy. What remains more generally valid is a basic algebraic relationship of analysis of variance,

$$\sum_j w_j (X_j - \mu)^2 = \sum_j w_j (X_j - X_. + X_. - \mu)^2$$

$$= \sum_j w_j (X_j - X_.)^2 + \sum_j w_j (X_. - \mu)^2, \qquad (28A.2)$$

for any variable X and weights w_j because we still have $\sum_j w_j (X_{j:k} - X_{.:k}) = 0$ by definition of $X_{.:k}$.

The method of analysis of variance is then based on the computation of weighted sums of squares related as follows:

$$\sum_{\substack{n_s \\ \text{samples}}} \sum_{\substack{n_i \\ \text{genes}}} w_i (X_{ij:k} - X_{..:k})^2 = \sum^{n_s} \sum^{n_i} w_i (X_{ij:k} - X_{i.:k})^2$$

$$+ \sum^{n_s} \sum^{n_i} w_i (X_{i.:k} - X_{..:k})^2 \qquad (28A.3)$$

$$\equiv SS_{\text{w[ithin]}} + SS_{\text{b[etween]}} \text{for allele } k. \qquad (28A.4)$$

For $w_i = 1$ these sums of squares will be expressed in terms of $S_1 \equiv \sum_i n_i$ and $S_2 \equiv \sum_i n_i^2$, of the observed frequency of pairs of different genes within samples which are both of type k, $\hat{Q}_{2:k} \equiv \sum_i \sum_{j \neq j'} X_{ij:k} X_{ij':k} / (S_2 - S_1)$, and of the observed frequency of pairs of different genes between samples which are both of type k, $\hat{Q}_{3:k} \equiv \sum_{i \neq i'} \sum_{j,j'} X_{ij:k} X_{i'j':k} / (S_1^2 - S_2)$. We first express the sums of squares using (28A.2), as follows:

$$SS_{\text{w}} = \sum_i \sum_j^{n_i} (X_{ij:k} - X_{i.:k})^2 = \sum_i \sum_j^{n_i} (X_{ij:k} - \pi_k)^2 - \sum_i n_i (X_{i.:k} - \pi_k)^2, \quad (28A.5)$$

and

$$SS_{\text{b}} = \sum_i \sum_j^{n_i} (X_{i.:k} - X_{..:k})^2 = \sum_i n_i (X_{i.:k} - \pi_k)^2 - S_1 (X_{..:k} - \pi_k)^2. \quad (28A.6)$$

The values of different variables that appear in these expressions, $(X_{j:k} - \pi_k)^2$ and $(X_{j:k} - \pi_k)(X_{j':k} - \pi_k)$, for pairs j, j' of different genes, are summarized in Table 28.1 where $Q_{.k}$ is the probability that both genes of a pair are of type k. Note that $E[(X_{j:k} - \pi_k)(X_{j':k} - \pi_k)]$ is the covariance between allele frequencies in the subpopulations from

Table 28.1 Values of variables in comparisons of pairs of genes.

Pair of gene	kk	k, not k	None k
Probability of each pair	$Q_{:k}$	$2(\pi_k - Q_{:k})$	$1 - 2\pi_k + Q_{:k}$
Frequency in total sample	$\hat{Q}_{:k}$	$2(\tilde{\pi}_k - \hat{Q}_{:k})$	$1 - 2\tilde{\pi}_k + \hat{Q}_{:k}$

Variable	Value of variable for each pair		
$(X_{j:k} - \pi_k)(X_{j':k} - \pi_k)$	$(1 - \pi_k)^2$	$-\pi_k(1 - \pi_k)$	π_k^2
$(X_{j:k} - \pi_k)^2 + (X_{j':k} - \pi_k)^2$	$2(1 - \pi_k)^2$	$(1 - \pi_k)^2 + \pi_k^2$	$2\pi_k^2$

which j and j' are sampled. From this table we see that

$$\sigma_a^2 + \sigma_e^2 = \mathrm{E}[(X_{j:k} - \pi_k)^2] = \pi_k(1 - \pi_k), \tag{28A.7}$$

$$\mathrm{E}[(X_{j:k} - \pi_k)(X_{j':k} - \pi_k)] = Q_{:k} - \pi_k^2. \tag{28A.8}$$

where $Q_{:k}$ is $Q_{w:k} = Q_{2:k}$ for two genes within a sample and $Q_{b:k} = Q_{3:k}$ for two genes between samples. In particular we have

$$Q_{2:k} - \pi_k^2 = \mathrm{Cov}[X_{ij}X_{ij'}] = \sigma_a^2 + \mathrm{E}[\varepsilon_{ij}\varepsilon_{ij'}], \tag{28A.9}$$

$$Q_{3:k} - \pi_k^2 = \mathrm{Cov}[X_{ij}X_{i'j'}] = \mathrm{E}[\alpha_i\alpha_{i'}]. \tag{28A.10}$$

The latter expression confirms that $\mathrm{E}[\alpha_i\alpha_{i'}] \neq 0$: in general two genes in different subpopulations are more likely to be identical than two independent genes, i.e. $Q_{3:k} - \pi_k^2 > 0$. In the present case one could consider a slightly different parameterization of the model, so that this positive component would appear as a variance (e.g. Cockerham and Weir, 1987), but more generally this would be confusing because we may also have to consider negative terms, as shown below for diploid data.

The table also shows that the sample averages of $(X_{j:k} - \pi_k)(X_{j':k} - \pi_k)$ and $(X_{j:k} - \pi_k)^2$ are $\hat{Q}_{:k} + \pi_k^2 - 2\pi_k\tilde{\pi}_k$ and $\tilde{\pi}_k + \pi_k^2 - 2\pi_k\tilde{\pi}_k$, respectively. Here $\tilde{\pi}_k$ is the average allele frequency among all *pairs of genes* for which the average is written. This is the observed allele frequency by gene counting (denoted $\hat{\pi}_k$ or $\hat{\pi}_{i:k}$) when all pairs in the total sample or in sample i are considered. Among all pairs of genes sampled without replacement within each sample, this is $\hat{\pi}_{w:k} \equiv \sum_i n_i(n_i - 1)\hat{\pi}_{i:k}/(S_2 - S_1)$. Among all pairs of genes from two different samples, $\hat{\pi}$ has value $\hat{\pi}_{b:k}$ given by $(S_1^2 - S_2)\hat{\pi}_{b:k} + (S_2 - S_1)\hat{\pi}_{w:k} = (S_1^2 - S_1)\hat{\pi}$.

Then

$$\sum_i \sum_j^{n_j} (X_{ij:k} - \pi_k)^2 = S_1(X_{..:k} + \pi_k^2 - 2\pi_k X_{..:k}). \tag{28A.11}$$

Next

$$(X_{..:k} - \pi_k)^2 = \left(\frac{\sum_{i,j} X_{ij:k} - \pi_k}{S_1}\right)^2 \tag{28A.12}$$

$$= \frac{1}{S_1^2}\left(\sum_{i,j}(X_{ij:k} - \pi_k)^2\right.$$

$$+ \sum_{i \neq i', j, j'} (X_{i'j':k} - \pi_k)(X_{i'j':k} - \pi_k)$$

$$+ \sum_{i, j \neq j'} (X_{ij:k} - \pi_k)(X_{ij':k} - \pi_k) \Bigg) \tag{28A.13}$$

$$= \frac{1}{S_1^2} \Bigg(S_1(\hat{\pi} + \pi_k^2 - 2\hat{\pi}\pi_k) + (S_1^2 - S_2)(\hat{Q}_{3:k} + \pi_k^2 - 2\hat{\pi}_{b:k}\pi_k)$$

$$+ (S_2 - S_1)(\hat{Q}_{2:k} + \pi_k^2 - 2\hat{\pi}_{w:k}\pi_k) \Bigg) \tag{28A.14}$$

$$= (\hat{\pi} + \pi_k^2 - 2\hat{\pi}\pi_k) + \frac{1}{S_1^2} \Big((S_1^2 - S_2)(\hat{Q}_{3:k} - \hat{\pi}) + (S_2 - S_1)(\hat{Q}_{2:k} - \hat{\pi}) \Big), \tag{28A.15}$$

by definition of $\hat{\pi}_{b:k}$. Likewise

$$(X_{i.:k} - \pi_k)^2 = \left(\frac{\sum_{j=1}^{n_i} X_{ij:k} - \pi_k}{n_i} \right)^2 \tag{28A.16}$$

$$= \frac{1}{n_i^2} \left(\sum_{j=1}^{n_i} \left(X_{ij:k} - \pi_k \right)^2 + \sum_{j \neq j'} \left(X_{ij:k} - \pi_k \right) \left(X_{ij':k} - \pi_k \right) \right) \tag{28A.17}$$

$$= \frac{1}{n_i^2} \Big(n_i(\hat{\pi}_{i:k} + \pi_k^2 - 2\hat{\pi}_{i:k}\pi_k) + n_i(n_i - 1)(\hat{Q}_{i2:k} + \pi_k^2 - 2\hat{\pi}_{i:k}\pi_k) \Big) \tag{28A.18}$$

$$= (\hat{\pi}_{i:k} + \pi_k^2 - 2\hat{\pi}_{i:k}\pi_k) + \frac{n_i - 1}{n_i}(\hat{Q}_{i2:k} - \hat{\pi}_{i:k}), \tag{28A.19}$$

hence

$$\sum_{i=1}^{n_s} n_i(X_{i.:k} - \pi_k)^2 = S_1(\hat{\pi}_k + \pi_k^2 - 2\hat{\pi}_k\pi_k) + \sum_i (n_i - 1)(\hat{Q}_{i2:k} - \hat{\pi}_{i:k}). \tag{28A.20}$$

Then from (28A.5), (28A.11) and (28A.20)

$$SS_w = (S_1 - n_s)(\hat{\pi}_k - \hat{Q}_{2:k}), \tag{28A.21}$$

and from (28A.6), (28A.15) and (28A.20)

$$SS_b = \sum_i (\hat{\pi}_{i:k} - \hat{\pi}_k) + \sum_i (n_i - 1)\hat{Q}_{i2:k} - (S_1 - S_2/S_1)\hat{Q}_{3:k} - (S_2/S_1 - 1)\hat{Q}_{2:k}. \tag{28A.22}$$

Note that as in (28A.20), allele frequencies terms do not reduce to a function of $\hat{\pi}$ only: the term $\sum_i (\hat{\pi}_{i:k} - \hat{\pi})$ will usually be nonzero when sample sizes are unequal. Taking expectations, one has

$$E[SS_w] = (S_1 - n_s)(\pi_k - Q_{2:k}),$$

$$E[SS_b] = (S_1 - S_2/S_1)(Q_{2:k} - Q_{3:k}) + (n_s - 1)(\pi_k - Q_{2:k})$$

$$= (S_1 - S_2/S_1)(\sigma_a^2 - E[\alpha_i\alpha_{i'}] + E[\varepsilon_{ij}\varepsilon_{ij'}]) + (n_s - 1)(\sigma_e^2 - E[\varepsilon_{ij}\varepsilon_{ij'}]).$$

(28A.23)

These relationships hold whatever the model considered (fixed or random, etc.). They are formally equivalent to a standard analysis of variance (e.g. Searle, 1971) on the indicator variables $X_{ij:k}$, except that (1) $E[\varepsilon_{ij}\varepsilon_{ij'}]$ and $E[\alpha_i\alpha_{i'}]$ are not assumed null, and (2) the sums of squares are themselves summed over alleles. When we write these two modifications as '$\overset{1}{\rightarrow}$' and '$\overset{2}{\rightarrow}$' the equivalence of expectations in the standard formulas of analysis of variance with expressions in terms of probabilities of identity is as follows:

$$\sigma_a^2 \overset{1}{\rightarrow} \sigma_a^2 - E[\alpha_i\alpha_{i'}] + E[\varepsilon_{ij}\varepsilon_{ij'}] \overset{2}{\rightarrow} Q_2 - Q_3 \equiv (1 - Q_3)F_{ST}$$

$$\sigma_e^2 \overset{1}{\rightarrow} \sigma_e^2 - E[\varepsilon_{ij}\varepsilon_{ij'}] \overset{2}{\rightarrow} 1 - Q_2 = (1 - Q_3)(1 - F_{ST}). \quad (28A.24)$$

Hence the 'intraclass covariance' $\sigma_a^2/(\sigma_a^2 + \sigma_e^2)$, often taken as a definition of F_{ST}, should be interpreted as

$$F_{ST} = \frac{\sigma_a^2}{\sigma_a^2 + \sigma_e^2} \overset{1}{\rightarrow} F_{ST} = \frac{\sigma_a^2 + E[\varepsilon_{ij}\varepsilon_{ij'}] - E[\alpha_i\alpha_{i'}]}{\sigma_a^2 + \sigma_e^2 - E[\alpha_i\alpha_{i'}]}, \quad (28A.25)$$

where the latter expression may be considered more general.

For diploid data, the model is $X_{ijl:k} = \mu + \alpha_i + \beta_{ij} + \varepsilon_{ijl}$ for gene l ($l = 1, 2$ for diploids) of individual j in population i. With $\sigma_a^2 \equiv E[\alpha_i^2]$, $\sigma_b^2 \equiv E[\beta_{ij}^2]$, $\sigma_e^2 \equiv E[\varepsilon_{ijl}^2]$, we have

$$\sigma_a^2 \overset{1}{\rightarrow} \sigma_a^2 - E[\alpha_i\alpha_{i'}] + E[\beta_{ij}\beta_{ij'}] \overset{2}{\rightarrow} Q_2 - Q_3 \equiv (1 - Q_3)F_{ST}$$

$$\sigma_b^2 \overset{1}{\rightarrow} \sigma_b^2 - E[\beta_{ij}\beta_{ij'}] + E[\varepsilon_{ijl}\varepsilon_{ijl'}] \overset{2}{\rightarrow} Q_1 - Q_2 \equiv (1 - Q_3)F_{IS}(1 - F_{ST})$$

$$\sigma_e^2 \overset{1}{\rightarrow} \sigma_e^2 - E[\varepsilon_{ijl}\varepsilon_{ijl'}] \overset{2}{\rightarrow} 1 - Q_1 = (1 - Q_3)(1 - F_{IS})(1 - F_{ST}),$$

(28A.26)

where Q_1 is the probability of identity of genes within a diploid individual, Q_2 is for genes between individuals within subpopulations, and Q_3 between subpopulations. Thus in both formalisms we see that the 'components of variance' actually translate into more general expressions that can be negative. When there is an heterozygote excess within demes ($Q_1 < Q_2$), $E[\varepsilon_{ijl}\varepsilon_{ijl'}]$ is negative, and $\sigma_b^2 - E[\beta_{ij}\beta_{ij'}] + E[\varepsilon_{ijl}\varepsilon_{ijl'}]$ is negative.

In the haploid case, (28A.23) implies that

$$\frac{(S_1 - n_s)E[SS_b] - (n_s - 1)E[SS_w]}{(S_1 - n_s)E[SS_b] + (W_c - 1)(n_s - 1)E[SS_w]} = \frac{Q_2 - Q_3}{1 - Q_3}, \quad (28A.27)$$

where SS_w and SS_b are now summed over alleles (e.g. $SS_w \equiv \sum^{n_s} \sum^{n_i} \sum_k (X_{ij:k} - X_{i.:k})^2$), and $W_c \equiv (S_1 - S_2/S_1)/(n_s - 1)$. Although we have related the expectation of the different terms to 'components of variance' in a model such as (28A.1), we note again that the last equality holds independently of such a model, because it is only based on the basic relationship (28A.2). Accordingly, an estimator of F_{ST} is the ratio of unbiased estimators

$$\frac{(S_1 - n_s)SS_b - (n_s - 1)SS_w}{(S_1 - n_s)SS_b + (W_c - 1)(n_s - 1)SS_w}. \tag{28A.28}$$

(see also Weir, 1996, p. 182). One could hope that this estimator is directly interpretable as $(\hat{Q}_2 - \hat{Q}_3)/(1 - \hat{Q}_3)$, where $\hat{Q}_j = \sum_k \hat{Q}_{j:k}$ are the frequencies of identical pairs of genes in the sample, computed by simple counting either within (for Q_2) or between (for Q_3) samples ($\hat{Q}_2$ being is an average over the different samples, weighted according to the number of pairs in each sample). But this is not so when sample sizes are unequal, because the term $\sum_i (\hat{\pi}_{i:k} - \hat{\pi})$ from (28A.22) remains in the above expression. The expression closest to $(\hat{Q}_2 - \hat{Q}_3)/(1 - \hat{Q}_3)$ that I have found for Weir and Cockerham's estimator is

$$\frac{\tilde{Q}_2 - \hat{Q}_3}{1 - \hat{Q}_3 + \sum_i (n_i - 1)(\hat{Q}_{i2} - \hat{Q}_2)\frac{S_2 - S_1}{(S_1^2 - S_2)(S_1 - n_s)}}, \tag{28A.29}$$

in terms of the weighted frequency

$$\tilde{Q}_2 = \frac{(S_1 - 1)\sum_i (n_i - 1)\hat{Q}_{i2} - (S_1 - n_s)(S_2/S_1 - 1)\hat{Q}_2}{(S_1 - n_s)(S_1 - S_2/S_1)}, \tag{28A.30}$$

and where $\hat{Q}_{i2}$ is the observed frequency of pairs of genes identical in state among all pairs taken without replacement within sample i. Compared to the analysis-of-variance estimator, the simple strategy of estimating any function of probabilities of identities by the equivalent function of frequencies of identical pairs of genes is equally 'unbiased', has no obvious drawback and is easily adaptable to different settings.

For multilocus data it is usual to compute the estimator as a sum of locus-specific numerators over a sum of locus-specific denominators; see e.g. Weir and Cockerham (1984) or Weir (1996) for details. Note that the sums are weighted differently in these two references. The numerator in Weir and Cockerham (1984), eqs. (2) and (10), is $\bar{n}/W_c$ times the one in Weir (1996), p.178–179. Parallel changes in the denominator ensure that the one-locus estimators are identical, but the multilocus estimators will be different if $\bar{n}/W_c$ varies between loci.

APPENDIX B: LIKELIHOOD ANALYSIS OF THE ISLAND MODEL

Sampling Formulas

Consider an infinite island model of haploid subpopulations where there are K alleles and n_s subpopulations are sampled. The following notation will be used: π_k is the frequency of allele k in the total population (which is not a random variable here) and p_{ki} is frequency of allele k in subpopulation i; n_{ki} is the number of genes of type k in the sample from subpopulation i; n_i is the size of sample i, $\bar{n}$ is the average n_i, $\tilde{\pi}_k \equiv \sum_i n_{ki}/\sum_i n_i$ and $\tilde{p}_{ki} \equiv n_{ki}/n_i$ are the observed frequencies of allele k in the total sample and in sample i, respectively.

The distribution of the p_{ki}s in population i follows a Dirichlet distribution,

$$L(p_{ki}, \ldots, p_{Ki}) = \Gamma(M) \prod_k \frac{p_{ki}^{M\pi_k - 1}}{\Gamma(M\pi_k)}. \tag{28B.1}$$

This type of distribution arises as a diffusion approximation to the discrete generation Wright–Fisher model, where M is twice the number of migrant genes per generation (Wright, 1949), e.g. $2Nm$ or $4Nm$, and more generally in any scenario that can be approximated by the n-coalescent.

It should be noted that this equation is valid for each subpopulation with its own size, N_i, and its own immigration rate, m_i. Thus the likelihood of samples may be given for an infinite island model only characterized by the homogeneous dispersal of individuals to other demes: the N_is and m_is need not be identical in all subpopulations. This result implies that, with a large number of subpopulations, one can only estimate the products $N_i m_i$ for each deme.

Consider the vector of counts n_{ki} of allele k in subpopulation i, $\mathbf{n}_i \equiv (n_{1i}, \cdots, n_{Ki})$, and the corresponding multinomial coefficient $C(\mathbf{n}_i) \equiv n_i! / \prod_{k=1}^{K} n_{ki}!$. The conditional probability distribution of the ith sample, given the subpopulation frequencies $\mathbf{p}_i \equiv (p_{1i}, \ldots, p_{Ki})$, is multinomial:

$$C(\mathbf{n}_i) \prod_k^K p_{ki}^{n_{ki}}. \tag{28B.2}$$

The probability distribution of a sample $\mathbf{n}_i$ in subpopulation i must be expressed as a function only of the parameters, M and of expected allele frequencies (expectations under stochastic model) $\boldsymbol{\pi} \equiv (\pi_1, \ldots, \pi_K)$, by combining (28B.1) and (28B.2) and summing over the set S of possible values of allele frequencies $\mathbf{p}_i$:

$$L(M, \boldsymbol{\pi}; n_{1i}, \ldots, n_{Ki}) = \int \overset{..}{\underset{S}{}} \int \Gamma(M) \prod_k^K \frac{p_{ki}^{M\pi_k - 1}}{\Gamma(M\pi_k)} C(\mathbf{n}_i) \prod_k^K p_{ki}^{n_{ki}} \, d\mathbf{p}_i \tag{28B.3}$$

$$= \frac{\Gamma(M)}{\Gamma(M + n_i)} C(\mathbf{n}_i) \prod_k^K \frac{\Gamma(M\pi_k + n_{ki})}{\Gamma(M\pi_k)}. \tag{28B.4}$$

This distribution is the Dirichlet-multinomial distribution. In the infinite island model, subpopulation frequencies are independent from each other in each subpopulation, so that the likelihood of a total sample from n_s subpopulations is

$$L(M, \boldsymbol{\pi}) = \left(\frac{\Gamma(M)}{\Gamma(M + n_i)} \right)^{n_s} \prod_{i=1}^{n_s} C(\mathbf{n}_i) \prod_k^K \frac{\Gamma(M\pi_k + n_{ki})}{\Gamma(M\pi_k)}. \tag{28B.5}$$

The pseudo–maximum likelihood estimator $\hat{M}_A$ of M had been previously defined by Chuang and Cox (1985) as the solution of $\partial \ln L / \partial M |_{\boldsymbol{\pi} = \tilde{\boldsymbol{\pi}}, M = \hat{M}_A} = 0$. From (28B.5), this is the solution of

$$0 = \left[\sum_k^K \sum_i^{n_s} \tilde{\pi}_k \left(\sum_{k=0}^{n_{ki}-1} \frac{1}{\tilde{\pi}_k M + k} \right) - \sum_i^{n_s} \left(\sum_{k=0}^{n_i-1} \frac{1}{M + k} \right) \right]. \tag{28B.6}$$

Efficiency in the Island Model

When $M \to \infty$ (high migration rates) the Dirichlet-multinomial distribution converges to a multinomial with parameter π, so the sampling distribution for the total sample is a product of multinomials with the same parameter π: this corresponds to the case of no population differentiation. Thus we can construct asymptotically efficient statistics for detecting weak differentiation from a study of the properties of the likelihood when $M \to \infty$. To that aim it is simpler to express it as a function of $\psi \equiv 1/M$ and compute the Taylor expansion near $\psi = 0$. From (28B.6), it may be shown that

$$\frac{\partial \ln L}{\partial \psi} = \sum_k^K \sum_i^{n_s} \frac{n_{ki}(n_{ki} - 1)}{2\pi_k} - \sum_i^{n_s} \frac{n_i(n_i - 1)}{2} + O(\psi), \qquad (28B.7)$$

and a statistic of interest (effectively a score statistic, Cox and Hinkley, 1974, Chapter 9) may be constructed as

$$\tilde{U} \equiv \lim_{\psi \to 0} \frac{\partial \ln L}{\partial \psi} \bigg|_{\pi = \tilde{\pi}} = \sum_k^K \sum_i^{n_s} \frac{n_{ki}(n_{ki} - 1)}{2\tilde{\pi}_k} - \sum_i^{n_s} \frac{n_i(n_i - 1)}{2}, \qquad (28B.8)$$

where the $\tilde{\pi}$s are the observed allele frequencies in the total sample, which are the MLEs of the πs in the case $\psi = 0$. This result draws a connection between the likelihood and the moment methods (see also Balding, 2003). Since the second sum in (28B.8) is fixed for given sample sizes n_i, the score statistic is essentially a sum of squares and can be considered in an analysis of variance framework. It shows that asymptotically efficient weights w_k of the sum of squares for the different alleles are proportional to $1/\tilde{\pi}_k$, and the weights of the sum of squares for the different samples are proportional to n_i^2 for the different samples, i.e. $w_i = n_i$ for each individual in (28A.3). The allele weighting is not new: it is implicit in the matrix formulations of Smouse and Williams (1982) and Long (1986) (see Weir and Cockerham, 1984; Chakraborty, 1992) and in standard test statistics such as the χ^2 or log-likelihood for multinomial models. Consistent with the above analysis assuming weak differentiation ($\psi \to 0$), it leads to estimators with efficient properties only for low differentiation (Raufaste and Bonhomme, 2000). By contrast the sample size weighting is odd and has not been previously considered in analysis of variance. But it may bring very little (F.R., unpublished data): the allele weighting is generally sufficient to turn moment statistics into efficient statistics when differentiation is low.

Weighting according to observed allele frequencies or to other measures of genetic diversity may have some drawbacks, particularly when one considers more general models than the island model. This weighting seems to imply that the ratio of expected sum of squares, conditional on genetic diversity, is independent of the value of the conditioning variable. As noted in the main Text, this is approximately so in the island model, and more generally under a separation of timescales when the slow process is an n-coalescent, but may not hold more generally. Then, the only consistent method would be to sum the $\hat{Q}$ terms directly in the numerator and denominator; any other method would introduce a bias. Further, selection of markers with specific levels of variability–as is often the case in practice–could also introduce an ascertainment bias.

29

Analysis of Population Subdivision

L. Excoffier

Zoological Institute, Department of Biology, University of Berne, Berne, Switzerland

This paper reviews the basic measures of intraspecific population subdivision, without assuming any particular spatial model. The structure of the paper follows from the historical development of the field, starting with the presentation of Wright's fixation indexes as correlations of gene frequencies, then Nei's extension of F statistics to multiple alleles as function of heterozygosities and his introduction of G statistics as function of gene diversities, to end with Cockerham's intraclass correlations obtained under an analysis of variance (ANOVA) framework. We also present an extension of the ANOVA approach to handle molecular data (DNA sequences, microsatellites) as well as dominant data. The relationships between different estimators, different approaches, and their potential limitations are discussed.

29.1 INTRODUCTION

29.1.1 Effects of Population Subdivision

Species or populations usually do not constitute a single panmictic unit where individuals breed at random over the whole species range. They are rather subdivided into smaller entities, which can be arranged in space, time, ecology, or otherwise. Different levels of population subdivisions may exist, which may be hierarchically arranged or not. A population can thus be divided into groups or regions, themselves divided into smaller units, and so on until we reach a basic unit of population that may be treated as a homogeneous group, a fundamental unit called a *deme* (Gilmour and Gregor, 1939) or a neighborhood (Wright, 1946) in the context of continuous habitats. Even within demes, individuals are often related to each other, due to some shared ancestry and finite deme sizes, which introduces some local levels of inbreeding often resulting in homozygote excess.

Population structure, if not properly recognized, can affect our interpretation of the observed pattern of genetic diversity, and can often mimic the effect of selection. Wahlund (1928) thus showed that hidden subdivision leads to an apparent excess of homozygotes (see also Nagylaki, 1985). When several loci are considered, population subdivision can

Handbook of Statistical Genetics, Third Edition. Edited by D.J. Balding, M. Bishop and C. Cannings.
© 2007 John Wiley & Sons, Ltd.   ISBN: 978-0-470-05830-5.

lead to observed patterns of linkage disequilibrium, even between physically unlinked loci (e.g. Ohta and Kimura, 1969; Nei and Li, 1973; Li and Nei, 1974; Ohta, 1982), and recently admixed populations showing high levels of linkage disequilibrium over long chromosomal segments can be interesting to locate disease genes (e.g. Chakraborty and Weiss, 1988; Chapman and Thompson, 2001). At the molecular level, the presence of hidden and unrecognized population subdivision can increase the number of segregating sites within samples relative to the average number of substitutions between pairs of sequences (Tajima, 1993; Wakeley, 1998), while local inbreeding due to selfing can result in a drastic reduction of genetic diversity (Fu, 1997; Nordborg and Donnelly, 1997). In a population made up of demes interconnected by migrations, the mean number of pairwise differences (or the number of heterozygous sites within individuals) only depends on the sum of deme sizes, irrespective of the migration pattern (e.g. Slatkin, 1987; Strobeck, 1987; Nagylaki, 1998a), but this property does not hold for the variance (Marjoram and Donnelly, 1994). It follows that statistical tests of selective neutrality can often be significant in absence of selection, when performed on samples drawn from subdivided populations (e.g. Tajima, 1993; Simonsen et al., 1995; Nielsen, 2001). Also, hidden subdivision or population stratification can severely affect tests of genetic association and case control studies (e.g. Pritchard et al., 2000b; Pritchard and Donnelly, 2001; Marchini et al., 2004). When selection is present, population subdivision can allow the distinction between several forms of selection that have different effects on diversity within and between demes (Charlesworth et al., 1997). The inference of population subdivision is also useful for its own sake. For instance, in conservation biology it may be crucial to understand phylogeographic patterns and to recognize subspecies or the effect of settlement history on genetic diversity (e.g. Taberlet et al., 1998; Hewitt, 2001; Falush et al., 2003b; Jaarola and Searle, 2004; Magri et al., 2006). Moreover, the presence and the assessment of genetic structure is also essential in studies of species adaptation under Wright (1932; 1982) shifting balance theory.

In statistical terms, population structure introduces some correlation or covariance between genes taken from different subdivision levels. We are interested in describing here how we can detect, measure and test for the presence of internal subdivisions, often on the basis of the observed correlations, without assuming any particular migration model, which should be treated in the section on spatial genetics (**Chapter 28**). However, we sometimes have to assume that subdivisions are independent (implying no migration between demes).

Owing to space limitation, this chapter is not a comprehensive review of the huge literature on population subdivision. Additional and sometimes more exhaustive reviews can be found elsewhere (e.g. Jorde, 1980; Weir and Cockerham, 1984; Nei, 1986; Chakraborty and Danker-Hopfe, 1991; Weir and Hill, 2002; Rousset, 2004). Instead, we focus on introducing readers to the basic concepts and measures of population subdivision, trying to preserve the spirit in which they have been initially presented. We show the foundations and the relationships between different approaches, as well as their potential limitations. While the treatment presented here assumes that populations are monoecious, the results apply quite well to species with separate sexes, and interested readers could consult Wang (1997b) for a specific treatment of dioecious species.

29.2   THE FIXATION INDEX F

The fact that a population is subdivided means in genetic terms that individuals do not mate at random, and that the population is thus not panmictic. This departure from panmixia translates into various levels of apparent inbreeding when considering the total population. The concept of fixation index has been introduced by Sewall Wright to describe the effect of population structure on the amount of inbreeding at a given level of subdivision. Departure from panmixia indeed creates a correlation between homologous genes in uniting gametes relative to a pair of genes taken at random from the population. Wright (1921; 1922) proposed to use this correlation to describe the genetic structure and to quantify the effect of inbreeding.

As we shall see, the fixation index F is precisely a measure of that correlation. We shall briefly see how this correlation arises in the simple case of a single locus with two alleles A and a of respective and unknown frequencies p and $1 - p$ in a total population subdivided into arbitrary breeding units. Gametes unite to form diploid individuals with genotype proportions as shown in Table 29.1, where H_o denotes observed proportion of heterozygotes.

Let us define an indicator variable y, such that, in an elementary experiment consisting in drawing a gamete at random from the population, $y = 1$ if a gamete is of allelic type A and 0 otherwise. We see immediately that over replicates of this sampling process, $E(y) = p.1 + (1 - p).0 = p$, $E(y^2) = p$, and $\mathrm{var}(y) = E(y^2) - E(y)^2 = p(1 - p)$. The correlation ρ_{12} between two uniting gametes G_1 and G_2 can be written as

$$\rho_{12} = \frac{\mathrm{cov}(y_1, y_2)}{\sqrt{\mathrm{var}(y_1)\mathrm{var}(y_2)}}. \tag{29.1}$$

From Table 29.1, it is clear that $\mathrm{cov}(y_1, y_2) = E(y_1 y_2) - E(y_1)E(y_2) = p - H_o/2 - p^2 = p(1 - p) - H_o/2$, leading to

$$\rho_{12} = 1 - \frac{H_o}{2p(1 - p)}. \tag{29.2}$$

Note that the quantity $2p(1 - p)$ is the expected heterozygosity H_e under Hardy–Weinberg equilibrium (HWE) (panmixia), and therefore,

$$\rho_{12} = 1 - \frac{H_o}{H_e}. \tag{29.3}$$

Table 29.1   Pattern of union of gametes in the total population. H_o and p are the observed proportion of heterozygotes and the frequency of allele A in the total population respectively. The observed frequencies of the homozygotes depend only on these two random variables.

		Gamete 2		
		A	a	Total
Gamete 1	A	$p - H_o/2$	$H_o/2$	p
	a	$H_o/2$	$1 - p - H_o/2$	$1 - p$
	Total	p	$1 - p$	1

For a (total) polymorphic population at an arbitrary time considered to be the beginning of an evolutionary process, H_o should be equal to H_e in the initial generation if there is complete panmixia, and therefore $\rho_{12} = 0$. If after an arbitrary period of genetic drift, there is complete fixation of one or the other allele in every breeding unit (whatever it is), then $H_o = 0$ and $\rho_{12} = 1$. Therefore, Wright proposed to call this correlation of uniting gametes the 'fixation index' F because it is a relative measure of where the demes are on their course to fixation due to random genetic drift. Note that this index was first introduced in studies of inbreeding (Wright, 1921) under the notation f. Later, Wright (1922) proposed to use f in a narrower sense as a 'coefficient of inbreeding', as it is still known until today.

On the course to fixation, p and H_o are random variables that change at each generation because of genetic drift and other potential evolutionary forces. Thus, in practice, both H_o and H_e fluctuate at each generation in the total population and F, defined as a function of observed and expected heterozygosities as in (29.3), is also a random variable with a potentially complex distribution. However, let us consider the simple case where the number of demes is infinite and where gene frequencies within subdivisions only change because of genetic drift. In that case, p and therefore H_e remain constant even though the size of each deme, say N, is finite. The frequency of heterozygotes H_o in any deme should decrease *on average* by a factor $[1 - 1/(2N)]$ each generation, such that

$$F^{t+1} = 1 - \frac{H_o^{t+1}}{H_e} = 1 - \frac{\left(1 - \dfrac{1}{2N}\right) H_o^t}{H_e}.$$

Therefore, the fixation index F should approximately change over time as

$$F^{t+1} = 1 - \frac{H_o^0}{H_e}\left(1 - \frac{1}{2N}\right)^t = 1 - (1 - F^0)\left(1 - \frac{1}{2N}\right)^t, \tag{29.4}$$

to asymptotically reach the value of 1.

Note that, in full generality, the observed proportion of heterozygotes H_o can be defined in terms of allele frequencies and fixation index by rearranging (29.2) as

$$H_o = 2p(1 - p)(1 - F). \tag{29.5}$$

This relationship holds for any level of subdivision, as is shown in the next section.

29.3 WRIGHT'S F STATISTICS IN HIERARCHIC SUBDIVISIONS

Suppose now that the total diploid population is subdivided into a finite number of demes d. Since the derivation of the fixation index F in the previous section does not require that the gametes are uniting to produce zygotes, it can apply to any pair of gametes, and in fact to any pair of genes at a given locus. F can therefore be seen as a correlation between homologous genes taken from a given subdivision level (individuals, demes, or any other higher level) relative to any higher subdivision level (Wright, 1943; 1951; 1965; 1969). The correlation between genes within individuals (I) relative to the genes of the total population (T) is represented by F_{IT}. The correlation between genes within a deme or subdivision (S) relative to the genes of the total population is represented by F_{ST}. Finally, the correlation between genes within individuals relative to those within a subdivision is Wright's local inbreeding coefficient f that is represented here by F_{IS}. The list of F statistics can be easily extended to further levels of subdivisions.

We first consider the correlation of genes within demes F_{ST}. In that case, (29.5) still holds, but H_o is not the observed heterozygosity anymore. Here, the probability is that two random genes *within a subdivision* are different, without need for these two genes to be on the homologous chromosomes of an individual. If we call this probability H_S, then we have

$$H_S = 2p(1 - p)(1 - F_{ST}).$$ (29.6)

Note that H_S can also be defined as an average probability over all d subdivisions as

$$H_S = \frac{1}{d} \sum_k^d H_{Sk},$$ (29.7)

where H_{Sk} is the probability that two genes randomly chosen from the kth deme are different, which is equivalent to the expected heterozygosity in the kth deme under HWE and $H_{Sk} = 2p_k(1 - p_k)$. It should be clear however that since these two genes do not necessarily come from the same individual, the assumption of HWE does not need to be made here. Therefore,

$$H_S = \frac{1}{d} \sum_k^d 2p_k(1 - p_k) = 2\left(\bar{p} - \frac{1}{d} \sum_k^d p_k^2\right) = 2(\bar{p} - \overline{p^2})$$

$$H_S = 2[\bar{p} - \text{var}_S(p) - \bar{p}^2] = 2\bar{p}(1 - \bar{p}) - 2\text{var}_S(p),$$ (29.8)

where the S subscript stresses the fact that the variance of p is computed over subdivisions. Note that this equation also describes the 'Wahlund effect' (Wahlund, 1928), which represents the deficit of observed heterozygotes in a subdivided population made up of random mating demes. Rearranging (29.8) with (29.6) leads to the classical relationship

$$F_{ST} = \frac{\text{var}_S(p)}{\bar{p}(1 - \bar{p})}.$$ (29.9)

Therefore, the fixation index F_{ST} can also be defined as the standardized variance of allele frequencies. It is the ratio of the observed variance of gene frequencies over subdivisions divided by the maximum possible variance $\bar{p}(1 - \bar{p})$ that can be reached when alleles have gone to fixation in all subdivisions, that is when the allele frequencies within each deme are as divergent as they can be from the average population frequency. Here again, the fixation index F_{ST} is a relative measure of where the subdivisions are in the process of allelic fixation. Note that F_{ST} has not been derived by assuming random mating within demes, and therefore it should not be affected by local inbreeding. Its definition only depends on local allele frequencies. However, if there is random mating within demes, the correlation of pairs of genes within or between individuals is equivalent and $F_{ST} = F_{IT}$.

Let us now consider subdivisions where mating is not at random, so that $F_{IT} \neq F_{ST}$, and

$$H_o = 2\bar{p}(1 - \bar{p})(1 - F_{IT}).$$ (29.10)

The total proportion of heterozygotes H_o can also be expressed as an average over subdivisions, as for H_S in (29.7), but allowing for local inbreeding within subpopulations it becomes

$$H_o = \frac{1}{d} \sum_{k=1}^{d} 2p_k(1 - p_k)(1 - F_{ISk}).$$

Assuming that the level of local inbreeding is independent of the allele frequencies, which has been rightly criticized (see e.g. Barrai, 1971), Wright (1951; 1965) obtains

$$H_o = 2(1 - \bar{F}_{IS})(\bar{p} - \frac{1}{d} \sum_{k}^{d} p_k^2) = 2(1 - \bar{F}_{IS})(\bar{p} - \mathrm{var}_S(p) - \bar{p}^2), \qquad (29.11)$$

where $\bar{F}_{IS}$ is the unweighted average F_{IS} value over subdivisions. From (29.10), we have

$$\mathrm{var}_S(p) = \frac{\bar{p}(1 - \bar{p})(F_{IT} - \bar{F}_{IS})}{(1 - \bar{F}_{IS})}. \qquad (29.12)$$

Combining (29.12) with (29.9) leads to the classical relationship between fixation indexes as

$$(1 - F_{IT}) = (1 - \bar{F}_{IS})(1 - F_{ST}). \qquad (29.13)$$

The bar over $\bar{F}_{IS}$ is often omitted in the literature, but we keep it to remind us that it is in fact the average inbreeding value over subpopulations. Note that F_{ST} has no meaning if the population is not subdivided and made up of only one deme. Note also that (29.11) and therefore (29.13) are valid even when local inbreeding depends on gene frequencies provided that $\bar{F}_{IS}$ is defined as a weighted average of the form (Wright, 1969, vol. 4, chap. 3)

$$\bar{F}_{IS_w} = \frac{\displaystyle\sum_{k}^{d} w_k p_k(1 - p_k) F_{ISk}}{\displaystyle\sum_{k}^{d} w_k p_k(1 - p_k)}, \qquad (29.14)$$

where w_k is the weight given to subpopulation k. A common choice is to set the weights equal to the relative deme effective size, with $\sum_k w_k = 1$. Alternatively, if these sizes are unknown, one usually assumes that all w_k,s are equal to $1/d$, which has the effect of giving more weight to subpopulations where allele A is of intermediate frequency.

29.3.1 Multiple Alleles

The extension of the above derivations to multiple alleles is straightforward (Nei, 1977). Consider the case where we have r alleles at a given locus. Within each subpopulation, $r(r - 1)/2$ inbreeding coefficients are needed for a complete specification of the genotype frequencies $P_{kij}(i = 1 \ldots r, j = 1 \ldots i)$, as

$$P_{kij} = 2p_{ki}p_{kj}(1 - F_{IS_{kij}}),$$

$$P_{kii} = p_{ki}^2(1 - F_{IS_{kii}}) + F_{IS_{kii}} p_{ki}. \qquad (29.15)$$

If we only consider the r homozygote genotypes (see Nagylaki, 1998b, for a complete treatment of all genotypes), a global inbreeding coefficient can be obtained as a weighted average of the form

$$
F_{IS_k} = \frac{\sum\limits_{i=1}^{r} p_{ki}(1 - p_{ki})F_{IS_{kii}}}{\sum\limits_{i=1}^{r} p_{ki}(1 - p_{ki})},
$$

which becomes

$$
F_{IS_k} = \frac{\sum\limits_{i=1}^{r}(P_{kii} - p_{ki}^2)}{\sum\limits_{i=1}^{r} p_{ki}(1 - p_{ki})} = \frac{H_{Sk} - H_{ok}}{H_{Sk}},
$$

where $H_{Sk} = 1 - \sum_i p_{ki}^2$ and $H_{ok} = 1 - \sum_i P_{kii}$ are the expected and the observed heterozygosities in subpopulation k respectively. Similar to (29.14), we can define an average inbreeding coefficient over all subpopulations as

$$
\bar{F}_{IS} = \frac{\sum\limits_{k}^{d} w_k H_{Sk} F_{ISk}}{\sum\limits_{k}^{d} w_k H_{Sk}} = \frac{H_S - H_o}{H_S}, \tag{29.16}
$$

where $H_S = 1 - \sum_k \sum_i w_k p_{ki}^2$ and $H_o = 1 - \sum_k \sum_i w_k P_{kii}$ are respectively the global expected and observed heterozygosities. Similarly, the other fixation indexes can be obtained as ratios of heterozygosities

$$
F_{IT} = \frac{H_T - H_o}{H_T}, \tag{29.17}
$$

$$
F_{ST} = \frac{H_T - H_S}{H_T}, \tag{29.18}
$$

where $H_T = 1 - \sum_i \bar{p}_i^2$ and $\bar{p}_i = \sum_k w_k p_{ki}$.

With these definitions, fixation indexes can be viewed as ratios of expected or observed heterozygosities (Nei, 1977), which allows arbitrary numbers of alleles at a locus. This simplicity, however, hides the fact that subpopulations have to be given weights.

29.3.2 Sample Estimation of F Statistics

Up to this point, F statistics have been defined from population allele and genotype frequencies (which are random variables if the number of demes and deme sizes are finite or if there are other random evolutionary forces acting within demes). Some corrections need to be performed when allele frequencies are estimated from population samples. An estimation of these fixation indexes has been developed in Nei and Chesser (1983). The estimators differ from the above definitions by replacing expected and observed population heterozygosities (H_o, H_S, and H_T) by unbiased sample estimates ($\hat{H}_o$, $\hat{H}_S$, and $\hat{H}_T$), taking into account sample sizes as (Nei and Chesser, 1983)

$$\hat{H}_o = 1 - \sum_{k=1}^{d} \sum_{i=1}^{r} w_k \hat{P}_{kii}$$

$$\hat{H}_S = \frac{\tilde{n}}{\tilde{n}-1} \left(1 - \sum_{i=1}^{d} \sum_{k=1}^{r} w_k \hat{p}_{ki}^2 - \frac{\hat{H}_o}{2\tilde{n}} \right),$$

$$\hat{H}_T = 1 - \sum_{i=1}^{r} \left(\sum_{k=1}^{d} w_k \hat{p}_{ki} \right)^2 + \frac{\hat{H}_S}{\tilde{n}d} - \frac{\hat{H}_o}{2\tilde{n}d} \tag{29.19}$$

with $\tilde{n}$ being the harmonic mean of the sample sizes, and d the number of sampled demes. Note that the estimates of $\hat{H}_S$ and $\hat{H}_T$ are only unbiased if one assumes that there is no correlation between sample size and observed or expected heterozygosity levels (Nei and Chesser, 1983, p. 255). Sample estimators of F_{IS}, F_{ST}, and F_{IT} are obtained by substituting (29.19) in (29.16), (29.17), and (29.18) respectively.

However, the above estimation procedure only takes into account the sampling process within demes. It does not consider the sampling of demes within the population and therefore assumes that all demes have been sampled. Pons and Chaouche (1995) have developed estimators of F statistics that remove this assumption by extending the approach of Nei and Chesser (1983) to an additional level of sampling. They considered that d demes (supposed to be independent from each other) had been sampled and obtained

$$\hat{H}_o^* = \frac{1}{d} \sum_{i=1}^{d} \hat{H}_{ok} = 1 - \sum_{k=1}^{d} \sum_{i=1}^{r} \hat{P}_{kii}$$

$$\hat{H}_S^* = \frac{1}{d} \sum_{k=1}^{d} \frac{n_k}{n_k - 1} \left(1 - \sum_{i=1}^{r} \hat{p}_{ki}^2 - \frac{\hat{H}_{ok}}{2n_k} \right),$$

$$\hat{H}_T^* = \frac{1}{d(d-1)} \sum_{k=1}^{d} \sum_{k' \neq k}^{d} \left(1 - \sum_{i=1}^{r} \hat{p}_{ki} \hat{p}_{k'i} \right) \tag{29.20}$$

where n_k is the size of the kth sample. Again, sample estimators of F_{IS}, F_{ST}, and F_{IT} are obtained by substituting (29.20) into (29.16), (29.17), and (29.18) respectively. $\hat{H}_o^*$ is identical to $\hat{H}_o$ defined in (29.19), except that Pons and Chaouche (1995) have preferred to weight demes equally. The difference between $\hat{H}_S^*$ and $\hat{H}_S$ is in the weighting scheme of the sample sizes, but the main difference between the two above approaches is that between $\hat{H}_T^*$ and $\hat{H}_T$, where we note that $\hat{H}_T^*$ only depends on the comparison of allele frequencies between samples.

29.3.3 G Statistics

G_{ST} has been defined by Nei (1973) as an analog of F_{ST} by extending Nei's genetic distance between a pair of populations (Nei, 1972) to the case of a hierarchical structure of populations. He showed that gene frequency variation at loci with multiple alleles could be analyzed directly in terms of gene diversity, which is defined as the probability that two randomly chosen genes from a population are different (Nei, 1973). For diploid organisms with random mating, this probability only depends on allele frequencies and should be equal to the average proportion of heterozygote individuals. However, there are many organisms or genetic systems where the concept of heterozygosity does not hold, but where gene diversity would always make sense (e.g. with mitochondrial DNA, haploid, or haplo-diploid organisms). By analogy with Wright's F_{ST}, Nei defined the coefficient of gene differentiation

$$ G_{ST} = \frac{D_{ST}}{H_T}, \tag{29.21} $$

expressing the magnitude of gene differentiation among subpopulations, where D_{ST} is the average gene diversity between subpopulations, including comparisons of subpopulations with themselves. However, for diploid random mating populations $D_{ST} = H_T - H_S$, and G_{ST} is identical to F_{ST} defined in (29.18) as function of heterozygosities.

Some confusion still prevails in the literature about the differences between F statistics and G statistics. Technically, G statistics are only function of allele frequencies (in gene diversities) and do not incorporate information on observed proportions of heterozygotes, unlike F statistics. They can thus only be analogs of F statistics that do not depend on population genotype frequencies, such as F_{ST}, or correlations of gene frequencies at higher levels of subdivisions.

The concept of gene identity is explicit in the definition of G statistics, where it has been introduced by Crow and Aoki (1984), but it is also implicit in the definition of F statistics as correlation coefficients. F statistics defined in (29.16–29.18) can also be defined in terms of identity coefficients (see **Chapter 28**) as

$$ \bar{F}_{IS} \equiv \frac{Q_I - Q_S}{1 - Q_S}; \quad F_{IT} \equiv \frac{Q_I - Q_T}{1 - Q_T}; \quad F_{ST} \equiv \frac{Q_S - Q_T}{1 - Q_T}, \tag{29.22} $$

where Q_I is the probability that two genes within individuals are identical, whereas Q_S and Q_T are the probabilities that two genes within subpopulations and within the total population are identical, respectively. Of course, the values taken by these probabilities will depend on a particular spatial and mutation model (see e.g. Crow and Aoki, 1984; Cockerham and Weir, 1993; Rousset, 2004).

Estimators of G statistics are obtained by estimating gene diversities similarly to heterozygosities in (29.19) and (29.20), except that they only depend on allele frequencies (Nei, 1987, p. 191; Pons and Chaouche, 1995).

29.4 ANALYSIS OF GENETIC SUBDIVISION UNDER AN ANALYSIS OF VARIANCE FRAMEWORK

Cockerham (1969; 1973) has shown how one could decompose the total variance of gene frequencies into variance components associated to different subdivision levels within the framework of an ANOVA of gene frequencies. He also showed that variance components, identity by descent measures (in absence of mutation and migration), and F statistics were just different ways to express correlations of genes (see Table 29.3). We present below his basic derivations in some detail, as it has become a method of choice for the analysis of population subdivision. We shall use Cockerham's notations, but we will mention how to relate them to F statistics, as defined previously.

Cockerham's ANOVA framework preserves Wright's definition of F statistics in terms of correlation of gene frequencies. The notion of demes used in the ANOVA framework is close to the notion of population in the statistical sense, as demes are supposed to be independent replicates of the same evolutionary process (drawn from the same distribution). Here the unknown distribution is a stochastic evolutionary process essentially shaped by random genetic drift of gene frequencies between generations. The mutation process will be introduced later.

29.4.1   The Model

We assume that several demes of finite sizes have diverged simultaneously from an ancestral population that was at HWE. Since that time, the demes have remained separate and have been exposed to the same conditions. Samples from different demes are thus expected to differ from each other because of the sampling process of individuals within each deme (statistical sampling), and because of the stochasticity in the evolutionary process between generations (genetic sampling). We will consider a hierarchically structured diploid population, with genes in individuals, individuals in demes, and demes in the population.

Let x_{kij} be the jth allele ($j = 1, 2$) in the ith individual in the kth deme, and y_{kij} be the indicator variable for any given gene: y_{kij} is equal to 1 if x_{kij} is of type A and 0 otherwise. If p is the population frequency of the allele A, then $E(y_{kij}) = p$, $E(y_{kij}^2) = p$, and thus $\mathrm{var}(y_{kij}) = p(1 - p)$. Cockerham (1969; 1973) considered a linear model of the form

$$y_{kij} = p + a_k + b_{ki} + w_{kij}. \tag{29.23}$$

The variable y_{kij} is assumed to differ from the mean p because of the additive, random, and uncorrelated effects due to demes (a), individuals within demes (b) and genes within individual (w).

We are interested in defining the relations (correlations, covariances) between genes found in different levels of the hierarchy. Formally, the expectations of the products of gene frequencies from different subdivision levels are

$$E(y_{kij} y_{k'i'j'}) = \mathrm{cov}(y_{kij}, y_{k'i'j'}) + E(y_{kij}) E(y_{k'i'j'})$$
$$= \sigma_T^2 + p^2 \qquad \text{if } k = k', i = i', j = j'$$
$$= \mathrm{cov}_{a \supset b} + p^2 \qquad \text{if } k = k', i = i', j \neq j'.$$
$$= \mathrm{cov}_a + p^2 \qquad \text{if } k = k', i \neq i'$$
$$= \mathrm{cov}_g + p^2 \qquad \text{if } k \neq k' \tag{29.24}$$

We have assumed that the demes are independent, implying that $\mathrm{cov}_g = 0$, but this simplification could be removed, for instance, to accommodate migration (see e.g. Cockerham and Weir, 1987). The ANOVA layout corresponding to this hierarchy can be found in Table 29.2.

The covariance of gene frequencies within individuals within demes, $\mathrm{cov}_{a \supset b}$, depends on the probability of genes being identical by descent $F_i = \mathrm{Pr}(x_{ki1} \equiv x_{ki2})$, a notion

Table 29.2 Analysis of variance layout. The total sums of square (SS(T)) is decomposed in a standard fashion into sums of square due to different sources of variation. The expected mean squares are expressed here as functions of variance components. Comparisons with (29.27), (29.30), and (29.34) show that these variance components are a parameterization of the model equivalent to the correlations $\bar{\theta}$ and $\bar{F}$.

Source of variation	d.f.	Sums of squares*	Mean squares[†]	Expected mean squares
Among demes (AD)	$d-1$	$SS(AD) = \sum_{k}^{d} 2n_k(y_{k..} - y_{...})^2$	$\dfrac{SS(AD)}{d-1}$	$\sigma_w^2 + 2\sigma_b^2 + 2n'\sigma_a^2$
Among individuals within demes (WD)	$n-d$	$SS(WD) = \sum_{k}^{d} \sum_{i}^{n_k} 2(y_{ki.} - y_{k..})^2$	$\dfrac{SS(WD)}{n-d}$	$\sigma_w^2 + 2\sigma_b^2$
Between genes within individuals (WI)	n	$SS(WI) = \sum_{k}^{d} \sum_{i}^{n_k} \sum_{j}^{2} (y_{kij} - y_{ki.})^2$	$\dfrac{SS(WI)}{n}$	σ_w^2
Total	$2n-1$	$SS(T) = \sum_{k}^{d} \sum_{i}^{n_k} \sum_{j}^{2} (y_{kij} - y_{...})^2$		

$$n = \sum_{k}^{d} n_k, n' = \frac{n - \sum_{k}^{d} \frac{n_k^2}{n}}{d-1},$$

* We use the conventional dot notation to specify different means of y (e.g. $y_{ki.} = \sum_j y_{kij}$).

[†] The mean squares among demes MS(AD), within demes MS(WD), and within individuals MS(WI) are obtained as usual by dividing the corresponding sum of squares by the appropriate degrees of freedom.

introduced by Malécot (1948), which is equivalent to Wright's total inbreeding coefficient F_{IT} here. The two alleles of an individual can be of type A because they are identical by descent with probability F_i, or not identical by descent with probability $(1 - F_i)p$. Therefore, $\Pr(x_{ki1} \equiv A, \; x_{ki2} \equiv A) = E(y_{ki1}y_{ki2}) = \Pr(x_{i2} \equiv A | x_{i1} \equiv A) \cdot \Pr(x_{i1} \equiv A) = [F_i + (1 - F_i)p]p$. It follows that $\text{cov}(y_{ki1}, \; y_{ki2}) = F_i p(1 - p)$. Taking the average F_i over all individuals, $\text{cov}_{a \supset b} = \bar{F}p(1 - p)$, which shows that Malécot's identity by descent measure is also equivalent to the correlation of gene frequencies within individuals.

The covariance of gene frequencies between individuals depends on another identity measure called the *coancestry coefficient* $\theta_{ii'}$ by Cockerham, which is the probability that two random genes from different individuals, i and i', are identical by descent. Note that this coancestry coefficient is indeed equivalent for most practical purposes to Malécot (1948) *coefficient of kinship* $\Phi_{ii'}$. We now have $\Pr(x_{kij} = A, x_{ki'j'} = A) = p[\theta_{ii'} + (1 - \theta_{ii'})p]$ and $\text{cov}(y_{kij}, y_{ki'j'}) = \theta_{ii'}p(1 - p)$, and again taking the average over all pairs of individuals within demes, $\text{cov}_a = \bar{\theta}p(1 - p)$ showing that $\bar{\theta}$ is also the average correlation between two genes from two *different* individuals within the same deme. It is worth stressing that $\bar{\theta}$ can only be estimated if we have samples from several independent demes, because this correlation is relative to the correlation of the least related individuals in the population, which are assumed to be in different demes and therefore independent. In that sense, it differs from the coefficient of kinship, which is supposed to be an absolute measure of relatedness between individuals. In other words, the relatedness

Table 29.3 Relationship between intraclass correlations, variance components, Cockerham's gene correlations, and Wright's fixation indexes.

Intraclass correlations	Expressed as covariances	Expressed as variance components	Cockerham gene correlations	Wright's fixation indexes
ρ_a	$\dfrac{\text{cov}_a}{p(1-p)}$	$\dfrac{\sigma_a^2}{\sigma_a^2+\sigma_b^2+\sigma_w^2}$	$\bar{\theta}$	F_{ST}
$\rho_{a\supset b}$	$\dfrac{\text{cov}_{a\supset b}}{p(1-p)}$	$\dfrac{\sigma_a^2+\sigma_b^2}{\sigma_a^2+\sigma_b^2+\sigma_w^2}$	$\bar{F}$	F_{IT}
ρ_b	$\dfrac{\text{cov}_b}{p(1-p)}$	$\dfrac{\sigma_b^2}{\sigma_b^2+\sigma_w^2}$	$\bar{f}=\dfrac{\bar{\theta}-\bar{F}}{1-\bar{F}}$	$\bar{F}_{IS}$
ρ_w	$\dfrac{\text{cov}_w}{p(1-p)}$	$\dfrac{\sigma_w^2}{\sigma_w^2}=1$		

between individuals within demes cannot be measured without a reference point, which in this case is built by comparing individuals between demes. Note that for F or G statistics estimated from heterozygosities or gene diversities, the reference point is the comparison of a random pair of genes, either within or between demes, as is made clearer below. See also Weir (1996, see Chapter 5) for further discussions on this distinction. Because it is the variability of the genetic process among demes that introduces a correlation of genes between individuals within demes, a measure of the relatedness between individuals within demes is also a measure of the degree of differentiation between demes.

In Table 29.3, we have expressed each intraclass correlation ρ as a ratio of variance components, and made explicit the relationship between Wright's fixation indexes and Cockerham's correlations. We have also introduced the correlation $\bar{f}$, which is an average correlation of genes within individuals, like $\bar{F}$, but it is relative to the least related pair of genes within the same deme, instead of being relative to the least related pair of genes between demes. It is a composite measure that depends on the two parameters of the model $\bar{\theta}$ and $\bar{F}$. $\bar{f}$ is the analog of Wright's $\bar{F}_{IS}$, the inbreeding coefficient within demes.

From Table 29.2, (29.27), (29.30), and (29.34) and Table 29.3, we see that

$$\sigma_a^2 = p(1-p)\rho_a \qquad = \text{cov}_a \qquad = p(1-p)\bar{\theta}$$
$$\sigma_b^2 = p(1-p)(\rho_{a\supset b}-\rho_a) \quad = \text{cov}_{a\supset b}-\text{cov}_a \quad = p(1-p)(\bar{F}-\bar{\theta}).$$
$$\sigma_w^2 = p(1-p)(1-\rho_{a\supset b}) \quad = \sigma_T^2 - \text{cov}_{a\supset b} \quad = p(1-p)(1-\bar{F}) \quad (29.25)$$

Thus, despite the conventional σ^2 notation, these variance components are indeed functions of gene covariances and can therefore take negative values. As explained by Rousset (2000), these variance components do not represent a decomposition of the total variance into a sum of variances as in conventional ANOVA.

29.4.2 Estimation Procedure

29.4.2.1 Expected Mean Squares

The correlation of genes at different levels of subdivision are estimated by the method of moments by equating the mean squares expressed in terms of allele and genotype frequencies to their expectation in terms of correlation parameters. The expected mean squares are derived below in terms of the parameters of the model. From Table 29.2, we have

$$E[SS(WI)] = E\left[\sum_k^d \sum_i^{n_k} \sum_j^2 (y_{kij} - y_{ki.})^2\right]$$

$$= \sum_k^d \sum_i^{n_k} \sum_j^2 \left[E(y_{kij})^2 + E(y_{ki.})^2 - 2E(y_{kij}y_{ki.})\right]$$

$$= 2n\left\{\left[p^2 + p(1-p)\right] + \left[p^2 + \frac{1}{2}p(1-p)(1+\bar{F})\right]\right.$$

$$\left. - 2\left[p^2 + \frac{1}{2}p(1-p)(1+\bar{F})\right]\right\}$$

$$= np(1-p)(1-\bar{F}) \tag{29.26}$$

and therefore

$$E[MS(WI)] = p(1-p)(1-\bar{F}). \tag{29.27}$$

The expectation of the sum of squares between individuals within demes is computed as

$$E[SS(WD)] = 2E\left[\sum_k^d \sum_i^{n_k} (y_{ki.} - y_{k..})^2\right]$$

$$= 2\sum_k^d \sum_i^{n_k} E(y_{ki.}^2) + E(y_{k..}^2) - 2E(y_{ki.}y_{k..}).$$

In order to derive this quantity, we need to compute the variance of gene frequencies within deme $\mathrm{var}(y_{k..})$, which is obtained as

$$\mathrm{var}(y_{k..}) = \mathrm{var}\left(\frac{1}{2n}\sum_{i=1}^{n_k}\sum_{j=1}^2 y_{kij}\right)$$

$$= \frac{1}{4n_k^2}\left[\sum_{i=1}^{n_k}\sum_{j=1}^2 \mathrm{var}(y_{kij}) + 2\sum_{i=1}^{n_k} \mathrm{cov}(y_{ki1}, y_{ki2})\right.$$

$$\left. + 2\sum_{i=1}^{n_k}\sum_{i'<i}^{n_k}\sum_{j=1}^2\sum_{j'=1}^2 \mathrm{cov}(y_{kij}, y_{ki'j'})\right]$$

$$= \frac{1}{4n_k^2}\left[2n_k(p(1-p)) + 2\sum_{i=1}^{n_k} F_i p(1-p) + 2\sum_{i=1}^{n_k}\sum_{i'<i}^{n_k} 4\theta_{ii'} p(1-p)\right]$$

$$= \frac{1}{4n_k^2}\left[2n_k p(1-p) + 2p(1-p)n_k\bar{F} + 4n_k(n_k-1)\bar{\theta} p(1-p)\right]$$

$$\mathrm{var}(y_{k..}) = \frac{p(1-p)}{2n_k}[1 + \bar{F} + 2(n_k-1)\bar{\theta}], \tag{29.28}$$

where $\bar{\theta}$ is the average correlation of genes between individuals (the $\theta_{ii'}$) or the average kinship coefficient between the individuals of the sample. The derivation of (29.28) is instructive, as it shows that the variance of gene frequencies within demes depends on the breeding structure of the population ($\bar{F}$) and the relatedness of the individuals within demes ($\bar{\theta}$). If the sample was noninbred and made up of unrelated individuals, the genetic variance would be removed and the variance would simply be equal to the binomial variance $p(1-p)/(2n_k)$.

The expected sum of squares within demes can now be computed as

$$E[SS(WD)] = 2\sum_k^d \sum_i^{n_k} \left\{ \left[p^2 + \frac{p(1-p)}{2}(1+\bar{F}) \right] \right.$$

$$+ \left[p^2 + \frac{p(1-p)}{2n_k}(1+\bar{F}+(n_k-1)\bar{\theta} \right]$$

$$\left. -2\left[p^2 + \frac{p(1-p)}{2n_k}(1+\bar{F}+(n_k-1)\bar{\theta} \right] \right\}$$

$$= (n-d)p(1-p)(1+\bar{F}-2\bar{\theta}), \qquad (29.29)$$

and therefore

$$E[MS(WD)] = p(1-p)(1-\bar{F}+2(\bar{\theta}-\bar{F})). \qquad (29.30)$$

Finally, the expectation of the sum of squares between demes is obtained as

$$E[SS(AD)] = E\left[\sum_k^d 2n_k(y_{k..}-y_{...})^2 \right] = \sum_k^d 2n_k[E(y_{k..}^2)+E(y_{...}^2)-2E(y_{k..}y_{...})]. \qquad (29.31)$$

In order to derive this quantity, we first need to get the total variance of gene frequencies:

$$\mathrm{var}(y_{...}) = \mathrm{var}\left(\frac{1}{2n} \sum_k^d \sum_i^{n_k} \sum_1^2 y_{kij} \right) = \frac{1}{4n^2}\left[\sum_k^d \mathrm{var}(2n_k y_{k..}) \right.$$

$$\left. +2\sum_{k'<k}^d 16n_k^2 n_{k'}^2 \mathrm{cov}(y_{k..}y_{k'..}) \right].$$

Because we have assumed that the demes were independent from each other, the covariance terms can be ignored, and the total variance reduces to

$$\mathrm{var}(y_{...}) = \left[\sum_k^d \frac{n_k^2}{n^2}\mathrm{var}(y_{k..}) \right] = \frac{p(1-p)}{2n}\left[1+\bar{F}+2\bar{\theta}\left(\frac{\left(\sum_k^d n_k^2\right)}{n}-1 \right) \right]. \qquad (29.32)$$

Then, the expectation of the product of the total mean $(y_{..})$ and the within population mean $(y_{k..})$ is obtained as

$$E(y_{k..}y_{...}) = E\left[y_{k..}\left(\frac{1}{2n}\sum_{k'=1}^{d}2n_{k}y_{k'..}\right)\right] = \frac{n_{k}}{n}\sum_{k'=1}^{d}E[y_{k'..}y_{k..}]$$

$$= \frac{n_{k}}{n}\left[E(y_{k..}^{2}) + \sum_{k'\neq k}^{d}E(y_{k'..}y_{k..})\right]$$

$$= \frac{n_{k}}{n}\left\{p^{2} + \frac{p(1-p)}{2n_{k}}[1 + \bar{F} + 2(n_{k}-1)\bar{\theta}] + (d-1)p^{2} + \sum_{k'\neq k}^{d}\text{cov}(y_{k'..},y_{k..})\right\}$$

$$= p^{2} + \frac{p(1-p)}{2n}[1 + \bar{F} + 2(n_{k}-1)\bar{\theta}]. \tag{29.33}$$

After some algebra, we obtain the expected mean squares between populations as

$$E[MS(AD)] = \frac{E[SS(AD)]}{d-1} = p(1-p)\left[1 - F + 2(\theta - F) + \frac{2\theta}{d-1}\left(n - \frac{\sum\limits_{k}^{d}n_{k}^{2}}{n}\right)\right], \tag{29.34}$$

which explains the weighting scheme of the variance components in Table 29.2.

29.4.2.2 *Moment Estimators*

Unbiased estimators of the variance components are obtained by the method of moments, equating the observed mean squares to their expectations in Table 29.2, and rearranging as

$$\hat{\sigma}_{w}^{2} \triangleq MS(WI)$$

$$\hat{\sigma}_{b}^{2} \triangleq \frac{1}{2}[MS(WD) - MS(WI)],$$

$$\hat{\sigma}_{a}^{2} \triangleq \frac{1}{2n'}[MS(AD) - MS(WD)] \tag{29.35}$$

from which the estimators of gene correlations are defined as

$$\hat{\bar{F}} = \frac{\hat{\sigma}_{a}^{2} + \hat{\sigma}_{b}^{2}}{\hat{\sigma}_{a}^{2} + \hat{\sigma}_{b}^{2} + \hat{\sigma}_{w}^{2}}$$

$$\hat{\bar{\theta}} = \frac{\hat{\sigma}_{a}^{2}}{\hat{\sigma}_{a}^{2} + \hat{\sigma}_{b}^{2} + \hat{\sigma}_{w}^{2}}.$$

$$\hat{\bar{f}} = \frac{\hat{\bar{F}} - \hat{\bar{\theta}}}{1 - \hat{\bar{\theta}}} = \frac{\hat{\sigma}_{b}^{2}}{\hat{\sigma}_{b}^{2} + \hat{\sigma}_{w}^{2}} \tag{29.36}$$

Note that these estimators do not depend on the unknown gene frequency p, and that they are not necessarily unbiased, as they are just ratios of unbiased quantities.

29.4.2.3 Weighted Averages for Several Alleles and Several Loci

The above derivations have been made for a single allele at a given locus. Similar computations could be made for each allele of a multiallelic (say r alleles) polymorphism, giving rise to $r - 1$ independent estimates of the same parameters $\bar{f}$, $\bar{F}$, and $\bar{\theta}$. Noting that estimators of correlations are ratios of variance components, weighted estimators can be defined as

$$\hat{\bar{F}}_w = \frac{\sum\limits_{u=1}^{r} \hat{\sigma}_{b_u}^2 + \sum\limits_{u=1}^{r} \hat{\sigma}_{w_u}^2}{\sum\limits_{u=1}^{r} \hat{\sigma}_{T_u}^2} = \frac{\bar{\hat{\sigma}}_b^2 + \bar{\hat{\sigma}}_w^2}{\bar{\hat{\sigma}}_T^2}$$

$$\hat{\bar{\theta}}_w = \frac{\sum\limits_{u=1}^{r} \hat{\sigma}_{a_u}^2}{\sum\limits_{u=1}^{r} \hat{\sigma}_{T_u}^2} = \frac{\bar{\hat{\sigma}}_a^2}{\bar{\hat{\sigma}}_T^2}.$$

$$\hat{\bar{f}}_w = \frac{\sum\limits_{u=1}^{r} \hat{\sigma}_{b_u}^2}{\sum\limits_{u=1}^{r} \hat{\sigma}_{b_u}^2 + \sum\limits_{u=1}^{r} \hat{\sigma}_{w_u}^2} = \frac{\bar{\hat{\sigma}}_b^2}{\bar{\hat{\sigma}}_b^2 + \bar{\hat{\sigma}}_w^2} \qquad (29.37)$$

From (29.25), it is clear that these estimators are weighted averages of the intraclass correlations ρ, with weight for allele u equal to the total variance $\sigma_{Tu}^2 = p_u(1 - p_u)$. Similarly, if estimators are available from l loci, they can be combined as

$$\hat{\bar{F}}_w = \frac{\sum\limits_{t=1}^{l} \sum\limits_{u=1}^{r_t} \hat{\sigma}_{b_{tu}}^2 + \sum\limits_{t=1}^{l} \sum\limits_{u=1}^{r_t} \hat{\sigma}_{w_{tu}}^2}{\sum\limits_{t=1}^{l} \sum\limits_{u=1}^{r_t} \hat{\sigma}_{T_{tu}}^2}, \quad \hat{\bar{\theta}}_w = \frac{\sum\limits_{t=1}^{l} \sum\limits_{u=1}^{r} \hat{\sigma}_{a_{tu}}^2}{\sum\limits_{t=1}^{l} \sum\limits_{u=1}^{r_t} \hat{\sigma}_{T_{tu}}^2}, \quad \hat{\bar{f}}_w = \frac{\sum\limits_{t=1}^{l} \sum\limits_{u=1}^{r_t} \hat{\sigma}_{b_{tu}}^2}{\sum\limits_{t=1}^{l} \sum\limits_{u=1}^{r_t} \hat{\sigma}_{b_{tu}}^2 + \sum\limits_{t=1}^{l} \sum\limits_{u=1}^{r_t} \hat{\sigma}_{w_{tu}}^2}.$$

Alternative ways of building average estimators such as unweighted or least-squares estimators are discussed in Reynolds *et al.* (1983), while other weighting schemes are discussed in Weir and Cockerham (1984) and in Robertson and Hill (1984). It appears that weighted estimators as in (29.37) have a very low bias and are rather insensitive to unequal sample sizes, even though they may have a slightly larger variance than other combined estimators (Goudet *et al.*, 1996).

29.4.2.4 Likelihood-based Methods

A number of authors have derived likelihood-based estimators of F statistics by specifying a given parametric distribution of the unknown allele frequencies (p_i) in subpopulations (e.g. Rannala and Hartigan, 1996; Holsinger *et al.*, 2002; Nicholson *et al.*, 2002; Weir and Hill, 2002; Balding, 2003; Kitada and Kishino, 2004; Foll and Gaggiotti, 2006; Kitakado *et al.*, 2006). While some authors have assumed a Normal distribution of these allele frequencies (e.g. Nicholson *et al.*, 2002; Weir and Hill, 2002), most others have chosen to assume that the r allele frequencies in subpopulations followed a β or a Dirichlet distribution (Wright, 1951; Balding, 2003) with parameters $\alpha p_i (i = 1, \ldots, r)$, where p_i is the unknown frequency of the ith allele. An interesting discussion on the validity of a Dirichlet distribution for describing allele frequencies in subpopulations under different evolutionary models or for different markers can be found in Balding (2003). Under this framework, the variance of p_i over populations is given by $\mathrm{var}(p_i) = p_i(1 - p_i)/(\alpha + 1)$, showing that $F_{ST} = 1/(\alpha + 1)$ from (29.9), and that $\alpha = F_{ST}^{-1} - 1$. It follows that the allelic counts (n_i) in each subpopulation follow a multinomial-Dirichlet distribution of the form

$$\mathrm{Pr}(n_i | \alpha, \mathbf{p}) = \frac{n!\Gamma(\alpha)}{\Gamma(n + a)} \prod_{i=1}^{r} \frac{\Gamma(n_i + \alpha p_i)}{n_i!\Gamma(\alpha p_i)}, \tag{29.38}$$

where $n = \sum_r n_i$, which provides a simple basis for maximum-likelihood or Bayesian estimation of F_{ST} (Balding, 2003). Note that (29.38) needs to be multiplied across loci and across subpopulations to get a global likelihood and a global F_{ST} estimate. This equation still depends on unknown allele frequencies, which can either be integrated out or coestimated with F_{ST} using an Markov chain Monte Carlo (MCMC) approach (Balding, 2003; Kitakado *et al.*, 2006). Interestingly, subpopulation specific or locus-specific F_{ST} can also be estimated, if one believes that some demes had distinct histories (e.g. Falush *et al.*, 2003a; Foll and Gaggiotti, 2006), or that selection has shaped genetic diversity at a given locus (Beaumont and Balding, 2004). Extensions of this approach have been proposed to take into account local inbreeding (Ayres and Overall, 1999) or linkage disequilibrium between loci (Kitada and Kishino, 2004).

29.4.3 Dealing with Mutation and Migration using Identity Coefficients

When mutations and migrations are considered, two major changes take place as compared to previous derivations: the global population allele frequencies cannot be considered as constant over time (due to mutations), and the subpopulations cannot be considered as independent (due to migration). Cockerham and Weir (1987; 1993) have overcome these difficulties by expressing correlation measures as functions of the identity coefficients Q. Note that a formalization of the effect of a population structure in terms of identity coefficients has led to the derivation of equilibrium values of the fixation indexes for different spatial models incorporating mutations (see e.g. Nagylaki, 1983; Crow and Aoki, 1984; Cockerham and Weir, 1987; Slatkin, 1991; 1993; Slatkin and Voelm, 1991; Rousset, 1996; 1997; 2004; Herbots, 1997; Wilkinson-Herbots, 1998; Weir and Hill, 2002; Wilkinson-Herbots and Ettridge, 2004).

The variance components now depend on migration and mutation parameters as well as on unknown allele frequencies, but an ANOVA can still be performed. In that case, however, Cockerham and Weir (1987) showed that the variance components corresponding to the linear model described in (29.23) were equal to

$$\sigma_a^2 = Q_1 - Q_2$$
$$\sigma_b^2 = Q_0 - Q_1$$
$$\sigma_w^2 = 1 - Q_0, \tag{29.39}$$

where Q_0 is the probability that two genes from the same individual are identical in state, Q_1 is the same identity probability for two genes from different individuals within the same deme, and Q_2 is the identity probability for two genes in different demes. Therefore, the intraclass correlations defined in the same way as in (29.36) are now given by

$$\rho_0 = \frac{Q_0 - Q_1}{1 - Q_1}; \rho_1 = \frac{1 - Q_1}{1 - Q_2}; \rho_2 = \frac{Q_1 - Q_2}{1 - Q_2}, \tag{29.40}$$

where $\rho_0 \equiv \bar{F}_{IS}$, $\rho_1 \equiv F_{IT}$, and $\rho_2 \equiv F_{ST}$. Rousset (2002) more recently showed that such definitions of inbreeding coefficients as ratios of differences of probability of identity in state are very general and do not depend on a particular evolutionary model.

29.5   RELATIONSHIP BETWEEN DIFFERENT DEFINITIONS OF FIXATION INDEXES

The use of identity coefficients provides a convenient way to understand the relationship between the different indexes used to quantify population subdivision. It is interesting to compare the quantities defined in (29.40) with those defined in (29.22) as functions of similar but different identity coefficients to understand a major difference between the gene diversity and the intraclass correlation approach to estimate F statistics. In Nei's (1973; 1977) approach, the correlations are always relative to a random pair of genes at a given subdivision level (Q_S or Q_T in (29.22)), whereas in Cockerham's approach (1969; 1973) it is always relative to the least related pair of genes for a given subdivision level (Q_1 or Q_2 in (29.40)). Note that when the number of sampled diploid individuals within demes (n_k) is large then $\hat{Q}_S \approx \hat{Q}_1$. This is because $n_k(n_k - 1)$ pairs of genes are compared when computing $\hat{Q}_S$, whereas only $n_k(n_k - 2)$ pairs are considered when computing $\hat{Q}_1$. In other words, the n_k pairs of genes within individuals that are not taken into account when computing $\hat{Q}_1$ (as compared to $\hat{Q}_S$) become negligible relative to the number of between-individual gene pairs. For a similar reason, $\hat{Q}_T \approx \hat{Q}_2$ when the number of sampled demes is large, and therefore F statistics defined from gene diversities or heterozygosities are close to those defined in terms of variance components. We study below the relationships between different estimators of F_{ST} in more detail.

   As usual now, let Q be the probability that two random genes are identical in state, but here we simply have Q_1 for genes within the same deme, and Q_2 for genes in different demes. The intraclass correlation within demes equivalent to ρ_2 in (29.40) is equal to (Cockerham and Weir, 1987)

$$\beta = \frac{Q_1 - Q_2}{1 - Q_2}, \tag{29.41}$$

while Crow and Aoki (1984) have redefined the G_{ST} statistic in (29.21) as a parameter corresponding to F_{ST} in (29.22). Following Cockerham and Weir (1993), we call this parameter G_{CA} (to differentiate it from the G_{ST} statistic defined in (29.21)). It is therefore equal to

$$G_{CA} = \frac{Q_1 - \bar{Q}}{1 - \bar{Q}}, \tag{29.42}$$

where $\bar{Q} = [Q_1 + (d-1)Q_2]/d$ is the probability corresponding to Q_T in (29.22) that two genes are alike when drawn randomly from the total population. Note that the two quantities are therefore related by the following relationships (Cockerham and Weir, 1987):

$$G_{CA} = \frac{(d-1)\beta}{d - \beta}; \quad \beta = \frac{d G_{CA}}{G_{CA} + d - 1}, \tag{29.43}$$

which corresponds to the relationship between F_{ST} and ρ_2 defined in (29.22) and (29.40) respectively. As explained above, $G_{CA} \approx \beta$ if the number of demes is large, but β should always be slightly larger than G_{CA} (Goudet, 1993).

If we now consider estimators based on these parameters, the relationship between $\hat{\beta}$ and $\hat{G}_{CA}$ would be identical to (29.43), but with d being here the number of sampled demes (Cockerham and Weir, 1993). For the estimator of G_{ST} as defined in (29.21) or F_{ST} as defined from (29.19), its relationship with $\hat{\beta}$ and $\hat{G}_{CA}$ depends on both the numbers of sampled demes and of sampled individuals per deme (see Cockerham and Weir, 1993, (29.4)). We note here that an estimator of F_{ST} based on the estimation of heterozygosities defined in (29.20) (Pons and Chaouche, 1995) should not depend on the number of demes.

Comparison between the actual values taken by different estimators has been achieved through simulations, or by comparing equilibrium values in different spatial models (e.g. Weir and Cockerham, 1984; Slatkin and Barton, 1989; Chakraborty and Danker-Hopfe, 1991; Rousset, 1997; Weir and Hill, 2002). The results of these comparisons vary in details, but the general trend is that estimators behave quite similarly and lead to nearly identical values provided that sample sizes and the number of sampled demes are large. For small samples (100 individuals or less), Chakraborty and Danker-Hopfe (1991) find that both heterozygosity and intraclass correlation approaches overestimate F_{ST} and underestimate $\bar{F}_{IS}$, while F_{IT} estimates are close to large samples values. From (29.43), it is clear that expectations and estimations based on heterozygosity are affected by the number of demes sampled, which is not the case for estimators based on intraclass correlation. In order to address this deficiency, Nei (1986; 1987, p. 163) has proposed to modify D_{ST} as $D'_{ST} = d D_{ST}/(d-1)$ to remove the comparison of samples with themselves. He also had to redefine H_T as $H'_T = D'_{ST} + H_S$, which results in a new definition of $G'_{ST} = D'_{ST}/H'_T$ similar to that of β in (29.41). Note that this modification is not entirely satisfactory, as the actual number of demes is usually unknown, and not necessarily equal to the number of demes sampled. Thus, even though different approaches lead to quite similar estimators in most practical situations, the ANOVA approach seems preferable for its clear statistical foundations (but see Nagylaki, 1998b for a statistically sound approach based on gene diversities). Another advantage is that the intraclass correlation approach can easily be applied to dominant data (Peakall et al., 1995; Stewart and Excoffier, 1996), as well as to molecular markers (Weir and Basten, 1990; Excoffier et al., 1992; Michalakis and Excoffier, 1996), while retaining exactly the same analytical design. Moreover, its connections with identity by descent and kinship coefficients make it suitable for analyses of regular systems of mating. However, several features of the ANOVA approach have been criticized. First, estimates of variance components with the highest subdivision levels can be negative. This can arise by the way they are extracted from the mean squares using the method of moments (see (29.35)). When the true value of the variance component is positive but small, slightly negative estimates can be obtained due to random fluctuations with small samples, but negative values can also be obtained because the true variance components are negative. This latter situation is not embarrassing, because they are not obtained by a decomposition of the total variance into variances. Rather, they are functions of covariances (see Table 29.3 and (29.25)). As explained in **Chapter 28** (Appendix 1), they differ from the strictly positive variance components in conventional ANOVA because we do not assume that the effects due to deme (a), individuals (b), and genes (w) are uncorrelated between demes, between individuals within demes, or between genes within individuals. Negative coefficients can indeed arise in real populations if alleles are more alike between individuals than within individuals (if there is self-mating avoidance or separate sexes), or between demes than within demes (if there is predominant outbreeding) (see e.g. Wright, 1969; Cockerham, 1973 for a discussion on this subject). Note, however, that heterozygosity-based estimates of F statistics can also become negative in similar situations if random mating is not assumed (Goudet, 1993). In that case, it is clear that the definition of F statistics as correlations still makes sense, whereas definitions in terms of identity probabilities do not. A more justified criticism made to the approach of Weir and Cockerham is that they consider the different subpopulations as replicates, and therefore as having identical sizes and being subjected to identical evolutionary forces, which is usually implausible in biological situations except for experimental populations (Chakraborty and Danker-Hopfe, 1991). On the contrary, approaches based on gene diversity and heterozygosity do not make any such assumptions and can in principle handle unequal deme sizes if properly estimated, which may be difficult (Gaggiotti and Excoffier, 2000). It seems therefore that intraclass correlation approaches should be preferred when the number of demes is small, supposed so, or totally unknown. On the contrary, in the case of an exhaustive sampling, or when demes are supposed to be of highly unequal size, gene diversity and heterozygosity approaches could be preferred due to lower variance consideration (Cockerham and Weir, 1993). However, maximum-likelihood-based methods (e.g. Beerli and Felsenstein, 2001; Balding, 2003; Beerli, 2006) should be superior in the context of gene flow estimations.

29.6 F STATISTICS AND COALESCENCE TIMES

Fixation indexes and related quantities can be conveniently described in terms of average coalescence times within and between subdivision levels assuming small mutation rates (Slatkin, 1991; Slatkin and Voelm, 1991). Using this formalization, equilibrium properties of different spatial models of population subdivision can be easily obtained (see e.g. Slatkin, 1995; Wilkinson-Herbots, 1998; Rousset, 2004). This approach gives results essentially identical to classical approaches based on recursions (see e.g. Crow and Aoki, 1984; Cockerham and Weir, 1987; 1993; Rousset, 1996).

Following Slatkin (1991), let us consider a set of demes that have diverged from an ancestral population some T generations ago. The demes have remained separate without exchanging migrants ever since (models with migration are reviewed in **Chapter 28**). We assume that the demes are made up of haploid or diploid individuals that randomly unite their gametes (no inbreeding).

Two genes are of the same allelic type if no mutation occurred since their divergence from a most recent common ancestor (since their coalescence time), say some t generations ago. Conditional on t, this event has a probability $Q(t) = (1 - \mu)^{2t}$, where μ is the mutation rate per generation. Then the unconditional expectation of Q is

$$Q = \sum_{t=1}^{\infty} (1 - \mu)^{2t} \Pr(t) \qquad (29.44)$$

where $\Pr(t)$ is the probability that there was a coalescence event t generations ago. For small μ, it becomes

$$Q \approx \sum_{t=1}^{\infty} (1 - 2\mu t) \Pr(t) = 1 - 2\mu \bar{t} \qquad (29.45)$$

where $\bar{t}$ is simply the average coalescence time. Note that (29.45) is a good approximation if $\mu \bar{t} \ll 1$ (Slatkin and Voelm, 1991), but see Wilkinson-Herbots (1998) for the exact result with large mutation rates. It follows that we can use (29.45) directly to express β defined in (29.41) as a function of average coalescence times as (Slatkin, 1991)

$$\beta = \frac{\bar{t}_1 - \bar{t}_0}{\bar{t}_1} \qquad (29.46)$$

where $\bar{t}_1$ is the average coalescence time of two genes from different demes and $\bar{t}_0$ is the average coalescence time of two genes drawn from the same deme. Similarly, Crow and Aoki's G_{CA} can be expressed as

$$G_{CA} = \frac{\bar{t} - \bar{t}_0}{\bar{t}}, \qquad (29.47)$$

where $\bar{t}$ is the average coalescence time of two genes drawn from the whole population, either in the same or in different demes. This parameterization of the fixation index in terms of average coalescence times can be extended to higher subdivision levels such as equivalents of F_{SC} (correlation of a pair of genes taken within demes relative to a pair of genes taken at random from a group of demes) or F_{CT} (correlation of a pair of genes taken within a group of demes relative to a pair of genes taken at random from the population). They are good approximations if migration rates are much smaller between demes from different subdivision levels than from the same subdivision level (Slatkin and Voelm, 1991). As more generally stated by Rousset (2002; 2004, p. 58–62), inbreeding coefficients or F statistics can be seen as the increased probability of recent coalescence of one pair of genes relative to another pair, the choice of pairs of genes to be compared defining the inbreeding coefficient.

We can use these relationships to study the rate of approach of these statistics to equilibrium. The average coalescent time within diploid demes of size N is simply $\bar{t}_0 = 2N$, while the average coalescence time between demes is given by $\bar{t}_1 = T + 2N$

(Tajima, 1993), since two genes have to be in the same ancestral population in order to coalesce, then it takes again on average $2N$ more generations. From (29.46), it follows that $\beta = T/(T + 2N)$, suggesting that

$$\frac{\beta}{1 - \beta} = \frac{T}{2N} \qquad (29.48)$$

can be used as an estimate of the divergence time (T) of the demes scaled by the population size. It is interesting to see that (29.48) is independent from the number of demes, which means that it can also be used as a genetic distance between populations (Slatkin, 1995) for loci with low mutation rates. Note that this is not the case for the same ratio of G_{CA} whose expectation is $G_{CA} = (d-1)T/[(d-1)T + 2Nd]$, implying

$$\frac{G_{CA}}{1 - G_{CA}} = \frac{T(d-1)}{2Nd}. \qquad (29.49)$$

If the true (but usually unknown) number of demes is two, then the ratio of G_{CA} values is only one half that for β. As G_{CA} depends on the unknown number of demes d, it implies that the scaled divergence time between demes is poorly estimated from gene diversity measures like G_{CA}.

29.7   ANALYSIS OF MOLECULAR DATA: THE AMOVA FRAMEWORK

Because conventional ANOVA compares average gene frequencies among populations, it is not best suited to deal with molecular markers for which we have information not only on allele frequencies, but also on the amount of differences (mutations) between alleles. We show hereafter how ANOVA can be modified to directly incorporate this molecular information and thus to become an analysis of molecular variance (AMOVA) (Excoffier *et al.*, 1992). AMOVA can be considered as an extension of Cockerham's ANOVA of gene frequencies, where intraclass correlation coefficients do depend on the mutational process. In this sense, it is analogous to Cockerham and Weir's (1987; 1993) extension of their correlation measures to identity coefficients to include mutation and migration. The common denominator between the two approaches is to estimate F statistics through intraclass correlations defined as ratios of variance components.

29.7.1   Haplotypic Diversity

Molecular information at a given locus can be considered as a particular combination of alleles at different but completely linked loci, which is one way to define a haplotype. The nature of the molecular information may vary (DNA sequence, restriction fragment length polymorphism (RFLP), microsatellites), but the format of the analysis remains the same. For sake of simplicity, let us assume that each mutation occurs at a different site (the infinite-site model), and consider a haploid genetic system or assume that the gametic phase of S-linked polymorphic loci can be determined without ambiguity (e.g. mitochondrial DNA, Y chromosome, X chromosome in males). In that case, each haplotype can be coded as a vector of biallelic states of dimension S, where each element

Table 29.4 Analysis of Molecular Variance (AMOVA) layout for the linear model defined in (29.50). The sums of squares are expressed here as functions of squared Euclidean distances between haplotypes i and j (δ_{ij}^2's).

Source of variation	d.f.*	Sums of squares	Mean squares	Expected mean squares
Among groups	$G-1$	$SS(AG) = SS(T)$ $-\sum\limits_{g}^{G}\dfrac{1}{2n_g}\sum\limits_{k}^{d}\sum\limits_{k'}^{d}\sum\limits_{i}^{n_{gk}}\sum\limits_{j}^{n_{gk}}\delta_{ij}^2$	$\dfrac{SS(AG)}{G-1}$	$\sigma_w^2 + n''\sigma_b^2 + n'''\sigma_a^2$
Among demes within groups	$d-G$	$SS(AD) = \sum\limits_{g}^{G}\dfrac{1}{2n_g}\sum\limits_{k}^{d_g}\sum\limits_{k'}^{d_g}\sum\limits_{i}^{n_{gk}}\sum\limits_{j}^{n_{gk'}}\delta_{ij}^2$ $-SS(WD)$	$\dfrac{SS(AD)}{d-G}$	$\sigma_w^2 + n'\sigma_b^2$
Among haplotypes within demes	$n-d$	$SS(WD) = \sum\limits_{g}^{G}\sum\limits_{k}^{d_g}\dfrac{1}{2n_{gk}}\sum\limits_{i}^{n_{gk}}\sum\limits_{j}^{n_{gk}}\delta_{ij}^2$	$\dfrac{SS(WD)}{n-d}$	σ_w^2
Total	$n-1$	$SS(T) = \dfrac{1}{2n}\sum\limits_{i}^{n}\sum\limits_{j}^{n}\delta_{ij}^2$		

$$d = \sum_{g}^{G} d_g, \quad n = \sum_{g}^{G}\sum_{k}^{d_g} n_{gk}, \quad n_g = \sum_{k}^{d_g} n_{gk}, \quad S_G = \sum_{g}^{G}\sum_{k}^{d_g}\frac{n_{gk}^2}{n_g}, \quad n' = \frac{n - S_G}{d - G}, \quad n'' = \frac{S_G - \sum\limits_{k}^{d}\dfrac{n_k^2}{n}}{G-1}, \quad n''' = \frac{n - \sum\limits_{g}^{G}\dfrac{n_g^2}{n}}{G-1},$$

* G is the number of groups of demes.

y_s is an indicator variable that may have the value 0 or 1. We show later how to handle multiallelic markers.

We consider here a hierarchical model where genes are within demes, demes within groups of demes, and groups within the population. In that case, $\mathbf{y}_{gki}$ represents the ith haplotype of the kth deme in the gth group. We shall use a linear model similar to that seen in conventional ANOVA (29.23),

$$\mathbf{y}_{gki} = \mathbf{p} + \mathbf{a}_g + \mathbf{b}_{gk} + \mathbf{w}_{gki}, \tag{29.50}$$

where $\mathbf{p}$ is the vector of the expected allelic states. The effects are here $\mathbf{a}$ for the group, $\mathbf{b}$ for the deme, and $\mathbf{w}$ for the individual, and, as usual, are assumed random, additive, and independent, with associated variance components σ_a^2, σ_b^2, and σ_c^2.

The ANOVA layout for this partitioning of the total variance is shown in Table 29.4. We introduce here an alternative but completely equivalent way of computing the sums of squares. They are expressed as functions of square Euclidean distances instead of squared deviations from some mean. As a matter of fact, a sum of squared deviations from the mean can be written as a sum of squared differences between observations, barring a given constant, as

$$SS(z) = \sum_{i=1}^{N}(z_i - \bar{z})^2 = \frac{1}{2N}\sum_{i=1}^{N}\sum_{j=1}^{N}(z_i - z_j)^2.$$

We can thus rewrite the sum of squares within demes

$$SS(WD) = \sum_g^G \sum_k^{d_g} \sum_i^{n_{gk}} (\mathbf{y}_{gki} - \mathbf{y}_{gk.})' \mathbf{W} (\mathbf{y}_{gki} - \mathbf{y}_{gk.}),$$

as

$$SS(WD) = \sum_g^G \sum_k^{d_g} \frac{1}{2n_{gk}} \sum_i^{n_{gk}} \sum_j^{n_{gk}} (\mathbf{y}_{gki} - \mathbf{y}_{gkj})' \mathbf{W} (\mathbf{y}_{gki} - \mathbf{y}_{gkj})$$

where $\mathbf{W}$ is a square weight matrix that can be used to specify the relationships between the different loci and/or their respective weights. The expression $(\mathbf{y}_{gki} - \mathbf{y}_{gkj})' \mathbf{W} (\mathbf{y}_{gki} - \mathbf{y}_{gkj})$ is nothing else but a squared Euclidean distance δ_{ij}^2 between haplotype i and j in a S-dimensional space. Note that if all loci are assumed independent and equally informative, then $\mathbf{W} = \mathbf{I}$, the identity matrix, and we can write the sum of squares in a simple way as

$$SS(WD) = \sum_g^G \sum_k^{d_g} \frac{1}{2n_{gk}} \sum_i^{n_{gk}} \sum_j^{n_{gk}} \delta_{ij}^2$$

$$= \sum_g^G \sum_k^{d_g} \frac{1}{2n_{gk}} \sum_i^{n_{gk}} \sum_j^{n_{gk}} \sum_s^S (y_{gki_s} - y_{gkj_s})^2. \tag{29.51}$$

In that case, the squared Euclidean distance is just the number of pairwise differences between two haplotypes, equivalent to a Hamming distance. The other sums of squares are given by

$$SS(AD) = SS(\text{Within Group}) - SS(WD)$$

$$= \sum_g^G \frac{1}{2n_g} \sum_k^{d_g} \sum_{k'}^{d_g} \sum_i^{n_{gk}} \sum_j^{n_{gk'}} \delta_{ij}^2 - \sum_g^G \sum_k^{d_g} \frac{1}{2n_{gk}} \sum_i^{n_{gk}} \sum_j^{n_{gk}} \delta_{ij}^2, \tag{29.52}$$

$$SS(AG) = SS(T) - SS(\text{Within Group})$$

$$= \frac{1}{2n} \sum_i^n \sum_j^n \delta_{ij}^2 - \sum_g^G \frac{1}{2n_g} \sum_k^{d_g} \sum_{k'}^{d_g} \sum_i^{n_{gk}} \sum_j^{n_{gk}} \delta_{ij}^2. \tag{29.53}$$

We can thus directly express the variance components and the intraclass correlations as a function of Euclidean distances between haplotypes (Table 29.4). By analogy with Wright's F statistics, we have called these intraclass correlations Φ statistics (Excoffier *et al.*, 1992). They are conventionally defined as

$$\Phi_{ST} = \frac{\sigma_a^2 + \sigma_b^2}{\sigma_T^2}; \quad \Phi_{SC} = \frac{\sigma_b^2}{\sigma_b^2 + \sigma_w^2}; \quad \Phi_{CT} = \frac{\sigma_a^2}{\sigma_T^2}. \tag{29.54}$$

If $\mathbf{W} = \mathbf{I}$, the identity matrix, the estimated Φ statistics are equivalent to Weir and Cockerham (1984) weighted correlations of gene frequencies over all loci (Michalakis and Excoffier, 1996) as defined in (29.37) for another hierarchical model. For other weighting schemes, the relationship between Φ statistics and Weir and Cockerham's statistics would be different, even though the underlying parameters (variance components) to estimate would remain identical.

29.7.2 Genotypic Data

Even though the AMOVA framework has been essentially developed for haplotypic data, data consisting of diploid genotypes can be accommodated to estimate correlations between genes in different individuals at any level of subdivision. This is possible because the correlation of genes between individuals (the coancestry coefficient $\bar{\theta}$ in Cockerham's terminology) does not depend on how genes are associated within individuals. Diploid molecular data can be transformed into pseudohaploid data by defining the arbitrary gametic phase in all individuals (Michalakis and Excoffier, 1996). If one assumes that the weighting matrix $\mathbf{W}$ has zero entries on off-diagonal elements (implying that the loci are independent), the Euclidean distances can be computed between pseudohaplotypes as if they were real haplotypes. Intraclass correlations are estimated as for the haploid case above, and the sums of squares within and between demes should be insensitive to the definition of the pseudohaplotypes. This procedure can thus be applied to loci with an arbitrary degree of linkage and even completely unlinked loci. This is because multilocus average estimators are constructed by summing up variance components on both the numerator and denominator, as in (29.37). Even though linkage information does not affect the way overall sums of squares and variance components are estimated, it should affect the variance of the estimates, which should be incorporated into the assessment of the significance of the inferred statistics, as discussed below.

29.7.3 Multiallelic Molecular Data

With the exception of RFLP and single nucleotide polymorphism (SNP) data, more than two allelic states are possible for most molecular data types like DNA sequences or microsatellites. Several options for the handling of such data are possible, depending on the underlying model of mutation we have in mind.

Under a strict infinite-site model, no more than two allelic states should coexist at any locus, even though the allelic repertoire can be much larger. In that case, the multiallelic data can be safely re-encoded into binary data.

29.7.3.1 DNA Sequences

Under a finite-site model, multiple allelic states can coexist at a given locus, and identity in state no longer means identity by descent. One can still re-encode the information into a binary vector (in an augmented space) if one can infer the phylogenetic tree underlying the observed data. Efficient techniques have been developed in the case of DNA sequence (e.g. Felsenstein, 1981; Swofford *et al.*, 1996; Piontkivska, 2004; Boussau and Gouy, 2006) as well as **Chapters 15** and **16**, often allowing the reconstruction of the actual number m of mutations having occurred in the ancestry of a sample of sequences. The binary vector used to compute the squared distances should then be of length m and encode the mutational status of each sequence instead of its nucleotide status. It should have an entry of 1 at a given position if the corresponding mutation has occurred on the lineage of the haplotype to the most recent common ancestor of all haplotypes and an entry of zero otherwise. An unweighted squared Euclidean distance computed on such vectors is therefore equivalent to counting the number of mutations having happened on the branches of a phylogeny separating two haplotypes (a patristic distance). Although we can still partition total genetic diversity into variance components and perform an AMOVA analysis on such data, it should be clear that the resulting Φ statistics are not average correlations of nucleotide frequencies anymore, but are rather average correlations of mutation frequencies at different subdivision levels. The weight matrix $\mathbf{W}$ could be used in such an analysis to reflect the weight of different types of mutations, like transitions, transversions, or insertion/deletions. If no phylogenetic tree is available, the number of pairwise differences between sequences can be used as a squared Euclidean distance, which would be equivalent to a patristic distance computed on minimum spanning network relating sequences (Excoffier and Smouse, 1994).

Note that corrections for multiple substitutions per site (see e.g. Nei and Kumar, 2000) could be incorporated into a conventional ANOVA framework (see Weir and Basten, 1990). Because these corrections aim at uncovering the true number of mutations having occurred in the ancestry of a pair of sequences, they could also be incorporated into the AMOVA framework by using corrected genetic distances as approximations to squared Euclidean distances. Further work is needed to establish exactly how multiple substitutions affect the partitioning of genetic variance within and among populations.

A number of other studies have attempted to estimate F statistics from DNA sequences (Takahata and Palumbi, 1985; Lynch and Crease, 1990; Weir and Basten, 1990; Hudson *et al.*, 1992a; 1992b; Slatkin, 1993), some of them being reviewed in Charlesworth (1998). They all attempt in partitioning the total diversity into within and between components, and can all be related either to gene diversity or variance components approaches, implying they should lead to similar results for large number of large samples.

29.7.3.2 *Microsatellite Data*

Microsatellite data cannot be properly recoded as a binary vector because it is quite difficult to reconstruct a phylogenetic tree from it. Microsatellite data are supposed to predominantly follow a stepwise mutation model (SMM) (e.g. Weber and Wong, 1993, but see Leblois *et al.*, 2003), where identity in state does not mean identity by descent. However, the difference between allelic states (numbers of repeats of a motif of a few base pairs) is an indicator of the evolutionary proximity between alleles.

Slatkin (1995) has shown that the square of this difference in allele size increased approximately linearly with time. He considered a single microsatellite locus, where μ is the mutation rate per generation. The mutation can lead either to an increase or to a decrease in repeat number, and the increase size is a random variable with mean 0 and variance σ_m^2, assumed to be independent of allele size. Note that this mutation model is more general than the single-step model, for which $\sigma_m^2 = 1$. Let y_i and y_j be as usual indicator variables, here equal to the number of repetitions of the microsatellite motif for the gene copies i and j respectively. If these gene copies had a most recent common ancestor t generations ago (i.e. their coalescence time is t) and we assume a Poisson distribution of mutations, then the expectation of $\delta_{ij}^2 = (y_i - y_j)^2$ increases linearly with coalescence time: $E(\delta_{ij}^2) = 2\mu t \sigma_m^2$. Slatkin (1995) then defined the sum of squares of the difference in allele size within d demes, from which samples of n diploid individuals are supposed to have been drawn, as

$$S_W = \frac{1}{d} \sum_{k=1}^{d} \frac{2}{2n(n-1)} \sum_{i=1}^{2n} \sum_{j<i} \delta_{ij}^2, \qquad (29.55)$$

and the average number of square differences in repeat size between demes as

$$S_B = \frac{2}{(2n)^2 d(d-1)} \sum_{k=1}^{d} \sum_{k'<k} \sum_{i=1}^{2n} \sum_{j=1}^{2n} \delta_{ij}^2. \qquad (29.56)$$

The average squared difference in repeat size in the total population is then obtained as

$$\bar{S} = \frac{2n-1}{2nd-1} S_W + \frac{2n(d-1)}{2nd-1} S_B. \qquad (29.57)$$

By analogy with Nei's G_{ST}, Slatkin (1995) defined an estimator of population subdivision

$$R_{ST} = \frac{\bar{S} - S_W}{\bar{S}}. \qquad (29.58)$$

Noting that the δ_{ij}^2s are indeed squared Euclidean distances, the AMOVA framework is a natural alternative to Slatkin (1995) approach for the analysis of microsatellite data. It also implies that Slatkin's analysis is close to an ANOVA of allele size frequencies. However, it should be clear from the comparison of (29.41) with (29.42) that the analog of R_{ST} should be (Michalakis and Excoffier, 1996; Rousset, 1996),

$$\Phi_{ST} = \frac{S_B - S_W}{S_B}. \qquad (29.59)$$

Therefore, assuming identical sample sizes in all demes, the correct expected relationships between R_{ST} and Φ_{ST} are

$$\Phi_{ST} = \frac{R_{ST}}{1 - c(1 - R_{ST})}, \qquad (29.60)$$

$$R_{ST} = \frac{(1-c)\Phi_{ST}}{1 - c\Phi_{ST}} \qquad (29.61)$$

where $c = (2n-1)/(2nd-1)$. (29.60) corrects a typo in (29.9) of Michalakis and Excoffier (1996). The equilibrium values of a statistic equivalent to Φ_{ST} under the SMM, but called ρ_{ST}, has been defined for an island model by Rousset (1996).

Even though the analysis of microsatellite data has exactly the same form as that of other data types, estimators based on such data have a much larger associated variance (Slatkin, 1995), which is essentially due to the SMM, and for which large samples do not necessarily produce more consistent estimates (Pritchard and Feldman, 1996). Several authors have underlined the need for a very large number of independent loci to get meaningful estimates of various genetic quantities from microsatellite data (Goldstein *et al.*, 1995; Zhivotovsky and Feldman, 1995; Bertorelle and Excoffier, 1998). Thus, even though it remains to be investigated in further detail, F statistics derived under the SMM could have the same limitations and requirements. As has been shown empirically (e.g. Takezaki and Nei, 1996; Gaggiotti *et al.*, 1999; Balloux and Goudet, 2002), it may be adequate to compute F_{ST} estimates only from allele frequencies if a few microsatellite loci are available because, even though such estimates would be biased under an SMM, they would have a smaller mean square error.

29.7.4 Dominant Data

An important fraction of newly generated molecular data are dominant such as randomly amplified polymorphic DNA (RAPD) markers or amplified fragment length polymorphism (AFLP) markers. For these markers, diploid individuals typically show, for each locus, either a positive or a negative scoring. A negative scoring implies that the individual is homozygote for the recessive allele, while a positive scoring is obtained for individuals that are either homozygote or heterozygote for the dominant allele. Lynch and Milligan (1994) have been the first authors to propose an estimation procedure for quantifying the amount of genetic structure based on such markers. Using second order Taylor series expansions, they have developed an asymptotically unbiased estimator of G_{ST} as defined in (29.21), that depends on estimated gene diversities within ($\hat{H}_W$) and between ($\hat{H}_B$) demes, as (Lynch and Milligan, 1994)

$$
\hat{G}_{ST} = \frac{\hat{H}_B}{\hat{H}_B + \hat{H}_W} \left(1 + \frac{\hat{H}_B \text{var}(\hat{H}_W) - \hat{H}_W \text{var}(\hat{H}_B) + (\hat{H}_B - \hat{H}_W)\text{cov}(\hat{H}_B, \hat{H}_W)}{\hat{H}_B (\hat{H}_B + \hat{H}_W)^2} \right)^{-1}.
$$
(29.62)

Gene diversities are conventionally estimated from allele frequencies. However, instead of using the maximum-likelihood estimator of the recessive allele frequency ($\hat{p}$) equal to $\sqrt{\hat{P}}$, where $\hat{P}$ is the fraction of individuals that do not exhibit the marker allele, Lynch and Milligan proposed to use an asymptotically unbiased estimator, namely,

$$
\hat{p} = \sqrt{\hat{P}} \left(1 - \frac{\text{var}(\hat{P})}{8\hat{P}^2} \right)^{-1},
$$
(29.63)

where $\text{var}(\hat{P}) = \hat{P}(1 - \hat{P})/(2n_k)$ is the sampling variance of $\hat{P}$ in deme k. While this estimator is less biased than the maximum-likelihood estimator, it is still biased for small sample sizes and low frequencies of the recessive allele. In order to address this problem, Zhivotovsky (1999) has recently proposed an elegant Bayesian approach to estimate recessive allele frequencies. He empirically shows that his approach leads to a slightly improved estimation of recessive allele frequencies and to a much better estimation of the fixation index.

An alternative and economical way to test for the presence of genetic structure with dominant data is to consider differences in genotype frequencies between demes. Thus, Peakall *et al.* (1995) proposed to analyze multilocus RAPD data under the AMOVA framework. They used as a Euclidean square distance the number of pairwise difference between RAPD profile, and obtained in this case Φ statistics equivalent to the correlation of *phenotype* frequencies within or between demes. Because these Φ statistics were not comparable to conventional fixation indexes, we (Stewart and Excoffier, 1996) developed a methodology that allows the computation of correlations of *gene* frequencies from dominant data under the AMOVA framework. This was done by realizing that the comparison of two diploid phenotypes corresponds to the comparisons of four pairs of genes. It is therefore possible to replace squared Euclidean distances between phenotypes by the sum of four expected square Euclidean distances between their constituent genes, and to adjust the degrees of freedom in order to get estimates of Φ statistics directly comparable to those defined from codominant markers. However, for each individual, the expected number of different genes given its phenotype depends on the frequency of the recessive allele, which thus needs to be estimated. It also depends quite strongly on a possible departure of HWE that is quite common in partially selfing plant communities where RAPD markers are most often used. If the selfing rate is S, then under inbreeding equilibrium, an asymptotically unbiased (and apparently unpublished) estimator of the recessive allele frequency equivalent to (29.63) can be obtained if $0 \leq S < 1$ as

$$\hat{p} = \frac{1}{4(1-S)} \left\{ (A\hat{P} + S^2)^{1/2} \left[1 - \frac{A^2 \text{var}(\hat{P})}{8(A\hat{P} + S^2)^2} \right]^{-1} - S \right\}, \qquad (29.64)$$

where $A = 8S^2 - 24S + 16$. Obviously, $\hat{p} = \hat{P}$ if $S = 1$. One can check that (29.64) reduces to (29.63) if $S = 0$. If departure from Hardy–Weinberg is not due to self-fertilization, but to other forms of inbreeding, (29.25) can still be used due to the relationship $S = 2\bar{F}_{IS}/(1 + \bar{F}_{IS})$. Note also that Zhivotovsky (1999) has proposed a Bayesian estimator of $\hat{p}$ that takes $\bar{F}_{IS}$ into account. More recently, Holsinger *et al.* (2002) have introduced a Bayesian approach to estimate F statistics from dominant markers allowing for unknown amount of local inbreeding. Under the assumption that, in populations related by migrations, the allele frequencies follow a β distribution (e.g. Crow and Kimura, 1970; Balding and Nichols, 1995), the parameters of which explicitly depend on F statistics (see (29.38)), Holsinger *et al.* (2002) are able to simultaneously estimate F_{ST} (as reported in (29.9)) and the local inbreeding coefficient F_{IS} (assumed to be identical in all subpopulations). Simulation results show that F_{ST} is usually well estimated but that a large number of loci and populations are necessary to get meaningful estimates of F_{IS} from dominant data. Further effect of selfing on the estimation of population subdivision can be found elsewhere (Maruyama and Tachida, 1992; Pollak and Sabran, 1992; Jarne, 1995; Nordborg, 1997; Wang, 1997a; 1997b).

29.7.5 Relation of AMOVA with other Approaches

The AMOVA framework is therefore fully equivalent to Cockerham's ANOVA framework. AMOVA is also closely related to Long *et al.* (1986) multivariate extension of Cockerham's ANOVA for loci with multiple alleles, which uses variance–covariance matrices of allele frequencies as weighting matrices similar to **W**. Explicit relationships between Long *et al.* (1986), Nei (1977; 1983), and Weir and Cockerham (1984) estimators can be found in Chakraborty and Danker-Hopfe (1991). For single-locus estimators, it is noted that when covariance of allele frequencies are neglected, Long's estimators are close to unweighted average F statistics $\hat{\theta}_U$ (Weir and Cockerham, 1984), while AMOVA produces estimates completely identical to Weir and Cockerham's weighted estimators $\hat{\theta}_W$ (Weir and Cockerham, 1984). For multilocus estimators, if some individuals are not typed at all loci, implying that there is some missing data, it is better to perform distinct analyses for each locus separately, and to combine variance components estimates for each locus as in (29.37). In that case, the AMOVA haplotypic framework would wrongly assume that the squared distances are based on the same number of characters for each pairwise comparison, which could bias the estimators, particularly if missing data are not equally balanced among demes. The advantage of the AMOVA framework is therefore to highly increase the speed of the computations by carrying out a single analysis for arbitrary numbers of alleles and loci, and by allowing the covariance of genetic information (i.e. linkage disequilibrium) across loci to be taken into account.

29.8   SIGNIFICANCE TESTING

Event though the variance of some F statistics estimates has been derived (see e.g. Nei et al., 1977; Cockerham and Weir, 1983; Weir and Cockerham, 1984), it seems of little use to build reliable significance tests for small samples. Instead, several traditional resampling techniques, like bootstrap, jackknife (Efron, 1982), or permutations are commonly used to assess the significance of the moment-estimated statistics. Likelihood-based estimations can lead to posterior distributions of the parameters of interest, from which credible interval can be constructed (e.g. Holsinger et al., 2002; Balding, 2003; Foll and Gaggiotti, 2006), which is not detailed here.

29.8.1   Resampling Techniques

Typically, bootstrap and jackknife methods aim at defining confidence intervals around an estimated value, implicitly assuming that genetic structure exists. Therefore, appropriate bootstrapping methods should try to mimic the action of evolutionary forces that have created the genetic structure (Weir, 1996, p. 174). The use of nonparametric bootstrap of sampled loci is advocated because the genetic diversity observed at unlinked loci should be the result of the replication of independent evolutionary (coalescent and mutational) processes, giving us some idea of the genetic component of the variance. This nonparametric bootstrapping approach would then satisfactorily explore the stochasticity of the evolutionary process only if a large number of unlinked loci are available. However, bootstrapping over linked nucleotide sites of a DNA sequence would certainly not be appropriate, as it would only give us some idea of the stochasticity of the mutation process, which is much smaller than the stochasticity of the coalescent process. Even though nonparametric bootstrap procedures do not always allow one to estimate the complete stochasticity of the (often unknown) evolutionary process, it may be justified to appreciate the sensitivity of a given statistic to the choice of alleles, loci, or demes as is implemented in various computer packages (e.g. Goudet, 1995; Raymond and Rousset, 1995b; Excoffier et al., 2005). Under an assumed spatial model, some parametric bootstrap procedure mimicking the coalescent and mutational process would be in order.

On the other hand, permutational approaches construct confidence intervals for a given statistic under the hypothesis of no genetic structure (e.g. Manly, 1991; Excoffier et al., 1992; Hudson et al., 1992a), and therefore do not require to specify the evolutionary forces at work that would lead to a genetic structure. However, note that mixed bootstrap and permutational approaches can be used to obtain power curves, for instance, as a function of the number of sampled loci (Excoffier et al., 1992). Some care should however be taken on how to perform permutations, depending on not only the statistic to test ($\bar{F}_{IS}$, F_{IT}, F_{ST}, or F_{CT}, as well as their corresponding variance components) but also on the nature of data (linked or independent loci, dominant data). For haplotypic data, with known gametic phase, all linked loci can be considered as a single locus, and it would be irrelevant to bootstrap loci. Differences between haplotypes within individuals within demes ($\bar{F}_{IS}$) can be tested by permuting haplotypes between individuals within demes. Differences between haplotypes within individuals within the total population (F_{IT}) can be tested by permuting haplotypes between individuals and between demes. Differences between demes (F_{ST}) are tested by permuting individuals between demes. If random mating is assumed or if no information is available on how haplotypes are grouped into individuals, F_{ST} can be tested by permuting haplotypes between demes. Under such permutation procedures, note that the total sums of squares and degrees of freedom remain unchanged in an ANOVA approach. If a higher level of subdivision is considered, such as groups of demes within populations, differences between groups (F_{CT}) are tested by permuting whole demes between groups, which changes the sums of square and the degrees of freedom.

With dominant data, $\bar{F}_{IS}$ or F_{IT} cannot be generally estimated (but see Holsinger *et al.*, 2002), and one has to estimate allele or haplotype frequencies (Lynch, 1991; Lynch and Milligan, 1994; Stewart and Excoffier, 1996; Zhivotovsky, 1999). Under the hypothesis of no population structure, genotypes can be permuted between demes only if frequencies have been estimated at the population level and not at the deme level because, otherwise, estimated frequencies would change with permutations. This feature of dominant data makes it much less powerful than codominant markers to study population subdivision (Lynch and Milligan, 1994; Stewart and Excoffier, 1996).

For data consisting of completely unlinked loci, permutations should be done as for haplotypic data, but independently for each locus. A problem that is not completely settled occurs for data consisting of incompletely linked loci. In that case, permutations cannot be done for each locus independently, as it would disrupt the potentially existing linkage disequilibrium pattern. Permuting all loci simultaneously as if they were completely linked seems, however, a conservative but not entirely satisfactory procedure (Michalakis and Excoffier, 1996).

29.8.2.9 *Exact Tests*

Exact tests of population subdivision have been proposed for testing differences in allele (Raymond and Rousset, 1995a) or genotype (Goudet *et al.*, 1996) frequencies among a set of demes. Contrary to previously described statistics, these tests do not aim at measuring directly correlations of genes within and among populations. They rather follow from the application of the classical Fisher exact test on R $\times$ C contingency tables. Here, rows are for demes, whereas columns are for alleles or genotypes. Under the null hypothesis of absence of genetic differentiation between samples, the conditional probability of the observed table Π_0 is computed as

$$\Pi_0 = \frac{\prod_{k=1}^{d} n_{k.}! \prod_{i=1}^{r} n_{.i}!}{n_{..}! \prod_{k=1}^{d} \prod_{i=1}^{r} n_{ki}!} \tag{29.65}$$

where, for allelic data, n_{ki} is the number of genes of allelic type i in the kth sample, $n_{.i}$ is the marginal count of the ith allele in the d samples, $n_{k.}$ is the marginal size of the kth sample, and $n_{..}$ is the total allelic counts over all samples. For genotype data, the n's are simply genotypic counts instead of allelic counts.

Instead of enumerating all possible contingency tables as in the Fisher exact test (Mehta and Patel, 1983), a Markov chain is used to efficiently explore the space of all possible tables by a random walk (see Guo and Thompson, 1992; Raymond and Rousset, 1995a). It is done in such a way that the probability to visit a particular table corresponds to its actual probability under the null hypothesis of no genetic structure. The P value of the test is then defined as the fraction of explored tables that are equally or less likely than the observed one. Simulations have confirmed that exact tests performed on genotype counts are less powerful than those performed on allelic counts (Goudet *et al.*, 1996), as they use less information. The incorporation of additional levels of subdivision could be envisioned by extending the analysis of two-dimensional tables to cubic ones, which could be quite challenging to implement in practice. Multilocus tests can be performed by combining probabilities of single-locus tests, assuming that all loci are independent.

29.9 RELATED AND REMAINING PROBLEMS

29.9.1 Testing Departure from Hardy–Weinberg Equilibrium

HWE can be disrupted by population subdivision or inbreeding and specific attempts have been made to detect discrepancy between observed and expected genotype frequencies. Recent attempts include procedures related to Fisher's exact test on R × C contingency tables, where alternative tables are generated either using permutations or by a random walk along a Markov chain (Guo and Thompson, 1992). Departure from HWE due to inbreeding can also be tested by the use of likelihood ratio statistics. Numerical methods to obtain minimum variance and maximum-likelihood estimators of $\bar{F}_{IS}$ have been proposed in Robertson and Hill (1984) and in Hill *et al.* (1995). Several Bayesian methods have been used to estimate inbreeding coefficients within demes in the case of biallelic loci (see Shoemaker *et al.*, 1998 for a review). For multiple alleles, a Bayesian MCMC method has been developed to test departure from HWE that is specifically due to inbreeding (Ayres and Balding, 1998). In that paper, a Bayesian procedure is described to estimate the posterior distribution of $\bar{F}_{IS}$ statistics associated with each genotype, or a global $\bar{F}_{IS}$ statistic, along with the distribution of allele frequencies. The use and effect of several priors are also discussed. Note that, in both maximum-likelihood and Bayesian procedures, care has to be taken to keep $\bar{F}_{IS}$ estimates within reasonable bounds (Robertson and Hill, 1984; Ayres and Balding, 1998). The power of different testing procedures is reviewed in Rousset and Raymond (1995). It is found that exact probability tests (Guo and Thompson, 1992) are more powerful than tests based on maximum likelihood (Hill *et al.*, 1995), minimum variance (Robertson and Hill, 1984), or variance component estimators of $\bar{F}_{IS}$ (Cockerham, 1969; 1973), when the departure from HWE is due to specific genotypes. Alternatively, Robertson and Hill's (1984) estimator is found to perform best when departure from HWE is due to regular inbreeding that can be accounted for by a single $\bar{F}_{IS}$ value for all genotypes.

29.9.2 Detecting Loci under Selection

It has since long been realized (Cavalli-Sforza, 1966; Lewontin and Krakauer, 1973) that gene frequencies at loci involved in adaptive events should show increased levels of differentiation among populations, while genes under balancing selection should present relatively uniform frequencies across populations, and thus lead to low F_{ST} values. Capitalizing on these ideas, Beaumont and Nichols (1996) proposed to detect loci under selection by comparing observed levels of genetic diversity within (heterozygosity) and between (F_{ST}) populations to those expected under a simple evolutionary scenario, like an infinite-island model. This approach does not seem to be sensitive to past population demography (Beaumont, 2005) and it seems applicable to a wide diversity of genetic markers assessed in several populations (e.g. Luikart et al., 2003, Table 29.2). It has been successfully applied to recent genome scans to detect outlier loci (e.g. Wilding et al., 2001; Storz et al., 2004; Murray and Hare, 2006), which can be considered as candidate loci for having had a history of selection. More recently, Beaumont and Balding (2004), proposed a Bayesian method to detect loci under selection based on (29.38), to identify signatures of selection, and compared its power to that of Beaumont and Nichols (1996) for loci under positive or balancing selection. They find that both methods can identify loci under positive selection if the selection coefficient is about five times larger than the migration rate between populations. However, for low levels of population differentiation ($F_{ST} = 0.1$), these approaches cannot really distinguish loci under balancing selection, even for quite strong selection coefficients. Because it is more and more widely recognized that our ability to evidence patterns of selection depends on a sound knowledge of past demographic history (e.g. Ometto et al., 2005; Williamson et al., 2005; Kelley et al., 2006), complex demographic inferences from observed patterns of genetic diversity will need to be developed to better understand selective processes.

29.9.3 What is the Underlying Genetic Structure of Populations?

The estimation procedures defined above all assumed that the (potentially hierarchical) genetic structure of the population was known and that the main issue was to estimate the degree of population subdivision. However, the history and therefore the true genetic structure of a population are very often unknown. Some attempts have been made to try to disentangle population history from population structure (e.g. Templeton et al., 1995; Templeton 2004), but these procedures have not been developed within a clear statistical framework. Moreover, individuals may not always be grouped into well-recognized sampling units, such that their assignment to a given subdivision may be a relevant issue. Quite recently, a number of studies have tried to address some of these problems. They attempted to partition sampled populations into distinguishable sets (e.g. Holsinger and Mason-Gamer, 1996; Dupanloup et al., 2002; Corander et al., 2003) to allocate individuals to specific or reconstructed populations (Pritchard et al., 2000a; Dawson and Belkhir, 2001; Falush et al., 2003a; Wilson and Rannala, 2003; Guillot et al., 2005; Corander and Marttinen, 2006), or to find the number of hidden subdivisions (Pritchard et al., 2000a; Guillot et al., 2005; Corander and Marttinen, 2006). Another complicating factor is that genetic structure estimates may differ depending on which chromosomal segments are studied (mitochondrial DNA, Y chromosome, or autosomes) because of sex-biased dispersal and sexual selection (Chesser, 1991; Chesser and Baker, 1996). In that case, the estimation of sex-specific F statistics has been proposed (e.g. Petit et al., 2001; Vitalis,

2002; Fontanillas *et al.*, 2004). It appears that the development of statistical procedures to uncover the demographic or selection history of a set of populations that best explains the observed genetic structure is certainly one of the most interesting challenges of population genetics.

Acknowledgments

I am grateful to Jérôme Goudet for providing access to his unpublished Ph.D thesis and for long discussions on various aspects of this work. I would also like to thank David Balding, Sergei Gavrilets, François Rousset, and Stefan Schneider for their careful reading of earlier versions of this manuscript and constructive suggestions. The AMOVA procedures as well as several exact tests described in the present paper are implemented in the Arlequin package (Excoffier *et al.*, 2005), freely available on `http://cmpg.unibe.ch/software/arlequin3`. This work and the development of the Arlequin package were supported by the Swiss National Science Foundation.

REFERENCES

Ayres, K.L. and Balding, D.J. (1998). Measuring departures from Hardy-Weinberg: a Markov chain Monte Carlo method for estimating the inbreeding coefficient. *Heredity* **80**, 769–777.

Ayres, K.L. and Overall, D.J. (1999). Allowing for within-subpopulation inbreeding in forensic match probabilities. *Forensic Science International* **103**, 207.

Balding, D.J. (2003). Likelihood-based inference for genetic correlation coefficients. *Theoretical Population Biology* **63**, 221–230.

Balding, D.J. and Nichols, R.A. (1995). A method, for quantifying differentiation between populations at multi-allelic loci and its implications for investigating identity and paternity. *Genetica* **96**, 3–12.

Balloux, F. and Goudet, J. (2002). Statistical properties of population differentiation estimators under stepwise mutation in a finite island model. *Molecular Ecology* **11**, 771–783.

Barrai, I. (1971). Subdivision and inbreeding. *American Journal of Human Genetics* **23**, 95–96.

Beaumont, M.A. (2005). Adaptation and speciation: what can F-st tell us? *Trends in Ecology and Evolution* **20**, 435–440.

Beaumont, M.A. and Balding, D.J. (2004). Identifying adaptive genetic divergence among populations from genome scans. *Molecular Ecology* **13**, 969–980.

Beaumont, M.A. and Nichols, R.A. (1996). Evaluating loci for use in the genetic analysis of population structure. *Proceedings of the Royal Society London, Series B* **263**, 1619–1626.

Beerli, P. (2006). Comparison of Bayesian and maximum-likelihood inference of population genetic parameters. *Bioinformatics* **22**, 341–345.

Beerli, P. and Felsenstein, J. (2001). Maximum likelihood estimation of a migration matrix and effective population sizes in n subpopulations by using a coalescent approach. *Proceedings of the National Academy of Sciences of the United States of America* **98**, 4563–4568.

Bertorelle, G. and Excoffier, L. (1998). Inferring admixture proportions from molecular data. *Molecular Biology and Evolution* **15**, 1298–1311.

Boussau, B. and Gouy, M. (2006). Efficient likelihood computations with nonreversible models of evolution. *Systematic Biology* **55**, 756–768.

Cavalli-Sforza, L.L. (1966). Population structure and human evolution. *Proceedings of the Royal Society London, Series B* **164**, 362–379.

Chakraborty, R. and Danker-Hopfe, H. (1991). Analysis of population structure: a comparative study of different estimators of Wright's fixation indices. In *Handbook of Statistics*, C.R. Rao and R. Chakraborty, eds. Elsevier Science Publishers, Amsterdam, pp. 203–254.

Chakraborty, R. and Weiss, K.M. (1988). Admixture as a tool for finding linked genes and detecting that difference from allelic association between loci. *Proceedings of the National Academy of Sciences of the United States of America* **85**, 9119–9123.

Chapman, N.H. and Thompson, E.A. (2001). Linkage disequilibrium mapping: the role of population history, size, and structure. *Advances in Statistics* **42**, 413–437.

Charlesworth, B. (1998). Measures of divergence between populations and the effect of forces that reduce variability. *Molecular Biology and Evolution* **15**, 538–543.

Charlesworth, B., Nordborg, M. and Charlesworth, D. (1997). The effects of local selection, balanced polymorphism and background selection on equilibrium patterns of genetic diversity in subdivided populations. *Genetical Research* **70**, 155–174.

Chesser, R.K. (1991). Influence of gene flow and breeding tactics on gene diversity within populations. *Genetics* **129**, 573–583.

Chesser, R.K. and Baker, R.J. (1996). Effective sizes and dynamics of uniparentally and diparentally inherited genes. *Genetics* **144**, 1225–1235.

Cockerham, C.C. (1969). Variance of gene frequencies. *Evolution* **23**, 72–83.

Cockerham, C.C. (1973). Analysis of gene frequencies. *Genetics* **74**, 679–700.

Cockerham, C.C. and Weir, B.S. (1983). Variance of actual inbreeding. *Theoretical Population Biology* **23**, 85–109.

Cockerham, C.C. and Weir, B.S. (1987). Correlations, descent measures: drift with migration and mutation. *Proceedings of the National Academy of Sciences of the United States of America* **84**, 8512–8514.

Cockerham, C.C. and Weir, B.S. (1993). Estimation of gene flow from F-statistics. *Evolution* **47**, 855–863.

Corander, J. and Marttinen, P. (2006). Bayesian identification of admixture events using multilocus molecular markers. *Molecular Ecology* **15**, 2833–2843.

Corander, J., Waldmann, P. and Sillanpaa, M.J. (2003). Bayesian analysis of genetic differentiation between populations. *Genetics* **163**, 367–374.

Crow, J. and Kimura, M. (1970). *An Introduction to Population Genetics Theory*. Harper and Row, New York.

Crow, J.F. and Aoki, K. (1984). Group selection for a polygenic behavioral trait: estimating the degree of population subdivision. *Proceedings of the National Academy of Sciences of the United States of America* **81**, 6073–6077.

Dawson, K.J. and Belkhir, K. (2001). A Bayesian approach to the identification of panmictic populations and the assignment of individuals. *Genetical Research* **78**, 59–77.

Dupanloup, I., Schneider, S. and Excoffier, L. (2002). A simulated annealing approach to define the genetic structure of populations. *Molecular Ecology* **11**, 2571–2581.

Efron, B. (1982). *The Jacknife, the Bootstrap and Other Resampling Plans. Regional Conference Series in Applied Mathematics*. Society for Industrial Applied Mathematics, Philadelphia, PA.

Excoffier, L., Laval, G. and Schneider, S. (2005). Arlequin (version 3.0): an integrated software package for population genetics data analysis. *Evolutionary Bioinformatics Online* **1**, 47–50.

Excoffier, L. and Smouse, P. (1994). Using allele frequencies and geographic subdivision to reconstruct gene genealogies within a species. Molecular variance parsimony. *Genetics* **136**, 343–359.

Excoffier, L., Smouse, P. and Quattro, J. (1992). Analysis of molecular variance inferred from metric distances among DNA haplotypes: application to human mitochondrial DNA restriction data. *Genetics* **131**, 479–491.

Falush, D., Stephens, M. and Pritchard, J.K. (2003a). Inference of population structure using multilocus genotype data: linked loci and correlated allele frequencies. *Genetics* **164**, 1567–1587.

Falush, D., Wirth, T., Linz, B., Pritchard, J.K., Stephens, M., Kidd, M., Blaser, M.J., Graham, D.Y., Vacher, S., Perez-Perez, G.I., Yamaoka, Y., Megraud, F., Otto, K., Reichard, U., Katzowitsch, E., Wang, X., Achtman, M. and Suerbaum, S. (2003b). Traces of human migrations in *Helicobacter pylori* populations. *Science* **299**, 1582–1585.

Felsenstein, J. (1981). Evolutionary trees from DNA sequences: a maximum likelihood approach. *Journal of Molecular Evolution* **17**, 368–376.

Foll, M. and Gaggiotti, O. (2006). Identifying the environmental factors that determine the genetic structure of populations. *Genetics* **174**, 875–891.

Fontanillas, P., Petit, E. and Perrin, N. (2004). Estimating sex-specific dispersal rates with autosomal markers in hierarchically structured populations. *Evolution* **58**, 886–894.

Fu, Y.X. (1997). Coalescent theory for a partially selfing population. *Genetics* **146**, 1489–1499.

Gaggiotti, O. and Excoffier, L. (2000). A simple method of removing the effect of a bottleneck and unequal population sizes on pairwise genetic distances. *Proceedings of the Royal Society London B* **267**, 81–87.

Gaggiotti, O.E., Lange, O., Rassmann, K. and Gliddon, C. (1999). A comparison of two indirect methods for estimating average levels of gene flow using microsatellite data. *Molecular Ecology* **8**, 1513–1520.

Gilmour, J.S.L. and Gregor, J.W. (1939). Demes: a suggested new terminology. *Nature* **144**, 333.

Goldstein, D.B., Ruiz-Linares, A., Cavalli-Sforza, L.L. and Feldman, M.W. (1995). Microsatellite loci, genetic distances, and human evolution. *Proceedings of the National Academy of Sciences of the United States of America* **92**, 6723–6727.

Goudet, J. (1993). The genetics of geographically structured populations. Ph.D., University of Wales, Bangor.

Goudet, J. (1995). Fstat version 1.2: a computer program to calculate F-statistics. *Journal of Heredity* **86**, 485–486.

Goudet, J., Raymond, M., De Meeüs, T. and Rousset, F. (1996). Testing differentiation in diploid populations. *Genetics* **144**, 1933–1940.

Guillot, G., Estoup, A., Mortier, F. and Cosson, J.F. (2005). A spatial statistical model for landscape genetics. *Genetics* **170**, 1261–1280.

Guo, S. and Thompson, E. (1992). Performing the exact test of Hardy-Weinberg proportion for multiple alleles. *Biometrics* **48**, 361–372.

Herbots, H.M. (1997). The structured coalescent. In *Progress in Population Genetics and Human Evolution*, P. Donnely and S. Tavaré, eds. Springer-Verlag, New York, pp. 231–255.

Hewitt, G.M. (2001). Speciation, hybrid zones and phylogeography – or seeing genes in space and time. *Molecular Ecology* **10**, 537–549.

Hill, W.G., Babiker, H.A., Ranford-Cartwright, L.C. and Walliker, D. (1995). Estimation of inbreeding coefficients from genotypic data on multiple alleles, and application to estimation of clonality in malaria parasites. *Genetical Research* **65**, 53–61.

Holsinger, K.E., Lewis, P.O. and Dey, D.K. (2002). A Bayesian approach to inferring population structure from dominant markers. *Molecular Ecology* **11**, 1157–1164.

Holsinger, K.E. and Mason-Gamer, R.J. (1996). Hierarchical analysis of nucleotide diversity in geographically structured populations. *Genetics* **142**, 629–639.

Hudson, R.R., Boos, D.D. and Kaplan, N.L. (1992a). A statistical test for detecting geographic subdivision. *Molecular Biology and Evolution* **9**, 138–151.

Hudson, R.R., Slatkin, M. and Maddison, W.P. (1992b). Estimation of levels of gene flow from DNA sequence data. *Genetics* **132**, 583–589.

Jaarola, M. and Searle, J.B. (2004). A highly divergent mitochondrial DNA lineage of *Microtus agrestis* in southern Europe. *Heredity* **92**, 228–234.

Jarne, P. (1995). Mating system, bottleneck, and genetic polymorphism in hermaphroditic animals. *Genetical Research* **65**, 193–207.

Jorde, L. (1980). The genetic structure of subdivided human populations. In *Current Developments in Anthropological Genetics*, J. Mielke and M. Crawford, eds. Plenum Press, New York, pp. 135–208.

Kelley, J.L., Madeoy, J., Calhoun, J.C., Swanson, W. and Akey, J.M. (2006). Genomic signatures of positive selection in humans and the limits of outlier approaches. *Genome Research* **16**, 980–989.

Kitada, S. and Kishino, H. (2004). Simultaneous detection of linkage disequilibrium and genetic differentiation of subdivided populations. *Genetics* **167**, 2003–2013.

Kitakado, T., Kitada, S., Kishino, H. and Skaug, H.J. (2006). An integrated-likelihood method for estimating genetic differentiation between populations. *Genetics* **173**, 2073–2082.

Leblois, R., Estoup, A. and Rousset, F. (2003). Influence of mutational and sampling factors on the estimation of demographic parameters in a "continuous" population under isolation by distance. *Molecular Biology and Evolution* **20**, 491–502.

Lewontin, R.C. and Krakauer, J. (1973). Distribution of gene frequency as a test of the theory of the selective neutrality of polymorphisms. *Genetics* **74**, 175–195.

Li, W.H. and Nei, M. (1974). Stable linkage disequilibrium without epistasis in subdivided populations. *Theoretical Population Biology* **6**, 173–183.

Long, J.C., Naidu, J.M., Mohrenweiser, H.W., Gershowitz, H., Johnson, P.L., Wood, J.W. and Smouse, P.E. (1986). Genetic characterization of Gainj- and Kalam-speaking peoples of Papua New Guinea. *American Journal of Physical Anthropology* **70**, 75–96.

Luikart, G., England, P.R., Tallmon, D., Jordan, S. and Taberlet, P. (2003). The power and promise of population genomics: from genotyping to genome typing. *Nature Reviews Genetics* **4**, 981–994.

Lynch, M. (1991). Analysis of population genetic structure by DNA fingerprinting. *Experientia Supplementum* **58**, 113–126.

Lynch, M. and Crease, T.J. (1990). The analysis of population survey data on DNA sequence variation. *Molecular Biology and Evolution* **7**, 377–394.

Lynch, M. and Milligan, B.G. (1994). Analysis of population genetic structure with RAPD markers. *Molecular Ecology* **3**, 91–99.

Magri, D., Vendramin, G.G., Comps, B., Dupanloup, I., Geburek, T., Gomory, D., Latalowa, M., Litt, T., Paule, L., Roure, J.M., Tantau, I., van der Knaap, W.O., Petit, R.J. and de Beaulieu, J.L. (2006). A new scenario for the quaternary history of European beech populations: palaeobotanical evidence and genetic consequences. *New Phytologist* **171**, 199–221.

Malécot, G. (1948). *Les Mathématiques de l'Hérédité*. Masson, Paris.

Manly, B.F.J. (1991). *Randomization and Monte Carlo methods in Biology*. Chapman and Hall, London.

Marchini, J., Cardon, L.R., Phillips, M.S. and Donnelly, P. (2004). The effects of human population structure on large genetic association studies. *Nature Genetics* **36**, 512–517.

Marjoram, P. and Donnelly, P. (1994). Pairwise comparisons of mitochondrial DNA sequences in subdivided populations and implications for early human evolution. *Genetics* **136**, 673–683.

Maruyama, K. and Tachida, H. (1992). Genetic variability and geographic structure in partially selfing populations. *Japanese Journal of Genetics* **67**, 39–51.

Mehta, C.R. and Patel, N.R. (1983). A network algorithm for performing Fisher's exact test in rxc contingency tables. *Journal of the American Statistical Association* **78**, 427–434.

Michalakis, Y. and Excoffier, L. (1996). A generic estimation of population subdivision using distances between alleles with special reference for microsatellite loci. *Genetics* **142**, 1061–1064.

Murray, M.C. and Hare, M.P. (2006). A genomic scan for divergent selection in a secondary contact zone between Atlantic and Gulf of Mexico oysters, *Crassostrea virginica*. *Molecular Ecology* **15**, 4229–4242.

Nagylaki, T. (1983). The robustness of neutral models of geographical variation. *Theoretical Population Biology* **24**, 268–294.

Nagylaki, T. (1985). Homozygosity, effective number of alleles, and interdeme differentiation in subdivided populations. *Proceedings of the National Academy of Sciences of the United States of America* **82**, 8611–8613.

Nagylaki, T. (1998a). The expected number of heterozygous sites in a subdivided population. *Genetics* **149**, 1599–1604.

Nagylaki, T. (1998b). Fixation indices in subdivided populations. *Genetics* **148**, 1325–1332.

Nei, M. (1972). Genetic distance between populations. *American Naturalist* **106**, 283–292.

Nei, M. (1973). Analysis of gene diversity in subdivided populations. *Proceedings of the National Academy of Sciences of the United States of America* **70**, 3321–3323.

Nei, M. (1977). F-statistics and analysis of gene diversity in subdivided populations. *Annals of Human Genetics* **41**, 225–233.

Nei, M. (1986). Definition and estimation of fixation indices. *Evolution* **40**, 643–645.

Nei, M. (1987). *Molecular Evolutionary Genetics*. Columbia University Press, New York.

Nei, M., Chakravarti, A. and Tateno, Y. (1977). Mean and variance of FST in a finite number of incompletely isolated populations. *Theoretical Population Biology* **11**, 291–306.

Nei, M. and Chesser, R.K. (1983). Estimation of fixation indices and gene diversities. *Annals of Human Genetics* **47**, 253–259.

Nei, M. and Kumar, S. (2000). *Molecular Evolution and Phylogenetics*. Oxford University Press, Oxford.

Nei, M. and Li, W.H. (1973). Linkage disequilibrium in subdivided populations. *Genetics* **75**, 213–219.

Nicholson, G., Smith, A.V., Jonsson, F., Gustafsson, O. and Stefansson, K. (2002). Assessing population differentiation and isolation from single-nucleotide polymorphism data. *Journal of the Royal Statistics Society, Series B* **64**, 695–715.

Nielsen, R. (2001). Statistical tests of selective neutrality in the age of genomics. *Heredity* **86**, 641–647.

Nordborg, M. (1997). Structured coalescent processes on different time scales. *Genetics* **146**, 1501–1514.

Nordborg, M. and Donnelly, P. (1997). The coalescent process with selfing. *Genetics* **146**, 1185–1195.

Ohta, T. (1982). Linkage disequilibrium due to random genetic drift in finite subdivided populations. *Proceedings of the National Academy of Sciences of the United States of America* **79**, 1940–1944.

Ohta, T. and Kimura, M. (1969). Linkage disequilibrium due to random genetic drift. *Genetical Research* **13**, 47–55.

Ometto, L., Glinka, S., De Lorenzo, D. and Stephan, W. (2005). Inferring the effects of demography and selection on *Drosophila melanogaster* populations from a chromosome-wide scan of DNA variation. *Molecular Biology and Evolution* **22**, 2119–2130.

Peakall, R., Smouse, P.E. and Huff, D.R. (1995). Evolutionary implications of allozyme and RAPD variation in diploid populations if dioecious buffalograss "Buchloe dactyloides". *Molecular Ecology* **4**, 135–147.

Petit, E., Balloux, F. and Goudet, J. (2001). Sex-biased dispersal in a migratory bat: a characterization using sex-specific demographic parameters. *Evolution* **55**, 635–640.

Piontkivska, H. (2004). Efficiencies of maximum likelihood methods of phylogenetic inferences when different substitution models are used. *Molecular Phylogenetics and Evolution* **31**, 865–873.

Pollak, E. and Sabran, M. (1992). On the theory of partially inbreeding finite populations. III. Fixation probabilities under partial selfing when heterozygotes are intermediate in viability. *Genetics* **131**, 979–985.

Pons, O. and Chaouche, K. (1995). Estimation, variance and optimal sampling of gene diversity. II. Diploid locus. *Theoretical and Applied Genetics* **91**, 122–130.

Pritchard, J.K. and Donnelly, P. (2001). Case-control studies of association in structured or admixed populations. *Theoretical Population Biology* **60**, 227–237.

Pritchard, J.K. and Feldman, M.W. (1996). Statistics for microsatellite variation based on coalescence. *Theoretical Population Biology* **50**, 325–344.

Pritchard, J.K., Stephens, M. and Donnelly, P. (2000a). Inference of population structure using multilocus genotype data. *Genetics* **155**, 945–959.

Pritchard, J.K., Stephens, M., Rosenberg, N.A. and Donnelly, P. (2000b). Association mapping in structured populations. *American Journal of Human Genetics* **67**, 170–181.

Rannala, B. and Hartigan, J.A. (1996). Estimating gene flow in island populations. *Genetical Research* **67**, 147–158.

Raymond, M. and Rousset, F. (1995a). An exact test for population differentiation. *Evolution* **49**, 1280–1283.

Raymond, M. and Rousset, F. (1995b). GENEPOP Version 1.2: population genetics software for exat tests and ecumenicism. *Journal of Heredity* **86**, 248–249.

Reynolds, J., Weir, B.S. and Cockerham, C.C. (1983). Estimation for the coancestry coefficient: basis for a short-term genetic distance. *Genetics* **105**, 767–779.

Robertson, A. and Hill, W.G. (1984). Deviations from Hardy-Weinberg proportions: sampling variances and use in estimation of inbreeding coefficients. *Genetics* **107**, 703–718.

Rousset, F. (1996). Equilibrium values of measures of population subdivision for stepwise mutation processes. *Genetics* **142**, 1357–1362.

Rousset, F. (1997). Genetic differentiation and estimation of gene flow from F-statistics under isolation by distance. *Genetics* **145**, 1219–1228.

Rousset, F. (2000). Inferences from spatial population genetics. In *Handbook of Statistical Genetics*, D. Balding, M. Bishop and C. Cannings, eds. John Wiley & Sons, Chichester.

Rousset, F. (2002). Inbreeding and relatedness coefficients: what do they measure? *Heredity* **88**, 371–380.

Rousset, F. (2004). *Genetic Structure and Selection in Subdivided Populations*. Princeton University Press, Princeton, NJ.

Rousset, F. and Raymond, M. (1995). Testing heterozygote excess and deficiency. *Genetics* **140**, 1413–1419.

Shoemaker, J., Painter, I. and Weir, B.S. (1998). A Bayesian characterization of Hardy-Weinberg disequilibrium. *Genetics* **149**, 2079–2088.

Simonsen, K.L., Churchill, G.A. and Aquadro, C.F. (1995). Properties of statistical tests of neutrality for DNA polymorphism data. *Genetics* **141**, 413–429.

Slatkin, M. (1987). The average number of sites separating DNA sequences drawn from a subdivided population. *Theoretical Population Biology* **32**, 42–49.

Slatkin, M. (1991). Inbreeding coefficients and coalescence times. *Genetical Research* **58**, 167–175.

Slatkin, M. (1993). Isolation by distance in equilibrium and non-equilibrium populations. *Evolution* **47**, 264–279.

Slatkin, M. (1995). A measure of population subdivision based on microsatellite allele frequencies. *Genetics* **139**, 457–462.

Slatkin, M. and Barton, N.H. (1989). A comparison of three indirect methods for estimating average levels of gene flow. *Evolution* **43**, 1349–1368.

Slatkin, M. and Voelm, L. (1991). FST in a hierarchical island model. *Genetics* **127**, 627–6629.

Stewart, N. and Excoffier, L. (1996). Assessing population genetic structure and variability with RAPD data: application to "vaccinium macrocarpon" (American cranberry). *Journal of Evolutionary Biology* **9**, 153–171.

Storz, J.F., Payseur, B.A. and Nachman, M.W. (2004). Genome scans of DNA variability in humans reveal evidence for selective sweeps outside of Africa. *Molecular Biology and Evolution* **21**, 1800–1811.

Strobeck, K. (1987). Average number of nucleotide differences in a sample from a single subpopulation: a test for population subdivision. *Genetics* **117**, 149–153.

Swofford, D.L., Olsen, G.J., Waddell, P.J. and Hillis, D.M. (1996). Phylogenetic inference. In *Molecular Systematics*, D.M. Hillis, C. Moritz and B.K. Mable, eds. Sinauer Associates, Inc., Sunderland, MA, pp. 407–514.

Taberlet, P., Fumagalli, L., Wust-Saucy, A.G. and Cosson, J.F. (1998). Comparative phylogeography and postglacial colonization routes in Europe. *Molecular Ecology* **7**, 453–464.

Tajima, F. (1993). Statistical analysis of DNA polymorphism. *Japanese Journal of Genetics* **68**, 567–595.

Takahata, N. and Palumbi, S.R. (1985). Extranuclear differentiation and gene flow in the finite island model. *Genetics* **109**, 441–457.

Takezaki, N. and Nei, M. (1996). Genetic distances and reconstruction of phylogenetic trees from microsatellite DNA. *Genetics* **144**, 389–399.

Templeton, A.R. (2004). Statistical phylogeography: methods of evaluating and minimizing inference errors. *Molecular Ecology* **13**, 789–809.

Templeton, A.R., Routman, E. and Phillips, C.A. (1995). Separating population structure from population history: a cladistic analysis of the geographic distribution of mitochondrial DNA haplotypes in the tiger salamander, Ambystoma tigrinum. *Genetics* **140**, 767–782.

Vitalis, R. (2002). Sex-specific genetic differentiation and coalescence times: estimating sex-biased dispersal rates. *Molecular Ecology* **11**, 125–138.

Wahlund, S. (1928). Zusammensetzung von populationen und korrelationerscheinungen vom standpunkt der vererbungslehre aus betrachtet. *Hereditas* **11**, 65–106.

Wakeley, J. (1998). Segregating sites in Wright's island model. *Theoretical Population Biology* **53**, 166–174.

Wang, J. (1997a). Effective size and F-statistics of subdivided populations. I. Monoecious species with partial selfing. *Genetics* **146**, 1453–1463.

Wang, J. (1997b). Effective size and F-statistics of subdivided populations. II. Dioecious species. *Genetics* **146**, 1465–1474.

Weber, J.L. and Wong, C. (1993). Mutation of human short tandem repeats. *Human Molecular Genetics* **2**, 1123–1128.

Weir, B.S. (1996). *Genetic Data Analysis II: Methods for Discrete Population Genetic Data*. Sinauer Associates, Inc., Sunderland, MA.

Weir, B.S. and Basten, C.J. (1990). Sampling strategies for distances between DNA sequences. *Biometrics* **46**, 551–582.

Weir, B.S. and Cockerham, C.C. (1984). Estimating F-statistics for the analysis of population structure. *Evolution* **38**, 1358–1370.

Weir, B.S. and Hill, W.G. (2002). Estimating F-statistics. *Annual Review of Genetics* **36**, 721–750.

Wilding, C.S., Butlin, R.K. and Grahame, J.W. (2001). Differential gene exchange between parapatric morphs of *Littorina saxatilis* detected using AFLP markers. *Journal of Evolutionary Biology* **14**, 611–619.

Wilkinson-Herbots, H.M. (1998). Genealogy and subpopulation differentiation under various models of population structure. *Journal of Mathematical Biolology* **37**, 535–585.

Wilkinson-Herbots, H.M. and Ettridge, R. (2004). The effect of unequal migration rates on FST. *Theoretical Population Biology* **66**, 185–197.

Williamson, S.H., Hernandez, R., Fledel-Alon, A., Zhu, L., Nielsen, R. and Bustamante, C.D. (2005). Simultaneous inference of selection and population growth from patterns of variation in the human genome. *Proceedings of the National Academy of Sciences of the United States of America* **102**, 7882–7887.

Wilson, G.A. and Rannala, B. (2003). Bayesian inference of recent migration rates using multilocus genotypes. *Genetics* **163**, 1177–1191.

Wright, S. (1921). Systems of mating. *Genetics* **6**, 111–178.

Wright, S. (1922). Coefficients of inbreeding and relationship. *American Naturalist* **56**, 330–338.

Wright, S. (1932). The roles of mutation, inbreeding, crossbreeding and selection in evolution In *Proceedings of the 6th International Congress of Genetics*. Vol. 1. Brooklyn Botanic Garden, Menasha, pp. 356–366.

Wright, S. (1943). Isolation by distance. *Genetics* **28**, 114–138.

Wright, S. (1946). Isolation by distance under diverse systems of mating. *Genetics* **31**, 39–59.

Wright, S. (1951). The genetical structure of populations. *Annals of Eugenics* **15**, 323–354.

Wright, S. (1965). The interpretation of population structure by F-statistics with special regard to systems of mating. *Evolution* **19**, 395–420.

Wright, S. (1969). *The Theory of Gene Frequencies*. The University of Chicago Press, Chicago, IL.

Wright, S. (1982). The shifting balance theory and macroevolution. *Annual Review of Genetics* **16**, 1–19.

Zhivotovsky, L.A. (1999). Estimating population structure in diploids with multilocus dominant DNA markers. *Molecular Ecology* **8**, 907–913.

Zhivotovsky, L.A. and Feldman, M.W. (1995). Microsatellite variability and genetic distances. *Proceedings of the National Academy of Sciences of the United States of America* **92**, 11549–11552.

30

Conservation Genetics

M.A. Beaumont

School of Biological Sciences, University of Reading, Reading, UK

This chapter reviews some of the statistical methodologies used in the genetic analysis of endangered and managed populations. The topics covered include the estimation of effective population size, N_e, the detection of past changes in population size, the estimation of admixture proportions, and the analysis of local population structure, kinship, and relatedness through genotypic methods. The reasons why it may be useful to measure N_e are discussed. Using genetic information it is possible to estimate N_e from changes in gene frequencies between samples taken at different times, or from genotypic disequilibria in one sample, or from differences in gene frequencies between two recently descended populations. The different statistical methodologies that have been applied are discussed in some detail. The ability to detect past changes in population size may also be useful in conservation management, and some of the more recent approaches are described. Another area of concern in conservation biology is introgressive hybridisation. At the population level it is possible to infer admixture proportions by comparing the gene frequencies in the admixed and non-admixed populations, providing they are available, and the different statistical approaches to analysing this problem are reviewed. Statistical methods have also been developed to detect immigrant and hybrid individuals from their multi-locus genotype, and the main methods developed in this area are also discussed.

30.1 INTRODUCTION

The use of genetic analysis in conservation falls into two main areas (Hedrick, 2001; DeSalle and Amato, 2004): examination of the genetic consequences of small population size for mean fitness and the probability of extinction (Frankham, 1995; 2005); and the use of data from genetic surveys to infer aspects of the demographic history of populations (Avise, 1994). This latter may then impinge on the former, e.g. in the estimation of 'effective population size', but may also be independent – for example, in the detection of hybrids. In many ways the first aspect, although historically motivating the use of genetics in conservation, is the most challenging. The theories describing the effect of recurrent deleterious mutation in small populations (Lynch *et al.*, 1995) depend on parameters whose

Handbook of Statistical Genetics, Third Edition. Edited by D.J. Balding, M. Bishop and C. Cannings.
© 2007 John Wiley & Sons, Ltd.   ISBN: 978-0-470-05830-5.

values are the subject of some debate even in model organisms (Keightley *et al.*, 1998; Lynch *et al.*, 1999; Whitlock *et al.*, 2003). The relative importance of genetic factors on extinction risk, in comparison with the ecological consequences of small population size, have been questioned (Lande, 1998). However, an increasing number of studies suggest that the loss of genetic variation can indeed contribute to extinction (Saccheri *et al.*, 1998; Spielman *et al.*, 2004), and this remains an area of active investigation in conservation biology.

This review will concentrate on the problems of statistical estimation that arise when using genetic data to infer aspects of the demographic history of populations, which can then be used to make informed decisions in the management of populations (Luikart and England, 1999). The statistical methodology is derived from the general area of inference in population genetics covered in Part 2 of this volume. To avoid overlap I have chosen four topics that have direct relevance to conservation biology – estimation of effective population size and detecting historical population bottlenecks, the analysis of hybridisation, and genotypic analysis of local population structure. One area, that of subdivision and gene flow from different populations, is only tangentially covered in the sections on hybridisation and genotypic analysis, because it is the subject of chapters by Excoffier and Rousset (**Chapters 28** and **29**). Methods for inferring relatedness and recovering pedigrees from genotypic data are briefly covered at the end of this review.

30.2   ESTIMATING EFFECTIVE POPULATION SIZE

It is useful to estimate the current rate of inbreeding in populations, and recent changes in this rate, for a number of reasons. First, although the effect of small population size (large inbreeding rate) on population viability is controversial, it is potentially significant, and therefore quantification of inbreeding rate in natural populations whose viability can also be monitored is an important component of any study to resolve these questions. Second, the inbreeding rate can be useful for calibrating timescales of recent inbreeding. For example, information on census size is often available, and if the rate of inbreeding for a given census size can be calibrated for one population then predictions may be made about other (comparable) populations on other (comparable) timescales (e.g. O'Ryan *et al.*, 1998).

The idealized 'effective' population size that gives rise to the observed rate of inbreeding is denoted N_e (Wright, 1931). The effect of different mating systems on N_e has been the subject of many studies (reviewed in Caballero (1994)), and, unsurprisingly, given the complexity of population genetic phenomena, it turns out that there are subtly different definitions of effective population size (Crow and Kimura, 1970), depending on the phenomena studied. For example, for the same theoretical 'inbreeding effective size' (Crow and Kimura, 1970), different mating systems can lead to slightly different predictions in the degree of linkage disequilibrium observed (Weir and Hill, 1980). Since these phenomena (e.g. linkage disequilibrium) can be used to infer N_e, a tacit assumption is that, given the uncertainty in estimation, these differences do not matter very much, and an estimate of N_e made by one method will lead to reasonable predictions of other phenomena that depend on N_e. Although a careful consideration of the mating system can give some estimate of N_e from the census size (Caballero, 1994; Nunney, 1995), it is often more practicable to estimate N_e directly, using genetic information, and listed below are some of the commoner approaches to this problem (see also Schwartz *et al.*, 1998). A final caveat is that implicit in the methods below, the estimate of N_e is made in a closed population over a short interval (with N_e the harmonic mean over this interval). On longer time scales, with mutation, immigration, and past population restructuring, estimates of (indeed, notions of) N_e obtained from genetic data become highly model dependent (Wakeley, 2001) and increasingly divorced from any possible measurement of local census size, mating systems or any factor with relevance to population management.

30.2.1   Estimating N_e Using Two Samples from the Same Population: The Temporal Method

Random genetic drift causes gene frequencies to change over an interval as a function of the population size and duration of the interval. Thus the gene frequencies in samples taken from a population at two time points can be compared, allowing N_e to be estimated. This approach is known as the *temporal method* and has a wide history of use with allozyme and microsatellite data (Krimbas and Tsakas, 1971; Nei and Tajima, 1981; Pollak, 1983). The bulk of the theory has been developed from the consideration of the Wright–Fisher model (see Jorde and Ryman, 1995, for a treatment with overlapping generations). Method-of-moments estimators were originally developed (Krimbas and Tsakas, 1971; Waples, 1989) but in recent years there has been a concentration on likelihood-based approaches. The general area has recently been reviewed in detail by Wang (2005), and a brief overview will be given here.

Assume a sample is taken at generation 0 from a diploid population of N individuals, comprising the total number of individuals available to be sampled. The frequency of a particular allele is p_0. The gene frequency in the next generation, p_1 is determined by taking a sample of size $2N_e$ with replacement from generation 0. With independent realisations of this sampling process, the mean, $E[p_1]$, is unchanged at p_0, and the variance, $E[(p_1 - p_0)^2]$, is $p_0(1 - p_0)/(2N_e)$ from binomial sampling. Extrapolating to t generations of such sampling, gives $E[p_t] = p_0$ and $E[(p_t - p_0)^2] = p_0(1 - p_0)(1 - (1 - 1/(2N_e))^t)$. Defining F to be $(p_t - p_0)^2/p_0(1 - p_0)$, $E[F] = 1 - (1 - 1/(2N_e))^t$ and hence, replacing $E[F]$ by an estimate $\hat{F}$ one can estimate $\hat{N}_e$ as $t/(2\hat{F})$. In general p_t and p_0 are unknown and are estimated by x_0 and x_t, with sample sizes S_0 and S_t, using some sampling scheme. The nature of the sampling scheme affects aspects of the estimation (Nei and Tajima, 1981; Pollak, 1983; Waples, 1989): e.g. whether the samples are taken with or without replacement. A commonly used method of moments estimator that takes into account uncertainty in the gene frequencies is

$$\hat{N}_e = \frac{t}{2[\hat{F} - 1/(2S_0) - 1/(2S_t)]}, \tag{30.1}$$

(Krimbas and Tsakas, 1971; Waples, 1989). A number of different estimators $\hat{F}$ have been proposed (Nei and Tajima, 1981; Pollak, 1983).

The likelihood-based method of Williamson and Slatkin (1999) uses the Wright–Fisher model. They wish to estimate $p(\mathbf{n}_0, \mathbf{n}_t | N)$ where $\mathbf{n}_0$ and $\mathbf{n}_t$ are vectors of counts of different allelic types scored at a locus at generation 0 and generation t, and N is the number of (diploid) individuals. At time 0 there is a vector $\mathbf{p}_0$ of the (unknown) allele frequencies in the entire population from which $\mathbf{n}_0$ is sampled with replacement. In this model the population allele frequencies are discrete and could be obtained as counts divided by $2N$. A tacit assumption is that N is always N_e in this model. Williamson and

Slatkin's method sums over the uncertainty in $\mathbf{p}_0$ in the following way. In the simplest case of diallelic loci, there are $2N + 1$ possible states of $\mathbf{p}_0$. Denote by $\mathbf{x}_0$ the vector of length $2N + 1$ giving the initial (prior) probability distribution of the ordered set of states of $\mathbf{p}_0$. Then

$$p(\mathbf{n}_0, \mathbf{n}_t | N) = (\mathbf{x}_0^T \mathbf{s}_0) \mathbf{x}_0^T \mathbf{M}^t \mathbf{s}_t, \tag{30.2}$$

where $\mathbf{s}_0$ and $\mathbf{s}_t$ are vectors of multinomial probabilities of sampling $\mathbf{n}_0$ and $\mathbf{n}_t$ from each of the possible population frequencies in generation 0 and t, and $\mathbf{M}$ is the standard Wright–Fisher transition matrix (see Ewens, 2004) where the element m_{ij} gives the probability of moving from state i to j by multinomial sampling. Equation (30.2) is evaluated for a range of possible value of N to obtain the maximum likelihood estimate. As Williamson and Slatkin (1999) point out, the likelihood can be evaluated straightforwardly only when the number of possible states is small. With diallelic loci there are only $2N + 1$ possible states and therefore the size of the computational problem will scale as $(2N)^2$. However with larger numbers of alleles estimation rapidly becomes unmanageable. An importance-sampling strategy has been proposed by Anderson et al. (2000) to overcome this.

An advantage of Williamson and Slatkin's method is that it is straightforward to make a composite estimate of N_e from samples taken at different time points, and also to make estimates of population growth rates using genetic samples. In this latter case, population size changes exponentially from initial size N_0 with rate r as $N_i = N_0 r^t$, rounded down to the nearest integer, and $\mathbf{M}^t$ in (30.2) is replaced by the multiplied sequence of matrices M_i with columns equal to the number of possible allelic states in the ith generation and rows equal to the number of possible states in the $(i - 1)$th generation. Using simulations, they demonstrate that with a large number of loci (150) it is possible to obtain reasonably accurate joint estimates of N_0 and r.

Wang (2001) has made two modifications to the approach of Williamson and Slatkin (1999), resulting in the ability to estimate likelihoods very rapidly, even for multi-allelic loci. These improvements have enabled detailed simulation testing to be carried out. As in the study of Williamson and Slatkin (1999), Wang is also able to extend the method to work with multiple samples and to estimate changes in N_e. One of the changes proposed by Wang (2001) is to suggest that the likelihoods for multi-allelic loci with K alleles can be well approximated by converting them into K biallelic loci, whose frequencies $(\mathbf{n}_{0k}, \mathbf{n}_{tk})$ consist of the frequencies of the kth allele and the sum of the frequencies of the others. The likelihood is then calculated as

$$p(\mathbf{n}_0, \mathbf{n}_t | N) = \left(\prod_{k=1}^{K} p(\mathbf{n}_{0k}, \mathbf{n}_{tk} | N) \right)^{\frac{k-1}{k}}.$$

There is no theory to say why and under what circumstances this approximation will work. However Wang (2001) extends the transition matrix approach to the three-allele case, which is feasible for small sample sizes and small N, and is able to demonstrate that the approximation and the exact method are comparable in precision and accuracy. The second modification is to use a variety of computational 'tricks' to greatly speed up evaluation of (30.2), the most important of which is to note that only the diagonal and adjacent off-diagonal elements of $\mathbf{M}$ contribute significantly to the calculation of the likelihood.

Like Williamson and Slatkin (1999) and Wang (2001) demonstrated modest gains in accuracy and precision when using the likelihood method. The gains are most substantial when initial allele frequencies are low. For alleles at intermediate frequencies, however, as noted by Williamson and Slatkin, the improvement on moment-based methods through the use of likelihood methods appears to be small. The accuracy and precision are approximately linearly proportional to sample size, number of generations between samples, and the number of independent alleles used (i.e. the sum of $K - 1$ over loci), and inversely proportion to N_e. The method of Wang (2001) has been further extended in Wang and Whitlock (2003) to allow for the estimation of immigration rates.

An alternative likelihood-based approach suitable for multiple alleles has been developed using coalescent theory (Berthier *et al.*, 2002; Beaumont, 2003a; Anderson, 2005). The use of a coalescent model assumes that the changes in gene frequencies can be accurately described by the diffusion approximation, and will therefore lead to discrepancies with those based on the Wright–Fisher model when the population size is low.

30.2.2   Estimating N_e from Two Derived Populations

It is commonly the case that a once large population occupying a contiguous area becomes fragmented into isolated populations. Also, managed populations are often divided into separate groups. In these cases, when there is knowledge of the time of splitting, it is possible to infer the N_e for each sub-population from the amount of genetic divergence between the two populations. Most of the statistical methods devised for this case appear to be based on the likelihood method (Cavalli-Sforza and Edwards, 1967; Felsenstein, 1981). These authors used a Brownian-motion approximation for genetic drift based on 2-allele models and extensions, discussed in a little more detail in a later section. More recently, methods based on coalescent theory have been developed which are directly applicable to multi-allelic models. This approach will be described here.

A method based on coalescent theory for estimating N_e in recently isolated, genetically diverging populations, was described in O'Ryan *et al.* (1998) and also by Nielsen *et al.* (1998) (see also Saccheri *et al.*, 1999). For a review of coalescent theory see Nordborg (**Chapter 25**), Hudson (1991) and Donnelly and Tavaré (1995). In the method described in O'Ryan *et al.* (1998), the genetic samples are taken from s isolated populations that are derived from some common ancestral population at time T in the past. In the discussion below N_e refers to the number of (diploid) individuals in the population, and sample sizes refer to the number of chromosomes. The populations are assumed to be have been isolated sufficiently recently that mutations arising since the time of isolation can be ignored. The genealogies for a particular locus of the s populations can be traced back to a set of founder lineages that have descendants in the samples. The founders are assumed to be sampled from an ancestral gene frequency vector $\mathbf{x}$ that is common to all the populations. The gene frequency configuration in the founder lineages is given by $\mathbf{a}_f$, and that in the sample is given by $\mathbf{a}_s$. The number of founder lineages, n_f, is a random variable given by

$$n_f = n_s - n_c,$$

where n_s is the sample size and n_c is the number of coalescences over the period T in the genealogy of the sample. The probability function $p(n_c | T/(2N_e), n_s)$ is given in Tavaré (1984; see also O'Ryan *et al.*, 1998). Thus the probability of a sample configuration for a particular population can be calculated as

$$p(\mathbf{a_s}|T/(2N_e), n_s, k, \mathbf{x}) = \sum_{\mathbf{a_f}} p(\mathbf{a_s}|\mathbf{a_f}, n_s)p(\mathbf{a_f}|\mathbf{x}, n_f)p(n_c|T/(2N_e), n_s), \qquad (30.3)$$

where the summation is over all possible $\mathbf{a_f}$. In the case of populations that split from a common ancestor, the likelihoods can be multiplied over populations, each with their own $T/(2N_e)$, but assuming a common $\mathbf{x}$. The first term in (30.3) is the probability of obtaining the sample configuration from the founder configuration by successively choosing a lineage at random and duplicating it n_c times, and is given by

$$p(\mathbf{a_s}|\mathbf{a_f}, n_s, n_f) = \frac{\prod_{i=1}^{k}\binom{a_{si}-1}{a_{fi}-1}}{\binom{n_s-1}{n_f-1}}.$$

This can be derived by considering it as an urn problem (Slatkin, 1996; Nielsen *et al.*, 1998): the denominator is the number of ways of allocating the n_s lineages in the sample to the n_f founder lineages, and the numerator is the number of different ways of allocating the sample lineages within each allelic class to the corresponding class in the founders. The second term in (30.3) is the multinomial probability of drawing $\mathbf{a_f}$ from $\mathbf{x}$ given n_f. Clearly the number of possible configurations $\mathbf{a_f}$ is very large, and evaluation of (30.3) can be problematic. An additional problem is that the evaluation of $p(n_c|T/(2N_e), n_s)$ can be numerically unstable. O'Ryan *et al.* (1998) use importance sampling (Griffiths and Tavaré, 1994) as an alternative method of evaluation, and this is embedded within a Markov chain Monte Carlo (MCMC) sampler to infer marginal posterior densities for individual parameters (see Beaumont, 2003a, for a detailed description of this general approach).

The general scheme for two populations is illustrated in Figure 30.1. Using the MCMC scheme, marginal posterior distributions for $T/(2N_e)$ can be obtained. In the study of O'Ryan *et al.* (1998) a number of isolated populations were modelled in this way and the effective population size could be estimated because the splitting times were known. As an example, one population, (St. Lucia), was seeded from another umfolozi-hluhluwe complex (UHC) 16 years prior to sampling. Cull data suggested that the generation time was 7.5 years in these park populations. Seven microsatellite loci were analysed in samples of around 20–30 individuals (the sample size varied among loci due to missing data) sampled from the two parks. Figure 30.2(a) shows the posterior distribution for N_e in the St. Lucia population. There was extensive census data available and the harmonic mean census size was estimated as 58 for the St. Lucia population, giving a modal estimate of effective to census size of 13%. In the UHC population, which was estimated to have a harmonic mean census size over this period of over 2000, the posterior density appeared to rise to an asymptote (a rectangular prior for N_e was used to make estimation practicable), and most values from $N_e > 500$ had similar posterior density (Figure 30.2b). Essentially, this means that an infinite population size has similar likelihood to most values of N_e greater than 500.

Obviously this modelling approach is only applicable when it can be assumed that the time scale is short enough that the effect of mutation can be ignored. Specifically it is assumed that all the descendent copies of alleles originally present in the founders are identical.

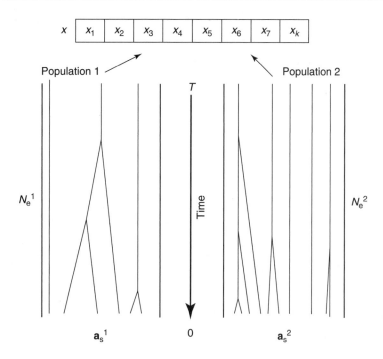

Figure 30.1 Illustration of the genealogy of samples taken from two different populations. The two populations are assumed to have been derived from a common ancestral population with allele frequencies $x_1, \ldots, x_k$. The gene frequencies in the two populations diverge over the time period of duration T. From a genealogical perspective this is illustrated as coalescences of lineages derived from the present day genetic samples with frequencies $\mathbf{a_s}^1$ in population 1 and $\mathbf{a_s}^2$ in population 2. The rates of coalescence depend on the population sizes N_e^1 and N_e^2.

It is possible to extend the methodology described here to test whether a model of population splitting and divergence through drift explains the data better than a model of gene flow. As discussed in (**Chapter 28**), the probability of the sample configuration $\mathbf{a_s}$ in a model of gene flow has a very simple description if an infinite island or continent-island model is assumed. In this case it is given by the multinomial Dirichlet

$$p(\mathbf{a_s}|M, n_s, k, \mathbf{x}) = \frac{\prod_{i=1}^{k} \binom{a_{si}+Mx_i-1}{a_{si}}}{\binom{n_s+M-1}{n_s}},$$

(Rannala and Hartigan, 1996; Balding and Nichols, 1995; 1997). As with the drift model, the assumption here is that the difference between the gene flow model and the isolation model is that in the latter case all lineages are available to coalesce until the time the population is founded whereupon the founder lineages have genotypes drawn at random from $\mathbf{x}$. By contrast in the gene flow model, looking backward in time, lineages emigrate out of the population (their genotypes being then drawn at random from $\mathbf{x}$) and are no longer available to coalesce. This corresponds to the 'scattering phase' of Wakeley (1999). Thus an assumption in this model is that the descendent lineages of each immigrant that

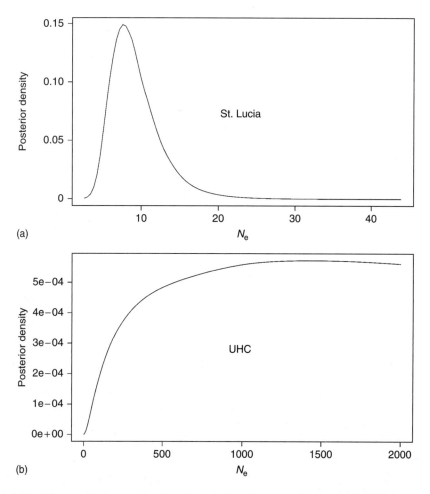

Figure 30.2 The marginal posterior distributions for N_e over a 16-year period, estimated from buffaloes sampled from the St. Lucia (a) and UHC (b) national parks in South Africa. Uniform priors are assumed for N_e. A rectangular prior was used for the UHC population, with an upper limit on N_e of 2000.

are present in the sample are all identical by descent (IBD), and that the population size of the external metapopulation, or continent is sufficiently large that $\mathbf{x}$ does not change between the times of the first and last immigration event. The differences between the two models are illustrated in Figure 30.3.

As described in Ciofi *et al.* (1999) it is possible to reparameterize both the drift model and gene flow model in terms of F, the probability that two lineages coalesce before the type of either lineage is drawn from $\mathbf{x}$ as $(1 - F)/F$ for M and $-\log(1 - F)$ for T/N_e. An indicator variable can then be introduced such that when $I = 0$ the probability of the sample configuration is calculated from the isolation-drift model, or when $I = 1$ the probability is calculated from the gene flow model. Thus it is possible to use $p(\mathbf{a_s}|I, F, n_s, k, \mathbf{x})$ to drive an MCMC scheme as described in Ciofi *et al.* to estimate the marginal posterior density $p(I|n, k, \mathbf{a_s})$. Simulations suggest that with 25 individuals,

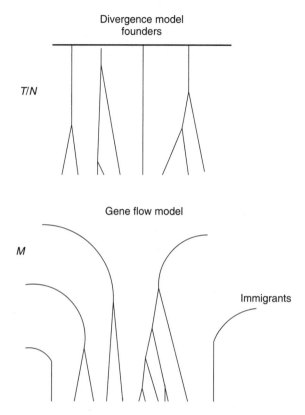

Figure 30.3 The genealogies of samples taken under a model of divergence (as in Figure 30.1) compared with those in a gene flow model. The symbols are defined in the text.

5 loci, 10 alleles per locus, and 5 populations, the two models can be distinguished relatively well (contribution by Beaumont in Balding *et al.* (2002)). Applications of the approach, implemented in the program *2mod*, are described in Ciofi *et al.* (1999), Goodman *et al.* (2001), Palo *et al.* (2001), and Hanfling and Weetman (2006). A method with related aims, but allowing for mutation as well as drift, is described in Nielsen and Wakeley (2001) and Hey and Nielsen (2004). In this case the model consists of a pair of populations that split from an ancestral population and there is gene flow between them. The data are assumed to consist of homologous copies of a single sequence (i.e. this is a one-locus model), and follow an infinite sites mutation model without recombination. Immigration and time of splitting can be jointly inferred using MCMC, thereby enabling the relative importance of each to be assessed.

Alternative methods for inference in these drift-based models are those by Wang (2003), using the Wright–Fisher model and the approximations discussed above, and also another approximation described by Nicholson *et al.* (2002), which is somewhat similar in spirit to that of Cavalli-Sforza and Edwards (1967). In the method of Nicholson *et al.* (2002), it is assumed that the data are frequencies of biallelic (SNP) markers, and the sample frequencies are a random draw from some unknown sub-population gene frequency, which is in turn a random draw from a distribution with mean corresponding to the baseline gene frequency, p (shared between populations), and variance $cp(1 - p)$. The distribution is assumed to be Gaussian, and probability in the tails that extend outside $(0,1)$ are given as atoms of probability on 0 and 1. The parameter c approximates T/N_e in the standard Wright–Fisher diffusion model, where T is the time of the split and N_e is the harmonic mean effective size over the interval. Using MCMC, Nicholson *et al.* (2002) are able to integrate out the unknown gene frequencies and obtain marginal posterior distributions for individual c_j for the jth population.

30.2.3 Estimating N_e Using One Sample

Often temporal information is not available for estimating short-term N_e and there have been attempts to estimate it using genotypic information from single samples. The essential idea here is that the gametes are sampled by a finite number of zygotes, which leads to small departures from the expected genotype frequencies both within and between loci. Thus, there are two possible approaches using genotypic data: to detect departures from Hardy–Weinberg equilibrium or to detect departures from linkage equilibrium. Both approaches have been studied and these will be briefly described here. A fundamental concern is that many other processes can cause departures from equilibrium. For example, admixture between two partially separated subpopulations can lead to departure from Hardy–Weinberg expectation (Wahlund effect) and cause substantial linkage disequilibrium. Furthermore, genotyping errors, particularly prevalent in microsatellites, can cause apparent substantial deviations from equilibrium proportions of genotypes, at least for particular loci, which can bias the results.

For a two-allele system with zygotes sampled from a base population with frequencies p and $1 - p$, the expected proportion of heterozygotes in a population of size N is $p(1 - p)/N$ in excess of that expected from an infinite population (Robertson, 1965; see also Cannings and Edwards, 1969). This result can be obtained by noting that N maternal gametes fuse with N paternal gametes to produce the zygote. The gene frequencies in these two classes of gamete will be slightly different, with frequencies p_f and p_m (having expectation p). Thus the expected proportion of heterozygotes is $p_f(1 + p_m) + p_m(1 + p_f)$. The expected proportion, based on the combined gene frequencies, $p_o = (p_f + p_m)/2$, is $2p_o(1 - p_o)$. Thus there is a heterozygote excess of $(p_f - p_m)^2/2$. Assuming that $E[p_f - p_m] = 0$, the expected value of the squared difference is given by the sum of binomial sampling variances, giving an expected heterozygote excess of $p(1 - p)/N$ (i.e. the expected frequency of heterozygotes is $2p(1 - p)(1 + 1/2N)$). Note that Crow and Kimura (1970), following Hogben (1946) and Levene (1949), suggest $2p(1 - p)(1 + 1/(2N - 1))$, instead, which can be obtained by assuming that all gametes contributing to the observed zygotes can potentially be joined together in a zygote.

Based on the Robertson (1965) result, Pudovkin *et al.* (1996) proposed an estimator that appeared to be unbiased. This has been investigated further by Luikart and Cornuet (1999) through simulation testing in which they aimed to identify the effect of sampling error and different mating systems on the robustness of the method. They concluded that in general the method slightly overestimated N_e, but this was generally less than 10 % for sample sizes of 30, and became smaller with larger sample sizes. The bias was largest (25 %) for extreme polygyny (1 male mating with 99 females), which is perhaps unsurprising, given that the method assumes equal male and female contributions. Overall, they note that relatively large numbers of loci are required because otherwise the confidence intervals are typically very wide. The method has been developed further by Balloux (2004), who extended its use to subdivided population, and corrected some errors in the original treatment by Pudovkin *et al.* (1996).

The use of linkage disequilibrium to estimate N_e has been proposed by Langley *et al.* (1978), Laurie-Ahlberg and Weir (1979), and Hill (1981). There is an extensive and technical literature deriving approximate expectations of linkage disequilibrium statistics in terms of demographic parameters, stemming from the work of Hill and Robertson (1968) and Ohta and Kimura (1969). These approximation are discussed by Hudson (1985) and McVean (**Chapter 27**). The squared correlation in gene frequencies between two loci is defined to be

$$r^2 = \frac{D^2}{(1 - F_A)(1 - F_B)},$$

where $D^2 = \sum \sum (f_{ij} - p^A{}_i p^B{}_j)^2$ for loci A and B with alleles i, j, and $F_A = \sum (p_i^A)^2$, $F_B = \sum (p_j^B)^2$ (Hudson, 1985; see also McVean, **Chapter 27**). These definitions are based on known population gene frequencies. For samples (where usually the gametic phase of heterozygotes is unknown) there are several estimators available (see Hill, 1981; Weir, 1996).

In a finite population of N_e diploid individuals at equilibrium $E[r^2]$ tends to a steady state value that is approximately independent of the mutation model and mutation rate when $4N_e c$ is large, for recombination rate c. An approximation for this value, $E[r^2] \approx 1/(4N_e c)$, was first obtained by Hill and Robertson (1968) based on observations of simulations, and analytically by Ohta and Kimura (1969). Following these authors, most analytical approaches have tended to solve $E[r^2] \approx E[D^2]/E[(1 - F_A)(1 - F_B)]$. A number of other approximations have been obtained (see also McVean, **Chapter 27**), and the most widely used approximation is

$$E[\tilde{r}^2] = \frac{(1 - c)^2 + c^2}{2N_e c(2 - c)} + \frac{1}{n},$$

where $\tilde{r}^2$ is a sample estimate (Laurie-Ahlberg and Weir, 1979; Hill, 1981, suggest using the Burrows method in Cockerham and Weir (1977); see Weir and Hill, 1980, for other approximations for different estimators and mating systems).

Laurie-Ahlberg and Weir (1979) used the equation above to provide an estimator for N_e. Their analysis was based on 17 enzyme loci surveyed in nine laboratory populations of *Drosophila melanogaster*, from which they selected unlinked pairs of loci. Some of these pairs involved the same loci, and were thus not independent. In total they analysed 4–18 pairs among eight of the nine populations. The sample sizes were generally between 50 and 100. The census sizes fluctuated in these populations, but never exceeded 500–700, and could reach values as low as 10–12. With $c = 0.5$, Laurie-Ahlberg and Weir estimated $\hat{N}_e = m/3 \sum_i[(\tilde{r}_i^2 - 1/n_i)]$. In four of the populations the denominator was negative, implying an estimated N_e of infinity. Of the remaining four populations $\hat{N}_e$ varied from 3 to 27. Laurie-Ahlberg and Weir made no attempt to give standard errors on these estimates.

Laurie-Ahlberg and Weir suggested that it is necessary for $n \gg \hat{N}_e$ to be able to estimate population size with some precision. For most taxa of conservation interest, linkage information will be unavailable, and, on the assumption that the markers are randomly scattered through the genome (as with microsatellites identified through cloning, for example) the assumption of $c = 0.5$ would generally be reasonable.

Hill (1981) performed a similar analysis on other data sets, including linked pairs of markers. Hill suggested using information on variances to provide a weighted estimator across pairs of loci. He suggested that pairs of loci could be treated as uncorrelated. Simulation studies indicated that $var[\tilde{R}_i^2] \approx 2(E[\tilde{R}_i^2])^2$ (i.e. approximating χ^2 with 1 df), and Hill therefore suggested a weighted estimator

$$\frac{1}{\hat{N}} = \frac{\sum_i \alpha_i / var[\alpha_i]}{\sum_i 1/var[\alpha_i]} = \frac{\sum_i \gamma_i (\tilde{r}_i^2 - 1/n_i) / \left(\frac{\gamma_i}{N} + \frac{1}{n_i}\right)^2}{\sum_i 1 / \left(\frac{1}{N} + \frac{1}{\gamma_i n_i}\right)^2},$$

where $\alpha_i = (\tilde{r}_i^2 - 1/n_i)/\gamma_i$ and $\gamma_i = ((1 - c_i)^2 + c_i^2)/(2c_i(2 - c_i))$. The variance of $\hat{N}$ is approximated as $\hat{N} = var[\hat{N}] = 2N^2/\sum_i 1/(1 + N/(\gamma_i n_i))^2$. It is necessary to substitute estimates of N into these equations.

Hill analysed two *Drosophila* data sets: one based on chromosomes extracted from a sample of 198 flies isolated from a wild population, and another using a sample of around 700 flies from a caged population. In the first sample, 11 enzyme loci were scored, all of which were linked with most $c_i < 0.1$, and $\tilde{r}_i^2$ was measured in 25 pair-wise comparisons. The estimate of $\hat{N}_e$ was negative for these data. In the caged populations Hill analysed data from seven loci. The estimated census figure was around 1000. He used 9 pair-wise comparisons of linked markers (with all c_i less than 0.15) and 12 pair-wise comparisons of unlinked loci. The estimate of $\hat{N}_e$ was 363 with standard deviation 170.

Thus these studies appear to indicate that information on linkage disequilibrium has limited power to estimate effective population size. Even with large sample sizes and linked loci the standard errors on the estimates are fairly large, and negative estimates of N_e are common. Furthermore, as Hill (1981) points out, with very tightly linked loci, estimates of population size will be influenced by earlier demographic events, including gene flow. Also linked loci are only readily available in model organisms.

Notwithstanding these problems, Waples (1991) pointed out that if many pairs of loci are used, estimates of $E[\tilde{r}^2]$ can be made with reasonable precision. His own simulations confirmed the observation of Hill (1981) that pair-wise comparisons among k loci appeared to behave as independent data points with $\sum_{i=1}^{m} \tilde{r}_i^2$, for $m = k(k-1)/2$, distributed as approximately χ^2 with m degrees of freedom. Recently it has been noted that the method of Hill (1981) appears to be quite biased when the sample sizes are low (England *et al.*, 2006) and Waples (2006) has developed a correction for this effect. Additional simulation-based tests by England *et al.* (2006) and Waples (2006) provide some encouragement that this general approach may be an effective method for estimating current population sizes.

The method has been used to estimate effective population sizes in managed or endangered populations (e.g. Bartley *et al.*, 1992; Bucci *et al.*, 1997). In the latter example, as part of a more detailed genetic survey, N_e was estimated in 5 Italian populations of the endangered pine *Pinus leucodermis* using 23 RAPD markers, assumed to be unlinked. The sample sizes were around 20–30 within each population. They estimated N_e using the method of Laurie-Ahlberg and Weir (1979), and estimated confidence limits for r^2 using the χ^2 approximation above, as suggested by Waples (1991), which were then transformed to give approximate limits on $\hat{N}_e$. Two of the 5 populations gave estimates of infinity for N_e and the remaining three estimates were: 17 (11–31); 17 (11–30); 31 (14–650).

Thus, in conclusion, there appears to be scope for using multi-locus genotype information for estimating N_e with reasonable precision, and it may well be worthwhile investigating this approach further, perhaps using likelihood-based approaches to extract as much information as possible from the genotypic information. The development of methods to calculate approximate likelihoods and posterior distributions based on summary statistics (Beaumont *et al.*, 2002; discussed further below) may be helpful in this regard. If unlinked markers are used, low levels of gene flow may not affect the estimates too greatly, or, alternatively, can be included in an enhanced model.

30.2.4 Inferring Past Changes in Population Size: Population Bottlenecks

In order to obtain a better understanding of the genetic consequences of small size for population viability, it is useful to be able to identify whether a population has undergone a recent reduction in population size. For example inbreeding depression is assumed to be a transient phenomenon. A reduction in population size can inflate the frequency of rare deleterious recessives at some loci, leading to a reduction in population fitness, which is then 'purged' by natural selection. Thus populations that have had small sizes for a long time are unlikely to have high frequencies for highly deleterious genes (although mildly deleterious genes may be at high frequencies because of the strength of drift in small populations; Lynch *et al.*, 1995). Furthermore, if it is possible to model the time scale of the reduction in population size, one may be able to distinguish among different ecological explanations for the reduction. For these reasons the development of statistical methods to infer past changes in population size is useful in conservation biology. Although the focus in this section is on past reductions in size, the same methodology can also be used to detect population expansions, and this case will also be addressed.

A number of methods for detecting historical changes in population size have relied on discrepancies between different summary statistics that are used to estimate $\theta = 4N_e\mu$ assuming a stable population model, where N_e is the diploid population size and μ is the mutation rate. When the data do not fit a stable population model, the different

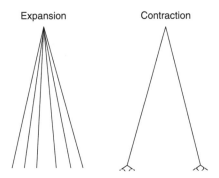

Figure 30.4 Two extreme examples illustrating how changes in population size affect the shape of genealogies. On the left is a genealogy that could only occur in the idealized case of a very small population that instantaneously increases to a size that is sufficiently large for no coalescences to occur during this interval. On the right is a genealogy that might be found in samples from a population undergoing a recent population contraction.

summary statistics give differing estimates of θ. This is best explained from a genealogical perspective (Figure 30.4) (see also **Chapter 25**). If populations have recently expanded, coalescences have a higher probability of occurring nearer the time of the most recent common ancestor than would be expected in a stable population. This is because the ancestral population size is smaller than the current population size, and therefore the coalescence rate is relatively higher. It should be noted that the schematic in Figure 30.4 is exaggerated: such a genealogy could only arise if there was a large change in population size over a very short period. For most reasonable demographic changes there are still many recent coalescences because the coalescence rate is almost quadratic in the number of lineages at any time (**Chapter 25**), and this tends to have a dominating effect on the shape of genealogies. This tendency for short recent coalescences is intensified when there is a population contraction. In this case there is a possibility that the most recent common ancestor occurs within the timescale of the contraction, or alternatively that some lineages 'escape' back into the ancestral population, which has longer coalescence times. In the example in Figure 30.4, two lineages are present in the ancestral population. Mutations occur along the lineages and thus for a sample from an expanding population most mutations have only a few descendants (a single descendent in the extreme example in Figure 30.4). Each gene in a pair differs by unique mutations. By contrast in a contracting population the majority of mutations have many descendants, and pairs of genes are either identical because of recent shared ancestry or differ by the same mutations. The differing distribution of mutations between these two scenarios alters the relationship, e.g. between summary statistics that are influenced by the total number of mutations in the genealogy and those that measure pair-wise differences between genes. Thus, conditioning on, say, measures of pair-wise divergence, there will be too many/too few alleles or segregating sites if the populations have expanded/contracted. Or, conversely, conditioning on the number of alleles or segregating sites, there will be too few/too many pair-wise differences if the populations have expanded/contracted.

There is an asymmetry in the detectable effects of population expansions and contractions. Contractions, by accelerating drift, produce instantaneous changes in the gene frequencies, whereas expansions do not. The only information about an expansion comes from mutations or recombination events that occur in the genealogical history over the time of expansion. Thus if the expansion has occurred very recently there will be little information in the data with which to detect it. Similarly if the population contraction is recent, although it should be easy to detect, without mutations there is no information on the shape of the contraction and the change in gene frequencies will depend only on the harmonic mean N_e over the interval. Thus the effects of a recent 'bottleneck' which traditionally means a contraction followed by an expansion, will be difficult to distinguish from any other model of recent population decline, and are considered together here.

For sequence data there are a large number summary statistics that have been devised to detect departures from the standard neutral model. Often these have been ascribed to the effects of selection rather than demographic history. This aspect is discussed in the chapters by Neuheuser (**Chapter 22**) and McVean (**Chapter 27**). A number of the tests are carried out by the Arlequin program (`http://lgb.unige.ch/arlequin/software/`) and the methodological section of the Arlequin manual provides a good description of the statistics, tests, and underlying assumptions.

One of the earliest tests for departures from neutrality (and also as a test for past changes in population size), is the test of Watterson (1978) based on the sampling theory of Ewens (1972). This test generates the distribution of sample homozygosity (average pair-wise similarity), $\sum x_i^2$, from sample gene frequencies x_i, conditional on the observed number of alleles, assuming the standard neutral model. In a survey of many loci the observation of homozygosities that are extreme under the theoretical sampling distribution would then imply a departure from the model assumptions. If only one of the loci studied was strongly affected, it might be reasonable to infer that departure from neutrality at that locus was a likely cause. Alternatively, if many loci appeared to show the same extreme heterozygosities, then it may be more likely that an historic change in population size is the explanation. If the homozygosity calculated from the real data is too high in comparison with the theoretical distribution this would suggest a population expansion, and if too low it would suggest a contraction.

For DNA sequences one of the most commonly used statistics for detecting past changes in population size is Tajima's D (Tajima, 1989), which is based on the difference in estimates of θ, assuming an infinite allele model (IAM), from average pair-wise differences between sequences and from the number of segregating sites, scaled by an estimate of the standard deviation of the difference:

$$D = \frac{\pi - \frac{S}{a_1}}{\sqrt{\widehat{\mathrm{var}}\left(\pi - \frac{S}{a_1}\right)}},$$

$$a_1 = \sum_{i=1}^{n-1} \frac{1}{i}.$$

π is the average number of pair-wise differences between sequences, which directly gives an estimate of θ (Tajima, 1983). S is the number of segregating sites in the sample, which is identical to the number of mutations in the genealogical history of the sample under the infinite sites assumption, and $\sum_{i=1}^{n-1} \frac{1}{i}$ is equal to half the expected total length of lineages in the genealogy, given the sample size, n, so that $E[S] = \theta \sum_{i=1}^{n-1} \frac{1}{i}$ (Watterson, 1975). The variance of the difference between the two estimators is given in Tajima (1989) as:

$$\widehat{\mathrm{Var}}\left(\pi - \frac{S}{a_1}\right) = \frac{S}{a_1}\left(\frac{n+1}{3(n-1)} - \frac{1}{a_1}\right) + S(S-1)\left(\frac{\frac{2(n^2+n+3)}{9n(n-1)} - \frac{n+2}{a_1 n} + \frac{a_2}{a_1^2}}{a_1^2 + a_2}\right),$$

where

$$a_2 = \sum_{i=1}^{n-1} \frac{1}{i^2}.$$

Negative values of Tajima's D indicate population growth (too many segregating sites/too few pair-wise differences) whereas positive values of Tajima's D indicate population contraction (too few segregating sites/too many pair-wise differences). A significance test can be developed by simulating data sets under the coalescent with the same sample size and number of segregating sites as in the real data. An obvious caveat is that any test based on summary statistics such as Tajima's D measures departures from the standard neutral model, assuming a particular mutation model, and not growth or decline *per se*. There are many ancestral processes, including variability in mutation rates among sites and selection, that can affect the value of the summary statistics.

Another summary statistic, similar in spirit to Tajima's D, is Fu's F_S (Fu, 1997). This is based on the discrepancy between the number of alleles observed in the sample and that predicted from Ewens' sampling theory for the IAM (Ewens, 1972), given $\hat{\theta}$ estimated as π, discussed above.

$$F_S = \log\left(\frac{S'}{1 - S'}\right),$$

where (from Ewens, 1972)

$$S' = p(k \geq k_O | \theta = \pi) = \sum_{k=k_O}^{n} \frac{S_k \pi^k}{S_n(\pi)},$$

and

$$S_n(\pi) = \pi(\pi + 1)\ldots(\pi + n - 1),$$

and S_k, an unsigned Stirling number of the first kind, is the coefficient of π^k in $S_n(\pi)$. If there has been population growth, there will be an excess of alleles (in particular, an excess of singleton haplotypes) given π, therefore S' will be small and F_S negative.

The statistical power of tests for population growth based on these summary statistics has been evaluated under a number of different scenarios using simulations by Ramos-Onsins and Rozas (2002). They found that tests based on F_S of Fu (1997), and another similar test developed by themselves, had a greatest power to detect population growth. Interestingly they found that tests based on the distribution of pair-wise differences (Rogers and Harpending, 1992) tended to be very conservative. The power of the tests to detect population decline has not been investigated.

A key point to reiterate is that these tests cannot in general be used to distinguish between departures from the standard neutral model caused by selection or by past changes in population size. A reasonable way to tackle this problem is to use many loci, on the assumption that selection will only affect a small proportion of loci, and if all loci are supporting the same tendency to depart from the standard neutral model it would be reasonable to ascribe this to the effect of past population history rather than selection. Until recently the markers that have been best suited for this are microsatellites because of their polymorphism and the relative ease of obtaining large numbers of loci. In conservation biology most of the tests for past population bottlenecks have been designed for microsatellites.

Based on these considerations Cornuet and Luikart (1996) proposed a test for bottlenecks using many loci in which coalescent simulations are performed to derive distributions of heterozygosities conditional on the observed number of alleles. Cornuet and Luikart consider two mutation models: the IAM, generally considered to be more appropriate for allozyme data; and the stepwise mutation model (SMM), often assumed for microsatellites (the distributed program, BOTTLENECK, Piry *et al.*, (1999), also incorporates the two-phase model (TPM) of Di Rienzo *et al.* (1994)). In the case of the IAM they used the method of Watterson (1978) described above, but looking at the distribution of heterozygosity (calculated as $\frac{n}{n-1}(1 - \sum x_i^2)$), conditional on the observed number of alleles. This is independent of θ (Watterson, 1978) and can be simulated by first creating a sample genealogy and then adding mutations at a frequency proportional to branch length until the required number of alleles are obtained. In the case of the SMM, where θ does affect the distribution of heterozygosities, a uniform prior distribution of θ is assumed and the conditional heterozygosity is simulated by sampling θ uniformly and then simulating a genealogy with stepwise mutations, rejecting those that do not give rise to samples with the required number of alleles (see Cornuet and Luikart, 1996, for specific details of how to sample θ efficiently).

Cornuet and Luikart propose two methods for using the simulated heterozygosities to test for significant departures of the observed data from the null hypothesis. Firstly they propose a type of sign test whereby each locus is scored as having a heterozygosity above or below its expected value. For each locus it is possible to estimate by simulation the probability of obtaining a heterozygosity greater than the expected value. From this one can obtain a probability distribution of observing $l = 0 \ldots L$ loci with heterozygosities greater than the expected value, and thereby estimate the probability of observing at least l_O loci with greater than expected heterozygosity. The other test Cornuet and Luikart propose is to calculate deviations of the observed heterozygosity from the simulated mean scaled by the simulated standard deviation in heterozygosity. Under the null hypothesis, the sum of these should approximate normal deviates with mean 0 and variance equal to the number of loci. Cornuet and Luikart (1996) carry out a number of tests of their method using simulated data sets. They suggest that their test is most powerful with loci evolving according to the IAM compared to the SMM, and find that the second test is somewhat more powerful than the first, although the difference is not large. If the test is applied to microsatellites, where there is abundant evidence from pedigree information that the IAM does not apply, it would seem reasonable never to cite results assuming the IAM because of the tendency to type-I error, and a case could be made for only citing results for the TPM model, using parameters that are reasonable, given current pedigree information.

Other summary statistics can also be monitored in microsatellites. For example, Luikart *et al.* (1998) suggest that the shape of the frequency spectrum can provide a useful graphical test of a bottleneck. In the case of microsatellites the distribution of lengths holds useful information on the past demographic history (Reich and Goldstein, 1998; Reich *et al.*, 1999). For example, as noted by Cornuet and Luikart (1996) the distribution of lengths becomes more uneven with gaps when the population has been subject to a bottleneck. This is exploited in the test of Garza and Williamson (2001), who monitor the distribution of the statistic $M = k/r$, where k is the number of alleles, and r is the difference in length between the longest and shortest microsatellite allele in the sample. In a population that has a history of recently reduced population size, or bottleneck, there are expected to be relatively few alleles for a given value of r, giving a small M. The mean value, $\bar{M}_O$, is calculated across loci and compared to relevant quantiles of simulated values of $\bar{M}$ obtained from coalescent simulations corresponding to the same number of loci, and assuming a TPM model. The critical values depend on the parameters chosen for the TPM model. If the mutation process allows larger than single-step changes in length then these can give rise to allele frequency distributions that are similar to those that arise from population contraction.

An additional summary statistic that has been used to detect population expansions from microsatellite is the expansion index of Kimmel *et al.* (1998). This has been shown by King *et al.* (2000) to be more sensitive to population growth than summary statistics based on the kurtosis of the length distribution or the variance among loci of the variances in microsatellite length (Di Rienzo *et al.*, 1998; Reich and Goldstein, 1998; Reich *et al.*, 1999). As with the summary statistics discussed earlier it is based on the discrepancy of two different estimators of θ in a standard neutral model. For microsatellites evolving according to a strict stepwise model one estimate of θ is $\theta_V = 2V$, where V is the sample variance in allele length. Another estimate is $\theta_f = (1/F^2 - 1)/2$ (Ohta and Kimura, 1973), where F is the sample homozygosity. The expansion index of Kimmel *et al.* (1998) is then

$$\log(\beta_1) = \log(\hat{\theta}_{\bar{V}}) - \log(\hat{\theta}_{\bar{F}}).$$

King *et al.* (2000) also suggest

$$\log(\beta_2) = \frac{1}{L} \sum_{i=1}^{L} \left((\log(\hat{\theta}_V))_i - (\log(\hat{\theta}_F))_i \right).$$

In the first case the two estimates of θ are calculated from the mean homozygosity and variance across or loci, and hence $\log(\beta_1)$ depends only on the marginal distribution of these statistics. In the second case the two estimates of θ are calculated for each locus separately, and $\log(\beta_2)$ depends on the joint distribution of homozygosity and variance across loci. King *et al.* (2000) show that the latter statistic is the most sensitive. Although devised to detect population growth, the imbalance index should also be useful for detecting population contractions or bottlenecks.

In addition to performing tests to detect departures from the null model of historically constant population size, it is possible to estimate parameters in specific models of population growth and/or decline. Primarily this has been done by the method of moments (e.g. Rogers, 1995; Reich and Goldstein, 1998; Reich *et al.*, 1999; Schneider and Excoffier, 1999). Increasingly, however, likelihood-based methods are being used. It is beyond the scope of this chapter to consider these in detail. Beaumont (2003b) gives a description of some of these approaches from the point of view of human demographic history, and general likelihood-based approaches are described in **Chapter 26.**

An example from an analysis of microsatellite data is given in Figure 30.5, which comes from Beaumont (1999), using a similar approach to that of Wilson and Balding (1998) and Wilson *et al.* (2003). The general idea is that the probability of the data and a genealogy consistent with the data, as a function of some unknown demographic and genetic parameters (such as mutation rate), can be derived from coalescent theory: $P(G, S|\Phi)$. MCMC is then used to obtain $P(\Phi, G|S)$ by sampling over genealogies and parameter values in the demographic model. In the example illustrated, the method has been applied to a sample from the last remaining population of the Northern Hairy Nosed Wombat, studied by Taylor *et al.* (1994). The species is believed to have undergone a sharp decline in the last 120 years. Sixteen loci were genotyped in a sample of 28 individuals. Seven loci were monomorphic and one locus had length differences that were not consistent multiples of two, leaving eight polymorphic loci. All the loci were polymorphic when samples from the Southern Hairy Nosed Wombat were included, suggesting that none of the monomorphic loci had unusually low mutation rates. Two analyses were performed with the data, one using the polymorphic data only and the other using the combined set of 15 loci. The results illustrated in Figure 30.5 strongly support a substantial population bottleneck. The population size was around 70 when the samples were taken, and even if one assumes that there were only 10 breeding adults, the data would suggest the

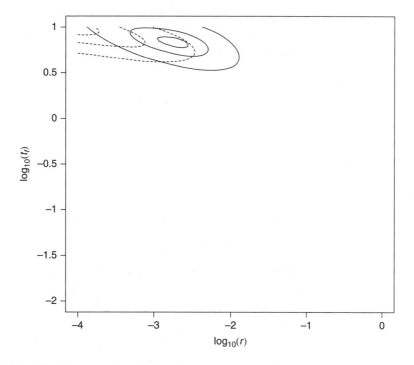

Figure 30.5 The joint posterior density for a parameter measuring the duration over which a population of wombats has been changing in size and a parameter measuring the degree to which it has changed. Specifically on the y axis is the logarithm to base 10 of the number of generations since the population started to change in size divided by the current population size, and on the x axis is the logarithm to base 10 of the ratio of current population size to ancestral population size. Contours giving the 10, 50, and 50 % highest posterior density intervals are shown. The dotted lines show the density for the polymorphic loci only.

decline began at least 1000 years ago, and that there has been a thousand-fold to ten thousand-fold decline in numbers. These conclusions will be highly model dependent, as discussed earlier, because of the difficulty in estimating the trajectory of a bottleneck without mutational information. An additional caveat is that the assumption of the SMM is restrictive, and inferences about reductions in population size will be affected by the presence of mutations that cause greater than single-step changes in allele length. Recently Calmet (2003) has modified the likelihood-based method of Beaumont (1999) to include the TPM, which corrects for the confounding effects of the mutation process.

It can be seen that there is a significant ascertainment bias associated with the use of the polymorphic markers alone. The effect is counter-intuitive in that the polymorphic markers suggest a more severe contraction over a longer time scale. Essentially, as discussed in Beaumont (1999), the polymorphic markers, while supporting a bottleneck effect imply a larger number of mutations in the ancestral population than the monomorphic loci. The number of mutations depends on $2N_1\mu = \theta/r$, and thus, for a given prior on θ, smaller values of r imply larger numbers of mutations in the pre-bottlenecked population. The ascertainment effect is much stronger for SNP markers, as discussed in Wakeley *et al.* (2001).

In conclusion, there appears to be considerable potential for the use of genetic data to make inferences about parameters in demographic models. However, a key caveat is whether one can have confidence in the models. In particular, as emphasised by Wakeley (1999), models with population structure can mimic many effects of past changes in population size without any change in census size. It is quite possible that most inferences about changes in population size are strongly confounded with phylogeographic effects.

30.2.5   Approximate Bayesian Computation

A problem with genealogical inference is that the parameter space is potentially very large and is not necessarily well explored by Monte Carlo methods, leading, e.g. to problems of convergence with MCMC. This will be exacerbated by the large volumes of genetic data that may soon be generated, even for non-model organisms. Motivated by difficulties in analysing human data a number of approximate approaches have been developed (**Chapter 26**). One of these, approximate Bayesian computation (ABC), has been used extensively for genetic analysis problems in conservation and management because it is relatively straightforward to model the complex scenarios that frequently arise in applied problems (as in, e.g. Miller *et al.* (2005)).

A number of related techniques have been developed (reviewed in Beaumont *et al.* (2002)). In particular Weiss and von Haeseler (1998) developed a method whereby a vector S_O of length d summary statistics are measured from the data and then n coalescent simulations are performed for a fixed parameter value Φ. For each simulation we measure summary statistics $S_{1,...,n}$, and the proportion of simulations that give rise to summary statistics sufficiently close to those measured from the data are recorded. This provides an approximate Monte Carlo estimate of the likelihood of $P(S = S_O|\Phi)$. If the parameter values are explored on a grid a likelihood surface can be estimated. Weiss and von Haeseler (1998) used this approach to detect patterns of population growth and also decline in different human populations from mtDNA sequence data.

A Bayesian approach taken by Pritchard *et al.* (1999) is to simulate the parameter values from the prior, $\Phi_i \sim P(\Phi)$, and then simulate data sets with these parameter values, and measure the summary statistics, $S_i \sim P(S|\Phi)$, to obtain samples from the joint distribution $P(S, \Phi)$. The posterior distribution $P(\Phi|S = S_O)$ is given by the conditional density

$$p(\Phi|S = S_O) = \frac{P(S_O, \Phi)}{P(S = S_O)}.$$

As with Weiss and von Haeseler (1998) this is estimated by choosing to accept simulations that give values of S_i that are close to S_O within some tolerance. Pritchard *et al.* (1999) apply the method to infer population growth from haplotype frequencies of linked microsatellite loci in the human Y chromosome. They used three summary statistics: the number of distinct haplotypes, mean heterozygosity, and mean variance in repeat length. The latter two were motivated by the study of Kimmel *et al.* (1998), discussed above. A problem for these methods is that only a limited number of summary statistics can be used for accurate inference because fewer points will be accepted for a given level of tolerance as the number of summary statistics is increased. The method of Pritchard *et al.* (1999) has been used by Estoup *et al.* (2001) to explore a relatively complex demographic and phylogeographic model of the history of introductions of cane toads to islands in the Caribbean and Pacific.

Beaumont *et al.* (2002) have pointed out that the method can be made more efficient by treating it as a problem of conditional density estimation, and using non-parametric regression methods to substantially increase the number of points used in the estimation of the posterior density, thereby allowing many more summary statistics to be used. Here, S_O and S are scaled so that each summary statistic in S has unit variance. The aim is to estimate the posterior mean $\alpha = E(\Phi|S = S_O)$ (see, e.g. Ruppert and Wand, 1994, for background to the approach). Beaumont *et al.* (2002) use least squares to minimize

$$\sum_{i=1}^{n} \left\{ \Phi_i - \alpha - \beta^{\mathrm{T}}(S_i - S_O) \right\}^2 K_\delta(\| S_i - S_0 \|),$$

where $\| x \| = \sqrt{\sum_{i=1}^{d} x_i^2}$, and $K_\delta()$ is a weight given by an Epanechnikov kernel. This is a quadratic function that has a maximum at $\| S_i - S_0 \| = 0$, and is zero for $\| S_i - S_0 \| \geq \delta$ Standard weighted regression gives the estimates $\hat{\alpha}$ and $\hat{\beta}$. In order to estimate posterior densities, Beaumont *et al.* (2002), took a heuristic approach, in which they make an assumption that the errors are constant in the interval and adjust the parameter values as

$$\Phi_i^* = \Phi_i - (S_i - S_O)^{\mathrm{T}} \hat{\beta}.$$

Scripts written in R are available to infer parameters using this method (`http://www.rubic.rdg.ac.uk/~mab/stuff/ABC_distrib.zip`).

An alternative ABC approach that does not require simulation from the prior, which will typically be rather inefficient, is that of Marjoram *et al.* (2003), which uses a rejection approach as in Pritchard *et al.* (1999) embedded within an MCMC sampler in which the likelihood ratio calculation is replaced by an accept/reject step depending on whether simulated summary statistics lie within a tolerance range (see also **Chapter 26**). The method of Beaumont *et al.* (2002) has been applied to a number of problems in population genetics, including some in conservation and population management (Estoup *et al.*, 2004; Chan *et al.*, 2006; Miller *et al.*, 2005).

30.3   ADMIXTURE

Often, closely related taxa that were previously isolated from each other, have been brought into contact as a consequence of recent habitat disturbance. Hybridisation occurs when individuals from different taxa mate together and produce offspring, and this may be followed by introgression when there is back-crossing into either population. Introgressive hybridisation, and its detection is a significant area of concern in conservation genetics (Haig, 1998), and can be studied at three levels. At the highest level one can ask questions based on gene frequencies in separate populations, the subject of this section. At the next level one can identify hybrid individuals through genotypic modelling, discussed in the following section. Lastly, at the lowest level, hybrid individuals can be identified by inferring degrees of relatedness between individuals and the reconstruction of pedigrees, described in the final section.

The early development of statistical methods to infer admixture proportions was stimulated by an interest in human populations (e.g. Glass and Li, 1953). Two populations diverge from a common ancestor and then hybridize in a single event (Figure 30.6). The current samples are taken some time after this event from the known descendants of the two parental populations and the hybrid population. The main aim of the statistical analyses is to estimate the admixture proportion μ. This has been called an *intermixture model* by Long (1991) to distinguish it from a gene flow model where the admixture occurs on a longer time scale. The various statistical methods that have been developed to estimate μ have made different simplifying assumptions about the parameters in the model. In general, other than the approach of Bertorelle and Excoffier (1998), the relationship between the parental populations has been ignored. Most approaches have modelled the gene frequency, x, of a particular allele in the hybrid population as $x_h = \mu x_1 + (1 - \mu)x_2$, where the indices refer to the hybrid, parental 1, and parental 2 populations respectively. Obviously, if there are many alleles at many loci, no single value of μ will explain all the allele frequencies and a maximum likelihood or least-squares estimate of μ is made where the variance in x_h from its expected value is modelled according to different assumptions. In the approaches of Glass and Li (1953) and Elston (1971) genetic drift is assumed absent, sampling error is assumed absent in the parentals, and all the variation is due to sampling error in the hybrids. The approach of Thompson (1973) takes into account drift in all populations as well as sampling error. That of Long (1991) separately estimates both drift and sampling error in the hybrid population, but assumes that the parental frequencies are known without drift or sampling error. Long suggests an approach to take into account sampling (but not drift) error in the parental frequencies, but this is not used in the most widely implemented version of Long's method, that of Chakraborty *et al.* (1992).

Thompson (1973) provides the framework for an analysis of the admixture model including drift in all populations. The gene frequencies are first arcsine square root transformed, $x' = \sin^{-1}(\sqrt{x})$. Under this transformation drift can be modelled as a Brownian-motion diffusion with variance $1/(8N_e)$ per generation, where N_e is the effective number of individuals, assumed to be diploid. In the case of k alleles at a locus, rather more complex transformations and projection to a $k - 1$ dimensional space can be used (see discussion in Thompson, 1975a). At the time of the admixture event the vector of transformed gene frequencies in the hybrid population is given as $\mathbf{x}'_h = \mu \mathbf{x}'_1 + (1 - \mu)\mathbf{x}'_2$. Following drift, the observed sample vectors are $\mathbf{a}_h$, $\mathbf{a}_1$, $\mathbf{a}_2$, the component elements of which are normal random variables with mean x_i and variance $T/8N_e + 1/8n$ for sample size n. Thus it is possible to estimate $p(\mathbf{a}_h, \mathbf{a}_1, \mathbf{a}_2|\mathbf{x}'_h, \mu, \mathbf{x}'_1, \mathbf{x}'_2)$. Thompson (1973) uses this to obtain conditional likelihood functions for μ. This model has also been studied using a coalescent-based method similar to that discussed in Section 30.2 (Chikhi *et al.*, 2001), and is described in slightly more detail later.

The method of Long is based on a weighted least squares solution of $\mathbf{a}_h = \mu \mathbf{x}_1 + (1 - \mu)\mathbf{x}_2 + \varepsilon$, where ε has zero mean and separate components of variance due to sampling and drift. The weighting accounts for the inhomogeneity of the variances, and is solved iteratively in Long (1991). Chakraborty *et al.* (1992) provide a closed form expression:

$$\hat{\mu} = \frac{\sum_{l=1}^{L} \sum_{i=1}^{r_l} (a_{1li} - a_{2li})(a_{hli} - a_{2li})/a_{hli}}{\sum_{l=1}^{L} \sum_{i=1}^{r_l} (a_{1li} - a_{2li})^2/a_{hli}},$$

for L loci with r_l alleles at each locus. This has mean square error

$$\text{MSE} = \frac{\sum_{l=1}^{L} \sum_{i=1}^{r_l} [(a_{hli} - a_{2li}) - (\hat{\mu} a_{1li} - a_{2li})]^2 / a_{hli}}{\sum_{l=1}^{L} r_l - 1}.$$

Long (1991) suggests that since the sample size is known, drift in the hybrid population can be separately estimated as $\hat{F} = (\text{MSE} - 1/2n)/(1 - 1/2n)$ where F has the usual definition (drift variance)/$(x_h(1 - x_h))$. An estimate of T/N_e can be made using the Wright's formula $F = 1 - \exp(-T/(2N_e))$.

Another approach to inferring admixture proportions, with relevance to conservation, is that by Bertorelle and Excoffier (1998). They consider a model where two parental populations, each of size N chromosomes, themselves diverged from a common population over a time period τ, hybridize in one event at time t_A from the present to produce a hybrid population of size N, with μN chromosomes from one population and $(1 - \mu)N$ chromosomes from the other.

The essential approach is to define the expected admixture proportion as a function of the ratios of expected coalescence times between groups of genes. Genetic distances defined for the genetic marker that is being studied have expectations that depend only on the product of coalescence time and mutation rate. Assuming a constant mutation rate, the ratios of coalescence times can be replaced by ratios of expected genetic distances. An estimator can be obtained by substituting the expected distances by the estimated distances. Bertorelle and Excoffier define two estimators, one of which is described here. They derive the expected coalescence times for a pair of genes, one sampled from the hybrid population and one sampled from a parental. In this case, there can be no coalescences since the time of admixture. There are two possibilities: with probability μ the hybrid lineage is derived from the same parental populations as that being compared, and therefore the expected coalescence time is $t_A + t_1$, where t_1 is the expected coalescent time in population 1; alternatively, with probability $1 - \mu$, the hybrid lineage comes from the other parental, in which case the expected coalescence time is t_{12}. Two equations can be obtained, one for each of the parental populations in the pair, and a least-squares estimate for μ can be obtained. The expected times, t_1, t_2 and t_{12} are replaced by the estimated genetic distances d_1, d_2 and d_{12}, on the assumption that for both microsatellites and sequences the typical genetic distances have the form $d = 2vt$, where v is the mutation rate and t is the expected coalescence time. Bertorelle and Excoffier denote this estimator m_Y.

It should be noted that this involves an estimate of t_A which is in general unknown. Bertorelle and Excoffier suggest replacing the estimate of $(2v)t_A$ by the minimum observed distance in pair-wise comparisons of genes between either parental and the hybrid population. However, in practice this is almost always equivalent to assuming $t_A = 0$. For multi-locus data, Bertorelle and Excoffier suggest that the distances should be calculated as averages over loci, rather than separately calculating estimates of μ for each locus before averaging.

Using molecular distances, they make comparisons between m_Y and the estimator of Roberts and Hiorns (1965), which they denote m_R, and that of Chakraborty et al. (1992; based on Long, 1991, discussed earlier), denoted m_C. Their overall finding is that in general m_Y is much less biased than either m_R or m_C, but tends to have a higher variance. When the parental populations have been separated for a substantial period and when the mutation rate is high the variance in m_Y tends to approach that of the other estimators.

However, with single microsatellite loci, the variance is still substantially larger than that of the other estimators. With many microsatellite loci (the situation commonly found in practice), the variance becomes close to that of the other estimators with negligible bias, while the other estimators remain appreciably biased towards admixture proportions of 0.5.

Bertorelle and Excoffier apply their method to microsatellite data from grey wolf, coyote, and red wolf populations obtained by Roy et al. (1994). They use the data from the red wolf, and use data from populations believed to be respectively pure wolf, pure coyote, wolf hybrid, coyote hybrid. They then infer admixture proportions in the wolf hybrid, coyote hybrid, and red wolf populations. They suggest that estimates of sampling error in m_Y, m_R and m_C can be made by bootstrap resampling chromosomes independently across loci. This is equivalent to adding an extra inbreeding step into each population, and it is more appropriate to resample among loci, conditioning on the observed frequencies at each locus. This procedure will take into account both the genealogical and sampling variance. For reasonable sample sizes the bulk of the variance in the estimators will be between loci.

For the coyote and wolf hybrid populations they obtain estimates of admixture that are consistent across estimators – ~ 0.5 in the case of the wolf hybrid populations and ~ 0.1–0.15 in the case of coyote hybrid populations. The bootstrap standard error in the case of the coyote hybrid population is sufficiently large that the estimate is consistent with the true admixture proportion of zero. Thus, for these known hybrid populations the results support the conclusions of Roy et al. (1994).

The method of Bertorelle and Excoffier (1998) has been extended to include multiple parental populations (Dupanloup and Bertorelle, 2001). Essentially the estimation procedure described above for m_Y, is generalized to allow estimates of multiple μ_i, where $\sum_{i=1}^{d} \mu_i = 1$. Standard errors for the m_{Yi} are estimated using the same bootstrapping procedure as described above. They evaluate the performance of the m_{Yi} using simulations and conclude that despite the increased number of parameters the level of precision in the estimates remains approximately the same as in the original study with two parental populations. A general improvement to the method of Bertorelle and Excoffier (1998) for sequence data, assuming an infinite sites model, has been described by Wang (2006). He derives formulae assuming an infinite sites model for the expected number of segregating sites in each population, in each pair of populations pooled together, and in all 3 populations pooled together. The actual numbers are counted in the data and the method of least squares is used to estimate parameters. This method appears to provide more accurate point estimates for the parameters in the general admixture model of Figure 30.6.

Chikhi et al. (2001) analyse the same model as that of Thompson (1973) using the coalescent method described in Section 30.2.2. Importance sampling can be used to estimate the likelihood $p(\text{data}|\mathbf{x}_1, \mathbf{x}_2, \mu, T/N_1, T/N_2, T/N_h)$ and then marginal posterior densities (e.g. $p(\mu|\text{data})$, $p(T/N_h|\text{data})$, etc.) are estimated with MCMC. Although the use of the likelihood method should allow for increased accuracy and precision the method is highly computer intensive and no simulation-based study of precision and accuracy has been published to allow comparison with earlier studies. On an example data set of gene frequencies obtained from Jamaican people, who are assumed to be of mixed European and African descent (Parra et al., 1998) to the point estimates for μ were similar to those obtained by the method of Long (1991), but with estimates of error that were more than twice as great (assuming the posterior distribution is sufficiently close to normal so that Bayesian and frequentist estimates are comparable). By estimating the degree of drift subsequent to admixture for all subpopulations it does provide extra detail in the analysis, unlike commonly used methods. The posterior distributions for T/N are, however, generally broad.

As discussed in Section 30.2.1, an alternative way of computing the likelihoods is to use the Wright–Fisher model. Wang (2003) has inferred the parameters discussed above using the same approach as in Wang (2001). In addition, like Bertorelle and Excoffier (1998), and unlike Chikhi *et al.* (2001), who used uniform Dirichlet priors for the parental gene frequencies, Wang (2003) explicitly models the genetic divergence between the parental populations by introducing additional scaled parameters T_A, N_{A1}, and N_{A2}, for the ancestral divergence time, and the ancestral population sizes, using a uniform prior for the common ancestral gene frequencies. Wang demonstrates that the assumption of independent uniform priors for the parental gene frequencies is quite problematic for most scenarios in which there is a high degree of divergence between the ancestral populations (the ideal case for detecting admixture). In this case there is tendency for estimates of the admixture proportion to be biased away from 0.5 towards 0 or 1. In a detailed comparison of the performance of different methods for inferring admixture, Choisy *et al.* (2004) noted a similar phenomenon, which could be partially corrected by using Dirichlet $(1/K, \ldots, 1/K)$ priors for the parental gene frequencies, as in Rannala and Mountain (1997), discussed further below. The prior of Wang (2003) is computationally intensive, and an alternative would be to use a Dirichlet distribution, parameterized by F_{ST} as discussed in Sections 30.2.2 and 30.4–i.e. Dirichlet $(\theta_j x_1, \ldots, \theta_j x_K)$, where $\theta_j = 1/F_{ST_j}$ for the jth parental population and $x_1, \ldots, x_K$ are the ancestral allele frequencies for alleles $1, \ldots, K$. The method of Wang (2003) is substantially faster than that of Chikhi *et al.* (2001), and, because of this, extensive simulation testing is possible, and Wang demonstrates that the likelihood-based approach has superior performance, generally, than moment-based estimators.

The likelihood-based models described thus far are models of drift without mutation. They assume the ancestral variation is partitioned among the descendent populations solely through drift and admixture. This is due to the convenience of modelling–e.g. it allows relatively straightforward analytical expressions for the likelihood in Wang (2003). In order to allow for mutations over the period since the common ancestral population, full genealogical models would have to be considered. The most straightforward implementation would be to use MCMC, or to compute likelihoods using importance sampling (**Chapter 26**). To date, no such methods have been developed for the specific class of intermixture models considered here. A Bayesian approach has been introduced for microsatellite data, however, using the ABC technique discussed above. In this case, 15 summary statistics are computed from the data and compared with those obtained from simulations. These comprise: for all 3 populations, numbers of alleles, heterozygosity, a modified version of the M statistic of Garza and Williamson, and pair-wise F_{ST} for all pairs; the $(\delta\mu)^2$ statistic of Goldstein *et al.* (1995) for the two parental populations; the linkage disequilibrium estimator, D' for the hybrid population; and the m_Y statistic (above). They use the SIMCOAL2 program (Laval and Excoffier, 2004) to infer the parameters in the full model of Figure 30.6. When the times since admixture are short (i.e. when the assumptions of drift-based models are best met) their method has only marginally weaker performance than that of Wang (2003), and substantially improved performance when they are longer.

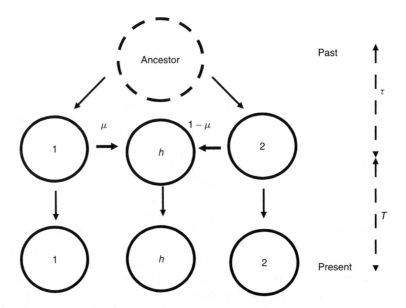

Figure 30.6 A figure illustrating the model of admixture described in the text. The two parental populations (1 and 2), diverge from a common ancestor over a period τ and then hybridize to produce a population h, with a proportion μ of alleles coming from population 1. These populations then remain distinct and diverge from each other over a period T without intervening migration.

30.4 GENOTYPIC MODELLING

In contrast to the approach described in the previous section, where gene frequencies are used to infer admixture, a relatively new area of modelling is to use multi-locus genotypic information to detect individuals that are immigrants or have immigrants in their recent ancestry. Thus, in effect, the method will only work if there is appreciable gametic disequilibrium within a population caused by gene flow. An influential paper in the context of conservation was that of Paetkau *et al.* (1995; see also Waser and Strobeck, 1998). They studied populations of polar bears in Canada and were interested in knowing whether populations managed as independent units were in fact connected through dispersal of individuals. The approaches that have been taken to address questions such as this have led to several overlapping areas of research that are outlined in the following sections.

30.4.1 Assignment Testing

In order to identify migrant individuals Paetkau *et al.* (1995) made an estimate of the gene frequencies at each locus in each of the populations from which they had samples. For each individual they calculated the joint probability of obtaining its multi-locus genotype from each of the populations by multinomial sampling, $\prod p(\mathbf{a}_j | \hat{\mathbf{x}}_j, n_j)$, with observed genotype at the jth locus, $\mathbf{a}_j$, and estimate of gene frequencies $\hat{\mathbf{x}}_j$. The individual was then 'assigned' to the population in which this probability was highest. The rationale behind this approach is that, in the absence of immigration, the genotype of each individual is a random draw from its own population gene frequencies and therefore, providing gene frequencies vary sufficiently among populations, individuals should be assigned to their own populations. Migrants can be detected because they have a higher probability of being drawn from some other population. In the study of Paetkau et al. (1995), 4 populations of polar bears were sampled, using eight microsatellite loci. They found that 60 % of individuals were assigned to the population in which they were sampled, 33 % to the nearest population and 7 % to more distant populations. In a later study on brown bears from seven populations in NW Canada using the same loci (Paetkau et al., 1998), 92 % were assigned to the population from which they were sampled, and those assigned to other populations illustrated biologically plausible patterns of dispersal.

Estimation of the population gene frequencies is an important part of the procedure, and in the method of Paetkau et al. in order to avoid zero elements in $\hat{\mathbf{x}}$, the genotype of each individual is incorporated into the estimate of $\hat{\mathbf{x}}$ for each population, when the assignment test is made. Alternative methods may be preferable, and, in particular, it would be useful to estimate $\hat{\mathbf{x}}$ jointly with the assignment parameters. In the initial study no attempt was made to estimate confidence intervals, but resampling schemes have latterly been considered (see http://www2.biology.ualberta.ca/jbrzusto/Doh.php for more details).

Some of these points are addressed in a study by Rannala and Mountain (1997). Their approach differs from that of Paetkau et al. in that they include Bayesian estimation of the allele frequencies in each population, and a likelihood ratio test to compare different hypotheses. They consider a number of populations with many loci sampled. For ease of exposition, one locus and two populations are considered here. They condition on the number of alleles, k, observed at a locus among all populations. The observed allele frequency counts are given by the vectors $\mathbf{a}_0$ and $\mathbf{a}_1$ for the two populations. The posterior distribution for the unknown gene frequency vector $\mathbf{x}$ is given by $p(\mathbf{x}|\mathbf{a}_0)$ and $p(\mathbf{x}|\mathbf{a}_1)$. To estimate this a Dirichlet prior is assumed with parameters all equal to $1/k$ and written $D(1/k, \ldots 1/k)$, giving a posterior $p(\mathbf{x}|\mathbf{a}_0) = D(\mathbf{a}_0 + 1/k)$ and $p(\mathbf{x}|\mathbf{a}_1) = D(\mathbf{a}_1 + 1/k)$. Rannala and Mountain describe their prior as assigning equal probability density to the frequencies of the alleles, although this only occurs with a Dirichlet $D(1, \ldots 1)$. Clearly the choice of prior may depend on the genetic model assumed.

Under a model where none of the individuals are immigrants or have immigrant ancestry, their genotypes are assumed to be multinomial samples of size 2 taken, e.g. from the posterior $p(\mathbf{x}|\mathbf{a}_0)$, defined above. Integrating out $\mathbf{x}$, the marginal probability of an individual having genotype $\mathbf{X} = (X_i, X_j)$ is then multinomial-Dirichlet,

$$p(\mathbf{X}|\mathbf{a}_0) = \int p(X|\mathbf{x}) p(\mathbf{x}|\mathbf{a}_0) \, \mathrm{d}\mathbf{x}.$$

The probabilities for each locus can then be multiplied together, under the assumption that they are independent.

Other hypotheses can be considered, e.g. whether the individual has ancestry from one immigrant d generations earlier. The probability of observing genotype $\mathbf{X} = (X_i, X_j)$ in an individual where one allele is drawn at random from one population and one is drawn at random from the other population, is $p(\mathbf{X}|\mathbf{a}_0, \mathbf{a}_1) = 1/2(p(X_i|\mathbf{a}_0)p(X_j|\mathbf{a}_1) + p(X_i|\mathbf{a}_1)p(X_j|\mathbf{a}_0))$, where the probabilities are calculated as single draws from the multinomial Dirichlet outlined above.

Rannala and Mountain (1997) extend the argument above to consider the probability of observing the genotype, $p(\mathbf{X}|\mathbf{a}_0, \mathbf{a}_1, d)$, given that an individual might have one ancestor,

d generations ago, from a different population (with all other ancestors coming from the same population). There is a probability of $1/2^{d-1}$ that the parent which is immigrant or has immigrant ancestry contributes an immigrant gene, in which case the probability is as given above, and there is a probability $1 - 1/2^{d-1}$ that both genes come from the same population in which case the probability is calculated as a sample of size 2 from the multinomial Dirichlet, $p(\mathbf{X}|\mathbf{a}_0)$ or $p(\mathbf{X}|\mathbf{a}_1)$.

Rannala and Mountain suggest that for particular individuals, hypotheses can be tested using likelihood ratio tests of the form

$$\Lambda = \frac{p(\mathbf{X}|\mathbf{a}_0)}{p(\mathbf{X}|\mathbf{a}_0, \mathbf{a}_1, d)}.$$

Critical regions of a given size can be estimated by parametric bootstrapping–i.e. Monte Carlo simulations of genotypes $\mathbf{X}$ with probability $p(X|\mathbf{a}_0)$ under the null hypothesis. The power of the tests can be estimated by additionally simulating under the alternative hypothesis–genotypes $\mathbf{X}$ with probability $p(\mathbf{X}|\mathbf{a}_0, \mathbf{a}_1, d)$.

A comparative analysis of the performance of the two approaches on test data sets has been carried out by Cornuet *et al.* (1999). They introduce the use of a distance-based method for assigning individuals to populations. The method is analogous to that of Paetkau *et al.* but individuals are assigned to populations with the smallest genetic distance. Cornuet *et al.* study the most commonly used genetic distances for microsatellite or allozyme data (**Chapter 29**), modified to take into account that individuals are compared with populations. In addition to the question of assigning individuals to particular populations, Cornuet *et al.* also consider the question whether an individual is likely to have come from the population in which it resides. In order to test the latter, which they term *testing for exclusions*, Cornuet *et al.* suggest simulating genotypes at random from an estimate of the sample population gene frequencies and comparing the likelihood (or genetic distance) of the individual's genotype with the distribution of likelihoods or distances from random sampling.

Cornuet *et al.* tested the methods using data simulated from a model of diverging populations with mutations according to either an IAM or SMM. Their overall conclusion was that the Rannala and Mountain method outperformed all other methods, and the genetic distance methods, in general, performed less well than the methods using likelihood. Since Rannala and Mountain's calculation of likelihoods differs from that of Paetkau *et al.* only in the estimation of population frequencies $\mathbf{x}$, this implies that, taking a Bayesian approach, the choice of a suitable prior distribution for $\mathbf{x}$ is an important consideration. Tests for exclusions also appear to work well using the Rannala and Mountain method to calculate likelihoods. However, Cornuet *et al.* noted that the method used for simulating null distributions was an important component of testing–with increasing numbers of loci there was a tendency to incorrectly exclude individuals from all populations.

30.4.2 Genetic Mixture Modelling and Clustering

One aspect of the method of Rannala and Mountain, discussed above, is the necessity for a large number of hypothesis tests, requiring some care in specifying the critical values for assessing significance. Also it would be better if the estimates of the population frequencies $\mathbf{x}$ could take into account the possibility that some individuals are immigrant rather than resident. An additional feature, common to these approaches (but examined by Cornuet *et al.*), is that hypothesis testing is limited to the populations that are actually included in the survey, whereas immigrants may come from other, unsurveyed, populations.

These aspects are addressed in a study by Pritchard *et al.* (2000), which has resulted in the very widely used program, STRUCTURE. They initially consider the problem in terms of one sample composed of individuals of possibly heterogenous origin, where the genotype of each individual is a mixture of alleles drawn from an unknown number of contributing populations with unknown gene frequencies in each population. The aim is to infer jointly: (1) the number of populations contributing to the genotypes of individuals; (2) the allele frequencies in each of these populations; (3) the proportions from each of these populations contributed to the genotype of each individual. This can be regarded as a standard mixture problem in statistics (Robert, 1996), often analysed using Bayesian MCMC methods, and Pritchard *et al.* use similar approaches to solve it.

In this approach, it is assumed that there are K populations contributing to the gene pool of the sample population. Pritchard *et al.* consider two main models: in one case each individual is drawn from one of the K populations; whereas in the more general case a proportion of the individual's alleles are drawn from each of the K populations. The former situation is a special case of the latter, and this will be described here. The K potential source populations each have an unknown gene frequency distribution at each locus, p_{kl} for the $k = 1 \ldots K$ populations and $l = 1 \ldots L$ loci. The p_{kl} are elements of some multidimensional vector $\mathbf{P}$. For the ith individual it is assumed that the proportion of its genotype that is drawn from population k is q_{ik}, an element of the vector $\mathbf{Q}$. For the purpose of analysis they introduce an indicator vector $\mathbf{Z}$ where the element z_{ail} gives the population from which allele copy $a = 1, 2$ at locus l in individual i originates, and can take any value from $1 \ldots K$. This use of indicator variables (missing data) is widely used in Bayesian mixture modelling with Gibbs sampling (e.g. Robert, 1996). The genotype of each individual is given by vector $\mathbf{X}$ with elements x_{ail}. It is assumed that the alleles at each locus in each individual are drawn independently. With this formulation, the likelihood $p(\mathbf{X}|\mathbf{Z}, \mathbf{P}, \mathbf{Q})$ can be calculated straightforwardly.

The interest is in estimating the posterior distribution $p(\mathbf{Z}, \mathbf{P}, \mathbf{Q}|\mathbf{X})$ for the parameters of interest. In order to estimate this, priors are required for $\mathbf{P}$, $\mathbf{Q}$, and $\mathbf{Z}$. They assume a Dirichlet prior for $\mathbf{P}$, as in Rannala and Mountain (1997), but a uniform $D(1, \ldots 1)$ rather than (for J alleles at a locus) the $D(1/J, \ldots 1/J)$ prior of the latter authors. They assume a prior of $p(z_{ail} = k) = 1/K$ for $\mathbf{Z}$. For $\mathbf{Q}$ they model the prior hierarchically as a Dirichlet with form $D(\alpha, \ldots \alpha)$. When α is small the genotype of an individual is drawn almost completely from one population, and when α is large the genotype is evenly mixed from all the populations. The prior for α is uniform on [0,10]. The posterior distribution of α gives evidence of the degree to which there is heterogeneity among individuals in their ancestry.

Generally they wish to estimate $p(\mathbf{P}, \mathbf{Q}, \alpha|\mathbf{X})$ (marginal to $\mathbf{Z}$, which is introduced to make the Gibbs sampling tractable, see Robert, 1996). This is performed by successively sampling from the full conditional distributions $p(\mathbf{P}, \mathbf{Q}|\mathbf{X}, \mathbf{Z})$, and $p(\mathbf{Z}|\mathbf{P}, \mathbf{X}, \mathbf{Q})$ and updating α using Metropolis–Hastings sampling. The conditional distributions and sampling methods are given in Pritchard *et al.* (2000).

The estimation of K, the number of populations contributing to genotype distributions among the samples, is more problematic. Pritchard *et al.* develop an *ad hoc* procedure for estimating $P(\mathbf{X}|K)$ using the sample mean and variance of the Bayesian Deviance $-2 \log p(\mathbf{X}|\mathbf{Z}, \mathbf{P}, \mathbf{Q})$ over values of $\mathbf{Z}, \mathbf{P}, \mathbf{Q}$ generated by the Gibbs sampler. An assumption of the method is that the sampled values of $p(\mathbf{X}|\mathbf{Z}, \mathbf{P}, \mathbf{Q})$ are lognormal, and Pritchard *et al.* argue that it should be viewed as an approximate guide, which appears to give satisfactory answers with test data. With a suitable prior for K it is then possible to estimate $P(K|\mathbf{X})$.

An additional point that Pritchard *et al.* note is that since the population labels are arbitrary, they can be switched in $\mathbf{Z}, \mathbf{P}, \mathbf{Q}$ without affecting the likelihood. Thus $p(\mathbf{Z}, \mathbf{P}, \mathbf{Q}|\mathbf{X})$ has $K!$ symmetric modes. If the Gibbs sampler explores these fully then, for example, a summary statistic such as the posterior mean of the q_{ik} will have expectation $1/K$ irrespective of the amount of substructure in the data. In their analyses the Gibbs sampler appears to explore only one mode, and thus they find that posterior mean values and 95 % credible intervals are satisfactory summaries of the posterior distribution. It is reasonable to assume that when there is less information on population structure label-switching will be more frequent, and this will in turn make detection of population structure more difficult if based on such summaries.

The method of Pritchard *et al.* can also be used to answer questions about the ancestry of individuals in a similar way to the method of Rannala and Mountain (1997). In order to avoid restructuring of the model, the approach they have taken is to specify that individuals have resident ancestry with fixed probability v and non-resident ancestry with probability $1 - v$. If they have non-resident ancestry, they can be themselves immigrants or have one immigrant parent, or one immigrant grandparent, etc. This is equivalent to using geographic information as a prior to restrict the possible values the q_{ik} can take. The Gibbs sampler can then be run as before, but the q_{ik} can only take a restricted set of values that can be mapped back to each of the above possibilities. In this way, by interpreting the q_{ik}, it is possible to estimate the posterior probability of each type of ancestry, whether immigrant or non-immigrant, and from which population. The method, however, relies on the user specifying in advance a particular value of v. The method offers an improvement on the approach of Rannala and Mountain in that it avoids the problems of multiple hypothesis testing.

Dawson and Belkhir (2001) have developed a method to partition a sample into subgroups within which there is random-mating, similar to the model without admixture analysed by Pritchard *et al.* (2000). There are, however, a number of differences between the two approaches. Following the notation of Pritchard *et al.* (2000), used above, Dawson and Belkhir calculate the likelihood function $p(\mathbf{X}|\mathbf{Z}, \mathbf{P}, K)$, and use Metropolis–Hastings sampling to estimate $p(\mathbf{Z}, \mathbf{P}, K|\mathbf{X})$. The prior for $\mathbf{Z}$ is such that all partitions have equal prior probability, and the prior for $\mathbf{P}$ is a Poisson–Dirichlet distribution, which is appropriate for the infinite alleles model, and is parameterized by θ, which is equivalent to the scaled mutation rate. A point prior value of θ was used in their examples. Unlike the method of Pritchard *et al.* (2000) the number of partitions, K, is allowed to vary from 1 to some upper value, and within this interval the prior for K is given by a power function $p(K) = Au^K$ for $0 < u \le 1$. The main concern in Dawson and Belkhir (2001) is the problem of label-switching discussed above. Although Pritchard *et al.* (2000) are careful to highlight this, it still remains the case that in their method the MCMC 'works' only because it does not converge. This seems somewhat unsatisfactory and Dawson and Belkhir attempt to avoid this by recording the proportion of times during the MCMC simulation in which groups of individuals co-occur together within the same partition. The posterior probability of co-assignment of d individuals to the same partition forms a measure of similarity among the d individuals, which can then be used in an agglomerative clustering algorithm to maximize the minimum probability of coassignment within clusters. In practice, Dawson and Belkhir use $d = 2$, which corresponds to the 'furthest-neighbour' or 'complete-linkage' clustering algorithm. They find that the resulting tree is a more useful graphical summary to identify cryptic population structure than the posterior distribution for K, which tends to overestimate the number of partitions, and appears to be affected by the value chosen for θ. They suggest that a future improvement would be to put a more diffuse prior on θ.

30.4.3   Hybridisation and the Use of Partially Linked Markers

The method of Pritchard *et al.* has been used to analyse hybridisation of wildcats with domestic cats in Scotland (Beaumont *et al.*, 2001). Applying the method, implemented in the program STRUCTURE, to wild-living cats in Scotland there is very strong evidence of two groups. In this analysis samples from known domestic cats were also included, but these were set to have $q_i = 0$ (where the subscript k has been dropped and q_i is the proportion of non-domestic ancestry of individual i). The point estimates of the q_i and their 95 % credible intervals are illustrated in Figure 30.7 for the wild-living cats. Museum specimens are shown as dotted lines in the figure. Interestingly, when logit transformed, these point estimates correlate strongly with the results from the first axis of classical metrical scaling of genetic distances between cats. As discussed in Beaumont *et al.* (2001), given that the linkage disequilibrium caused by admixture will decay rapidly for unlinked genes, the existence of a second group of cats does not necessarily mean that a 'pure' wildcat population with no introgression has been identified. It only suggests that this group of cats shows little evidence of very recent domestic cat ancestry.

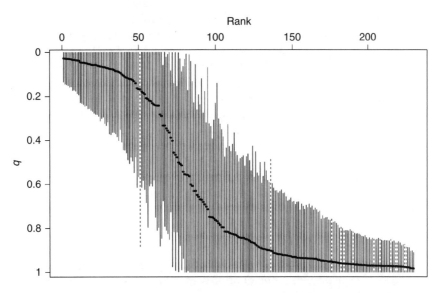

Figure 30.7   The figure shows the means and 95 % credible intervals for estimates of the probability that individual wild-living cats have purely non-domestic ancestry. These are plotted against the rank of the mean estimate. Vertical dashed lines refer to museum specimens.

Motivated by considerations such as this, Falush *et al.* (2003) have extended the basic approach in the program STRUCTURE to allow the use of closely linked (but recombining) markers. A further improvement lies in the priors used for the parental gene frequencies, which follow the Dirichlet approximation for modelling immigration-drift equilibrium. This allows additionally for F_{ST} to be inferred in each population, as in, e.g. the *2mod* program of Ciofi *et al.* (1999) discussed above. To allow for recombination, in their extension of the method, they modify $\mathbf{Z}$ so that the z_{ail} are no longer independent at each locus, but form a Markov chain, with neighbouring loci tending to originate from the same population with probability inversely related to the map distance between markers and the total rate of recombination events, r, per unit distance. In this model r can be loosely equated to the time since admixture. The idea is that the indicator for the first locus is drawn according to $\mathbf{Q}$ above and then the indicator for the second locus is either the same as this if there is no recombination event, or is drawn randomly from $\mathbf{Q}$ if there is an event. Although this approach has not been used to analyse the Scottish wildcats it has been used to look at wildcats in Hungary (Lecis *et al.*, 2006). In this case the results suggested that the admixture was probably ancient, but could not be dated accurately. More definite results come from an analysis of wolf populations in Italy (Verardi *et al.*, 2006). In wolves there is the prospect of historical admixture with domestic dogs, and the aim of this study was to elucidate how much had occurred, and over what period. In this case they were able to obtain clear estimates of admixture time of 70($\pm$20) generations (around 140–200 years), although the overall amount of admixture was quite low. This study illustrates the potential power of genotypic methods to enhance frequency based analysis of population structure and hybridisation.

The use of multilocus genotypic information to make inferences about recent migration has great potential, and a number of papers have used methods similar to those of Pritchard *et al.* (2000) to study different aspects of population structure. For example, Anderson and Thompson (2002) have developed a method for inferring the proportion of hybrids in a sample. In their method they consider hybridisation over n generations. It is then possible to partition genotype frequencies into classes corresponding to the pedigree of the individuals–e.g. when $n = 2$, whether they are purely of one species or the other, or F1 hybrids, or F2 hybrids, or back-crosses. For each individual the posterior probability of belonging to one of these genotype frequency classes can be computed using MCMC, and hence non-admixed individuals can be detected (at least, given the n generations of admixture). There are similarities between these methods and those used for genetic stock identification (GSI), where often the population affinities of individuals is of less interest than some estimate of the degree of admixture. A Bayesian method specifically for GSI, using MCMC, has also been developed by Pella and Masuda (2001). Corander and Marttinen (2006) discuss the general problem of quantifying the extent of hybridisation and admixture in populations, and propose an alternative approach.

30.4.4 Inferring Current Migration Rates

As noted above, the method of Pritchard *et al.* (2000) also allows calculation of the posterior probability that an individual has a specified degree of immigrant ancestry. In order to calculate this a point prior is specified for v, the probability that an individual has resident ancestry. Taking this approach further, it is possible to use genotypic methods to infer parameters in a demographic model, such as the current levels of gene flow between populations. The model of Wilson and Rannala (2003) assumes that all populations in the system have been sampled, and that the individuals each have at most one immigrant ancestor. The data consist of the genotypes (X) and sampling locations of individuals (S), and the parameters are the population source of the immigrant ancestor for each individual (M), the number of generations back in the past that the immigrant ancestor occurred (t), the probability that two alleles in an individual are IBD from a recent ancestor (F), leading to a departure from Hardy–Weinberg equilibrium, and the population frequencies (p). The prior for M and t for each individual can then be written as a function of the immigration rate, m_{lq}, the proportion of individuals in population q that are immigrant from population l. It is this hierarchical parameter that is of most interest. MCMC is then used to obtain the posterior distribution of the parameters. Thus the method can identify the immigrant ancestry of particular individuals, but it also, for example, constructs a matrix with point estimates of current migration rates between populations. An example where this approach has been used is in the study of movement between Orang-utan populations in Borneo (Goossens et al., 2006). In this case the model of Wilson and Rannala (2003) was used to demonstrate that there was very little current gene flow between populations on two sides of a river. Goosens et al note that in more general comparisons the method appeared to converge well when the migration rate is low, but convergence was more problematic on data sets with higher migration rates.

The method of Wilson and Rannala (2003) need not be closely correlated with, e.g. pair-wise F_{ST}, both because the latter takes some time to equilibrate and also because in a migration model F_{ST} is a function of the product of the effective size, N, and migration rate m. Thus it should be possible to obtain estimates of N given information on the genetic divergence between populations. Interestingly it is not necessary to use a likelihood framework to make such estimates. Vitalis and Couvet (2001) have developed method-of-moments estimators to infer effective size and immigration rate using single locus and two locus measures of identity by state.

30.4.5 Spatial Modelling

A common application of some of these genotypic methods has been to discover clusters of individuals, with a view to better understand population structure (see Excoffier and Heckel, 2006, for a general review of software for performing these analyses). Typically the data are based on individuals of known origin, and the idea is to then apply, e.g. the program STRUCTURE of Pritchard et al. (2000) to see if it identifies interesting groupings (see Corander et al., 2003, for an alternative approach to this problem). Thus inference of parameter K, the number of clusters, becomes a focus of interest, and this is rather problematic, since it is only estimated through an approximation that may not always be valid, although further modifications have been suggested (Evanno et al., 2005). These considerations have motivated the development of methods that identify the number of groups through model-selection and also incorporate explicit spatial information into the genotypic models. These spatial models tend to borrow heavily from the techniques of geostatistics.

One example of this is the study by Guillot et al. (2005), who propose a model similar to that of Pritchard et al. (2000), with K populations, and with the aim of inferring K. However the prior for K is structured around a Voronoi tessellation. Here, a series of points, the number, m, of which is drawn from a Poisson distribution with parameter λ are located uniformly at random over a rectangle, covering the geographic area of interest. Around each point it is possible to draw a convex polygon that contains the region of the rectangle that is closer to that point than to any other point. The rectangle can be divided into m such non-overlapping regions. This is a Voronoi tessellation, and, given K populations, each tile of the tessellation is assumed to be assigned uniformly at random to one of the K populations. One implication of this model is that, although the tessellation provides a spatial element, it only specifies loosely, via the interaction of K and λ, how the space is subdivided. Thus, e.g. if K is 2 and λ is high the two populations are distributed as a mosaic through the region, whereas if λ is low the region is likely to be dominated by two main blocks. In contrast to the method of Pritchard *et al.* (2000), reversible jump MCMC (Green, 1995) is used to infer the posterior distribution of K. The method appears able to be able to recover accurately spatial genetic discontinuities in simulated data sets. In an application of this method (implemented in the program GENELAND) to roe deer populations in France (Coulon *et al.*, 2006) the method could find weak evidence of spatial discontinuity into two populations, north and south of a region including a highway, canal, and river running close together. In contrast, analysis with *Structure* suggested that there was only one population.

An alternative spatial model, specifically for assigning individuals to populations (also possible with *Geneland*), is presented in Wasser *et al.* (2004), which includes an application of their method to the problem of locating the geographic origin in Africa of elephants from their DNA samples. Theirs is primarily an assignment method, similar to that of Rannala and Mountain (1997). However, they incorporate a spatial model for the population frequencies. Importantly also they explicitly allow for genotyping errors in the microsatellite data that they analyse. The frequencies f_{jlk} for allele j at locus l in population k are modelled logistically $f_{jlk}(\theta) = \exp(\theta_{jlk})/(\sum_{j'} \exp(\theta_{j'lk}))$, and the θ is modelled by a Gaussian process. A Gaussian process is one that generates random variables, any sample n of which are drawn from a n dimensional multivariate Gaussian distribution (and hence any linear combination of these random variables is also Gaussian). The mean vector and covariance matrix of this distribution are determined by the problem under consideration. For spatial problems, as here, this is more generally called a *Gaussian random field*. In the case of the spatial model here, for the same allele at the same locus in two different locations k and k' the covariance depends on d the distance between the two locations and they use the function given by $(1/\alpha_0) \exp[-(\alpha_1 d)^{\alpha_2}]$. Wasser *et al.* (2004) further extend the model by allowing a location W to be specified so that genetic samples can be assigned to locations that are not specified in advance. This is implemented using MCMC, allowing the posterior distribution of W to be computed. With this method Wasser *et al.* analysed samples from 399 individuals, and demonstrated that they were able to make accurate spatial assignments. The implication of this study is that in future, DNA from ivory samples could be used to determine their provenance, and thereby help control the trade in ivory.

30.5   RELATEDNESS AND PEDIGREE ESTIMATION

It is tempting to predict that the genotypic methods discussed above will in future become less distinct from those based on the kin structure of populations. The latter aim to estimate the relatedness of pairs of individual, test hypotheses about degrees of relationships between individuals, and more generally uncover the recent pedigree of a sample. If such methods could be embedded into a population model, then it would be feasible to unite frequency-based methods for uncovering demographic history with the genotypic approaches. For example, recently methods have been developed for partitioning a sample of individuals into groups according to their degrees of relationship (e.g. sib, half-sib, or non-sib groups), and in this regard the methods are similar to the multilocus techniques described above. Currently, analysis of relatedness has been used to estimate genetic components of variance of phenotypic trait in outbred populations (e.g. Mousseau *et al.*, 1998), and thus can potentially be used in conservation to quantify the amount of phenotypically important genetic variation in endangered populations (Storfer, 1996; Carvajal-Rodríguez *et al.*, 2005; Thomas, 2005). Furthermore information about the kin structure of populations can help inform conservation decisions (Regnaut *et al.*, 2006).

There have been a number of recent reviews in this area (Blouin, 2003; Thomas, 2005; Oliehoek *et al.*, 2006; see also **Chapters 23** and **24**), and further relevant information is available in the chapters on pedigree analysis in the current volume. The aim of this section is to briefly introduce the general concepts and mainly follows the treatment in Thomas (2005) and Oliehoek *et al.* (2006).

An early interest has been to estimate the coefficient of relatedness, r. In a large random mating population

$$r = 2\theta = \phi/2 + \Delta, \tag{30.4}$$

where θ is the coefficient of coancestry or kinship between two individuals, and is the probability that an allele taken at random from a locus in one individual is IBD with an allele taken at random in another individual, ϕ is the probability that at a particular locus the two individuals have exactly one allele each i.e. IBD, and Δ is the probability that the two individuals have exactly two alleles that are IBD. The only other possibility is that no alleles are IBD. With inbreeding, it is of course possible that the alleles within individuals are IBD, leading to more complex possibilities, but those are typically not considered, by making the assumption of a large random mating population. Different degrees of relationship have different expected values for r, ϕ and Δ. The coefficient of relatedness can be viewed as the proportion of alleles in one individual that are IBD with those in another.

There are a number of different ways to estimate r. For example a simple method of moments estimator can be constructed from the same considerations that are used in the estimation of F_{ST} (Ritland, 1996): a pair of randomly chosen alleles, one from individual x and the other from individual y, has a probability of being identical in state (IIS), s_{xy}, given by $s_{xy} = \theta + (1 - \theta)s$, where s is the probability of choosing two identical alleles at random from the population (i.e. the sample homozygosity). This can be rewritten as

$$\theta = \frac{s_{xy} - s}{1 - s}. \tag{30.5}$$

As with estimators of F_{ST}, this invites the substitution of s_{xy} and s by estimates from the data. Various versions of this type of estimator, with different weighting schemes, are discussed in Ritland (1996), Lynch and Ritland (1999), and Oliehoek *et al.* (2006). It should be noted that Lynch and Ritland (1999) suggest that the specific estimator derived above (their equation 10) has poor sampling variance in comparison with others. It is given here solely because of its simplicity of derivation. As with F_{ST}, less biased estimates will be obtained by averaging the numerator and denominator across loci before evaluating the fraction. A more sophisticated class of moment estimators uses a similar approach to that above to estimate the parameters in (30.4) (Lynch and Ritland, 1999; Wang, 2002). A widely-used method of moments estimator, derived using rather different considerations, is that of Queller and Goodnight (1989), which has often been used to estimate the relatedness of groups of individuals. It can, however, be used to estimate r for a pair of individuals, but in this case it is necessary to use multi-allelic loci because the estimator is undefined for heterozygotes when there are biallelic loci. A recent comparison of methods based on empirical data with known pedigrees and also on simulated data suggests that the method of Lynch and Ritland has generally superior performance among these estimators (Csillery et al., 2006). This is supported, at least for panmictic populations, by the study of Oliehoek et al. (2006), but they also observed that for structured populations a modified version of (30.5) was more accurate.

Likelihood-based methods have also been developed (Thompson, 1975b). Maximum likelihood estimates tend to have marginally improved mean square error performance but can be biased with small sample sizes (Milligan, 2003; Thomas, 2005). The basic form of the likelihood calculation that has been used (following Thomas, 2005) is:

$$P(G|\phi, \Delta) = \prod_l \left\{ (1 - \phi - \Delta) P_{[0]}(g_l|x) + \phi P_{[1]}(g_l|x) + \Delta P_{[2]}(g_l|x) \right\},$$

where g_l is the vector of the four allele types at a particular locus (in this example, which assumes independence across loci, it is assumed that the alleles in an individual are unordered with respect to which parent they were inherited from), G is the genotype across loci, x is the baseline frequency. Note that this will typically be estimated from the data in a separate computation, although in principle it could be imputed by MCMC, as in the program STRUCTURE. The indices [0], [1], [2] refer to the number of alleles that are IBD, giving, e.g. $P_{[1]}((\{a, b\}, \{a, c\})|x) = 2p_a p_b p_c$. Point estimates can then be obtained by maximising the likelihood (Milligan, 2003). One use of the likelihood approach has been to estimate heritability in outbred populations (Mousseau et al. 1998). For each pair of individuals the likelihood of the joint genotypes and phenotypes can be calculated as a function of the degree of relationship and parameters in a quantitative genetic model, which can then be inferred. Typically, since the computations are based on all pairs of individuals, these give composite likelihood estimates and the confidence intervals have to be computed through bootstrapping.

Recently, moving beyond these pair-wise analyses, there has been an increase in likelihood-based approaches for uncovering the kin-structure of populations. Methods have been developed for calculating the likelihood for groups of individuals that share two parents or none (Painter, 1997). In this case it is equivalent to finding a simple partition of the sample into full sib families and single individuals. Alternatively it is possible to identify groups that share two parents (full-sibs), one parent (half-sibs) or none (Thomas and Hill, 2002; Wang, 2004). The *Parentage* program of Wilson (in Emery et al. (2001)) directly assigns individuals to parents. It is possible to specify the genotypes of known potential parents, and impute the genotypes of missing parents. The methods of Wang (2004) and Emery et al. take care to deal with the possibility of genotyping error, which can have serious consequences for such analyses. Hadfield et al. (2006) have developed a method, extending that of Emery et al. that allows for assignment of parents but also the estimation of a number of parameters relating to social structure, such as the size of male territories.

An ideal goal is to be able to reconstruct the pedigree of a sample from genotypic information (Steel and Hein, 2006). Parentage assignment (Emery *et al.*, 2001; Hadfield *et al.*, 2006) can be regarded as construction of a pedigree back one generation in a discrete-generation model, with a population size given by the sum of the number of observed and missing parents, and with a prior distribution for the genotypes of the missing parents, as well as for family sizes and mating structure as in Hadfield *et al.* (2006). Such a system could then in principle be taken back an arbitrary number of generations. The pedigree can be regarded as a more detailed genealogical model for a diploid sexual organism than the coalescent, and, as with the coalescent, it may often be a nuisance parameter to be integrated out while inferring parameters related to demography and life-history. The most complete approach to date in this regard is that of Gasbarra *et al.* (2007), which is based on a population-based model of pedigrees constructed backwards in time (Gasbarra *et al.*, 2005). The method of Gasbarra *et al.* (2007) uses a highly complex MCMC updating scheme, and can be applied to both unlinked and partially linked genetic data. The pedigree is constructed back to some pre-specified point in time. Pedigrees are sampled from their posterior distribution, given the genotypic data. It is also possible to use the method to infer the phase of multi-locus data. Such a method has enormous potential in conservation genetics, both in being able to infer social structure as well as demographic parameters, such as immigration rates.

Acknowledgments

I am very grateful to Eric Anderson, Jonathan Pritchard, and David Balding for their helpful comments on manuscripts for the first edition of this chapter, and to David Balding for comments on the current edition.

Related Chapters

Chapter 22;   **Chapter 23**;   **Chapter 24**;   **Chapter 25**;   **Chapter 26**;   **Chapter 27**; **Chapter 28**; and **Chapter 29**.

Websites

Below is a list of websites containing programs that perform some of the analyses described in this chapter.

- Estimation of effective size and migration using temporally spaced samples; admixture; relatedness:
  http://www.zoo.cam.ac.uk/ioz/software.htm
- Estimation of effective size; changes in population size; migration and isolation; ABC methods:
  http://www.rubic.rdg.ac.uk/~mab/
- Estimation of effective size; hybridization:
  http://swfsc.noaa.gov//staff.aspx?Division=FED&id=740
- Assignment testing; bottleneck detection:
  http://www.montpellier.inra.fr/URLB/
- General population genetics software (ARLEQUIN); general-purpose programs for simulating gene-frequency data under a wide variety of conditions; programs for performing ABC analysis:
  http://cmpg.unibe.ch/software.htm

- Bottleneck detection:
  `http://swfsc.noaa.gov/textblock.aspx?Division=FED&id=3298`
- Estimation of migration rates:
  `http://www.rannala.org`
- Analysis of population structure; population assignment; hybridization:
  `http://pritch.bsd.uchicago.edu/software.html`
- Analysis of population structure; population assignment: hybridization:
  `http://www.rni.helsinki.fi/~jic/bapspage.html`

REFERENCES

Anderson, E.C. (2005). An efficient Monte Carlo method for estimating N_E from temporally spaced samples using a coalescent-based likelihood. *Genetics* **170**, 955–967.

Anderson, E.C. and Thompson, E.A. (2002). A model-based method for identifying species hybrids using multilocus genetic data. *Genetics* **160**, 1217–1229.

Anderson, E.C., Williamson, E.G. and Thompson, E.A. (2000). Monte Carlo evaluation of the likelihood for N_e from temporally spaced samples. *Genetics* **156**, 2109–2118.

Avise, J.C. (1994). *Molecular Markers, Natural History and Evolution*. Chapman & Hall, London.

Balding, D.J., Carothers, A.D., Marchini, J.L., Cardon, L.R., Vetta, A., Griffiths, B., Weir, B.S., Hill, W.G., Goldstein, D., Strimmer, K., Myers, S., Beaumont, M.A., Glasbey, C.A., Mayer, C.D., Richardson, S., Marshall, C., Durrett, R., Nielsen, R., Visscher, P.M., Knott, S.A., Haley, C.S., Ball, R.D., Hackett, C.A., Holmes, S., Husmeier, D., Jansen, R.C., ter Braak, C.J.F., Maliepaard, C.A., Boer, M.P., Joyce, P., Li, N., Stephens, M., Marcoulides, G.A., Drezner, Z., Mardia, K., McVean, G., Meng, X.L., Ochs, M.F., Pagel, M., Sha, N., Vannucci, M., Sillanpaa, M.J., Sisson, S., Yandell, B.S., Jin, C.F., Satagopan, J.M., Gaffney, P.J., Zeng, Z.B., Broman, K.W., Speed, T.P., Fearnhead, P., Donnelly, P., Larget, B., Simon, D.L., Kadane, J.B., Nicholson, G., Smith, A.V., Jonsson, F., Gustafsson, O., Stefansson, K., Donnelly, P., Parmigiani, G., Garrett, E.S., Anbazhagan, R. and Gabrielson, E. (2002). Discussion on the meeting on 'statistical modelling and analysis of genetic data'. *Journal of the Royal Statistical Society, Series B: Statistical Methodology* **64**, 737–775.

Balding, D.J. and Nichols, R.A. (1995). A method for quantifying differentiation between populations at multi-allelic loci and its implications for investigating identity and paternity. *Genetica* **96**, 3–12.

Balding, D.J. and Nichols, R.A. (1997). Significant genetic correlations among Caucasians at forensic DNA loci. *Heredity* **78**, 583–589.

Balloux, F. (2004). Heterozygote excess in small populations and the heterozygote-excess effective population size. *Evolution* **58**, 1891–1900.

Bartley, D., Bagley, M., Gall, G. and Bentley, B. (1992). Use of linkage disequilibrium data to estimate effective size of hatchery and natural fish populations. *Conservation Biology* **6**, 365–375.

Beaumont, M.A. (1999). Detecting population expansion and decline using microsatellites. *Genetics* **153**, 2013–2029.

Beaumont, M.A. (2003a). Estimation of population growth or decline in genetically monitored populations. *Genetics* **164**, 1139–1160.

Beaumont, M.A. (2003b). Recent developments in genetic data analysis: what can they tell us about human demographic history? *Heredity* **92**, 365–379.

Beaumont, M.A., Barratt, E.M., Gottelli, D., Kitchener, A.C., Daniels, M.J., Pritchard, J.K. and Bruford, M.W. (2001). Genetic diversity and introgression in the Scottish wildcat. *Molecular Ecology* **10**, 319–336.

Beaumont, M.A., Zhang, W. and Balding, D.J. (2002). Approximate Bayesian computation in population genetics. *Genetics* **162**, 2025–2035.

Berthier, P., Beaumont, M.A., Cornuet, J.M. and Luikart, G. (2002). Likelihood-based estimation of the effective population size using temporal changes in allele frequencies: a genealogical approach. *Genetics* **160**, 741–751.

Bertorelle, G. and Excoffier, L. (1998). Inferring admixture proportions from molecular data. *Molecular Biology and Evolution* **15**, 1298–1311.

Blouin, M.S. (2003). DNA-based methods for pedigree reconstruction and kinship analysis in natural populations. *Trends in Ecology and Evolution* **18**, 503–511.

Bucci, G., Vendramin, G.G., Lelli, L. and Vicario, F. (1997). Assessing the genetic divergence of *Pinus leucodermis* Ant. endangered populations: use of molecular markers for conservation purposes. *Theoretical and Applied Genetics* **95**, 1138–1146.

Caballero, A. (1994). Developments in the prediction of effective population size. *Heredity* **73**, 657–679.

Calmet, C. (2003). Inférences sur l'histoire des populations à partir de leur diversité genetique: étude de séquences démographiques de type fondation-explosion. Ph.D. thesis, University of Paris.

Cannings, C. and Edwards, A.W.F. (1969). Expected genotypic frequencies in a small sample: deviation from Hardy-Weinberg equilibrium. *American Journal of Human Genetics* **21**, 245–247.

Carvajal-Rodríguez, A., Rolan-Alvarez, E. and Caballero, A. (2005). Quantitative variation as a tool for detecting human-induced impacts on genetic diversity. *Biological Conservation* **124**, 1–13.

Cavalli-Sforza, L.L. and Edwards, A.W.F. (1967). Phylogenetic analysis: models and estimation procedures. *Evolution* **32**, 550–570.

Chakraborty, R., Kamboh, M.I., Nwankwo, M. and Ferrell, R.E. (1992). Caucasian genes in American blacks: new data. *American Journal of Human Genetics* **50**, 145–155.

Chan, Y.L., Anderson, C.N.K. and Hadly, E.A. (2006). Bayesian estimation of the timing and severity of a population bottleneck from ancient DNA. *PLoS Genetics* **2**, 451–460.

Chikhi, L., Bruford, M.W. and Beaumont, M.A. (2001). Estimation of admixture proportions: a likelihood-based approach using Markov chain Monte Carlo. *Genetics* **158**, 1347–1362.

Choisy, M., Franck, P. and Cornuet, J.M. (2004). Estimating admixture proportions with microsatellites: comparison of methods based on simulated data. *Molecular Ecology* **13**, 955–968.

Ciofi, C., Beaumont, M.A., Swingland, I.R. and Bruford, M.W. (1999). Genetic divergence and units for conservation in the Komodo Dragaon *Varanus komodoensis*. *Proceedings of the Royal Society of London, Series B* **266**, 2269–2274.

Cockerham, C.C. and Weir, B.S. (1977). Digenic descent measures for finite populations. *Genetical Research* **30**, 121–147.

Corander, J. and Marttinen, P. (2006). Bayesian identification of admixture events using multilocus molecular markers. *Molecular Ecology* **15**, 2833–2843.

Corander, J., Waldmann, P. and Sillanpaa, M.J. (2003). Bayesian analysis of genetic differentiation between populations. *Genetics* **163**, 367–374.

Cornuet, J.M. and Luikart, G. (1996). Description and power analysis of two tests for detecting recent population bottlenecks from allele frequency data. *Genetics* **144**, 2001–2014.

Cornuet, J.M., Piry, S., Luikart, G., Estoup, A. and Solignac, M. (1999). New methods employing multilocus genotypes to select or exclude populations as origins of individuals. *Genetics* **153**, 1989–2000.

Coulon, A., Guillot, G., Cosson, J.F., Angibault, J.M.A., Aulagnier, S., Cargnelutti, B., Galan, M. and Hewison, A.J.M. (2006). Genetic structure is influenced by landscape features: empirical evidence from a roe deer population. *Molecular Ecology* **15**, 1669–1679.

Crow, J.F. and Kimura, M. (1970). *An Introduction to Population Genetics Theory.* Harper & Row, New York.

Csillery, K., Johnson, T., Beraldi, D., Clutton-Brock, T., Coltman, D., Hansson, B., Spong, G. and Pemberton, J.M. (2006). Performance of marker-based relatedness estimators in natural populations of outbred vertebrates. *Genetics* **173**, 2091–2101.

Dawson, K.J. and Belkhir, K. (2001). A Bayesian approach to the identification of panmictic populations and the assignment of individuals. *Genetical Research, Cambridge* **78**, 59–77.

DeSalle, R. and Amato, G. (2004). The expansion of conservation genetics. *Nature Reviews Genetics* **5**, 702–712.

Di Rienzo, A., Donnelly, P., Toomajian, C., Sisk, B., Hill, A., Petzl-Erler, M.L., Haines, G.K. and Barch, D.H. (1998). Heterogeneity of microsatellite mutations within and between loci and implications for human demographic histories. *Genetics* **148**, 1269–1284.

Di Rienzo, A., Peterson, A.C., Garza, J.C., Valdes, A.M., Slatkin, M. and Freimer, N.B. (1994). Mutational processes of simple sequence repeat loci in human populations. *Proceedings of the National Academy of Sciences of the United States of America* **91**, 3166–3170.

Donnelly, P. and Tavaré, S. (1995). Coalescents and genealogical structure under neutrality. *Annual Review of Genetics* **29**, 410–421.

Dupanloup, I. and Bertorelle, G. (2001).. Inferring admixture proportions from molecular data: extension to any number of parental populations. *Molecular Biology and Evolution* **18**, 672–675.

Elston, R.C. (1971). The estimation of admixture in racial hybrids. *Annals of Human Genetics* **35**, 9–17.

Emery, A.M., Wilson, I.J., Craig, S., Boyle, P.R. and Noble, L.R. (2001). Assignment of paternity groups without access to parental genotypes: multiple mating and developmental plasticity in squid. *Molecular Ecology* **10**, 1265–1278.

England, P.R., Cornuet, J.M., Berthier, P., Tallmon, D.A. and Luikart, G. (2006). Estimating effective population size from linkage disequilibrium: severe bias in small samples. *Conservation Genetics* **7**, 303–308.

Estoup, A., Beaumont, M.A., Sennedot, F., Moritz, C. and Cornuet, J.M. (2004). Genetic analysis of complex demographic scenarios: the case of spatially expanding populations in the cane toad, *Bufo marinus*. *Evolution* **58**, 2021–2036.

Estoup, A., Wilson, I.J., Sullivan, C., Cornuet, J.-M. and Moritz, C. (2001). Inferring population history from microsatellite and enzyme data in serially introduced cane toads *Bufo marinus*. *Genetics* **159**, 1671–1687.

Evanno, G., Regnaut, S. and Goudet, J. (2005). Detecting the number of clusters of individuals using the software STRUCTURE: a simulation study. *Molecular Ecology* **14**, 2611–2620.

Ewens, W.J. (1972). The sampling theory of selectively neutral alleles. *Theoretical Population Biology* **3**, 87–112.

Ewens, W.J. (2004). *Mathematical Population Genetics: Theoretical Introduction*, vol. 1. Springer-Verlag, New York.

Excoffier, L. and Heckel, G. (2006). Computer programs for population genetics data analysis: a survival guide. *Nature Reviews Genetics* **7**, 745–758.

Falush, D., Stephens, M. and Pritchard, J.K. (2003). Inference of population structure using multilocus genotype data: linked loci and correlated allele frequencies author(s). *Genetics* **164**, 1567–1587.

Felsenstein, J. (1981). Evolutionary trees from gene frequencies and quantitative characters: finding maximum likelihood estimates. *Evolution* **35**, 1229–1242.

Frankham, R. (2005). Genetics and extinction. *Biological Conservation* **126**, 131–140.

Frankham, R. (1995). Conservation genetics. *Annual Review of Genetics* **29**, 305–327.

Fu, Y.-X. (1997). Statistical tests of neutrality of mutations against population growth, hitchhiking and background selection. *Genetics* **147**, 915–925.

Garza, J.C. and Williamson, E.G. (2001). Detection of reduction in population size using data from microsatellite loci. *Molecular Ecology* **10**, 305–318.

Gasbarra, D., Sillanpaa, M.J. and Arjas, E. (2005).. Backward simulation of ancestors of sampled individuals. *Theoretical Population Biology* **67**, 75–83.

Gasbarra, D., Pirinen, M., Sillanpää, M.J., Salmela, E. and Arjas, E. (2007). Estimating genealogies from unlinked marker data: a Bayesian approach. *Theoretical Population Biology* (Submitted: preprint at http://www.rni.helsinki.fi/dag/genealogy.tar).

Glass, B. and Li, C.C. (1953). The dynamics of racial intermixture – an analysis based on the American Negro. *American Journal of Human Genetics* **5**, 1–20.

Goldstein, D.B., Ruiz Linares, A., Cavalli-Sforza, L.L. and Feldman, M.W. (1995).. Genetic absolute dating based on microsatellites and the origin of modern humans. *Proceedings of the National Academy of Sciences of the United States of America* **92**, 6723–6727.

Goodman, S.J., Tamate, H.B., Wilson, R., Nagata, J., Tatsuzawa, S., Swanson, G.M., Pemberton, J.M. and McCullough, D.R. (2001). Bottlenecks, drift and differentiation: the population structure and demographic history of sika deer (Cervus nippon) in the Japanese archipelago. *Molecular Ecology* **10**, 1357–1370.

Goossens, B., Chikhi, L., Ancrenaz, M., Lackman-Ancrenaz, I., Andau, P. and Bruford, M.W. (2006). Genetic signature of anthropogenic population collapse in orang-utans. *PLoS Biology* **4**, 285–291.

Green, P.J. (1995). Reversible jump Markov chain Monte Carlo and Bayesian model determination. *Biometrika* **82**, 711–7732.

Griffiths, R.C. and Tavaré, S. (1994). Simulating probability distributions in the coalescent. *Theoretical Population Biology* **46**, 131–159.

Guillot, G., Estoup, A., Mortier, F. and Cosson, J.F. (2005). A spatial statistical model for landscape genetics. *Genetics* **170**, 1261–1280.

Hadfield, J.D., Richardson, D.S. and Burke, T. (2006). Towards unbiased parentage assignment: combining genetic, behavioural and spatial data in a Bayesian framework. *Molecular Ecology* **15**, 3715–3730.

Haig, S.M. (1998). Molecular contributions to conservation. *Ecology* **79**, 413–425.

Hanfling, B. and Weetman, D. (2006). Concordant genetic estimators of migration reveal anthropogenically enhanced source-sink population structure in the River Sculpin, *Cottus gobio*. *Genetics* **173**, 1487–1501.

Hedrick, P.W. (2001). Conservation genetics: where are we now? *Trends in Ecology and Evolution* **16**, 629–636.

Hey, J. and Nielsen, R. (2004). Multilocus methods for estimating population sizes, migration rates and divergence time, with applications to the divergence of *Drosophila pseudoobscura* and *D. persimilis*. *Genetics* **167**, 747–760.

Hill, W.G. (1981). Estimation of effective population size from data on linkage disequilibrium. *Genetical Research* **38**, 209–216.

Hill, W.G. and Robertson, A. (1968). Linkage disequilibrium in finite populations. *Theoretical and Applied Genetics* **38**, 226–231.

Hogben, L. (1946). *An Introduction to Mathematical Genetics*. W. W. Norton, New York.

Hudson, R.R. (1985). The sampling distribution of linkage disequilibrium under an infinite allele model without selection. *Genetics* **109**, 611–631.

Hudson, R.R. (1991). Gene genealogies and the coalescent process. In *Oxford Surveys in Evolutionary Biology*, K.J. Futuyama and J. Antonovics, eds. Oxford University Press, Oxford, pp. 1–44.

Jorde, P.E. and Ryman, N. (1995). Temporal allele frequency change and estimation of effective size in populations with overlapping generations. *Genetics* **139**, 1077–1090.

Kimmel, M., Chakraborty, R., King, J.P., Bamshad, M., Watkins, W.S. and Jorde, L.B. (1998). Signatures of population expansion in microsatellite repeat data. *Genetics* **148**, 1921–1930.

King, J.P., Kimmel, M. and Chakraborty, R. (2000). A power analysis of microsatellite-based statistics for inferring past population growth. *Molecular Biology and Evolution* **17**, 1859–1868.

Keightley, P.D., Caballero, A. and GarciaDorado, A. (1998). Population genetics: surviving under mutation pressure. *Current Biology* **8**, R235–R239.

Krimbas, C.B. and Tsakas, S. (1971). The genetics of *Dacus oleae*. V. Changes of esterase polymorphism in a natural population following insecticide control – selection or drift? *Evolution* **25**, 454–460.

Lande, R. (1998). Anthropogenic, ecological and genetic factors in extinction and conservation. *Research in Population Economics* **40**, 259–269.

Langley, C.H., Smith, D.B. and Johnson, F.M. (1978). Analysis of linkage disequilibrium between allozyme loci in natural populations of Drosophila melanogaster. *Genetical Research* **32**, 215–229.

Laurie-Ahlberg, C.C. and Weir, B.S. (1979). Allozyme variation and linkage disequilibrium in some laboratory populations of *Drosophila melanogaster*. *Genetics* **92**, 1295–1314.

Laval, G. and Excoffier, L. (2004).. SIMCOAL 2.0: a program to simulate genomic diversity over large recombining regions in a subdivided population with a complex history. *Bioinformatics* **20**, 2485–2487.

Lecis, R., Pierpaoli, M., Biro, Z.S., Szemethy, L., Ragni, B., Vercillo, F. and Randi, E. (2006). Bayesian analyses of admixture in wild and domestic cats (*Felis silvestris*) using linked microsatellite loci. *Molecular Ecology* **15**, 119–131.

Levene, H. (1949). On a matching problem arising in genetics. *Annals of Mathematical Statistics* **20**, 91–94.

Long, J.C. (1991). The genetic structure of admixed populations. *Genetics* **127**, 417–428.

Luikart, G., Allendorf, F.W., Cornuet, J.M. and Sherwin, W.B. (1998). Distortion of allele frequency distributions provides a test for recent population bottlenecks. *The Journal of Heredity* **89**, 238–247.

Luikart, G. and Cornuet, J.M. (1999). Estimating the effective number of breeders from heterozygote excess in progeny. *Genetics* **151**, 1211–1216.

Luikart, G. and England, P.R. (1999). Statistical analysis of microsatellite DNA data. *Trends in Ecology and Evolution* **14**, 253–256.

Lynch, M., Blanchard, J., Houle, D., Kibota, T., Schultz, S., Vassivlieva, L. and Willis, J. (1999). Perspective: spontaneous deleterious mutation. *Evolution* **53**, 645–663.

Lynch, M., Conery, J. and Burger, R. (1995). Mutation accumulation and the extinction of small populations. *The American Naturalist* **146**, 489–518.

Lynch, M. and Ritland, K. (1999). Estimation of pairwise relatedness with molecular markers. *Genetics* **152**, 1753–1766.

Marjoram, P., Molitor, J., Plagnol, V. and Tavaré, S. (2003).. Markov chain Monte Carlo without likelihoods. *Proceedings of the National Academy of Sciences of the United States of America* **100**, 15324–15328.

Miller, N., Estoup, A., Toepfer, S., Bourguet, D., Lapchin, L., Derridj, S., Kim, K.S., Reynaud, P., Furlan, F. and Guillemaud, T. (2005). Multiple transatlantic introductions of the western corn rootworm. *Science* **310**, 992.

Milligan, B.G. (2003). Maximum-likelihood estimation of relatedness. *Genetics* **163**, 1153–11167.

Mousseau, T.A., Ritland, K. and Heath, D.D. (1998). A novel method for estimating heritability using molecular markers. *Heredity* **80**, 218–224.

Nei, M. and Tajima, F. (1981). Genetic drift and estimation of effective population size. *Genetics* **98**, 625–640.

Nicholson, G., Smith, A.V., Jonsson, F., Gustafsson, O., Stefansson, K. and Donnelly, P. (2002). Assessing population differentiation and isolation from single-nucleotide polymorphism data. *Journal of the Royal Statistical Society, Series B: Statistical Methodology* **64**, 695–715.

Nielsen, R., Mountain, J.L., Huelsenbeck, J.P. and Slatkin, M. (1998). Maximum-likelihood estimation of population divergence times and population phylogeny in models without mutation. *Evolution* **52**, 669–677.

Nielsen, R. and Wakeley, J. (2001). Distinguishing migration from isolation: a Markov chain Monte Carlo approach. *Genetics* **158**, 885–896.

Nunney, L. (1995). Measuring the ratio of effective population size to adult numbers using genetic and ecological data. *Evolution* **49**, 389–392.

Ohta, T. and Kimura, M. (1969). Linkage disequilibrium due to random genetic drift. *Genetical Research* **13**, 47–55.

Ohta, T. and Kimura, M. (1973). A model of mutation appropriate to estimate the number of electrophoretically detectable alleles in a finite population. *Genetical Research, Cambridge* **22**, 201–204.

Oliehoek, P.A., Windig, J.J., van Arendonk, J.A.M. and Bijma, P. (2006). Estimating relatedness between individuals in general populations with a focus on their use in conservation programs. *Genetics* **173**, 483–496.

O'Ryan, C., Harley, E.H., Bruford, M.W., Beaumont, M., Wayne, R.K. and Cherry, M.I. (1998). Microsatellite analysis of genetic diversity in fragmented South African buffalo populations. *Animal Conservation* **1**, 85–94.

Paetkau, D., Calvert, W., Stirling, I. and Strobeck, C. (1995). Microsatellite analysis of population structure in Canadian polar bears. *Molecular Ecology* **4**, 347–354.

Paetkau, D., Shields, G.F. and Strobeck, C. (1998). Gene flow between insular, coastal and interior populations of brown bears in Alaska. *Molecular Ecology* **7**, 1283–1292.

Painter, I. (1997). Sibship reconstruction without parental information. *Journal of Agricultural Biological and Environmental Statistics* **2**, 212–229.

Palo, J.U., Makinen, H.S., Helle, E., Stenman, O. and Vainola, R. (2001). Microsatellite variation in ringed seals (*Phoca hispida*): genetic structure and history of the Baltic Sea population. *Heredity* **86**, 609–617.

Parra, E.J., Marcini, A., Akey, J., Martinson, J., Batzer, M.A., Cooper, R., Forrester, T., Allison, D.B., Deka, R., Ferrell, R.E. and Shriver, M.D. (1998). Estimating African American admixture proportions by use of population-specific alleles. *American Journal of Human Genetics* **63**, 1839–1851.

Pella, J. and Masuda, M. (2001). Bayesian methods for analysis of stock mixtures from genetic characters. *Fishery Bulletin* **99**, 151–167.

Piry, S., Luikart, G. and Cornuet, J.M. (1999). BOTTLENECK: a computer program for detecting recent reductions in the effective population size using allele frequency data. *The Journal of Heredity* **90**, 502–503.

Pollak, E. (1983). A new method for estimating the effective population size from allele frequency changes. *Genetics* **104**, 531–548.

Pritchard, J.K., Seielstad, M.T., Perez-Lezaun, A. and Feldman, M.W. (1999). Population growth of human Y chromosomes: a study of Y chromosome microsatellites. *Molecular Biology and Evolution* **16**, 1791–1798.

Pritchard, J.K., Stephens, M. and Donnelly, P. (2000). Inference of population structure using multilocus genotype data. *Genetics* **155**, 945–959.

Pudovkin, A.I., Zaykin, D.V. and Hedgecock, D. (1996). On the potential for estimating the effective number of breeders from heterozygote-excess in progeny. *Genetics* **144**, 383–387.

Queller, D.C. and Goodnight, K.F. (1989). Estimating relatedness using genetic markers. *Evolution* **43**, 258–275.

Ramos-Onsins, S.E. and Rozas, J. (2002). Statistical properties of new neutrality tests against population growth. *Molecular Biology and Evolution* **19**, 2092–2100.

Rannala, B. and Hartigan, J.A. (1996). Estimating gene flow in island populations. *Genetical Research, Cambridge* **67**, 147–158.

Rannala, B. and Mountain, J.L. (1997). Detecting immigration by using multilocus genotypes. *Proceedings of the National Academy of Sciences of the United States of America* **94**, 9197–9201.

Regnaut, S., Christe, P., Chapuisat, M. and Fumagalli, L. (2006). Genotyping faeces reveals facultative kin association on capercaillie's leks. *Conservation Genetics* **7**, 665–674.

Reich, D.E., Feldman, M.W. and Goldstein, D.B. (1999). Statistical properties of two tests that use multilocus data sets to detect population expansions. *Molecular Biology and Evolution* **16**, 453–466.

Reich, D.E. and Goldstein, D.B. (1998). Genetic evidence for a Paleolithic human population expansion in Africa. *Proceedings of the National Academy of Sciences of the United States of America* **95**, 8119–8123.

Ritland, K. (1996). Estimators for pairwise relatedness and individual inbreeding coefficients. *Genetical Research* **67**, 175–185.

Robert, C.P. (1996). Mixtures of distributions: inference and estimation. In *Markov Chain Monte Carlo in Practice*, W.R. Gilks, S. Richarson and D.J. Spiegelhalter, eds. Chapman & Hall, London, pp. 441–464.

Roberts, D. and Hiorns, R. (1965). Methods of analysis of the genetic composition of a hybrid population. *Human Biology* **37**, 38–43.

Robertson, A. (1965). The interpretation of genotypic ratios in domestic animal populations. *Animal Production* **7**, 319–324.

Rogers, A.R. (1995). Genetic evidence for a Pleistocene population explosion. *Evolution* **49**, 608–615.

Rogers, A.R. and Harpending, H. (1992). Population growth makes waves in the distribution of pairwise genetic differences. *Molecular Biology and Evolution* **9**, 552–569.

Roy, M.S., Geffen, E., Smith, D., Ostrander, E.A. and Wayne, R.K. (1994). Patterns of differentiation and hybridization in North American wolf-like canids, revealed by analysis of microsatellite loci. *Molecular Biology and Evolution* **11**, 553–570.

Ruppert, D. and Wand M.P., (1994).. Multivariate locally weighted least squares regression. *Annals of Statistics*, **22**, 1346–1370.

Saccheri, I., Kuussaari, M., Kankare, M., Vikman, P., Fortelius, W. and Hanski, I. (1998). Inbreeding and extinction in a butterfly metapopulation. *Nature* **392**, 491–494.

Saccheri, I.J., Wilson, I.J., Nichols, R.A., Bruford, M.W. and Brakefield, P.M. (1999). Inbreeding of bottlenecked butterfly populations: estimation using the likelihood of changes in marker allele frequencies. *Genetics* **151**, 1053–1063.

Schneider, S. and Excoffier, L. (1999). Estimation of past demographic parameters from the distribution of pairwise differences when the mutation rates vary among sites: application to human mitochondrial DNA. *Genetics* **152**, 1079–1089.

Schwartz, M.K., Tallmon, D.A. and Luikart, G. (1998). Review of DNA-based census and effective population size estimators. *Animal Conservation* **1**, 293–299.

Slatkin, M. (1996). Gene genealogies within mutant allelic classes. *Genetics* **145**, 579–587.

Spielman, D., Brook, B.W. and Frankham, R. (2004). Most species are not driven to extinction before genetic factors impact them. *Proceedings of the National Academy of Sciences of the United States of America* **101**, 15261–15264.

Steel, M. and Hein, J. (2006). Reconstructing pedigrees: a combinatorial perspective. *Journal of Theoretical Biology* **240**, 360–367.

Storfer, A. (1996). Quantitative genetics: a promising approach for the assessment of genetic variation in endangered species. *Trends in Ecology and Evolution* **11**, 343–348.

Tajima, F. (1983). Evolutionary relationship of DNA sequences in finite populations. *Genetics* **105**, 437–460.

Tajima, F. (1989). Statistical method for testing the neutral mutation hypothesis by DNA polymorphism. *Genetics* **123**, 585–595.

Tavaré, S. (1984). Lines-of-descent and genealogical processes, and their application in population genetics models. *Theoretical Population Biology* **26**, 119–164.

Taylor, A.C., Sherwin, W.B. and Wayne, R.K. (1994). Genetic variation of microsatellite loci in a bottlenecked species: the northern hairy-nosed wombat. *Molecular Ecology* **3**, 277–290.

Thomas, S.C. (2005). The estimation of genetic relationships using molecular markers and their efficiency in estimating heritability in natural populations. *Philosophical Transactions of the Royal Society of London, Series B* **360**, 1457–1467.

Thomas, S.C. and Hill, W.G. (2002). Sibship reconstruction in hierarchical population structures using Markov chain Monte Carlo techniques. *Genetical Research* **79**, 227–234.

Thompson, E.A. (1973). The Icelandic admixture problem. *Annals of Human Genetics* **37**, 69–80.

Thompson, E.A. (1975a). *Human Evolutionary Trees*. Cambridge University Press, Cambridge.

Thompson, E.A. (1975b). The estimation of pairwise relationships. *Annals of Human Genetics* **39**, 173–188.

Verardi, A., Lucchini, V. and Randi, E. (2006). Detecting introgressive hybridization between free-ranging domestic dogs and wild wolves (*Canis lupus*) by admixture linkage disequilibrium analysis. *Molecular Ecology* **15**, 2845–2855.

Vitalis, R. and Couvet, D. (2001). Estimation of effective population size and migration rate from one- and two-locus identity measures. *Genetics* **157**, 911–925.

Wakeley, J. (1999). Nonequilibrium migration in human history. *Genetics* **153**, 1863–1871.

Wakeley, J. (2001). The coalescent in an island model of population subdivision with variation among demes. *Theoretical Population Biology* **59**, 133–144.

Wakeley, J., Nielsen, R., Liu-Cordero, S.N. and Ardlie, K. (2001). The discovery of single-nucleotide polymorphisms – and inferences about human demographic history. *American Journal of Human Genetics* **69**, 1332–1347.

Wang, J. (2001). A pseudo-likelihood method for estimating effective population size from temporally spaced samples. *Genetical Research, Cambridge* **78**, 243–257.

Wang, J. (2002). An estimator for pairwise relatedness using molecular markers. *Genetics* **160**, 1203–1215.

Wang, J.L. (2003). Maximum-likelihood estimation of admixture proportions from genetic data. *Genetics* **164**, 747–765.

Wang, J.L. (2004). Sibship reconstruction from genetic data with typing errors. *Genetics* **166**, 1963–1979.

Wang, J.L. (2005). Estimation of effective population sizes from data on genetic markers. *Philosophical Transactions of the Royal Society of London, Series B* **360**, 1395–1409.

Wang, J.L. (2006). A coalescent-based estimator of admixture from DNA sequences. *Genetics* **173**, 1679–1692.

Wang, J.L. and Whitlock, M.C. (2003). Estimating effective population size and migration rates from genetic samples over space and time. *Genetics* **163**, 429–446.

Waples, R.S. (1989). A generalized approach for estimating effective population size from temporal changes in allele frequency. *Genetics* **121**, 379–392.

Waples, R.S. (1991). Genetic methods for estimating the effective size of Cetacean populations. *Genetic Ecology of Whales and Dolphins*, (special issue 13). *Report of the International Whaling Commission*, pp. 279–300.

Waples, R.S. (2006). A bias correction for estimates of effective population size based on linkage disequilibrium at unlinked gene loci. *Conservation Genetics* **7**, 167–184.

Waser, P.M. and Strobeck, C. (1998). Genetic signatures of interpopulation dispersal. *Trends in Ecology and Evolution* **13**, 43–44.

Wasser, S.K., Shedlock, A.M., Comstock, K., Ostrander, E.A., Mutayoba, B. and Stephens, M. (2004). Assigning African elephant DNA to geographic region of origin: applications to the ivory trade. *Proceedings of the National Academy of Sciences of the United States of America* **101**, 14847–14852.

Watterson, G.A. (1975). On the number of segregating sites in genetical models without recombination. *Theoretical Population Biology* **7**, 256–276.

Watterson, G.A. (1978). The homozygosity test of neutrality. *Genetics* **88**, 405–417.

Weir, B.S. (1996). *Genetic Data Analysis*, 2nd edition. Sinauer Associates, Sunderland.

Weir, B.S. and Hill, W.G. (1980). Effect of mating structure on variation in linkage disequilibrium. *Genetics* **95**, 477–488.

Weiss, G. and von Haeseler, A. (1998). Inference of population history using a likelihood approach. *Genetics* **149**, 1539–1546.

Whitlock, M.C., Griswold, C.K. and Peters, A.D. (2003). Compensating for the meltdown: the critical effective size of a population with deleterious and compensatory mutations. *Annales Zoologici Fennici* **40**, 169–183.

Williamson, E.G. and Slatkin, M. (1999). Using maximum likelihood to estimate population size from temporal changes in allele frequencies. *Genetics* **152**, 755–761.

Wilson, I.J. and Balding, D.J. (1998). Genealogical inference from microsatellite data. *Genetics* **150**, 499–510.

Wilson, G.A. and Rannala, B. (2003). Bayesian inference of recent migration rates using multilocus genotypes. *Genetics* **163**, 1177–1191.

Wilson, I.J., Weale, M.E. and Balding, D.J. (2003). Inferences from DNA data: population histories, evolutionary processes, and forensic match probabilities. *Journal of the Royal Statistical Society, Series A* **166**, 155–188.

Wright, S. (1931). Evolution in Mendelian populations. *Genetics* **16**, 97–159.

31

Human Genetic Diversity and its History

G. Barbujani

Dipartimento di Biologia ed Evoluzione, Università di Ferrara, Ferrara, Italia

and

L. Chikhi

Laboratoire Evolution et Diversité Biobgique, Université Paul Sabatier, Toulouse, France

Aspects of the evolutionary and historical processes that shaped present-day human diversity can be inferred from patterns of genetic variation within and between populations. The main questions currently being addressed include the evolutionary relationships among contemporary humans and the different human forms documented in the fossil record, the extent and causes of the genetic differences among modern populations, and the implications of such differences for applied research in fields such as medical genetics, pharmacogenetics, and forensic science. In this chapter, we outline the main current models of human evolution and review the available ancient and modern DNA evidence in the light of these models. We suggest that, despite ongoing controversy, most data are easier to reconcile with a model in which the ancestors of modern populations dispersed recently (<200 000 years ago) from Africa and essentially or completely replaced previously settled human forms. However, it has become increasingly evident that models assuming the expansion of a small African group are oversimplified, and more complex scenarios need be envisaged, including genetic substructuring in the expanding population. How much substructure is needed to explain existing data is unclear, and whether this apparent substructure could be partly due to admixture between modern and archaic humans is debated. The available studies among modern populations show that the differences tend to be patterned in geographical space, but that variation tends to be continuous and strong genetic boundaries rarely occur, suggesting a major role of isolation by distance in shaping human diversity. As a consequence, global human diversity is not well described by a small set of well-defined races. The extent to which racial classifications might be useful in some applications remains to be demonstrated. Throughout the chapter, we emphasise the large amount of uncertainty surrounding the interpretation of genetic data, when they are used either to make inferences on our origins or to predict disease risk.

Handbook of Statistical Genetics, Third Edition. Edited by D.J. Balding, M. Bishop and C. Cannings.
© 2007 John Wiley & Sons, Ltd.   ISBN: 978-0-470-05830-5.

31.1   INTRODUCTION

Individuals of all species form phenotypic and genotypic clusters that are generally structured in space and represent the outcome of the species' evolutionary history. Identifying the geographical patterns of human genetic diversity, and understanding their evolutionary and historical causes are the goals of anthropological genetics. Until recently, to address questions in this field, scientists had only measures of anatomical traits available and, from the mid twentieth century, a handful of polymorphic genetic markers. With the development of fast and inexpensive methods for DNA typing, the amount of relevant information available has increased dramatically, and so has the possibility to test hypotheses about the evolutionary factors underlying human diversity. In this way, some traditional questions found an apparently satisfactory answer, but many others emerged, so that the field seems no less controversial now than it used to. The main controversies concern the degree of genetic differentiation among human populations, whether the existing differences can be regarded as discontinuous and, if so, whether it is possible to agree on a non-subjective list of distinct human groups, each of which could potentially be associated with specific disease risks. A lesser, but by no means less debated problem, is whether such groups can or cannot be legitimately called *races*.

Interpreting the data bearing on human diversity and on its history is not straightforward for a number of reasons, including the obvious difficulty for all, including the authors of this paper, to study our species with the same emotional detachment with which we study *Escherichia coli* or *Drosophila melanogaster*. Words such as diversity and race do not immediately correspond to obvious biological concepts, and take different meanings in science from their non-scientific usage. Even in the technical literature, a high degree of ambiguity is present, so that according to Templeton (1999) the genetic differences among human populations are among the lowest in mammals, whereas according to Sarich (2000) they are highest of all mammals. Therefore, it seems useful to briefly review the available evidence, to outline the main open questions, and to consider the potentials and the limitations of the approaches chosen to address them.

In this chapter, we shall deal with some of the many studies leading to our current understanding of the patterns of human biological diversity and the evolutionary and demographic processes that generated them. Given the increasing role of genetic data in the debate, we describe in some detail the main methods used to interpret them. We also outline, the way some concepts took shape in the course of time. In the next section, we deal with historical inference, outlining the main lines of evidence that suggest a recent African origin of humankind. We include a short description of the palaeontological and archaeological data and of the models that provide the framework necessary for interpreting the biological evidence. In Section 31.3, we move to the studies describing how human genetic diversity is structured in space. This leads to a final section in which we deal with a number of recently emerged practical issues, in fields such as medicine, pharmacogenetics, and forensics, for which assumptions on the structure of human diversity appear to be important.

31.2   HUMAN GENETIC DIVERSITY: HISTORICAL INFERENCES

31.2.1   Some Data on Fossil Evidence

It is currently accepted that hominins, i.e. bipedal primates (Lewin and Foley, 2004), separated from the other African apes (*Gorilla* and *Pan* genera) some 5–8 million years (MY) ago. On the basis of fossil evidence, the first small-brained clearly bipedal primates are documented in Africa from at least 6 MY ago in what are now Chad and Kenya. Between 6 and 4 MY ago, the data are scanty, and it is only from approximately 4 MY ago that the hominin fossil record starts to be wider, with the appearance of the australopithecines (genus *Australopithecus*, meaning the 'southern ape man'), the most famous being 'Lucy' (*A. afarensis* 3.2 MY old, Figure 31.1). There is no consensus on the number of australopithecine species that existed between 4 and 1 MY ago (e.g. Klein, 1999), but many believe that there may have been as many as seven, some of them living at the same time (Toth and Schick, 2005; Klein, 2005; Scarre, 2005).

Australopithecines are usually divided into robust (sometimes referred to as *Paranthropus*) and gracile (genus *Australopithecus*) forms, and it is from the latter that the genus *Homo* (more precisely, *H. habilis* and *H. rudolfensis*) probably evolved, perhaps from east African *A. garhi* (Figure 31.1). This split probably took place around 2.5 or 2.0 MY ago, when the first stone tools appear in the African archaeological record. The early *Homo* forms include *H. habilis, H. rudolfensis*, and *H. ergaster*, the latter appearing sometime around 1.8 MY. It is from *H. ergaster* that *H. heidelbergensis* and *H. erectus* are thought to have derived, with *H. heidelbergensis* being regarded as the most likely ancestor of Neandertals and modern humans. *H. erectus* separated from *H. ergaster* less than 2 MY ago (Figure 31.1). With *H. ergaster, H. heidelbergensis, H. neandertalensis*, and *H. erectus*, we enter the heart of the controversies surrounding the models of modern human emergence, in particular, the contribution of so-called archaic forms to the ancestry of modern *H. sapiens*.

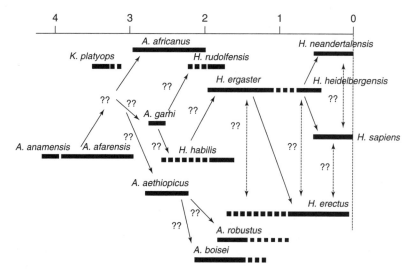

Figure 31.1   A schematic representation of hominin evolution. *X* axis: million years ago.

Until recently, *H. erectus* was thought to be the first *Homo* species to have left Africa and reached eastern Asia more than 1.6 MY ago (Antón and Swisher, 2004). The recent Dmanisi finds in the Caucasus, attributed to *H. ergaster* or to a closely related species (Vekua *et al.*, 2002; Toth and Schick, 2005) and dated about 1.77–1.95 MY ago, suggested that other *Homo* species also left Africa, which would imply that the three species potentially involved in the *H. sapiens* ancestry developed from *H. ergaster* ancestors. It thus appears that between 1 and 2 MY ago these potential ancestors were already geographically separated, namely *H. erectus* in eastern Asia, *H. neandertalensis* in western Asia and Europe, and *H. sapiens* in Africa. The first European settlers (ancestors of *H. neandertalensis*) are often pooled with their African contemporaries (ancestors of *H. sapiens*) under the name *H. heidelbergensis*, mentioned above, and may have lived around 1.1–1.3 MY ago.

Many details of this history are unknown and, among those that are known, just a few can be properly discussed here. But, before moving further to the models of modern human origins, we note that many fossils are still not assigned with certainty to a genus. As a consequence, there is still incomplete agreement on how many species of humans (or australopithecines, for that matter) have lived in the past. In fact, simply deciding which fossils were part of the human lineage and which were not depends largely on what are considered the key human features (Tattersall, 1995; Toth and Schick, 2005; Klein, 1999). We have only a vague idea about variation in time and space within fossil species, and hence the attribution to species and the reconstruction of their evolutionary relationships contain elements of arbitrariness (see Figure 31.1). For example, some authors do not consider the differences between *H. erectus* and *H. ergaster* large enough to warrant the need for a separate *H. ergaster* species. Moreover, proposed hominid species increased in numbers during the early twentieth century, decreased when the tendency prevailed to group together individuals in a limited number of species, and then increased again. Many early controversies originated from Darwin's intuition that bipedal locomotion, tool making, and a large brain case size developed simultaneously. We now know that these features appeared at different times, respectively around circa 6, 2.5, and 2 MY ago (Klein, 1999; Toth and Schick, 2005). Dating can also be problematic, as has been suggested, for instance, for the Dmanisi specimen, who might be as recent as 0.79 MY ago, or as old as 2.4 MY due to stratigraphic uncertainty (Klein, 2005). Finally, technologies are also not always easily linked to one species, or even one genus. In brief, there is currently no uncontroversial *Homo* fossil older than 1.8 MY and no uncontroversial data suggesting sustained occupation of temperate areas before 1 MY ago (Klein, 2005). The first dispersal of *Homo* species outside Africa probably started as early as ~2 MY, but more recent dates (~1 MY) cannot be ruled out (Antón and Swisher, 2004; Klein, 2005).

31.2.2 Models of Modern Human Origins

Despite the many uncertainties, most palaeoanthropologists would agree that *H. ergaster* is the ancestor from which Asian *H. erectus* and African *H. heidelbergensis* evolved, and that *H. neandertalensis* and *H. sapiens* then evolved from *H. heidelbergensis*. However, the relative contributions of these three forms, which may (Tattersall, 1997) or may not (Wolpoff, 1999) be regarded as separate species, to modern human morphologies remains controversial, as well as the exact timing (Grine *et al.*, 2007) and route (Anikovich *et al.*, 2007) of their dispersal in the Near East and Eurasia.

In recent decades, the debate has focused on two main models of human evolution. Under the Out of Africa (OOA), or Recent African Origin (Howells, 1976; Stringer and Andrews, 1988; Stringer, 1989) or Replacement (e.g. Relethford, 2003) model, present-day populations are regarded as deriving from a major spatial and demographic expansion of African populations of *H. sapiens*, starting less than 200 KY ago. These populations colonised the whole planet, replacing, essentially with no admixture, previously settled *H. erectus* and *H. neandertalensis* populations. On the other hand, the Multi-Regional (MR) model posits that Eurasian populations of *H. erectus* or *H. neandertalensis* are part of the genealogy of present-day humans (Wolpoff *et al.*, 1988). This means that only one human species existed in the last MY or so, encompassing all these different morphologies, which should then be simply classified as archaic and modern forms of a polytypic *H. sapiens* species (Relethford, 2003). These basic models really represent the extremes of a range of possible models, whose classification, in turn, is not completely clear (but see Aiello, 1993). For instance, Bräuer (1984) and Smith (1985) proposed that both admixture and replacement occurred between expanding African and previously settled Eurasian populations. Despite their similarities, these two hypotheses are considered, respectively, as versions of the OOA and of the MR models because of the greater weight they give to either admixture (Bräuer, 1984) or continuous gene flow (Smith, 1985) over long periods.

The MR model stems from work by Weidenreich (1943; 1947) and others, who described morphological similarities between Asian people and the fossil *Sinanthropus* (Peking Man), and between modern Australians and *Pithecanthropus* (Java Man) fossils, the so-called regional continuity traits (both Peking and Java man are now classified as *H. erectus*). These findings led to the proposal that anatomically modern Africans, Asians, Australians, and Europeans had descended from anatomically archaic ancestors who largely occupied the same regions. On the contrary, a comparatively recent origin of our species was proposed to explain the similarities between the first anatomically modern Europeans of the Cro-Magnon type and 130-KY-old fossils from the Omo Kibish region of Ethiopia (Stringer, 1978). In large-scale comparisons of modern and ancient bone specimens, the data appeared to be in significantly better agreement with the OOA rather than the MR model (Waddle, 1994; Lahr, 1994).

Further support for the OOA model comes from studies showing little or no evidence for hybrid populations displaying both archaic and modern traits, which suggests that the fossil record exhibits a discontinuous process of demographic replacement rather than continuity (Stringer and Andrews, 1988; Lahr and Foley, 1998). The controversy is still going on (Zilhão, 2006), and different authors appear to interpret the apparent lack of continuity in very different ways regarding the admixture process itself. For instance, it has been claimed that, even with significant admixture, complete continuity is not expected for morphological traits (Relethford, 1999; 2001), and hence lack of continuity may not necessarily play against the MR model. If all models involving any contribution of archaic populations to the modern gene pool are to be considered as versions of the MR model, rejecting it becomes in effect impossible.

31.2.3 Methods for Inferring Past Demography

Genetic data have the advantages over other types of data that their transmission mode is well known and that theory describing the evolution of genes in populations is well developed. In the following, we discuss from a biologist's point of view the manner in which genetic data have been used in recent decades to infer the demographic history of populations. For more statistical and mathematical treatments, see **Chapters 26**, **28**, **29** and **30**.

31.2.3.1 Summary Statistics

Population genetics inference is possible only if past demographic events leave specific genetic signatures in present-day populations. In particular, the effect of changes in population size on different summary statistics has been extensively studied. For diploids, it is customary to define the scaled mutation rate $\theta = 4N_e\mu$, where N_e is the effective population size of a stationary (or demographically stable) population and μ the locus mutation rate.

For neutral markers, evolving under an infinite-allele model in a Wright–Fisher population, Kimura and Crow (1964) showed that at mutation-drift equilibrium $E[H] = 1/(\theta + 1)$, where $E[.]$ represents expectation and H is the probability that two alleles drawn at random from the population are the same, which can be estimated by $\Sigma(n_i/n)^2$, where n_i is the observed count of the ith allele, and n is the sample size. Ewens (1972) derived from what is now known as *Ewens' sampling formula*, an estimator of H given n_A, the number of distinct alleles in the sample, which depends on and n and θ but not the observed n_i values. Significant differences between these two H estimates can be interpreted in terms of departure from any of the neutrality, size constancy, or mutation model assumptions Watterson (1978). For instance, when large populations go through a bottleneck (Nei *et al.*, 1975), rare alleles are lost first, which only mildly influences $\Sigma(n_i/n)^2$, whereas n_A is substantially reduced, and so the former estimate of H is significantly lower than that expected given n_A until a new equilibrium is approached. Conversely, expanding populations tend to accumulate new (and hence rare) alleles, and $\Sigma(n_i/n)^2$ is initially significantly higher than the estimate of H based on n_A.

The rationale behind this approach, and Watterson's (1978) related approach, has since been extended to different genetic markers by assuming different mutation models and using different summaries of the data (Tajima, 1989a; 1989b; Fu and Li, 1993). For DNA sequence data, the mean number of nucleotide differences between sequences, π, and the number of segregating sites, S (i.e. single nucleotide polymorphisms (SNPs)), provide two estimators, θ_S and θ_π, of the scaled mutation parameter, θ, at mutation-drift equilibrium that are differently affected by demographic events and selection (Tajima, 1989a; 1989b). Tajima (1989a) hence suggested the use of $D = (\theta_\pi - \theta_S)/[\text{var}(\theta_\pi - \theta_S)]^{1/2}$, now known as *Tajima's D* (see also **Chapter 26**), as a measure of departure from equilibrium conditions. Demographic bottlenecks and balancing selection tend to reduce S without much affecting π, and hence lead to positive D values, whereas population expansions and positive selection (selective sweeps) lead to negative values. For microsatellites, equilibrium equations have not been derived analytically, but simulation-based approaches have been developed (Cornuet and Luikart, 1996).

Beyond the use of summary statistics, theoretical work has also focused on other properties of allele-frequency distributions, or, for sequence data, on the properties of the so-called mismatch distributions. The mismatch distributions are obtained by plotting a histogram of all pairwise comparisons between sequences in a sample, and have been shown to be sensitive to demographic changes (Slatkin and Hudson, 1991; Rogers and Harpending, 1992; Excoffier and Schneider, 1999; Ray *et al.*, 2004). For other markers,

Marth *et al.* (2004) studied the properties of the full allele-frequency spectrum under different demographic scenarios, and found computationally quick ways to estimate these frequency spectra. Garza and Williamson (2001) have shown that bottlenecked populations have 'gappy' allele-frequency distributions at microsatellite loci because the number of alleles is strongly reduced, whereas the allelic range is only marginally affected.

Summary statistics based on genetic data may be easy to compute, but may discard most of the genealogical information in the data (Felsenstein, 1992). As a consequence, different sets of evolutionary factors can exert similar effects on summary statistics and cannot be separated unless detailed archaeological or historical information is available. Furthermore, these methods can *detect* demographic events, but usually provide poor *quantification* or *dating* of such events.

Indeed, the magnitude of a departure from equilibrium may not be a monotonic function of the time since the demographic event. Whereas mismatch distributions appear to follow simple temporal dynamics, this is not the case of Tajima's D (Fay and Wu, 1999). After a population expansion from a small population size, a unimodal mismatch distribution is expected whose mode will move from low to high values as mutations accumulate, whereas its height will decrease (Rogers and Harpending, 1992). On the contrary, simulations have shown that, after a bottleneck, Tajima's D is first positive, and then becomes negative when the population recovers from the bottleneck, before tending towards equilibrium (Fay and Wu, 1999). Thus, Tajima's D, one of the most popular statistics used to detect population size changes, may be transiently positive or negative, depending on the severity of the change and on the number of generations since the original demographic event, and hence it can behave differently for haploid and diploid genomes. The mutation model assumed can also generate significant departures from equilibrium (Aris-Brosou and Excoffier, 1996), but in many studies rejection of equilibrium is directly taken as evidence of demographic changes or selection. Future work will probably need to separate the effects of the mutation model from those of population structure and selection. Because crucial evolutionary information is lost as summary statistics are estimated, such an endeavour will require methods that are more efficient at retrieving information on demographic parameters from genetic data.

31.2.3.2 *Likelihood-based and Bayesian Approaches*

Following Felsenstein (1992), several authors have developed statistical methods that aim at extracting information from the full allelic distribution (Griffiths and Tavaré, 1994; Kuhner *et al.*, 1995; Wilson and Balding, 1998; Beaumont and Rannala, 2004). The aim of likelihood-based approaches is to compute or approximate the probability $P_M(D|\theta)$ of the observed data D under some demographical model M, defined by a set of parameters $\theta = (\theta_1, \ldots, \theta_k)$. This probability is the likelihood $L_M(\theta|D)$, or simply $L_M(\theta)$. Some methods focus on point estimators, e.g. maximum likelihood estimates (MLE), others take a Bayesian perspective and compute a posterior probability distribution for θ. Using Bayes formula, we can write the following:

$$P_M(\theta|D) = P_M(\theta) \times P_M(D|\theta)/P_M(D) = P_M(\theta) \times L_M(\theta|D)/P_M(D)$$

$$= P_M(\theta) \times L_M(\theta)/P_M(D). \tag{31.1}$$

Since the denominator is constant, *given* the data, $P_M(\theta|D)$ is proportional to $P_M(\theta) \times L_M(\theta)$. In the Bayesian framework, $P_M(\theta)$ summarises knowledge (or lack thereof) about

θ *before* the data are observed and is referred to as the *prior*. $P_M(\theta|D)$ is the posterior and represents updated knowledge about θ *after* the data have been observed. Equation (31.1) asserts that the posterior is obtained by weighting the prior with the likelihood function. The use of priors involves subjectivity, but this is introduced via clear and explicit assumptions about parameters. It also allows one to explicitly account for uncertainty in all parameters, rather than assuming point values for, say, mutation rates or generation times (Goldstein *et al.*, 1995; Chikhi *et al.*, 1998).

Computing the likelihood can present a substantial difficulty for these approaches, even for simple demographic models involving one population (Wilson and Balding, 1998; Beaumont, 1999; Storz and Beaumont, 2002; Wilson *et al.*, 2003). For complex models, there is often no solution (but see Kuhner *et al.*, 1995; Chikhi *et al.*, 2001; Nielsen and Wakeley, 2001; Beerli and Felsenstein, 2001; Rannala and Yang, 2003; Hey and Nielsen, 2004). Whereas population data can be rapidly simulated using coalescent methods (see **Chapters 25, 26, 29** or **30**), the probabilities involved are too low for classical Monte Carlo (MC) integration even for reasonable sample sizes and multilocus data sets.

In order to improve the efficiency of MC integration and significantly reduce the computation time, importance sampling (IS) schemes have been introduced (Griffiths and Tavaré, 1994; Stephens and Donnelly, 2000). The idea is to sample coalescent trees from distributions that are as close as possible to the conditional distribution, given the data $P(T|D)$, where T represents the trees. One way of doing that is to construct coalescent trees starting from the data so that trees that cannot possibly produce the data are not explored (see Griffiths and Tavaré, 1994; Stephens and Donnelly, 2000). Under the classical frequentist approach, some optimisation algorithm is used to find the MLE. Alternatively, the likelihood profile is obtained using a grid across the parameter space. On the contrary, most Bayesian methods of the last decade use Markov chain Monte Carlo (MCMC) algorithms to obtain samples from the required posterior distribution. This approach has proven useful to analyse various demographic models, including migration between splitting populations of variable size (Nielsen and Wakeley, 2001; Hey and Nielsen, 2004; an application in Hey, 2005) and admixture (Chikhi *et al.*, 2001; 2002).

The application of full-likelihood methods to real data is often difficult, and limited to specific demographic models. Analyses may run for weeks with no result because the MCMC algorithm fails to reach equilibrium, due to the high dimensionality of the problems addressed. This has favoured a renewed interest in approximate methods (Tavaré *et al.*, 1997; Weiss and von Haeseler, 1998; Excoffier *et al.*, 2005). One line of research consists in simplifying the calculation of the full likelihood into the product of simpler components that are treated as if they were independent, sometimes called a *composite likelihood approach* (Wang, 2003; McVean *et al.*, 2004). The use of product of approximate conditionals (PAC) likelihood by Li and Stephens (2003) follows a similar rationale and was used to simplify inference with recombining sequence data (see also **Chapter 26**).

Another possibility is to simulate a parameter value θ under the prior, but accept it as an approximate simulation from the posterior $P_M(\theta|D)$ if a dataset D^* simulated under the model given this value of θ is sufficiently close to the real data. How close D^* must be from the real data and how to measure 'closeness' are the focus of ongoing research. Different authors suggest either the use of summary statistics (e.g. Pritchard *et al.*, 1999; Beaumont *et al.*, 2002; Marjoram *et al.*, 2003), or the data directly (Marjoram *et al.*, 2003) to obtain a distance measure $d(D^*, D)$ between the observed and simulated data. Not all methods using this approach are Bayesian (e.g. Weiss and von Haeseler, 1998), but most are, and these have been dubbed Approximate Bayesian Computation (ABC) methods (Beaumont *et al.*, 2002). Their main advantages are that they are extremely flexible, suitable for any demographic model under which data can be simulated, and, contrary to MCMC-based methods, computations do not need to reach an equilibrium. Non-genetic information, such as that deriving from archaeology, can also be incorporated into the priors. Figure 31.2 shows a simple example where posterior distribution obtained using a full-likelihood method is compared to posteriors obtained using an ABC approach with different summary statistics or combinations thereof (see also **Chapters 26** and **30**).

Summary statistics could also be used in simulation-based approaches not requiring likelihoods or posterior distributions to be estimated. Akey *et al.* (2004) analysed the properties of four summary statistics, namely Tajima's D, Fu and Li's D^* and F^*, and Fay and Wu's H, both under the standard neutral model and four alternative demographic scenarios (an exponential expansion, a bottleneck, a two-island model, and an ancient population split). They then compared average summary statistics estimated from the data and from the simulations, choosing parameter values that minimised the difference. By analysing variation at 132 genes sequenced in African-American and European-American samples in this way, the authors could (1) reject implausible evolutionary hypotheses, (2) estimate parameter values under the likely hypotheses, and (3) detect loci potentially under selection (outliers). Eight loci appeared as outliers under all the simulated scenarios and were thus termed *demographically robust selection genes*.

Detection of loci under selection has again come to be a major focus of research (see **Chapters 22**, **27** and **30**, Beaumont and Balding, 2004; Beaumont, 2005). Many methods analysing genomic data derive from the suggestion of Cavalli-Sforza (1966) that all demographic parameters being equal, very high or very low values of the standardised allele-frequency variance, F_{ST}, can only be due, respectively, to diversifying or stabilising selection. The main problem with this approach is that the null distribution of F_{ST} is unknown (Nei *et al.*, 1977). Possible solutions include simulating the data under simple models such as Wright's island model (Beaumont and Nichols, 1996), incorporating different mutation models (Flint *et al.*, 1999), or avoiding assumptions on the demographic history by using many random loci (Goldstein and Chikhi, 2002; Akey *et al.*, 2002). Akey *et al.* (2002) and Storz *et al.* (2004) chose this approach to study variation of, respectively, 26 000 SNPs in Africans, Europeans, and Asians, and microsatellite loci in Europeans and Africans. In both cases, all outlier loci showed reduced variability in non-African populations. Assuming the OOA model, these results might point to adaptive selective sweeps following expansion from Africa. However, the choice of the populations sampled could also generate spurious signals of selection. For instance, complex demographic processes, such as the wave of advance and admixture proposed by Eswaran (2002) and Eswaran *et al.* (2005) could also produce unusual patterns in genomic data, with some neutral loci showing drastically extreme behaviours. Recent work indicates that the fate of mutations during spatial expansions, whether to increase in frequency or disappear, is not easily predicted (Edmonds *et al.*, 2004; Klopfstein *et al.*, 2006).

31.2.4　Reconstructing Past Human Migration and Demography

In the last 25 years, genetic data have been increasingly used to investigate aspects of the human past that were previously studied only by paleontologists and archaeologists.

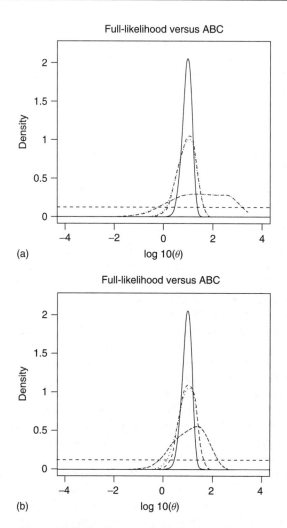

(a)

(b)

Figure 31.2 A comparison of full-likelihood and ABC approaches. The posterior densities for $\log(\theta)$ are compared, where $\theta = 2Nu$ and u is the mutation rate. The horizontal dashed lines represent the flat prior distributions for $\log(\theta)$. The data analysed were simulated using a value of $\theta = 10$. The solid lines in each panel represent the posterior obtained using Beaumont's (1999) full-likelihood method as implemented in the msvar software. Three summary statistics were used in the ABC approaches: expected heterozygosity (H_e), variance in allele size (va), and number of alleles (n_A). The three curves in each panel were obtained by using H_e only (dash-dotted line), H_e and va (dotted) or H_e, va and n_A (dashed). The upper and lower panels were obtained by changing the tolerance level. In the upper and lower panel, the summary statistics were allowed to differ from the real data by 50 and 30 % respectively. Altogether 10 000 coalescent simulations were performed using an R code originally written by M. Beaumont.

The first large-scale analyses of human DNA diversity were descriptions of mitochondrial DNA and showed that, contrary to what was expected at that time, geographical structuring is weak in humans. The estimated evolutionary trees are rather shallow, with long terminal branches and an excess of low-frequency polymorphisms (Cann *et al.*,

1987; Vigilant *et al.*, 1991; Ingman *et al.*, 2000), all features expected in populations that expanded rapidly or were subjected to diversifying selection. Because this signal was also detected in several autosomal regions (Goldstein *et al.*, 1995; Marth *et al.*, 2003; Voight *et al.*, 2005), it was interpreted as resulting from a demographic process, thus supporting a comparatively recent origin of our species from a small group of founders. Various independent lines of evidence suggest that these founders lived in Africa (Reich and Goldstein, 1998; Takahata *et al.*, 2001). The deepest branches of the mitochondrial trees consistently separated African sequences from heterogeneous clusters containing both African and non-African sequences, a pattern also observed in successive Y-chromosome studies (Underhill *et al.*, 2000; Ke *et al.*, 2001). Thus, both mitochondrial and Y-chromosome gene genealogies contained a significant signal of a recent (<200 KY ago) population bottleneck followed by a demographic (e.g. Slatkin and Hudson, 1991; Rogers and Harpending, 1992) and spatial (Ray *et al.*, 2004) expansion.

These results are not easy to reconcile with the predictions of the MR model. In addition, the OOA model seems to better account not only for highest genetic variation in Africa but also for two other findings, namely the limited human genetic diversity in general and the limited diversity among continental populations, both greater in all species of great apes despite their smaller geographical range and census sizes (Gagneux *et al.*, 1999; Kitano *et al.*, 2003; Fischer *et al.*, 2006, but see Yu *et al.*, 2003). Therefore, in the last decade of the twentieth century, a relatively clear picture of human demographic history seemed to be emerging. In synthesis, human genomic diversity appears to result from the recent expansion of a single African population. This population replaced the pre-existing human forms encountered in its dispersal across the other continents, in the course of a process of which the spatial details and the exact timing needed to be worked out.

In the last 5–10 years, however, the resequencing of autosomal regions provided evidence that it is necessary to envisage more complex sets of evolutionary pressures, not all of them fully understood yet, to account for the current patterns and levels of DNA diversity. Indeed, studies of autosomal and X-chromosome variation showed no general excess of rare polymorphisms (Hey, 1997; Harding *et al.*, 1997; Harris and Hey, 1999; Hammer *et al.*, 2004). That points to diversifying selection, rather than to a sudden demographic growth, as the cause of the excess polymorphism observed in non-recombining DNA regions (Przeworski *et al.*, 2000). In addition, Tajima's D shows different values in Africans and non-Africans (reviewed in Garrigan and Hammer (2006)), which may be explained by different selective pressures in the different continents, and/or different demographic histories in African versus Eurasian populations.

At present, the conclusion that the human gene pool is entirely derived from a single and small ancestral population that expanded from East Africa in the last 200 KY is being questioned anew. Opinions also differ on whether data support a single expansion, or rather a more complex set of demographic changes, during which distinct archaic African human forms came in contact and mixed, before or after migrating to other continents (Garrigan and Hammer, 2006). The latter view is supported by the observation of deep haplotype divergence and long-range linkage disequilibrium in the X chromosome.

On the other hand, explicit modelling of evolutionary processes in the geographical space, incorporating population structure, gives results in good agreement with the hypothesis that it was a single group of African hunter–gatherers who expanded dramatically in numbers. The most recent of such studies concluded that humans started expanding very recently (<60 KY ago) from a population whose effective size N_e was close to 1000 individuals (Liu *et al.*, 2006). The likely origin of the expansion was located in East Africa (Liu *et al.*, 2006), i.e. the region where the oldest modern *Homo* fossils have been found (Trinkaus, 2005). An analysis of the correlation between genetic and geographic distances concluded that current human diversity reflects a series of founder effects starting from central or western Africa, but paucity of samples from East Africa may have influenced this result (Ramachandran *et al.*, 2005). Ray *et al.* (2005) also explicitly modelled the spatial population structure and found a likely expansion centre in north Africa, but argued that this result is probably an artefact due to ascertainment bias; once the bias is removed, the most likely origin of modern humans returns to be East Africa.

A certain degree of inconsistency across studies is not surprising. The contrasting conclusions drawn from analyses of variation at, respectively, autosomal loci and uniparentally transmitted markers, reflect, at least in part, the different effective population sizes for these classes of markers, which in turn have an impact on the temporal change of various statistical indexes, including Tajima's D (Fay and Wu, 1999). And, more generally, current genetic diversity is the result of a history spanning hundreds of thousand years, of which we can infer from genetic data only some trends through time and place, with no guarantee that these trends affected in the same way all geographical regions and time periods. However, at present, the patterns observed in studies of different markers and different population samples are not easily combined in a coherent and comprehensive picture (Harris and Hey, 1999; Wall and Przeworski, 2000; Chikhi and Beaumont, 2005). All factors considered, most authors currently maintain that the OOA model is in clearer agreement with genetic data, but more elaborate versions of the model seem necessary to account for the way different genome regions seem to vary (Excoffier 2002; Goldstein and Chikhi, 2002). Efforts are currently directed at better understanding the effects of ancient population structuring (Sherry *et al.*, 1998; Goldstein and Chikhi, 2002; Excoffier, 2002; Harding and McVean, 2004), which may have generated genetic patterns of ambiguous interpretation. Indeed, in the absence of detailed knowledge of ancient gene flow, isolation, and extinctions, models must consider a broad range of possibilities, and then the data may not contain sufficient information for us to discriminate between alternative hypotheses.

Two examples, among the many possible, illustrate this problem. High coalescence times were estimated for some nuclear genes (Harding *et al.*, 1997; Templeton, 2002; Garrigan *et al.*, 2005; Garrigan and Hammer, 2006), and interpreted as resulting from admixture between modern and archaic humans, and hence in favour of the MR model. But with an effective size (N_e) of around 10 000 and a generation time of 20 years, the expected time to the most recent common ancestor (T_{MRCA}) is close to 800 KY (Goldstein and Chikhi, 2002), i.e. close to the values estimated by Harding *et al.* (1997). Even higher coalescence times such as those found by Garrigan *et al.* (2005) can be accounted for by assuming a larger N_e, as has been found by some authors (e.g. Hammer *et al.*, 2004) and a larger generation length. However, subdivision of the ancestral human population in genetically differentiated subpopulations or, in other words, ancient population structure, seems to more simply explain the available data. On the other hand, population structure and extinctions can lead to low N_e estimates, even in populations whose census size was actually large. Thus, the low N_e value, around 10 000, often estimated and used for humans, could have been caused by extinctions of early communities that were spread across Africa and Eurasia, rather than by a really small founding human population (Eller, 2002).

The second example comes from the study of the geographic distribution of derived haplotypes (the 'D' haplogroups) at microcephalin and abnormal spindle-like microcephaly (ASPM), two loci involved in the determination of brain size, and hence potential determinants of some cognitive abilities. The D haplogroups are recent (their estimated ages are 37 and 5.8 KY, with large standard errors) and widespread outside Africa. These two findings led to the conclusion that the D haplogroups carried a selective advantage, possibly related with the emergence of modern human cognition, and hence increased in frequency through a selective sweep (Evans *et al.*, 2005; Mekel-Bobrov *et al.*, 2005). This conclusion was supported by neutrality tests and by simulations, in which neither a population expansion from Africa nor presence of ancient population subdivision in Africa could generate similar patterns in short evolutionary times. However, simulations considering *at the same time* the presence of ancestral differences among African subpopulations and expansion from one subpopulation, did indeed reproduce under neutral conditions the pattern observed (Currat *et al.*, 2006). Studies currently in progress suggest that non-pathologic variation at the microcephalin and ASPM loci is not associated with variation in IQ tests (Mekel-Bobrov *et al.*, 2007); to establish whether or not a selective sweep played a role in the diffusion of the D haplogroups, more robust demographic models are still needed.

Data from other fields, such as archaeology and linguistics, provide potentially useful information to refine demographic models. However, language change may or may not be associated with population changes, and, similarly, changes in the material culture documented in the archaeological record are not necessarily caused by demographic changes (Collard and Shennan, 2000). Therefore, predicting evolutionary change from linguistic and archaeological information remains a complicated task. Multidisciplinary, comparative studies have enormous potential, but a shared methodology that everybody might be satisfied with has not yet emerged (cf Cavalli-Sforza *et al.*, 1988; Bateman *et al.*, 1990).

A substantial leap forward in our ability to reconstruct at least aspects of our evolutionary past is represented by the technical possibility to directly study fragments of DNA coming from fossil bones. The main drawback of the polymerase chain reaction (PCR)-based methods for the study of ancient DNA is the risk of amplifying contaminating DNA (Serre *et al.*, 2004). In the study of non-human species, it is simple to tell contaminating, generally human, from endogenous DNA. Much greater problems of authenticity arise in the study of human fossil DNA, which cannot differ much from those of potential contaminators (Pääbo *et al.*, 2004).

The risk to mistake modern human sequences for the sequences endogenous to the ancient sample has not affected the study of the mitochondrial relationships between anatomically modern and archaic Europeans, the Neandertals. At present, seven sequences of the Neandertal mtDNA covering >300 bp have been published, spanning a time range between 50 and 29 KY ago. All these Neandertal sequences differ from those of modern individuals by at least 16 fixed nucleotide substitutions (Caramelli *et al.*, 2006), and hence cannot possibly result from contamination by modern people who manipulated the specimens. In addition, the Neandertal sequences bear no special resemblance to those of the Europeans, who, under the MR but not the OOA model, should be their direct descendants (Krings *et al.*, 2000).

Comparison of the first Neandertal sequence with modern European sequences under reasonable assumptions on population sizes, and hence on the impact of genetic drift, led Nordborg (1998) to conclude that a direct genealogical link between them is unlikely, but still cannot be ruled out with statistical confidence. However, later work on the increasing number of sequences available shows that the Neandertal contribution to the modern European mtDNA gene pool is very limited (Serre *et al.*, 2004) and probably does not exceed 0.1 % (Currat and Excoffier, 2004). Conversely, Plagnol and Wall (2006), analysing modern patterns of linkage disequilibrium, concluded that even a value greater than 5 % is compatible with the data, and suggested that admixture took place outside Europe, namely in west Africa, a place, however, for which no fossils are currently available. Admixture between archaic and modern Europeans has also been proposed to account for the results of a study of the microcephalin locus (Evans *et al.*, 2006). Evidence against a significant Neandertal contribution to the mtDNA of modern Europeans comes from the analysis of two sequences from 24-KY old anatomically modern humans of the Cro-Magnoid type. Despite the limited chronological distance from Neandertals, these sequences fall in the cluster of modern European sequences, and are clearly distinct from the cluster formed by the Neandertal sequences, thus suggesting that Neandertals are unlikely to be the genetic ancestors of anatomically modern Palaeolithic Europeans (Caramelli *et al.*, 2003).

Because DNA is generally degraded in ancient samples, so far applications have been essentially limited to the study of mtDNA, which is present in multiple copies in each cell, and hence has a higher probability to be retrieved in reasonably good, amplifiable, conditions. Hopefully, technological advances may allow this to change in the near future (Binladen *et al.*, 2006; Green *et al.*, 2006). However, at present, there is no way to look directly at past nuclear DNA variation in humans.

The main feature of modern nuclear DNA variation at the global level are the continent-wide gradients of allele frequencies, spanning much of Eurasia and extending into the Americas and Oceania (Cavalli-Sforza *et al.*, 1994; Barbujani *et al.*, 1994; Chikhi *et al.*, 1998; Serre and Pääbo, 2004; Mulligan *et al.*, 2004; Ray *et al.*, 2005; Ramachandran *et al.*, 2005; Liu *et al.*, 2006; Amos and Manica, 2006), with a more complicated pattern in Africa, probably reflecting complex gene-flow patterns within the continent, and selection (Reed and Tishkoff, 2006). Also, in good agreement with processes of gene flow originating in Africa are measures of genetic diversity estimated from microsatellite data, with a clear trend from minimal values in the Americas to maximum values in Africa, and populations known to be of particularly large or small size behaving as outliers (Conrad *et al.*, 2006).

Even in the absence of data convincing everybody that all modern human genealogies can be traced back to recent African ancestors, it is clear that Africa played a special role in human evolution, as shown by patterns of genetic diversity (Watkins *et al.*, 2001) and of linkage disequilibrium (Tishkoff *et al.*, 1996). Two main migration routes have been proposed on the basis of the available genetic data. An exit and dispersion through the Near East into Eurasia is commonly accepted and in good agreement with fossil evidence (Lahr and Foley, 1998) and nuclear diversity (Ramachandran *et al.*, 2005; Liu *et al.*, 2006) but another exit through the so-called southern route is currently receiving increasing attention (Luis *et al.*, 2004; Mellars, 2006). Indeed, mitochondrial DNA diversity in Asian genetic isolates has been interpreted as suggesting an early (65 KY ago) expansion through the Red Sea and India, directly into southeast Asia and Oceania (Macaulay *et al.*, 2005). Levels of linkage disequilibrium are maximal in the Americas and Oceania, lower in Eurasia and minimal in southern and central Africa, consistent with repeated founder effects in the course of an expansion from Africa (Tishkoff and Verrelli, 2003; Tishkoff *et al.*, 1998; 2000; Sawyer *et al.*, 2005; Montpetit *et al.*, 2006).

31.3   HUMAN GENETIC DIVERSITY: GEOGRAPHICAL STRUCTURE

Humans have been described for centuries as subdivided in discrete and easily identifiable genetic subgroups or races. During much of the twentieth century, most geneticists believed that these races differed profoundly in their physical and mental characteristics and that these differences have a biological basis and are hereditary (Provine, 1986). At the end of the century, anthropologists and population geneticists had mostly abandoned this idea, which was perceived to conflict with the available genetic evidence. However, genetic epidemiologists brought it back to the centre of the attention with the beginning of the twenty-first century.

Indeed, our probability to develop a disease reflects the interplay between two main classes of risk factors, genetic and environmental. Both are at present impossible to exactly measure, either because they are unknown (e.g. all the loci involved in the onset of a complex disease, and the way they interact) or because they are difficult to quantify (e.g. the effects on health of our diet, of the exposure to chemicals, smoke, alcohol, and of our lifestyle in general). As a consequence, epidemiologists are forced to use approximations in their effort to prevent disease occurrence, and it has been proposed that a way to approximate genetic disease risks is to classify people in traditional racial categories (Risch *et al.*, 2002; Burchard *et al.*, 2003; Mountain and Risch, 2004; Tang *et al.*, 2005; Risch, 2006). The task may not be easy because in the course of history no consensus has emerged on the number and definition of human races, with catalogues including from two to hundreds of proposed races.

31.3.1   Catalogues of Humankind

Only in the last few centuries could questions about human diversity be addressed scientifically, but in fact these questions are as old as humankind. Differences among humans were evident to the earliest naturalists; descriptions and classifications of human types can be found in many ancient texts. A central question was whether we belong to one or to several species. Such polygenistic theories flourished up to the nineteenth century (Cohen, 1991), but the development of evolutionary studies, and the demonstration that there is no reduction of fertility in crosses between members of very distant human populations (Chung *et al.*, 1967) led to their dismissal.

Science has long left creationism and polygenism behind, but the idea that a scientific study of humans necessarily starts from their racial classification was long unchallenged (Cohen, 1991) and still remains attractive to some. People differ in their aspect; all of us recognise variation in facial traits, hair textures, height, body structure, and skin colour. The typological approach consists in identifying some basic human types, defined on the basis of such traits, and then attributing individuals to one of those types, or races. Accordingly, starting with Linnaeus and for at least two centuries, analyses of human biological diversity were essentially race catalogues, differing from each other for the number and definition of the various items (Bernasconi, 2001). In turn, racial types were thought to be associated with other characteristics such as temperament, behaviour, and intellectual abilities (Cole, 1965). The main difficulty was represented by the fact that it is fairly simple to list typical anatomical features of a region or a population, but each human group, however defined, includes variable proportions of people who do not resemble much the typical individual. Even in the early years, it was evident that single morphological traits do not allow a stable classification of humans into discrete groups. Therefore, races were defined by a combination of several traits, often both biological and non-biological, the latter including language, house-building and tool-making techniques, mating systems, and other cultural variables (see Cohen, 1991, where reference to the original eighteenth and nineteenth century sources can be found).

The consequences of these difficulties are evident in the lists of races described in Table 31.1. Starting from Linnaeus' six races and going through Buffon's, Blumenbach's, and Cuvier's systems into the twentieth century, the number of races increased. In his *Systema naturae,* Linnaeus first defined the species *Homo sapiens* within the order primates and divided it in six varieties, four corresponding to continents (Australia was

Table 31.1 An incomplete scheme of proposed catalogs of human races. Reference to the original works can be found in Cohen (1991) and Barbujani (2005).

Author	Number of races	Races proposed
Linnaeus (1735)	6	Europaeus, Asiaticus, Afer, Americanus, Ferus, Monstruosus
Buffon (1749)	6	European, Tartar, Laplander, south Asian, Ethiopian, American
Blumenbach (1795)	5	Caucasian, Mongolian, Ethiopian, American, Malay
Cuvier (1828)	3	Caucasoid, Negroid, Mongoloid
Deniker (1900)	29	–
Weinert (1937)	17	–
Von Eickstedt (1937)	38	–
Biasutti (1959)	53	–
Coon (1962)	5	Congoid, Capoid, Caucasoid, Mongoloid, Australoid
Garn (1965)	9	African, European, Asian, Indian, Amerind, Melanesian, Polynesian, Micronesian, Australian
US Office of Management and Budget (1997)	5	African-American, White, American Indian or Alaska Native, Asian, Native Hawaiian or Pacific Islander
Metropolitan Police Service, London, UK (2005) (http://en.wikipedia.org/wiki/Race)	16	White-British, White-Irish, any other white background, Asian-Indian, Asian-Pakistani, Asian-Bangladeshi, any other Asian background, Black Caribbean, Black African, any other black background, Chinese, White and Black Caribbean, White and Black African, White and Asian, any other mixed background, any other race
Risch *et al.* (2002), Figure 1	5	African, Caucasian, Pacific islanders, east Asian, Native American
Risch *et al.* (2002), Table 3	5	African Americans, Caucasians, Hispanic Americans, east Asians, Native Americans

missing) and two, *H.s. ferus* and *H.s. monstruosus* (savage and monstrous), designating respectively wild human forms whose existence was not confirmed by successive work, and carriers of congenital malformations. These two races were eliminated from the list and Australia was added by Buffon, who considered laplanders as an additional, separate race. At the end of the eighteenth century, it was the German anatomist Blumenbach who, while refusing a relationship between humans and the other primates, proposed that humankind be composed of five races, broadly corresponding to the inhabitants of the five continents, four of them regarded as more or less serious degenerations from the European race, which he first termed *Caucasian*.

In the nineteenth and early twentieth centuries, as anthropologists came in contact with more and more populations, fitting all of them into pre-existing races proved difficult. The catalogues became broader, crossing in some instances the 100-mark (Molnar, 1975), and the borders between races therein ambiguous. In this way, the need emerged to explicitly define what a race is. Although published definitions differ, many converge around the idea that races are subspecies, namely collections of individuals associated with a geographical region, and separated by biological boundaries from other groups that differ from them in some measurable features (Mayr, 1947; Coon *et al.*, 1950). Classical population-genetics theory and empirical data show that genetic variation is shaped by a combination of factors affecting the genome (such as mutation, recombination, and selection) and acting at the population level (such as subdivision, admixture, migration, and population-size changes). Under reproductive isolation, genetic drift affects independently each reproductive group. This tends to reduce the group's internal variation, while groups diverge from each other, both phenomena causing the onset of zones of rapid genetic change, i.e. boundaries (Barbujani *et al.*, 1989). Conversely, when groups or populations exchange migrants, the effects of drift are opposed by those of gene flow, and genetic variation between populations and groups is continuous (Wright, 1943). Therefore, genetically differentiated groups that one can legitimately regard as races are surrounded by biological boundaries, which are the product of genetic drift affecting populations connected by little (or no) migratory exchanges.

By the second half of the twentieth century, a growing number of investigators had become unhappy with the idea that the discontinuous entities thus far defined give a faithful representation of human variation. This assumption appeared to conflict with the continuous change observed for most morphological (and, later, genetic) traits. An additional problem was the fact that different traits yield discordant human taxonomies (King and Wilson, 1975; Brown and Armelagos, 2001). Individuals can be clustered on the basis of, say, skin colour, but the clusters are not the same if a different trait, say skull shape, β-globin, or lactose intolerance, is considered (Relethford and Harding, 2001). With the introduction of the concept of cline (Huxley, 1938), namely a gradient of morphological measures or allele frequencies in the geographical space, continuous models of human population structure became conceivable, which ultimately led to the proposal that the concept of race is misleading for describing human biological diversity (Montagu, 1941; Livingstone, 1962). Dobzhansky (1967) maintained that human races could nevertheless be defined, if not by fixed allelic differences, at least as open genetic systems, each differing to some extent from its neighbours for some allele frequencies. However, it has been objected that according to this definition any human population would be a distinct race, which does not correspond to the general use of the concept in evolutionary biology. This debate is still open, with different authors siding with either Dobzhansky or Livingstone.

In what can be regarded as the last attempt to base a racial classification on anatomy, Coon (1963) proposed that the apparent complexity of the human population structure disappears if one disregards the effects of recent migration. In that way, humans can be subdivided into five major races, two in Africa, and one in Europe, Australia, and Asia, the latter including native American populations. However, by the early 1960s, the typological approach had shown its drawbacks, and the number and definition of human races had become extremely uncertain. In the meantime, genetic data had begun to accumulate, as well as quantitative methods for their analysis (Edwards and Cavalli-Sforza, 1965; Cavalli-Sforza, 1966; Sokal *et al.*, 1989). Most studies of the last decades of the twentieth century focused, then, on the levels and patterns of genetic variation in the geographical space, summarised in the atlas of Cavalli-Sforza *et al.* (1994).

31.3.2   Methods for Describing Population Structure

In this section, we shall focus on two main sets of methods for assessing the strength of population structure from multilocus genotypes given a tentative classification based on some non-genetic criteria such as physical morphology or geography. Under the first approach, similar to an analysis of variance, genetic diversity is compared within and among the pre-defined groups. The alternative approach is to define a criterion of similarity among genotypes and hence classify them into groups ignoring the available group labels. Later, the group labels are revealed in order to quantify the accuracy of the assignments. We now discuss these two approaches in more detail.

31.3.2.1   Genetic Diversity within and among Populations

Classical analysis of variance is unsuitable for assessing genetic variance within and between groups because the distribution of allele frequencies is not expected to be normal, but is better approximated by the beta or Dirichlet distributions (Wright, 1931). An ideal statistic for comparing variation within and between groups should have the following properties: (1) take value 0 when there is no polymorphism; (2) increase with the number of alleles; (3) reach a maximum when the frequencies of all alleles are equal, for any number of alleles; and (4) be convex, i.e. it should increase when groups are pooled, unless these averages are identical (Lewontin, 1972). Wright's (1943) F_{st} is an example of a convex statistics. These criteria led Lewontin (1972) to apportion the global species diversity at three hierarchical levels using the Shannon information measure, defined as

$$H = -\Sigma p_i \ln_2 p_i,$$

where p_i is the frequency of the ith allele at a locus, and summation is over all alleles. H is calculated independently for each locus and each population, and its average is calculated respectively within populations (H_{pop}), among populations attributed to one race (H_{race}), and in the entire species ($H_{species}$). The global genetic diversity was then partitioned at the three levels as follows:

Variance within populations: $V_{wp} = H_{pop}/H_{species}$

Variance between populations, within races: $V_{bp} = (H_{race} - H_{pop})/H_{species}$

Variance between races: $V_{br} = (H_{species} - H_{race})/H_{species}$.

Although it is customary to refer to these V indexes as variances, this is not strictly correct. In particular, since V_{bp} and V_{br} are estimated by subtraction, when there is little geographical structure these statistics can take negative, if small, values (see e.g. Jorde *et al.*, 2000).

With the advent of techniques for the direct study of DNA, measures of molecular differentiation between alleles were incorporated into the statistics. Much like in Lewontin's approach, analysis of molecular variance (AMOVA) is a non-parametric method for the analysis of variance that subdivides genetic diversity estimated from molecular data into hierarchical components (Excoffier *et al.*, 1992; see also **Chapter 29**).

AMOVA tests by a permutational procedure the significance of each variance component, whether estimated from allele-frequency differences or considering molecular information also. For this purpose, the null hypothesis is that all samples come from an unstructured population, without differences among the groups (populations or races) defined within it, so that all variation is due to the random sampling of individuals or populations. The three variances (which one can refer to as *pseudovariances*) are recalculated by assigning individuals and populations to random geographic locations, according to three independent resampling schemes; the procedure is repeated many times, so as to obtain empirical null distributions for the three pseudovariances. The observed variances are eventually compared with these distributions, and observed values falling in the upper percentiles are judged significant at the appropriate level.

31.3.2.2 Classification Methods

In discriminant analysis (Ripley, 1996), also known as *supervised classification*, each individual's group allocation is ignored in sequence, and their genotype is assigned to the most likely source population using the pre-defined allocations of all other individuals as the reference data set. The similarity between genotypes is usually measured as the minimum (and most likely) number of mutational events separating them. Both parametric and non-parametric methods of discriminant analysis are available. However, the standard parametric methods, namely linear, logistic, and quadratic discriminant analysis, assume that the variables are at least approximately normal. Rannala and Mountain (1997) proposed an assignment method specifically designed for genetic data, which has also been used for discriminant analysis (Romualdi *et al.*, 2002). Under the assumptions of Hardy–Weinberg and linkage equilibria, this method computes the probability of multilocus genotypes originating from different potential reference/source populations and allocates them to the most likely group of origin. Among non-parametric methods, the simplest approach is to assign each genotype to the group to which the majority of the k nearest genotypes belong; k is chosen arbitrarily in a range of reasonable values. Under more complex approaches, such as kernel methods, the density of the genotype frequencies is initially estimated in each group, and then a new observation (genotype) is assigned to the group for which its estimated density is maximal.

Unsupervised clustering methods ignore any pre-defined group assignments and infer from the data groups of genetically similar individuals. The underlying idea is to use a model-based approach to define the probability of generating the data assuming K hidden partitions. Typically, the basic model assumes that, within the partitions, alleles are independent within and between loci. Once this probability can be calculated or approximated, it is possible to construct an MCMC algorithm that will explore the parameter space in such a way that the parameter values sampled during an MCMC run are visited in proportion to their posterior probability of generating the data. In the general problem of uncovering hidden structure, parameters may include the number of partitions (Dawson and Belkhir, 2001) or the proportion of genes (in a population or in an individual) coming from any of K partitions conditional on K (Pritchard *et al.*, 2000; Falush *et al.*, 2003).

The most popular such approach is the one originally developed by Pritchard *et al.* (2000), later extended by Falush *et al.* (2003) and implemented in the STRUCTURE software (hereafter named the Pritchard and Falush method; see also **Chapter 30**). This method allows data analysis under two ancestry models, with or without admixture. In the simplest case of no admixture, the relevant parameters are (1) the population of origin of each individual and (2) the allele frequencies of each population. When admixture is assumed, an extra parameter is considered, (3) the proportion of each individual's ancestry from each population. The number of partitions K is not estimated and so must be specified, but Pritchard *et al.* (2000) proposed a criterion for choosing from multiple runs of the algorithm that providing the optimal K value. Recent simulation work on complex population structure where patterns of gene flow are not homogeneous has shown that this *ad hoc* method can be misleading (Evanno *et al.*, 2005).

Another approach is the Dawson and Belkhir (2001) method, based on an MCMC algorithm implemented in the PARTITION software, in which the parameter space is defined by (1) K, the number of possible partitions; (2) the distribution of individuals in the K partitions; and (3) the allele frequencies within partitions. During the exploration of the parameter space, the likelihood of the data is estimated for the different K values and for possible assignments of individuals to the K partitions. Thus, when the chain reaches equilibrium, the different values of K have been sampled in proportion to their probability of generating the data, and so it is possible to estimate the posterior probability distribution of K. The Dawson and Belkhir method also estimates the probability that sets of individuals lie in the same partition. Only the maximum number of partitions allowed has to be specified beforehand, whereas in the Pritchard and Falush method a pre-specified number of populations, K, is given and the algorithm uses a Gibbs sampler to obtain the posterior distribution of the parameters, i.e. the values of each parameter conditional on the observed individual's genotypes and on the number of populations K considered.

The Pritchard and Falush approach has some drawbacks, but it is also the most flexible. Besides assigning individuals to their population of origin and identifying possible migrants or admixed individuals, STRUCTURE can also incorporate information about the geographic origin of individuals, which is treated as a prior in the cluster analysis. The algorithm can also allow for allele frequencies to be correlated across populations (Falush *et al.*, 2003). This can be a crucial point when the groups being studied are not genetically distinct.

Other recent methods are suitable to detect and quantify population structure (Corander *et al.*, 2004; 2007; Corander and Marttinen, 2006). Corander and colleagues have developed Bayesian model–based clustering methods implemented in the different versions of the BAPS software. In the most recent version, Corander *et al.* (2007) proposed a method analogous to Dawson and Belkhir's in the sense that it tries to find the partition that best fits the data, treating as random variables the allele frequencies and the number of genetically divergent populations. However, contrary to the alternative approaches, BAPS is based on *a priori* information provided by the geographic location of sampled individuals. Given a maximum value of partitions, the algorithm uses a stochastic optimisation procedure (rather than an MCMC approach) to find the clustering solution with the highest marginal likelihood of K (i.e. the most probable number of differentiated populations conditional on observed data). For more extensive, recent reviews, the interested reader should refer to Beaumont (2004) and Chikhi and Bruford (2005).

31.3.3 Identifying the Main Human Groups

In the first global analysis of human genetic diversity, Lewontin (1972) quantified variation within and among seven groups, namely Caucasians (including western Asians and north

Table 31.2 Estimated percentages of the global human diversity at three hierarchical levels of population subdivision.

Polymorphism, number of loci	References	Within population	Between populations, within race or continent	Between races or continents
Protein, 17	Lewontin (1972)	85.4	8.3	6.3
Protein, 18	Latter (1980)	84.0	5.6	10.4
Protein, 18	Latter (1980)	87.0	5.5	7.5
Protein, 18	Latter (1980)	83.8	6.6	9.6
Protein, 25	Ryman et al. (1983)	86.0	2.8	11.2
mtDNA	Excoffier et al. (1992)	75.4	3.5	21.1
Autosomal DNA, 109	Barbujani et al. (1997)	84.4	4.7	10.8
MtDNA (non-coding)	Seielstad et al. (1998)	81.4	6.1	12.5
Y chromosome, 10	Seielstad et al. (1998)	35.5	11.8	52.7
Autosomal DNA, 90	Jorde et al. (2000)	84.8	1.6	13.6
MtDNA (non-coding)	Jorde et al. (2000)	71.5	6.1	23.4
Y chromosome, 10	Jorde et al. (2000)	83.3	18.5	−1.8
Autosomal DNA, 21	Romualdi et al. (2002)	82.9	8.2	8.9
Y chromosome, 10	Romualdi et al. (2002)	42.6	17.3	40.1
β-Globin	Romualdi et al. (2002)	79.4	2.8	17.8
Autosomal DNA, 377	Rosenberg et al. (2002)	93.2	2.5	4.3
Autosomal DNA, 377	Excoffier and Hamilton (2003)	87.6	3.1	9.2
Y chromosome, 13	Wilder et al. (2004)	64.3	17.0	18.7
MtDNA (coding)	Wilder et al. (2004)	59.9	21.1	18.9
X chromosome, 17	Ramachandran et al. (2004)	90.4	4.6	4.9
Autosomal indels, 40	Bastos-Rodrigues et al. (2006)	85.7	2.3	12.1
Median, all loci[a]		82.8	6.1	11.1
Median, autosomal[a]		85.7	4.6	9.7

[a] These values were obtained by treating all studies equally. The medians at the three levels of population subdivision do not sum up to 1, and hence they were normalised by dividing them by 100.6 for all loci, 99.0 for autosomal loci.

Africans), Black Africans, Mongoloids, south Asian aborigines (dark-skinned populations of India and the Asian southeast), Amerinds, Oceanians, and Australians (Table 31.2). The estimated variances differed across the 17 loci, as well as the sample sizes but, on average, 85.4 % of the total was attributed to differences between members of the same population, 8.3 % to differences between populations of the same group, and 6.3 % to differences between groups. Lewontin concluded that racial classification has no genetic or taxonomic significance.

Nei and Roychoudhury (1972) compared three groups, white, black, and Japanese, and found that allele frequencies in the three groups are remarkably similar, despite the conspicuous phenotypic differences. They suggested that the genes controlling pigmentation and facial structures evolved under stronger natural selection than 'average genes' (Nei and Roychoudhury, 1972), a notion that gained broad acceptance (see, e.g. Rees, 2003). On the other hand, Lewontin's bold conclusions also prompted a number of criticisms for more than three decades (see Edwards, 2003) discussed below. However, reanalyses of Lewontin's data by three alternative methods, based on the proportion of alleles that two random multilocus genotypes have in common, gave only slightly different results, with variances within populations estimated around 84.0 % of the total (Latter, 1980). Genetic variances largely overlapping with Lewontin's and Latter's estimates were also inferred from enzyme and serum protein allele frequencies, and from blood groups, in three populations from Africa, Europe, and Asia (Ryman et al., 1983).

The first apportionment of human genetic variances at the DNA level was an illustration of the AMOVA algorithm. Excoffier et al. (1992) found that the differences between continents in mitochondrial DNA RFLP diversity were larger than estimated in protein studies, and populations contained 75 % of the global mtDNA diversity. Conversely, in large-scale analyses of autosomal DNA, genetic variances were very close to those inferred from protein polymorphisms, with differences among individuals of the same population accounting for 84.5 % of the overall variance for both microsatellites and RFLPs (Barbujani et al., 1997). Genetic differences among continental populations were small in absolute terms, between 8 and 11.7 % of the total, but yet significantly greater than zero at most loci.

Similar figures have been inferred from studies of other autosomal polymorphisms (Rosenberg et al., 2002), including Alu insertions (Romualdi et al., 2002). Actually, the variances among continents estimated in the former study, by far the largest so far for the number of screened loci, are less than 5 %, although a reanalysis of the same data based on a more appropriate mutational model suggests a twice-as-large value (Excoffier and Hamilton, 2003). Despite these inconsistencies, the impression one receives from studies of autosomal polymorphisms (Table 31.2) is that a large share of the global genetic diversity, very close to Lewontin's 85 %, is present in each human population, with small differences between non-coding DNA regions (e.g. Excoffier and Hamilton, 2003) and coding regions subjected to selection, such as the β-globin locus (Romualdi et al., 2002).

Greater differences between populations and continents are usually inferred from markers transmitted by one parent only. In studies of biallelic Y-chromosome markers, between 40 % (Romualdi et al., 2002) and more than half of the total variance (Seielstad et al., 1998) was found to differentiate continents, but that component was zero in a study of short tandem repeats (STRs) (Jorde et al., 2000). The absence of detectable differences among continents in the study of STRs may depend, at least in part, on the high mutation rate for STRs and on constraints on the number of repeats, both phenomena leading, in the long run, populations of all continents to converge towards similar allelic distributions (Jorde *et al.*, 2000).

In general, given the smaller population sizes for uniparentally transmitted loci (in principle, one-fourth of the autosomal population size), higher diversity between populations and groups thereof should be expected for these markers, under a simple model of population differentiation by genetic drift. Therefore, the high variances between populations and groups inferred from the Y chromosome are easier to reconcile with theory than the lower variances inferred from mtDNA. This finding may reflect either an increased average impact of gene flow on females, or of drift on males, or both. Increased female migration is a consequence of patrilocality, i.e. a higher tendency to move to the spouse's place of origin for women than for men (Seielstad *et al.*, 1998); increased drift in males may mean that their effective population sizes are reduced by their higher variance in reproductive success, associated with the practice of polygyny (Dupanloup *et al.*, 2003).

All the above factors may have, or have had, an impact on the patterns of human diversity, but a part of the differences in the apportionment of genetic variances inferred from different markers may be due to biases in the choice of the markers. After identifying areas of high polymorphism on the Y chromosome, Wilder *et al.* (2004) sequenced more than 6 Kb in individuals from four continents, compared the observed sequence variation with variation in mtDNA sequences, and actually observed a slight excess of between-continent diversity for mtDNA than for the Y chromosome. Therefore, when sequences of random Y-chromosome regions are compared with mtDNA sequences, (1) similar distributions of variances emerge, and (2) the differences between populations (of the same or of different continents) is larger than inferred from autosomal DNA, which makes sense in the light of the different effective population sizes, fourfold larger for autosomal than for mitochondrial and Y-chromosome polymorphisms.

In all these studies, the global species diversity is probably overestimated because of (1) the unavoidable selection of populations that are far removed in the geographical space, to the exclusion of populations that may be of ambiguous classification, such as those at the boundaries between continents and (2) the over-representation of populations of anthropological interest, often reproductively isolated and hence presumably highly differentiated, rather than admixed urban populations. Because the internal diversity also tends to be low in genetic isolates, one can also expect the fraction of within-population variance to be underestimated. A random sample of humankind, say based on individuals collected at the nodes of a regular grid superimposed on the world map and taking population densities into account, would contain a greater proportion of admixed individuals and recent immigrants, presumably resulting in even higher diversity within populations.

On the other hand, genetic differences between putative races or continents, albeit small, are significantly greater than zero. This prompts the question whether these differences are large enough to allow one to distinguish discrete racial group, as suggested, among others, by Edwards (2003).

In evolutionary trees inferred from microsatellites (Bowcock *et al.*, 1994), or combining information from microsatellites, insertion/deletion and restriction polymorphisms (Jorde and Wooding, 2004), people of similar geographical origin tend to form clusters, and plausible evolutionary relationships among populations are apparent. As is commonly observed in humans (starting from Cann *et al.*, 1987), these trees have little deep structure and long terminal branches.

The question whether the limited genetic variances among populations and groups thereof are large enough for individuals to be accurately classified in continental groups was addressed considering Alu insertion/deletion polymorphisms and Y-chromosome SNPs. Using several discriminant analysis methods, Romualdi et al. (2002) could correctly assign to the right continent up to 73 % of their multilocus genotypes, with Y-chromosome data allowing a more accurate assignment. Accuracy decreased drastically as the number of partitions increased, i.e. when genotypes were assigned to subcontinental regions, and the misclassification rate decreased only slightly by adding new markers, once some 20 polymorphisms were considered (Romualdi et al., 2002).

The choice of the markers and of the populations deeply affects the results of the analysis, suggesting that discordant variation is not only typical of morphological traits but also of genetic polymorphisms. Analysis of two sets of Alu polymorphisms by STRUCTURE showed that, depending on the loci considered, either three clusters (two worldwide distributed and one Eurasian) or four clusters (two in Eurasia and two corresponding respectively to Africa with Oceania and Asia with the Americas) were identified. These clusters did not overlap (Romualdi et al., 2002), neither did they correspond to the clusters obtained by STRUCTURE in studies of X-chromosome STR markers, loci of pharmacogenomic relevance (Wilson et al., 2001), and autosomal STR markers (Rosenberg et al., 2002). Rosenberg et al. (2002) inferred the existence of six clusters from the analysis of a 377 STR markers from the Centre pour l'Etude des Polymorphismes Humains (CEPH) Diversity Panel, five of them corresponding to the main continents (with central Asia clustering with Europe rather than with east Asia), plus the Kalash population of Pakistan, presumably a genetic isolate. In a successive analysis of a dataset of almost 1000 STR markers by the same authors, the Kalash no longer seemed to be a genetic outlier, and appeared to be part of the European-Western-Central Asian cluster, but things became more complicated in the Americas, where two new clusters emerged (Rosenberg et al., 2005).

The 377-marker CEPH dataset has been reanalysed by several authors. Using a Bayesian MCMC approach, Corander et al. (2003) found that more than six groups are needed to represent global human genetic diversity, with evidence for subdivision in South America). Barbujani and Belle (2006) used a method for recognising genetic boundaries in maps of genomic variation (Manni et al., 2004), finding evidence for several distinct clusters in the Americas, and for three clusters separated by zones of rapid genomic change in Africa, a result consistent with extensive diversity known to exist in that continent (Kaessmann et al., 1999; Yu et al., 2003; Watkins et al., 2003). Evidence for substructuring in Africa had also been detected by STRUCTURE in one analysis where the inferred clusters were four (Bamshad et al., 2003). On the other hand, Serre and Pääbo (2004) investigated how study design can influence the inferred clustering. By resampling the CEPH dataset through a scheme devised to reduce the effects of the necessary discontinuous geographical collection of samples, they came to the conclusion that only genetic gradients, and not boundaries, are apparent when individuals are sampled homogeneously from all continents, and concluded that there is no evidence for major genetic discontinuities in the human species.

Is there a way to reconcile all these results? And which is the real structure of the human population? Some inconsistencies may be due to differences in the distribution of the studied populations. However, several of the above analyses were based on the same 377-STR CEPH diversity dataset, and yet their results are discordant as for the number and geographical span of the genetic clusters identified. Studies of different dataset yield an even more complicated picture (Wilson *et al.*, 2001; Romualdi *et al.*, 2002; Bamshad *et al.*, 2003; Tang *et al.*, 2005; Barbujani, 2005), and it is legitimate to say that the main common element one can recognise is that each study is inconsistent with all the others. One plausible conclusion is then that this incongruence represents a basic feature of human diversity, and that racial categories are imposed on data rather than inferred from them (Cooper, 2005). Different genetic polymorphisms show discordant geographic patterns, i.e. are differently distributed over the planet, and their distributions are generally weakly correlated or uncorrelated. Genetic clusters can be inferred from the data, but the fact that two populations fall in the same cluster (or in different clusters) when described at loci A, B, C does not imply that they will fall in the same cluster (or in different clusters) based on loci X, Y, Z.

31.3.4   Continuous versus Discontinuous Models of Human Variation

Continuous genetic variation in the geographic space, such as that commonly observed in humans, may result from three classes of evolutionary phenomena, one of which, selection along a gradient, is expected to affect single loci, but not the genome as a whole (Cavalli-Sforza, 1966; Kayser *et al.*, 2003). When clines are observed over a number of loci, they likely reflect population dispersal, or isolation by distance, or both. Isolation by distance can be loosely defined as the process whereby the effects of drift are independent in each population, but gene flow rates are higher between close than between distant populations, and so genetic similarity tends to decline with the spatial distances (Wright, 1943). Population dispersal differs from isolation by distance in that it entails directional migration, and can generate clinal genetic change either by demic diffusion or by founder effects. Demic diffusion (Menozzi *et al.*, 1978) means that the geographical expansion is accompanied by demographic growth. If the expanding population mixes and forms hybrids with previously settled populations, the fraction of members of the expanding population that will contribute to the hybrids decreases with the distance from the place of origin, and so does their contribution to the gene pool of the hybrid populations. Repeated founder effects may also result in clines because, as small numbers of individuals disperse, alleles are lost from their gene pool, and occasionally reintroduced by local gene flow.

Application of a formal model of isolation by distance to the analysis of worldwide patterns of protein and DNA polymorphisms, and of craniometric measures, showed an excellent fit of the model for both genetic and anatomical data, suggesting that patterns of human diversity can largely be accounted for by the simple interaction of drift and geographically structured gene flow (Relethford, 2004). These results suggest that, as a rule, migrational networks left a deep mark in the observed patterns of human biodiversity. The exceptions, namely zones of rapid genetic change or genetic boundaries, point to migrational or reproductive barriers. In the few studies so far attempting to map them at the DNA level, genetic boundaries have been shown to occur mostly between small genetic isolates, sometimes separated by just a few kilometres, such as the Suruì and Karitiana populations of Brazil (Barbujani and Belle, 2006). This suggests that, at the small geographical scale, chance and possibly adaptation to local conditions may lead to even sharp population differentiation, thus causing local departures from the general clinal pattern.

31.4 FINAL REMARKS

As we have seen, the interpretation of the data on human genetic diversity is controversial in several areas. However, it is at least well established that the main fraction of the global human genetic diversity is found between individuals of the same population. We also know that the remaining fraction, representing differences among populations and population groups, is small but not zero. Indeed, it is large enough for most of us to form an opinion, if rough, on the likely origin of the people we meet just by looking at them, and for population geneticists to reconstruct aspects of population history. However, different loci show uncorrelated, and sometimes very different, patterns of variation, a phenomenon termed *discordant variation* (Brown and Armelagos, 2001). Therefore, predictions of epidemiological relevance based on the patient's physical aspect or supposed racial affiliation may be severely inaccurate (Wilson *et al.*, 2001; Cooper *et al.*, 2003; Cooper, 2005; Orduñez *et al.*, 2005; Bamshad, 2005). Countries or regions where several subpopulations coexist in reproductive isolation, with little or no genetic exchange, may represent exceptions. Examples include India, where genetic differences are apparent among castes (Cordaux *et al.*, 2004; Watkins *et al.*, 2005); the Yanomama (Neel, 1978) and other Amazonian tribes (Crawford, 1998), in which small population sizes combined with the tendency of tribes to split along family lines led to extreme genetic drift effects; and various urban areas of the United States (Shriver *et al.*, 2004).

DNA variation has probably been studied among US population groups more extensively than in any other area of the world, and US ethnic groups show different allele frequencies at many loci (Shriver *et al.*, 2005; Redd *et al.*, 2006). The problem is to what extent and for what purposes the results of these studies can be generalised to other populations. Characteristic alleles, present at high frequencies in one US population while rare or absent in the others, were initially defined population-specific alleles (PSA), and proposed for studies of admixture and DNA-based personal identification (Shriver *et al.*, 1997; Parra *et al.*, 1998). Later, the set of markers was extended and refined, but also relabelled as AIM (ancestry informative markers), thus suggesting that by using these markers it is possible to make one more step, namely to correlate skin colour with biological ancestry (Shriver *et al.*, 2003). This view was echoed by authors who concluded that human genomic diversity is discontinuously distributed according to groups that correspond well to 'common concepts of race' (Rosenberg *et al.*, 2002; 2005). However, in a study of Brazilian individuals, Parra *et al.* (2003) found no correlation between skin colour and frequency of 10 AIM, selected as those that had shown the maximum differentiation between Africans and Europeans. Thus, genetic markers that are population specific, and perhaps even ancestry informative in certain populations, are neither population specific nor ancestry informative elsewhere. Local genetic differentiation may be sharp for certain loci and useful for certain purposes, but studies of single countries are unlikely to give us a sensible, all-purpose description of human diversity in general, especially if these countries are inhabited by people whose ancestors evolved elsewhere for millennia.

Overall, the global patterns of human diversity are the likely product of a comparatively short evolutionary history. During less than 200 KY, a population of six billion individuals has quickly developed, largely or exclusively, from a presumably small number of recent African founders. In their dispersal across the planet, these ancestors are unlikely to have mixed with the descendants of people who had previously left Africa. The main evolutionary processes affecting them were probably repeated bottlenecks, dispersals, and isolation by distance, all processes accounting for the broad gradients of genetic diversity observed (Cavalli-Sforza *et al.*, 1994). However, migration and genetic drift were not the only forces shaping human genetic diversity. Recent admixture episodes and selection have left a mark in current genetic diversity. Genomic regions of intense, and often environment specific, selection have been identified analysing both low-resolution and high-resolution single nucleotide polymorphism data (Kayser *et al.*, 2003; Akey *et al.*, 2002; 2004; Storz *et al.*, 2004; Voight *et al.*, 2006), and the different selection regimes, along with chance, may explain why patterns of genetic variation are discordant across loci.

Discordant variation has practical implications. Calculating how many polymorphic markers would be necessary to confidently assign individuals to racial groups, Risch *et al.* (2002) estimated that, with average allele-frequency differences between racial groups = 0.6, 17 markers would be enough, with an error rate as low as 10^{-5}. More markers would be needed if average allele-frequency differences are = 0.1, but with 474 loci the error rate would still be a comfortable 10^{-3}, and with 901 markers an excellent 10^{-5}. Because allele-frequency differences ≥ 0.4 are not uncommon between US ethnic groups, Risch *et al.* (2002) concluded that statistically robust racial clusters can be obtained by typing a few hundred markers. A similar rationale underlies a study suggesting that the limited genetic differences among putative races are sufficient to define well-distinct genetic clusters, provided one is willing to combine information at many loci (Edwards, 2003). The problem is that, for the calculations proposed in both such studies, genetic diversity is assumed to be concordant across loci; the more the loci in the analysis, the stronger the signal of differentiation. As we have seen, in humans that is not the case.

An additional problem is that, to test whether genetic differences confirm common concepts of race (as stated by Risch, 2006), one needs a list of biological races that most people are reasonably happy with, but so far there has been no agreement on such a list. This can only mean that, in practice, the task of compiling it is either impossible, or so difficult that nobody has been able to succeed. In particular, classical attempts to infer racial subdivision from skin colour have been remarkably unsuccessful (Cohen, 1991), and now we know why (Jablonski, 2006). Tens of loci affect pigmentation in humans, and they have epistatic effects. Most likely, these loci evolved under contrasting selective pressures, so that their current diversity represents a trade-off between protection against UV solar radiation (greater with darker skins) and the synthesis of vitamin D (more efficient with lighter skins) (Makova and Norton, 2005). There is evidence of relaxation of selection in non-African populations (Harding *et al.*, 2000), and different loci were probably selected at different times and places, so that some candidate genes show evidence of selection in the ancestral African population(s), others during the exit from Africa, and others in response to more recent local pressures (McEvoy *et al.*, 2006). Women have, on average, lighter skins than men of the same population, and this is probably accounted for by the higher vitamin D requirements during pregnancy and lactation (Madrigal and Kelly, 2006).

The importance of selection in shaping human pigmentation diversity had been already pointed out in the seventies (Nei and Roychoudhury, 1972), and is indirectly confirmed by the fact that skin-colour differences between populations and continents are fourfold greater than genetic differences estimated from autosomal loci (Relethford, 2002). Analogous cases can be made for traits such as lactase persistence and thalassemia. Because of decreasing levels of the enzyme lactase in the intestine, most people of the world lose the ability to digest lactose after weaning. However, lactase persists through lifetime at high frequency in populations from Europe and Africa, and remains at substantial frequencies in other populations sharing the habit to drink fresh milk (Campbell *et al.*, 2005). The alleles for lactose tolerance are different in Africa and Europe (Tishkoff *et al.*, 2007), and in Europe there is a clear geographic parallelism among variation in lactose tolerance, cattle milk genes, and locations of the Neolithic farming sites, implying a gene-culture coevolution between cattle and humans (Beja-Pereira *et al.*, 2003). Similarly, various pathologic variants of the haemoglobins including those responsible for thalassemias are common in areas of Africa, Asia, and Europe, where malaria is endemic. These alleles are maintained by balancing selection, because the heterozygotes are protected against the malarial parasite (Weatherall, 2001). In all these cases, common selective pressures have led to evolutionary convergence of populations that had only comparatively weak evolutionary relationships, and so should not be expected to resemble each other as much in other genome districts.

Recent analyses of genomic variation have confirmed this broad picture, at the same time adding intriguing details. Within the context of the International HapMap project, 51 autosomal regions, spanning 13 MB of the human genome, were resequenced (Gabriel *et al.*, 2002). The study of African, European, and Asian individuals shows that half of the blocks are cosmopolitan, another quarter is observed only in Africa, and the rest are essentially shared between continents, with only 4% of them being restricted to either Europe or Asia. Similar results were obtained in another study of the HapMap genotype data, estimating allele sharing for 1536 randomly selected individual SNPs (Montpetit *et al.*, 2006). In brief, the study of a sizable fraction of the genome in geographically distant populations shows that both single SNPs and large genomic regions in linkage disequilibrium are, with few exceptions, either specifically African, or generically human. Considering the expression level of various genes as variable phenotypes, Spielman *et al.* (2007) found substantial population differences, suggesting that variation at a few regulatory loci may account for even extensive population differences in liability to diseases.

This distribution of human genetic diversity suggests that the best way to predict whether certain individuals will have certain health risks or will benefit from pharmaceutical treatment is to study their genes, rather than relying on ill-defined racial stereotypes. This seems particularly true in the study of individual response to drug treatment, i.e. pharmacogenomics. Drug response is determined by many genetic and non-genetic factors, but variation at loci coding for drug-metabolizing enzymes accounts for a large fraction of the hereditary differences among normal, fast, and slow metabolisers, i.e. respectively those who benefit from a certain drug at the standard dosage, and those who do not, or even show negative side effects. Different populations show different allele frequencies at loci of pharmacogenomic interest (Meyer, 2004). However, most populations contain the whole spectrum of genotypes, which leaves little hope to develop different drugs, or drug dosages, specific for the Asian, African, or European market. Rather, it is now technically conceivable to concentrate the efforts on the genetic dissection of individual drug metabolism pathways, which in time may lead to tailoring-specific pharmacological treatment for different classes of metabolisers (Weber, 2001), no matter what their skin colour is and where in the world they live. Similarly, a serious understanding of the causes of disease requires an analysis of both their genetic and social causes, the latter being poorly approximated by ethnic and racial labels (Collins, 2004).

In the last 10 years, human genetic variation has also been actively studied for forensic purposes. The police authorities of many countries currently identify crime suspects by means of DNA profiling, and use the word race to summarise the general appearance of people. Association between race and patterns of DNA variation has gained a broad acceptance in courts (Cho and Sankar, 2004) and claims have been made that all scientific issues associated with DNA-based personal identification are resolved (Lander and Budowle, 1994). However, comparison of the racial categories used by the American federal bureau of investigations (FBI) and by the British police (Table 31.1) shows that these catalogs differ in the number and definition of races. People from the Indian subcontinent are either classified as whites or blacks in the United States, whereas they are attributed to three different Indian races under the UK system; the Irish and the British enjoy a special status in the UK system; and the possibility of mixed origins is contemplated in the United Kingdom, but not in the US system, where, instead, we find two races, 'Hispanics' and 'American Indian or Alaska natives', matching only with the 'Any other' race in the UK system. As a matter of fact, these differences do not matter much; Alaskan natives hardly cause criminal problems in the United Kingdom. However, these examples show that such race lists include categories that are considered useful for the practical needs of the local police, and are not *all-purpose* scientific descriptions of human variation.

The differences between UK and US forensic race catalogues reflect that common concepts of race are culture specific. In particular, there is no reason to believe that people know the race they belong to, and hence that self-assessed racial classification is a reasonable basis for categorizing genotypes. Under the one-drop rule that has been used in the United States to define the border between whites and blacks (everybody with one drop of black blood is black), many are considered (and consider themselves) as African Americans, because they have even a small documented fraction of black ancestors. Similarly, the hispanics or latinos (Risch, 2006), namely the Spanish-speaking community of the United States, include individuals whose origins represent very diverse mixtures, including native American, African, and European. They are defined as a group based on two non-biological factors, language and immigration, and would not refer to themselves as Hispanics in their country of origin. It may make sense to estimate disease prevalence and perhaps even allele frequencies among Hispanics, but only in very specific and local contexts. Moreover, disease prevalence also differs according to diverse social factors such as education level, often more strongly than is reflected in 'racial' labels such as Hispanic (Collins, 2004).

Genomic data are accumulating rapidly, and some or many of the views expressed in this chapter may have to change in the future. While we wait for the future to come, however, we have to conclude that the available studies of human biological diversity have not made it possible yet (1) to agree on a race list; (2) to place races on the world's map; and (3) to associate each race with diagnostic alleles or haplotypes. Humans differ genetically, but their differences do not seem to come in well-defined and consistent racial packages. After more than a century, it is hard to disagree with Darwin (1871) who (using the words race and species as synonymous) wrote: 'But the most weighty of all the arguments against treating the races of man as distinct species, is that they graduate into each other, independently in many cases, as far as we can judge, of their having intercrossed. Man has been studied more carefully than any other animal, and yet there is the greatest possible diversity amongst capable judges whether he should be classed as a single species or race, or as two (Virey), as three (Jacquinot), as four (Kant), five (Blumenbach), six (Buffon), seven (Hunter), eight (Agassiz), eleven (Pickering), fifteen (Bory de St-Vincent), sixteen (Desmoulins), twenty-two (Morton), sixty (Crawfurd), or as sixty-three, according to Burke.'

Acknowledgments

We thank Mark Beaumont, Lorena Madrigal, Rosalind Harding, and David Balding for critical reading of this chapter and for many insightful comments. While writing this chapter, G.B. was visiting the Department of Anthropology at the University of South Florida, Tampa, United States and L.C. was visiting the Instituto Gulbenkian de Ciências, Oeiras, Portugal.

REFERENCES

Aiello, L.C. (1993). The fossil evidence for modern human origins in Africa: a revised view. *American Anthropologist* **95**, 73–96.

Akey, J.M., Eberle, M.A., Rieder, M.J., Carlson, C.S., Shriver, M.D., Nickerson, D.A. and Kruglyak, L. (2004). Population history and natural selection shape patterns of genetic variation in 132 genes. *PLoS Biology* **2**, e286.

Akey, J.M., Zhang, G., Zhang, K., Jin, L. and Shriver, M.D. (2002). Interrogating a high-density SNP map for signatures of natural selection. *Genome Research* **12**, 1805–1814.

Amos, W. and Manica, A. (2006). Global genetic positioning: evidence for early human population centers in coastal habitats. *Proceedings of the National Academy of Sciences of the United States of America* **103**, 820–824.

Anikovich, M.V., Sinitsyn, A.A., Hoffecker, J.F., Holliday, V.T., Popov, V.V., Lisitsyn, S.N., Forman, S.L., Levkovskaya, G.M., Pospelova, G.A., Kuz'mina, I.E., Burova, N.D., Goldberg, P., Macphail, R.I., Giaccio, B. and Praslov, N.D. (2007). Early upper paleolithic in eastern Europe and implications for the dispersal of modern humans. *Science* **315**, 223–226.

Antón S.C., and Swisher C.C. III (2004). Early dispersals of Homo from Africa. *Annual Review of Anthropology* **33**, 271–296.

Aris-Brosou, S. and Excoffier, L. (1996). The impact of population expansion and mutation rate heterogeneity on DNA sequence polymorphism. *Molecular Biology and Evolution* **13**, 494–504.

Bamshad, M. (2005). Genetic influences on health: does race matter? *JAMA* **24**, 937–946.

Bamshad, M., Wooding, S., Salisbury, B.A. and Stephens, J.C. (2003). Deconstructing the relationship between genetics and race. *Nature Reviews Genetics* **5**, 598–609.

Barbujani, G. (2005). Human races: classifying people vs. Understanding diversity. *Current Genomics* **6**, 215–226.

Barbujani, G. and Belle, E.M.S. (2006). Genomic boundaries between human populations. *Human Heredity* **61**, 15–21.

Barbujani, G., Magagni, A., Minch, E. and Cavalli-Sforza, L.L. (1997). An apportionment of human DNA diversity. *Proceedings of the National Academy of Sciences of the United States of America* **94**, 4516–4519.

Barbujani, G., Oden, N.L. and Sokal, R.R. (1989). Detecting areas of abrupt change in maps of biological variables. *Systematic Zoology* **38**, 376–389.

Barbujani, G., Pilastro, A., De Domenico, S. and Renfrew, C. (1994). Genetic variation in North Africa and Eurasia: neolithic demic diffusion versus Paleolithic colonisation. *American Journal of Physical Anthropology* **95**, 137–154.

Bastos-Rodrigues, L., Pimenta, J.R. and Pena, S.D.J. (2006). The genetic structure of human populations studied through short insertion-deletion polymorphisms. *Annals of Human Genetics* **70**, 658–665.

Bateman, R., Goddard, I., O'Grady, R., Funk, V.A., Mooi, R., Kress, W.J. and Cannell, P. (1990). Speaking of forked tongues. *Current Anthropology* **31**, 1–24.

Beaumont, M.A. (1999). Detecting population expansion and decline using microsatellites. *Genetics* **153**, 2013–2029.

Beaumont, M.A. (2004). Recent developments in genetic data analysis: what can they tell us about human demographic history? *Heredity* **92**, 365–379.

Beaumont, M.A. (2005). Adaptation and speciation: what can Fst tell us? *Trends in Ecology and Evolution* **20**, 435–440.

Beaumont, M.A. and Balding, D.J. (2004). Identifying adaptive genetic divergence among populations from genome scans. *Molecular Ecology* **13**, 969–980.

Beaumont, M.A. and Nichols, R. (1996). Evaluating loci for use in the genetic analysis of population structure. *Proceedings of the Royal Society of London* **B 263**, 1619–1626.

Beaumont, M.A. and Rannala, B. (2004). The Bayesian revolution in genetics. *Nature Reviews Genetics* **5**, 251–261.

Beaumont, M.A., Zhang, W. and Balding, D.J. (2002). Approximate Bayesian computation in population genetics. *Genetics* **162**, 2025–2035.

Beerli, P. and Felsenstein, J. (2001). Maximum likelihood estimation of a migration matrix and effective population sizes in n subpopulations by using a coalescent approach. *Proceedings of the National Academy of Sciences of the United States of America* **98**, 4563–4568.

Beja-Pereira, A., Luikart, G., England, P.R., Bradley, D.G., Jann, O.C., Bertorelle, G., Chamberlain, A.T., Nunes, T.P., Metodiev, S., Ferrand, N. and Erhardt, G. (2003). Gene-culture coevolution between cattle milk protein genes and human lactase genes. *Nature Genetics* **35**, 311–313.

Bernasconi, R. (2001). *Concepts of Race in the Eighteenth Century*. Thoemmes Press, Bristol.

Binladen, J., Wiuf, C., Gilbert, M.T., Bunce, M., Barnett, R., Larson, G., Greenwood, A.D., Haile, J., Ho, S.Y., Hansen, A.J. and Willerslev, E. (2006). Assessing the fidelity of ancient DNA sequences amplified from nuclear genes. *Genetics* **172**, 733–741.

Bowcock, A.M., Ruiz-Linares, A., Tomfohrde, J., Minch, E., Kidd, J.R. and Cavalli-Sforza, L.L. (1994). High resolution of human evolutionary history trees with polymorphic microsatellites. *Nature* **368**, 455–457.

Bräuer, G. (1984). A craniological approach to the origin of anatomically modern human *Homo sapiens* and its implications for the appearance of modern Europeans. In *The Origins of Modern Humans*, F.H. Smith and F. Spencer, eds. Wenner-Gren Foundation for Anthropological Research New York, pp. 327–410.

Brown, R.A. and Armelagos, G.J. (2001). Apportionment of racial diversity: a review. *Evolutionary Anthropology* **10**, 24–20.

Burchard, E.G., Ziv, E., Coyle, N., Gomez, S.L., Tang, H., Karter, A.J., Mountain, J.L., Perez-Stable, E.J., Sheppard, D. and Risch, N. (2003). The importance of race and ethnic background in biomedical research and clinical practice. *The New England Journal of Medicine* **348**, 1170–1175.

Campbell, A.K., Waud, J.P. and Matthews, S.B. (2005). The molecular basis of lactose intolerance. *Science in Progress* **88**, 157–202.

Cann, R., Stoneking, M. and Wilson, A.J. (1987). Mitochondrial DNA and human evolution. *Nature* **325**, 31–36.

Caramelli, D., Lalueza-Fox, C., Condemi, S., Longo, L., Milani, L., Manfredini, A., de Saint Pierre, M., Adoni, F., Lari, M., Giunti, P., Ricci, S., Casoli, A., Calafell, F., Mallegni, F., Bertranpetit, J., Stanyon, R., Bertorelle, G. and Barbujani, G. (2006). A highly divergent mtDNA sequence in a neandertal individual from italy. *Current Biology* **16**, R630–R632.

Caramelli, D., Lalueza-Fox, C., Vernesi, C., Lari, M., Casoli, A., Mallegni, F., Chiarelli, B., Dupanloup, I., Bertranpetit, J., Barbujani, G. and Bertorelle, G. (2003). Evidence for a genetic discontinuity between Neandertals and 24,000-year-old anatomically modern Europeans. *Proceedings of the National Academy of Sciences of the United States of America* **100**, 6593–6597.

Cavalli-Sforza, L.L. (1966). Population structure and human evolution. *Proceedings of the Royal Society Section B* **164**, 362–379.

Cavalli-Sforza, L.L., Menozzi, P. and Piazza, A. (1994). *The History and Geography of Human Genes*. Princeton University Press, Princeton, NJ.

Cavalli-Sforza, L.L., Menozzi, P., Piazza, A. and Mountain, J. (1988). Reconstruction of human evolution: bringing together genetic, archaeological, and linguistic data. *Proceedings of the National Academy of Sciences of the United States of America* **85**, 6002–6006.

Chikhi, L. and Beaumont, M.A. (2005). Modelling human genetic history. *Encyclopaedia of Genetics, Genomics, Proteomics and Bioinformatics*. John Wiley & Sons, Chichester.

Chikhi, L. and Bruford, M.B. (2005). Mammalian population genetics and genomics. In *Mammalian Genomics*, A. Ruvinsky and J. Marshall Graves, eds. Chapter 21. CAB International Publishing, pp. 539–584.

Chikhi, L., Bruford, M.W. and Beaumont, M.A. (2001). Estimation of admixture proportions: a likelihood-based approach using Markov chain Monte Carlo. *Genetics* **158**, 1347–1362.

Chikhi, L., Destro-Bisol, G., Bertorelle, G., Pascali, V. and Barbujani, G. (1998). Clines of nuclear DNA markers suggest a largely neolithic ancestry of the European gene pool. *Proceedings of the National Academy of Sciences of the United States of America* **95**, 9053–9058.

Chikhi, L., Nichols, R.A., Barbujani, G. and Beaumont, M.A. (2002). Y genetic data support the Neolithic demic diffusion model. *Proceedings of the National Academy of Sciences of the United States of America* **99**, 10008–10013.

Cho, M.K. and Sankar, P. (2004). Forensic genetics and legal, ethic and social implications beyond the clinic. *Nature Genetics Supplement* **36**, S8–S12.

Chung, C.S., Mi, M.P. and Morton, N.E. (1967). *Genetics of Interracial Crosses in Hawaii*. Karger, Basel.

Cohen, C. (1991). *Aux Origines d'Homo Sapiens*, J.-J. Hublin and A.M. Tillier, eds. Presses Universitaries de France, Paris, pp. 9–47.

Cole, S. (1965). *Races of Man*. British Museum (Natural History), London.

Collard, M. and Shennan, S. (2000). Processes of culture change in prehistory: a case study from the European Neolithic. In *Archaeogenetics: DNA and the Population Prehistory of Europe*, C. Renfrew and K. Boyle, eds. MacDonald Institute for Archaeological Research, Cambridge, pp. 89–97.

Collins, F. (2004). What we do and don't know about 'race', 'ethnicity', genetics and health at the dawn of the genome era. *Nature Genetics Supplement* **36**, S13–S15.

Conrad, D.F., Jakobsson, M., Coop, G., Wen, X., Eall, J.D., Rosenberg, N.A. and Pritchard, J.K. (2006). A worldwide survey of haplotype variation and linkage disequilibrium in the human genome. *Nature Genetics* **38**, 1251–1260.

Coon, C.S. (1963). *The Origin of Races*. Alfred A. Knopf, New York.

Coon, C.S., Garn, S.M. and Birdsell, J.B. (1950). *A Study of the Problem of Race Formation in Man*. Charles Thomas, Springfield, Ill.

Cooper, R.S. (2005). Race and IQ: molecular genetics as deus ex machina. *The American Psychologist* **60**, 71–76.

Cooper, R.S., Kaufman, J.S. and Ward, R.W. (2003). Race and genomics. *The New England Journal of Medicine* **348**, 1166–1170.

Corander, J. and Marttinen, P. (2006). Bayesian identification of admixture events using multilocus molecular markers. *Molecular Ecology* **15**, 2833–2843.

Corander, J., Marttinen, P. and Mäntyniemi, S. (2007). Bayesian identification of stock mixtures from molecular marker data. *Fishery Bulletin* (in press).

Corander, J., Waldmann, P., Marttinen, P. and Sillanpaa, M.J. (2004). BAPS 2: enhanced possibilities for the analysis of genetic population structure. *Bioinformatics* **20**, 2363–2369.

Corander, J., Waldmann, P. and Sillanpaa, M.J. (2003). Bayesian analysis of genetic differentiation between populations. *Genetics* **163**, 367–374.

Cordaux, R., Aunger, R., Bentley, G., Nasidze, I., Sirajuddin, S.M. and Stoneking, M. (2004). Independent origins of Indian caste and tribal paternal lineages. *Current Biology* **14**, 231–235.

Cornuet, J.-M. and Luikart, G. (1996). Description and power analysis of two tests for detecting recent population bottlenecks from allele frequency data. *Genetics* **144**, 2001–2014.

Crawford, M. (1998). *The Origin of Native Americans: Evidence from Archeological Genetics.* Cambridge University Press, Cambridge.

Currat, M. and Excoffier, L. (2004). Modern humans did not admix with Neanderthals during their range expansion into Europe. *PLoS Biology* **2**, e421.

Currat, M., Excoffier, L., Maddison, W., Otto, S.P., Ray, N., Whitlock, M.C. and Yeaman, S. (2006). Comment on "Ongoing adaptive evolution of ASPM, a brain size determinant in Homo sapiens" and "Microcephalin, a gene regulating brain size, continues to evolve adaptively in humans". *Science* **313**, 172.

Darwin, C.R. (1871). *The Descent of Man, and Selection in Relation to Sex*, Chapter VII. John Murray, London.

Dawson, K.J. and Belkhir, K. (2001). A Bayesian approach to the identification of panmictic populations and the assignment of individuals. *Genetical Research* **78**, 59–77.

Dobzhansky, T. (1967). *Genetic Diversity and Human Behavior*, J.N. Spuhler, ed. Wenner-Gren Foundation for Anthropological Research, New York, pp. 1–19.

Dupanloup, I., Pereira, L., Bertorelle, G., Calafell, F., Prata, M.J., Amorim, A. and Barbujani, G. (2003). A recent shift from polygyny to monogamy in humans is suggested by the analysis of worldwide Y-chromosome diversity. *Journal of Molecular Evolution* **57**, 85–97.

Edmonds, C.A., Lillie, A.S. and Cavalli-Sforza, L.L. (2004). Mutations arising in the wave front of an expanding population. *Proceedings of the National Academy of Sciences of the United States of America* **101**, 975–979.

Edwards, A.W.F. (2003). Human genetic diversity: Lewontin's fallacy. *BioEssays* **25**, 798–801.

Edwards, A.W.F. and Cavalli-Sforza, L.L. (1965). A method for cluster analysis. *Biometrics* **21**, 362–375.

Eller, E. (2002). Population extinction and recolonisation in human demographic history. *Mathematical Biosciences* **177–178**, 1–10.

Eswaran, V. (2002). A diffusion wave out of Africa: the mechanism of the modern human revolution? *Current Anthropology* **43**, 749–774.

Eswaran, V., Harpending, H. and Rogers, A.R. (2005). Genomics refutes an exclusively African origin of humans. *Journal of Human Evolution* **49**, 1–18.

Evanno, G., Regnaut, S. and Goudet, J. (2005). Detecting the number of clusters of individuals using the software STRUCTURE: a simulation study. *Molecular Ecology* **14**, 2611–2620.

Evans, P.D., Gilbert, S.L., Mekel-Bobrov, N., Vallender, E.J., Anderson, J.R., Vaez-Azizi, L.M., Tishkoff, S.A., Hudson, R.R. and Lahn, B.T. (2005). Microcephalin, a gene regulating brain size, continues to evolve adaptively in humans. *Science* **309**, 1717–1720.

Evans, P.D., Mekel-Bobrov, N., Vallender, E.J., Hudson, R.R. and Lahn, B.T. (2006). Evidence that the adaptive allele of the brain size gene microcephalin introgressed into Homo sapiens from an archaic Homo lineage. *Proceedings of the National Academy of Sciences of the United States of America* **108**, 18178–18183.

Ewens, W.J. (1972). The sampling theory of selectively neutral alleles. *Theoretical Population Biology* **3**, 87–112.

Excoffier, L. (2002). Human demographic history: refining the recent African origin model. *Current Opinion in Genetics and Development* **12**, 675–682.

Excoffier, L., Estoup, A. and Cornuet, J.-M. (2005). Bayesian analysis of an admixture model with mutations and arbitrarily linked markers. *Genetics* **169**, 1727–1738.

Excoffier, L. and Hamilton, G. (2003). Comment on "Genetic structure of human populations". *Science* **300**, 1877.

Excoffier, L. and Schneider, S. (1999). Why hunter-gatherer populations do not show signs of Pleistocene demographic expansions. *Proceedings of the National Academy of Sciences of the United States of America* **96**, 10597–10602.

Excoffier, L., Smouse, P.E. and Quattro, J.M. (1992). Analysis of molecular variance inferred from emtric distances among DNA haplotypes: application ti human mitochondrial DNA restriction data. *Genetics* **131**, 479–491.

Falush, D., Stephens, M. and Pritchard, J.K. (2003). Inference of population structure using multilocus genotype data: linked loci and correlated allele frequencies. *Genetics* **164**, 1567–1587.

Fay, J.C. and Wu, C.I. (1999). A human population bottleneck can account for the discordance between patterns of mitochondrial versus nuclear DNA variation. *Molecular Biology and Evolution* **16**, 1003–1005.

Felsenstein, J. (1992). Estimating effcctive population size from samples of sequences: inefficiency of pairwise and segregating sites as compared to phylogenetic estimates. *Genetical Research* **59**, 139–147.

Fischer, A., Pollack, J., Thalmann, O., Nickel, B. and Pääbo, S. (2006). Demographic history and genetic differentiation in apes. *Current Biology* **16**, 1133–1138.

Flint, J., Bond, J., Rees, D.C., Boyce, A.J., Roberts-Thompson, J.M., Excoffier, L., Clegg, J.B., Beaumont, M.A., Nichols, R.A. and Harding, R.M. (1999). Minisatellite mutational processes reduce F_{ST} estimates. *Human Genetics* **105**, 567–576.

Fu, Y.X. and Li, W.H. (1993). Statistical tests of neutrality of mutations. *Genetics* **133**, 693–709.

Gabriel, S.B., Schaffner, S.F., Nguyen, H., Moore, J.M., Roy, J., Blumenstiel, B., Higgins, J., DeFelice, M., Lochner, A., Faggart, M., Liu-Cordero, S.N., Rotimi, C., Adeyemo, A., Cooper, R., Ward, R., Lander, E.S., Daly, M.J. and Altshuler, D. (2002). The structure of haplotype blocks in the human genome. *Science* **296**, 2225–2229.

Gagneux, P., Wills, C., Gerloff, U., Tautz, D., Morin, P.A., Boesch, C., Fruth, B., Hohmann, G., Ryder, O.A. and Woodruff, D.S. (1999). Mitochondrial sequences show diverse evolutionary histories of African hominoids. *Proceedings of the National Academy of Sciences of the United States of America* **96**, 5077–5082.

Garn, S.M. (1965). *Human Races*. Thomas, Springfield, Ill.

Garrigan, D. and Hammer, M.F. (2006). Reconstructing human origins in the genomic era. *Nature Reviews Genetics* **7**, 669–680.

Garrigan, D., Mobasher, Z., Kingan, S.B., Wilder, J.A. and Hammer, M.F. (2005). Deep haplotype divergence and long-range linkage disequilibrium at Xp21.1 provide evidence that humans descend from a structured ancestral population. *Genetics* **170**, 1849–1856.

Garza, J.C. and Williamson, E. (2001). Detection of reduction in population size using data from microsatellite DNA. *Molecular Ecology* **10**, 305–318.

Goldstein, D.B. and Chikhi, L. (2002). Human migrations and population structure: what we know and why it matters. *Annual Review of Genomics and Human Genetics* **3**, 129–152.

Goldstein, D.B., Ruiz Linares, A., Cavalli-Sforza, L.L. and Feldman, M.W. (1995). Genetic absolute dating based on microsatellites and the origin of modern humans. *Proceedings of the National Academy of Sciences of the United States of America* **92**, 6723–6727.

Green, R.E., Krause, J., Ptak, S.E., Briggs, A.W., Ronan, M.T., Simons, J.F., Du, L., Egholm, M., Rothberg, J.M., Paunovic, M. and Pääbo, S. (2006). Analysis of one million base pairs of Neanderthal DNA. *Nature* **444**, 330–336.

Griffiths, R.C. and Tavaré, S. (1994). Simulating probability distributions in the coalescent. *Theoretical Population Biology* **46**, 131–159.

Grine, F.E., Bailey, R.M., Harvati, K., Nathan, R.P., Morris, A.G., Henderson, G.M., Ribot, I. and Pike, A.W. (2007). Late Pleistocene human skull from Hofmeyr, South Africa, and modern human origins. *Science* **315**, 226–229.

Hammer, M.F., Garrigan, D., Wood, E., Wilder, J.A., Mobasher, Z., Bigham, A., Krenz, J.G. and Nachman, M.W. (2004). Heterogeneous patterns of variation among multiple human x-linked Loci: the possible role of diversity-reducing selection in non-Africans. *Genetics* **167**, 1841–1853.

Harding, R.M., Fullerton, S.M., Griffiths, R.C., Bond, J., Cox, M.J., Schneider, J.A., Moulin, D.S. and Clegg, J.B. (1997). Archaic African and Asian lineages in the genetic ancestry of modern humans. *American Journal of Human Genetics* **60**, 772–789.

Harding, R.M., Healy, E., Ray, A.J., Ellis, N.S., Flanagan, N., Todd, C., Dixon, C., Sajantila, A., Jackson, I.J., Birch-Machin, M.A. and Rees, J.L. (2000). Evidence for variable selective pressures at MC1R. *American Journal of Human Genetics* **66**, 1351–1361.

Harding, R.M. and McVean, G.A. (2004). A structured ancestral population for the evolution of modern humans. *Current Opinion in Genetics and Development* **14**, 667–674.

Harris, E.E. and Hey, J. (1999). Human demography in the Pleistocene: do mitochondrial and nuclear genes tell the same story? *Evolutionary Anthropology* **8**, 81–86.

Hey, J. (1997). Mitochondrial and nuclear genes present conflicting portraits of human origins. *Molecular Biology and Evolution* **14**, 166–172.

Hey, J. (2005). On the number of new world founders: a population genetic portrait of the peopling of the Americas. *PLoS Biology* **3**, 965–975.

Hey, J. and Nielsen, R. (2004). Multilocus methods for estimating population sizes, migration rates and divergence time, with applications to the divergence of *Drosophila pseudoobscura* and *D. persimilis*. *Genetics* **167**, 747–760.

Howells, W.W. (1976). Explaining modern man: evolutionists versus migrationists. *Journal of Human Evolution* **5**, 477–496.

Huxley, J. (1938). Clines: an auxiliary taxonomic principle. *Nature* **142**, 219–220.

Ingman, M., Kaessmann, H., Pääbo, S. and Gyllensten, U. (2000). Mitochondrial genome variation and the origin of modern humans. *Nature* **408**, 708–713.

Jablonski, N. (2006). *Skin. A Natural History*. University of California Press, Berkeley, CA.

Jorde, L.B., Watkins, W.S., Bamshad, M.J., Dixon, M.E., Ricker, C.E., Seielstad, M.T. and Batzer, M.A. (2000). The distribution of human genetic diversity: a comparison of mitochondrial, autosomal and Y-chromosome data. *American Journal of Human Genetics* **66**, 979–988.

Jorde, L.B. and Wooding, S.P. (2004). Genetic variation, classification, and 'race'. *Nature Genetics* **36**, S28–S33.

Kaessmann, H., Heissig, F., von Haeseler, A. and Pääbo, S. (1999). DNA sequence variation in a non-coding region of low recombination on the human X-chromosome. *Nature Genetics* **22**, 78–81.

Kayser, M., Brauer, S. and Stoneking, M. (2003). A genome scan to detect candidate regions influenced by local natural selection in human populations. *Molecular Biology and Evolution* **20**, 893–900.

Ke, Y., Su, B., Song, X., Lu, D., Chen, L., Li, H., Qi, C., Marzuki, S., Deka, R., Underhill, P., Xiao, C., Shriver, M., Lell, J., Wallace, D., Wells, R.S., Seielstad, M., Oefner, P., Zhu, D., Jin, J., Huang, W., Chakraborty, R., Chen, Z. and Jin, L. (2001). African origin of modern humans in East Asia: a tale of 12,000 Y chromosomes. *Science* **292**, 1151–1153.

Kimura, M. and Crow, J. (1964). The number of alleles that can be maintained in a finite population. *Genetics* **49**, 725–738.

Kitano, T., Schwarz, C., Nickel, B. and Pääbo, S. (2003). Gene diversity patterns at 10 X-chromosomal loci in humans and chimpanzees. *Molecular Biology and Evolution* **20**, 1281–1289.

King, M.C. and Wilson, A.C. (1975). Evolution at two levels in humans and chimpanzees. *Science* **188**, 107–116.

Klein, R.G. (1999). *The Human Career: Human Biological and Cultural Origins*. University of Chicago Press, Chicago, Ill.

Klein, R.G. (2005). Hominin dispersals in the old world. In *The Human Past*, C. Scarre, ed. Thames and Hudson Ltd, London, pp. 84–123.

Klopfstein, S., Currat, M. and Excoffier, L. (2006). The fate of mutations surfing on the wave of a range expansion. *Molecular Biology and Evolution* **23**, 482–490.

Krings, M., Capelli, C., Tschentscher, F., Geisert, H., Meyer, S., von Haeseler, A., Grossschmidt, K., Possnert, G., Paunovic, M. and Paabo, S. (2000). A view of Neandertal genetic diversity. *Nature Genetics* **26**, 144–146.

Kuhner, M., Yamoto, J. and Felsenstein, J. (1995). Estimating effective population size and mutation rate from sequence data using Metropolis-Hastings sampling. *Genetics* **140**, 1421–1430.

Lahr, M.M. (1994). The multiregional model of modern human origins: a reassessment of its morphological basis. *Journal of Human Evolution* **26**, 23–56.

Lahr, M.M. and Foley, R.A. (1998). Towards a theory of modern human origins: geography, demography, and diversity in recent human evolution. *Yearbook of Physical Anthropology* **41**, 137–176.

Lander, E. and Budowle, B. (1994). DNA fingerprinting dispute laid to rest. *Nature* **371**, 735–737.

Latter, B.D.H. (1980). Genetic differences within and between populations of the major human subgroups. *American Naturalist* **116**, 220–237.

Lewin, R. and Foley, R.A. (2004). *Principles of Human Evolution*. Blackwell Science, Oxford.

Lewontin, R.C. (1972). The apportionment of human diversity. *Evolutionary Biology* **6**, 381–398.

Li, N. and Stephens, M. (2003). Modelling linkage disequilibrium and identifying recombination hotspots using single nucleotide polymorphism data. *Genetics* **165**, 2213–2233.

Liu, H., Prugnolle, F., Manica, A. and Balloux, F. (2006). A geographically explicit genetic model of worldwide human-settlement history. *American Journal of Human Genetics* **79**, 230–137.

Livingstone, F.B. (1962). On the nonexistence of human races. *Current Anthropology* **3**, 279–281.

Luis, J.R., Rowold, D.J., Regueiro, M., Caeiro, B., Cinnioglu, C., Roseman, C., Underhill, P.A., Cavalli-Sforza, L.L. and Herrera, R.J. (2004). The Levant versus the Horn of Africa: evidence for bidirectional corridors of human migrations. *American Journal of Human Genetics* **74**, 532–544.

Macaulay, V., Hill, C., Achilli, A., Rengo, C., Clarke, D., Meehan, W., Blackburn, J., Semino, O., Scozzari, R., Cruciani, F., Taha, A., Shaari, N.K., Raja, J.M., Ismail, P., Zainuddin, Z., Goodwin, W., Bulbeck, D., Bandelt, H.J., Oppenheimer, S., Torroni, A. and Richards, M. (2005). Single, rapid coastal settlement of Asia revealed by analysis of complete mitochondrial genomes. *Science* **308**, 1034–1036.

Madrigal, L. and Kelly, W. (2006). Human skin-color sexual dimorphism: a test of the sexual selection hypothesis. *American Journal of Physical Anthropology* **132**, 470–482.

Makova, K. and Norton, H. (2005). Worldwide polymorphism at the MC1R locus and normal pigmentation variation in humans. *Peptides* **26**, 1901–1908.

Manni, F., Guerard, E. and Heyer, E. (2004). Geographic patterns of (genetic, morphologic, linguistic) variation: how barriers can be detected by using Monmonier's algorithm. *Human Biology* **76**, 173–190.

Marjoram, P., Molitor, J., Plagnol, V. and Tavaré, S. (2003). Markov chain Monte Carlo without likelihoods. *Proceedings of the National Academy of Sciences of the United States of America* **100**, 15324–15328.

Marth, G.T., Czabarka, E., Murvai, J. and Sherry, S.T. (2004). The allele frequency spectrum in genome-wide human variation data reveals signals of differential demographic history in three large world populations. *Genetics* **166**, 351–372.

Marth, G.T., Schuler, G., Yeh, R., Davenport, R., Agarwala, R., Church, D., Wheelan, S., Baker, J., Ward, M., Kholodov, M., Phan, L., Czabarka, E., Murvai, J., Cutler, C., Wooding, S., Rogers, A.R., Chakravarti, A., Harpending, H.C., Kwok, P.-Y. and Sherry, S.T. (2003). Sequence variations in the public human genome data reflect a bottlenecked population history. *Proceedings of the National Academy of Sciences of the United States of America* **100**, 376–381.

Mayr, E. (1947). *Systematics and the Origin of Species*, 3rd edition. Columbia University Press, New York.

McEvoy, B., Beleza, S. and Shriver, M.D. (2006). The genetic architecture of normal variation in human pigmentation: an evolutionary perspective and model. *Human Molecular Genetics* **15**, R176–R181.

McVean, G.A.T., Myers, S.R., Hunt, S., Deloukas, P., Bentley, D.R. and Donnelly, P. (2004). The fine-scale structure of recombination rate variation in the human genome. *Science* **304**, 581–584.

Mekel-Bobrov, N., Gilbert, S.L., Evans, P.D., Vallender, E.J., Anderson, J.R., Hudson, R.R., Tishkoff, S.A. and Lahn, B.T. (2005). Ongoing adaptive evolution of ASPM, a brain size determinant in Homo sapiens. *Science* **309**, 1720–1722.

Mekel-Bobrov, N., Posthuma, D., Gilbert, S.L., Lind, P., Gosso, M.F., Luciano, M., Harris, S.E., Bates, T.C., Polderman, T.J., Whalley, L.J., Fox, H., Starr, J.M., Evans, P.D., Montgomery, G.W., Fernandes, C., Heutink, P., Martin, N.G., Boomsma, D.I., Deary, I.J., Wright, M.J., de Geus, E.J. and Lahn, B.T. (2007). The ongoing adaptive evolution of ASPM and Microcephalin is not explained by increased intelligence. *Human Molecular Genetics* **16**, 600–608.

Mellars, P. (2006). Going east: new genetic and archaeological perspectives on the modern human colonization of Eurasia. *Science* **313**, 796–800.

Menozzi, P., Piazza, A. and Cavalli-Sforza, L.L. (1978). Synthetic maps of human gene frequencies in Europeans. *Science* **201**, 786–792.

Meyer, U.A. (2004). Pharmacogenetics – five decades of therapeutic lessons from genetic diversity. *Nature Reviews Genetics* **5**, 669–676.

Molnar, S. (1975). *Human Variation. Races, Types and Ethnic Groups*. Prentice Hall, Upper Saddle River.

Montagu, M.F.A. (1941). The concept of race in the human species in the light of genetics. *The Journal of Heredity* **32**, 243–247.

Montpetit, A., Nelis, M., Laflamme, P., Magi, R., Ke, X., Rmm, M., Cardon, L., Hudson, T.J. and Metspalu, A. (2006). An evaluation of the performance of TAG SNPs derived from HapMap in a Caucasian population. *PLoS Genetics* **2**, e27.

Mountain, J.L. and Risch, N. (2004). Assessing genetic contributions to phenotypic differences among 'racial' and 'ethnic' groups. *Nature Genetics Supplement* **36**, S48–S53.

Mulligan, C.J., Hunley, K., Cole, S. and Long, J.C. (2004). Population genetics, history, and health patterns in native americans. *Annual Review of Genomics and Human Genetics* **5**, 295–315.

Neel, J.V. (1978). The population structure of an Amerindian tribe, the Yanomama. *Annual Review of Genetics* **12**, 365–418.

Nei, M., Chakravarti, A. and Tateno, Y. (1977). Mean and variance of F_{st} in a finite number of incompletely isolated populations. *Theoretical Population Biology* **11**, 291–306.

Nei, M., Maruyama, T. and Chakraborty, R. (1975). The bottleneck effect and genetic variability in populations. *Evolution* **29**, 1–10.

Nei, M. and Roychoudhury, A.K. (1972). Gene differences between Caucasian, Negro and Japanese populations. *Science* **177**, 434–436.

Nielsen, R. and Wakeley, J. (2001). Distinguishing migration from isolation: a Markov chain Monte Carlo approach. *Genetics* **158**, 885–896.

Nordborg, M. (1998). On the probability of Neanderthal ancestry. *American Journal of Human Genetics* **63**, 1237–1240.

Orduñez, P., Bernal Munoz, J.L., Espinosa-Brito, A., Silva, L.C. and Cooper, R.S. (2005). Ethnicity, education, and blood pressure in Cuba. *American Journal of Epidemiology* **162**, 49–56.

Pääbo, S., Poinar, H., Serre, D., Jaenicke-Despres, V., Hebler, J., Rohland, N., Kuch, M., Krause, J., Vigilant, L. and Hofreiter, M. (2004). Genetic analyses from ancient DNA. *Annual Review of Genetics* **38**, 645–679.

Parra, F.C., Amado, R.C., Lambertucci, J.R., Rocha, J., Antunes, C.M. and Pena, S.D. (2003). Color and genomic ancestry in Brazilians. *Proceedings of the National Academy of Sciences of the United States of America* **100**, 177–182.

Parra, E.J., Marcini, A., Akey, J., Martinson, J., Batzer, M.A., Cooper, R., Forrester, T., Allison, D.B., Deka, R., Ferrell, R.E. and Shriver, M.D. (1998). Estimating African American admixture proportions by use of population-specific alleles. *American Journal of Human Genetics* **63**, 1839–1851.

Plagnol, V. and Wall, J.D. (2006). Possible ancestral structure in human populations. *PLoS Biology* **2**, 972–979.

Pritchard, J.K., Seielstad, M.T., Perez-Lezaun, A. and Feldman, M.W. (1999). Population growth of human Y chromosomes: a study of Y chromosome microsatellites. *Molecular Biology and Evolution* **16**, 1791–1798.

Pritchard, J.K., Stephens, M. and Donnelly, P. (2000). Inference of population structure using multilocus genotype data. *Genetics* **155**, 945–959.

Provine, W.B. (1986). Geneticists and race. *American Zoologist* **26**, 857–888.

Przeworski, M., Hudson, R.R. and Di Rienzo, A. (2000). Adjusting the focus on human variation. *Trends in Genetics* **16**, 296–302.

Ramachandran, S., Deshpande, O., Roseman, C.C., Rosenberg, N.A., Feldman, M.W. and Cavalli-Sforza, L.L. (2005). Support from the relationship of genetic and geographic distance in human populations for a serial founder effect originating in Africa. *Proceedings of the National Academy of Sciences of the United States of America* **102**, 15942–15947.

Ramachandran, S., Rosenberg, N.A., Zhivotovsky, L.A. and Feldman, M.W. (2004). Robustness of the inference of human population structure: a comparison of X-chromosomal and autosomal microsatellites. *Human Genomics* **1**, 87–97.

Rannala, B. and Mountain, J.L. (1997). Detecting immigration by using multilocus genotypes. *Proceedings of the National Academy of Sciences of the United States of America* **94**, 9197–9201.

Rannala, B. and Yang, Z. (2003). Bayes estimation of species divergence times and ancestral population sizes using DNA sequences from multiple loci. *Genetics* **164**, 1645–1656.

Ray, N., Currat, M., Berthier, P. and Excoffier, L. (2005). Recovering the geographic origin of early modern humans by realistic and spatially explicit simulations. *Genome Research* **15**, 1161–1167.

Ray, N., Currat, M. and Excoffier, L. (2004). Intra-deme molecular diversity in spatially expanding populations. *Molecular Biology and Evolution* **20**, 76–86.

Redd, A.J., Chamberlain, V.F., Kearney, V.F., Stover, D., Karafet, T., Calderon, K., Walsh, B. and Hammer, M.F. (2006). Genetic structure among 38 populations from the United States based on 11 U.S. core Y chromosome STRs. *Journal of Forensic Sciences* **51**, 580–585.

Reed, F.A. and Tishkoff, S.A. (2006). African human diversity, origins and migrations. *Current Opinion in Genetics and Development* **16**, 597–605.

Rees, J.L. (2003). Genetics of hair and skin color. *Annual Review of Genetics* **37**, 67–90.

Reich, D.E. and Goldstein, D.B. (1998). Genetic evidence for a Paleolithic human population expansion in Africa. *Proceedings of the National Academy of Sciences of the United States of America* **95**, 8119–8123.

Relethford, J.H. (1999). Models, predictions, and the fossil record of modern human origins. *Evolutionary Anthropology* **8**, 7–10.

Relethford, J.H. (2001). Absence of regional affinities of Neandertal DNA with living humans does not reject multiregional evolution. *American Journal of Physical Anthropology* **115**, 95–98.

Relethford, J.H. (2002). Apportionment of global human genetic diversity based on craniometrics and skin color. *American Journal of Physical Anthropology* **118**, 393–398.

Relethford, J.H. (2003). Genetic history of the human species. In *Handbook of Statistical Genetics*, 2nd edition, D. Balding, M. Bishop and C. Cannings, eds. John Wiley & Sons, Chichester, pp. 793–829.

Relethford, J.H. (2004). Global patterns of isolation by distance based on genetic and morphological data. *Human Biology* **76**, 499–513.

Relethford, J.H. and Harding, R.M. (2001). Population genetics of modern human evolution. *Encyclopedia of Life Sciences.* John Wiley & Sons, Chichester. http://www.els.net/.

Ripley, B.D. (1996). *Pattern Recognition and Neural Networks.* Cambridge University Press, Cambridge.

Risch, N. (2006). Dissecting racial and ethnic differences. *The New England Journal of Medicine* **354**, 408–411.

Risch, N., Burchard, E., Ziv, E. and Tang, H. (2002). Categorization of humans in biomedical research: genes, race and disease. *Genome Biology* **3**, 2007.1–2007.12.

Rogers, A.R. and Harpending, H. (1992). Population growth makes waves in the distribution of pairwise genetic differences. *Molecular Biology and Evolution* **9**, 552–569.

Romualdi, C., Balding, D., Nasidze, I.S., Risch, G., Robichaux, M., Sherry, S.T., Stoneking, M., Batzer, M.A. and Barbujani, G. (2002). Patterns of human diversity, within and among continents, inferred from biallelic DNA polymorphisms. *Genome Research* **12**, 602–612.

Rosenberg, N.A., Mahajan, S., Ramachandran, S., Zhao, C., Pritchard, J.K. and Feldman, M.W. (2005). Clines, clusters, and the effect of study design on the inference of human population structure. *PLoS Genetics* **1**, e70.

Rosenberg, N.A., Pritchard, J.K., Weber, J.L., Cann, H.M., Kidd, K.K., Zhivotovsky, L.A. and Feldman, M.W. (2002). Genetic structure of human populations. *Science* **298**, 2381–2385.

Ryman, N., Chakraborty, R. and Nei, M. (1983). Differences in the relative distribution of human gene diversity between electrophoretic and red and white cell antigen loci. *Human Heredity* **33**, 93–102.

Sarich, V. (2000). The final taboo. Race differences in ability. *Skeptic* **8**, 38–43.

Sawyer, S.L., Mukherjee, N., Pakstis, A.J., Feuk, L., Kidd, J.R., Brookes, A.J. and Kidd, K.K. (2005). Linkage disequilibrium patterns vary substantially among populations. *European Journal of Human Genetics* **13**, 677–686.

Scarre, C. (2005). *The Human Past*. Thames and Hudson Ltd, London.

Seielstad, M.T., Minch, E. and Cavalli-Sforza, L.L. (1998). Genetic evidence for a higher female migration rate in humans. *Nature Genetics* **20**, 278–280.

Serre, D., Langaney, A., Chech, M., Teschler-Nicola, M., Paunovic, M., Mennecier, P., Hofreiter, M., Possnert, G. and Pääbo, S. (2004). No evidence of Neandertal mtDNA contribution to early modern humans. *PLoS Biology* **2**, 313–317.

Serre, D. and Pääbo, S. (2004). Evidence for gradients of human genetic diversity within and among continents. *Genome Research* **14**, 1679–1685.

Sherry, S.T., Batzer, M.A. and Harpending, H. (1998). Modeling the genetic architecture of modern populations. *Annual Review of Anthropology* **27**, 153–169.

Shriver, M.D., Kennedy, G.C., Parra, E.J., Lawson, H.A., Sonpar, V., Huang, J., Akey, J.M. and Jones, K.W. (2004). The genomic distribution of population substructure in four populations using 8,525 autosomal SNPs. *Human Genomics* **1**, 274–286.

Shriver, M.D., Mei, R., Parra, E.J., Sonpar, V., Halder, I., Tishkoff, S.A., Schurr, T.G., Zhadanov, S.I., Osipova, L.P., Brutsaert, T.D., Friedlaender, J., Jorde, L.B., Watkins, W.S., Bamshad, M.J., Gutierrez, G., Loi, H., Matsuzaki, H., Kittles, R.A., Argyropoulos, G., Fernandez, J.R., Akey, J.M. and Jones, K.W. (2005). Large-scale SNP analysis reveals clustered and continuous patterns of human genetic variation. *Human Genomics* **2**, 81–89.

Shriver, M.D., Parra, E.J., Dios, S., Bonilla, C., Norton, H., Jovel, C., Pfaff, C., Jones, C., Massac, A., Cameron, N., Baron, A., Jackson, T., Argyropoulos, G., Jin, L., Hoggart, C.J., McKeigue, P.M. and Kittles, R.A. (2003). Skin pigmentation, biogeographical ancestry and admixture mapping. *Human Genetics* **112**, 387–399.

Shriver, M.D., Smith, M.W., Jin, L., Marcini, A., Akey, J.M., Deka, R. and Ferrell, R.E. (1997). Ethnic-affiliation estimation by use of population-specific DNA markers. *American Journal of Human Genetics* **60**, 957–964.

Slatkin, M. and Hudson, R.R. (1991). Pairwise comparisons of mitochondrial DNA sequences in stable and exponentially growing populations. *Genetics* **129**, 555–562.

Smith, F.H. (1985). Continuity and change in the origin of modern *Homo sapiens*. *Zeitschrift fur Morphologie und Anthropologie* **75**, 197–222.

Sokal, R.R., Jacquez, G.M. and Wooten, M.C. (1989). Spatial autocorrelation analysis of migration and selection. *Genetics* **121**, 845–855.

Spielman, R.S., Bastone, L.A., Burdick, J.T., Morley, M., Ewens, W.J. and Cheung, V.G. (2007). Common genetic variants account for differences in gene expression among ethnic groups. *Nature Genetics* **39**, 226–231.

Stephens, M. and Donnelly, P. (2000). Inference in molecular population genetics. *Journal of the Royal Statistical Society* **B 62**, 605–635.

Storz, J.F. and Beaumont, M.A. (2002). Testing for genetic evidence of population expansion and contraction: an empirical analysis of microsatellite DNA variation using a hierarchical Bayesian model. *Evolution* **56**, 154–166.

Storz, J.F., Payseur, B.A. and Nachman, M.W. (2004). Genome scans of DNA variability in humans reveal evidence for selective sweeps outside of Africa. *Molecular Biology and Evolution* **21**, 1800–1811.

Stringer, C.B. (1978). Some problems in Middle Pleistocene hominid relationships. In *Recent Advances in Primatology*, D. Chivers and K. Joysey, eds. Academic Press, London, pp. 395–418.

Stringer, C.B. (1989). The origin of modern humans: a comparison of European and non-European evidence. In *The Human Revolution: Behavioural and Biological Perspectives on the Origins of Modern Humans*, P. Mellars and C. Stringer, eds. Edinburgh University Press, Edinburgh, pp. 123–154.

Stringer, C.B. and Andrews, P. (1988). Genetic and fossil evidence for the origin of modern humans. *Science* **239**, 1263–1268.

Tajima, F. (1989a). Statistical method for testing the neutral mutation hypothesis by DNA polymorphism. *Genetics* **123**, 585–595.

Tajima, F. (1989b). The effect of change in population size on DNA polymorphism. *Genetics* **123**, 597–601.

Takahata, N., Lee, S.H. and Satta, Y. (2001). Testing multiregionality of human origins. *Molecular Biology and Evolution* **18**, 172–183.

Tang, H., Quertermous, T., Rodriguez, B., Kardia, S.L., Zhu, X., Brown, A., Pankow, J.S., Province, M.A., Hunt, S.C., Boerwinkle, E., Schork, N.J. and Risch, N.J. (2005). Genetic structure, self-identified race/ethnicity, and confounding in case-control association studies. *American Journal of Human Genetics* **76**, 268–275.

Tattersall, I. (1995). *The Fossil Trail: How We Know What We Think We Know About Human Evolution*. Oxford University Press, Oxford.

Tattersall, I. (1997). Out of Africa again ... and again? *Scientific American* **276**, 60–67.

Tavaré, S., Balding, D.J., Griffiths, R.C. and Donnelly, P. (1997). Inferring coalescence times from DNA sequence data. *Genetics* **145**, 505–518.

Templeton, A.R. (1999). Human races: a genetic and evolutionary perspective. *American Anthropologist* **100**, 632–650.

Templeton, A.R. (2002). Out of Africa again and again. *Nature* **416**, 45–51.

Tishkoff, S.A., Dietzsch, E., Speed, W., Pakstis, A.J., Kidd, J.R., Cheung, K., Bonne-Tamir, B., Santachiara-Benerecetti, A.S., Moral, P. and Krings, M. (1996). Global patterns of linkage disequilibrium at the CD4 locus and modern human origins. *Science* **271**, 1380–1387.

Tishkoff, S.A., Goldman, A., Calafell, F., Speed, W.C., Deinard, A.S., Bonne-Tamir, B., Kidd, J.R., Pakstis, A.J., Jenkins, T. and Kidd, K.K. (1998). A global haplotype analysis of the myotonic dystrophy locus: implications for the evolution of modern humans and for the origin of myotonic dystrophy mutations. *American Journal of Human Genetics* **62**, 1389–1402.

Tishkoff, S.A., Pakstis, A.J., Stoneking, M., Kidd, J.R., Destro-Bisol, G., Sanjantila, A., Lu, R.B., Deinard, A.S., Sirugo, G., Jenkins, T., Kidd, K.K. and Clark, A.G. (2000). Short tandem-repeat polymorphism/alu haplotype variation at the PLAT locus: implications for modern human origins. *American Journal of Human Genetics* **67**, 901–925.

Tishkoff, S.A., Reed, F.A., Ranciaro, A., Voight, B.F., Babbitt, C.C., Silverman, J.S., Powell, K., Mortensen, H.M., Hirbo, J.B., Osman, M., Ibrahim, M., Omar, S.A., Lema, G., Nyambo, T.B., Ghori, J., Bumpstead, S., Pritchard, J.K., Wray, G.A. and Deloukas, P. (2007). Convergent adaptation of human lactase persistence in Africa and Europe. *Nature Genetics* **39**, 31–40.

Tishkoff, S.A. and Verrelli, B.C. (2003). Patterns of human genetic diversity: implications for human evolutionary history and disease. *Annual Review of Genomics and Human Genetics* **4**, 293–340.

Toth, N. and Schick, K. (2005). African origins. In *The Human Past*, C. Scarre, ed. Thames and Hudson Ltd, London, pp. 46–83.

Trinkaus, E. (2005). Early modern humans. *Annual Review of Anthropology* **34**, 207–230.

Underhill, P.A., Shen, P., Lin, A.A., Jin, L., Passarino, G., Yang, W.H., Kauffman, E., Bonne-Tamir, B., Bertranpetit, J., Francalacci, P., Ibrahim, M., Jenkins, T., Kidd, J.R., Mehdi, S.Q., Seielstad, M.T., Wells, R.S., Piazza, A., Davis, R.W., Feldman, M.W., Cavalli-Sforza, L.L. and Oefner, P.J. (2000). Y chromosome sequence variation and the history of human populations. *Nature Genetics* **26**, 358–361.

Vekua, A., Lordkipanidze, D., Rightmire, G.P., Agusti, J., Ferring, R., Maisuradze, G., Mouskhe-lishvili, A., Nioradze, M., Ponce de Leòn, M.S., Tappen, M., Tvalchrelidze, M. and Zollikofer, C.P.E. (2002). A new skull of early *Homo* from Dmanisi, Georgia. *Science* **297**, 85–89.

Vigilant, L., Stoneking, M., Harpending, H., Hawkes, K. and Wilson, A.C. (1991). African populations and the evolution of human mitochondrial DNA. *Science* **253**, 1503–1507.

Voight, B.F., Adams, A.M., Frisse, L.A., Qian, Y., Hudson, R.R. and Di Rienzo, A. (2005). Interrogating multiple aspects of variation in a full resequencing data set to infer human population size changes. *Proceedings of the National Academy of Sciences of the United States of America* **102**, 18508–18513.

Voight, B.F., Kudaravalli, S., Wen, X. and Pritchard, J.K. (2006). A map of recent positive selection in the human genome. *PLoS Biology* **4**, e72.

Waddle, D.M. (1994). Matrix correlation tests support a single origin for modern humans. *Nature* **368**, 452–454.

Wall, J.D. and Przeworski, M. (2000). When did the human population size start increasing? *Genetics* **155**, 1865–1874.

Wang, J.L. (2003). Maximum-likelihood estimation of admixture proportions from genetic data. *Genetics* **164**, 747–765.

Watkins, W.S., Prasad, B.V., Naidu, J.M., Rao, B.B., Bhanu, B.A., Ramachandran, B., Das, P.K., Gai, P.B., Reddy, P.C., Reddy, P.G., Sethuraman, M., Bamshad, M.J. and Jorde, L.B. (2005). Diversity and divergence among the tribal populations of India. *Annals of Human Genetics* **69**, 680–692.

Watkins, W.S., Ricker, C.E., Bamshad, M.J., Carroll, M.L., Nguyen, S.V., Batzer, M.A., Harpending, H.C., Rogers, A.R. and Jorde, L.B. (2001). Patterns of ancestral human diversity: an analysis of Alu-insertion and restriction-site polymorphisms. *American Journal of Human Genetics* **68**, 738–752.

Watkins, W.S., Rogers, A.R., Ostler, C.T., Wooding, S., Bamshad, M.J., Brassington, A.M., Carroll, M.L., Nguyen, S.V., Walker, J.A., Prasad, B.V., Reddy, P.G., Das, P.K., Batzer, M.A. and Jorde, L.B. (2003). Genetic variation among world populations: inferences from 100 Alu insertion polymorphisms. *Genome Research* **13**, 1607–1618.

Watterson, G.A. (1978). The homozygosity test of neutrality. *Genetics* **88**, 405–417.

Weatherall, D.J. (2001). Phenotype-genotype relationships in monogenic disease: lessons from the thalassaemias. *Nature Reviews Genetics* **2**, 245–255.

Weber, W.W. (2001). Effect of pharmacogenetics on medicine. *Environmental and Molecular Mutagenesis* **37**, 179–184.

Weidenreich, F. (1943). The skull of *Sinanthropus pekinensis*: A comparative study of a primitive hominid skull. *Palaeontologica Sinica* **D10**, 1–485.

Weidenreich, F. (1947). Are human races in the taxonomic sense races or species? *American Journal of Physical Anthropology* **5**, 369–371.

Weiss, G. and von Haeseler, A. (1998). Inference of population history using a likelihood approach. *Genetics* **149**, 1539–1546.

Wilder, J.A., Kingan, S.B., Mobasher, Z., Metni Pilkington, M. and Hammer, M.F. (2004). Global patterns of human mtDNA and Y chromosome structure are not Influenced by higher rates of female migration. *Nature Genetics* **36**, 1122–1125.

Wilson, I.J. and Balding, D.J. (1998). Genealogical inference from microsatellite data. *Genetics* **150**, 499–510.

Wilson, I.J., Weale, M.E. and Balding, D.J. (2003). Inferences from DNA data: population histories, evolutionary processes and forensic match probabilities. *Journal of the Royal Statistical Society A* **166**, 155–188.

Wilson, J.E., Weale, M.E., Smith, A.C., Gratrix, F., Fletcher, B., Thomas, M.G., Bradman, N. and Goldstein, D.B. (2001). Population genetic structure of variable drug response. *Nature Genetics* **29**, 265–269.

Wolpoff, M.H. (1999). The systematics of *Homo*. *Science* **284**, 1774–1775.

Wolpoff, M.H., Spuhler, J.N., Smith, F.H., Radovcic, J., Pope, G., Frayer, D.W., Eckhardt, R. and Clark, G. (1988). Modern human origins. *Science* **241**, 772–774.

Wright, S. (1931). Evolution in Mendelian populations. *Genetics* **16**, 97–159.

Wright, S. (1943). Isolation by distance. *Genetics* **28**, 139–156.

Yu, N., Chen, F.C., Ota, S., Jorde, L.B., Pamilo, P., Patthy, L., Ramsay, M., Jenkins, T., Shyue, S.K. and Li, W.H. (2003). Larger genetic differences within Africans than between Africans and Eurasians. *Genetics* **161**, 269–274.

Zilhão, J. (2006). Neandertals and moderns mixed, and it matters. *Evolutionary Anthropology* **15**, 183–195.

Part 6

Genetic Epidemiology

32

Epidemiology and Genetic Epidemiology

P.R. Burton, J.M. Bowden and M.D. Tobin

Departments of Health Sciences and Genetics, University of Leicester, Leicester, UK

This chapter focuses on descriptive analysis in genetic epidemiology and some of its links with mainstream epidemiology. It starts with incidence and prevalence, briefly reviewing some of the relevant statistical methods, and then moves on to consider the analysis of correlated responses in mainstream epidemiology. The core of the chapter describes methods used in descriptive analysis in genetic epidemiology. These methods are primarily used to describe and draw inferences about the structure of phenotypic dependencies within families. The appropriate focus of traditional descriptive epidemiology on representative sampling frames and high response rates is paralleled by the equal importance of sampling schemes in descriptive genetic epidemiology and a full section is therefore devoted to non-random ascertainment.

32.1 INTRODUCTION

Epidemiology may be defined as 'the study of the distribution and determinants of health-related states and events in *populations*' (Last, 2001). The term *genetic epidemiology* is less clearly defined. Neel and Schull (1954) referred to *epidemiological genetics* almost 50 years ago and, since then, different authors have variously emphasised the focus of genetic epidemiology on familial aggregation (King *et al.*, 1984), inherited disease in populations (Morton and Chung, 1978), the genetic structure of populations (Roberts, 1985) and on gene–environment interactions (Cohen, 1980). Perhaps the most satisfactory single definition is that given by Morton when he defined genetic epidemiology as 'a science which deals with the aetiology, distribution and control of disease *in groups of relatives* and with *inherited causes of disease in populations*' (Morton, 1982). The italics in Last's definition of epidemiology and Morton's definition of genetic epidemiology are after Hopper (1992). Firstly, they draw attention to the fact that both disciplines have a focus on being able to draw inferences at the level of a population rather than at the level of an individual (or a single family). This is not to say that some of the analytical methods

Handbook of Statistical Genetics, Third Edition. Edited by D.J. Balding, M. Bishop and C. Cannings.
© 2007 John Wiley & Sons, Ltd.   ISBN: 978-0-470-05830-5.

used in genetic epidemiology cannot also be used to draw inferences in individual families, e.g. in genetic counselling, but that, like mainstream epidemiology, genetic epidemiology is primarily about the pooling of information across many individuals (or families) with the aim of drawing more powerful inferences about potentially weak effects at the level of a population. Secondly, they emphasise that a key difference between epidemiology and genetic epidemiology is that while the former usually (but not always) focuses on individuals, the latter often (though again not always) focuses on families and that interest often centres upon inherited causes of disease. In this regard, it is important to note that the term *inherited* does not necessarily imply 'genes', but that it also subsumes non-genetic causes of disease clustering within families and this includes cultural and environmental mechanisms of inheritance (Burton *et al.*, 2005).

In general terms, the studies conducted by mainstream epidemiological research groups may be divided into three broad categories: (1) those that aim to describe the distribution of a disease or a determinant at the level of a population of interest (often the general population); (2) those that attempt to investigate a potential aetiological link between one or more specific determinants and a disease of interest; and (3) those aimed at formally evaluating the effectiveness and/or cost of an intervention applied to individuals or groups of individuals in the general population. Historically, much of the work of genetic epidemiology may be viewed as having been of the first type and it is upon this category that this chapter will concentrate. However, many of the contemporary developments in genetic epidemiology address studies of the second type, and this category will be briefly discussed in Section 32.4. It is likely that, in the future, genetic epidemiologists will become increasingly involved in studies of the third type as they are required to evaluate interventions developed from the new genetics. However, the issues relevant to the undertaking of such studies go beyond the remit of this particular chapter.

We make no claim that this chapter represents a definitive review of all of the many issues it attempts to address. Its primary aim is to consider descriptive analysis in genetic epidemiology in the light of the analogous aims of mainstream epidemiology, and to point the reader to some of the highlights of a broad and complex literature. Some of the issues we raise are also addressed in other chapters in this handbook, and for reviews of much of the theory and practice of genetic epidemiology, we recommend the book *Biostatistical Genetics and Genetic Epidemiology* edited by Elston *et al.* (2002) and also the *Lancet Series in Genetic Epidemiology* (Burton *et al.*, 2005; Cordell and Clayton, 2005; Davey Smith *et al.*, 2005; Hattersley and McCarthy, 2005; Hopper *et al.*, 2005; Palmer and Cardon, 2005; Teare and Barrett, 2005).

32.2 DESCRIPTIVE EPIDEMIOLOGY

Descriptive studies in the traditional epidemiological setting typically focus upon the distribution of an outcome of interest (a disease, a health care event or possibly an aetiological exposure) in the general population, or in some defined subsection of the general population. A typical example of such a study might be a survey investigating the prevalence of symptoms related to asthma in children up to five years of age (Luyt *et al.*, 1993). Such a study might be undertaken for the purpose of determining appropriate health care provision, or alternatively as a lead into a potential aetiological study by asking a question such as, 'has there been a change in prevalence over time?' (Kuehni *et al.*, 2000). In practice, such studies are difficult to conduct. This is because it is essential that the study population is representative of the target population to be described and careful study design and conduct are therefore crucial. It is essential that the sampling frame (the formal list of potential sampling units) is clearly defined and that it is appropriate, given the specific purpose of the study. It is also crucial that the response rate to the study (if it involves seeking some form of response from individuals) is high and that it is unbiased with respect to the outcome of interest. It is important that the outcome of interest is assessed in a valid and rigorous manner (Hennekens and Buring, 1987; Rothman and Greenland, 1998). Finally, analysis has to be undertaken with care, in particular, it is important to ensure that any potential correlation between observational units is recognised, investigated and taken into appropriate account. Such correlation is occasionally of interest in its own right and, even if it is not, failure to model it correctly will lead to incorrect standard errors and sometimes to biased parameter estimates.

All of these concepts apply equally to descriptive studies in genetic epidemiology (see Section 32.3), but for a variety of reasons, the issues are even more important in this latter setting. Firstly, there is a wider class of genetic epidemiological studies that may reasonably be viewed as being primarily descriptive in nature than in traditional epidemiology. Indeed, if one looks at the history of research in genetic epidemiology, much of the work that has been undertaken *has* been descriptive in nature. Historically, the key question that genetic epidemiologists often addressed was: '*is the distribution of disease D between and within families, consistent with there being an inherited component that might be consequent upon genes?*' Secondly, the option of rendering a study unbiased at the level of the general population in relation to the outcome of interest, while desirable, is sometimes not possible in genetic epidemiology. Many conditions of interest to genetic epidemiologists are simply too rare to be efficiently studied using random samples. Instead genetic epidemiologists often have to work with samples that are deliberately biased (by oversampling families with affected individuals) and must then deal with the implications of this non-random ascertainment in the analysis (Fisher, 1934; Elston and Sobel, 1979; Burton *et al.*, 2000). Thirdly, the correlation structure between family members, which is in part consequent upon powerful biological mechanisms, is complex. Furthermore, it is often of primary interest to genetic epidemiologists and must therefore be modelled carefully.

Figure 32.1 is a flow chart illustrating a 'stylised' overview of the sequence of tasks that might be taken on by a genetic epidemiologist in trying to determine whether one or more genes might exist that are likely to have a large enough effect upon the phenotype of interest to make it worthwhile to invest substantial resources to identify these genes. Several points should be made. Firstly, this is a rather simplified perspective of the way that research is (or has ever been) conducted in the real world. Secondly, Figure 32.1 emphasises the fact that this form of descriptive analysis should not be viewed as an end in itself; its primary objective is to help prioritise the expensive research that then needs to be undertaken in collaboration with bioscientists. Thirdly, the increasing focus on complex diseases and the plethora of biological knowledge and data that have followed in the wake of the human genome project have rendered such a stepped approach less common in recent times. The most expensive part of many contemporary studies is the collection of families, and the phenotyping (Thompson, 2002). Therefore once such data have been collected, there is a natural tendency to jump straight to tests of

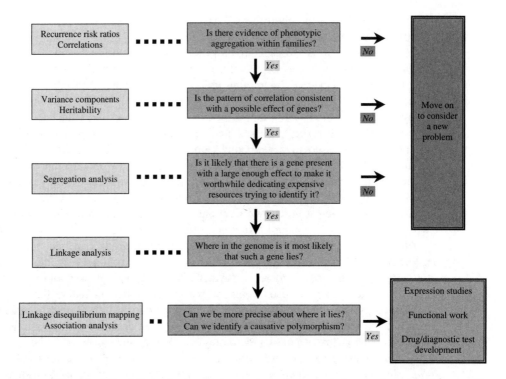

Figure 32.1 A stylised flow chart outlining a stepped approach in genetic epidemiology. (The failure to find evidence of genetic variant(s) with large enough effects will not necessarily impede contemporary aetiological studies shown in subsequent steps. See (Burton *et al.*, 2005) for a discussion of this issue.)

associations between phenotype and specific biomarkers (including genotypic markers) rather than investing extensive time and resources examining the familial distribution of the phenotype. Nevertheless, such analyses continue to be undertaken in an attempt to provide us with some initial guidance in our first attempts to understand the aetiology of the complex diseases (Duffy *et al.*, 1990; Palmer *et al.*, 2001). Fourthly, there are important overlaps with equivalent perspectives in agricultural genetics (see **Chapter 20**).

32.2.1   Incidence and Prevalence

Much of the descriptive analysis in epidemiology involves the investigation of prevalence or incidence.

Assuming that disease status is uncorrelated between individuals, the estimated prevalence of disease at time t in a population under study is defined as

$$\hat{\pi}_t = \frac{D_t}{N_t}, \tag{32.1}$$

where D_t is the number of individuals who are diseased at time t, and N_t is the total number of individuals who were evaluated at that same point in time. Because it relates to a specific point in time, prevalence is sometimes referred to as *point prevalence*. The statistic $\hat{\pi}_t$ is a proportion and exact or asymptotic inferences may be based on the binomial log likelihood which (up to a constant) is given by

$$\ell = D_t \log(\pi_t) + (N_t - D_t) \log(1 - \pi_t). \tag{32.2}$$

Standard errors for $\hat{\pi}_t$ may be obtained in a variety of ways (Clayton and Hills, 1993). One of the most convenient methods (Clayton and Hills, 1993) is to use the Gaussian approximation to the binomial log likelihood and to treat logit($\hat{\pi}_t$) as being asymptotically normally distributed with standard error:

$$\sqrt{\frac{1}{D_t} + \frac{1}{N_t - D_t}}. \tag{32.3}$$

Incidence is a rate (λ) rather than a proportion. It is estimated as the number of new cases of disease (failures) occurring over a period of time (D) divided by the total number of person-years of at-risk exposure (Y) experienced by individuals in the study over that same period of time (Clayton and Hills, 1993):

$$\hat{\lambda} = \frac{D}{Y}. \tag{32.4}$$

The person-years of exposure is obtained by summing the subject-specific periods of at-risk exposure across all individuals. For most purposes, individuals are not viewed as being at risk of developing the disease once they have already developed the disease and/or following an event such as death. These events therefore abbreviate the total person-years of exposure. Analysis is typically based on the Poisson log likelihood which (up to a constant) can be expressed as

$$\ell \approx D \log(\lambda) - \lambda Y. \tag{32.5}$$

On the basis of a Gaussian approximation to the Poisson log likelihood, a standard error for $\log(\hat{\lambda})$ can conveniently be obtained (Clayton and Hills, 1993) as

$$\sqrt{\frac{1}{D}}. \tag{32.6}$$

Analyses of incidence and prevalence are easily extended to incorporate observed covariates by using generalised linear modelling (McCullagh and Nelder, 1989) in the form of logistic or Poisson regression (see also **Chapter 36**) (Breslow and Day, 1980; 1987; Clayton and Hills, 1993). Incidence data can also be modelled by treating the analysis as a failure time problem and using conventional survival time methods (Miller *et al.*, 1981) including Cox proportional hazards regression to incorporate covariates (Cox, 1972).

32.2.2 Modelling Correlated Responses

Correlation, covariance or association between the discrete outcomes in individual subjects in descriptive mainstream epidemiology usually reflects one of the following: (1) cluster sampling by design; (2) a multi-centre study design; (3) longitudinal repeated measurements over time; (4) a natural clustering structure in the study population, e.g. data related to patients registered with general practitioners (family physicians); or (5) latent determinants of disease which have a tendency to cluster geographically or in time. Depending upon the purpose of analysis, and the structure of the data, a variety of different classes of models may be used to analyse such data. A comprehensive review of these models, and their implications for model interpretation would warrant a full chapter on its own (e.g. see: Neuhaus, 1992; Breslow and Clayton, 1993; Pendergast *et al.*, 2002) and we will restrict ourselves to a few brief comments.

The classes of model that may be used include (Neuhaus, 1992; Diggle *et al.*, 1994; Pendergast *et al.*, 2002): (1) conditional models in which the probability of response of an individual observational unit is conditioned on all other responses in a cluster; (2) transition models in which there is a logical ordering to the responses in a cluster and conditioning can be restricted to those responses that pre-date a response of interest; (3) marginal models in which one jointly estimates parameters underpinning the marginal mean and others reflecting the correlation or association between units in a cluster but, crucially, the model for the marginal mean does not include cluster-specific effects, and this results in parameters having a 'population-averaged' interpretation; (4) cluster-specific (random effects) models, which differ from marginal models in that they *do* have cluster-specific effects in the model for the mean, and parameters therefore have a 'subject-specific' interpretation. Many of the models that can be used to analyse correlated data may be viewed as being different types of generalised linear mixed model (GLMM) (Breslow and Clayton, 1993) and there are many different approaches to model fitting (Breslow and Clayton, 1993; Pendergast *et al.*, 2002). We will restrict our comments to some of the methods that are used most commonly in mainstream epidemiology.

If primary scientific interest focuses on the fixed effects and the correlation structure is essentially a nuisance, marginal models based on conventional (first order) generalised estimating equations (GEEs) provide a quick and convenient way to analyse the required data (Liang and Zeger, 1986; Zeger and Liang, 1986; 1992; Burton *et al.*, 1998; Pendergast *et al.*, 2002). In a conventional GEE the equations for the fixed effect parameters and the correlation parameters are assumed orthogonal and estimation is based solely on the mean and second order correlation components only. Provided the model for the mean is correctly specified and a robust sandwich-based estimator of the covariance matrix for the fixed parameter estimates is used (Huber, 1967), and provided inferences are then based on Wald tests and estimation intervals, the conventional GEE approach generates consistent (though not necessarily fully efficient) inferences for the fixed regression coefficients even if the covariance structure is incorrect or varies somewhat from cluster to cluster. Models based on GEEs can easily be fitted in many standard software packages including Stata (xtgee) (Stata Corporation, 2005) and SAS (Proc Genmod) (SAS Institute Inc., 2001).

Unfortunately, models fitted using first-order GEEs will often not provide good estimators of the parameters underlying the correlation/covariance structure itself (Pickles, 2002). In consequence, if the correlation is of primary interest in its own right, approaches based on multi-level modelling (Goldstein, 1986; 1991; Breslow and Clayton, 1993), structural equation modelling (Wright, 1921; Neale and Cardon, 1992) or higher order GEEs (Prentice and Zhao, 1991; Pendergast *et al.*, 2002; Pickles, 2002) are commonly used. In practice, models such as these are most often fitted using general purpose software such as MLWin (Rasbash *et al.*, 1999), Splus (lme, nlme) (MathSoft, 2000), Lisrel (Jöreskog and Sörbom, 1996) or Mx (Neale *et al.*, 2002) or specialist software such as GEE4 (Hanfelt, 1993). In recent times there has been an increasing use of Markov chain

Monte Carlo (MCMC) based approaches (Breslow and Clayton, 1993) particularly Gibbs sampling (Zeger and Karim, 1991; Breslow and Clayton, 1993; Spiegelhalter *et al.*, 2000).

Although most traditional epidemiological analysis is based upon the binomial and Poisson distributions, some quantitative outcomes (such as blood pressure) and exposures are more naturally modelled assuming normality either directly or following transformation. Standard statistical methods for Gaussian responses may then be used. These again include univariate methods (Armitage and Berry, 2002), generalised regression-based techniques (McCullagh and Nelder, 1989) and extensions to deal with correlated data (Goldstein, 1987; Breslow and Clayton, 1993; Burton *et al.*, 1998). Many mainstream epidemiologists use SAS (Proc Mixed) (SAS Institute Inc., 2001) or SPSS (GLM repeated measures) (SPSS Inc., 2002) to model correlated continuous responses.

32.3 DESCRIPTIVE GENETIC EPIDEMIOLOGY

Like any other epidemiologists, genetic epidemiologists sometimes need to undertake simple descriptive analyses based on prevalence, incidence or quantitative outcomes. Such analyses may be based on the methods outlined above. In addition, however, genetic epidemiologists need to be able to carry out the more specialised types of analysis that permit them to address the questions addressed in the stylised flow chart in Figure 32.1.

32.3.1 Is There Evidence of Phenotypic Aggregation within Families?

Simple familial aggregation can be studied in a number of different ways. For a binary phenotype (disease absent/present), one of the simplest forms of analysis is to estimate the recurrence risk ratio in relatives of type R of affected individuals (Risch, 1990). This risk ratio is defined as

$$\frac{\Pr(\text{Disease in relative of type } R | \text{ affected case})}{\hat{\pi}}, \tag{32.7}$$

where $\hat{\pi}$ is the estimated prevalence of the disease in the general population (Risch, 1990; Kopciuk and Bull, 2002). Most commonly it is the recurrence risk ratio for siblings ($\hat{\lambda}_S$) that is estimated. As well as evaluating the evidence in favour or against familial aggregation, Risch (1990) has shown that this statistic is an important determinant of the power of affected relative pair studies to detect linkage (see **Chapter 34**). Despite the simplicity of the λ_S statistic, some comments are warranted. Firstly, it reflects aggregation within a sibship regardless of its cause. It does *not* specifically estimate aggregation due to genes. Secondly, particularly in diseases with a late onset, any systematic differences between the age distribution in the siblings of the affected cases, and the age distribution in the population sample from which $\hat{\pi}$ was estimated could distort the statistic. Thirdly, in relation to diseases with a high population prevalence, such as asthma (Palmer *et al.*, 2001), $\hat{\lambda}_S$ has a theoretical upper bound at $\frac{1}{\hat{\pi}}$ which means that one should be cautious in comparing values between populations that have different prevalences.

A variety of other risk ratios for relatives that may be used to assess familial aggregation have also been described (Kopciuk and Bull, 2002).

Familial aggregation can also be assessed by estimating appropriate intra-class and inter-class correlation coefficients. For quantitative phenotypes, this approach dates back more than a century to Galton (1877; 1886) and Pearson (1896). From a modern perspective, the assessment of familial aggregation in this manner is no different to the more general problem of modelling clustered data with a simple correlation structure in mainstream epidemiology (see Section 32.2.2). All the same issues therefore apply and the same models and model fitting approaches may be used. It should however be noted that certain statistical packages, such as the fortran-based program FISHER (Lange *et al.*, 1988) which fits maximum-likelihood-based linear mixed models for normally distributed phenotypes, and Lisrel (Jöreskog and Sörbom, 1996) and Mx (Neale *et al.*, 2002) which may be used to fit structural equation models have been used much more commonly by genetic epidemiologists than by mainstream epidemiologists.

32.3.2 Is the Pattern of Correlation Consistent with a Possible Effect of Genes?

If and when it has been demonstrated that there is evidence of simple familial aggregation, the next logical step is to investigate the structure of the correlation underpinning that aggregation. The use of variance components models to explore complex correlation or covariance structures has a long and distinguished history in genetic epidemiology (Fisher, 1918; Wright, 1921; Jinks and Fulker, 1970; Neale and Cardon, 1992; Hopper, 1993; Khoury *et al.*, 1993a; Hopper, 2002b; Neale, 2002; Rao and Rice, 2002). More recently, extensive work has been carried out based upon mixed effects models fitted using MCMC methods (Guo and Thompson, 1992; Sobel and Lange, 1993; Gauderman and Thomas, 1994; Burton *et al.*, 1999; Scurrah *et al.*, 2000).

As in mainstream epidemiology, many of the relevant models may helpfully be viewed as being GLMMs (Breslow and Clayton, 1993). In order to facilitate a discussion of these models and the uses to which they may be put, we will now consider the structure of one such GLMM in more detail.

32.3.2.1 *Model Structure*

Combining the notation of several authors (Khoury *et al.*, 1993a; Burton, *et al.*, 1999; Hopper, 2002b), a general model with wide applicability may be written as

$$g(\mu_{ij}) = \eta_{ij} = \alpha + \boldsymbol{\beta}^{\mathrm{T}} \mathbf{z}_{ij} + \xi_{ij}, \tag{32.8}$$

$$Y_{ij} \sim f(\mu_{ij}, \varpi),$$

$$\xi_{ij} \sim N(0, [\sigma^2_{\mathrm{A}} + \sigma^2_{\mathrm{D}} + \sigma^2_{\mathrm{C}}]),$$

$$\mathrm{Cov}(\xi_{ij}, \xi_{ik})[j \neq k] = 2\phi_{ij,ik}\sigma^2_{\mathrm{A}} + \Delta_{ij,ik}\sigma^2_{\mathrm{D}} + \gamma_{ij,ik}\sigma^2_{\mathrm{C}},$$

where Y_{ij} is the observed phenotype in the jth member of the ith family, μ_{ij} is its expected value, $f(.)$ denotes an error distribution (from the exponential family) which may incorporate a nuisance parameter denoted ϖ. The expected value of the phenotype is predicted via a link function $g(.)$ applied (in inverse form) to a linear predictor (η_{ij}) comprising a baseline mean (α), a vector of observed covariates ($\mathbf{z}_{ij}$), a corresponding vector of unknown regression parameters ($\boldsymbol{\beta}$) and subject-specific random effects ξ_{ij} with an appropriate covariance structure. The components σ^2_{A}, σ^2_{D} and σ^2_{C} represent, respectively, the variances arising from polygenic additive effects, polygenic dominance effects and shared environmental effects (Fisher, 1918; Khoury *et al.*, 1993a; Hopper, 2002b). The term $\phi_{ij,ik}$ denotes the kinship coefficient between individuals ij and ik: the probability of randomly drawing a single allele in individual ij that is identical by descent (*ibd*) to a single allele at the same locus randomly drawn from individual ik. $\Delta_{ij,ik}$ is the probability that *both* alleles at a locus are shared *ibd* by individuals ij and ik. Table 32.1 details the $\phi_{ij,ik}$ and $\Delta_{ij,ik}$ values for selected relative pairs and the total genetic variances that these imply.

In many models, the elements $\gamma_{ij,ik}$ are simply binary indicators denoting whether two individuals live together ($\gamma_{ij,ik} = 1$) or apart ($\gamma_{ij,ik} = 0$). However, the effect of shared environment may be modelled in a more sophisticated manner. That is by allowing the impact of shared environment to increase over time living together and then to decline with time living apart (Hopper, 1993). Furthermore, one may choose to discriminate between the effects of living together *per se* from those of living together *as a child* by introducing an additional variance component to model shared sibling environment (σ^2_{Cs}) (Hopper, 1993; Burton *et al.*, 1999). However, σ^2_{Cs} is completely confounded with σ^2_D under most simple family designs (except those including monozygous (MZ) twins, double first cousins, adoptees or siblings reared apart). Given appropriate data, the genetic component of variance may be extended to incorporate gene:gene (epistatic) and/or gene:environment interaction effects.

Under model (32.8) (as it is specified) and regardless of the structure of the residual error, the distribution of the ξ_{ij} is multivariate normal within families. This parameterisation is often chosen for reasons of analytic and interpretational tractability (Burton *et al.*, 1999). However, depending how the model is fitted, alternative assumptions may be both possible and preferable. For example, γ frailties are often used in survival time models (Clayton, 1991; Hougaard and Thomas, 2002).

Table 32.2 lists combinations of error structures and link functions that are commonly used to model a number of important classes of trait (McCullagh and Nelder, 1989) (see also **Chapter 20**).

A variance components GLMM, such as model (32.8), invokes a number of critical assumptions (see Box 32.1). These include assumptions about the statistical model, assumptions about the underlying biology and assumptions about the epidemiological

Table 32.1 Genetic components of variance assuming random mating.

Relationship	ϕ	Δ	Genetic covariance
Same person	1/2	1	$\sigma^2_A + \sigma^2_D$
Parent–child	1/4	0	$1/2\sigma^2_A$
Full-siblings	1/4	1/4	$1/2\sigma^2_A + 1/4\sigma^2_D$
Half-siblings	1/8	0	$1/4\sigma^2_A$
Monozygous twins	1/2	1	$\sigma^2_A + \sigma^2_D$
Grandparent–grandchild	1/8	0	$1/4\sigma^2_A$
Uncle/aunt–nephew/niece	1/8	0	$1/4\sigma^2_A$
First cousins	1/16	0	$1/8\sigma^2_A$
Double first cousins	1/8	1/16	$1/4\sigma^2_A + 1/16\sigma^2_D$
Spouses	0	0	0

Table 32.2 Error-link combinations for common classes of phenotype.

Class of trait	$f(\mu_{ij}, \varpi)$	g(.)	Notes
Continuous normally distributed	$N(\mu_{ij}, \sigma^2_E)$	identity(.)	—
Binary	Bernoulli(μ_{ij})	logit(.)	probit(.) is a common alternative link
Count	Poisson(μ_{ij})	log(.)	—
Censored survival time	Poisson(μ_{ij})	log(.)	Response is the binary censoring indicator, log(survival time) is used as an offset

---

Mathematical assumptions

Conditional on fixed and random effects, all responses are independent
Random effects are mutually independent and are independent of the fixed effects
The structure of the model is appropriate (e.g. random effects *are* multivariate normal, and the assumed distribution of observed phenotype about its modelled predictions is correct).

Biological assumptions

The population is in Hardy–Weinberg equilibrium
The biological model is complete
The recorded family structures are correct (e.g. no unrecognised non-paternity).

Epidemiological assumptions

All recorded information (outcomes and covariates) is recorded faithfully
Ascertainment is either random (as assumed in model 1), or else the ascertainment model is appropriately reflected in the likelihood underpinning the model.

Box 32.1 **Modelling assumptions for variance components model.**

design. The risk of violation of each assumption and the relevance of a violation should it occur, will vary from trait to trait and will be influenced by the focus of scientific interest (Hopper, 1993).

32.3.2.2 *Model Fitting*

A GLMM equivalent to model (32.8), may be fitted using a range of different techniques. This includes many of the model-based approaches to analysing correlated data considered in Section 32.2.2. However, first-order GEE models are not ideal if the covariance structure is of primary interest. Model fitting using MCMC (for complex likelihoods in either a

Bayesian or non-Bayesian setting) also offers great flexibility and generalises relatively easily to a wide variety of non-normal traits and to complex extended pedigrees (Guo and Thompson, 1992; Hopper, 1993; Gauderman and Thomas, 1994; Burton *et al.*, 1999; Scurrah *et al.*, 2000). On the other hand, the use of MCMC can be demanding in terms of time commitment to setting up the data, undertaking the analysis and checking the MCMC process for convergence (Gilks *et al.*, 1996). Furthermore, it is often considered undesirable to use an MCMC approach when an exact likelihood-based solution is available.

32.3.2.3 The Interpretation of Parameters

Model (32.8) is a cluster-specific model that generates subject-specific inferences. This is because the random effects (ξ_{ij}) generating the covariance structure appear on the linear predictor. Under this model, $\hat{\beta}_x$ estimates the change in η_{ij}, or equivalently $g(\mu_{ij})$, associated with a one unit change in covariate $\mathbf{z}_x$ *in an individual*. The variance components also have a subject-specific interpretation (Burton *et al.*, 1999; Scurrah *et al.*, 2000). For example, in a model with a logit link, if $\hat{\sigma}_A^2 = 0.5$, then it suggests that a subject at the upper 95 % percentile for additive polygenic effects will have a predicted log(odds) of disease that will be $\Phi(0.95) \times \sqrt{0.5} = 1.645 \times \sqrt{0.5} \approx +1.16$ higher than it would have been if the same subject had been at the median for additive polygenic effects but nothing else had changed (i.e. observed covariates and other random effects were unaltered). This corresponds to an odds ratio of exp(1.16) $\approx$ 3.19.

In a marginal model, the variance components reflect the marginal covariance between the predicted means based solely on the estimated fixed effects (Breslow and Clayton, 1993; Pendergast *et al.*, 2002). Furthermore, $\tilde{\beta}_x$ (estimated under the marginal model) has a population-averaged interpretation reflecting the difference in the marginal expectation of η_{ij} associated with a one unit increase in $\mathbf{z}_x$. In effect, $\tilde{\beta}_x$ is averaged over the full range of frailties in the particular population under study (Diggle *et al.*, 1994). Under an identity link $\hat{\beta}_x = \tilde{\beta}_x$. Under logit and probit links $\hat{\beta}_x \geq \tilde{\beta}_x$. The difference increases as the variance of the frailties rises, and equality only holds if the frailty variance is zero (Neuhaus, 1992; Diggle *et al.*, 1994). The estimated variance components obtained from a marginal model are also shrunk relative to their equivalents in a cluster-specific model in GLMMs with logit and probit links. For a logarithmic link, there is an offset in the estimated marginal mean (Zeger *et al.*, 1988).

32.3.2.4 Identifiability of Variance Components

The joint estimation of the different variance components depends upon the availability of a data set containing a suitable range of different classes of familial relationship. For example, with no further information, a data set based on standard nuclear families allows the modelling of different covariances between two spouses, a parent and a child and between two siblings. This allows the simultaneous resolution of three variance components such as σ^2_A, σ^2_C and σ^2_{Cs} or σ^2_A, σ^2_C and σ^2_D (Burton *et al.*, 1999). Extension to three or more generations will not necessarily increase the number of variance components that may be identified. For example, despite the provision of an extra class of relation (grandparent:grandchild) addition of the grandparental generation will still not permit simultaneous resolution of $\sigma^2_A, \sigma^2_C, \sigma^2_{Cs}$ and σ^2_D because σ^2_{Cs} and σ^2_D will still be linearly dependent (see Table 32.1). On the other hand, such an extension *will* increase the power to resolve σ^2_A, σ^2_C and σ^2_{Cs} and will markedly improve mixing in the MCMC setting (Scurrah *et al.*, 2000). In order to increase the number of components that may be identified, one must either incorporate individuals with different environmental exposures (e.g. some siblings living together and some living apart) or use designs including additional genetically informative relative types such as adoptees, half-siblings or MZ twins. In this context, a particularly attractive design is the twin-family study (Hansen *et al.*, 2000) which enrols MZ and dizygous (DZ) twin pairs with their parents and siblings. Even when all members of every family unit cohabit, such a design permits simultaneous identification of $\sigma^2_A, \sigma^2_C, \sigma^2_{Cs}$ and σ^2_D.

The classical twin design, MZ and DZ twins reared together (with no information on other family members) (Merriman, 1924; Siemans, 1924), provides a robust test of $H_0 : \sigma^2_G = 0$. This is because under the null hypothesis and *assuming an equal environmental covariance*, the covariance between MZ twins and DZ twins should be the same and any genetic determinants, regardless of their nature, will tend to increase the covariance between MZ twins more than DZ twins. However, there are important interpretational limitations (Neale and Cardon, 1992; Neale, 2002). For example, in the absence of additional information (e.g. some twins are reared together and some apart) only two variance components may be resolved. This is because there are only two different covariances: that between a pairs of MZ twins and that between a pairs of DZ twins. Furthermore, the assumption that the shared environmental covariance between pairs of MZ and DZ twins is the same may be viewed as suspect. In contrast, if one is prepared to restrict the model to three covariance terms, the twin-family design (see above) enables this key assumption to be formally tested (Hansen *et al.*, 2000).

32.3.2.5   Twin Studies

There is no doubt that twin studies have been of great historical importance in the development of genetic epidemiology. This is principally because they provide a natural experiment that provides a straightforward means of separating the 'nature' and 'nurture' underlying phenotypic traits (Galton, 1875; Neale and Cardon, 1992; Spector *et al.*, 1999; Spector, 2000; Neale, 2002). That said, the classical twin design does have significant inferential limitations (see above), and it is important that the value of twin studies is not overstated. The potential role of a twin-based design in answering a scientific question of interest must be considered just as carefully as the corresponding role of any alternative design (Neale and Cardon, 1992; Neale, 2002).

From an epidemiological perspective, one of the great advantages of twin studies is that, as a population subgroup, twins are generally very proud of their status and are very motivated to join biomedical research studies. In consequence, there are many large twin registries around the world that provide an important opportunity for genomic and genetic epidemiological study (Strachan, 2002). A far from exhaustive list of major twin registries includes: the Australian Twin Register (http://www.twins.org.au/); GenomEUtwin, a biobank that pulls together twin registers across Europe (http://www.genomeutwin.org/); and the US Veterans Twins Study (http://www.iom.edu/CMS/3795/4907.aspx).

In the post-genomic era, the ability to work directly with characterised genotypes rather than having to infer genetic etiology indirectly from the covariance structure of a trait (Burton *et al.*, 2005) has meant that the role of twin registries is changing. They continue to provide an invaluable source of well-characterised epidemiological information on large numbers of subjects. But their role in providing a natural experiment to explore nature and nurture is becoming less crucial. In contrast increasing use is likely to be made of the special inferential opportunities that arise from twin-based studies. For example, MZ twins provide an ideal group in which to study gene–environment interactions. This is because they provide a 'natural experiment' in which the impact of different environments can be compared and contrasted between twins with exactly the same genetic background. Another opportunity arises from the fact that there is a clear need to increase understanding of the various sources of error in transcriptomics, proteomics and metabolomics. Twin studies, and in particular twin-family studies, provide a powerful design upon which to base such research and, in particular, to discriminate random measurement errors from variability arising from the systematic effect of modulating genes and environmental determinants.

32.3.2.6 *Negative Variance Components*

There is a debate as to whether a variance component such as σ^2_A should be allowed to take negative values. The biological motivation for variance components in genetic epidemiology usually invokes unobserved determinants (genetic or environmental) that are shared more commonly among close relatives. Under such a model, unless determinants are shared systematically *less* commonly among close relatives, truly negative correlations or covariances should not arise. However, unless one believes absolutely in the biological model motivating the variance components model, it can reasonably be argued that the only constraint that should be applied to the model is the fundamental requirement that a variance–covariance matrix should be positive definite; this requires, e.g. that $\sigma^2_A \geq -\sigma^2_{Cs}$ (Burton *et al.*, 1999) and is therefore less constraining that the requirement that it be non-negative. Further, it may be argued that it is only by permitting the variance components to take legal negative values that one can properly test the biological assumption that they *are* individually positive. The use of a less restrictive parameterisation (Burton *et al.*, 1999; Scurrah *et al.*, 2000) is particularly relevant for models fitted using an MCMC approach with inferences being based on the posterior mean, because one is then averaging the posterior distribution across iterations. If a variance component is in truth positive but close to zero, a conventional likelihood-based solution will be valid. On the other hand, unless the MCMC chain is allowed to traverse those legal parts of the parameter space associated with negative values, the MCMC generated estimate of a variance component will be positively biased (Zeger and Karim, 1991; Burton *et al.*, 1999).

32.3.2.7 *Modelling Discrete Traits*

Variance components models for discrete traits are often motivated by assuming an underlying normally distributed latent trait with one or more thresholds (Falconer, 1965; Hopper, 1993; Khoury *et al.*, 1993a; Todorov and Suarez, 2002). In the simple case of a binary disease state, the unobserved value of the latent trait in the jth subject in the ith family may be denoted X_{ij} and the threshold in that same individual as τ_{ij}. The subject is modelled as 'affected' if $X_{ij} \geq \tau_{ij}$ and 'unaffected' otherwise. It has been argued that this model is fundamentally incoherent (Edwards, 1969; Todorov and Suarez, 2002) because very similar latent trait values on either side of a threshold imply a discordant affection status, while potentially very different values on the same side imply concordance. However, this concern has been overstated (Todorov and Suarez, 2002). In its full generality, the latent trait in the ijth individual could be viewed as being drawn from a $N(\theta_{ij}, \sigma^2_{ij})$ distribution. But, a model with subject-specific values for $\theta_{ij}, \sigma^2_{ij}$ and τ_{ij} would be overparameterised given that the X_{ij} are unobserved. Constraints must therefore be applied. If one arbitrarily specifies $\tau_{ij} = 0$ and $\sigma^2_{ij} = 1$ for all i and j, and sets $\theta_{ij} = \eta_{ij}$ (from model 32.8) then the latent trait model can be shown to be mathematically equivalent to a GLMM with a probit link. Under this model, $\mu_{ij} = \Phi(\eta_{ij})$, where $\Phi(.)$ is the cumulative distribution function for a standard normal random variable and μ_{ij} is the expected probability of disease in the ijth subject. Equivalently, under the latent trait model, the probability that X_{ij}, a random drawn from a $N(\eta_{ij}, 1)$ distribution, equals or exceeds $\tau_{ij} = 0$ is also $\Phi(\eta_{ij})$. This indicates that the latent threshold model provides a convenient motivation of what is fundamentally a coherent GLMM, with a probit link, under which the expected probability of affectation increases monotonically with the value of the linear predictor (η_{ij}).

32.3.2.8 *Heritability*

A starting point for many scientists investigating disease aetiology has often been to study the heritability of a health-related trait. Formally, the heritability of a continuous trait is defined as the proportion of its total variance (σ^2_{TOT}) that is attributable to genetic factors in a particular population. Narrow sense heritability is defined as $\sigma^2_A/\sigma^2_{TOT}$ and broad sense heritability as $\sigma^2_G/\sigma^2_{TOT}$, where σ^2_G includes all genetic components of variance.

Heritability is not about cause *per se*, but is about the causes of variation in a particular trait *in a particular population* (Hopper, 2002a). Heritability varies from study to study depending on the population being investigated (environmental exposures vary between populations), the structure of the analytic model and measurement error (Hopper, 1992; 2002a). Fisher (1951) pointed out that, in calculating heritability, whilst the numerator has a simple genetic meaning, the denominator does not. Scientists and the media sometimes treat heritability as meaning the proportion of a disease 'caused' by genetic factors. This is incorrect. If a disease process is entirely dependent upon the presence of a particular allele of a particular gene, but *everybody* in the population is homozygotic for that allele, variation at that locus will play no role in variation of the disease phenotype and it will therefore make no contribution to heritability. On the other hand, the gene is clearly implicated in the causal architecture of the disease. Equivalently, a near ubiquitous environmental exposure will make little or no contribution to the denominator, σ^2_{TOT}. Interpretational ambiguities are also introduced by decisions about the covariates to be included in a model and by interactions between genes and environmental determinants. For example, failure to include an important environmental covariate may well increase σ^2_E (the residual variance) and therefore σ^2_{TOT}, while leaving σ^2_A unchanged. This will apparently decrease the narrow sense heritability. For all of these reasons, it is often preferable to quote the magnitude of the variance components (such as σ^2_A) individually, rather than relying solely on the overall value of the heritability itself (Fisher, 1951). The individual variance components each have a direct scientific interpretation, and their estimates can be compared meaningfully within and between populations (Hopper and Carlin, 1992; Burton *et al.*, 1999).

Because heritability is formally defined in terms of variation in a quantitative trait (Hopper, 2002a) there are specific interpretational problems that pertain to binary traits (Lichtenstein *et al.*, 2000; Spector, 2000). Although a binary trait can be modelled using a latent threshold model or equivalently a GLMM with a probit link (see above), this does not imply that the concept of heritability generalises directly from quantitative to binary traits. For example, in the setting of a binary disease state, the key assumption that the liability is normally distributed (Hopper, 2002a) cannot be tested. Furthermore, an important component of the variability in a binary problem is Bernoulli variation on the native scale of the trait. Mapping this to the scale of the linear predictor is a non-linear problem, and the required approximations are at their worst with binary responses when the binomial denominator is one (Breslow and Clayton, 1993). For a quantitative trait, a principal reason for calculating heritability, rather than quoting a raw value for the variance attributable to additive genetic effects, is that heritability 'standardises' σ^2_A for the 'intrinsic variability' of the trait under study. This is of obvious scientific utility. But for a binary phenotype, the intrinsic variability of the trait is determined principally by the Bernoulli error, which is fixed mathematically. Furthermore, for a typical complex trait, the variability will also include random unshared error terms that contribute directly to the assumed liability; e.g. those reflecting the impact of unmeasured aetiological determinants that are specific to an individual. In designs involving a single measurement of the trait, the additional variability arising from these error terms is generally non-identifiable for a binary phenotype (Neale and Cardon, 1992). In consequence, the 'variance of the liability' is fixed by assumption – it is not estimated from the data. There would appear to be little value in standardising σ^2_A for a quantity that is fixed by assumption, and this suggests that the calculation of 'heritability' for a binary trait in this setting is of rather limited utility.

Given the many pitfalls in the interpretation of heritability, why bother to calculate it at all? The power of most studies for discovering genes is positively associated with the heritability of the trait of interest; so, all else being equal and if the option exists, analytic efficiency may be enhanced by selecting a study population in which the heritability of the trait of interest is thought to be high. Furthermore, subject to all of the caveats above, knowledge that a trait of interest has a high heritability provides support for a study that proposes to investigate the genetic determinants of that trait. Equally, if heritability is low, the investigators and funders of the proposed study are forewarned that genetic effects may be difficult to find. In either case, interpretation demands expert understanding of the nature of the trait.

32.3.2.9 Transition Models

Most of the preceding discussion has focused on the use of marginal and cluster-specific models. However, in genetic epidemiology, transition models are also used to model within-family correlation structures. In particular, Bonney (1984) described a series of models that account for within-family dependencies by specifying a regression relationship between an individual's phenotype and those of certain ancestors and older relatives. In special cases such a model reflects a simple first-order Markov process, but a range of alternative biologically sensible autoregressive dependency structures may also be specified. Classically, and although his nomenclature has changed somewhat over time, (Bonney, 1984; 2002; Hopper, 1995) described four major classes of 'regressive model': (Class A) regression on parental phenotypes and on spousal phenotype if spouse correlations are non-zero; (Class B) add regression on the oldest sibling; (Class C) add the immediately preceding sibling and (Class D) add all preceding siblings. The advantages of the regressive approach include the fact that the models can be fitted using standard regression software (simply by extending the design matrix), and that they may be applied quite naturally both to continuous normally distributed phenotypes (Bonney, 1984) and to binary traits (Bonney, 1986; 1987) by choosing an appropriate regression model. Furthermore, the properties of the models have been investigated extensively and specific analytical structures have been devised that reflect a wide range of different biological scenarios (Bonney, 2002). However, the models also have some disadvantages. These include a potential loss of efficiency arising from the conditioning on earlier responses (Diggle *et al.*, 1994), occasional ambiguities in how to order a pedigree and how to deal with missing data (Bonney, 1984), and the indirect manner in which inferences about the correlation structure are obtained. Regressive models for specialist applications in genetic epidemiology have been built into the software package SAGE (Elston, 2001).

32.3.3   Segregation Analysis

Having determined that the data are consistent with the presence of genetic effects, one may then move on to address the third question in Figure 32.1: 'Is it likely that there is a gene present with a large enough effect to make it worthwhile dedicating expensive resources trying to identify it?' This falls under the heading of segregation analysis and entails assessing whether there are important variants in one or a small number of major genes whose segregation (usually in a Mendelian manner) explains all or a substantial part of the observed variation in the trait of interest. Segregation analysis is already considered elsewhere in this handbook (**Chapter 19** and **Chapter 20**) and we will keep our comments brief.

Elston (1981) defines segregation analysis as: 'The statistical methodology used to determine from family data the mode of inheritance of a particular phenotype, especially with a view to elucidating single gene effects'. He goes on to emphasise that segregation analysis has four major components: (1) the genotypic distribution of mating individuals; (2) the relationship between phenotype and genotype; (3) the mode of inheritance and (4) the sampling scheme. He considers each of these components in detail (Elston, 1981).

Segregation analysis can conveniently be split into classical methods (Majumder, 2002) and more sophisticated model-based approaches which deal with complex segregation problems, particularly those in which more than one mating type is possible for a given sibship (Blangero, 2002). Classical segregation analysis often involves estimation of the *segregation ratio*: that is the probability that an offspring is affected given parental genotypes. That this is often not straightforward is generally a reflection of the complications introduced by non-random ascertainment. Classical estimators have been designed to deal with a number of simple ascertainment schemes (see Section 32.3.4). Weinberg (1912) proposed the *proband method* and derived a simple estimator for use under single ascertainment. The Li–Mantel estimator (Li and Mantel, 1968) may be used under complete ascertainment and Davie (1979) proposed an efficient estimator for intermediate situations with incomplete ascertainment. In contrast, complex segregation methods reflect the fact that reality is more complicated and generally aim to build an analytical component reflecting one or more major genes into a more sophisticated model that simultaneously reflects other determinants of the phenotype of interest. Two approaches to complex segregation analysis that have been widely used are Morton and MacLean's mixed model (1974) and transition (regressive) models (see Section 32.3.2). Both of these approaches permit the modelling of a major gene against a background of polygenic and environmental effects.

Model fitting is often based upon regressive models or variance components models, sometimes incorporating a component of finite mixtures to reflect the major gene (Palmer *et al.*, 2001) (see also **Chapter 18**). Estimation may be based upon maximum likelihood, higher order GEEs or MCMC. Blangero provides an excellent contemporary review of complex segregation models (Blangero, 2002).

32.3.4 Ascertainment

Any descriptive study in mainstream epidemiology must invest considerable time and resource ensuring that its sampling frame is representative of the required target population, and that there is a high response rate. This emphasis highlights the importance of the recruitment process in descriptive epidemiology. Under many circumstances, the ideal sampling frame is one that permits the ascertainment of a random sample of the population to be studied. This is very convenient because it means that the ascertainment process can then be ignored in the analysis.

Unfortunately, genetic epidemiologists do not often have this luxury. If one is interested in the prevalence of disease Q amongst the offspring of parents who have mutually at-risk genotypes, then unless the at-risk genotype is relatively common, it is prohibitively expensive to recruit a random sample of the general population because most families will not be at risk and will therefore be uninformative. Consequently, it would be very useful to identify those families at risk and to restrict recruitment to this subset. If the at-risk genotype is expressed in the phenotype of the parents (e.g. if Q is caused by a dominant polymorphism with high penetrance at an early age) then it may be cheap and easy to identify relevant families through the parents. However, if the causative polymorphism (q) is rare and recessive, then almost all at-risk families will consist of doubly heterozygotic parents (Qq × Qq) who will both be unaffected. Consequently, in the absence of biological knowledge of the relevant genotypes, the only direct way to identify at-risk families is to sample families with affected offspring. From a traditional epidemiological perspective, this appears counter-intuitive: it would not normally be recommended that one should deliberately oversample affected individuals if one is trying to estimate the prevalence of a disease! It might be argued that if one obtained all or a representative sample of families with offspring affected by Q the problem would be circumvented. But this is not so. One group of at-risk families consists of parents who are both Qq heterozygotes, but (by chance) none of their offspring are qq homozygotes and, thus, none are affected by disease Q. By restricting recruitment to families with at least one affected child, this subset of families will be missed, and because amongst all those at risk these particular families have an unusually low prevalence (0 %), their exclusion will seriously bias a naïve estimate of the prevalence of Q in offspring. Ascertainment bias is a recurring problem throughout genetic epidemiology and must always be considered both in study design and analysis.

Non-random ascertainment refers to a mode of sampling that depends systematically on the outcome being analysed as the dependent variable. Most commonly, families containing individuals affected by a disease of interest are preferentially oversampled. Non-random ascertainment has two important implications. Firstly, it distorts the distribution of the outcome in the sample being studied (see above). This may be referred to as *classical ascertainment bias* and its importance has been recognised for many years (Weinberg, 1928; Fisher, 1934; Morton, 1959). Secondly, in situations of aetiological heterogeneity, non-random ascertainment leads to the oversampling of families that truly have an unusually high risk of disease (Burton *et al.*, 2000; Epstein *et al.*, 2002; Glidden and Liang, 2002; Burton, 2003).

Classically, ascertainment has been addressed using the π model (Weinberg, 1928; Morton, 1959). This approach is based upon the concept of a proband who may be defined as: 'an affected person who at any time was detected independently of the other members of the family, and who would therefore be sufficient to assure selection of the family in the absence of other probands' (Morton, 1959). The ascertainment probability (π) is then defined as the probability that an affected individual becomes a proband and the ascertainment probability for a family as the probability that that family contains at least one proband. Thus, in an ascertainment scheme based upon offspring, a family containing s children where p_j denotes the probability of disease in the jth child under the current model, the probability that the family is ascertained (assuming that ascertainment events within sibships are conditionally independent) is given by

$$1 - \prod_{j=1}^{s} (1 - \pi p_j). \tag{32.9}$$

There are two important 'limiting' scenarios: (1) complete ascertainment ($\pi = 1$) under which *all* affected individuals are ascertained and the ascertained sample will include all families with at least one affected child; and (2) Single ascertainment ($\pi \to 0$) when the probability of ascertainment is very low, and the probability that any single family contains more than one proband becomes vanishingly small. Incomplete ascertainment refers to any intermediate situation ($0 < \pi < 1$).

A variety of classical estimators (see Section 32.3.3) have been devised to address the estimation of segregation ratios under these various ascertainment schemes (Hodge, 2002). However, the generality of the relevant theory goes beyond this. If one ignores the ascertainment process, the contribution of an ascertained family F to the likelihood under a statistical model (M) may be denoted $\Pr(\mathbf{Y_F}|M)$ where $\mathbf{Y_F}$ is the response vector for the family. Under non-random ascertainment, this should be replaced by the conditioned likelihood $\Pr(\mathbf{Y_F}|A_F^+, M)$ where A_F^+ denotes the event that family F has been ascertained. Standard probability theory implies

$$\Pr(\mathbf{Y_F}|A_F^+, M) = \frac{\Pr(A_F^+|\mathbf{Y_F}, M)\, \Pr(\mathbf{Y_F}|M)}{\Pr(A_F^+|M)}. \tag{32.10}$$

Furthermore, if the probability of ascertainment given the response vector is unaffected by the parameters in the current model, which it will be under many ascertainment schemes, this simplifies to

$$\Pr(\mathbf{Y_F}|A_F^+, M) \propto \frac{\Pr(\mathbf{Y_F}|M)}{\Pr(A_F^+|M)}.$$

In words: *Up to a multiplicative constant, the likelihood component for family F condi-tioned for the fact that family F has been ascertained may be obtained as the likelihood assuming random ascertainment divided by the probability under the current model (for affection and for ascertainment given affection) that family F would have been ascertained.*

The ability to condition any likelihood for the probability of ascertainment has wide application and is an important feature of many analyses in genetic epidemiology. Of course, it is not always necessary. For example, non-random ascertainment is generally treated as being unimportant in linkage analysis provided ascertainment is based *either* on the trait of interest, *or* the marker, but not *both* (Hodge, 2002). On the other hand, there are other circumstances in which a failure to take proper account of non-random ascertainment can lead to an analysis that is so biased that conclusions are qualitatively incorrect. For example, Hodge (2002) describes an illustrative example consisting of a simple sibling-based design under complete ascertainment in which a true segregation ratio of 25 % (as expected for a rare recessive disease) is incorrectly estimated at 57.1 % (completely inconsistent with a recessive mode of inheritance) when the ascertainment scheme is ignored (Burton *et al.*, 2000; Kraft and Thomas, 2000). Failure to condition on the correct probability of ascertainment may also be due to reasons of computational infeasibility, rather than ignorance. A prime example is when classical bias and aetiological heterogeneity bias need to be accounted for in a complex variance component model of a binary response (Noh *et al.*, 2005). With this in mind, any method that can partially correct for ascertainment bias, that is also practically feasible should not be dismissed outright. Burton (Burton *et al.*, 2000; Burton, 2003) shows, e.g. that it is possible to remove bias due to classical ascertainment, without correcting for the effects of aetiological heterogeneity. However, after this partial correction, estimates for the prevalence of disease or the value of any variance component will pertain to the *ascertained sample* and not the *general population*. This method does not therefore provide a full correction, as noted by Epstein *et al.* (2002) and Glidden and Liang (2002), but it has been recently shown that parameter estimates for the ascertained population can be used to estimate general population parameters via a two-stage approach (Bowden *et al.*, 2006).

Under single ascertainment, adjustment for the sampling scheme can be based upon removal of the probands from all families (Khoury *et al.*, 1993b). However, this will generate inconsistencies under any ascertainment model in which the probability that an individual family contains a proband does *not* increase linearly with the number of affected members. Nevertheless, this is obviously related to the common practice of dividing the full likelihood without adjustment for ascertainment, by the likelihood (under the model) of the observed phenotype in each proband (Hopper, 2002b). Although the option to do this is built into a number of software applications, it is important that the true ascertainment mechanism is considered properly, and simple conditioning on the phenotype of the proband should only be undertaken if it makes sense in the given situation being considered.

In the real world ascertainment schemes are often poorly defined and may not involve unambiguous probands (Morton, 1959; Hodge, 2002). This means that although conditioning the likelihood for the probability that each family is ascertained remains theoretically possible, it may, in practice, be impossible to express this in terms of the parameters in M and this may thwart adjustment. Instead, Ewens and Shute (Ewens and Shute, 1986) proposed an alternative 'resolution of the ascertainment sampling problem'. This involved 'conditioning the likelihood of the sample on that part of the data relevant to ascertainment'. For example, the number of affected children may represent one such component (Ewens and Shute, 1986). Provided one conditions for all such components of the data, it is '... clear that this method will lead to estimates of genetic parameters free of any specific mathematical ascertainment assumption' (Ewens and Shute, 1986). The problem is that conditioning in this manner can discard much of the information driving the analysis in the first place. It can therefore be inefficient (Ewens and Shute, 1986).

As in mainstream epidemiology, the importance of study sampling mechanisms to genetic epidemiology cannot be overstated (Fisher, 1934). Particularly for descriptive studies, it is essential that the ascertainment strategy is considered carefully during study design, and is then taken into proper account at the time of analysis. Where possible it is better for the design to include an explicitly stated sampling scheme (including any sequential rules for pedigree extension once a family has been ascertained (Hodge, 2002) as it is generally easier to correct for the impact of a precisely defined mechanism than for an informal process. It is also important to properly understand the scientific question being addressed and the mathematical model that is to be used, in order to ensure that the analysis is conditioned in such a way that inferences are valid and yet valuable information is not wasted.

We do not believe that there can ever be a single answer to the 'ascertainment problem' and it is likely that most genetic epidemiologists will continue to encounter circumstances in which ascertainment bias is severe (and hence cannot be ignored) and yet the true sampling model is obscure (and hence it is impossible to identify the ideal approach to adopt). As a practical solution in such circumstances, it is our belief that one has no option but to try a variety of tractable approaches to ascertainment adjustment that are consistent with what is known about the problem. This should potentially include adjusting for sampling schemes that are more extreme than single or complete (e.g. quadratic schemes under which the probability of ascertainment of a family increases with the square of the number of affected children (Ewens and Shute, 1986)). If these produce qualitatively different answers to one another, then there is true uncertainty and no firm conclusions can be drawn. On the other hand, if all approaches generate comparable results then one can at least feel more secure in taking action on the basis of those results. Nevertheless, one can never completely exclude the possibility that the true ascertainment mechanism was different, and that if one had been able to properly adjust for it, different conclusions would have resulted.

32.4   STUDIES INVESTIGATING SPECIFIC AETIOLOGICAL DETERMINANTS

To finish, we return briefly to the second class of studies that is fundamental to mainstream epidemiology: 'studies that attempt to investigate a potential aetiological link between one or more specific determinants and a disease of interest'. Most such studies are observational and are based upon cross-sectional, case–control or cohort designs, with analysis being based upon unconditional or conditional, binomial or Poisson likelihoods

(Breslow and Day, 1980; 1987; Clayton and Hills, 1993; Greenland and Rothman, 1998). That we have chosen not to focus our chapter on their equivalents in genetic epidemiology does not in any sense reflect a belief that such studies are unimportant. Indeed, in the wake of the human genome project, studies that investigate the relationships between a phenotype and genomic or proteomic biomarkers are fast becoming the dominant areas of research in genetic epidemiology. Our reason for choosing to focus elsewhere is because most of the key issues pertaining to such studies are already dealt with thoroughly in other chapters in this handbook. **Chapter 18** deals with association studies which investigate 'the detection and analysis of statistical association, at the population level, between a trait and a genotype'. Linkage analysis investigates 'the dependence in inheritance of genes at different genetic loci, on the basis of phenotype observations on individuals' (see **Chapter 33**), and model-free methods are addressed in **Chapter 34**. **Chapter 38** describes the classical transmission disequilibrium test. The TDT is a form of matched case pseudocontrol design, wherein the cases are alleles observed to have been transmitted to the offspring within familial trios consisting of two parents and an affected child, and the pseudocontrols are their unobserved counterparts that were not transmitted. One of the most useful characteristics of the TDT is that it permits the drawing of inferences pertaining to linkage disequilibrium from an, otherwise rather restrictive, family trio design in which an affected child has been genotyped and his/her parents have been genotyped but not necessarily phenotyped. Unfortunately, the TDT approach so rapidly gained widespread popularity that it is sometimes recommended and used unthinkingly and without proper regard to all of the cautionary notes originally described (Spielman *et al.*, 1993).

Contemporary methodological developments in genetic epidemiology are largely focusing upon ways to investigate specific aetiological determinants. Major areas of current methodological research and development include issues pertaining to: (1) the efficient and robust use of single nucleotide polymorphisms in both candidate regions and in genome-wide association studies (Balding, 2006); (2) haplotype analysis; (3) population admixture and stratification (Freedman *et al.*, 2004; Marchini *et al.*, 2004); (4) the use of data from genome-wide SNP genotyping arrays (Clayton *et al.*, 2005; de Bakker *et al.*, 2005; Barrett and Cardon, 2006), comparative genome hybridisation and precise genotyping technologies such as dynamic allele specific hybridisation (DASH) to infer copy number variation and ultimately to relate this to disease (Fredman *et al.*, 2004; Freeman *et al.*, 2006; Redon *et al.*, 2006) and (5) methods for genetic meta-analysis and cross-study synthesis. There is also increasing interest in the use of graphical models to provide a flexible and powerful analytic environment within which to develop the sophisticated models ultimately required to disentangle the complex aetiological architectures of current interest to international medical science. Much of the relevant theory relating to many of these developments is discussed in various places throughout this handbook.

32.5 THE FUTURE

There is a plethora of biological information pertaining to the genome and to the proteome that is following in the wake of the human genome project (Lander *et al.*, 2001; Venter *et al.*, 2001). There is also a growing recognition of the fundamental importance (and serious resource implications) of ensuring that modern biomedical research is based upon studies that are not only large enough but also incorporate first-class phenotyping and a rigorous assessment of co-exposures and co-morbidities. As a result, there are a number of large national initiatives addressing epidemiology and genetics in several countries (Davey Smith, Ebrahim *et al.*, 2005) including the United Kingdom (BioBank (Vogel, 2002; Wellcome Trust and Medical Research Council, 2002)), Estonia (Frank, 2000) and Latvia (Abbott, 2001). Furthermore, many large pre-existing cohort studies such as the Framingham Study (Joost *et al.*, 2002) and the Nurses Health Study (Haiman *et al.*, 2002) in the United States, the Busselton Study in Australia (Palmer *et al.*, 2001), and european prospective investigation of cancer (EPIC) (Basham *et al.*, 2001), the avon longitudinal study of parents and children (ALSPAC) (Jones *et al.*, 2000; Golding *et al.*, 2001) and the 1958 National Birth Cohort in Britain (Jefferis *et al.*, 2002) all now have a significant focus on genetics. In addition, special family-based initiatives such as the deCODE project in Iceland (Gulcher *et al.*, 2001; Hakonarson *et al.*, 2002), and twin registries around the world (Hopper, 1995; Hansen, de Klerk *et al.* 2000; Strachan, 2002) will continue to produce important information. It is clear that those developing statistical methods in genetic epidemiology are unlikely to be underemployed. Nevertheless, it is crucial that genetic epidemiologists continue to work closely with mainstream epidemiologists and biostatisticians in order to ensure that we do not lose sight of the fact that we not only share methods but a fundamental scientific philosophy.

Acknowledgments

The methodological research program in genetic and genomic epidemiology at the University of Leicester is supported in part by a framework 6 Coordination Action (European Union - 518148), MRC Cooperative Grant #G9806740, Program Grant #003209 from the National Health and Medical Research Council (NHMRC) of Australia and by Leverhulme Research Interchange Grant # F/07134/K.; MRC Clinical Training Fellowship G106/1008 and; MRC Clinician Scientist Fellowship G0501942.

REFERENCES

Abbott, A. (2001). Hopes of biotech interest spur Latvian population genetics. *Nature* **412**, 468.

Armitage, P. and Berry, G. (2002). *Statistical Methods in Medical Research*. Blackwell Scientific, Oxford.

de Bakker, P.I., Yelensky, R., Pe'er, I., Gabriel, S.B., Daly, M.J. and Altshuler, D. (2005). Efficiency and power in genetic association studies. *Nature Genetics* **37**(11), 1217–1223.

Balding, D.J. (2006). A tutorial on statistical methods for population association studies. *Natutre Reviews Genetics* **7**(10), 781–791.

Barrett, J.C. and Cardon, L.R. (2006). Evaluating coverage of genome-wide association studies. *Nature Genetics* **38**(6), 659–662.

Basham, V.M., Pharoah, P.D., Healey, C.S., Luben, R.N., Day, N.E., Easton, D.F. Ponder, B.A.J. and Dunning, A.M. (2001). Polymorphisms in CYP1A1 and smoking: no association with breast cancer risk. *Carcinogenesis* **22**(11), 1797–1800.

Blangero, J. (2002). Segregation analysis, complex. In *Biostatistical Genetics and Genetic Epidemiology*, R. Elston, J. Olson and L. Palmer, eds. John Wiley & Sons, Chichester, pp. 696–708.

Bonney, G.E. (1984). On the statistical determination of major gene mechanisms in continuous human traits: regressive models. *American Journal of Medical Genetics* **18**, 731–749.

Bonney, G.E. (1986). Regressive logistic models for familial disease and other binary traits. *Biometrics* **42**(3), 611–625.

Bonney, G.E. (1987). Logistic regression for dependent binary observations. *Biometrics* **43**(4), 951–973.

Bonney, G.E. (2002). Regressive models. In *Biostatistical Genetics and Genetic Epidemiology*, R. Elston, J. Olson and L. Palmer, eds. John Wiley & Sons, Chichester, pp. 666–673.

Bowden, J.M., Thompson, J.R. and Burton, P.R. (2006). A two-step approach to ascertainment bias correction in complex genetic models with variance components. *Annals of Human Genetics* **71**, 220–229.

Breslow, N.E. and Clayton, D.G. (1993). Approximate inference in generalized linear mixed models. *Journal of the American Statistical Association* **88**, 9–25.

Breslow, N.E. and Day, N.E. (1980). *Statistical Methods in Cancer Research. Volume 1 – The Analysis of Case-Control Studies.* International Agency for Research on Cancer, Lyon.

Breslow, N.E. and Day, N.E. (1987). *Statistical Methods in Cancer Research. Volume 2 – The Design and Analysis of Cohort Studies.* International Agency for Research on Cancer, Lyon.

Burton, P., Gurrin, L. and Sly, P. (1998). Extending the simple linear regression model to account for correlated responses: an introduction to generalized estimating equations and multi-level mixed modelling. *Statistics in Medicine* **17**(11), 1261–1291.

Burton, P.R. (2003). Correcting for non-random ascertainment in generalized linear mixed models (GLMMs) fitted using Gibbs sampling. *Genetic Epidemiology* **24**(1), 24–35.

Burton, P.R., Palmer, L.J., Jacobs, K., Keen, K.J., Olson, J.M. and Elston, R.C. (2000). Ascertainment adjustment: where does it take us? *American Journal of Human Genetics* **67**(6), 1505–1514. Erratum American Journal Human Genetics 69:672.

Burton, P.R., Tiller, K.J., Gurrin, L.C., Cookson, W.O., Musk, A.W. and Palmer, L.J. (1999). Genetic variance components analysis for binary phenotypes using generalized linear mixed models (GLMMs) and Gibbs sampling. *Genetic Epidemiology* **17**(2), 118–140.

Burton, P.R., Tobin, M.D. and Hopper, J.L. (2005). Key concepts in genetic epidemiology. *Lancet* **366**, 941–951.

Clayton, D. and Hills, M. (1993). *Statistical Models in Epidemiology.* Oxford University Press, Oxford.

Clayton, D.G. (1991). A Monte Carlo method for Bayesian inference in frailty models. *Biometrics* **47**(2), 467–485.

Clayton, D.G., Walker, N.M., Smyth, D.J., Pask, R., Cooper, J.D., Maier, L.M., Smink, L.J., Lam, A.C., Ovington, N.R., Stevens, H.E., Nutland, S., Howson, J.M., Faham, M., Moorhead, M., Jones, H.B., Falkowski, M., Hardenbol, P., Willis, T.D. and Todd, J.A. (2005). Population structure, differential bias and genomic control in a large-scale, case-control association study. *Nature Genetics* **37**(11), 1243–1246.

Cohen, B.H. (1980). Chronic obstructive pulmonary disease: a challenge in genetic epidemiology. *American Journal of Epidemiology* **112**(2), 274–288.

Cordell, H.J. and Clayton, D.G. (2005). Genetic association studies. *Lancet* **366**(9491), 1121–1131.

Cox, D.R. (1972). Regression models and life tables (with discussion). *Journal of the Royal Statistical Society Series B* **34**, 187–220.

Davey Smith, G., Ebrahim, S., Lewis, S., Hansell, A.L., Palmer, L.J. and Burton, P.R. (2005). Genetic epidemiology and public health: hope, hype, and future prospects. *Lancet* **366**(9495), 1484–1498.

Davie, A.M. (1979). The 'singles' method for segregation analysis under incomplete ascertainment. *Annals of Human Genetics* **42**(4), 507–512.

Diggle, P.J., Liang, K.Y. and Zeger, S.L. (1994). *Analysis of Longitudinal Data.* Oxford University Press, New York.

Duffy, D.L., Martin, N.G., Battistutta, D., Hopper, J.L. and Mathews, J.D. (1990). Genetics of asthma and hay fever in Australian twins. *American Review of Respiratory Disease* **142**(6 Pt 1), 1351–1358.

Edwards, J.H. (1969). Familial predisposition in man. *British Medical Bulletin* **25**(1), 58–64.

Elston, R.C. (1981). Segregation analysis. *Advances in Human Genetics* **11**, 63–120, 372–373.

Elston, R.C. (2001). *S.A.G.E. Statistical Analysis for Genetic Epidemiology, S.A.G.E. 4.1*. Case Western Reserve University.

Elston, R.C., Olson, J.M. and Palmer, L.J. (2002). *Biostatistical Genetics and Genetic Epidemiology*. John Wiley & Sons, Chichester.

Elston, R.C. and Sobel, E. (1979). Sampling considerations in the gathering and analysis of pedigree data. *American Journal of Human Genetics* **31**(1), 62–69.

Epstein, M.P., Lin, X. and Boehnke, M. (2002). Ascertainment-adjusted parameter estimates revisited. *American Journal of Human Genetics* **70**(4), 886–895.

Ewens, W.J. and Shute, N.C. (1986). The limits of ascertainment. *Annals of Human Genetics* **50**(Pt 4), 399–402.

Falconer, D.S. (1965). The inheritance of liability to certain diseases, estimated from the incidence among relatives. *Annals of Human Genetics* **29**, 51–71.

Fisher, R. (1918). The correlation between relatives on the supposition of Mendelian inheritance. *Transactions of the Royal Society of Edinburgh* **52**, 399–433.

Fisher, R.A. (1934). The effect of methods of ascertainment upon the estimation of frequencies. *Annals of Eugenics* **6**, 13–25.

Fisher, R.A. (1951). Limits to intensive production in animals. *British Agricultural Bulletin* **4**, 217–218.

Frank, L. (2000). Estonia prepares for national DNA database. *Science* **290**(5489), 31.

Fredman, D., White, S.J., Potter, S., Eichler, E.E., Den Dunnen, J.T. and Brookes, A.J. (2004). Complex SNP-related sequence variation in segmental genome duplications. *Nature Genetics* **36**(8), 861–866.

Freedman, M.L., Reich, D., Penney, K.L., McDonald, G.J., Mignault, A.A., Patterson, N., Gabriel, S.B., Topol, E.J., Smoller, J.W., Pato, C.N., Pato, M.T., Petryshen, T.L., Kolonel, L.N., Lander, E.S., Sklar, P., Henderson, B., Hirschhorn, J.N. and Altshuler, D. (2004). Assessing the impact of population stratification on genetic association studies. *Nature Genetics* **36**(4), 388–393.

Freeman, J.L., Perry, G.H., Feuk, L., Redon, R., McCarroll, S.A., Altshuler, D.M., Aburatani, H., Jones, K.W., Tyler-Smith, C., Hurles, M.E., Carter, N.P., Scherer, S.W. and Lee, C. (2006). Copy number variation: new insights in genome diversity. *Genome Research* **16**(8), 949–961.

Galton, F. (1875). The history of twins, as criterion of the relative powers of nature and nurture. *Fraser's magazine* Nov, 566–576.

Galton, F. (1877). Typical laws of heredity. *Proceedings of the Royal Institution* **8**, 282–301.

Galton, F. (1886). Family likeness in stature. *Proceedings of the Royal Society* **40**, 42–73.

Gauderman, W.J. and Thomas, D.C. (1994). Censored survival models for genetic epidemiology: a Gibbs sampling approach. *Genet Epidemiol* **11**(2), 171–188.

Gilks, W., Richardson, S. and Spiegelhalter, D. (1996). *Markov Chain Monte Carlo in Practice*. Chapman and Hall, London.

Glidden, D.V. and Liang, K.Y. (2002). Ascertainment adjustment in complex diseases. *Genetic Epidemiology* **23**(3), 201–208.

Golding, J., Pembrey, M. and Jones, R. (2001). ALSPAC – the Avon longitudinal study of parents and children. I. Study methodology. *Paediatric and Perinatal Epidemiology* **15**(1), 74–87.

Goldstein, H. (1986). Multilevel mixed linear model analysis using iterative generalised least squares. *Biometrika* **73**(1), 43–56.

Goldstein, H. (1987). *Multilevel Models in Educational and Social Research*. Charles Griffin & Company, Ltd, London.

Goldstein, H. (1991). Nonlinear multilevel models with an application to discrete response data. *Biometrika* **78**(1), 45–51.

Greenland, S. and Rothman, K. (1998). Measures of effect and measures of association. *Modern epidemiology*, K. Rothman and G.S. Philadelphia, Lippincott-Raven, 47–64.

Gulcher, J., Kong, A. and Stefansson, K. (2001). The genealogic approach to human genetics of disease. *Cancer Journal* **7**(1), 61–68.

Guo, S.W. and Thompson, E.A. (1992). A Monte Carlo method for combined segregation and linkage analysis. *American Journal of Human Genetics* **51**(5), 1111–1126.

Haiman, C.A., Brown, M., Hankinson, S.E., Spiegelman, D., Colditz, G.A., Willett, W.C., Kantoff, P.W. and Hunter, D.J. (2002). The androgen receptor CAG repeat polymorphism and risk of breast cancer in the Nurses' Health Study. *Cancer Research* **62**(4), 1045–1049.

Hakonarson, H., Bjornsdottir, U.S., Halapi, E., Palsson, S., Adalsteinsdottir, E., Gislason, D., Finnbogason, G., Gislason, T., Kristjansson, K., Arnason, T., Birkisson, I., Frigge, M.L., Kong, A., Gulcher, J.R. and Stefansson, K. (2002). A major susceptibility gene for asthma maps to chromosome 14q24. *American Journal of Human Genetics* **71**(3), 483–491.

Hanfelt, J. (1993). *GEE4 Documentation, Department of Biostatistics*. Johns Hopkins University.

Hansen, J., de Klerk, N., Croft, M., Alessandri, P. and Burton, P. (2000). The Western Australian Twin Child Health (WATCH) study: work in progress. *Australian Epidemiologist* **7**(2), 16–20.

Hattersley, A.T. and McCarthy, M.I. (2005). What makes a good genetic association study? *Lancet* **366**(9493), 1315–1323.

Hennekens, C.H. and Buring, J.E. (1987). Descriptive studies. In *Epidemiology in medicine*, S.L. Mayrent, ed. Brown, Boston, Little, pp. 101–131.

Hodge, S.E. (2002). Ascertainment. In *Biostatistical Genetics and Genetic Epidemiology*, R. Elston, J. Olson and L. Palmer, eds. John Wiley & Sons, Chichester, pp. 20–28.

Hopper, J. (1993). Variance components for statistical genetics: applications in medical research to characteristics related to human diseases and health. *Statistical Methods in Medical Research* **2**, 199–223.

Hopper, J.L. (1992). The epidemiology of genetic epidemiology. *Acta Geneticae Medicae et Gemellologiae* **41**(4), 261–273.

Hopper, J.L. (1995). Australian NHMRC twin registry: a resource for pediatric research. *Pediatric Cardiology* **16**(2), 100.

Hopper, J.L. (2002a). Heritability. In *Biostatistical Genetics and Genetic Epidemiology*, R. Elston, J. Olson and L. Palmer, eds. John Wiley & Sons, Chichester, pp. 371–372.

Hopper, J.L. (2002b). Variance component analysis. In *Biostatistical Genetics and Genetic Epidemiology*, R. Elston, J. Olson and L. Palmer, eds. John Wiley & Sons, Chichester, pp. 778–788.

Hopper, J.L., Bishop, D.T. and Easton, D.F. (2005). Population-based family studies in genetic epidemiology. *Lancet* **366**(9494), 1397–1406.

Hopper, J.L. and Carlin, J.B. (1992). Familial aggregation of a disease consequent upon correlation between relatives in a risk factor measured on a continuous scale. *American Journal of Epidemiology* **136**, 1138–1147.

Hougaard, P. and Thomas, D. (2002). Frailty. In *Biostatistical Genetics and Genetic Epidemiology*, R. Elston, J. Olson and L. Palmer, eds. John Wiley & Sons, Chichester, pp. 279.

Huber, P.J. (1967). The behaviour of maximum likelihood estimates under non-standard conditions. *Proceedings of the Fifth Berkeley Symposium on Mathematical Statistics and Probability* **1**, 221–233.

Jefferis, B.J., Power, C. and Hertzman, C. (2002). Birth weight, childhood socioeconomic environment, and cognitive development in the 1958 British birth cohort study. *British Medical Journal* **325**(7359), 305.

Jinks, J.L. and Fulker, D.W. (1970). Comparison of the biometrical genetical, MAVA, and classical approaches to the analysis of human behaviour. *Psychological Bulletin* **73**(5), 311–349.

Jones, R.W., Ring, S., Tyfield, L., Hamvas, R., Simmons, H., Pembrey, M., and Golding, J. (2000). A new human genetic resource: a DNA bank established as part of the Avon longitudinal study of pregnancy and childhood (ALSPAC). *European Journal of Human Genetics* **8**(9), 653–660.

Joost, O., Wilk, J.B., Cupples, L.A., Harmon, M., Shearman, A.M., Baldwin, C.T., O'Connor, G.T., Myers, R.H. and Gottlieb, D.J. (2002). Genetic loci influencing lung function: a genome-wide scan in the Framingham study. *Americal Journal of Respiratory and Critical Care Medicine* **165**(6), 795–799.

Jöreskog, K.G. and Sörbom, D. (1996). *LISREL 8 User's Reference Guide*. Scientific Software International, Lincolnwood.

Khoury, M.J., Beaty, T.H. and Cohen, B.H. (1993a). Genetic approaches to familial aggregation: I. Analysis of heritability. *Fundamentals of Genetic Epidemiology* 200–232.

Khoury, M.J., Beaty, T.H. and Cohen, B.H. (1993b). Genetic approaches to familial aggregation: II. Segregation analysis. *Fundamentals of Genetic Epidemiology* 233–283.

King, M.C., Lee, G.M., Spinner, N.B., Thomson, G. and Wrensch, M.R. (1984). Genetic epidemiology. *Annual Review of Public Health* **5**, 1–52.

Kopciuk, K.A. and Bull, S.B. (2002). Risk Ratios. In *Biostatistical Genetics and Genetic Epidemiology*, R. Elston, J. Olson and L. Palmer, eds. John Wiley & Sons, Chichester, pp. 687–691.

Kraft, P. and Thomas, D.C. (2000). Bias and efficiency in family-based gene-characterization studies: conditional, prospective, retrospective, and joint likelihoods. *American Journal of Human Genetics* **66**(3), 1119–1131.

Kuehni, C.E., Brooke, A.M. and Silverman, M. (2000). Prevalence of wheeze during childhood: retrospective and prospective assessment. *European Respiratory Journal* **16**(1), 81–85.

Lander, E.S., Linton, L.M., Birren, B., Nusbaum, C., Zody, M.C., Baldwin, J., Devon, K., Dewar, K., Doyle, M., FitzHugh, W., Funke, R., Gage, D., Harris, K., Heaford, A., Howland, J., Kann, L., Lehoczky, J., LeVine, R., McEwan, P., McKernan, K., Meldrim, J., Mesirov, J.P., Miranda, C., Morris, W., Naylor, J., Raymond, C., Rosetti, M., Santos, R., Sheridan, A., Sougnez, C., N. Stange-Thomann, Stojanovic, N., Subramanian, A., Wyman, D., Rogers, J., Sulston, J., Ainscough, R., Beck, S., Bentley, D., Burton, J., Clee, C., Carter, N., Coulson, A., Deadman, R., Deloukas, P., Dunham, A., Dunham, I., Durbin, R., French, L., Grafham, D., Gregory, S., Hubbard, T., Humphray, S., Hunt, A., Jones, M., Lloyd, C., McMurray, A., Matthews, L., Mercer, S., Milne, S., Mullikin, J.C., Mungall, A., Plumb, R., Ross, M., Shownkeen, R., Sims, S., Waterston, R.H., Wilson, R.K., Hillier, L.W., McPherson, J.D., Marra, M.A., Mardis, E.R., Fulton, L.A., Chinwalla, A.T., Pepin, K.H., Gish, W.R., Chissoe, S.L., Wendl, M.C., Delehaunty, K.D., Miner, T.L., Delehaunty, A., Kramer, J.B., Cook, L.L., Fulton, R.S., Johnson, D.L., Minx, P.J., Clifton, S.W., Hawkins, T., Branscomb, E., Predki, P., Richardson, P., Wenning, S., Slezak, T., Doggett, N., Cheng, J.F., Olsen, A., Lucas, S., Elkin, C., Uberbacher, E., Frazier, M., Gibbs, R.A., Muzny, D.M., Scherer, S.E., Bouck, J.B., Sodergren, E.J., Worley, K.C., Rives, C.M., Gorrell, J.H., Metzker, M.L., Naylor, S.L., Kucherlapati, R.S., Nelson, D.L., Weinstock, G.M., Sakaki, Y., Fujiyama, A., Hattori, M., Yada, T., Toyoda, A., Itoh, T., Kawagoe, C., Watanabe, H., Totoki, Y., Taylor, T., Weissenbach, J., Heilig, R., Saurin, W., Artiguenave, F., Brottier, P., Bruls, T., Pelletier, E., Robert, C., Wincker, P., Smith, D.R., L. Doucette-Stamm, Rubenfield, M., Weinstock, K., Lee, H.M., Dubois, J., Rosenthal, A., Platzer, M., Nyakatura, G., Taudien, S., Rump, A., Yang, H., Yu, J., Wang, J., Huang, G., Gu, J., Hood, L., Rowen, L., Madan, A., Qin, S., Davis, R.W., Federspiel, N.A., Abola, A.P., Proctor, M.J., Myers, R.M., Schmutz, J., Dickson, M., Grimwood, J., Cox, D.R., Olson, M.V., Kaul, R., Shimizu, N., Kawasaki, K., Minoshima, S., Evans, G.A., Athanasiou, M., Schultz, R., Roe, B.A., Chen, F., Pan, H., Ramser, J., Lehrach, H., Reinhardt, R., McCombie, W.R., M. de la Bastide, Dedhia, N., Blocker, H., Hornischer, K., Nordsiek, G., Agarwala, R., Aravind, L., Bailey, J.A., Bateman, A., Batzoglou, S., Birney, E., Bork, P., Brown, D.G., Burge, C.B., Cerutti, L., Chen, H.C., Church, D., Clamp, M., Copley, R.R., Doerks, T., Eddy, S.R., Eichler, E.E., Furey, T.S., Galagan, J., Gilbert, J.G., Harmon, C., Hayashizaki, Y., Haussler, D., Hermjakob, H., Hokamp, K., Jang, W., Johnson, L.S., Jones, T.A., Kasif, S., Kaspryzk, A., Kennedy, S., Kent, W.J., Kitts, P., Koonin, E.V., Korf, I., Kulp, D., Lancet, D., Lowe, T.M., McLysaght, A., Mikkelsen, T.,

Moran, J.V., Mulder, N., Pollara, V.J., Ponting, C.P., Schuler, G., Schultz, J., Slater, G., Smit, A.F., Stupka, E., Szustakowski, J., D. Thierry-Mieg, J. Thierry-Mieg, Wagner, L., Wallis, J., Wheeler, R., Williams, A., Wolf, Y.I., Wolfe, K.H., Yang, S.P., Yeh, R.F., Collins, F., Guyer, M.S., Peterson, J., Felsenfeld, A., Wetterstrand, K.A., Patrinos, A., Morgan, M.J., Szustakowki, J., P. de Jong, Catanese, J.J., Osoegawa, K., Shizuya, H., Choi, S. and Chen, Y.J. (2001). Initial sequencing and analysis of the human genome. *Nature* **409**(6822), 860–921.

Lange, K., Weeks, D. and Boehnke, M. (1988). Programs for pedigree analysis: MENDEL, FISHER, and dGENE. *Genetic Epidemiology* **5**(6), 471–472.

Last, J. (2001). *A Dictionary of Epidemiology*. Oxford University Press, New York.

Li, C.C. and Mantel, N. (1968). A simple method of estimating the segregation ratio under complete ascertainment. *American Journal of Human Genetics* **20**(1), 61–81.

Liang, K.-Y. and Zeger, S.L. (1986). Longitudinal data analysis using generalized linear models. *Biometrika* **73**(1), 13–22.

Lichtenstein, P., Holm, N.V., Verkasalo, P.K., Iliadou, A., Kaprio, J., Koskenvuo, M., Pukkala, E. Skytthe, A. and Hemminki, K. (2000). Environmental and heritable factors in the causation of cancer–analyses of cohorts of twins from Sweden, Denmark, and Finland. *New England Journal of Medicine* **343**(2), 78–85.

Luyt, D.K., Burton, P.R. and Simpson, H. (1993). Epidemiological study of wheeze, doctor diagnosed asthma, and cough in preschool children in Leicestershire. *British Medical Journal* **306**(6889), 1386–1390.

Majumder, P.P. (2002). Heritability. In *Biostatistical Genetics and Genetic Epidemiology*, R. Elston, J. Olson and L. Palmer, eds. John Wiley & Sons, Chichester, pp. 693–696.

Marchini, J., Cardon, L.R., Phillips, M.S. and Donnelly, P. (2004). The effects of human population structure on large genetic association studies. *Nature Genetics* **36**(5), 512–517.

MathSoft, I. (2000). *Splus*. MathSoft, Inc., Cambridge, MA.

McCullagh, P. and Nelder, J. (1989). *Generalized Linear Models*. Chapman and Hall, London.

Merriman, C. (1924). The intellectual resemblance of twins. *Psychological Monographs* **33**, 1–58.

Miller, R.G., Gong, G. and Munoz, A. (1981). *Survival analysis*. John Wiley & Sons, New York.

Morton, N.E. (1959). Genetic tests under incomplete ascertainment. *American journal of Human Genetics* **11**, 1–16.

Morton, N.E. (1982). *Outline of Genetic Epidemiology*. Karger, London.

Morton, N.E. and Chung, C.S. (1978). *Genetic epidemiology*. Academic Press, New York.

Morton, N.E. and MacLean, C.J. (1974). Analysis of family resemblance. 3. Complex segregation of quantitative traits. *American Journal of Human Genetics* **26**(4), 489–503.

Neale, M.C. (2002). Twin analysis. In *Biostatistical Genetics and Genetic Epidemiology*, R. Elston, J. Olson and L. Palmer, eds. John Wiley & Sons, Chichester, pp. 743–756.

Neale, M.C., Boker, S.M., Xie, G. and Maes, H.H. (2002). *Mx: Statistical Modeling (6th Edition)*. Department of Human Genetics, Virginia Institute for Psychiatric and Behavioral Genetics, Virginia Commonwealth University, Richmond, Virginia, U.S.A.

Neale, M.C. and Cardon, L.R. (1992). *Methodology for Genetic Studies of Twins and Families*. Kluwer, Boston.

Neel, J.V. and Schull, W.J. (1954). *Human Heredity*. The University of Chicago Press, Chicago.

Neuhaus, J. (1992). Statistical methods for longitudinal and clustered designs with binary responses. *Statistical Methods in Medical Research* **1**, 249–273.

Noh, M., Lee, Y. and Pawitan, Y. (2005). Robust ascertainment-adjusted parameter estimation. *Genetic Epidemiology* **29**(1), 68–75.

Palmer, L.J. and Cardon, L.R. (2005). Shaking the tree: mapping complex disease genes with linkage disequilibrium. *Lancet* **366**(9492), 1223–1234.

Palmer, L.J., Cookson, W.O., James, A.L., Musk, A.W. and Burton, P.R. (2001). Gibbs sampling-based segregation analysis of asthma-associated quantitative traits in a population-based sample of nuclear families. *Genetic Epidemiology* **20**(3), 356–372.

Palmer, L.J., Knuiman, M.W., Divitini, M.L., Burton, P.R., James, A.L., Bartholomew, H.C., Ryan, G. and Musk, A.W. (2001). Familial aggregation and heritability of adult lung function: results from the Busselton health study. *European Respiratory Journal* **17**(4), 696–702.

Palmer, L.J., Rye, P.J., Gibson, N.A., Burton, P.R., Landau, L.I. and Lesouef, P.N. (2001). Airway responsiveness in early infancy predicts asthma, lung function, and respiratory symptoms by school age. *American Journal of Respiratory Cell and Molecular Biology* **163**(1), 37–42.

Pearson, K. (1896). Mathematical contributions to the theory of evolution – IIRegression, I., heredity, and panmixia. *Philosophical Transactions of the Royal Society, Series A* **187**, 253–318.

Pendergast, P.F., Gange, S.J. and Lindstrom, M.J. (2002). Correlated binary data. In *Biostatistical Genetics and Genetic Epidemiology*, R. Elston, J. Olson and L. Palmer, eds. John Wiley & Sons, Chichester, pp. 159–171.

Pickles, A. (2002). Generalized estimating equations. In *Biostatistical Genetics and Genetic Epidemiology*, R. Elston, J. Olson and L. Palmer, eds. John Wiley & Sons, Chichester, pp. 299–310.

Prentice, R.L. and Zhao, L.P. (1991). Estimating equations for parameters in means and covariances of multivariate discrete and continuous responses. *Biometrics* **47**(3), 825–839.

Rao, D.C. and Rice, T. (2002). Path analysis. In *Biostatistical Genetics and Genetic Epidemiology.*, R. Elston, J. Olson and L. Palmer, eds. John Wiley & Sons, Chichester, pp. 606–619.

Rasbash, J., Browne, W., Goldstein, H., Yang, M., Plewis, I., Draper, D., Healy, M. and Woodhouse, G. (1999). *A User's Guide to MLwiN*. Institute of Education, London.

Redon, R., Ishikawa, S., Fitch, K.R., Feuk, L., Perry, G.H., Andrews, T.D., Fiegler, H., Shapero, M.H., Carson, A.R., Chen, W., Cho, E.K., Dallaire, S., Freeman, J.L., Gonzalez, J.R., Gratacos, M., Huang, J., Kalaitzopoulos, D., Komura, D., MacDonald, J.R., Marshall, C.R., Mei, R., Montgomery, L., Nishimura, K., Okamura, K., Shen, F., Somerville, M.J., Tchinda, J., Valsesia, A., Woodwark, C., Yang, F., Zhang, J., Zerjal, T., Armengol, L., Conrad, D.F., Estivill, X., C. Tyler-Smith, Carter, N.P., Aburatani, H., Lee, C., Jones, K.W., Scherer, S.W. and Hurles, M.E. (2006). Global variation in copy number in the human genome. *Nature* **444**(7118), 444–454.

Risch, N. (1990). Linkage strategies for genetically complex traits. II. The power of affected relative pairs. *American Journal of Human Genetics* **46**(2), 229–241.

Roberts, D.F. (1985). A definition of genetic epidemiology. In *Diseases of Complex Etiology in Small Populations: Ethnic Differences and Research Approaches*, R. Chakraborty and E.J.E. Szathmary, eds. Alan R Liss, New York, pp. 9–20.

Rothman, K. and Greenland, S. (1998). Accuracy considerations in study design. In *Modern epidemiology*, K. Rothman and S. Greenland, eds. Lippincott-Raven, Philadelphia, pp. 135–145.

SAS Institute Inc. (2001). *SAS Release 8.02.*, SAS Institute Inc, Cary, NC.

Scurrah, K.J., Palmer, L.J. and Burton, P.R. (2000). Variance components analysis for pedigree-based censored survival data using generalized linear mixed models (GLMMs) and Gibbs sampling in BUGS. *Genetic Epidemiology* **19**(2), 127–148.

Siemans, H.W. (1924). *Zwillingspathologie: Ihre Bedeutung; Ihre Methodik, Ihre Bisherigen Ergebnisse*. Springer Verlag, Berlin.

Sobel, E. and Lange, K. (1993). Metropolis sampling in pedigree analysis. *Statistical Methods in Medical Research* **2**(3), 263–282.

Spector, N. (2000). Cancer, genes, and the environment. *New England Journal of Medicine* **343**(20), 1494. Discussion 1495-6.

Spector, T.D., Sneider, H. and MacGregor, A.J. (1999). *Advances in twin and sib pair analysis*. Greenwich Medical Media Ltd, London.

Spiegelhalter, D., Thomas, A. and Best, N. (2000). *WinBUGS Version 1.3 – User Manual*. MRC Biostatistics Unit, Cambridge.

Spielman, R.S., McGinnis, R.E. and Ewens, W.J. (1993). Transmission test for linkage disequilibrium: the insulin gene region and insulin-dependent diabetes mellitus (IDDM). *American Journal of Human Genetics* **52**(3), 506–516.

SPSS Inc. (2002). *SPSS 11.0 Guide to Data Analysis*. SPSS Inc Headquarters, Chicago.

Stata Corporation. (2005). *Stata User's Guide. Version 9.0*. Stata Press, College Station, Texas.

Strachan, D.P. (2002). Twin registers. In *Biostatistical Genetics and Genetic Epidemiology*, R. Elston, J. Olson and L. Palmer, eds. John Wiley & Sons, Chichester, pp. 759–761.

Teare, M.D. and Barrett, J.H. (2005). Genetic linkage studies. *Lancet* **366**(9490), 1036–1044.

Thompson, E.A. (2002). Human Genetics, Overview. In *Biostatistical Genetics and Genetic Epidemiology*, R. Elston, J.M. Olson and L.J. Palmer, eds. Wiley, Chichester, pp. 386–390.

Todorov, A.A. and Suarez, B.K. (2002). Liability model. In *Biostatistical Genetics and Genetic Epidemiology*, R. Elston, J. Olson and L. Palmer, eds. John Wiley & Sons, Chichester, pp. 430–435.

Venter, J.C., Adams, M.D., Myers, E.W., Li, P.W., Mural, R.J., Sutton, G.G., Smith, H.O., Yandell, M., Evans, C.A., Holt, R.A., Gocayne, J.D., Amanatides, P., Ballew, R.M., Huson, D.H., Wortman, J.R., Zhang, Q., Kodira, C.D., Zheng, X.H., Chen, L., Skupski, M., Subramanian, G., Thomas, P.D., Zhang, J., G. L. Gabor Miklos, Nelson, C., Broder, S., Clark, A.G., Nadeau, J., McKusick, V.A., Zinder, N., Levine, A.J., Roberts, R.J., Simon, M., Slayman, C., Hunkapiller, M., Bolanos, R., Delcher, A., Dew, I., Fasulo, D., Flanigan, M., Florea, L., Halpern, A., Hannenhalli, S., Kravitz, S., Levy, S., Mobarry, C., Reinert, K., Remington, K., J. Abu-Threideh, Beasley, E., Biddick, K., Bonazzi, V., Brandon, R., Cargill, M., Chandramouliswaran, I., Charlab, R., Chaturvedi, K., Deng, Z., V. Di Francesco, Dunn, P., Eilbeck, K., Evangelista, C., Gabrielian, A.E., Gan, W., Ge, W., Gong, F., Gu, Z., Guan, P., Heiman, T.J., Higgins, M.E., Ji, R.R., Ke, Z., Ketchum, K.A., Lai, Z., Lei, Y., Li, Z., Li, J., Liang, Y., Lin, X., Lu, F., Merkulov, G.V., Milshina, N., Moore, H.M., Naik, A.K., Narayan, V.A., Neelam, B., Nusskern, D., Rusch, D.B., Salzberg, S., Shao, W., Shue, B., Sun, J., Wang, Z., Wang, A., Wang, X., Wang, J., Wei, M., Wides, R., Xiao, C., Yan, C., Yao, A., Ye, J., Zhan, M., Zhang, W., Zhang, H., Zhao, Q., Zheng, L., Zhong, F., Zhong, W., Zhu, S., Zhao, S., Gilbert, D., Baumhueter, S., Spier, G., Carter, C., Cravchik, A., Woodage, T., Ali, F., An, H., Awe, A., Baldwin, D., Baden, H., Barnstead, M., Barrow, I., Beeson, K., Busam, D., Carver, A., Center, A., Cheng, M.L., Curry, L., Danaher, S., Davenport, L., Desilets, R., Dietz, S., Dodson, K., Doup, L., Ferriera, S., Garg, N., Gluecksmann, A., Hart, B., Haynes, J., Haynes, C., Heiner, C., Hladun, S., Hostin, D., Houck, J., Howland, T., Ibegwam, C., Johnson, J., Kalush, F., Kline, L., Koduru, S., Love, A., Mann, F., May, D., McCawley, S., McIntosh, T., McMullen, I., Moy, M., Moy, L., Murphy, B., Nelson, K., Pfannkoch, C., Pratts, E., Puri, V., Qureshi, H., Reardon, M., Rodriguez, R., Rogers, Y.H., Romblad, D., Ruhfel, B., Scott, R., Sitter, C., Smallwood, M., Stewart, E., Strong, R., Suh, E., Thomas, R., Tint, N.N., Tse, S., Vech, C., Wang, G., Wetter, J., Williams, S., Williams, M., Windsor, S., E. Winn-Deen, Wolfe, K., Zaveri, J., Zaveri, K., Abril, J.F., Guigo, R., Campbell, M.J., Sjolander, K.V., Karlak, B., Kejariwal, A., Mi, H., Lazareva, B., Hatton, T., Narechania, A., Diemer, K., Muruganujan, A., Guo, N., Sato, S., Bafna, V., Istrail, S., Lippert, R., Schwartz, R., Walenz, B., Yooseph, S., Allen, D., Basu, A., Baxendale, J., Blick, L., Caminha, M., J. Carnes-Stine, Caulk, P., Chiang, Y.H., Coyne, M., Dahlke, C., Mays, A., Dombroski, M., Donnelly, M., Ely, D., Esparham, S., Fosler, C., Gire, H., Glanowski, S., Glasser, K., Glodek, A., Gorokhov, M., Graham, K., Gropman, B., Harris, M., Heil, J., Henderson, S., Hoover, J., Jennings, D., Jordan, C., Jordan, J., Kasha, J., Kagan, L., Kraft, C., Levitsky, A., Lewis, M., Liu, X., Lopez, J., Ma, D., Majoros, W., McDaniel, J., Murphy, S., Newman, M., Nguyen, T., Nguyen, N., Nodell, M., Pan, S., Peck, J., Peterson, M., Rowe, W., Sanders, R., Scott, J., Simpson, M., Smith, T., Sprague, A., Stockwell, T., Turner, R., Venter, E., Wang, M., Wen, M., Wu, D., Wu, M., Xia, A., Zandieh, A. and Zhu, X. (2001). The sequence of the human genome. *Science* **291**(5507), 1304–1351.

Vogel, G. (2002). Population studies. U.K.'s mass appeal for disease insights. *Science* **296**(5569), 824.

Weinberg, W. (1912). Weitere Beiträge zur Theorie der Vererbung. 5. Zur Vererbung der Anlage zur Blutenkranheit mit methodologischen Ergänzungen meiner Geschwistermethode. *Archiv für Rassen- und Gesellschafts-Biologie* **2**, 694–709.

Weinberg, W. (1928). Mathematische grundlagen der probandenmethode. *Zeitschrift für Induktive Abstammungs- und Vererbungslehre* **48**, 179–228.

Wellcome Trust and Medical Research Council (2002). *Draft Protocol for BioBank UK: A study of genes, environment and health.* Medical Research Council, London.

Wright, S. (1921). Correlation and causation. *Journal of Agricultural Research* **20**, 557–585.

Zeger, S.L. and Karim, M.R. (1991). Generalized linear models with random effects; a Gibbs sampling approach. *Journal of the American Statistical Association* **86**(413), 79–86.

Zeger, S.L. and Liang, K.Y. (1986). Longitudinal data analysis for discrete and continuous outcomes. *Biometrics* **42**(1), 121–130.

Zeger, S.L. and Liang, K.Y. (1992). An overview of methods for the analysis of longitudinal data. *Statistics in Medicine* **11**(14–15), 1825–1839.

Zeger, S.L., Liang, K.Y. and Albert, P.S. (1988). Models for longitudinal data: a generalized estimating equation approach. *Biometrics* **44**(4), 1049–1060.

33

Linkage Analysis

E.A. Thompson

Department of Statistics, University of Washington, Seattle, WA, USA

Linkage analysis is the analysis of the dependence in inheritance of genes at different genetic loci, on the basis of phenotypic observations on individuals. The methods of linkage analysis that evolved over the years 1920 to 1970 closely followed more general developments in approaches to statistical inference. In recent years, the wealth of DNA markers and the near completion of the human genome project have led to changes in approaches to linkage analysis. New computational methods have been developed to make optimal use of available multilocus marker data. With the mapping of most simple Mendelian traits and the consequent emphasis on mapping the genes contributing to more complex traits, methods involving analysis of genome identity by descnt are increasingly used. In view of the limited numbers of meioses often available for inference, fine-scale mapping and estimation of accurate meiotic maps remain open challenges. As data on genetic variation becomes available at more detailed levels, sources of variation in the processes underlying meiosis can be investigated and more accurate genetic maps obtained. Genomic data at high resolution can allow models for meiosis outcomes to be tested and more accurate models developed. These may then be used in the mapping of trait genes.

33.1 INTRODUCTION

The last century has seen enormous change in methods of inference from genetic data, from the rediscovery of Mendel's laws in 1900, to near completion of Phase I of the Human Genome Project 100 years later. However, the scientific questions remain surprisingly constant: Where are the genes? What do they do? Genetics is the science of heritable variation, meiosis is the biological process whereby genetic information is transmitted from parent to offspring, and linkage analysis is inference concerning the outcomes of meioses from data on the genetic characteristics of individuals.

For the purposes of this chapter, linkage analysis is defined as the analysis of the dependence in inheritance of genes at different genetic loci, as evidenced in phenotypic observations on individuals in known pedigree relationships. A pedigree is a specification of the father and mother of each (nonfounder) individual. Individuals whose parents are

Handbook of Statistical Genetics, Third Edition. Edited by D.J. Balding, M. Bishop and C. Cannings.
© 2007 John Wiley & Sons, Ltd.   ISBN: 978-0-470-05830-5.

unspecified are *founders*: relationships are defined only relative to the specified pedigree. Dependence in inheritance of genes at different genetic loci reflects synteny of the loci, and the degree of dependence provides a measure of the genetic distance between them. When a genetic marker map is available, dependence between the inheritances of the trait and that at certain marker loci enables the locations of trait loci to be inferred.

Of the biological sciences, genetics is the one with the most clearly defined probability models, and hence the one in which classical parametric statistical inference has had the biggest role. Moreover, linkage analysis is the area of genetics which is most dependent on probability models and parametric inference. From the work of Fisher (1922b) onward, each major development in parametric statistical inference has been adopted by the developers of linkage analysis methods, and questions of genetic analysis have prompted new statistical developments. In many ways, statistical inference and genetic analysis have developed in parallel over the last 100 years.

33.2 THE EARLY YEARS

Modern genetics started with Mendel (1866), who postulated his two laws as a probability model. In modern terminology, they can be stated as follows:

- Every diploid organism has two factors [or *genes*] controlling a given trait, one from the mother, one from the father.
- When a (diploid) individual has an offspring, a copy of a randomly chosen one of his two genes is transmitted [or *segregates*] to the offspring.
- This transmission occurs independently of the other parent, independently for each child, and independently for each trait [or *locus*].

Mendel's work was rediscovered in 1900 and not long thereafter geneticists realized that independence of segregations for different traits is not always true. Instead, there are groups of traits, which are *linked*, the genes controlling them tending to be inherited by the offspring as a group, not independently. That is, the same combinations of alleles transmitted from grandparent to parent are transmitted to an offspring. The basic framework for linkage analysis was quickly established. The genetic material that underlies inherited characteristics consists of the chromosomes: linear structures of DNA in the form of a double helix contained within each cell nucleus. In a diploid organism, chromosomes come in pairs, one deriving from the genetic material of the mother and the other from the father. In *meiosis*, the process of formation of a gamete (sperm or egg cell), a parent provides one chromosome from each chromosome pair. In the formation of this chromosome, several *crossover* events may occur between the two parental chromosomes, such that the transmitted chromosome consists of alternating segments of the two parental chromosomes. Simple genetic characteristics are determined by the two segments of DNA sequence (*alleles*) at a specific location (*locus*) on a pair of chromosomes. Geneticists soon associated this dependence in inheritance (*linkage*) with the chromosomes. Sturtevant (1913) showed the patterns of coinheritance of genes for different sex-linked traits in *Drosophila* were best explained by a linear arrangement of genes and made the first *gene ordering* inference using methods analogous to those still used today.

Mathematically, we define

$$S_{i,j} = 0 \text{ or } 1 \tag{33.1}$$

to denote that in meiosis i at locus j the maternal or paternal gene (respectively) of the parent is transmitted to the offspring. For ease of notation we define the two sets of vectors

$$S_{\bullet,j} = (S_{i,j}, i = 1, ..., m) \text{ for locus } j, j = 1, ..., l$$

$$S_{i,\bullet} = (S_{i,j}, j = 1, ..., l) \text{ at meiosis } i, i = 1, ..., m. \tag{33.2}$$

These *meiosis indicators* or *switches* (Donnelly, 1983) specify the descent of genes in pedigrees. In experimental organisms, experiments can be designed such that the grandparental origin of each offspring allele is clear, and often large numbers of offspring can be observed. For two distinct locations on a chromosome, the *recombination frequency* θ is the probability that the alleles derive from different parental chromosomes; that is, have different grandparental origins. Between two loci, j and j', the recombination frequency is

$$\theta_{jl} = \Pr(S_{i,j} \neq S_{i,j'}).$$

The recombination frequency is often assumed not to depend on the specific meiosis i, or to depend only on the sex of the parent in which meiosis i occurs. In reality, many factors may affect the frequency of recombination.

For recombination to occur, there must be an odd number of crossovers between the two loci in the formation of the offspring gamete. For loci that are very close together on the chromosome, θ is close to 0, and alleles at the two loci will show strong dependence in their grandparental origins. Under most meiosis models, and apparently in nature, the value of θ increases with increasing length of chromosome intervening between the two loci, until for loci that are far apart on a chromosome, or are on different pairs of chromosomes, $\theta = \frac{1}{2}$ and the grandparental origins of alleles at the two loci are independent. Loci for which $\theta < \frac{1}{2}$ are said to be *linked*. Linkage analysis is the statistical analysis of genetic data with the goals of detecting whether $\theta < \frac{1}{2}$, of estimating θ, of ordering a set of genetic loci, and ultimately of placing the loci determining genetic traits of interest at correct locations in a genetic map.

An understanding of the population–genetic issues was also fundamental to the early development of statistical genetics. In 1908, G. Hardy and W. Weinberg independently established the idea of *Hardy–Weinberg equilibrium* showing that Mendelian segregation provided for the maintenance of stable genetic variation within a population. In the absence of selection, in an infinite population, allele frequencies remain constant. Moreover, for loci on the 22 pairs of autosomes, constant 'equilibrium' genotype frequencies are established by a single generation of random mating (see Hartl and Clark, 1997). For joint frequencies at two loci, this is not so. Denoting alleles at two diallelic loci A and B by a_i and b_i ($i = 1, 2$), there are four *haploypes*: a_1b_1, a_1b_2, a_2b_1, and a_2b_2. If the population frequencies of these are $p_k, k = 1, ..., 4$ the standard measure of *linkage disequilibrium* or *allelic association* is

$$D = \Pr(a_1b_1) - \Pr(a_1) \Pr(b_1)$$

$$= p_1 - (p_1 + p_2)(p_1 + p_3)$$

$$= p_1 p_4 - p_2 p_3,$$

since $p_1 + p_2 + p_3 + p_4 = 1$. Robbins (1918) showed that, after one generation of random mating, in an infinite population, the allelic association is decreased to

$$D^* = D(1 - \theta). \tag{33.3}$$

Thus equilibrium haplotype frequencies are not obtained in a single generation even for unlinked loci ($\theta = \frac{1}{2}$). If linkage is tight ($\theta \approx 0$), allelic associations persist over many generations.

Haldane (1919) defined genetic map distance between two loci as the expected number of crossover events between them; this distance measure is additive regardless of the meiosis model. One Morgan is the length of chromosome in which one crossover event is expected; map distances are normally given in centiMorgans (1 cM = 0.01 M). He also related this additive *map distance* to recombination frequencies between loci on the assumption that crossovers occur as a homogeneous Poisson process. (This is the model of 'no genetic interference'; a given crossover event does not affect the probability distribution of the number and locations of other crossover events.) This sparked an exploration that still continues, on the relationship between physical distance (actual DNA length in base pairs (bp)) and map distance (the statistical measure based on the degree of dependence between genetic loci). There is no simple relationship; in terms of physical distance recombination rates vary widely over the genome. The early statistical geneticists made clear that only recombination frequencies, and hence genetic map distances, could be estimated from segregation data alone.

Fisher (1922b) used a different relationship between recombination frequencies and map distance, assuming one crossover precluded any other crossover in the region of chromosome he considered. Under this model of 'complete interference', recombination frequencies are themselves additive. With this simplifying assumption, he used the multinomial distribution of recombination counts in offspring as one of his first examples of maximum likelihood estimation, citing, and drawing on his theoretical work published in the same year Fisher (1922a). Thus, by 1925, questions of gene ordering, genetic maps, linkage disequilibrium, and the relationship between genetic and physical maps had all been raised.

33.3 THE DEVELOPMENT OF HUMAN GENETIC LINKAGE ANALYSIS

In the 1930s, three British mathematicians, statisticians, and geneticists laid the foundations of most of modern linkage analysis, and did so through the use of statistical models. Haldane (1934) and Fisher (1934) recognized that the same methods and ideas that were being increasingly applied to designed crosses of experimental organisms could be applied also to observed patterns of inheritance in human families. At the same time, Penrose (1935) pointed out that, with a sufficiently large sample, linkage could be inferred from associated sharing of traits in sib pairs (see **Chapter 34**). Haldane (1934) considered use of likelihood to detect linkage, while Fisher (1934) addressed the estimation problem more generally, applying ideas of information and efficiency. Just as the work of the early years raised many still-topical map issues, the work of the 1930s raised some human trait issues. Trait heterogeneity in linkage detection was first addressed by Fisher (1936). While Fisher (1934) focused on the statistical issues of information and power to detect linkage, he also foresaw both the potential and the problems of using linked genetic markers in genetic counseling.

Statistically, the problem is one of missing data or latent variables. In human families, individuals may be unavailable for typing, the traits that are observed may have no direct correspondence with the underlying alleles, and even where they do the sharing of common alleles may make grandparental origins of alleles unclear. Since Fisher (1922b) and Haldane (1934), the usual approach to linkage analysis is through computation of likelihoods. Evaluation of the likelihood involves summing over all the latent events that could have led to the observed data. In the early approaches, the genotypes of individuals are considered the latent variables:

$$\Pr(\mathbf{Y}) = \sum_{\mathbf{G}} \Pr(\mathbf{Y}|\mathbf{G}) \, \Pr(\mathbf{G}) = \sum_{\mathbf{G}} \left(\prod_{\text{observed } i} \Pr(Y_{i,\bullet}|G_{i,\bullet}) \right) \Pr(\mathbf{G}), \tag{33.4}$$

where

$$\Pr(\mathbf{G}) = \prod_{\text{founders } i} \Pr(G_{i,\bullet}) \prod_{\text{nonfounders } i} \Pr(G_{i,\bullet}|G_{M_i,\bullet}, G_{F_i,\bullet}). \tag{33.5}$$

Here, and throughout, $\mathbf{Y}$ denotes phenotypic data at trait and marker loci: $Y_{i,\bullet}$ denotes all data on individual i. The bullet subscript denotes that multiple genetic loci may be involved. For ease of presentation, we assume that phenotypes $Y_{i,\bullet}$ are conditionally independent given the multilocus genotype $G_{i,\bullet}$ of individuals i. The population allele frequencies and allelic associations determine the probabilities $\Pr(G_{i,\bullet})$ of founder haplotypes, while the laws of Mendelian segregation and linkage relationships among the loci determine the transmission probabilities $\Pr(G_{i,\bullet}|G_{M_i,\bullet}, G_{F_i,\bullet})$ for the genotypes of nonfounder individuals i conditional on those of the parents M_i and F_i of i. Genotypes of individuals are conditionally independent given those of their parents. In the context of linkage analysis, recombination frequencies θ are the unknown parameters of the transmission probabilities, and $\Pr(\mathbf{Y})$ is the likelihood $L(\theta)$ for θ.

The next big advances came again with application to linkage analysis of new statistical approaches. Haldane and Smith (1947) developed the uses of the likelihood ratio and maximum likelihood estimation, including the computation (by hand) of likelihoods on extended pedigrees, while Smith (1953) investigated the properties of these likelihood methods and the information for linkage in various pedigree designs. Smith (1953) also introduced the *lod score* of Barnard (1949) to linkage analysis. The likelihood ratio for linkage at a given recombination frequency θ relative to that for unlinked loci $\theta = \frac{1}{2}$ is $L(\theta)/L\left(\frac{1}{2}\right)$. The *lod* score for linkage between two loci is

$$\max_{0 \le \theta \le \frac{1}{2}} \log \left(L(\theta)/L\left(\frac{1}{2}\right) \right).$$

Haldane and Smith (1947) also proposed use of a Bayesian prior probability for synteny of two randomly chosen loci, and for the locations of linked loci. The Bayesian approach was developed much further by Renwick (1969).

Statistical developments in hypothesis testing also influenced linkage analysis, continuing the distinction between testing and estimation first emphasized by Fisher (1934). Using logarithms to base 10, the *lod* score bounds for declaration of linkage (+3) and exclusion of linkage (−2) proposed by Morton (1955) derived from a consideration of type 1 and type II error following a sequential probability ratio test (SPRT) of Wald (1947). This base-10 *lod* score has become the standard criterion for assessing evidence of linkage. The use of base-10 logarithms also became standard, and is deeply embedded in current methodology in this area, although in some recent multipoint linkage approaches natural logarithms of likelihoods are also used (see *location score* below).

In the framework of testing, Smith (1963) formalized the ideas of Fisher (1936) to develop likelihood-based tests for linkage heterogeneity. Edwards (1971) also followed the likelihood and testing approach to linkage detection, referring to absence of linkage as 'the only true null hypothesis in biology'. He also considered questions of the amount of information in a set of observations, converting it to an 'equivalent number' of informative meioses on the basis of the curvature of the likelihood in the neighborhood of the maximum – in effect using the usual statistical measure of (Fisher) information.

33.4 THE PEDIGREE YEARS; SEGREGATION AND LINKAGE ANALYSIS

The issues facing linkage analysis in 1970 were primarily computational and statistical. Data had been collected on pedigrees segregating many Mendelian traits, some exhibiting more complex patterns of inheritance, and some quantitative traits. Markers remained few and mostly unmapped, but linkage analysis was increasingly attempted as digital computers began to provide the necessary computational power. Around 1970, Hilden (1970), Elston and Stewart (1971), Heuch and Li (1972) laid the basis for what would remain the primary approach to segregation and linkage analysis computations for the next 20 years. Using very similar ideas, they independently produced algorithms and programs for the computation of likelihoods on extended pedigrees. Very soon thereafter, Ott (1974) produced the linkage analysis program LIPED which is still in use today. This program used the Elston–Stewart algorithm to compute probabilities of data at two loci jointly, and hence likelihoods for linkage. Over the next several years, computers improved and computational algorithms were generalized to more complex pedigrees and more complex trait models (Cannings *et al.*, 1978; 1980).

Although more genetic markers that could be typed in human individuals were gradually accumulated, these methodological and computational advances would have meant little without the advent of DNA markers (Botstein *et al.*, 1980). Since 1980, new technology has made available a wealth of new types of genetic markers, on the basis of characteristics of the DNA sequence itself rather than on proteins and enzymes determined by the DNA. These markers rely primarily on length variations in DNA, either in terms of numbers of copies of a short repeat sequence, or lengths between occurrences of a given short motif. These genetic markers must first be mapped relative to each other. Then they can be used to localize the genes contributing to traits of interest. Now, there are thousands of such markers available, and the Human Genome Project goal of a marker map at 1 cM density has been achieved (Murray *et al.*, 1994). However, even a genetic length of

1 cM is approximately 1 million base pairs (bp) of DNA. More recent maps contain even more markers, including single-nucleotide-polymorphisms (SNPs). These may occur as frequently as 1 per 500 bp.

With the discovery of the first DNA markers, the restriction fragment-length polymorphisms (RFLPs), suddenly there was the prospect of markers throughout the genome. Instead of linkage groups, there would be genetic maps on which the loci contributing to traits could be placed. Questions of study design and map density arose: Who should be typed? For what markers? For many simple Mendelian traits, trait data on pedigrees either already existed or could be relatively easily collected. The marker typing was the expensive component. Again from the statistics arena, the *elod* entered the literature on genetic linkage (Thompson *et al.*, 1978). This is simply the expected log-likelihood difference or base-10 version of the Kullback–Leibler information (Kullback and Leibler, 1951)

$$elod(\theta) = \mathrm{E}_\theta \left(\log_{10} L(\theta) - \log_{10} L\left(\frac{1}{2}\right) \right),$$

where the expectation is over the probability distribution of trait and marker data under a recombination frequency θ between the trait and marker loci. The *elod* became the key measure of the information to be expected from data on a set of pedigrees, given a trait model. Since trait data were often already available, a more relevant measure of expected information for linkage is the *elod* conditional on these specific trait data. Ploughman and Boehnke (1989) solved the problem of providing these conditional expected *lod* scores in their program SIMLINK which became a mainstay for those involved in practical linkage analyses.

Another consequence of the existence of genetic marker maps, was the development of map-specific multipoint linkage analyses (Lathrop *et al.*, 1984). That is, the marker map is assumed known and a log-likelihood difference (or *location score*) is computed for each hypothesized location of the trait locus relative to the hypothesis that the trait locus is not linked to this segment of the marker map. Use of multiple marker loci increases the power of the analysis, combining information from markers that are informative in different meioses of the pedigree. Even *interval mapping*, in which a locus is mapped using data on two hypothesized flanking markers, provides much more information than mapping with each marker locus separately. These principles are embodied in programs such as VITESSE (O'Connell and Weeks, 1995) and FASTLINK (Cottingham *et al.*, 1993) which remain the standards for exact computation of lod scores on extended pedigrees, using a small number of markers.

With the use of data at larger numbers of marker loci, it becomes simpler to consider the *meiosis indicators* of (33.1) as the latent variables, rather than genotypes **G**. Then (33.4) becomes

$$\Pr(\mathbf{Y}) = \sum_{\mathbf{S}} \Pr(\mathbf{Y}|\mathbf{S})\Pr(\mathbf{S}),$$

where **S** is the total set of meiosis indicators $\{S_{i,j}; i = 1, \ldots, m, \ j = 1, \ldots, l\}$ for the m meioses of the pedigree and each of the l loci. For ease of presentation, we assume that the data are specific to known or putative genetic loci, and for convenience we suppose these to be ordered $1, 2, \ldots, l$ along a chromosome (or chromosomes). Then the data $Y_{\bullet,j}$, attributable to genotypes at locus j, depends only on the descent of genes at locus j. The analogue of (33.4) and (33.5) is then

$$\Pr(\mathbf{Y}) = \sum_{\mathbf{S}} \left(\prod_{j=1}^{l} \Pr(Y_{\bullet,j}|S_{\bullet,j}) \right) \left(\prod_{i=1}^{m} \Pr(S_{i,\bullet}) \right), \qquad (33.6)$$

since different meioses i are *a priori* independent. Although summation over the space of all $\mathbf{S}$ values is required, this is a smaller space than the space of all multilocus genotype configurations $\mathbf{G}$ on a pedigree, and has a simpler dependence structure. Moreover, probabilities $\Pr(\mathbf{S})$ depend only on the the pedigree structure and on the meiosis (chromosome) structure. Recombination parameters enter only into this term, while all other parameters relating to the genetic model for trait and marker loci affect only $\Pr(\mathbf{Y}|\mathbf{S})$.

In considering a location score curve for linkage of a particular trait, it is convenient to partition the data into the marker data $\mathbf{Y}_M$ and trait data Y_T, and likewise partition the meiosis indicators into those at the marker loci, $\mathbf{S}_M$, and those at the putative trait locus S_T. Often the genetic model for the marker phenotypes $\mathbf{Y}_M$ is assumed known. That is, the marker map Γ_M determining $\Pr(\mathbf{S}_M)$ and marker allele frequencies determining $\Pr(\mathbf{Y}_M|\mathbf{S}_M)$ are fixed. Also parameters that are often assumed to be known are the parameters β relating trait genes and trait phenotypes: $\Pr(Y_T|S_T)$ is a function of β. The only parameter which is varied is the location γ of the trait locus, with $\gamma = \infty$ corresponding to absence of linkage. Log likelihoods are plotted as a function of γ. The location score curve is thus twice the (natural) log of $L(\gamma)/L(\infty)$, where

$$L(\gamma) = \Pr(Y_T, \mathbf{Y}_M; \Gamma_M, \beta, \gamma) = \sum_{\mathbf{S}_M, S_T} \Pr(Y_T, \mathbf{Y}_M, \mathbf{S}_M, S_T; \Gamma_M, \beta, \gamma) \qquad (33.7)$$

and

$$\Pr(Y_T, \mathbf{Y}_M, \mathbf{S}_M, S_T; \Gamma_M, \beta, \gamma) = \Pr(\mathbf{Y}_M|\mathbf{S}_M)\Pr(\mathbf{S}_M; \Gamma_M)$$
$$\times \Pr(Y_T|S_T; \beta)\Pr(S_T|\mathbf{S}_M; \gamma). \qquad (33.8)$$

On the one hand, analysis is simplified by the fact that γ enters only into the final term $\Pr(S_T|\mathbf{S}_M; \gamma)$. On the other hand, analysis is complicated by the need to compute across the whole range of positions γ, and by the fact that there is often strong evidence that the trait locus is not coincident with any marker locus, leading to 0 or infinitesimal values of $L(\gamma)$ at some marker locations.

The development of genetic maps also initiated the era of genome scans, in which a test can be made for linkage of a trait with each available marker. The interpretation of *lod* scores under these multiple dependent tests is not straightforward. Even when the trait gene is in some map interval, data on a sufficient number of informative meioses will provide strong evidence for recombination between the trait locus and an adjacent marker. Thus, the log-likelihood curve for the location of the trait locus will normally have multiple peaks with sharp decreases at the marker locations.

The 1980s were the golden years of traditional linkage analysis. By 1990 simple Mendelian traits had been localized by linkage analysis, except for those so rare that too few data are available. Methods for fine-scale mapping were developed; many genes for simple traits were identified. Ott (1999) covers many of the developments of the last two sections in his text. He also says:

> Linkage analysis is sometimes perceived as a matter of simply using the proper computer program, so that anyone with sufficient computer expertise could 'do the linkage analysis' after family data and marker typing have been obtained.... It is dangerous to have linkage analyses carried out by individuals without the necessary theoretical background.

A full understanding of linkage likelihoods, and particularly of location score curves, requires that many complex statistical questions be addressed.

33.5   LIKELIHOOD AND LOCATION SCORE COMPUTATION

The basis of the Elston–Stewart algorithm, and related algorithms, is to sum latent variables sequentially over a pedigree but jointly over loci. Equation (33.4) leads to computational algorithms that are linear in pedigree size, but exponential in pedigree complexity, as measured by the number of interconnected pedigree loops. Moreover, algorithms are linear in the number of potential multilocus genotype triples of a father–mother–child trio, and hence exponential in the number of genetic loci. Although the efficiency of computer algorithms for exact likelihood evaluation have been much improved (Cottingham *et al.*, 1993), and various approximate methods have been proposed (Curtis and Gurling, 1993), exact computation is intrinsically limited in the number of loci that can be considered. With multiple linked multiallelic loci, the approach rapidly becomes computationally infeasible.

A new approach to likelihood computation on pedigrees was provided by Lander and Green (1987). Their algorithm was based on using $S_{i,j}$ (33.1) as the latent variables (33.6). They called $S_{\bullet,j}$ the *inheritance vector* at locus j. Instead of proceeding sequentially over the pedigree and jointly over loci, the computation proceeds sequentially along a chromosome, jointly over all meioses. Under the assumption of no genetic interference, the random vectors $S_{\bullet,j}$ have a first-order Markov property, and the likelihood (33.6) may be rewritten as

$$\Pr(\mathbf{Y}) = \sum_{\mathbf{S}} \Pr(\mathbf{S}, \mathbf{Y}) = \sum_{\mathbf{S}} \Pr(\mathbf{Y}|\mathbf{S})\Pr(\mathbf{S})$$

$$= \sum_{\mathbf{S}} \left(\Pr(S_{\bullet,1}) \prod_{j=2}^{l} \Pr(S_{\bullet,j}|S_{\bullet,j-1}) \prod_{j=1}^{l} \Pr(Y_{\bullet,j}|S_{\bullet,j}) \right). \qquad (33.9)$$

Under this Markov assumption, the framework is that of a hidden Markov model (HMM) and the Baum algorithm provides a computational method which is linear in the number of loci (Baum, 1972), but exponential in pedigree size. Thus, the method allows computation of probabilities of marker phenotypes for any number of loci, but on small pedigrees.

The dependence structure of data on a pedigree is shown in Figure 33.1. This shows the conditional independencies which underlie both the Elston–Stewart and Lander–Green

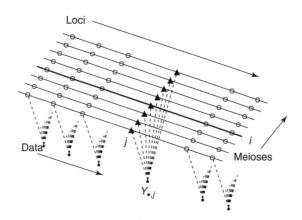

Figure 33.1   The dependence structure of data on a pedigree.

algorithms for likelihood computation on pedigrees. A pedigree is a collection of *a priori* independent meioses. In a given meiosis, transmissions are dependent along the chromosome. In the absence of interference this dependence is first-order Markov. Data are determined by the founder alleles and meioses at a given locus; observation of data at a locus creates *a posteriori* dependence of the meioses at that locus. The Elston–Stewart algorithm uses the *a priori* independence of meioses, which leads to conditional independence of genotypes (33.4). The Lander–Green uses the first-order Markov dependence of meioses along a chromosome (that is, of $S_{\bullet,j}$) in the absence of genetic interference.

Again the methods have been further developed, allowing for more efficient computation as implemented in the programs GENEHUNTER (Kruglyak *et al.*, 1996), ALLEGRO (Gudbjartsson *et al.*, 2000) and MERLIN (Abecasis *et al.*, 2002). A key recognition comes from placing the problem in the context of general graphical models (Lauritzen and Spiegelhalter, 1988), so that all dependencies in Figure 33.1 are treated equally. Due to the *a priori* independence of meioses, the structure is actually that of a *factored* HMM (Fischelson and Geiger, 2004). Thus performance is much improved by summing over the transitions of each meiosis in turn (from $S_{i,j-1}$ to $S_{i,j}$ for each $i = 1, ..., m$), rather than over the entire inheritance vector jointly (from $S_{\bullet,j-1}$ to $S_{\bullet,j}$). Regardless of the values of these transition probabilities (for example, using sex-specific maps), this factorization reduces the computation from order $2^m \times 2^m$ to order $m \times 2^m$. These and other computational advances are embodied in the program SUPERLINK (Fischelson and Geiger, 2004; Silberstein *et al.*, 2006), which has greatly increased the size of the pedigree on which exact computation of multimarker location score curve is feasible. However, Figure 33.1 also shows that computation is intrinsically exponential in the number of meioses m, so pedigree size is a limiting factor in this approach.

A disadvantage of the form (33.6) or (33.9) is the requirement to determine $\Pr(Y_{\bullet,j}|S_{\bullet,j})$, which is potentially more complex than the individual penetrance probabilities $\Pr(Y_{i,\bullet}|G_{i,\bullet})$ of (33.4). Genes are *identical-by-descent* (ibd) if they are copies of the same gene in some common ancestor. Disregarding mutation, such genes are necessarily of the same allelic type. The indicators $S_{\bullet,j} = (S_{i,j}; i = 1, ..., m)$ determine the descent of genes from the founders of the pedigree, and hence the pattern of gene *ibd* among observed individuals at locus j. If k denotes a founder gene, $a(k)$ its allelic type, and $q_j(a)$ the population frequency of allele a at locus j, then

$$\Pr(Y_{\bullet,j}|S_{\bullet,j}) = \sum_{\mathcal{A}(j)} \left(\prod_k q_j(a(k)) \right). \tag{33.10}$$

(Thompson, 1974), where $\mathcal{A}(j)$ denotes an allocation of allelic types to the founder genes labels (FGL) k at locus j, which, given $S_{\bullet,j}$, is consistent with the data $Y_{\bullet,j}$. Computation requires an efficient algorithm for the summation over all feasible allocations $\mathcal{A}(j)$ at locus j. For marker genotype data observed without error, one such algorithm is given by Sobel and Lange (1996). Using this algorithm, computation of $\Pr(Y_{\bullet,j}|S_{\bullet,j})$ is readily accomplished, even on large pedigrees. For more general marker models or quantitative traits, the problem is again one of peeling, this time over FGL k on an *FGL graph*. The nodes of this graph are the FGL and there is a dependency link between each pair of FGL that are, given $S_{\bullet,j}$, both present in an observed individual. Details are given by Thompson (2005).

33.6 MONTE CARLO MULTIPOINT LINKAGE LIKELIHOODS

The structure of dependence shown in Figure 33.1 also lends itself to a variety of Monte Carlo approaches, in which the latent variables are realized from some importance-sampling distribution. Even when $\Pr(\mathbf{Y})$ cannot be computed exactly, due to the size or complexity of the pedigree and the number of linked genetic loci, Monte Carlo estimates of the likelihood or of posterior probabilities can be made.

Since, in linkage analysis, the parameter θ is normally reserved for recombination frequencies, we use ξ for the full set of parameters of the genetic model. For example, in the case of the location score curves of (33.7), $\xi = (\Gamma_M, \beta, \gamma)$. The likelihood of a genetic model indexed by parameters ξ, given data on a pedigree, can be written

$$L(\xi) = P_\xi(\mathbf{Y}) = \sum_{\mathbf{X}} P_\xi(\mathbf{X}, \mathbf{Y}) = \sum_{\mathbf{X}} P_\xi(\mathbf{Y}|\mathbf{X}) P_\xi(\mathbf{X}), \tag{33.11}$$

where $\mathbf{X}$ are latent variables, either the genotypes $\mathbf{G}$ of (33.4) or the meiosis indicators $\mathbf{S}$ of (33.6). Thus, for fixed observed data $\mathbf{Y}$,

$$L(\xi) = \mathrm{E}_\xi(P_\xi(\mathbf{Y}|\mathbf{X})), \tag{33.12}$$

where the expectation is over $\mathbf{X}$ having the distribution $P_\xi(\mathbf{X})$. Where the variable $\mathbf{X}$ is the set of latent genotypes $\mathbf{G}$, this is the form given by Ott (1979). In principle, $L(\xi)$ could be estimated by simulating $\mathbf{G}$ from the genotype distribution under model ξ and averaging the value of the penetrance probabilities $P_\xi(\mathbf{Y}|\mathbf{G})$ for the realized values of $\mathbf{G}$. This does not work well, except on very small pedigrees, since each realized $\mathbf{G}$ is almost certain to be inconsistent with data $\mathbf{Y}$, or at best to make an infinitesimal contribution to the likelihood.

Better ideas normally involve some form of importance sampling:

$$L(\xi) = \mathrm{E}_{P^*} \left(\frac{P_\xi(\mathbf{X}, \mathbf{Y})}{P^*(\mathbf{X})} \right), \tag{33.13}$$

where now realizations are made from $P^*(\mathbf{X})$. An advantage of this approach is that a single set of realizations from $P^*(\mathbf{X})$ can be used to provide a Monte Carlo estimate of $L(\xi)$ over a range of models ξ. However, the importance sampling will be effective only if $P^*(\mathbf{X})$ is approximately proportional to $P_\xi(\mathbf{X}, \mathbf{Y})$. Since

$$P_\xi(\mathbf{X}|\mathbf{Y}) = \frac{P_\xi(\mathbf{X}, \mathbf{Y})}{P_\xi(\mathbf{Y})} \propto P_\xi(\mathbf{X}, \mathbf{Y}), \tag{33.14}$$

the ideal choice of sampling distribution is $P_\xi(\mathbf{X}|\mathbf{Y})$. However the constant of proportionality is $L(\xi) = P_\xi(\mathbf{Y})$. If $P_\xi(\mathbf{X}|\mathbf{Y})$ can be computed, so also can $L(\xi) = P_\xi(\mathbf{Y})$. Thus the choice of sampling distribution is a balance between approximation to $P_\xi(\mathbf{X}|\mathbf{Y})$ and computational feasibility.

The following approach to choice of importance-sampling distribution $P^*(\mathbf{G})$ is due to Kong *et al.* (1994) and Irwin *et al.* (1994). Suppose, as before, there are data at l genetic loci (say a disease and $l-1$ markers) on a chromosome, and assume absence of genetic interference. Note that genotypes $G_{\bullet,j}$ satisfy the same first-order Markov dependence over loci as do the meiosis indicators $S_{\bullet,j}$ (Figure 33.1). Let $Y_{\bullet,j}$ again denote the data for locus j and $G_{\bullet,j}$ the underlying genotypes at that locus for all members of the pedigree. Let $Y^{(j)} = (Y_{\bullet,1}, \ldots, Y_{\bullet,j})$, and $G^{(j)}$ be analogously defined. For any specified ξ of interest, a realization $G_{\bullet,j}^*$ is obtained for each locus in turn from the sequential imputation distribution

$$P_\xi(G_{\bullet,j}|G^{(j-1)}, Y^{(j)}) = P_\xi(G_{\bullet,j}|G_{\bullet,1}, \ldots G_{\bullet,j-1}, Y_{\bullet,1}, \ldots, Y_{\bullet,j-1}, Y_{\bullet,j})$$

$$= w_j^{-1} P_\xi(Y_{\bullet,j}|G_{\bullet,j}) P_\xi(G_{\bullet,j}|G_{\bullet,j-1}),$$

where the predictive weight w_j is $P_\xi(Y_{\bullet,j}|G_{\bullet,j-1})$. Computation of w_j is a single-locus computation which may be done using the Elston–Stewart algorithm. Then it is readily shown that the joint simulation distribution for $\mathbf{G}^* = (G_{\bullet,1}^*, \ldots, G_{\bullet,l}^*)$ is

$$P^*(\mathbf{G}^*) = \frac{P_\xi(\mathbf{Y}, \mathbf{G}^*)}{W_l(\mathbf{G}^*)}, \text{ where } W_l(\mathbf{G}^*) = \prod_{j=1}^{l} w_j. \tag{33.15}$$

Thus

$$\mathrm{E}_{P^*}(W_l(\mathbf{G}^*)) = \sum_{\mathbf{G}^*} W_l(\mathbf{G}^*) P^*(\mathbf{G}^*) = P_\xi(\mathbf{Y}) = L(\xi). \tag{33.16}$$

A Monte Carlo estimate of $L(\xi)$ is given by the mean value of $W_l(\mathbf{G}^*)$ over repeated independent repetitions of the sequential imputation process. Repeating the process for different trait-locus positions γ on the chromosome, keeping other components of ξ fixed, one can obtain an estimated location score curve (33.7). This procedure is implemented in the program SIMPLE (Skrivanek *et al.*, 2003), and works well for moderate numbers of loci.

Alternative methods use Markov chain Monte Carlo (MCMC) to sample directly from the conditional distribution of latent variables $\mathbf{X}$ given the data $\mathbf{Y}$. We shall not detail here the various MCMC methods that have been used to realize latent variables for location score estimation in linkage analysis. The early methods were primarily single-site methods using either genotypes $\mathbf{G}$ (Lange and Sobel, 1991) or meiosis indicators $\mathbf{S}$ (Thompson, 1994). More recently a variety of block-updating MCMC algorithms have been developed, including the locus-sampler (L-sampler) of Heath (1997) and the meiosis-sampler (M-sampler) of Thompson and Heath (1999), or their combination (Heath and Thompson, 1997). More recent methods allow the sampling of multiple meioses (Thomas *et al.*, 2000), and the use of sequential sampling distribution (33.15) as a Metropolis–Hastings proposal distribution (George and Thompson, 2003). These samplers require the combination of exact and Monte Carlo methods of analysis. Either the generalized Elston–Stewart algorithm (Cannings *et al.*, 1978) is used to perform the required single-locus computations conditional on neighboring loci, or the Lander–Green version of the Baum algorithm (Baum *et al.*, 1970) is used to resample some subset of the meioses.

The first MCMC-based linkage location score approach is due to Lange and Sobel (1991), who, using our current notation, write the likelihood (33.7) in the form

$$
\begin{aligned}
L(\beta, \gamma, \Gamma_M) = P_{\beta, \gamma, \Gamma_M}(Y_T, \mathbf{Y}_M) &\propto P_{\beta, \gamma, \Gamma_M}(Y_T | \mathbf{Y}_M) \\
&= \sum_{\mathbf{X}_M} P_{\beta, \gamma}(Y_T | \mathbf{X}_M) P_{\Gamma_M}(\mathbf{X}_M | \mathbf{Y}_M) \\
&= \mathrm{E}_{\Gamma_M}(P_{\beta, \gamma}(Y_T | \mathbf{X}_M) | \mathbf{Y}_M).
\end{aligned}
\tag{33.17}
$$

Now latent variables $\mathbf{X}_M$ are sampled from their conditional distribution given the marker data $\mathbf{Y}_M$. Provided exact computation of $P_{\beta, \gamma}(Y_T | \mathbf{X}_M)$ is possible for alternative trait models (β) and locations (γ), we have a Monte Carlo estimate of $L(\beta, \gamma, \Gamma_M)$, while comparison to the unlinked base point requires only $P_\beta(Y_T)$. Since Γ_M is fixed the Monte Carlo requires only a single set of N realizations $\mathbf{X}_M^{(\tau)}$, $\tau = 1, \ldots, N$. This approach is implemented in the program SIMWALK (Sobel and Lange, 1996).

The disadvantage of using (33.17) is that $P_{\beta, \gamma}(Y_T | \mathbf{X}_M^{(\tau)})$ must be computed for each such realization; this requires a single-locus peeling computation for the trait-locus data under the trait model. Further, this computation must be done not only for each realization $\mathbf{X}_M^{(\tau)}$ but also for each β and γ at which a likelihood estimate is required. In many cases, however, the gains outweigh the costs except when the simulation distribution $P_{\Gamma_M}(\mathbf{X}_M | \mathbf{Y}_M)$ is not close to proportional to the ideal importance-sampling target distribution $P_{\beta, \gamma, \Gamma_M}(\mathbf{X}_M | Y_T, \mathbf{Y}_M)$. This is particularly so for models (trait locations) γ which are not close to the truth, and for a trait which provides substantial information about the inheritance patterns of genes at the underlying trait locus, and hence also at linked marker loci (Thompson, 2000).

An alternative MCMC approach is based on the estimation of likelihood ratios between models that impose similar distributions on the latent variables conditional on the data, $\mathbf{Y}$. In this case MCMC is used to sample, not from $P_\xi(\mathbf{X}|\mathbf{Y})$ but from $P_{\xi_0}(\mathbf{X}|\mathbf{Y})$, where $\xi_0 \approx \xi$.

Then

$$P_\xi(\mathbf{Y}) = \sum_{\mathbf{X}} P_\xi(\mathbf{Y}, \mathbf{X}) = \sum_{\mathbf{X}} \frac{P_\xi(\mathbf{Y}, \mathbf{X})}{P_{\xi_0}(\mathbf{X}|\mathbf{Y})} P_{\xi_0}(\mathbf{X}|\mathbf{Y})$$

$$= \mathrm{E}_{\xi_0}\left(\frac{P_\xi(\mathbf{Y}, \mathbf{X})}{P_{\xi_0}(\mathbf{X}|\mathbf{Y})}\,\bigg|\,\mathbf{Y}\right) = P_{\xi_0}(\mathbf{Y})\mathrm{E}_{\xi_0}\left(\frac{P_\xi(\mathbf{Y}, \mathbf{X})}{P_{\xi_0}(\mathbf{Y}, \mathbf{X})}\,\bigg|\,\mathbf{Y}\right).$$

Hence in genetic analysis, or in any latent-variable context, we have

$$\frac{L(\xi)}{L(\xi_0)} = \frac{P_\xi(\mathbf{Y})}{P_{\xi_0}(\mathbf{Y})} = \mathrm{E}_{\xi_0}\left(\frac{P_\xi(\mathbf{Y}, \mathbf{X})}{P_{\xi_0}(\mathbf{Y}, \mathbf{X})}\,\bigg|\,\mathbf{Y}\right). \tag{33.18}$$

(Thompson and Guo, 1991). In this expectation, $\mathbf{Y}$ is fixed, and the distribution of $\mathbf{X}$ is $P_{\xi_0}(\cdot|\mathbf{Y})$. If $\mathbf{X}^{(\tau)}$, $\tau = 1, ..., N$, are realized from this distribution then the likelihood ratio can be estimated by

$$\frac{1}{N} \sum_{\tau=1}^{N} \left(\frac{P_\xi(\mathbf{Y}, \mathbf{X}^{(\tau)})}{P_{\xi_0}(\mathbf{Y}, \mathbf{X}^{(\tau)})}\right).$$

For estimation a location score curve (33.7), the form that follows directly from (33.18) is

$$\frac{L(\beta, \gamma_1, \Gamma_M)}{L(\beta, \gamma_0, \Gamma_M)} = \mathrm{E}_{\xi_0}\left(\frac{P_{\xi_1}(Y_T, \mathbf{Y}_M|X_T, \mathbf{X}_M)P_{\xi_1}(X_T, \mathbf{X}_M)}{P_{\xi_0}(Y_T, \mathbf{Y}_M|X_T, \mathbf{X}_M)P_{\xi_0}(X_T, \mathbf{X}_M)}\,\bigg|\,Y_T, \mathbf{Y}_M\right).$$

for two trait-locus positions γ_1 and γ_0. Since only the position of the trait locus differs between numerator and denominator, the above equation reduces to

$$\frac{L(\beta, \gamma_1, \Gamma_M)}{L(\beta, \gamma_0, \Gamma_M)} = \mathrm{E}_{\xi_0}\left(\frac{P_{\gamma_1}(X_T|\mathbf{X}_M)}{P_{\gamma_0}(X_T|\mathbf{X}_M)}\,\bigg|\,Y_T, \mathbf{Y}_M\right). \tag{33.19}$$

Thus only the conditional probability of trait-locus latent variables given marker-loci latent variables appears explicitly in the estimator. Although realization of the latent variables is complex, and requires MCMC methods, computation of the estimate from the realizations is straightforward (Thompson and Guo, 1991).

For an accurate Monte Carlo estimator based on (33.18) the distribution of $\mathbf{X}$ given $\mathbf{Y}$ should be similar under ξ and ξ_0. In the context of the location score curve estimator (33.19), this means that trait locations γ_1 and γ_0 should lie in the same marker interval. While effective methods have been developed for combining local likelihood ratio estimates from (33.18) into a smooth likelihood surface, multipoint location score curves are far from smooth across marker positions. Attempts to use the estimator (33.19) directly for locations score curves have thus had only limited success (Thompson, 2000). However, see also Section 33.7, for another use of (33.18).

Another approach to estimating location score curves has been proposed by George and Thompson (2003). Since γ is a single scalar parameter, we may regain the likelihood from a posterior probability distribution:

$$L(\beta, \gamma, \Gamma_M) = \frac{P_{(\beta, \Gamma_M)}(\gamma \mid \mathbf{Y})}{\pi(\gamma)},$$

where (β, Γ_M) are held fixed, and $\pi(\gamma)$ is any chosen prior on trait location γ. Since $\pi(\gamma)$ is a pseudoprior in the sense of Geyer and Thompson (1995), it may be arbitrarily chosen to improve MCMC performance. In practice, it is chosen as a discrete prior on points at which the location score curve is to be estimated. To improve MCMC performance, a preliminary run is used to provide a prior approximately inversely proportional to the likelihood, so that in the main MCMC run the marginal posterior $P_{(\beta, \Gamma_M)}(\gamma \mid \mathbf{Y})$ is approximately uniform.

In Monte Carlo estimation, Rao-Blackwellization (Gelfand and Smith, 1990) can play an important role in improving estimators. This procedure replaces an estimator by its conditional expectation given some subset of the sampled variables. Thus the Rao-Blackwellized estimator requires more intensive exact computations, but benefits from reduced Monte Carlo sample size to achieve the same accuracy. For dependent realizations, Rao–Blackwellization does not guarantee reduced Monte Carlo variance (Liu *et al.*, 1994), but in practice it normally provides significant gains in computational efficiency. For example, suppose we have realizations $(\gamma^{(\tau)}, \mathbf{X}^{(\tau)})$, $\tau = 1, ..., N$ from $P_{(\beta, \Gamma_M)}(\gamma, \mathbf{X} \mid \mathbf{Y})$. Rather than estimating the marginal posterior $P_{(\beta, \Gamma_M)}(\gamma \mid \mathbf{Y})$ by

$$\frac{1}{N} \sum_{\tau=1}^{N} I(\gamma^{(\tau)} = \gamma),$$

we instead use the estimators

$$\frac{1}{N} \sum_{\tau=1}^{N} \mathrm{E}(I(\gamma^{(\tau)} = \gamma) \mid \mathbf{X}_M^{(\tau)}, \mathbf{Y}).$$

In fact, it is easily shown that

$$\mathrm{E}(I(\gamma^{(\tau)} = \gamma) \mid \mathbf{X}_M^{(\tau)}, \mathbf{Y}) = \frac{P(Y_T \mid \gamma, \mathbf{X}_M^{(\tau)}) \pi(\gamma)}{\sum_{\gamma^*} P(Y_T \mid \gamma^*, \mathbf{X}_M^{(\tau)}) \pi(\gamma^*)}, \qquad (33.20)$$

leading to an estimator similar in appearance to (33.17) in that both depend on the probabilities of trait data Y_T given sampled marker latent variables $\mathbf{X}_M$ computed for each desired trait location γ. However, note that here the trait data and trait model β enter into the sampling of variables $\mathbf{X}_M$.

The new MCMC methods can provide accurate location scores on large pedigrees with substantial missing data for two or three marker loci many times faster than the exact computational methods, and can estimate multipoint location score curves when exact computation is infeasible (George and Thompson, 2003). The MCMC lod score estimation methods of (33.17), (33.19), and (33.20) are all implemented in programs in the MORGAN package (Thompson, 2005).

33.7 LINKAGE ANALYSIS OF COMPLEX TRAITS

As the traits considered in linkage analyses become increasingly complex, there are two alternative approaches to analysis. One is to avoid explicit modeling of the trait, using so-called 'nonparametric' genome-sharing computations for linkage detection (see **Chapter 34**). The other is to develop more complex models for more complex phenotypes. Several factors enhance the detection and resolution of genes contributing to complex genetic traits. These include joint analysis of data on members of extended pedigrees, joint analysis of data at multiple genetic markers, and selective sampling of pedigrees (Wijsman and Amos, 1997). Although data are increasingly available at multiple tightly linked genetic markers, often a high proportion of the individuals in an extended pedigree is not observed. Data may be available only on affected individuals, or even on only a few of their relatives. Moreover, selectively sampled extended pedigrees are often also complex. Often the traits of interest are not simple genetic traits. They may have delayed onset, or there may be incomplete penetrance; individuals carrying the disease allele may never show symptoms. Conversely, there may be individuals who apparently have the disease but do not carry the gene, there may be several different genetic causes of a disease, or even several genes interacting to produce the observed characteristics. In some cases, the traits of interest may be quantitative.

Because of difficulties of likelihood computation, trait model uncertainties, and the mass of marker data, often with different markers typed or informative in different families, a number of researchers moved away from likelihood analyses of linkage in the 1980s, and instead developed a variety of association tests for linkage detection. There are two main classes of such tests, those at the population level and those at the family or pedigree level. Genome-sharing methods are based on analysis of marker data, and rely on the fact that, regardless of the trait model provided it has some genetic component, related affected individuals or related individuals exhibiting extreme trait values will share genes *ibd* at trait loci with some increased probability. Hence also they will share genes *ibd* with increased probability at marker loci linked to those trait loci. While genome-sharing methods on individuals with known relationship derive first from Penrose (1935), and are best known in the form of sib pair analyses (Suarez *et al.*, 1978) and affected-relative methods (Weeks and Lange, 1988) they can also involve probability computations on a pedigree structure. For example, the method of *homozygosity mapping* for rare recessive traits developed by Lander and Botstein (1987) can be viewed either as computation of a linkage likelihood (Smith, 1953) or as an inference based on the genome sharing between the two haplotypes of inbred affected individuals.

Another area of linkage analysis which uses gene *ibd* is fine-scale localization of genes. The resolution of pedigree-based methods of linkage analysis is limited by the number of meioses in the pedigrees. However, when a new trait allele arises by mutation, it does so on some specific chromosome with a specific collection of alleles at nearby markers; that is, there is a specific marker *haplotype* that carries the new trait mutation. Where the loci are very tightly linked, associations of the disease allele with an allele at a linked marker locus may be maintained for many generations before decaying due to recombination (33.3). Thus, there will be population associations due to chance historical associations. The study of association can be very useful in narrowing the region in which a gene is located. For a general discussion of the use of association tests in linkage analysis, see **Chapter 34**. Here we focus on aspects relating to analysis of marker data on extended pedigrees.

There is much more information about genome-sharing patterns in the genome if all the available data on the pedigree are used to compute *ibd* probabilities, and if *ibd* is scored jointly among the several affected individuals of an extended pedigree, rather than pair-wise. Exact computation of gene *ibd* patterns jointly among relatives in an extended pedigree, using data at multiple linked markers, is equivalent to likelihood computation for these marker data, as given by (33.4) or (33.9). Note that the gene *ibd* at any locus j is a function of the inheritance vector $S_{\bullet,j}$ (Section 33.5). For data at multiple marker loci, exact computation is thus feasible only on a small pedigree. Moreover, although likelihoods for relationship and the estimates of gene *ibd* patterns do incorporate the linkage information and data at all loci, the Baum algorithm (Baum *et al.*, 1970) gives only estimates of the probabilities $\Pr(S_{\bullet,j} \mid \mathbf{Y})$ marginally at each locus j. At best, on small pedigrees, it is possible to obtain the probabilities $\Pr(S_{\bullet,j-1}, S_{\bullet,j} \mid \mathbf{Y})$ for pairs of loci.

However, Monte Carlo sampling of the meiosis indicators $\mathbf{S}_M$ given marker data $\mathbf{Y}_M$ provides direct estimates of posterior probabilities of patterns of gene *ibd*. These realizations can be scored jointly over loci, and jointly over individuals, and hence provides for more general analyses of genome sharing in relatives (Thompson, 2000). Further, given each realization $\mathbf{S}_M^{(\tau)}$, realizations of the inheritance vector $S(\gamma)$ at location γ may be readily obtained. If γ lies between markers $j - 1$ and j

$$\Pr(S(\gamma) \mid \mathbf{S}_M, \mathbf{Y}_M) = \Pr(S(\gamma) \mid S_{\bullet,j-1}^{(\tau)}, S_{\bullet,j}^{(\tau)}).$$

That is, $S(\gamma)$ depends only on the realized inheritance vector at the two neighboring markers. On small pedigrees or for certain *ibd* patterns of interest exact probability computations given the realized marker inheritance vectors are often possible.

For linkage analysis, the statistical problem becomes, first, one of development of appropriate test statistics to detect linkage on the basis of estimated posterior *ibd* probabilities (for example, see McPeek, 1999). Second a way of assessing statistical significance of estimated *ibd* statistics is needed, taking into account both uncertainties in the *ibd* given marker data, that *ibd* patterns at linked locations with a region of the genome are highly dependent, and the fact that numerous regions of the genome may be tested for linkage to the trait of interest. Under the hypothesis of no linkage to the trait, resimulation of marker data is always possible, but analysis of multiple resimulated data sets is computationally intensive and this approach is sensitive to marker model misspecification. Permutation tests (Churchill and Doerge, 1994) are robust, but on an extended pedigree the set of available valid permutations may be small. A compromise is provided by the approach of Thompson and Geyer (2007), which, like a permutation test conditions on the observed marker data, and uses the probability distribution of the underlying latent variables $\mathbf{S}_M$ to provide a test of significance.

The alternative general approach for analysis of complex traits uses an explicit trait model. An early such model was the *mixed model* (Morton and MacLean, 1974), in which the phenotypic value is dependent both on the genotype at a 'major' Mendelian locus and on a heritable *polygenic* Gaussian random effect. While the polygenic component can be a convenient way to model additional heritable variation, likelihood computation for the mixed model is problematic, and linkage analysis more so. Increasingly, methods are being developed for *oligogenic models* in which several major genes contribute to a trait, and a Bayesian MCMC analysis is used to detect and localize these genes and estimate their effects (Heath, 1997). Although these methods have been directed primarily toward quantitative traits (see **Chapter 19**), they can also be used to analyze a qualitative trait controlled by a latent quantitative liability. Age-of-onset of a disease can also be analyzed, using these methods, by treating it as a censored quantitative trait (Daw *et al.*, 1999).

Robust genome-sharing methods for linkage detection and Bayesian methods of linkage estimation are gaining popularity in part because of the difficulties of both computation and interpretation of linkage likelihoods for complex traits. These difficulties of interpretation result from uncertainties in the trait model, heterogeneity in the genetic causes of the trait, and interpretation of location score curves. In terms of assessing sensitivity to trait, marker, or map model assumptions, (33.18) provides a useful MCMC approach. Recall that for any probability models, ξ and ξ_0

$$\frac{L(\xi)}{L(\xi_0)} = \frac{P_\xi(\mathbf{Y})}{P_{\xi_0}(\mathbf{Y})} = \mathrm{E}_{\xi_0}\left(\frac{P_\xi(\mathbf{Y}, \mathbf{X})}{P_{\xi_0}(\mathbf{Y}, \mathbf{X})} \;\middle|\; \mathbf{Y}\right),$$

provided only that ξ_0 assigns positive probability to any latent values $\mathbf{X}$ that have positive probability under ξ. Thus a single set of Monte Carlo realizations at some assumed model ξ_0 can be used to obtain an estimate of the likelihood ratio $L(\xi)/L(\xi_0)$ for any model that is close to ξ. The models ξ may differ from ξ_0 in any, or several, parameters.

As data on more complex traits are analyzed, the problems of trait model misspecification and trait heterogeneity have assumed increasing importance. The excellent recent text by Ott (1999) gives a very thorough review of human genetic linkage, and methods of linkage analysis. The Bayesian approach, with a prior probability distribution for the location of a trait gene, avoids some of the problems of multiple tests for linkage. These methods can also be used to test for heterogeneity, providing a posterior probability that a particular pedigree is segregating a putative trait gene Ott (1999). Although Bayesian methods have a long history in linkage analysis (Haldane and Smith, 1947) and have recently attracted more attention, the problems of trait model misspecification and heterogeneity are no less with a Bayesian approach. Inference based on the *lod* score or location score curve is still the accepted practical approach. Further, new Monte Carlo–based approaches to *lod* score estimation can allow for more complex trait models, two such major loci, and a polygenic component all contributing jointly to a quantitative trait value (Sung *et al.*, 2006).

33.8   MAP ESTIMATION, MAP UNCERTAINTY, AND THE MEIOSIS MODEL

In the previous sections, we assumed the marker map Γ_M to be known. However, the usefulness of multipoint linkage analyses is dependent on the accuracy of both the marker map and of the trait model. Uncertainties about the marker map can have a strong impact on the location log-likelihood curve (Halpern and Whittemore, 1999; Daw *et al.*, 2000). Although there is now a wealth of DNA markers, their exact positions and the frequencies of their alleles in the study population are often uncertain. Thus, there is a need to compute location scores under alternative assumptions about the marker loci and to compute likelihoods relating to estimation of the marker map. Evaluation of multipoint lod scores is extremely computationally intensive in human pedigrees where there are often missing data and many alternative patterns of gene descent that are compatible with the observed data.

Estimation of an accurate meiotic map becomes harder as maps become denser. The same markers will not be typed in all studies; even if typed, the same will not be informative for linkage. Moreover, rather than seeking to estimate recombination rates of perhaps 20 %, it is necessary to order markers between which recombination rates are less than 1 %. To analyze these small recombination frequencies, far more data are required. Typically, the amounts of data available in pedigree studies will not permit a finer scale of resolution than about 1 cM (Boehnke, 1994). Even 1 cM distance is approximately 1 million DNA base pairs – too great a length for current methods of physical mapping to be practical.

Many factors influence recombination frequencies. A major one is the sex of the parent; in humans the total female map length of the 22 chromosome pairs (not including the sex chromosomes) is 39 M, about 1.5 times the male map length (26.5 M). However, this ratio is not constant over the genome. In many linkage analyses, male and female recombination frequencies are estimated jointly, although sometimes they are constrained to be equal, or a fixed relationship between them may be assumed. Differences in male and female genetic map distances have an impact on multipoint location scores. However, even the best current meiotic maps (NIH-CEPH Collaborative Mapping Group, 1992) are based on very limited numbers of meioses particularly when sex-specific maps are desired. Often a large study may have as many meioses available for analysis as these standard maps. In fact, several recent large-scale genotyping studies, such as that of the Icelandic population by deCODE Genetics, provide large numbers of informative meioses, and will become a future resource for investigating the relationship between physical and genetic distance and sources of variation in genetic maps. For example a study of stroke by Gretarsdottir et al. (2002) yielded over 1000 informative meioses, over five times as many as in the standard CEPH panel. Map estimation is needed to refine maps in critical regions, where valid likelihoods for linkage detection or association analyses are desired.

The problem of map estimation is well suited to an EM algorithm approach (Dempster et al., 1977). As earlier, suppose, we have l marker loci along a chromosome, now with recombination frequencies $\theta_{m,j-1}$ and $\theta_{f,j-1}$ in male and female meioses, respectively, between locus $j-1$ and locus j. With data $\mathbf{Y}$ and latent variables $\mathbf{S}$ consider the complete-data log-likelihood

$$\log \Pr(\mathbf{S}, \mathbf{Y}) = \log(\Pr(S_{\bullet,1})) + \sum_{j=2}^{l} \log(\Pr(S_{\bullet,j}|S_{\bullet,j-1})) + \sum_{j=1}^{l} \log(\Pr(Y_{\bullet,j}|S_{\bullet,j})) \quad (33.21)$$

(see equation (33.9)). Now, in the absence of interference, the recombination probabilities $\theta_{m,j-1}$ and $\theta_{f,j-1}$ enter only into the term $\log(\Pr(S_{\bullet,j} \mid S_{\bullet,j-1}))$ which takes the form

$$\log\left(\Pr\left(S_{\bullet,j}|S_{\bullet,j-1}\right)\right) = R_{m,j-1} \log\left(\theta_{m,j-1}\right) + \left(M_m - R_{m,j-1}\right) \log\left(1 - \theta_{m,j-1}\right)$$
$$+ R_{f,j-1} \log\left(\theta_{f,j-1}\right) + \left(M_f - R_{f,j-1}\right) \log\left(1 - \theta_{f,j-1}\right),$$

where $R_{m,j-1}$ is the number of recombinations in interval $(j-1, j)$ in male meioses, and M_m is the total number of male meioses scored in the pedigree. The recombination counts $R_{f,j-1}$, for $j = 2, \ldots, l$, and total meioses M_f are similarly defined for the female meioses. Thus computation of the expected complete-data log-likelihood requires only computation of

$$\tilde{R}_{m,j-1} = \mathrm{E}(R_{m,j-1}|\mathbf{Y}).$$

Since this is a simple binomial log likelihood, the M-step sets the new estimate of $\theta_{m,j-1}$ to $\tilde{R}_{m,j-1}/M_m$, and similarly for all intervals $j = 2, 3, \ldots, l$ and for both the male and female meioses. On a small pedigree, expectations can be computed exactly, and the EM algorithm is thus readily implemented to provide estimates of recombination frequencies for all intervals and for both sexes.

An alternative is Monte–Carlo EM (Guo and Thompson, 1994). Instead of computing the bivariate distributions of $(S_{\bullet,j-1}, S_{\bullet,j})$ (Baum et al., 1970) in order to estimate the expected proportions of recombinations in interval $(j - 1)$, N realizations $\{S^{(\tau)}; \tau = 1, \ldots, N\}$ are obtained from the conditional distribution of $\Pr(S \mid Y)$ under the current parameter values. Recombinations and nonrecombinations are scored in the realized $S^{(\tau)}$, and Monte Carlo estimates of the expectations are obtained. This Monte Carlo EM is readily implemented, and, like many Monte Carlo EM procedures, performs as well as the deterministic version. Initially the Monte Carlo sample size N need not be large, although for the final EM steps it should be increased to gain precision in the final Monte Carlo estimates of the maximum likelihood estimates (MLEs) of the recombination frequencies. A generalization of this basic approach has been developed by Stewart and Thompson (2006), permitting map estimation on extended pedigrees with extensive missing data. These methods facilitate assessment of the accuracy and uncertainty in meiotic maps and provide an approach to combine data sets from several sources, including the large pedigrees used for disease–gene mapping.

MCMC realization of meiosis indicators S, or of multilocus genotypes G, on a pedigree under a genetic map given genetic marker data Y also leads directly to methods for detecting genotype or pedigree error. The meiosis indicators determine imputed multilocus haplotypes which can be used to detect probable genotyping errors that are apparent only when multilocus data are analyzed jointly. Methods for detection of such pedigree and typing errors are important as preliminaries to other multipoint analysis methods and have been a focus of research in recent years (McPeek and Sun, 2000; Epstein et al., 2000; Sieberts et al., 2002).

Another aspect of meiotic maps is the presence of genetic interference. In experimental organisms, genetic interference has been well known and well studied since Haldane (1919). For example, Weinstein (1936) investigated estimation of multilocus recombination pattern frequencies using data he had previously collected on seven linked loci in a sample of 28 239 meioses of Drosophila X chromosomes. Although Bailey (1961) developed the mathematical theory and models for the process of interference, only now do we have, in principle, the information to investigate interference on the basis of data on humans or other mammalian species. For example, King et al. (1991) have shown almost complete interference in mice over regions up to $10 \, \text{cM}$. Genetic interference in meiosis is a more complex issue than that of sex-specific maps, since it destroys the first-order Markov conditional-independence-structure of the inheritance vectors along a chromosome. Although genetic interference exists, it is seldom incorporated into multilocus linkage computations. This can reduce the power to detect linkage (Goldstein et al., 1995). In an analysis of data at multiple tightly linked markers from actual maternal meioses in loblolly pine, Thompson and Meagher (1998) have shown that interference can have a significant impact on patterns of cosegregation of genes and gene ibd at distances of $30–50 \, \text{cM}$.

Almost all likelihood computations on pedigrees assume absence of interference, and hence the independence of recombination in different marker intervals. An earlier practice was to transform estimated recombination fractions using some mapping function relating genetic distance d (in Morgans) to recombination fraction θ. For example, in the absence of interference

$$\theta(d) = \frac{1}{2}(1 - \exp(-2d)),$$

(Haldane, 1919), while for the Kosambi map function which exhibits a moderate amount of interference

$$\theta(d) = \frac{1}{2}\tanh(2d) = \frac{1}{2}\left(\frac{\exp(4d) - 1}{\exp(4d) + 1}\right).$$

Since for all natural map functions, $\theta \approx d$ when both are small, this transformation practice becomes increasingly futile as maps become denser. Interference must be accommodated within the multilocus computation itself. The question of the impact of interference on linkage analysis and linkage detection can only be resolved through likelihood computation and data analysis under both interference and noninterference models. Two methods of likelihood evaluation under interference have been proposed. Weeks *et al.* (1993) provides an approach for models of count interference (Liberman and Karlin, 1984), while Lin and Speed (1996) provides a method for the renewal process χ-square models of position interference (Zhao *et al.*, 1995). Both methods are computationally intensive, and limited to small numbers of loci and/or simple pedigrees.

MCMC and Monte Carlo likelihood methods greatly extend the feasibility of likelihood evaluation under interference models. If more complex meiosis models are to be considered, a meiosis-based inheritance-vector M-sampler has advantages over a locus-based genotypic L-sampler. Using the M-sampler, all the inheritance indicators for all the linked loci in an entire meiosis are resampled jointly and incorporation of an interference model is feasible. For a fixed marker map, it is feasible to precompute and store probabilities of all patterns of recombination and nonrecombination in a set of marker intervals for up to about 12 markers ($2^{11} = 2048$). However, as for any multilocus problem, exact likelihood computation on an extended or complex pedigree remains computationally infeasible. The MCMC M-sampler can, however, be extended. In that sampler, given marker data $\mathbf{Y}$ at loci $j = 0, \ldots, K$, meiosis indicators at meiosis i, $S_{i,\bullet} = (S_{i,0}, \ldots, S_{i,K})$ are realized from

$$P^{(H)}(S_{i,\bullet}|S_{k,\bullet}, k \neq i, \mathbf{Y}) \propto P(\mathbf{Y}|\mathbf{S})P^{(H)}(\mathbf{S}), \tag{33.22}$$

where $\mathbf{S}$ is the total set of meiosis indicators for all loci at all meioses of the pedigree, and the superscript (H) denotes the Haldane (no-interference) model. Using (33.22) as the proposal distribution, a Metropolis acceptance step (Metropolis *et al.*, 1953) can be added into the process to provide the correct conditional distribution of $S_{i,\bullet}$ under interference (denoted $P^{(I)}$). The required Hastings-ratio α (Hastings, 1970) for the acceptance probability $\min(1, \alpha)$ for current $\mathbf{S}$ and proposed $\mathbf{S}^\dagger$ is

$$\alpha = \frac{P^{(I)}(\mathbf{S}^\dagger, \mathbf{Y})}{P^{(I)}(\mathbf{S}, \mathbf{Y})} \frac{P^{(H)}(S_{i,\bullet}|S_{k,\bullet}, k \neq i, \mathbf{Y})}{P^{(H)}(S_{i,\bullet}^\dagger|S_{k,\bullet}^\dagger, k \neq i, \mathbf{Y})}$$

$$= \frac{P^{(I)}(S_{i,\bullet}^\dagger)}{P^{(I)}(S_{i,\bullet})} \frac{P^{(H)}(S_{i,\bullet})}{P^{(H)}(S_{i,\bullet}^\dagger)}. \tag{33.23}$$

This considerable reduction, and consequent ease of computation of the acceptance probability $\min(1, \alpha)$ relies on the independence of segregation patterns at different meioses i (when not conditioned on data $\mathbf{Y}$), and of the fact that the probability of data $\mathbf{Y}$ given meiosis pattern $\mathbf{S}$ does not depend on the interference process giving rise to $\mathbf{S}$. Thompson (2000) gives an example of the use of this Metropolis–Hastings sampler to investigate the impact of interference on probabilities of patterns of gene identity by descent conditional on observed marker data on a pedigree.

33.9   THE FUTURE

Just as the questions of linkage analysis have remained unchanged, so also have the basic approaches. The ideas developed by Sturtevant (1913), Robbins (1918), Haldane (1919), and Fisher (1922b) are still the basis of analysis of the outcomes of meiosis, and hence of linkage analysis. With the completion of the Human Genome Project, the practical relevance of linkage analysis is sometimes questioned, particularly in view of the variable relationship between the physical map (DNA sequence lengths in base pairs) and the genetic map (recombination frequencies). However, this issue dates back to (Fisher, 1922b), and despite changing markers and changing traits of interest, analysis of the cotransmission of genes is still the only route to a fuller understanding of the meiotic map, and sources of variation in that map. Even when our markers become the DNA sequence, and traits are levels of gene expression in a given tissue or organ, patterns of similarity and difference among relatives still provide the basic information of genetic inheritance.

As data on multiple linked markers have become available, the computational paradigm has shifted from one based on genotypes of individuals in a pedigree structure (Elston and Stewart, 1971) to one based on the meiosis indicators along a chromosome (Lander and Green, 1987). As DNA data become available at even higher resolution, it may become impractical to study the huge number of marker loci available. Instead of considering recombination events between discrete marker loci, it becomes possible to analyze the precise crossover points in a segregation, or the segments of genome shared by relatives, or by individuals having a trait in common. Such models date back to Haldane (1919) but became more formalized with Fisher's theory of junctions (Fisher, 1949). In the context of data on pedigrees, Donnelly (1983) first developed continuous genome models in terms of the meiosis indicators of (33.1). As for classical linkage analysis, likelihoods must be computed, and the computational issues must become more complex. Monte Carlo estimation of likelihoods will have a role here also (Browning, 2000), in conjunction with other computational approaches.

From the situation in 1980, when many trait data were available but marker typing was hard and expensive, we have moved to an era in which the marker typing of available individuals is relatively cheap, fast, and easy. The major cost of a study of a complex trait is now in the family collection and trait phenotyping. The limits to resolution in linkage analysis now lies in the trait data rather than in the markers. Data on extended pedigrees are necessary to resolve questions of locus and allelic heterogeneity, but in an extended pedigree a high proportion of individuals may be unavailable. The robustness of linkage inferences not only to trait model deviations but also to genetic map and meiosis model variations remains to be fully investigated. These questions raise further computational and statistical challenges to be met in the twenty-first century.

Acknowledgments

This research was supported in part by NIH grant GM-46255. I am grateful to Mike Badzioch, Nicola Chapman, and Hao Liu for comments on an earlier version of this chapter, to Professor A.W.F. Edwards for discussion of the linkage analysis contributions of Smith (1953), and to Professor Dan Geiger for discussion of graphical model computations.

REFERENCES

Abecasis, G.R., Cherny, S.S., Cookson, W.O. and Cardon, L.R. (2002). Merlin – rapid analysis of dense genetic maps using sparse gene flow trees. *Nature Genetics* **30**, 97–101.

Bailey, N.T.J. (1961). *Introduction to the Mathematical Theory of Genetic Linkage*. Clarendon Press, Oxford.

Barnard, G.A. (1949). Statistical inference. *Journal of the Royal Statistical Society Series B* **11**, 115–139.

Baum, L.E. (1972). An inequality and associated maximization technique in statistical estimation for probabilistic functions on Markov processes. In *Inequalities-III; Proceedings of the Third Symposium on Inequalities. University of California Los Angeles, 1969*, O. Shisha, ed. Academic Press, New York, pp. 1–8.

Baum, L.E., Petrie, T., Soules, G. and Weiss, N. (1970). A maximization technique occurring in the statistical analysis of probabilistic functions on Markov chains. *Annals of Mathematical Statistics* **41**, 164–171.

Boehnke, M. (1994). Limits of resolution of genetic linkage studies: implications for the positional cloning of human disease genes. *American Journal of Human Genetics* **55**(2), 379–390.

Botstein, D., White, R.L., Skolnick, M.H. and Davis, R.W. (1980). Construction of a linkage map in man using restriction fragment polymorphism. *American Journal of Human Genetics* **32**, 314–331.

Browning, S. (2000). A Monte Carlo approach to calculating probabilities for continuous identity by descent data. *Journal of Applied Probability* **37**, 850–864.

Cannings, C., Thompson, E.A. and Skolnick, M.H. (1978). Probability functions on complex pedigrees. *Advances of Applied Probability* **10**, 26–61.

Cannings, C., Thompson, E.A. and Skolnick, M.H. (1980). Pedigree analysis of complex models. In *Current Developments in Anthropological Genetics*, J. Mielke and M. Crawford, eds. Plenum Press, New York, pp. 251–298.

Churchill, G.A. and Doerge, R.W. (1994). Empirical threshold values for quantitative trait mapping. *Genetics* **138**(3), 963–71.

Cottingham, R.W., Idury, R.M. and Schäffer, A.A. (1993). Faster sequential genetic linkage computations. *American Journal of Human Genetics* **53**, 252–263.

Curtis, D. and Gurling, H. (1993). A procedure for combining two-point lod scores into a summary multipoint map. *Human Heredity* **43**, 173–185.

Daw, E.W., Kumm, J., Snow, G.L., Thompson, E.A. and Wijsman, E.M. (1999). MCMC methods for genome screening. *Genetic Epidemiology* **17**(Suppl. 1), S133–S138.

Daw, E.W., Thompson, E.A. and Wijsman, E.M. (2000). Bias in multipoint linkage analysis arising from map misspecification. *Genetic Epidemiology* **19**, 366–380.

Dempster, A.P., Laird, N.M. and Rubin, D.B. (1977). Maximum likelihood from incomplete data via the EM algorithm (with discussion). *Journal of the Royal Statistical Society Series B* **39**, 1–37.

Donnelly, K.P. (1983). The probability that related individuals share some section of genome identical by descent. *Theoretical Population Biology* **23**(1), 34–63.

Edwards, J.H. (1971). The analysis of X-linkage. *Annals of Human Genetics* **34**, 229–259.

Elston, R.C. and Stewart, J. (1971). A general model for the analysis of pedigree data. *Human Heredity* **21**, 523–542.

Epstein, M., Duren, W. and Boehnke, M. (2000). Improved inference of relationship for pairs of individuals. *American Journal of Human Genetics* **67**, 1219–1231.

Fischelson, M. and Geiger, D. (2004). Optimizing exact linkage computations. *Journal of Computational Biology* **11**, 263–275.

Fisher, R.A. (1922a). On the mathematical foundations of theoretical statistics. *Philosophical Transactions of the Royal Society of London Series A* **222**, 309–368.

Fisher, R.A. (1922b). The systematic location of genes by means of crossover observations. *American Naturalist* **56**, 406–411.

Fisher, R.A. (1934). The amount of information supplied by records of families as a function of the linkage in the population sampled. *Annals of Eugenics* **6**, 66–70.

Fisher, R.A. (1936). Heterogeneity of linkage data for Friedreich's ataxia and the spontaneous antigens. *Annals of Eugenics* **7**, 17–21.

Fisher, R.A. (1949). *The Theory of Inbreeding*. Oliver and Boyd, Edinburgh.

Gelfand, A.E. and Smith, A.F.M. (1990). Sampling based approaches to calculating marginal densities. *Journal of the American Statistical Association* **85**, 398–409.

George, A.W. and Thompson, E.A. (2003). Multipoint linkage analyses for disease mapping in extended pedigrees: a Markov chain Monte Carlo approach. *Statistical Science* **18**, 515–531.

Geyer, C.J. and Thompson, E.A. (1995). Annealing Markov chain Monte Carlo with applications to ancestral inference. *Journal of the American Statistical Association* **90**, 909–920.

Goldstein, D.R., Zhao, H. and Speed, T.P. (1995). Relative efficiencies of chi-square models of recombination for exclusion mapping and gene ordering. *Genomics* **27**(2), 265–273.

Gretarsdottir, S., Sveinbjornsdottir, S., Jonsson, H.H., Jakobsson, F., Einarsdottir, E., Agnarsson, U., Shkolny, D., Einarsson, G., Gudjonsdottir, H.M., Valdimarsson, E.M., Einarsson, O.B., Thorgeirsson, G., Hadzic, R., Jonsdottir, S., Reynisdottir, S.T., Bjarnadottir, S.M., Gudmundsdottir, T., Gudlaugsdottir, G.J., Gill, R., Lindpaintner, K., Sainz, J., Hannesson, H.H., Sigurdsson, G.T., Frigge, M.L., Kong, A., Gudnason, V., Stefansson, K. and Gulcher, J.R. (2002). Localization of a susceptibility gene for common forms of stroke to 5q12. *American Journal of Human Genetics* **70**, 593–603.

Gudbjartsson, D., Jonasson, K., Frigge, M. and Kong, A. (2000). Allegro, a new computer program for multipoint linkage analysis. *Nature Genetics* **25**, 12–13.

Guo, S.W. and Thompson, E.A. (1994). Monte Carlo estimation of mixed models for large complex pedigrees. *Biometrics* **50**(2), 417–432.

Haldane, J.B.S. (1919). The combination of linkage values and the calculation of distances between the loci of linked factors. *Journal of Genetics* **8**, 229–309.

Haldane, J.B.S. (1934). Methods for the detection of autosomal linkage in man. *Annals of Eugenics* **6**, 26–65.

Haldane, J.B.S. and Smith, C.A.B. (1947). A new estimate of the linkage between the genes for colour-blindness and haemophilia in man. *Annals of Eugenics* **14**, 10–31.

Halpern, J. and Whittemore, A.S. (1999). Multipoint linkage analysis. A cautionary note. *Human Heredity* **49**, 194–196.

Hartl, D.L. and Clark, A.G. (1997). *Principles of Population Genetics*, 3rd edition. Sinauer, Sunderland, MA.

Hastings, W.K. (1970). Monte Carlo sampling methods using Markov chains and their applications. *Biometrika* **57**, 97–109.

Heath, S.C. (1997). Markov chain Monte Carlo segregation and linkage analysis for oligogenic models. *American Journal of Human Genetics* **61**(3), 748–760.

Heath, S. and Thompson, E.A. (1997). MCMC samplers for multilocus analyses on complex pedigrees. *American Journal of Human Genetics* **61**(Suppl.), A278.

Heuch, I. and Li, F.M.H. (1972). PEDIG – A computer program for calculation of genotype probabilities, using phenotypic information. *Clinical Genetics* **3**, 501–504.

Hilden, J. (1970). GENEX – An algebraic approach to pedigree probability calculus. *Clinical Genetics* **1**, 319–348.

Irwin, M., Cox, N. and Kong, A. (1994). Sequential imputation for multilocus linkage analysis. *Proceedings of the National Academy of Sciences of the United States of America* **91**, 11684–11688.

King, T.R., Dove, W.F., Guénet, J., Hermann, B.G. and Shedlovsky, A. (1991). Meiotic mapping of murine chromosome 17: the string of loci around l(17)-2-Pas. *Mammalian Genome* **1**, 37–46.

Kong, A., Liu, J. and Wong, W.H. (1994). Sequential imputations and Bayesian missing data problems. *Journal of the American Statistical Association* **89**, 278–288.

Kruglyak, L., Daly, M.J., Reeve-Daly, M.P. and Lander, E.S. (1996). Parametric and nonparametric linkage analysis: a unified multipoint approach. *American Journal of Human Genetics* **58**(6), 1347–1363.

Kullback, S. and Leibler, R.A. (1951). On information and sufficiency. *Annals of Statistics* **22**, 79–86.

Lander, E.S. and Botstein, D. (1987). Homozygosity mapping: a way to map human recessive traits with the DNA of inbred children. *Science* **236**, 1567–1570.

Lander, E.S. and Green, P. (1987). Construction of multilocus genetic linkage maps in humans.*Proceedings of the National Academy of Sciences of the United States of America* **84**(8), 2363–2367.

Lange, K. and Sobel, E. (1991). A random walk method for computing genetic location scores. *American Journal of Human Genetics* **49**, 1320–1334.

Lathrop, G.M., Lalouel, J.M., Julier, C. and Ott, J. (1984). Strategies for multilocus linkage analysis in humans. *Proceedings of the National Academy of Sciences of the United States of America* **81**, 3443–3446.

Lauritzen, S.L. and Spiegelhalter, D.J. (1988). Local computations with probabilities on graphical structures and their application to expert systems. *Journal of the Royal Statistical Society Series B* **50**, 157–224.

Liberman, U. and Karlin, S. (1984). Theoretical models of genetic map functions. *Theoretical Population Biology* **25**(3), 331–346.

Lin, S. and Speed, T.P. (1996). Incorporating crossover interference into pedigree analysis using the χ^2 model. *Human Heredity* **46**, 315–322.

Liu, J., Wong, W.H. and Kong, A. (1994). A covariance struture of the Gibbs sampler with applications to the comparisons of estimators and augmentation schemes. *Biometrika* **81**, 27–40.

McPeek, M.S. (1999). Optimal allele-sharing statistics for genetic mapping using affected relatives. *Genetic Epidemiology* **16**, 225–249.

McPeek, M. and Sun, L. (2000). Statistical tests for detection of misspecified relationships by use of genome-screen data. *American Journal of Human Genetics* **66**, 1076–1094.

Mendel, G. (1866). Experiments in plant hybridisation. In *English Translation and Commentary by R. A. Fisher*, J.H. Bennett, ed. Oliver and Boyd, Edinburgh 1965.

Metropolis, N., Rosenbluth, A.W., Rosenbluth, M.N., Teller, A.H. and Teller, E. (1953). Equations of state calculations by fast computing machines. *Journal of Chemical Physics* **21**, 1087–1092.

Morton, N.E. (1955). Sequential tests for the detection of linkage. *American Journal of Human Genetics* **7**, 277–318.

Morton, N.E. and MacLean, C.J. (1974). Analysis of family resemblance.III. Complex segregation of quantitative traits. *American Journal of Human Genetics* **26**, 489–503.

Murray, J.C., Buetow, K.H., Weber, J.L., Ludwigson, S., Scherpier-Heddema, T., Manion, F., Quillen, J., Sheffield, V., Sunden, S., Duyk, G.M., Weissenbach, J., Gyapay, G., Dib, C., Morrissette, J., Lathrop, G.M., Vignal, A., White, R., Matsunami, N., Gerken, S., Melis, R., Albertsen, H., Plaetke, R., Odelberg, S., Ward, D., Dausset, J., Cohen, D. and Cann, H. (1994). A comprehensive human linkage map with centimorgan density. *Science* **265**, 2049–2064.

NIH-CEPH Collaborative Mapping Group (1992). A comprehensive genetic linkage map of the human genome, *Science* **258**, 67–86.

O'Connell, J.R. and Weeks, D.E. (1995). The algorithm for rapid exact multilocus linkage analysis via genotype set-recoding and fuzzy inheritance. *Nature Genetics* **11**(4), 402–408.

Ott, J. (1974). Estimation of the recombination frequency in human pedigrees: efficient computation of the likelihood for human linkage studies. *American Journal of Human Genetics* **26**, 588–597.

Ott, J. (1979). Maximum likelihood estimation by counting methods under polygenic and mixed models in human pedigrees. *American Journal of Human Genetics* **31**, 161–175.

Ott, J. (1999). *Analysis of Human Genetic Linkage*, 3rd edition. The Johns Hopkins University Press, Baltimore, MD.

Penrose, L.S. (1935). The detection of autosomal linkage in data which consist of pairs of brothers and sisters of unspecified parentage. *Annals of Eugenics* **6**, 133–138.

Ploughman, L.M. and Boehnke, M. (1989). Estimating the power of a proposed linkage study for a complex genetic trait. *American Journal of Human Genetics* **44**, 543–551.

Renwick, J.H. (1969). Progress in mapping the human autosomes. *British Medical Bulletin* **25**, 65–73.

Robbins, R.B. (1918). Some applications of mathematics to breeding problems.III. *Genetics* **3**, 375–389.

Sieberts, S.K., Wijsman, E.M. and Thompson, E.A. (2002). Relationship inference from trios of individuals in the presence of typing error. *American Journal of Human Genetics* **70**, 170–180.

Silberstein, M., Tzemach, A., Dovgolesky, N., Fishelson, M., Schuster, A. and Geiger, D. (2006). Online system for faster multipoint linkage analysis via parallel execution on thousands of personal computers. *American Journal of Human Genetics* **78**, 922–935.

Skrivanek, Z., Lin, S. and Irwin, M.E. (2003). Linkage analysis with sequential imputation. *Genetic Epidemiology* **25**, 25–35.

Smith, C.A.B. (1953). Detection of linkage in human genetics. *Journal of the Royal Statistical Society Series B* **15**, 153–192.

Smith, C.A.B. (1963). Testing for heterogeneity of recombination fractions in human genetics. *Annals of Human Genetics* **27**, 175–182.

Sobel, E. and Lange, K. (1996). Descent graphs in pedigree analysis: applications to haplotyping, location scores, and marker-sharing statistics. *American Journal of Human Genetics* **58**, 1323–1337.

Stewart, W.C.L. and Thompson, E.A. (2006). Improving estimates of genetic maps: a maximum likelihood approach. *Biometrics* **62**, 728–734.

Sturtevant, A.H. (1913). The linear association of six sex-linked factors in *Drosophila*, as shown by their mode of association. *Journal of Experimental Zoology* **14**, 43–59.

Suarez, B.K., Rice, J. and Reich, T. (1978). The generalized sib pair IBD distribution: its use in the detection of linkage. *Annals of Human Genetics* **42**, 87–94.

Sung, Y.J., Thompson, E.A. and Wijsman, E.M. (2006). MCMC-based multipoint linkage analysis for two loci plus a polygenic component and general pedigrees. *Genetic Epidemiology* **31**, 103–114.

Thomas, A., Gutin, A. and Abkevich, V. (2000). Multilocus linkage analysis by blocked Gibbs sampling. *Statistics and Computing* **10**, 259–269.

Thompson, E.A. (1974). Gene identities and multiple relationships. *Biometrics* **30**, 667–680.

Thompson, E.A. (1994). Monte Carlo likelihood in genetic mapping. *Statistical Science* **9**, 355–366.

Thompson, E.A. (2000). *Statistical Inferences from Genetic Data on Pedigrees, Vol. 6 of NSF-CBMS Regional Conference Series in Probability and Statistics*, Institute of Mathematical Statistics, Beachwood, OH.

Thompson, E.A. (2005). Chapter 4: MCMC in the analysis of genetic data on pedigrees. In *Markov Chain Monte Carlo: Innovations and Applications*, F. Liang, J.-S. Wang and W. Kendall, eds. World Scientific Co Pvt Ltd, Singapore, pp. 183–216.

Thompson, E.A. and Geyer, C.J. (2007). Fuzzy p-values in latent variable problems. *Biometrika* **94**, 49–60.

Thompson, E.A. and Guo, S.W. (1991). Evaluation of likelihood ratios for complex genetic models. *I.M.A. Journal of Mathematics Applied in Medicine and Biology* **8**(3), 149–169.

Thompson, E.A. and Heath, S.C. (1999). Estimation of conditional multilocus gene identity among relatives. In *Statistics in Molecular Biology and Genetics: Selected Proceedings of a 1997 Joint AMS-IMS-SIAM Summer Conference on Statistics in Molecular Biology, IMS Lecture Note–Monograph Series*, Vol. 33, F. Seillier-Moiseiwitsch, ed. Institute of Mathematical Statistics, Hayward, CA, pp. 95–113.

Thompson, E.A., Kravitz, K., Hill, J. and Skolnick, M.H. (1978). Linkage and the power of a pedigree structure. In *Genetic Epidemiology*, N.E. Morton, ed. Academic Press, New York, pp. 247–253.

Thompson, E.A. and Meagher, T.R. (1998). Genetic linkage in the estimation of pairwise relationship. *Theoretical and Applied Genetics* **97**, 857–864.

Wald, A. (1947). *Sequential Analysis*. John Wiley & Sons, New York.

Weeks, D.E. and Lange, K. (1988). The affected pedigree member method of linkage analysis. *American Journal of Human Genetics* **42**, 315–326.

Weeks, D.E., Lathrop, G.M. and Ott, J. (1993). Multipoint mapping under genetic interference. *Human Heredity* **43**(2), 86–97.

Weinstein, A. (1936). The theory of multiple-strand crossing over. *Genetics* **21**, 155–199.

Wijsman, E.M. and Amos, C.I. (1997). Genetic analysis of simulated oligogenic traits in nuclear and extended pedigrees: summary of GAW10 contributions. *Genetic Epidemiology* **14**, 719–735.

Zhao, H., Speed, T.P. and McPeek, M.S. (1995). Statistical analysis of crossover interference using the chi-square model. *Genetics* **139**(2), 1045–1056.

34

Non-parametric Linkage

P. Holmans

Department of Psychological Medicine, Cardiff University, Cardiff, UK

Model-free methods are commonly used to detect linkage to both dichotomous and quantitative traits. This review begins by discussing the principles underlying model-free methods for detecting linkage, and their merits relative to traditional lod-score methods (see also **Chapter 33**). Affected sib-pair methods and their extensions to include other typed unaffected relatives and extra affected siblings are discussed, together with model-free methods for the analysis of dichotomous traits on larger pedigrees. Extensions of the methods for the analysis of covariates, multiple marker loci and disease loci are reviewed, together with an investigation of methods for meta-analysis of genome scans. Finally, methods for the analysis of quantitative traits are briefly discussed.

34.1 INTRODUCTION

The basis of the initial (Penrose, 1935) and all subsequent model-free tests for linkage is that individuals concordant for a given genetic trait should show greater than expected concordance for another trait (or marker) if the two traits are linked. This is because concordant pairs are likely to have inherited the same alleles at the first trait locus, and are therefore likely to have inherited the same alleles at the second trait locus, if this is linked to the first locus. They are thus likely to be concordant for the second trait also. The expected degree of concordance if the traits are not linked can be calculated, and a statistical test performed.

The most commonly used measure of concordance of two individuals at a locus is the number of alleles they share *identical by descent* (*ibd*). This is illustrated for a pair of siblings in Figure 34.1. If the marker is not linked to a disease susceptibility locus, then, provided the Mendelian laws of inheritance hold, the probabilities of a sib pair sharing 0, 1 or 2 alleles ibd are 1/4, 1/2, 1/4 respectively. If the marker is linked to a disease locus, the probability of an affected sib pair sharing alleles ibd is increased. This is because an affected sib pair is likely to share (disease) alleles at the disease susceptibility locus, and therefore also at a linked marker locus. This is illustrated in Figure 34.2 for a recessive trait. The expected ibd sharing probabilities depend on the mode of inheritance. For a

---

Handbook of Statistical Genetics, Third Edition. Edited by D.J. Balding, M. Bishop and C. Cannings.
© 2007 John Wiley & Sons, Ltd.   ISBN: 978-0-470-05830-5.

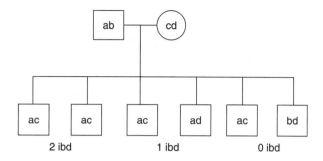

Figure 34.1   Two individuals are said to share an allele identical by descent if they both inherit a copy of that allele from a common ancestor.

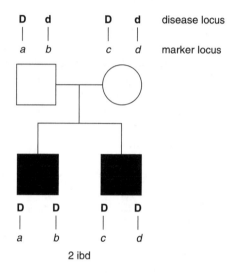

Figure 34.2   Two related affected individuals will exhibit increased ibd sharing at a marker locus linked to a disease susceptibility locus.

dominant or additive trait, the probability of 2 ibd increases above 1/4, but the probability of 1 ibd stays close to 1/2. For a recessive trait, the probability of 2 ibd increases, but the probability of 1 ibd decreases from 1/2 (Suarez, 1978). Concordant unaffected individuals should also exhibit increased ibd sharing. However, the increase is much smaller for such pairs, making them relatively uninformative (Suarez *et al.*, 1978), so in practice, such pairs are not included in linkage analyses of dichotomous traits. For a review of sib-pair methods for dichotomous traits, see also Holmans (1998). A more general overview of model-free methods is provided by Elston and Cordell (2001).

34.2   PROS AND CONS OF MODEL-FREE METHODS

A major advantage of model-free methods over the traditional lod-score approach (Ott, 1974) is that they do not require specification of the disease model. Lod-score methods are more powerful if the model is known accurately, but their power is reduced if an incorrect model is used, and the resulting estimate of the recombination fraction may be biased (Clerget-Darpoux *et al.*, 1986). A common procedure is to perform the analysis under a number of different disease models, which involves multiple testing and necessitates adjustment of the significance level of the lod score (MacLean *et al.*, 1993; Risch, 1991). An alternative is to maximise the lod score over (single-locus) models. For example, Curtis and Sham (1995) vary the penetrance of the heterozygote in their program MFLINK. Greenberg *et al.* (1998) recommend performing analyses with a dominant and a recessive model, each with 50 % penetrance, and taking the larger lod score. However, the performance of such methods when the true disease model involves several interacting loci is unclear. Some model-free methods also make implicit assumptions regarding the disease model (Whittemore, 1996). For example, the mean test has been shown to be equivalent to a lod-score analysis assuming a recessive model (Knapp *et al.*, 1994). However, they require only one test to be performed, avoiding multiple testing and making it easier to interpret the significance of the test statistics.

Another advantage is that the large multiply affected pedigrees for which traditional lod-score approaches are most informative are relatively rare, particularly for complex traits with late onset. However, large samples of relatively small families may still be available. Indeed, it has been suggested (Risch, 1994) that, for multilocus traits, sib pairs may be even more informative for linkage than multiply affected pedigrees. This is because the presence of a large number of affecteds in a pedigree increases the chance that one or more of the founders was homozygous for the disease allele at the locus being analysed, which will reduce the amount of ibd sharing. In practice, the decision regarding the correct method of analysis will be influenced by the structures of the available families. For complex multifactorial traits, where model-free methods may be advantageous, the majority of the sample will be affected sib pairs or other small pedigrees. For single-locus traits, however, large pedigrees containing multiple affecteds will be more likely, suggesting the use of parametric analyses. A major advantage of model-free methods in the analysis of complex traits is that it is relatively easy to include covariates, since their effect on disease penetrance need not be explicitly specified.

Model-free linkage methods also have some drawbacks – they do not give estimates of the recombination fraction, and are thus of limited use for mapping genes, although this defect may be overcome to some extent by the use of multipoint analyses. Neither do they explicitly allow for heterogeneity in the analysis, resulting in greatly reduced power when heterogeneity is present. Heterogeneity in linkage between *samples* (e.g. samples collected in different countries) can be overcome to some extent. For example, the ibd sharing probability can be modelled as a logistic function including a covariate with different levels for each sample (Rice, 1997; Levinson *et al.*, 2000). However, it is not possible to model heterogeneity in linkage between *individual pedigrees*, unlike model-based methods.

The conclusion seems to be that model-free analyses such as affected sib-pair methods are most advantageous in analysing complex multilocus traits where the mode of inheritance is unclear, particularly if the genetic model involves environmental or clinical covariates. For single-locus, high-penetrance traits, a lod-score analysis on large pedigrees will be more powerful. Issues raised in the analysis of complex traits are more fully reviewed by Lander and Schork (1994) and Weeks and Lathrop (1995), and the relative merits of model-free and model-based linkage analyses are discussed by Elston (1998).

It is important to note that the power of all linkage methods applied to dichotomous traits depends on the frequency of the trait in the population, decreasing as this frequency increases (Blangero *et al.*, 1998). For a common trait, it might be advantageous to determine a related quantitative measure and analyse that instead. In particular, one should not attempt to dichotomise a quantitative trait by applying a threshold, since this may greatly reduce power (Blangero *et al.*, 1998; Duggirala *et al.*, 1997).

34.3 MODEL-FREE METHODS FOR DICHOTOMOUS TRAITS

34.3.1 Affected Sib-pair Methods

The simplest approach is to count up the number of pairs sharing 0, 1 and 2 ibd and compare these to the expected frequencies under the hypothesis of no linkage using a χ^2 test (Cudworth and Woodrow, 1975). Alternatively, one could compare the average number of alleles shared ibd by the affected pairs (the 'mean test') or the number of pairs sharing 2 alleles ibd with their expected values. In each case, an excess of ibd sharing is taken as evidence for linkage. The powers of these 'counting' methods were investigated by Blackwelder and Elston (1985). Although the relative efficiencies of the methods depend on the disease model, the mean test was found to perform well in many situations.

All the above methods assume that the ibd status of the sib pairs can be determined unequivocally. This is often impossible, except for highly polymorphic markers such as the HLA system, even when the parents are typed. For an example, see Figure 34.3. Simply omitting pairs for which the ibd status is not known with certainty is likely to bias the results against detecting linkage, since it is easier to unequivocally determine pairs sharing 0 alleles ibd than pairs sharing 1 or 2 ibd (for sib pairs with no typed relatives, it is *only* possible to determine 0 sharers unequivocally). The mean test has been adapted to deal with this problem by estimating the ibd score as the average of the ibd scores under the various possible parental genotype combinations, weighted by their probabilities, and

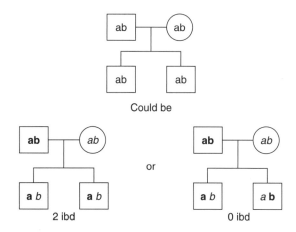

Figure 34.3 Two alleles which look the same are not necessarily shared ibd.

is implemented in packages such as SAGE (2006). Whilst the mean test is optimal under an additive genetic model, it can lose power in certain situations (e.g. under a recessive model). Whittemore and Tu (1998) considered the range of 1 degree of freedom (df) tests defined by varying the weight assigned to pairs sharing one allele ibd, and proposed the 'minimax test', which counts the 1 sharers as 0.55, rather than 1 as in the means test.

Another commonly used technique is the likelihood-ratio method proposed by Risch (1990b; 1990c). Here, the likelihood of the observed marker data is expressed as a linear function of the (unknown) ibd probabilities. The likelihood is then maximised with respect to these probabilities, and a likelihood-ratio test performed. The power of this method can be increased by restricting the maximisation to the set of ibd probabilities consistent with possible genetic models (Holmans, 1993), i.e. $\Pr(0\,\text{ibd}) \leq 1/4$, $2\Pr(0\,\text{ibd}) \leq \Pr(1\,\text{ibd})$. For a more detailed description, see Risch (1990b; 1990c) and Holmans (1993). If the dominance variance component of the genetic model is small, holding $\Pr(1\,\text{ibd}) = 0.5$ may improve the power still further (Lunetta and Rogus, 1998). The method is implemented in the packages SPLINK (Holmans and Clayton, 1995), MAPMAKER/SIBS (Kruglyak and Lander, 1995) and ASPEX (Hinds and Risch, 1996). Risch has also extended the method to other pairs of relatives (Risch, 1990b).

Another approach to the problems caused by incomplete polymorphic marker loci and/or untyped parents is to consider the number of alleles shared *identical by state* (*ibs*). Two individuals are said to share an allele ibs if they both have a copy of that allele, regardless of from whom it was inherited. It follows that the ibs status of a pair of typed individuals can always be calculated with certainty, no matter how uninformative the marker. A number of ibs methods have been suggested, of which the most notable is the affected-pedigree-member (APM) method of Weeks and Lange (1988). However, they are generally less powerful than ibd methods, especially when parental data is available (Davis and Weeks, 1997), and are thus not in common usage nowadays.

Note that the methods described above are designed for use on *autosomal* chromosomes. Variations of the methods are available for X-linked data (e.g. Cordell *et al.*, 1995a) and are implemented in most analysis software. Particular care must be taken for *pseudoautosomal* data (Dupuis and Van Eerdewegh, 2000).

34.3.2 Parameter Estimation and Power Calculation Using Affected Sib Pairs

The ibd sharing probabilities may be expressed in terms of the parameters of the genetic model in a number of different ways. The observed ibd sharing probabilities can then be used to obtain estimates of the parameters, or to infer the mode of inheritance. A major drawback of using affected sib pairs for parameter estimation is that at most two parameters may be estimated uniquely, since there are only 2 df in the data (3 ibd probabilities which are constrained to sum to 1). It is therefore necessary to make assumptions about the other parameters. For example, Thomson and Bodmer (1977) assumed that the penetrance of the normal trait genotype was zero, in order to test whether the trait exhibited recessive inheritance. This assumption was also made by Thomson (1986) to test recessive vs additive inheritance, and to estimate disease gene frequency. Day and Simons (1976) used sib pairs to estimate both disease gene frequency and relative risk, but had to specify the mode of inheritance in advance. With data from larger groups of affected siblings, it is possible to estimate more parameters. Payami *et al.* (1985) used affected trios to estimate disease gene frequency in addition to the relative risks associated with both heterozygous and homozygous disease genotypes, and to test various modes of inheritance.

The ability of sib-pair data to yield reliable estimates of genetic parameters may also be influenced by other factors, such as the method of ascertainment (Risch, 1983; Olson and Cordell, 2000), selection (Payami *et al.*, 1984) and the presence of other susceptibility loci (Risch, 1983). For these reasons, it is advisable to be cautious when using affected sib pairs for estimating genetic model parameters, although it is possible to correct for ascertainment bias (Cordell and Olson, 2000).

Although it may not be possible to estimate model parameters from the sib-pair sample itself, they may be available from previous studies. In such situations, the expected values of the ibd sharing may be calculated, and the power of the study estimated. Formulae which express the ibd probabilities in terms of quantities which may be estimated from population data are particularly useful, since power calculations may then be made without the need for prior segregation analyses. Such formulae were provided by Suarez *et al.* (1978), who expressed the ibd probabilities in terms of the prevalence and the additive and dominance variances, and Risch (1990a), who used the recurrence risks in various types of relatives of a proband. Power calculations using Risch's parameterisation are given by Weeks and Lathrop (1995). Caution should be exercised when using estimates of ibd sharing or sibling recurrence risks estimated directly from genome-wide scans since these are prone to upward bias (Goring *et al.*, 2001). Correcting fully for this bias is difficult, although partial solutions are available (Sun and Bull, 2005).

34.3.3 Typing Unaffected Relatives in Sib-pair Analyses

The power of any sib-pair method is increased by increasing the amount of information available regarding the ibd status of the affected pairs. For this reason, typing the parents increases power, especially for relatively unpolymorphic markers (Risch, 1990c; 1992; Holmans and Clayton, 1995). If parents are unavailable, a degree of information about their genotypes may be obtained by typing unaffected siblings. Again, the increase in power is greatest when the information content of the marker is low. For the likelihood-ratio method, typing both parents was found to reduce the sample size needed for a given power by at most one-third, while doubling the amount of genotyping. Typing just one parent or one unaffected sib reduced the sample size by at most one-sixth while increasing the amount of genotyping by one-half (Holmans and Clayton, 1995). It can therefore be seen that typing unaffected relatives actually *increases* the number of individuals who must be genotyped to obtain a given power. It should be noted that genotyping unaffected siblings will be necessary if one wishes to analyse discordant sibling pairs (e.g. Xing *et al.*, 2006).

Typing unaffected relatives has some other major advantages. If there are a limited number of affected pairs, it may be important to get the maximum amount of information out of each pair. Most importantly, if both parents are typed, the probability of obtaining false-positive linkage evidence will not be increased by population stratification, Hardy–Weinberg disequilibrium at the marker locus, consanguinity in the population or incorrect specification of marker allele frequencies. Typing unaffected relatives increases the chance of detecting genotyping errors in the family (Douglas *et al.*, 2002), thereby increasing power (since genotyping errors tend to reduce power). Inappropriate correction for genotyping errors may also inflate Type I error rate (Seaman and Holmans, 2005). Typing unaffected siblings enables the assumption of Mendelian inheritance to be tested, by investigating the distribution of alleles shared ibd in discordant pairs (Khoury *et al.*, 1991). This depends on the disease penetrance; for low penetrance one would expect ibd sharing close to the expected values, while for high penetrance one might observe an excess of pairs sharing 0 ibd. In any case, the proportion of alleles shared ibd by discordant pairs should be less than for affected pairs. If discordant pairs are found to show an excess of ibd sharing, then the Mendelian assumption may be false (e.g due to transmission-ratio distortion), and the usual sib-pair tests invalid. In this situation, one possibility is to compare haplotype sharing in concordant and discordant pairs (Flanders and Khoury, 1991; Wang and Elston, 2005). Typing unaffected siblings may also increase power substantially if there is assortative mating (Sribney and Swift, 1992).

For these reasons, it is desirable to genotype unaffected relatives, particularly both parents. However, affected pairs without relatives available for typing should not be discarded (Holmans and Clayton, 1995). If some of the parents in the sample are missing, the analysis should be repeated on the subset of pairs with both parents typed, to check that any positive result is not a result of incorrect assumptions about the marker locus.

If parental data is missing, all ibd methods require the specification of marker allele frequencies. It is possible to use frequencies obtained from a previous study, or a database such as CEPH; however, if these are incorrect for the population being studied, the false-positive rate may be increased (Ott, 1992; Freimer *et al.*, 1993). A safer procedure is to estimate the marker allele frequencies from the sample itself. This eliminates the risk of using frequencies estimated from a different population, and does not lose much power compared to the situation where the true frequencies are known (Holmans, 1993).

34.3.4   Application of Sib-pair Methods to Multiplex Sibships

The methods described above were originally designed for use on affected sibling *pairs*. In practice, it is likely that the sample collected for an affected sib-pair study will contain a number of sibships containing three or more affected siblings. There are a number of ways to analyse such sibships. For some methods, sibships of any size can be analysed without modification. An example is the binomial maximum-likelihood approach of Abel *et al.* (1998), which models the number of affected siblings inheriting a given parental allele as a binomial distribution, enabling a likelihood-ratio test to be performed.

However, many commonly used methods, such as the likelihood-ratio methods of Risch (1990b; 1990c), restrict the analysis to sib pairs only. Data from multiply affected sibships must therefore be broken up. The most common ways to do this, as implemented in GENEHUNTER (Kruglyak *et al.*, 1996) are: (1) to choose just one affected pair from each sibship (2) to choose one affected sibling and include all pairs containing this sibling in the analysis (3) to include all possible sib pairs from the sibship in the analysis. For example, an affected trio of sibs A, B and C would contribute the affected pair $A - B$ under (1), the two pairs $A - B$ and $A - C$ under (2), and the three pairs $A - B$, $A - C$ and $B - C$ under (3). The main problem with choosing just one sib pair from each sibship is that a large proportion of the data is discarded, reducing power. Furthermore, the result may depend on which pair is chosen, giving rise to the possibility, in extreme cases, of two researchers applying the same analysis to the same dataset, but reaching different conclusions. The same drawbacks apply to the method of analysing all pairs containing a given individual.

It is therefore usual to include all possible affected pairs from each sibship in the analysis. However, pairs obtained in this way from the same sibship are no longer independent. A number of weighting schemes for such pairs have been suggested (e.g. Hodge, 1984) to deal with this problem. The most commonly used weighting scheme is that of Suarez and Hodge (1979), in which pairs formed from a sibship containing n affected individuals are weighted by a factor of $2/n$. This scheme has the intuitively appealing result that the total weight assigned to pairs formed from a n-affected sibship is $n - 1$ ($= n(n - 1)/2$ pairs $\times 2/n$), which is the number of *independent* pairs which could be formed from the sibship.

In fact, the lack of independence does not inflate the false-positive rate for 'counting' methods such as the mean test. This is because the number of alleles shared ibd by two pairs of affected sibs are independent, even if the sib pairs share a common member (Blackwelder and Elston, 1985). The variance of the test statistics is therefore unaltered by the dependence among pairs from the same sibship. For the likelihood-ratio method, however, there is evidence that analysing pairs from the same sibship as if they were independent may alter the false-positive rate. If the genotypes of the affected sibs in the sibship which are not part of the pair being analysed are used to infer missing parental genotypes (as is the case for programs such as MAPMAKER/SIBS), the test is conservative even without weighting (Holmans, 2001). Otherwise, the absence of parental genotypes increased the false-positive rate if no weighting was applied. However, the $2/n$ weighting scheme of Suarez and Hodge is too conservative. These effects become more pronounced as marker informativeness decreases. If parents are genotyped, there is a slight inflation of false-positive rate without weighting, most pronounced for small nominal p values (≤ 0.001). The $2/n$ weighting scheme of Suarez and Hodge was found to be conservative in all situations.

The behaviour of likelihood-ratio methods when applied to multiplex sibships depends on both family structure and marker informativeness, so no single weighting scheme can be correct in all circumstances. Thus, it is advisable to estimate the significance of observed results by simulation, rather than relying on pre-defined test criteria based on asymptotic theory, when the sample contains a high proportion of multiply affected sibships.

Even when the method of analysis does not require the use of weighting to ensure the correct false-positive rate, it may still be possible to increase power by assigning different weights to pairs formed from sibships containing different numbers of affected siblings (Suarez and Van Eerdewegh, 1984; Sham *et al.*, 1997).

34.3.5 Methods for Analysing Larger Pedigrees

The APM method was proposed by Weeks and Lange (1988) for the model-free linkage analysis of general pedigrees. It compares the sum of the observed number of alleles shared ibs by each pair of affected relatives, weighted according to the frequency of the shared allele(s), to its expected value in the absence of linkage. The method was extended to include unaffected relatives by Ward (1993).

One drawback of the method was that the results were dependent on the (arbitrary) choice of the weighting function. Another was that, since the APM method used only ibs information, it was less powerful than methods which inferred ibd, especially when other relatives were typed (Goldin and Weeks, 1993). Therefore, Davis *et al.* (1996) developed SimIBD, which counts the ibd between affected pairs and computes an empirical p value using conditional simulation. This method was found to perform poorly when analysing sibships without typed parents (Davis and Weeks, 1997), although it may be useful for larger pedigrees.

An alternative approach, proposed by Kruglyak *et al.* (1996), performs a non-parametric test (NPL) based on the observed inheritance patterns of the affected individuals, and is implemented in the program GENEHUNTER. The significance levels of the NPL statistic calculated by GENEHUNTER assume complete ibd information, and the test is conservative when this is not the case (Kruglyak *et al.*, 1996), resulting in a loss of power. This may make the NPL statistics less efficient than other methods for analysing sib-pair data (Davis and Weeks, 1997), although it appears to be a powerful method of non-parametric analysis on large pedigrees. The problem of the conservativeness of the NPL statistics has been addressed by Kong and Cox (1997) in their program GENEHUNTER-PLUS, and is also implemented in ALLEGRO (Gudbjartsson *et al.*, 2005). Two scoring functions are commonly available for NPL-type statistics: S_{pairs}, which is based on numbers of alleles shared ibd by *pairs* of affected relatives, and S_{all}, which is based on ibd patterns among all affecteds in a pedigree. S_{pairs} is equivalent to the mean test when applied to sibship data. These and other scoring functions were reviewed by McPeek (1999). The optimal scoring function depends on the (unknown) disease model, but S_{pairs} was found to perform well over all models tested.

The weighted pairwise correlation (WPC) method (Commenges, 1994) is also designed to perform non-parametric analysis on arbitrarily large pedigrees. Its basic premise is that, under linkage, the correlation of the residuals (i.e. the observed trait value minus its expected value) of a pair of related individuals will increase with the number of alleles shared ibd. The WPC method can be applied to either quantitative or dichotomous traits provided the form of the residuals is suitably chosen.

34.3.6   Extensions to Multiple Marker Loci

The power of all methods for detecting linkage increases with the informativeness of the marker, since this enables the ibd status of the affected individuals to be determined more accurately. Extra information regarding ibd sharing may be obtained by extending the methods to analyse two or more linked marker loci simultaneously, thereby increasing power in both dichotomous (Holmans and Clayton, 1995; Olson, 1995) and quantitative (Fulker and Cardon, 1994) traits. A further advantage of analysing multiple linked markers is that a greater area of the chromosome can be studied, and an estimate of the location of the disease locus provided, although this may not always be very precise, even for large samples, when the genetic effect is small for complex traits (Roberts *et al.*, 1999, Cordell, 2001). The likelihood-ratio method of Risch and Holmans has been extended to multiple linked markers in the program GENEHUNTER (Kruglyak *et al.*, 1996), which also allows multipoint NPL and parametric analyses of larger pedigrees. The maximum size of pedigree that GENEHUNTER can analyse is determined by the quantity $2n - f$, where n is the number of individuals in the pedigree and f the number of founders, and is limited by the memory of the computer. If the pedigree is too large, either individuals must be removed or the pedigree must be split, and in some cases the way this is done may lose power (Romero-Hidalgo *et al.*, 2005). The ALLEGRO package (Gudbjartsson *et al.*, 2005) performs GENEHUNTER and GENEHUNTER-PLUS analyses, but is faster and capable of handling larger pedigrees, as is MERLIN (Abecasis *et al.*, 2002). However, very large pedigrees must still be split, or model-based analyses performed using programs such as FASTLINK (Cottingham *et al.*, 1993), although these are restricted in the number of marker loci they can analyse. Alternatively, Monte-Carlo techniques, such as those implemented in SIMWALK2 (Sobel and Lange, 1996; Sobel *et al.*, 2002) may be used.

34.3.7   Multipoint Analysis with Tightly Linked Markers

It should be noted that all programs for multipoint analysis assume the marker loci to be in linkage equilibrium when calculating haplotype frequencies. The presence of linkage disequilibrium might therefore lead to spurious results in the absence of typed parents. This is only likely to be a problem if the marker loci are close together (< 1 cM), which was rarely the case for grids of microsatellite markers. Recent advances in genotyping techniques have made it feasible to perform genome-wide linkage studies using dense maps of SNPs. These give more linkage information than the microsatellite grids, and thus can increase power (Schaid *et al.*, 2004). However, linkage disequilibrium (LD) between the SNPs can also result in false-positive linkages (Schaid *et al.*, 2004), and the extent to which Type I error rate is inflated by inter-marker LD has been investigated by several authors (Huang *et al.*, 2004; 2005; Boyles *et al.*, 2005; Levinson and Holmans, 2005).

One approach to dealing with inter-marker LD is to measure the LD between all pairs of SNPs, and then remove SNPs until the LD between all remaining pairs of SNPs is below a pre-set criterion. The choice of criterion differs from study to study: Schaid *et al.* used $|D'| < 0.7$, whereas Levinson and Holmans used $r^2 < 0.06$. The optimal criterion for controlling Type I error rate while losing as little linkage information as possible is still unclear (and is likely to depend on the structure of the pedigrees – in particular, the number of typed parents).

An alternative approach has been implemented in recent versions of MERLIN (Abecasis and Wigginton, 2005). Here, consecutive SNPs are grouped into clusters such that all pairs of SNPs within a cluster satisfy a user-selected criterion based either on LD (e.g. $r^2 < 0.1$) or genetic distance (e.g. less than 0.1 cM). Haplotype frequencies are calculated within clusters without assuming linkage equilibrium, while different clusters are assumed to be in linkage equilibrium. Recombination within clusters is assumed to be zero. If a recombination is observed, data from the entire cluster in that pedigree is dropped from the analysis, which could reduce power. The relative performances of these approached under varying LD structures have not as yet been assessed.

If unaffected siblings are typed, an analysis comparing ibd sharing in affected sib pairs to that in discordant sib pairs can be performed, and this is robust to inter-marker LD (Xing *et al.*, 2006). However, such an approach is less powerful than a study of affected sib pairs alone (Blackwelder and Elston, 1985).

34.3.8   Inclusion of Covariates

Since there is often epidemiological evidences showing that measurable environmental or clinical factors influence disease risk, it is desirable to allow for the effects of these factors as covariates in a model-free linkage analysis. One approach is to use logistic regression. Greenwood and Bull (1997; 1999) propose modelling the probability that a sib pair shares 0, 1 or 2 ibd as a multinomial logistic regression, into which parameters modelling covariate effects can be incorporated. A similar approach has been taken by Olson (1999). How best to apply constraints to the ibd probabilities when maximising the likelihood, as the constraints may not necessarily hold for all covariate values is one issue with this approach. Olson *et al.* (2001; 2002) recommend a linear constraint based on the results of Whittemore and Tu (1998). Alternatively, one can model the probability that an affected sib pair share a given parental allele ibd as a logistic regression, and to assume that the ibd probabilities of maternal and paternal alleles are independent (Rice, 1997; Rice et al., 1999). This avoids the constraint problems, and also requires fewer degrees of freedom, but may lose power if the true genetic model violates the independence assumption.

Another approach is to consider the sample as a mixture of linked and unlinked sib pairs, in a similar way to heterogeneity analysis in parametric linkage (Devlin et al., 2002). The probability that a sib pair is linked can be modelled in terms of the covariates (e.g. by logistic regression). Whilst this approach is efficient for modelling linkage heterogeneity, it does not allow pairs to share fewer alleles ibd than expected under the null (as might happen under certain gene – environment interaction models where the pairs are discordant for the environmental factor).

All of the methods described above operate on one sib pair at a time, which may cause problems when applying them to multiplex sibships, as noted earlier. The binomial likelihood method of Abel et al. (1998) has been extended (Alcais and Abel, 2001) to incorporate covariates by performing a logistic regression of disease status on the covariates, then analysing the residuals as a quantitative trait (Alcais and Abel, 1999). This method deals easily with multiplex sibships, but would not be able to analyse covariates defined only on *pairs* of sibs (e.g. ibd sharing at a second locus).

A final approach, applicable to general pedigrees, is ordered subsets analysis (Hauser et al., 2004). Here, pedigrees are ranked in order of some covariate (e.g. earliest age at onset). Pedigrees are added to the sample in rank order, and the linkage analysis repeated after each. The largest of the resulting linkage statistics is taken as the overall test statistics, with significance estimated by randomly permuting the order in which pedigrees are added. Again, this method would not be able to analyse covariates defined on pairs of sibs. Also, it is limited to the analysis of one covariate at a time, unlike the other approaches. However, it makes no assumptions (other than monotonicity) on the relationship of linkage evidence with covariate, so may be advantageous if this is non-(log)linear.

Power comparisons of these methods were carried out by Tsai and Weeks (2006) under a variety of models incorporating gene – environment interactions. Pairwise/familywise covariates were defined as the average of the covariates of the individuals involved. This raises the question of whether other definitions of pairwise covariate (e.g. differences) might be superior. This has not yet been studied.

In theory, one can analyse several covariates simultaneously in the logistic-regression and mixture model approaches. In practice, power may be reduced due to the large number of degrees of freedom corresponding to the regression parameters. An interesting way of dealing with this problem is the propensity score method (Doan et al., 2006). Here, logistic regression of disease status is performed on all covariates simultaneously. Residuals for the affecteds are used as covariates in a logistic-regression linkage analysis. A drawback of this approach is that sufficient numbers of unaffected siblings must be genotyped and measured for the covariates.

Issues regarding the inclusion of covariates in linkage analysis are further discussed by Schaid et al. (2003).

34.3.9 Multiple Disease Loci

For many complex traits, there is evidence that more than one locus is involved in the aetiology, e.g.: schizophrenia (Risch, 1990a), bipolar disorder (Craddock *et al.*, 1995) and type 1 diabetes (Davies *et al.*, 1994). Schork *et al.* (1993) extended the lod-score approach to analyse two trait loci simultaneously, including two marker loci, one linked to each trait locus, in the analysis. In certain situations, they found a substantial increase in power over standard analysis involving just one trait locus. However, if just one marker locus is used, including a second trait locus in the analysis does not greatly increase power (Goldin, 1992; Vieland *et al.*, 1992). Knapp *et al.* (1994) found that sib-pair analysis applied to two marker loci, each linked to a different trait locus, could be much more powerful than analysing the loci separately. The maximum-likelihood method used by Knapp *et al.* was refined by Cordell *et al.* (1995b) to include the restrictions on ibd sharing proposed by Holmans (1993), and used not only to test for linkage with type 1 diabetes, but also to distinguish between models of interaction between the disease loci. This approach was extended to multiple *linked* trait loci by Farrall (1997). Cordell *et al.* (2000) extended the method further to allow likelihood models based on variance components to be fitted to ibd data from pairs of affected relatives from extended pedigrees at an arbitrary number of loci, linked or unlinked. This enables the evidence for a disease locus to be tested while conditioning on the presence of other loci, and for hypotheses regarding epistatic interactions to be tested, but it is unclear how other covariates could be included. The logistic-regression method of Olson (1999) is capable of performing two-locus analyses (Olson *et al.*, 2002). Alternatively, the proportion of alleles shared ibd by each pair at a second locus may be included as a covariate in the Rice model (Holmans, 2002).

Another approach to testing for epistasis was proposed by Cox *et al.* (1999). The principle underlying this approach is that linkage statistics (e.g. NPL score, ibd proportion) for each sampling unit (pedigree, affected sib pair) at two loci A and B should be positively correlated if A and B interact epistatically, but negatively correlated if the loci conform to a heterogeneity model. Then, one can select the families whose test statistics at locus A is greater (if testing for epistasis) or less (if testing for heterogeneity) than a pre-defined criterion, and perform analysis at locus B using these families only. Significance is estimated by randomly sampling replicate sets containing the same number of families used in the actual analysis without reference to their linkage score at locus A. This method was used by Cox *et al.* to detect interactions between susceptibility loci for diabetes. This is an intuitively appealing procedure, but it is often by no means clear *a priori* what criterion should be used in the analysis. Whilst the issue of multiple testing caused by the use of several criteria giving non-independent tests was addressed by Buhler *et al.* (1997), allowing for this may reduce power. Furthermore, it is not clear how to apply this method when loci A and B are linked, since one would then expect the linkage scores to be positively correlated whatever the underlying disease model.

A problem with using a two-trait-locus approach is that the number of possible pairs of marker loci rapidly becomes large, leading to multiple testing problems. For this reason, such approaches probably ought not to be used in an initial genome scan, but rather to test sets of candidate loci, or to follow-up loci to which (possibly weak) evidence of linkage has been found using single trait-locus analysis (Schork *et al.*, 1993). Furthermore, when only linkage data (rather than a candidate gene) is available, it is unlikely that a two-trait-locus analysis could give a significant lod without some evidence of linkage being visible from single-locus analyses (Holmans, 2002).

34.3.10   Significance Levels for Genome Scans

Advances in genotyping technology mean that genome scans for linkage are now very common. However, they still involve a large number of linkage tests, and it is thus necessary to apply a correction for multiple testing to reduce the number of false positives. Simple corrections such as Bonferroni are inappropriate due to the non-independence of these tests. Lander and Kruglyak (1995) devised theoretical criteria for linkage statistics to be deemed genome-wide 'significant' (i.e. obtained in fewer than 5 % of genome scans in the absence of disease loci) or 'suggestive' (obtained less than once per genome scan). These criteria depend on the analysis method. Whilst these criteria are useful rules of thumb, they are based on the assumption of complete linkage information, and are conservative in more realistic situations (Sawcer et al., 1997). In addition, as noted earlier, the distribution of linkage statistics may also depend on the number of affected sibs, the number of typed parents, and the inclusion of covariates. Thus, significance levels for observed results should be obtained by simulation.

34.3.11   Meta-analysis of Genome Scans

Genetic effects underlying complex traits are likely to be small. Thus, large-sample sizes will be required to achieve power, often more than can be collected by any single group. Ideally, several groups would pool their raw genotype and phenotype data and perform a single analysis on the combined sample (e.g. Levinson et al., 2000). Often, however, raw data is unavailable, but summary measures of individual studies (e.g. linkage statistics) can be obtained. In such situations, studies can be combined via a meta-analysis.

One method for performing meta-analysis (Badner and Goldin, 1999; Badner and Gershon, 2002) is based on Fisher's method for combining p values (i.e. $-2 \sum_{i=1}^{n} \ln p_i \sim \chi^2_{2n}$ if none of the studies is a true positive). Zaykin et al. (2002) extended the Fisher method to use only p values below a certain threshold the truncated product method (TPM), which may be useful if studies only report significant regions. Some linkage analysis methods (e.g. likelihood-ratio sib-pair analysis) truncate the test statistics at zero, which can bias the Fisher method. This bias was overcome by Province (2001).

Another commonly used meta-analysis method is the genome search meta-analysis (GSMA) proposed by Wise et al. (1999). Here, the genome is split into a number of 'bins' of approximately equal width. Within each study, bins are ranked according to the most significant test statistics occurring in that bin. Ranks for each bin are summed across studies and compared to the expected distribution in the absence of susceptibility genes (i.e. ranks assigned randomly). This gives a bin-wise p value p_{bin}. Correction for testing multiple bins may be performed by randomly permuting ranks within each study. This yields a p value p_{ord} depending on the position of the bin in the ranked p_{bin} values (e.g. most significant, second-most significant etc.) Combining p_{bin} and p_{ord} was shown to increase power to detect true linkages, as was weighting by sample size (Levinson et al., 2003). The GSMA has been applied to numerous complex traits, e.g. schizophrenia (Lewis et al., 2003), and software is now available (Pardi et al., 2005).

The final class of methods involves combining Z scores (i.e. linkage statistics with a standard normal distribution, such as NPL scores) across studies (Loesgen et al., 2001). Statistics which do not have a standard normal distribution (such as lod scores) can be transformed into Z scores (Etzel et al., 2005). Weighting schemes can be applied, such as by study size or linkage information content (Dempfle and Loesgen, 2004).

All of these methods are generally applicable, since they do not require that the studies use the same linkage analysis method, or the same set of markers (although clearly the different maps must be combined such that their locations agree). The performance of the Z-score, Fisher, TPM and GSMA methods was compared on simulated data by Dempfle and Loesgen (2004). The Z-score method was found to be most powerful when all studies were simulated under the same linkage model. However, if there is linkage heterogeneity, the TPM method may be preferable (Zaykin *et al.*, 2002), while the power of the GSMA is reduced (Levinson *et al.*, 2003). Methodolgy has been developed to test for linkage heterogeneity in the GSMA (Zintzaras and Ioannidis, 2005), although its power is questionable (Lewis and Levinson, 2006).

34.4 MODEL-FREE METHODS FOR ANALYSING QUANTITATIVE TRAITS

This topic is also covered in **Chapter 19**, and thus only a brief overview of the most commonly used methods for linkage analysis of quantitative traits in humans will be given.

A commonly used method for analysing quantitative traits is the Haseman–Elston method (Haseman and Elston, 1972). In this method, the squared trait differences between pairs of siblings is regressed on the proportion of alleles that the pair is estimated to share their ibd with. This method retains large-sample validity if all distinct sib pairs from sibships with three or more members are included in the analysis as independent pairs (Amos *et al.*, 1989; Collins and Morton, 1995), does not require the trait to be normally distributed, and allows the testing of multiple trait loci, including interactions, by multiple linear regression. It was pointed out (Wright, 1997) that using the sib-pair difference ignores information available in the sib-pair trait sum. The method was therefore 'revisited' to use the mean-corrected cross-product of trait values instead (Elston *et al.*, 2000), thereby increasing power in some situations. This method has been implemented in the program SIBPAL2, part of the SAGE package. Power may be increased further by regressing the sum of the squared trait sum and squared trait difference, each weighted inversely proportional to their variance, of each pair on ibd. This is equivalent to variance-components analysis on sib-pair data (Sham and Purcell, 2001).

The other major approach to analysing quantitative traits is to use variance-components analysis (Goldgar, 1990; Amos, 1994; Amos *et al.*, 1996; Blangero and Almasy, 1996). Here, the covariance matrix for the trait is partitioned into components attributable to additive genetic variance at each of the trait loci, residual genetic effects and random environmental effects. The size of each effect is estimated, and its significance tested by means of a likelihood-ratio test. The main advantage of this approach is its flexibility: it is applicable to general pedigrees, it enables the effects of covariates, epistasis between multiple trait loci and gene – environment interactions to be modelled (Blangero and Almasy, 1997). Multiple traits may be analysed simultaneously, both quantitative and qualitative (Almasy *et al.*, 1997; Williams *et al.*, 1999). The method appears to be more powerful than the original (Williams and Blangero, 1997) and 'revisited' (Xu *et al.*, 2000). Haseman–Elston methods, but requires that the traits being analysed have a normal distribution, with non-normality inflating Type I error rates (Allison *et al.*, 1999).

Applying transformations (e.g. Box – Cox) to the data in an attempt to approximate normality appears not to be efficient at correcting the Type I error rate, so estimating significance by simulation is recommended (Deutsch *et al.*, 2005). Variance-components analysis is implemented in the SOLAR package (Almasy and Blangero, 1998), and in MERLIN.

An interesting alternative approach (Sham *et al.*, 2002) is to regress the estimated ibd sharing of affected relative pairs on the squared sums and differences of their trait values (i.e. reversing the direction of regression used by Haseman–Elston and variance-components methods). This has the potential to combine the power of the variance-components approach with robustness to non-normality (e.g. it can be applied to samples selected on their trait values). However, the mean and variance of the trait in the population must be specified. The method is implemented in MERLIN-REGRESS.

34.5 CONCLUSIONS

When the mode of inheritance is known, such as for high-penetrance Mendelian traits, model-based maximum-likelihood methods applied to multigenerational pedigrees are the preferred form of analysis, due to their extra power. However, model-free methods do have some advantages over traditional model-based parametric analyses for the detection of linkage to complex traits. Firstly, they do not require specification of a disease model, thereby evading the problem of multiple testing caused by analysing a number of different (incorrect) models. Secondly, large, multigenerational pedigrees containing multiple affected members are usually rare for complex traits, especially those with late onset. Small pedigrees, such as nuclear families or affected sib pairs are relatively common, enabling large samples to be collected. Model-free methods may be advantageous in the analyses of such pedigrees. Whilst model-free methods do not allow direct estimation of the recombination fraction, estimates of the location of the disease locus are available by using multipoint analysis.

Given the small locus-specific genetic effects observed in complex traits, together with likely genetic heterogeneity, it is important that future linkage analyses incorporate clinical subphenotypes and/or environmental risk factors as covariates. Model-free methods are very suitable for this, as they do not require the relationship between the covariate and the disease model to be explicitly specified. Numerous methods have been proposed, but optimal procedures have yet to be determined.

Small genetic effect sizes also require large-sample sizes to achieve power. This, in turn, increases the importance of efficient meta-analysis methods, and there is scope for future work on these.

To conclude: carefully designed linkage studies making appropriate use of clinical and/or epidemiological data, and information from other studies, still have an important role to play in genetic epidemiology. Their results can be useful in informing the design of follow-up association studies, and samples of multiply affected pedigrees can be a powerful resource for such studies.

Related Chapters

Chapter 19; **Chapter 33**.

REFERENCES

Abecasis, G.R., Cherny, S.S., Cookson, W.O. and Cardon, L.R. (2002). Merlin-rapid analysis of dense genetic maps using sparse gene flow trees. *Nature Genetics* **30**, 97–101.

Abecasis, G.R. and Wigginton, J.E. (2005). Handling marker-marker linkage disequilibrium: pedigree analysis with clustered markers. *American Journal of Human Genetics* **77**, 754–767.

Abel, L., Alcais, A. and Mallet, A. (1998). Comparison of four sib-pair linkage methods for analysing sibships with more than two affecteds: interest of the binomial maximum likelihood approach. *Genetic Epidemiology* **15**, 371–390.

Alcais, A. and Abel, L. (1999). Maximum-Likelihood-Binomial method for genetic model-free linkage analysis of quantitative traits in sibships. *Genetic Epidemiology* **17**, 102–117.

Alcais, A. and Abel, L. (2001). Incorporation of covariates in multipoint model-free linkage analysis of binary traits: how important are unaffecteds? *European Journal of Human Genetics* **9**, 613–620.

Allison, D.B., Neale, M.C., Zannolli, R., Schork, N.J., Amos, C.I. and Blangero, J. (1999). Testing the robustness of the likelihood-ratio test in a variance-component quantitative-trait loci-mapping procedure. *American Journal of Human Genetics* **65**, 531–544.

Almasy, L. and Blangero, J. (1998). Multipoint quantitative trait linkage analysis in general pedigrees. *American Journal of Human Genetics* **62**, 1198–1211.

Almasy, L., Dyer, T.D. and Blangero, J. (1997). Bivariate quantitative trait linkage analysis: pleiotropy versus co-incident linkages. *Genetic Epidemiology* **14**, 953–958.

Amos, C.I. (1994). Robust variance-components approach for assessing genetic linkage in pedigrees. *American Journal of Human Genetics* **54**, 535–543.

Amos, C.I., Elston, R.C., Wilson, A.F. and Bailey-Wilson, J.E. (1989). A more powerful robust sib-pair test of linkage for quantitative traits. *Genetic Epidemiology* **6**, 435–449.

Amos, C.I., Zhu, D.K. and Boerwinkle, E. (1996). Assessing genetic linkage and association with robust components of variance approaches. *Annals of Human Genetics* **60**, 143–160.

Badner, J.A. and Gershon, E.S. (2002). Meta-analysis of whole-genome linkage scans of bipolar disorder and schizophrenia. *Molecular Psychiatry* **7**, 405–411.

Badner, J.A. and Goldin, L.R. (1999). Meta-analysis of linkage studies. *Genetic Epidemiology* **17**(Suppl. 1), S485–S490.

Blackwelder, W.C. and Elston, R.C. (1985). A comparison of sib-pair linkage tests for disease susceptibility loci. *Genetic Epidemiology* **2**, 85–97.

Blangero, J. and Almasy, L. (1997). Multipoint oligogenic linkage analysis of quantitative traits. *Genetic Epidemiology* **14**, 959–964.

Blangero, J., Almasy, L. and Williams, J.T. (1998). Tragedy of the commons: common misconceptions about mapping genes for common diseases. *American Journal of Human Genetics* **63**, A45.

Boyles, A.L., Scott, W.K., Martin, E.R., Schmidt, S., Li, Y.J., Ashley-Koch, A., Bass, M.P., Schmidt, M., Pericak-Vance, M.A., Speer, M.C. and Hauser, E.R. (2005). Linkage disequilibrium inflates type I error rates in multipoint linkage analysis when parental genotypes are missing. *Human Heredity* **59**, 220–227.

Buhler, J., Owerbach, D., Schaffer, A.A., Kimmel, M. and Gabbay, K.H. (1997). Linkage analyses in type I diabetes using CASPAR, a software and statistical program for conditional analysis of polygenic diseases. *Human Heredity* **47**, 211–222.

Clerget-Darpoux, F., Bonaiti-Pellie, C. and Hochez, J. (1986). Effects of misspecifying genetic parameters in lod score analysis. *Biometrics* **42**, 393–399.

Collins, A. and Morton, N.E. (1995). Nonparametric tests for linkage with dependent sib pairs. *Human Heredity* **47**, 333–353.

Commenges, D. (1994). Robust genetic linkage analysis based on a score test of homogeneity: the weighted pairwise correlation statistic. *Genetic Epidemiology* **11**, 189–200.

Cordell, H.J. (2001). Sample size requirements to control for stochastic variation in magnitude and location of allele-sharing linkage statistics in affected sibling pairs. *Annals of Human Genetics* **65**, 491–502.

Cordell, H.J., Kawaguchi, Y., Todd, J.A. and Farrall, M. (1995a). An extension of the maximum lod score method to X-linked loci. *Annals of Human Genetics* **59**, 435–449.

Cordell, H.J. and Olson, J.M. (2000). Correcting for ascertainment bias of relative-risk estimates obtained using affected-sib-pair linkage data. *Genetic Epidemiology* **18**, 307–321.

Cordell, H.J., Todd, J.A., Bennett, S.T., Kawaguchi, Y. and Farrall, M. (1995b). Two-locus maximum lod score analysis of a multifactorial trait: joint consideration of IDDM2 and IDDM4 with IDDM1 in type 1 diabetes. *American Journal of Human Genetics* **57**, 920–934.

Cordell, H.J., Wedig, G.C., Jacobs, K.B. and Elston, R.C. (2000). Multilocus linkage tests based on affected relative pairs. *American Journal of Human Genetics* **66**, 1273–1286.

Cottingham, R.W. Jr., Idury, R.M. and Schaffer, A.A. (1993). Faster sequential genetic linkage computations. *American Journal of Human Genetics* **53**, 252–263.

Cox, N.J., Frigge, M., Nicolae, D.L., Concannon, P., Hanis, C.L., Bell, G.I. and Kong, A. (1999). Loci on chromosomes 2 (NIDDM1) and 15 interact to increase susceptibility to diabetes in Mexican Americans. *Nature Genetics* **21**, 213–215.

Craddock, N., Khodel, V., Van Eerdewegh, P. and Reich, T. (1995). Mathematical limits of multilocus models: the genetic transmission of bipolar disorder. *American Journal of Human Genetics* **57**, 690–702.

Cudworth, A.G. and Woodrow, J.C. (1975). Evidence for HLA-linked genes in "juvenile" diabetes mellitus. *British Medical Journal* **3**, 133–135.

Curtis, D. and Sham, P.C. (1995). Model-free linkage analysis using likelihoods. *American Journal of Human Genetics* **57**, 703–716.

Davies, J.L., Kawaguchi, Y., Bennett, S.T., Copeman, J.B., Cordell, H.J., Pritchard, L.E., Reed, P.W., Gough, S.C.L., Jenkins S.C., Palmer, S.M., Balfour, K.M., Rowe, B.R., Farrall, M., Barnett, A.H., Bain, S.C., Todd, J.A. (1994). A genome-wide search for human type 1 diabetes susceptibility genes. *Nature* **371**, 130–136.

Davis, S., Schroeder, M., Goldin, L.R. and Weeks, D.E. (1996). Nonparametric simulation-based statistics for detecting linkage in general pedigrees. *American Journal of Human Genetics* **58**, 867–880.

Davis, S. and Weeks, D.E. (1997). Comparison of nonparametric statistics for detection of linkage in nuclear families: single-marker evaluation. *American Journal of Human Genetics* **61**, 1431–1444.

Day, N.E. and Simons, M.J. (1976). Disease susceptibility genes-their identification by multiple case family studies. *Tissue Antigens* **8**, 109–117.

Dempfle, A. and Loesgen, S. (2004). Meta-analysis of linkage studies for complex diseases: an overview of methods and a simulation study. *Annals of Human Genetics* **68**, 69–83.

Deutsch, S., Lyle, R., Dermitzakis, E.T., Attar, H., Subrahmanyan, L., Gehrig, C., Parand, L., Gagnebin, M., Rougemont, J., Jongeneel, C.V. and Antonarakis, S.E. (2005). Gene expression variation and expression quantitative trait mapping of human chromosome 21 genes. *Human Molecular Genetics* **14**, 3741–3749.

Devlin, B., Jones, B.L., Bacanu, S.A. and Roeder, K. (2002). Mixture models for linkage analysis of affected sibling pairs and covariates. *Genetic Epidemiology* **22**, 52–65.

Doan, B.Q., Sorant, A.J., Frangakis, C.E., Bailey-Wilson, J.E. and Shugart, Y.Y. (2006). Covariate-based linkage analysis: application of a propensity score as the single covariate consistently improves power to detect linkage. *European Journal of Human Genetics* **14**, 1018–1026.

Douglas, J.A., Skol, A.D. and Boehnke, M. (2002). Probability of detection of genotyping errors and mutations as inheritance inconsistencies in nuclear-family data. *American Journal of Human Genetics* **70**, 487–495.

Dupuis, J. and Van Eerdewegh, P. (2000). Multipoint linkage analysis of the pseudoautosomal regions, using affected sibling pairs. *American Journal of Human Genetics* **67**, 462–475.

Duggirala, R., Williams, J.T., Williams-Blangero, S. and Blangero, J. (1997). A variance component approach to dichotomous trait linkage analysis using a threshold model. *Genetic Epidemiology* **14**, 987–992.

Elston, R.C. (1998). Linkage and association. *Genetic Epidemiology* **15**, 565–576.

Elston, R.C., Buxbaum, S., Jacobs, K.B. and Olson, J.M. (2000). Haseman and Elston revisited. *Genetic Epidemiology* **19**, 1–17.

Elston, R.C. and Cordell, H.J. (2001). Overview of model-free methods for linkage analysis. *Advances in Genetics* **42**, 135–150.

Etzel, C.J., Liu, M. and Costello, T.J. (2005). An updated meta-analysis approach for genetic linkage. *BMC Genetics* **6**(Suppl. 1), S43.

Farrall, M. (1997). Affected sibpair linkage tests for multiple linked susceptibility genes. *Genetic Epidemiology* **14**, 103–115.

Flanders, W. and Khoury, M. (1991). Extensions to methods of sib-pair linkage analysis. *Genetic Epidemiology* **8**, 399–408.

Freimer, N., Sandkuijl, L. and Blower, S. (1993). Incorrect specification of marker allele frequencies: effects on linkage analysis. *American Journal of Human Genetics* **52**, 1102–1110.

Fulker, D.W. and Cardon, L.R. (1994). A sib-pair approach to interval mapping of quantitative trait loci. *American Journal of Human Genetics* **54**, 1092–1103.

Goldgar, D.E. (1990). Multipoint analysis of human quantitative genetic variation. *American Journal of Human Genetics* **47**, 957–967.

Goldin, L.R. (1992). Detection of linkage under heterogeneity: comparison of two-locus vs. admixture models. *Genetic Epidemiology* **9**, 61–66.

Goldin, L.R. and Weeks, D.E. (1993). Two-locus models of disease: comparison of likelihood and nonparametric linkage methods. *American Journal of Human Genetics* **53**, 908–915.

Goring, H.H., Terwilliger, J.D. and Blangero, J. (2001). Large upward bias in estimation of locus-specific effects from genomewide scans. *American Journal of Human Genetics* **69**, 1357–1369.

Greenberg, D.A., Abreu, P. and Hodge, S.E. (1998). The power to detect linkage in complex disease by means of simple LOD-score analyses. *American Journal of Human Genetics* **63**, 870–879.

Greenwood, C.M.T. and Bull, S.B. (1997). Incorporation of covariates into genome scanning using sib pair analysis in bipolar affective disorder. *Genetic Epidemiology* **14**, 635–640.

Greenwood, C.M.T. and Bull, S.B. (1999). Analysis of affected sib pairs, with covariates – with and without constraints. *American Journal of Human Genetics* **64**, 871–885.

Gudbjartsson, D.F., Thorvaldsson, T., Kong, A., Gunnarsson, G. and Ingolfsdottir, A. (2005). Allegro version 2. *Nature Genetics* **37**, 1015–1016.

Haseman, J.K. and Elston, R.C. (1972). The investigation of linkage between a quantitative trait and a marker locus. *Behavior Genetics* **2**, 3–19.

Hauser, E.R., Watanabe, R.M., Duren, W.L., Bass, M.P., Langefeld, C.D. and Boehnke, M. (2004). Ordered subset analysis in genetic linkage mapping of complex traits. *Genetic Epidemiology* **27**, 53–63.

Hinds, D.A., Risch, N. (1996). *The ASPEX package: affected sib-pair exclusion mapping.* http://aspex.sourceforge.net/.

Hodge, S.E. (1984). The information contained in multiple sibling pairs. *Genetic Epidemiology* **1**, 109–122.

Holmans, P. (1993). Asymptotic properties of affected sib-pair linkage analysis. *American Journal of Human Genetics* **52**, 362–374.

Holmans, P. and Clayton, D. (1995). Efficiency of typing unaffected relatives in an affected sib-pair linkage study with single locus and multiple tightly-linked markers. *American Journal of Human Genetics* **57**, 1221–1232.

Holmans, P. (1998). Affected sib-pair methods for detecting linkage to dichotomous traits: a review of the methodology. *Human Biology* **70**, 1025–1040.

Holmans, P. (2001). Likelihood-ratio affected sib-pair tests applied to multiply affected sibships: issues of power and type I error rate. *Genetic Epidemiology* **20**, 44–56.

Holmans, P. (2002). Detecting gene-gene interactions using affected sib pair analysis with covariates. *Human Heredity* **53**, 92–102.

Huang, Q., Shete, S. and Amos, C.I. (2004). Ignoring linkage disequilibrium among tightly linked markers induces false-positive evidence of linkage for affected sib pair analysis. *American Journal of Human Genetics* **75**, 1106–1112.

Huang, Q., Shete, S., Swartz, M. and Amos, C.I. (2005). Examining the effect of linkage disequilibrium on multipoint linkage analysis. *BMC Genetics* **6**(Suppl. 1), S83.

Khoury, M., Flanders, W., Lipton, R. and Dorman, J. (1991). The affected sib-pair method in the context of an epidemiologic study design. *Genetic Epidemiology* **8**, 277–282.

Knapp, M., Seuchter, S.A. and Baur, M.P. (1994). Linkage analysis in nuclear families. II. Relationship between affected sib-pair tests and lod score analysis. *Human Heredity* **44**, 44–51.

Kong, A. and Cox, N.J. (1997). Allele sharing models: LOD scores and accurate linkage tests. *American Journal of Human Genetics* **61**, 1179–1188.

Kruglyak, L., Lander, E.S. (1995). Complete multipoint sib-pair analysis of qualitative and quantitative traits. *American Journal of Human Genetics* **57**, 439–454.

Kruglyak, L., Daly, M.J., Reeve-Daly, M.P. and Lander, E.S. (1996). Parametric and nonparametric linkage analysis: a unified multipoint approach. *American Journal of Human Genetics* **58**, 1347–1363.

Lander, E.S. and Kruglyak, L. (1995). Genetic dissection of complex traits: guidelines for reporting linkage results. *Nature Genetics* **11**, 241–247.

Lander, E.S. and Schork, N.J. (1994). Genetic dissection of complex traits. *Science* **265**, 2037–2048.

Levinson, D.F. and Holmans, P. (2005). The effect of linkage disequilibrium on linkage analysis of incomplete pedigrees. *BMC Genetics* **6**(Suppl. 1), S6.

Levinson, D.F., Holmans, P., Straub, R.E., Owen, M.J., Wildenauer, D.B., Gejman, P.V., Pulver, A.E., Laurent, C., Kendler, K.S., Walsh, D., Norton, N., Williams, N.M., Schwab, S.G., Lerer, B., Mowry, B.J., Sanders, A.R., Antonarakis, S.E., Blouin, J.L., DeLeuze, J.F. and Mallet, J. (2000). Multicenter linkage study of schizophrenia candidate regions on chromosomes 5q, 6q, 10p and 13q: schizophrenia linkage collaborative group III. *American Journal of Human Genetics* **67**, 652–663.

Levinson, D.F., Levinson, M.D., Segurado, R. and Lewis, C.M. (2003). Genome scan meta-analysis of schizophrenia and bipolar disorder, part I: methods and power analysis. *American Journal of Human Genetics* **73**, 17–33.

Lewis, C.M. and Levinson, D.E. (2006). Testing for genetic heterogeneity in the genome search meta-analysis method. *Genetic Epidemiology* **30**, 348–355.

Lewis, C.M., Levinson, D.F., Wise, L.H., DeLisi, L.E., Straub, R.E., Hovatta, I., Williams, N.M., Schwab, S.G., Pulver, A.E., Faraone, S.V., Brzustowicz, L.M., Kaufmann, C.A., Garver, D.L., Gurling, H.M., Lindholm, E., Coon, H., Moises, H.W., Byerley, W., Shaw, S.H., Mesen, A., Sherrington, R., O'Neill, F.A., Walsh, D., Kendler, K.S., Ekelund, J., Paunio, T., Lonnqvist, J., Peltonen, L., O'Donovan, M.C., Owen, M.J., Wildenauer, D.B., Maier, W., Nestadt, G., Blouin, J.L., Antonarakis, S.E., Mowry, B.J., Silverman, J.M., Crowe, R.R., Cloninger, C.R., Tsuang, M.T., Malaspina, D., Harkavy-Friedman, J.M., Svrakic, D.M., Bassett, A.S., Holcomb, J., Kalsi, G., McQuillin, A., Brynjolfson, J., Sigmundsson, T., Petursson, H., Jazin, E., Zoega, T. and Helgason, T. (2003). Genome scan meta-analysis of schizophrenia and bipolar disorder, part II: Schizophrenia. *American Journal of Human Genetics* **73**, 34–48.

Loesgen, S., Dempfle, A., Golla, A. and Bickeboller, H. (2001). Weighting schemes in pooled linkage analysis. *Genetic Epidemiology* **21**(Suppl. 1), S142–S147.

Lunetta, K.L. and Rogus, J.J. (1998). Strategy for mapping minor histocompatibility genes involved in graft-versus-host disease: a novel application of discordant sib pair methodology. *Genetic Epidemiology* **15**, 595–607.

MacLean, C., Bishop, D.T., Sherman, S. and Diehl, S. (1993). Distribution of lod scores under uncertain mode of inheritance. *American Journal of Human Genetics* **52**, 354–361.

McPeek, M.S. (1999). Optimal allele-sharing statistics for genetic mapping using affected relatives. *Genetic Epidemiology* **16**, 225–249.

Olson, J.M. (1995). Multipoint linkage analysis using sib pairs: an interval mapping approach for dichotomous outcomes. *American Journal of Human Genetics* **56**, 788–798.

Olson, J.M. (1999). A general conditional-logistic model for affected relative pair linkage studies. *American Journal of Human Genetics* **65**, 1760–1769.

Olson, J.M. and Cordell, H.J. (2000). Ascertainment bias in the estimation of sibling genetic risk parameters. *Genetic Epidemiology* **18**, 217–235.

Olson, J.M., Goddard, K.A. and Dudek, D.M. (2001). The amyloid precursor protein locus and very-late-onset Alzheimer disease. *American Journal of Human Genetics* **69**, 895–899.

Olson, J.M., Goddard, K.A. and Dudek, D.M. (2002). A second locus for very-late-onset Alzheimer disease: a genome scan reveals linkage to 20p and epistasis between 20p and the amyloid precursor protein region. *American Journal of Human Genetics* **71**, 154–161.

Ott, J. (1974). Estimation of the recombination fraction in human pedigrees: efficient computation of the likelihood for human linkage studies. *American Journal of Human Genetics* **26**, 588–597.

Ott, J. (1992). Strategies for characterizing highly polymorphic markers in human gene mapping. *American Journal of Human Genetics* **51**, 283–290.

Pardi, F., Levinson, D.F. and Lewis, C.M. (2005). GSMA: software implementation of the genome search meta-analysis method. *Bioinformatics* **21**, 4430–4431.

Payami, H., Thomson, G. and Louis, E. (1984). The affected sib method. III. Selection and recombination. *American Journal of Human Genetics* **36**, 352–362.

Payami, H., Thomson, G., Motro, U., Louis, E. and Hudes, E. (1985). The affected sib method. IV. Sib trios. *Annals of Human Genetics* **49**, 303–314.

Penrose, L. (1935). The detection of autosomal linkage in data which consist of pairs of brothers and sisters of unspecified parentage. *Annals of Eugenics* **6**, 133–138.

Province, M.A. (2001). The significance of not finding a gene. *American Journal of Human Genetics* **69**, 660–663.

Rice, J.P. (1997). The role of meta-analysis in linkage studies of complex traits. *American Journal of Medical Genetics* **74**, 112–114.

Rice, J.P., Rochberg, N., Neuman, R.J., Saccone, N.L., Liu, K.Y., Zhang, X. and Culverhouse, R. (1999). Covariates in linkage analysis. *Genetic Epidemiology* **17**(Suppl. 1), S703–S708.

Risch, N. (1983). The effects of reduced fertility, method of ascertainment, and a second unlinked locus on affected sib-pair marker allele sharing. *American Journal of Medical Genetics* **16**, 243–259.

Risch, N. (1990a). Linkage strategies for genetically complex traits: I. Multilocus models. *American Journal of Human Genetics* **46**, 222–228.

Risch, N. (1990b). Linkage strategies for genetically complex traits: II. The power of affected relative pairs. *American Journal of Human Genetics* **46**, 229–241.

Risch, N. (1990c). Linkage strategies for genetically complex traits: III. The effect of marker polymorphism on analysis of affected relative pairs. *American Journal of Human Genetics* **46**, 242–253.

Risch, N. (1991). A note on multiple testing procedures in linkage analysis. *American Journal of Human Genetics* **48**, 1058–1064.

Risch, N. (1992). Corrections to "Linkage strategies for genetically complex traits. III. The effect of marker polymorphism on analysis of affected relative pairs". *American Journal of Human Genetics* **51**, 673–675.

Risch, N. (1994). Mapping genes for psychiatric disorders. In *Genetic Approaches to Mental Disorders*, E.S. Gershon, C.R. Cloninger and J.E. Barrett, eds. American Psychiatric Press, Washington, DC, pp. 47–61.

Roberts, S.B., MacLean, C.J., Neale, M.C., Eaves, L.J. and Kendler, K.S. (1999). Replication of linkage studies of complex traits: an examination of variation in location estimates. *American Journal of Human Genetics* **65**, 876–884.

Romero-Hidalgo, S., Rodrigues, E.R., Gutierrez-Pena, E., Riba, L. and Tusie-Luna, M.T. (2005). GENEHUNTER versus SimWalk2 in the context of an extended kindred and a qualitative trait locus. *Genetica* **123**, 235–244.

SAGE (2006). *Statistical Analysis for Genetic Epidemiology* 5.2.0 Computer package available from the Department of Epidemiology and Biostatistics, Rammelkamp Center for Education and Research, MetroHealth campus. Case Western Reserve University, Cleveland, OH. `http://darwin.cwru.edu`.

Sawcer, S., Jones, H.B., Judge, D., Visser, F., Compston, D.A.S., Goodfellow, P.N. and Clayton, D. (1997). Empirical genomewide significance levels established by whole genome simulations. *Genetic Epidemiology* **14**, 223–229.

Schaid, D.J., Guenther, J.C., Christensen, G.B., Hebbring, S., Rosenow, C., Hilker, C.A., McDonnell, S.K., Cunningham, J.M., Slager, S.L., Blute, M.L., Thibodeau, S.N. (2004). Comparison of microsatellites versus single-nucleotide polymorphisms in a genome linkage screen for prostate cancer-susceptibility Loci. *American Journal of Human Genetics* **75**, 948–965.

Schaid, D.J., Olson, J.M., Gauderman, W.J. and Elston, R.C. (2003). Regression models for linkage: issues of traits, covariates, heterogeneity, and interaction. *Human Heredity* **55**, 86–96.

Schork, N.J., Boehnke, M., Terwilliger, J.D. and Ott, J. (1993). Two-trait-locus linkage analysis: a powerful strategy for mapping complex genetic traits. *American Journal of Human Genetics* **53**, 1127–1136.

Seaman, S.R. and Holmans, P. (2005). Effect of genotyping error on type-I error rate of affected sib pair studies with genotyped parents. *Human Heredity* **59**, 157–164.

Sham, P.C. and Purcell, S. (2001). Equivalence between Haseman-Elston and variance-components linkage analyses for sib pairs. *American Journal of Human Genetics* **68**, 1527–1532.

Sham, P.C., Purcell, S., Cherny, S.S. and Abecasis, G.R. (2002). Powerful regression-based quantitative-trait linkage analysis of general pedigrees. *American Journal of Human Genetics* **71**, 238–253.

Sham, P.C., Zhao, J.H. and Curtis, D. (1997). Optimal weighting scheme for affected sib-pair analysis of sibship data. *Annals of Human Genetics* **61**, 61–69.

Sobel, E. and Lange, K. (1996). Descent graphs in pedigree analysis: applications to haplotyping, location scores, and marker sharing statistics. *American Journal of Human Genetics* **58**, 1323–1337.

Sobel, E., Papp, J.C. and Lange, K. (2002). Detection and integration of genotyping errors in statistical genetics. *American Journal of Human Genetics* **70**, 496–508.

Sribney, W.M. and Swift, M. (1992). Power of sib-pair and sib-trio linkage analysis with assortative mating and multiple disease Loci. *American Journal of Human Genetics* **51**, 773–784.

Suarez, B.K. (1978). The affected sib pair IBD distribution for HLA-linked disease susceptibility genes. *Tissue Antigens* **12**, 87–93.

Suarez, B.K. and Hodge, S.E. (1979). A simple method to detect linkage for rare recessive diseases: an application to juvenile diabetes. *Clinical Genetics* **15**, 126–136.

Suarez, B.K., Rice, J. and Reich, T. (1978). The generalized sib pair IBD distribution: its use in the detection of linkage. *Annals of Human Genetics* **42**, 87–94.

Suarez, B.K. and Van Eerdewegh, P. (1984). A comparison of three affected sib-pair scoring methods to detect HLA-linked disease susceptibility genes. *American Journal of Medical Genetics* **18**, 135–146.

Sun, L. and Bull, S.B. (2005). Reduction of selection bias in genomewide studies by resampling. *Genetic Epidemiology* **28**, 352–367.

Thomson, G. (1986). Determining the mode of inheritance of RFLP-associated diseases using the affected sib-pair method. *American Journal of Human Genetics* **39**, 207–221.

Thomson, G. and Bodmer, W. (1977). The genetic analysis of HLA and disease associations. In *HLA and Disease*, J. Dausset and A. Svejgaard, eds. Munksgaard, Copenhagen, pp. 84–93.

Tsai, H.J. and Weeks, D.E. (2006). Comparison of methods incorporating quantitative covariates into affected sib pair linkage analysis. *Genetic Epidemiology* **30**, 77–93.

Vieland, V.J., Hodge, S.E. and Greenberg, D.A. (1992). Adequacy of single-locus approximations for linkage analysis of oligogenetic traits. *Genetic Epidemiology* **9**, 45–59.

Wang, T. and Elston, R.C. (2005). The bias introduced by population stratification in IBD based linkage analysis. *Human Heredity* **60**, 134–142.

Ward, P. (1993). Some developments on the affected pedigree-member method of linkage analysis. *American Journal of Human Genetics* **52**, 1200–1215.

Weeks, D.E. and Lange, K. (1988). The affected-pedigree-member method of linkage analysis. *American Journal of Human Genetics* **42**, 315–326.

Weeks, D.E. and Lathrop, G.M. (1995). Polygenic disease: methods for mapping complex disease traits. *Trends in Genetics* **11**, 513–519.

Whittemore, A.S. (1996). Genome scanning for linkage: an overview. *American Journal of Human Genetics* **59**, 704–716.

Whittemore, A.S. and Tu, I.-P. (1998). Simple, robust linkage tests for affected sibs. *American Journal of Human Genetics* **62**, 1228–1242.

Williams, J.T. and Blangero, J. (1997). Comparison of variance-components and sibpair methods for quantitative trait linkage analysis in sibships and nuclear families. *Genetic Epidemiology* **14**, 543.

Williams, J.T., Van Eerdewegh, P., Almasy, L. and Blangero, J. (1999). Joint multipoint linkage analysis of multivariate quantitative traits. I. Likelihood formulation and simulation results. *American Journal of Human Genetics* **65**, 1134–1147.

Wise, L.H., Lanchbury, J.S. and Lewis, C.M. (1999). Meta-analysis of genome searches. *Annals of Human Genetics* **63**, 263–272.

Wright, F.A. (1997). The phenotypic difference discards sib-pair QTL linkage information. *American Journal of Human Genetics* **60**, 740–742.

Xing, C., Sinha, R., Xing, G., Lu, Q. and Elston, R.C. (2006). The affected-/discordant-sib-pair design can guarantee validity of multipoint model-free linkage analysis of incomplete pedigrees when there is marker-marker disequilibrium. *American Journal of Human Genetics* **79**, 396–340.

Xu, X., Weiss, S., Xu, X. and Wei, L.J. (2000). A unified Haseman-Elston method for testing linkage with quantitative traits. *American Journal of Human Genetics* **67**, 1025–1028.

Zaykin, D.V., Zhivotovsky, L.A., Westfall, P.H. and Weir, B.S. (2002). Truncated product method for combining P-values. *Genetic Epidemiology* **22**, 170–185.

Zintzaras, E. and Ioannidis, J.P. (2005). Heterogeneity testing in meta-analysis of genome searches. *Genetic Epidemiology* **28**, 123–137.

35

Population Admixture and Stratification in Genetic Epidemiology

P.M. McKeigue

Conway Institute, University College Dublin, Dublin, Ireland

Population admixture and stratification generally occur together. Admixture between subpopulations generates allelic associations that decay with map distance, whereas stratification generates allelic associations that are independent of map distance. The autocorrelation of ancestry on gametes inherited from parents of mixed descent can be exploited to localize genes in which the pool of disease risk alleles is differentially distributed between subpopulations. Tests for linkage are based on testing for association of the disease or outcome of interest with locus ancestry, conditioning on parental admixture proportions to eliminate confounding by genetic background. This approach, known as *admixture mapping*, is an extension of the principles underlying linkage analysis of an experimental cross between inbred strains. Most current approaches to modelling admixture are based on a standard model in which the stochastic variation of ancestry on gametes inherited from admixed parents is generated by independent Poisson arrival processes. For such models, the posterior distribution of locus ancestry and parental admixture proportions can be generated by Markov chain Monte Carlo simulation, given a sample of individuals typed at ancestry-informative marker loci. Tests for linkage are constructed by averaging over this posterior distribution. The same statistical model can be used with unselected marker loci to detect and control for population stratification in ordinary genetic association studies. For studies using arrays of closely spaced tag SNPs, this approach has limitations as it requires that the marker loci be spaced far enough apart for all allelic association to be attributable to admixture and stratification. An alternative approach is to use principal components analysis to infer a few underlying axes of variation that summarize the allelic associations in the dataset. This approach is computationally efficient, and can be extended to datasets with closely spaced markers using a simple adjustment for short-range allelic associations. Tests of the number of underlying latent variables can be constructed; for a given F_{ST} distance between subpopulations, there is a threshold size of dataset at which stratification can be detected. With either modelling approach, confounding by stratification can be controlled by adjusting for

Handbook of Statistical Genetics, Third Edition. Edited by D.J. Balding, M. Bishop and C. Cannings.
© 2007 John Wiley & Sons, Ltd.   ISBN: 978-0-470-05830-5.

genetic background when testing for allelic association; alternative 'genomic control' approaches based on correcting the variance of the test statistic has serious limitations.

35.1 BACKGROUND

Population admixture and stratification are often considered together, because they usually occur together and because the two phenomena can be modelled with similar statistical methods. Admixture between subpopulations with different allele frequencies generates gametes that consist of a mosaic of segments inherited from each of the ancestral subpopulations. On a gamete inherited from a parent of mixed descent, there will be autocorrelation of ancestry states: the shorter the distance between two loci, the higher the probability that the subpopulation of ancestry will be the same at these two loci. Where allele frequencies vary between ancestral subpopulations, this autocorrelation of ancestry gives rise to allelic association that decays with distance. This makes it possible to infer recent admixture from data on a sample of individuals typed at linked marker loci. Where admixture is recent, the distances over which this association is detectable are far longer than the few hundred kilobases (kb) over which haplotype block structure is detectable.

Genetic stratification, on the other hand, generates allelic associations that are independent of map distance. In principle, it is possible for genetic stratification to occur without admixture: one possible scenario is a total population consisting of discrete endogamous subpopulations, with strong social barriers to mating between subpopulations. Another possible scenario for stratification without admixture is a geographic cline of allele frequencies, with no migration between regions. Such extreme situations would usually be obvious to a researcher, without the need to infer stratification from genetic data. In a stratified population with no admixture, the strength of allelic associations is independent of map distance beyond the very short distances (typically a few hundred kb in human populations) over which haplotype structure is detectable. At the other extreme, it is possible in principle for an admixed population to have no stratification, so that admixture proportions – the proportions of the individual's genome that have ancestry from each ancestral subpopulation – are the same in all individuals. A possible scenario for this would be an island population where an initial pulse of admixture has been followed by isolation, with no new gene flow, and mating that has not been assortative with respect to individual admixture proportions. Usually, stratification and recent admixture occur together. Stratification in admixed populations can be maintained by continuing gene flow from one or more of the unadmixed ancestral subpopulations, or by assortative mating resulting from socioeconomic stratification on the basis of genetic background. See Excoffier (**Chapter 29**) and Beaumont (**Chapter 30**) for further discussion of population stratification and admixture.

Admixture and stratification present both opportunities and challenges to genetic epidemiology. There are opportunities to exploit the genetic structure of admixed and stratified populations for more efficient approaches to mapping and identification of disease susceptibility loci. The challenges are to control for hidden population stratification as a confounder in genetic association studies, and to control for the long-range associations generated by admixture when undertaking fine mapping of a disease locus. This chapter begins by discussing the theory of admixture mapping, which exploits the genetic structure of admixed populations to localize genes that underlie variation in disease risk between ethnic groups.

35.2 ADMIXTURE MAPPING

35.2.1 Basic Principles

Because the allelic associations generated by admixture decay with map distance, admixture generates information about linkage. Early suggestions for exploiting this information were based on testing the allelic associations generated by admixture in a manner similar to classical linkage disequilibrium mapping of a Mendelian disease locus in an isolated population (Chakraborty and Weiss, 1988; Stephens *et al.*, 1994). With this approach, the information available is limited by the magnitude of allelic associations, which depend upon the size of the allele frequency differentials between the ancestral subpopulations. Admixture mapping, in contrast, uses the observed allelic associations to infer the underlying states of locus ancestry. This can be viewed as an extension, to outbred admixed populations, of the principles underlying linkage analysis of an experimental cross between inbred strains (see **Chapter 18**). In family linkage studies, it is the segregation indicators at each locus, not the marker alleles, that convey information about linkage (see **Chapter 33**). In admixture mapping, it is the locus ancestry states, not the marker alleles, that convey information about linkage. When exploiting admixture to localize genetic effects, the marker genotypes are irrelevant once all available information about locus ancestry has been extracted.

To exploit the information about linkage that is generated by admixture, it is necessary to model the underlying stochastic variation of ancestry on chromosomes inherited from admixed parents. To extract information about ancestry, it is desirable to select markers that are informative for ancestry in that they have large allele frequency differentials between the ancestral populations.

Admixture mapping is complementary to association studies and family linkage studies: each has the objective of identifying genes that underlie between-population variation in disease risk. Where admixture mapping can be applied, it has several advantages in comparison with linkage or association study designs.

1. To detect loci that generate between-population variation in disease risk, statistical power is higher for admixture mapping than for family linkage studies. For example, to detect a locus that underlies a twofold risk ratio between populations by admixture mapping, the required sample size is much lower than that required to detect a locus that underlies a twofold sibling recurrence risk ratio in an affected sib-pair study. The low statistical power of family linkage studies to detect loci of modest effect is a fundamental property of this study design (Lander and Schork, 1994). To explain this, consider a study in which the disease locus itself has been typed. The most direct test for an effect at this locus is to test for association of outcome with genotype. Family linkage analysis of a single pedigree can be viewed as an association study in which each gene copy in each founder has been coded as a separate allele, and the association of outcome with the genotypes defined by these 'alleles' is examined within that pedigree. The direction of association (defined by the coding of founder

copies as 'alleles') is fixed within pedigrees, but random between pedigrees. Unless the pedigrees are very large and the number of founders is small, tests that combine information across pedigrees have low statistical power because the direction of association is integrated out of the likelihood function. In admixture mapping, by contrast, the direction of association between outcome and locus ancestry is fixed across all individuals; analysis of an experimental cross between inbred strains can be viewed as linkage analysis of a single large sibship.

2. The required marker density is much lower for admixture mapping than for genome-wide association studies. Genome-wide admixture mapping studies typically require about 2000 ancestry-informative markers, compared with at least 300 K for genome-wide association studies. Now that high-density genotyping arrays are relatively inexpensive, this is no longer a key advantage.

3. Admixture mapping, like family linkage studies, is not dependent upon the assumption of low allelic heterogeneity, as Terwilliger and Weiss (1998) have pointed out. Association studies using haplotype–tagging SNPs depend critically upon the common disease–common variant hypothesis (Reich and Lander, 2001): if there are many disease alleles distributed over all the modal haplotypes in each gene, SNP-based association studies are unlikely to detect them. In contrast, admixture mapping studies can detect a locus as long as the total pool of risk alleles is differentially distributed between the ancestral populations that have contributed to the admixed population under study, whether there are many rare risk alleles or only a few common ones in the disease susceptibility gene.

In a classic linkage analysis of an experimental cross, inbred strains produced by brother–sister mating over many generations are crossed over two generations to generate a sample of $F2$ intercrosses or $F1$ backcrosses. These animals are typed at a set of marker loci at which different alleles have become fixed in each strain, and tested for association between the trait and the marker genotypes at each locus. In extending this approach to admixed human populations, three main problems arise.

1. The history of admixture is not under experimental control or even known. In an experimental cross, the design can be specified so that, for instance, we can study a sample of individuals that consists entirely of $F2$ intercrosses. In human admixed populations, admixture proportions – the proportions of the genome that have ancestry from each continental group – vary between individuals. This confounds any associations between a trait and locus ancestry. This confounding can be controlled by conditioning on parental admixture proportions when testing for association of the trait with locus ancestry. In an affected-only design, we can compare the observed ancestry state frequencies at the locus under study with the expected frequencies given parental admixture proportions of each individual. In a cross-sectional study, we can fit a regression model for the dependence of the trait upon parental admixture proportions, and test for association of adjusted trait values with ancestry state frequencies.

2. Human subpopulations are not inbred strains. In a cross between two inbred strains, the strains of ancestry of the two gene copies at each locus can be inferred directly from the observed genotypes, as the marker alleles are differentially fixed in the two strains. In humans, the F_{ST} distance between continental populations is typically of

the order of 0.1–0.15 (Cavalli-Sforza *et al.*, 1994); another way of expressing this is that human continental groups have lost only 10–15 % of their shared ancestral allelic diversity. Loci at which alleles have become differentially fixed in each continental group are rare: one of the few examples is the *FY* (Duffy antigen) locus, at which a null allele is fixed in populations originating from sub-Saharan Africa and rare in other populations. Even when hundreds of thousands of markers are screened to identify a subset with extreme allele frequency differentials between continental groups, the average information content for ancestry of these ancestry-informative markers is typically no more than 40 %. Thus, we cannot directly infer locus ancestry from the genotype at that locus. As in classical linkage analysis, this problem can be overcome by a multipoint analysis that combines all genotype information on each chromosome to infer the posterior distribution of ancestry at each marker locus.

3.   The ancestral populations are not available for study. In an experimental cross, the marker allele frequencies in the ancestral strains are known, or can be estimated directly in the case of a cross between outbred populations. In human admixture, this is generally not the case: for instance, we cannot sample the exact mix of West African subpopulations that contributed genes to the modern African-American population. This problem can be overcome by re-estimating ancestry-specific allele frequencies within the admixed population under study, as described below.

35.2.2 Statistical Power and Sample Size

To examine the construction of hypothesis tests for admixture mapping, and the statistical power of studies based on these tests, initially we assume that parental admixture proportions, and ancestry at each locus can be inferred without error. For statistical power calculations, this assumption is reasonable, as any region of putative linkage detected in an initial analysis can be saturated with additional markers to reduce any uncertainty in the inference of locus ancestry.

In a study of a binary trait, the parameter under test at each locus is the ancestry risk ratio r, defined as the risk ratio between individuals with 2 and 0 gene copies with ancestry from the high-risk population at the locus under study. For simplicity, we assume a multiplicative model in which there is a linear relationship between the log risk and the number a of gene copies at the disease susceptibility locus that have ancestry from the high-risk population. For an individual whose parents' genomes have proportions μ_1 and μ_2 of ancestry from the high-risk population, the probabilities p_0, p_1, p_2 of 0, 1, 2 gene copies from the high-risk population at any given locus are $(1 - \mu_1)(1 - \mu_2), \mu_1(1 - \mu_2) + (1 - \mu_1)\mu_2, \mu_1\mu_2$, respectively. For an affected individual, given that we observe x gene copies at the locus under study that have ancestry from the high-risk population, the likelihood L as a function of r, conditional on parental admixture proportions is

$$L(r) = \frac{p_x r^{x/2}}{p_0 + p_1\sqrt{r} + p_2 r},$$

where p_x is the probability of x gene copies with ancestry from the high-risk population at the locus under study, defined as above in terms of the parental admixture proportions μ_1, μ_2. For calculations of statistical power, and construction of score tests, it is convenient to evaluate the likelihood as a function of $\theta = \log r$, which makes the asymptotic approximation of the log likelihood by a quadratic function more accurate.

In a cross-sectional study design, the parameter under test is the regression slope θ for the dependence of the trait value upon locus ancestry, adjusted for parental admixture proportions μ. Given data on a single individual with observed trait value y and x gene copies at the locus under study that have ancestry from the high-risk population, the log likelihood as a function of θ is

$$L(\theta) = -\frac{1}{2}\nu(y - \alpha - x\theta)^2,$$

where α is the expected value of y under the null hypothesis $\theta = 0$, and ν is the residual precision (inverse variance) in the regression model, after including parental admixture and any other covariates that are adjusted for.

We can approximate the statistical power of any study design that tests the null hypothesis that a scalar parameter θ has value θ_0 from the Fisher information V, defined as the expectation over repeated experiments of $-d \log L(\theta)/d\theta$ (Clayton and Hills, 1993). This calculation assumes that the second derivative of the log likelihood is approximately constant over the plausible range of the parameter under test (equivalent to approximating the log likelihood by a quadratic function of θ). For one-sided Type 1 error probability α, Type 2 error probability β, and effect size θ, the required sample size n is given by

$$n = \left(\frac{Z_{1-\alpha} + Z_\beta}{\theta}\right)^2 / V,$$

where Z_p is the pth quantile of the standard normal distribution.

For 90% power ($Z_\beta = 1.28$) to detect an effect of size $\theta = 1$ at a one-sided threshold p-value of 10^{-5} ($Z_{1-\alpha} = 4.27$), we require the Fisher information V to be 30.8. The required sample size for a given study design and effect size $\theta = 1$ can be calculated by dividing 30.8 by the average Fisher information contributed by a single individual. The statistical power for any other effect size can be calculated by dividing by the square of the effect size.

For an affected-only design testing the null hypothesis that $\theta = 0$ (where θ is the log ancestry risk ratio), the Fisher information contributed by a single individual whose parental gametes have proportionate admixture μ is $\frac{1}{2}\mu(1 - \mu)$. Thus, with $\mu = 0.2$ (a typical value for mean European admixture in African-American populations), a single affected individual contributes information of 0.08 and 385 individuals are required to detect a locus that accounts for an effect size $\theta = 1$. The number of individuals required to detect an ancestry risk ratio $r = 2$ (equivalent to $\theta = 0.69$) is about 800. A practical lower limit for the ancestry risk ratio generated by a locus to be detectable with realistic sample sizes (no more than a few thousand affected individuals) by admixture mapping is about $r = 1.5$.

Where cases and controls for a rare disease have been sampled, almost all information about linkage with locus ancestry is contributed by the cases. If we can assume that ancestry state frequencies do not vary systematically across the genome within the admixed population under study, we can compare observed with expected ancestry state frequencies (calculated from parental admixture proportions) in affected individuals only, and ignore the controls. If the number of founders of the admixed population is large, and the effects of selection and drift since admixture can be ignored, the assumption of no systematic variation of ancestry state frequencies within the admixed population under study should be valid. Alternatively, we can relax this assumption, and compare ancestry state frequencies in cases and controls, adjusting for expected ancestry state frequencies. This case–control test, however, has lower statistical power (for a given number of cases) because it is testing for a difference between the observed minus expected ancestry state frequency in the cases and the observed minus expected frequency in the controls, rather than comparing the observed ancestry state frequency in the cases to the expected frequency (which can usually be inferred quite accurately). For a case–control comparison of locus ancestry to have the same statistical power as an affected-only design, four times as many individuals must be genotyped as in an affected-only study. It is, however, useful to sample at least some controls to allow a check on the assumption of no systematic variation of ancestry state frequencies, and more generally as a sanity check. Controls may also be required to obtain additional information about ancestry-specific allele frequencies.

The statistical power of a test for association with a quantitative trait can be calculated by similar arguments (Hoggart *et al.*, 2004). The null hypothesis is that $\theta = 0$ (where θ is the slope of regression of the trait upon locus ancestry (coded as the proportion of copies $(0, 1/2, 1)$ from the high-risk population). The Fisher information contributed by a single individual whose parental gametes have proportionate admixture μ, for a trait with residual precision λ, is $\frac{1}{2}\lambda\mu(1 - \mu)$. Thus for a cross-sectional study of a quantitative trait, the sample size required to detect a locus that accounts for an effect size equal to the residual standard deviation $(1/\sqrt{\lambda})$ is the same as that required to detect an effect size $\theta = 1$ equivalent to ancestry risk ratio $r = 2.7$ in an affected-only study. In practice, effects of locus ancestry on a quantitative trait are unlikely to be as large as this; with the exception of skin melanin content, there are few quantitative traits for which the mean difference between continental populations exceeds one standard deviation. Thus admixture mapping studies of quantitative traits, using unselected cross-sectional samples from the admixed population under study, typically require larger sample sizes than studies of extreme phenotypes for which affected-only and case–control designs can be used. For instance, to detect genes underlying ethnic differences in blood pressure, a study based on ascertainment of individuals with the extreme phenotype (hypertension) via clinical records is likely to be more efficient than a cross-sectional study of blood pressure.

In practice, individual admixture proportions and locus ancestry cannot be inferred without error from marker genotypes. This is a 'missing-data' problem, similar to the problem of inferring segregation indicators from marker genotypes in classical linkage studies (Thompson **Chapter 33**). A general approach to such problems is to generate the posterior distribution of the missing data – admixture proportions and locus ancestry – given the observed data – marker genotypes and trait values – under the null hypothesis of no effect of any locus on the outcome. We can then construct tests based on the likelihood functions given above by averaging over the posterior distribution of the missing data under the null hypothesis, as described later.

35.2.3 Distinguishing between Genetic and Environmental Explanations for Ethnic Variation in Disease Risk

Admixture mapping is most likely to work where there are large ethnic differences in disease risk and these differences are at least partly attributable to genetic factors. The classical epidemiological approach to distinguishing between genetic and environmental explanations for between-population differences in disease rates is to study migrants (Reid, 1971). When migrant groups living in the same environment have different disease rates that are not accounted for by adjusting for known environmental determinants of disease risk, it is reasonable to consider genetic explanations. Genetic explanations are most likely where differences in disease rates persist even in migrants who have been settled outside the home country for several generations, and where such differences are consistently found in all countries where the migrant group has settled. On these criteria, genetic factors are likely to underlie the high rates of coronary heart disease and non-insulin-dependent diabetes, which have been reported in people of South Asian (Indian, Pakistani, Bangladeshi and Sri Lankan) descent settled overseas (McKeigue et al., 1989). In contrast, migrant studies suggest that genetic factors are unlikely to account for the low rates of coronary heart disease, colorectal cancer and breast cancer in native Japanese compared with European-ancestry populations in the United States; although initial studies showed that the rates in first-generation Japanese migrants to the United States were similar to Japanese in Japan, in second- and third-generation migrants the rates of these diseases are close to those in European-ancestry individuals in the United States (Dunn, 1975).

A more definitive approach to distinguishing between environmental and genetic explanations for ethnic differences in disease risk is to study populations in which the proportionate admixture of genes from high-risk and low-risk populations varies between individuals. If disease risk is dependent upon the proportion of the individual's genome that has ancestry from the high-risk population, independently of environmental factors, this suggests genetic influences. However, as with any epidemiological study design, inference of an independent relationship depends on the ability to control for confounding factors such as socioeconomic status. The shape of the relationship between disease risk and individual admixture proportions may also be informative. Even if the difference in disease risk is accounted for by a single locus with an autosomal recessive or dominant mode of action, the relationship of disease risk to individual admixture proportions will be approximately linear. It is possible to construct genetic models in which the relationship of disease risk to admixture proportions is non-linear: for instance, we could postulate a model in which the risk of disease is highest in heterozygous individuals. This might be biologically plausible for an autoimmune disease, as for instance in experimental models of lupus (Rudofsky and Lawrence, 1999). Under such a model, the risk of disease would show an inverse U-shaped relationship to admixture, with highest risk in those who have one parent from each of the two ancestral subpopulations. If a model that allows admixture proportions to differ between the two parents is fitted, such a relationship can be modelled and tests for linkage can be easily constructed.

Some examples of ethnic differences in diseases or traits for which epidemiological evidence points to a genetic explanation are listed in Table 35.1. It is remarkable that non-insulin-dependent diabetes, obesity and hypertension should be so prominent among the conditions for which genetic differentiation has apparently led to ethnic differences in disease risk. This may be because genes influencing risk of these diseases also influence traits on which there has been differential selection pressure, such as the ability to survive famine (Neel et al., 1998). Whatever the explanation, these diseases are likely to be the ones most amenable to admixture mapping.

Table 35.1 Ethnic differences in disease rates that are likely to have a genetic component.

Disease or trait	High-risk groups	Low-risk groups	Typical risk ratios between high-risk and low-risk groups
Non-insulin-dependent diabetes	Pacific islanders, Native Americans, Native Australians, South Asians (Zimmet, 1992)	Europeans	5–12
Hypertension	West Africans (Prineas and Gillum, 1985)	Europeans	2–5
Hypertensive renal disease	West Africans (Qualheim *et al.*, 1991)	Europeans	6–20
Generalised obesity	Native Americans, Pacific islanders, West African women (Hodge and Zimmet, 1994)	Europeans	1–2 SD (continuous scale
Central adiposity	South Asians (McKeigue *et al.*, 1991), Native Australians (O'Dea *et al.*, 1993)	Europeans	1 SD (continuous scale)
Coronary heart disease	South Asians (McKeigue *et al.*, 1989)	West Africans (Miller *et al.*, 1989)	2–3
Systemic lupus erythematosus	West Africans (Hopkinson *et al.*, 1994)	Europeans	10
Prostate cancer	West Africans (Merrill and Brawley, 1997)	Europeans	2–3
Angle-closure glaucoma	Alaskan Natives and other Inuit (Congdon *et al.*, 1992)	Europeans	5–10

35.3 STATISTICAL MODELS

35.3.1 Modelling Admixture

To combine information from linked marker loci to infer admixture proportions and locus ancestry, we have to model the stochastic variation of ancestry on gametes generated by parents of mixed descent. Most current programs for modelling admixture, including

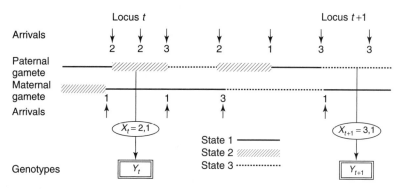

Figure 35.1 Structure of the standard statistical model of admixture with three ancestral populations. Genotypes at each locus (two are shown) depend on the ancestral states, which change along gametes according to a Poisson arrival process.

ADMIXMAP, STRUCTURE and ANCESTRYMAP, specify a model of admixture between K subpopulations in which the stochastic variation of ancestry on each gamete is generated by K independent Poisson arrival processes. In this chapter, this is referred to as the 'standard statistical model' of admixture.

In this model (Figure 35.1), the K arrival processes have intensities $(\mu_1 \rho, \ldots, \mu_K \rho)$, where $(\mu_1, \ldots, \mu_K)$ is the vector of proportions of the parent's genome that have ancestry from each of the K subpopulations, and ρ is the sum of intensities of the K arrival processes. The probability of a transition to ancestry state j at locus $t + 1$, given probability p_j of ancestry state j at locus t, is then $\left[e^{-\rho d_t} p_j + \left(1 - e^{-\rho d_t}\right) \mu_j \right]$, where d_t is the distance in morgans from locus t to locus $t + 1$. If admixture has occurred in a single pulse, the arrival rate parameter ρ can be interpreted as the number of generations back to unadmixed ancestors (Falush *et al.*, 2003). More generally, we can interpret ρ as the effective number of generations since admixture. The assumption of independent Poisson arrival processes is convenient for modelling purposes, but has no basis in any biological model of recombination. Even if we assume a Haldane mapping function – in which meiotic crossovers occur as a Poisson process with intensity 1 per morgan – the stochastic variation of ancestry on admixed gametes does not in general follow a Markov process, except on gametes produced by an $F1$ individual who has one parent from each of two unadmixed ancestral populations (McKeigue, 1998). However, this simple model has been found to perform well in inferring admixture proportions, locus ancestry, and number of generations since admixture. A key assumption of the model is that there is no allelic association between markers loci within any of the ancestral subpopulations, and thus that all allelic association between linked loci can be attributed to admixture. This assumption is generally valid as long as the spacing between marker loci is at least 1 cM.

35.3.2 Modelling Stratification

To model stratification as well as admixture, we allow the admixture proportion vector μ to vary between individuals. If we assume that mating in the parental generation is completely assortative with respect to admixture, we can specify that the admixture proportion vector μ has the same value on both parental gametes. Alternatively, if we assume random mating within the parental generation, we can specify a model in which

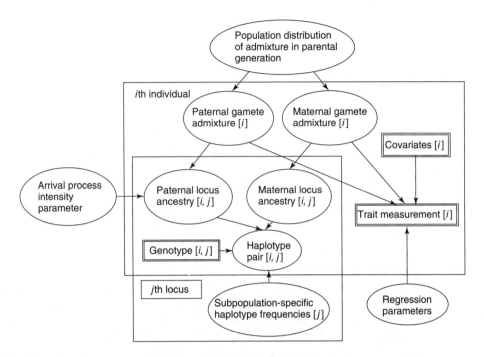

Figure 35.2 Graph representing the statistical model for individual admixture and locus ancestry. This figure does not show the dependence of locus ancestry between linked loci.

the admixture proportions of the two parents are drawn independently from the same population distribution. As the conjugate prior for a probability vector is a Dirichlet distribution, it is natural to use this distribution to model the distribution of admixture proportions over individuals (or gametes) in the population. We place a prior on the parameters of this Dirichlet distribution, and infer the posterior distribution of these parameters from the data. If all elements of the Dirichlet parameter vector α are greater than 1, the distribution of individual admixture proportions in the population is unimodal. If all elements are less than 1, the distribution of individual admixture proportions is U-shaped: that is, most individuals have admixture proportions in which one element of the proportion vector is close to 1 and the others are 0. This is a hierarchical model, as shown in Figure 35.2; inference about each individual's admixture proportions uses not only the information from that individual's genotype data, but also the inferred distribution of admixture proportions based on other individuals sampled from the same population. Where a trait or outcome variable has been measured, the standard model of admixture can be extended to model the possible dependence of the trait value upon individual admixture proportions; this extension is implemented in the ADMIXMAP program. The dependence of trait value on individual admixture can be modelled with logistic regression (for a binary trait) or linear regression (for a quantitative trait). This is useful not only when we want to examine the relationship of an outcome variable to individual admixture proportions (for instance, when attempting to distinguish between genetic and environmental contributions to ethnic variation in disease risk) but also because the likelihood from the regression model feeds back into the inference of individual admixture proportions, enhancing the ability to detect subtle degrees of population stratification.

In principle, it is possible to specify more complex models for the distribution of individual admixture proportions in the population – for instance, we could specify a mixture of two Dirichlet distribution – but this is not implemented in programs currently available. Where enough information is available from marker data to infer the arrival rate parameter ρ on each gamete, we can similarly extend the model to allow the arrival rate to vary across gametes, using a γ distribution to model the distribution of arrival rates over gametes in the population.

35.3.3 Modelling Allele Frequencies

For admixture mapping, the ancestry-specific allele frequencies are usually estimated from samples of modern unadmixed descendants of the populations that contributed genes to the admixed population under study. To allow for sampling error in these estimates, we can specify Dirichlet priors based on the posterior that would be obtained by combining the observed counts with a 'reference' prior on allele frequencies. For a diallelic locus, the 'reference' prior is $\beta(0.5, 0.5)$.

More generally, we may be uncertain as to whether the unadmixed modern descendants are representative of the subpopulations that contributed to the admixed population: for instance, it is not possible to sample the exact mix of Native American populations that contributed to the modern Mexican population. To allow for this uncertainty, we can specify a 'dispersion' model in which both the ancestry-specific allele frequencies in the admixed population and the allele frequencies in the corresponding modern unadmixed population are drawn from a Dirichlet distribution with parameter vector $(\alpha_1 \eta, \ldots, \alpha_K \eta)$, where the mean parameter α is a proportion vector, and η is the precision parameter. The mean parameter vector is allowed to vary across loci, but the precision parameter is assumed to be constant across loci. The biological basis for this is the Wright–Fisher model for drift of allele frequencies in subpopulations that have split from an ancestral total population; the precision parameter η is related to the F_{ST} distance between the two subpopulations by $F_{ST} = 1/(1 + \eta)$ (Lockwood et al., 2001). Thus, for instance, if we assume that the dispersion of allele frequencies between modern West African populations and the African-ancestry gene pool is of similar magnitude to the F_{ST} distance between West African subpopulations (estimated to be about 0.02 (Cavalli-Sforza et al., 1994)) we might specify a prior mean of about 50 ($0.98/0.02 = 49$) for the African-specific allele frequency precision parameter in African-Americans. For the European-specific allele frequency precision parameter, we might specify a prior mean of about 500 based on estimates that the F_{ST} distance between European subpopulations is typically about 0.002 (Cavalli-Sforza et al., 1994).

With large samples from both the admixed population under study and the modern unadmixed descendants, and a highly informative marker panel, the allele frequency precision parameter η can be estimated for each of the continental groups under study. In the few studies that have estimated this parameter for admixed populations such as African-Americans, the posterior mode for the African-specific precision parameter is large, implying that the dispersion of allele frequencies between modern unadmixed West Africans and the African gene pool in the African-American population is fairly small (Patterson et al., 2004). One reason for this may be that markers that show large frequency variation between West African subpopulations are typically excluded from panels of ancestry-informative markers. The practical implication of these results is that it may only rarely be necessary to specify a dispersion model for allele frequencies, as the additional uncertainty generated by dispersion of allele frequencies is usually small compared with the uncertainty in the Dirichlet prior based on the observed allele counts.

Once we have studied one sample from an admixed population, we can re-use the posterior distribution of allele frequencies obtained from that study to specify priors on allele frequencies for any study of new samples from the same admixed population. For simplicity, we can approximate the posterior distribution of allele frequencies by a Dirichlet distribution with the same mean and variance, and use the parameters of this Dirichlet distribution to specify the prior. From this stage onwards, it is unnecessary to specify a dispersion model, because the posterior distribution from the initial study has already taken into account our uncertainty about the extent to which we can rely on modern unadmixed descendants to estimate ancestry-specific allele frequencies in the admixed population under study. In effect, the ancestry-specific allele frequencies have been re-estimated within the admixed population under study. To re-use samples from the posterior distributions of allele frequencies, we can approximate these posteriors by Dirichlet distributions, equating the means and variances of the Dirichlet distribution with the mean and variance of the sample.

35.3.4 Fitting the Statistical Model

Two general approaches to model fitting have been used: a fully Bayesian approach in which the posterior distribution of model parameters is generated by Markov chain Monte Carlo (MCMC) simulation, and a maximum likelihood approach in which the model parameters are held at their maximum likelihood values. The Bayesian approach is more computationally intensive, but has the advantage that samples from the posterior distribution (under the null) can be used to construct hypothesis tests as outlined below. As the transitions of ancestry are generated by Poisson arrival processes, the variation of ancestry along the chromosome is a Markov process. If we assume no allelic association between loci other than that attributable to the autocorrelation of locus ancestry, this is a hidden Markov model (HMM). For such models, standard 'message-passing' algorithms are available to calculate the likelihood of the model parameters μ and ρ given the data, to sample the joint distribution of hidden states (ancestry) at all loci, and to calculate the marginal distribution of hidden states at each locus, conditional on the model parameters (MacDonald and Zucchini, 1997). In a Bayesian MCMC approach, we can use the likelihood calculated from the HMM forward recursion (combined with the prior specified by the population-level parameters) in a Metropolis algorithm to sample the model parameters for each individual. If we also calculate the HMM backward recursions, we can calculate the marginal distribution of ancestry states at each locus, conditional on the realized model parameters. Alternatively, in the maximum likelihood approach, we can use the HMM forward–backward algorithm to calculate the marginal expectations of the numbers of 'arrivals' of each state of ancestry, and maximize the likelihood by an expectation-maximization (EM) algorithm.

The HMM recursions account for most of the computational burden of the MCMC simulations. When standard HMM algorithms are used to compute the forward and backward recursions by matrix multiplication, these recursions have complexity $O(K^4)$ for diploid individuals, as the transition matrix for each interval between adjacent loci is of order K^2 and each element of this matrix has to be multiplied by the corresponding element of the vector of 'emission' probabilities. As in other situations where the transition probabilities in the HMM arise from a simpler stochastic process, the algorithms for the forward and backward recursions can be rewritten in a more efficient form that does not require matrix multiplication, and has complexity $O(K^2)$ for diploid individuals. Updating the population-level parameters at each iteration requires less computation as the number of these parameters does not depend upon the number of individuals. The parameters of the Dirichlet distribution of individual admixture proportions can be updated with a simple Metropolis random walk. After the ancestry states have been sampled on each gamete, and the ordered genotypes or haplotype pairs have been sampled conditional on the realized ancestry states, the allele frequency parameters can be updated with a conjugate Dirichlet update.

35.3.5   Model Comparison

For any given problem, we have a wide choice of the model to be specified. For instance, we can specify more subpopulations, we can allow admixture proportions to differ on the two parental gametes, or we can specify a 'dispersion model' for the ancestry-specific allele frequencies. Two general approaches are available to model choice: construction of diagnostic tests for specific inadequacies of the model, or evaluating the marginal likelihood of the model given the data.

35.3.5.1   Model Diagnostics

One approach to model choice is to construct diagnostic tests for specific inadequacies of the model. For instance, we can construct a test for residual allelic association between unlinked loci, implying residual population stratification not explained by the model. To account for this residual stratification, we might have to specify a model with more subpopulations. Another example would be to test for lack of fit of the observed allele frequencies to the prior, implying that a dispersion model should be fitted. Where the diagnostic test can be set up in the form of a test of a null hypothesis for a continuous parameter, it is possible to evaluate a score test of this null hypothesis by averaging over the posterior distribution. For example, to test for departure from Hardy–Weinberg equilibrium that is not explained by the fitted model, we can set up a test of the null hypothesis of zero inbreeding coefficient. For any realization of the complete data, the score is then the observed frequency of heterozygotes at a locus minus the expected frequency given the realized locus ancestry states and ancestry-specific allele frequencies. Where the diagnostic test cannot be set up as a null hypothesis for a scalar parameter, an alternative approach is to construct a test based on the posterior predictive distribution of some scalar variable T that captures the type of deviation from the model that we are looking for. For instance, to test for dispersion of allele frequencies when the model specifies no dispersion between ancestry-specific allele frequencies and the corresponding allele frequencies in unadmixed modern descendants, we can calculate at each realization of the model parameters a scalar variable T_{obs} as the log likelihood of the prior parameters (of the Dirichlet distribution of allele frequencies) given the realized allele frequencies. Although no obvious sampling distribution (over repeated experiments) is available for the posterior mean of T_{obs}, we can compare the posterior distribution of T_{obs} with the posterior distribution of a quantity T_{rep} calculated by the same method from a replicate dataset drawn conditional on the model parameters at each iteration (Rubin, 1984). We can calculate a posterior predictive check probability ("Bayesian p value") as the frequency with which T_{rep} is more extreme than T_{obs}. Under the null hypothesis, T_{obs} and T_{rep} are exchangeable, and the posterior predictive check probability has expectation 0.5 over repeated experiments. This provides a useful diagnostic check for bad loci: at loci where the observed genotype data are incompatible with the prior on ancestry-specific allele frequencies, the posterior predictive check probability is usually close to 0.

35.3.5.2 *Evaluation of the Marginal Likelihood*

In a Bayesian framework, formal model comparison is based on the marginal likelihood $P(Y \mid M)$ of each model M, given the observed data Y. The ratio of marginal likelihoods between models M_1 and M_0 is the Bayes' factor, which by Bayes theorem is the ratio of posterior to prior odds in favour of M_1 versus M_0. The marginal likelihood can be evaluated by subtracting a measure of 'complexity' (the information conveyed by the data, defined as the decrease in entropy when passing from the prior to the posterior) from a measure of 'fit' (posterior mean of the log likelihood of the model parameters, given the data). The Bayesian framework for hypothesis testing thus incorporates a penalty for complexity (Mackay, 2003). It is straightforward to evaluate the posterior mean of the log likelihood of the model parameters, given the data, from MCMC simulation. Evaluation of the information is more computationally demanding. In the program STRUCTURE, the information is approximated by the posterior variance of the log likelihood of the model parameters (Falush *et al.*, 2003). Although this has been widely used to evaluate how many subpopulations are required to model admixed and stratified subpopulations, it may give answers that are rather different from direct calculation of the marginal likelihood. It is possible, although computationally intensive, to calculate the information from a series of MCMC runs at different temperatures by the method of thermodynamic integration (Neal, 1993).

35.3.6 Assembling and Evaluating Panels of Ancestry-informative Marker Loci

Admixture mapping depends critically upon the availability of markers that are informative for ancestry. Simulations and empirical studies show that to extract about 70 % of information about locus ancestry across the genome in a population where the effective number of generations back to unadmixed ancestors is 5, markers with average information content for ancestry of 40 % are required at average spacing of 1–2 cM (Hoggart *et al.*, 2004). For admixture between populations originating in different continents, where the F_{ST} distance between these populations is at least 0.1, it is feasible to assemble such marker panels by screening large numbers of SNPs and selecting those with the most extreme allele frequency differentials. Before such markers can be used in an admixture mapping study, it is necessary to re-estimate the allele frequencies in independent samples of unadmixed individuals. Marker panels informative for West African/European admixture are already available (Smith *et al.*, 2004), and marker panels for admixture between other continental populations are being assembled.

In principle, it is possible to model admixture with unselected marker panels at higher density. Thus, for African/European admixture, using unselected SNP markers (for which the average marker information content for ancestry is about 0.08) would require only about five times higher density than using markers informative for ancestry (average information content for ancestry about 0.4). With arrays that can score tens of thousands of markers in parallel, the additional genotyping costs of typing more markers may not be a barrier. In practice, however, with such a dense panel of markers, it is difficult to ensure that there is no allelic association between adjacent markers in any of the ancestral subpopulations, which is a key modelling assumption. Another disadvantage, of course, is the greater computational burden.

35.4 TESTING FOR LINKAGE WITH LOCUS ANCESTRY

The model of admixture and locus ancestry is based on the null hypothesis that the effect size parameter θ is 0 at any locus. Such a null hypothesis can be tested by averaging over the posterior distribution generated by MCMC simulation. Two general approaches to constructing tests are available: score tests based on the gradient of the log likelihood at the null value of the parameter, and direct calculation of the likelihood ratio. To calculate a score test, we calculate for each realization of the missing data X the gradient $d \log P(Y, X)/d\theta$ and second derivative $d^2 \log P(Y, X)/d\theta^2$ of the log-likelihood function. We evaluate the score U as the posterior mean of the realized score, the missing information as the posterior variance of the realized score, and the complete information as the posterior mean of the realized information. The observed information V is calculated by subtracting the missing information from the complete information (Louis, 1982). A chi-square test statistic can then be evaluated as $UV^{-1}U$. An attractive feature of this algorithm for calculating score tests by averaging over the posterior distribution is that the ratio of observed information to complete information – the 'proportion of information extracted' – can be interpreted as a measure of the efficiency of the study design, relative to one in which individual admixture proportions and locus ancestry are inferred without uncertainty.

Alternatively, we can calculate, for each locus and for specified values of the effect size parameter θ, the likelihood ratio $P(Y \mid \theta)/P(Y \mid \theta_0)$, where Y is the observed data. We make use of a standard result that $P(Y \mid \theta)/P(Y \mid \theta_0)$ is the expectation of the complete-data likelihood ratio $P(Y, X \mid \theta)/P(Y, X \mid \theta_0)$ over the posterior distribution of the missing data X under the null hypothesis $\theta = \theta_0$ (Thompson and Guo, 1991). In linkage analysis of a pedigree, the missing data X are the segregation indicators at the locus under test; in an admixture mapping study, the missing data for a single individual are the ancestry states at the locus under test and the parental admixture proportions. In a multipoint linkage analysis with known marker map, there is no need to include the recombination fraction as an unknown parameter in the disease model as we can test each position on the genome and specify a disease model in which the effect is mediated only through the locus under test. Under this assumption, $P(Y \mid X, \theta)$ is independent of θ. The likelihood ratio then simplifies to

$$\frac{P(Y \mid \theta)}{P(Y \mid \theta_0)} = \left\langle \frac{P(Y, X \mid \theta)}{P(Y, X \mid \theta_0)} \right\rangle_{P(X \mid Y, \theta_0)} = \left\langle \frac{P(Y \mid X, \theta) P(X \mid \theta)}{P(Y \mid X, \theta_0) P(X \mid \theta_0)} \right\rangle_{P(X \mid Y, \theta_0)}$$

$$= \left\langle \frac{P(X \mid \theta)}{P(X \mid \theta_0)} \right\rangle_{P(X \mid Y, \theta_0)},$$

where angled brackets denote that expectation is evaluated over the distribution specified in the subscript.

In an admixture mapping study, we can treat the individual's admixture proportions μ as fixed by the design, and evaluate the likelihood ratio as

$$\left\langle \frac{P\left(x, \mid \boldsymbol{\mu}, \theta\right)}{P\left(x \mid \boldsymbol{\mu}, \theta_0\right)} \right\rangle_{P(x, \boldsymbol{\mu} \mid Y, \theta_0)},$$

where x is the individual's ancestry at the locus under study (Patterson *et al.*, 2004). Averaging this ratio over the prior on θ yields a Bayes factor (ratio of integrated likelihoods) comparing the hypothesis of an effect at the locus under test with the null hypothesis of no disease locus. The log to base 10 of the Bayes factor can be interpreted as equivalent to a log score in a classical linkage analysis in which the genetic model parameters are fixed. An extension of this argument is to average the Bayes factor over all positions on the genome: equivalent to averaging the likelihood over a diffuse prior on location. This yields a Bayes factor comparing the hypothesis of a single disease locus at an unknown location with the null hypothesis of no disease locus (Patterson *et al.*, 2004). With this Bayesian approach to hypothesis testing, it is unnecessary to correct for multiple testing: averaging over a diffuse prior on location imposes the correct penalty for not specifying the position of the disease locus.

In an affected-only study, the parameter under test is the ancestry risk ratio r and the complete-data likelihood can be calculated at each iteration from the realized values of x and μ using the expression given earlier. For a score test, the asymptotic properties of the test statistic are improved by evaluating the score and information as functions of $\theta = \log r$. For a single individual, the realized score at $\theta = 0$ is $x - \left(\mu_1 + \mu_2\right)$, where x is the number of gene copies at the locus under test that have ancestry from the high-risk population, and μ_1, μ_2 are the parental admixture proportions. This is simply the observed minus expected proportion of gene copies that have ancestry from the high-risk population. The complete information is $\frac{1}{4}\left(\mu_1\left[1 - \mu_1\right] + \mu_2\left(1 - \mu_2\right)\right]$.

In a case–control or cross-sectional study, we can calculate a conventional test for association with locus ancestry, conditional on parental admixture proportions. This can be implemented as a test for the effect size parameter in a regression model, or simply as a test for dependence of the trait on the observed minus expected proportion of gene copies that have ancestry from the high-risk population. In comparison with the affected-only test, the case–control test has lower statistical power but does not depend upon the assumption of no heterogeneity of ancestry state across the genome within the admixed population under study.

With either test, the computational burden can be reduced by using the conditional distribution of locus ancestry given the model parameters λ and averaging over the posterior distribution of model parameters, rather than using the realized states of locus ancestry. The conditional distribution of hidden states (locus ancestry) is calculated from the forward and backward recursions of the hidden Markov model, as described elsewhere (MacDonald and Zucchini, 1997). To calculate the posterior variance of the score from posterior samples of the conditional mean $\langle U \mid \lambda \rangle$ and variance $V\left(U \mid \lambda\right)$ of the realized score, we use the identity

$$\mathrm{V}\left(U\right) = \left\langle \mathrm{V}\left(U \mid \lambda\right) \right\rangle_{P(\lambda)} + \mathrm{V}\left(\langle U \mid \lambda \rangle\right)_{P(\lambda)},$$

where angled brackets denote taking expectations. This calculation provides a useful breakdown of the missing information into two components: the expectation $\langle V(U \mid \lambda) \rangle_{P(\lambda)}$ of the score variance conditional on model parameters λ is a measure of how much missing information is attributable to uncertainty about locus ancestry. To reduce this component of missing information, we can increase the density of markers in the genomic region containing the locus under study. The variance $V(\langle U \mid \lambda \rangle)_{P(\lambda)}$ of the conditional expectation of the score is a measure of how much missing information is attributable to uncertainty about model parameters such as parental admixture proportions. To reduce this component of missing information, we would have to increase the density or information content of the marker panel in other regions unlinked to the locus under study.

35.4.1 Modelling Population Stratification

Population stratification is characterized by the presence of allelic associations between unlinked loci; to model it we have to specify the underlying latent variables that generate these associations. Where no prior information is available about demographic background or allele frequencies in ancestral subpopulations, this is an unsupervised learning problem. Given a dataset in which a sample of individuals from a possibly heterogeneous subpopulation have been typed at marker loci, one possible modelling approach is to use principal components analysis to infer a few uncorrelated latent variables that explain most of the observed allelic associations and to calculate the coordinates of each individual with respect to these latent variables (Patterson *et al.*, 2006). An alternative approach is to fit the standard statistical model of admixture and stratification described above and infer the admixture proportions of each individual (Falush *et al.*, 2003). These two approaches are described below.

35.4.1.1 Modelling Stratification with Principal Components Analysis

Principal components analysis is a transformation of the data that does not depend upon any specific statistical model. The theory of this approach is briefly outlined below. Given a data matrix in which M variables have been measured on N units of observation, principal components analysis performs an orthogonal rotation of the axes defined by the M observed variables, so as to define M new variables that are linear combinations of the original variables and that maximize the proportion of residual variance explained by each successive latent variable. Algebraically, this reduces to finding the eigenvalues and eigenvectors of the $M \times M$ covariance matrix, and ranking them in descending order of magnitude of the eigenvalues. The first eigenvalue λ_1 measures the variance explained by the first principal component, and the corresponding eigenvector represents the loadings (weights) of each variable on that component. Where the M observed variables have been measured on different scales, it is usual to standardize them to zero mean and unit variance. If we specify a statistical model in which the latent variables and measurement errors have Gaussian distributions, it is not possible to infer anything about how the axes should be rotated, because orthogonal rotations do not change the multivariate Gaussian likelihood. It is, however, possible to test hypotheses about the number of independent latent variables that underlie the observed covariance between observed variables. The null hypothesis of no latent variable specifies that the M observed variables correspond to M independent Gaussian variables. This can be compared with alternative hypotheses such as, for instance, a model in which a single latent variable gives rise to covariance between the observed variables. If, for instance, a single latent variable underlies the observed covariances, we expect that the first eigenvalue λ_1 will be large compared with the other $M - 1$ eigenvalues; we can thus use the ratio of the first eigenvalue to the sum of the M eigenvalues $\lambda_1 / \sum_m \lambda_m$ as a test statistic.

In the context of an MCMC simulation, we can calculate at each realization a test statistic T_{obs} as the ratio of the first eigenvalue λ_1 to the sum of the M eigenvalues, and compare it with a test statistic calculated from a replicate dataset drawn from the distribution defined by the realized model parameters. This yields a posterior predictive check probability (Rubin, 1984). More recently it has been shown that if M and N are large, the distribution of the the test statistic $\lambda_1 / \sum_m \lambda_m$ (appropriately normalized for the size of the dataset) under the null hypothesis of no latent variable has has a density discovered by Tracy and Widom from which tests of significance can be obtained (Johnstone, 2001; Patterson et al., 2006).

Where the observed variables are allele frequencies or allele counts, and all loci are unlinked, inferring that the number of latent variables is greater than 0 is equivalent to detecting population stratification. Where a model for stratification has already been fitted, the expected values can be subtracted from the observed values before calculating the covariance matrix for the residuals. We can thus test for residual stratification not explained by the fitted model. In the first applications of principal components analysis to infer genetic structure, the data matrices consisted of allele frequency estimates in geographically defined subpopulations. The object was to study geographic clines in genetic background (Cavalli-Sforza et al., 1994). Such analyses demonstrated that a high proportion of geographic variation of allele frequencies within the European continent could be explained by the first principal component, which varies from South-East to North-West Europe. More recently, principal components analysis has been applied to individual genotype data, such that each row of the data matrix contains allele counts (scored as 0, 1, 2), standardized so that each column has zero mean and unit variance (Price et al., 2006). To allow tightly linked loci to be included as if they were unlinked, the observed allele counts at locus t can be regressed on the observed counts at loci $t - 1, \ldots, t - T$, where T is chosen to be the smallest number that makes the residual allelic associations independent of map distance.

Patterson et al. (2006) has demonstrated how tests for stratification can be constructed using the asymptotic distribution of the magnitude of the first eigenvalue under the null. If the first eigenvalue is declared to be significant, the test procedure can be applied to the remaining eigenvalues $\lambda_2, \ldots, \lambda_M$. This procedure is repeated until $K - 1$ significant eigenvalues have been retained and the Kth eigenvalue is not significant. The number K inferred by this test for stratification corresponds to the number of subpopulations in a mixture model of population stratification (Patterson et al., 2006). Patterson et al. (2006) exploit recent work on the distribution of the largest eigenvalue of a sample covariance matrix to examine the number of individuals and marker loci required to detect significance. They show that where the total population consists of two equally sized subpopulations, the ability to detect stratification with a sample of N individuals typed at M marker loci depends critically upon whether the fixation index F_{ST} between the two subpopulations is greater than $1/\sqrt{MN}$. If F_{ST} is less than the threshold defined by the data size, the test statistic will have the distribution expected under the null, and stratification will be undetectable. Above this threshold, evidence for stratification will accumulate very rapidly as the data size is increased. They denote this as a phase-change phenomenon. Thus for a sample of 10^3 individuals typed at 10^5 marker loci, stratification into two equally sized subpopulations will be detectable if the F_{ST} distance between subpopulations is greater than 10^{-4}. Although this result was derived for detecting stratification by principal components analysis, it appears to apply to any method for detecting stratification from marker genotypes alone (Patterson *et al.*, 2006).

A key advantage of using principal components analysis to model stratification is that the approach is computationally feasible even for very large datasets. Another advantage is that it is possible to adjust for allelic association between tightly linked loci simply by regressing the genotype on the last few loci as outlined above. This makes it possible to analyse data from whole-genome tag SNP arrays. A limitation of principal components analysis is that it ignores the additional information that may be available from marker positions. In admixed populations, the decay of allelic associations with map distance yields additional information about the latent variables that underlie these associations. In admixed populations, the presence of long-range allelic associations that are not attributable to stratification will invalidate tests for stratification unless the analysis is restricted to unlinked loci, although principal components analysis may still be effective in defining a few axes that summarize most variation in genetic background. In principle, it is possible to augment the genotype data matrix with phenotypic traits or demographic variables relevant to genetic stratification, so that inference of genetic structure is based on 'supervised' learning.

35.4.1.2 Modelling Stratification with a Mixture Model

The standard model of admixture and stratification described above can also serve as a model for population stratification, and programs such as STRUCTURE and ADMIXMAP can thus be used to infer stratification. In contrast to a principal components analysis, where the number of 'dimensions' to be retained is not fixed at the start of the analysis, for the admixture/stratification model it is necessary to specify the number K of discrete subpopulations in advance. Where the demographic background of the population is not known, or we do not want to make assumptions about the ancestral subpopulations, the subpopulation-specific allele frequencies can be specified with uninformative priors. Stratification is inferred from allelic associations between unlinked loci, and admixture from the decay of allelic associations with map distance. Thus, in the extreme situation where a population consists of discrete endogamous strata, the strength of allelic association will be independent of map distance (except over very short distances where allelic associations are generated by haplotype structure). In the admixture model, this implies that all individuals have admixture proportions such that one element of the proportion vector is close to 1 and the others are close to 0. This distribution can be represented as a Dirichlet distribution with low precision parameter; the mean vector represents the proportion of the population assigned to each of the K subpopulations. The problem of inferring the number K of ancestral subpopulations, and the posterior assignment probabilities of each individual to one of these subpopulations, then resembles the classical 'K-means' clustering problem. To infer K, we can evaluate the marginal likelihood of models with different values of K as described above, or we can construct diagnostic tests for residual stratification not explained by the model, using principal component analyses of the residuals as described above.

Other plausible models can also be subsumed within the standard model of admixture and stratification. Thus, a gradient of allele frequencies across a geographic cline can be represented by an admixture model in which the distribution of individual admixture proportions in the population is specified by a Dirichlet distribution with moderately high precision, and the arrival rate parameter is large (so that allelic association does not decay with distance). Thus, for instance, the north–south gradient of allele frequencies in the native British population could be modelled as a gradient of admixture between two ancestral subpopulations corresponding to 'southern' and 'northern' .

Where the population under study is admixed, and its demographic history is known, as for African-American and Hispanic-American populations, this information can be taken into account when specifying the model, so that for instance we can specify two or three ancestral continental populations and can set informative priors on the ancestry-specific allele frequencies. If the demographic history of the population is unknown or we do not wish to make any assumptions about demographic history, uninformative 'reference' priors on the allele frequencies can be specified. In this situation, the K subpopulations are not identifiable, and inference that relies on assuming that the subpopulation labels do not switch during a sampling run may be unreliable. When using unselected markers to infer stratification and admixture between less genetically distant subpopulations, with no prior information about allele frequencies, it is necessary to specify in the model our prior expectation that the allele frequencies at each locus will be correlated between subpopulations. This improves the ability to detect subtle degrees of stratification (Falush et al., 2003). Specifying correlated allele frequencies is equivalent to specifying at each locus a Dirichlet-multinomial model with Dirichlet parameter vector $\boldsymbol{\alpha} = \eta\boldsymbol{\mu}$ for the distribution of allele frequencies across subpopulations. The proportion parameter vector $\boldsymbol{\mu}$ is sampled from a flat Dirichlet prior independently for each locus, with a dispersion parameter η that is the same across all loci. The Dirichlet parameters are given by $\alpha_i = \eta\mu_i$. Where the allelic associations between unlinked loci are weak, the ability to detect stratification may be improved further by specifying an informative prior on the allele frequency precision parameter. With randomly chosen markers, the average correlation of allele frequencies between subpopulations is given by the fixation index F_{ST} between those subpopulations. Values of F_{ST} in the range 0.1–0.2 are typical for subpopulations originating in different continents, and values less than 0.05 are typical for subpopulations originating in the same continent. This implies that plausible prior values for η, in studies of stratification within populations originating from the same continent, are at least 20.

Where a trait that is strongly dependent upon genetic background has been modelled, it is possible to allow for this dependence in the model, for instance by specifying a linear or logistic regression model as implemented in ADMIXMAP. Any demographic variable, such as self-reported ethnicity or geographic origin, can be included in the model as a phenotypic trait. This improves the ability to detect subtle degrees of stratification, because information from the observed trait value is fed back into the admixture model via the contribution of the regression model to the likelihood.

For modelling population stratification, the standard model of admixture/stratification has two main limitations: it is computationally intensive, and the assumption of no allelic association other than that generated by admixture or stratification requires a minimum spacing between markers: typically about 1 cM within European-ancestry populations. Advantages of this approach to modelling are that both marker–trait and marker–marker associations can be used simultaneously to infer stratification, and that it allows tests of association based on averaging over the posterior distribution to be constructed as described below.

35.4.1.3 Controlling for Population Stratification as a Confounder in Genetic Association Studies

With either of the two modelling approaches described above, we can obtain an estimate of the genetic background of each individual: the coordinates with respect to the retained components of a principal components analysis, or the posterior means of individual admixture proportions in a model of admixture with K subpopulations. To control for possible confounding by genetic background in genetic association studies, we can simply adjust for estimated genetic background in regression models of the effect of genotype on outcome. This has been denoted the 'structured association' approach, and is implemented in the program STRUCTURE/STRAT (Pritchard and Donnelly, 2001). A variant of this approach, implemented in the program ADMIXMAP (Hoggart *et al.*, 2003), is to construct score tests for association based on averaging over the joint posterior distribution of individual admixture proportions and parameters of the model for regression of outcome on admixture proportions. In theory, this one-step procedure has the advantage that it allows correctly for uncertainty in the inference of individual admixture proportions, where two-step procedures may fail to allow for residual confounding where the information about genetic background available from marker loci is limited and the confounding effect of genetic background is strong. Uncertainty in the inference of genetic background is quantified as missing information in the score test, and problems of lack of identifiability of the K subpopulations are eliminated. Where many loci are to be tested for association, stored samples from the posterior distribution of individual admixture proportions and regression parameters (inferred from genotypes at a subset of ancestry-informative marker loci) can be used without having to include all the tested loci in the model of admixture. In association studies that use DNA pools from multiple individuals, it is possible to control for stratification if an initial panel of markers has been typed on individuals to obtain estimates of genetic background, simply by forming pools that are stratified by genetic background.

A somewhat different method for using marker genotype data is the 'genomic control' approach proposed by (Devlin *et al.*, 2003). In this approach, confounding by population stratification is treated as a random effect that inflates the variance of the χ-square test statistic. The magnitude of this random effect is inferred from trait–marker associations: the information available from marker–marker associations is ignored. In effect, genomic control rescales the crude p values so that the observed Type 1 error rate (over all loci) approximates the target Type 1 error rate. The re-scaling of the crude χ-square statistic is equivalent to multiplying the log p values by a constant factor, because the right tail probability of a chi-square distribution approximates an exponential curve. A serious limitation of the genomic control approach is that it assumes that the locus under study and the marker loci are exchangeable with respect to F_{ST}. If, for instance, we are studying a candidate gene that influences a trait such as drug metabolism or immune response, we might expect allele frequency differentials to be larger for functional polymorphisms in this candidate gene than for randomly chosen neutral SNPs. In this situation, the genomic control method may fail to control for confounding. A recent example is the demonstration of an association between height and a lactase polymorphism in European-Americans (Campbell *et al.*, 2005). Adjustment for marker–trait associations at 178 SNPs by the genomic control method did not reduce the size of the crude effect. However, the association appeared to be at least partly explained by confounding by demographic origin (defined on an axis from north-western to south-eastern Europe).

Another limitation of the genomic control approach is that adjustment necessarily increases the Type 2 error rate. Where the confounding effect of population stratification and the true effect of genotype at the locus under study on the outcome variable are opposite in direction, adjustment of the test statistic by the genomic control approach will mask the true effect, whereas adjustment by the structured association approach will unmask the true effect. These limitations of the genomic control approach are consequences of modelling confounding as a random effect, rather than measuring the confounder (genetic background) and modelling its effect on the outcome variable.

35.5   CONCLUSIONS

The statistical problems of admixture mapping with samples of unrelated individuals have now been largely solved. Panels of ancestry-informative markers are available for West African/European admixture, and will soon be available for other continental groups also. Several programs are available to model individual admixture and locus ancestry, using either Bayesian approaches to generate the posterior distribution of model parameters (Falush *et al.*, 2003; Patterson *et al.*, 2004; Hoggart *et al.*, 2004) or fitting of model parameters by maximizing the likelihood. An unsolved problem is how to combine modelling admixture with other types of genetic model: for instance, modelling pedigree data, or modelling extended haplotypes using genotypes at tag SNPs scored on arrays. Modelling admixture and inheritance in pedigrees simultaneously is possible in theory, by extending the standard HMM algorithms. Thus, we could specify a model in which the hidden state space is defined by the ancestry of the gametes in each founder in the pedigree, and by the segregation indicators in each meiosis. Even for a sib pair, the four founder gametes and two meioses would define a hidden state vector of length $2K^4$ at each locus, for which the HMM recursions would be computationally demanding. Modelling admixture and extended haplotypes simultaneously is even less computationally feasible with current methods. With more than a few thousand marker loci, it becomes increasingly difficult to maintain the assumption of no allelic association between marker loci other than that generated by admixture and stratification. To exploit the information about locus ancestry and individual admixture that is available from ancestry-informative markers, one possible approach would be a two-stage analysis in which estimates of individual admixture or locus ancestry are 'plugged in' to a second step in which extended haplotypes are modelled.

For inferring population stratification with tag SNP arrays that include tens of thousands or hundreds of thousands of markers, ordination methods such as principal components analysis are the most feasible approach. The recent development of formal methods to test for residual stratification unexplained by the retained components strengthens this approach. Statistical modelling of admixture and stratification may, however, yield results that are more directly interpretable in terms of genetic structure and population history:

for instance, the model parameters can be interpreted in terms of the length of time over which admixture has occurred. One possible approach would be to use ordination methods to preselect for each continental group, a subset of markers that are highly informative for stratification, and then to use these markers to model stratification and admixture.

REFERENCES

Campbell, C.D., Ogburn, E.L., Lunetta, K.L., Lyon, H.N., Freedman, M.L., Groop, L.C., Altshuler, D., Ardlie, K.G. and Hirschhorn, J.N. (2005). Demonstrating stratification in a European American population. *Nature Genetics* **37**, 868–872.

Cavalli-Sforza, L.L., Menoozz, P. and Piazzi, A. (1994). *The History and Geography of Human Genes*. Princeton University Press, Princeton, NJ.

Chakraborty, R. and Weiss, K.M. (1988). Admixture as a tool for finding linked genes and detecting that difference from allelic association between loci. *Proceedings of the National Academy of Sciences of the United States of America* **85**, 9119–9123.

Clayton, D. and Hills, M. (1993). *Statistical Models in Epidemiology*. Oxford University Press.

Congdon, N., Wang, F. and Tielsch, J.M. (1992). Issues in the epidemiology and population-based screening of primary angle-closure glaucoma. *Survey of Ophthalmology* **36**, 411–423.

Devlin, B., Roeder, K. and Wasserman, L. (2003). Analysis of multilocus models of association. *Genetic Epidemiology* **25**, 36–47.

Dunn, J.E. (1975). Cancer epidemiology in populations of the United States–with emphasis on Hawaii and California–and Japan. *Cancer Research* **35**, 3240–3245.

Falush, D., Stephens, M. and Pritchard, J.K. (2003). Inference of population structure using multilocus genotype data: linked loci and correlated allele frequencies. *Genetics* **164**, 1567–1587.

Hodge, A.M. and Zimmet, P.Z. (1994). The epidemiology of obesity. *Bailliere's Clinical Endocrinology and Metabolism* **8**, 577–599.

Hoggart, C.J., Parra, E.J., Shriver, M.D., Bonilla, C., Kittles, R.A., Clayton, D.G. and McKeigue, P.M. (2003). Control of confounding of genetic associations in stratified populations. *American Journal of Human Genetics* **72**, 1492–1504.

Hoggart, C.J., Shriver, M.D., Kittles, R.A., Clayton, D.G. and McKeigue, P.M. (2004). Design and analysis of admixture mapping studies. *American Journal of Human Genetics* **74**, 965–978.

Hopkinson, N.D., Doherty, M. and Powell, R.J. (1994). Clinical features and race-specific incidence/prevalence rates of systemic lupus erythematosus in a geographically complete cohort of patients. *Annals of the Rheumatic Diseases* **53**, 675–680.

Johnstone, I. (2001). On the distribution of the largest principal component. *Annals of Statistics* **29**, 295–327.

Lander, E.S. and Schork, N.J. (1994). Genetic dissection of complex traits. *Science* **265**, 2037–2048.

Lockwood, J.R., Roeder, K. and Devlin, B. (2001). A Bayesian hierarchical model for allele frequencies. *Genetic Epidemiology* **20**, 17–33.

Louis, T.A. (1982). Finding the observed information matrix when using the EM algorithm. *Journal of the Royal Statistical Society, Series A* **44**, 226–232.

MacDonald, I.L. and Zucchini, W. (1997). *Hidden Markov and Other Models for Discrete-Valued Time Series – Monographs on Statistics & Applied Probability*. Chapman and Hall/CRC.

Mackay, D.J. (2003). *Information Theory, Inference and Learning Algorithms*. Cambridge University Press, Cambridge.

McKeigue, P.M. (1998). Mapping genes that underlie ethnic differences in disease risk: methods for detecting linkage in admixed populations, by conditioning on parental admixture. *American Journal of Human Genetics* **63**, 241–251.

McKeigue, P.M., Miller, G.J. and Marmot, M.G. (1989). Coronary heart disease in south Asians overseas: a review. *Journal of Clinical Epidemiology* **42**, 597–609.

McKeigue, P.M., Shah, B. and Marmot, M.G. (1991). Relation of central obesity and insulin resistance with high diabetes prevalence and cardiovascular risk in South Asians. *Lancet* **337**, 382–386.

Merrill, R.M. and Brawley, O.W. (1997). Prostate cancer incidence and mortality rates among white and black men. *Epidemiology* **8**, 126–131.

Miller, G.J., Beckles, G.L., Maude, G.H., Carson, D.C., Alexis, S.D., Price, S.G. and Byam, N.T. (1989). Ethnicity and other characteristics predictive of coronary heart disease in a developing community: principal results of the St James Survey, Trinidad. *International Journal of Epidemiology* **18**, 808–817.

Neal, R. (1993). Probabilistic inference using Markov chain Monte Carlo methods. Technical report CRG-TR-93-1, Department of Computer Science, University of Toronto.

Neel, J.V., Weder, A.B. and Julius, S. (1998). Type II diabetes, essential hypertension, and obesity as "syndromes of impaired genetic homeostasis": the "thrifty genotype" hypothesis enters the 21st century. *Perspectives in Biology and Medicine* **42**, 44–74.

O'Dea, K., Patel, M., Kubisch, D., Hopper, J. and Traianedes, K. (1993). Obesity, diabetes, and hyperlipidemia in a central Australian aboriginal community with a long history of acculturation. *Diabetes Care* **16**, 1004–1010.

Patterson, N., Hattangadi, N., Lane, B., Lohmueller, K.E., Hafler, D.A., Oksenberg, J.R., Hauser, S.L., Smith, M.W., O'Brien, S.J., Altshuler, D., Daly, M.J. and Reich, D. (2004). Methods for high-density admixture mapping of disease genes. *American Journal of Human Genetics* **74**, 979–1000.

Patterson, N., Price, A.L. and Reich, D. (2006). Population structure and eigenanalysis. *PLoS Genetics* **2**, e190.

Price, A.L., Patterson, N.J. Plenge, R.M., Weinblatt, M.E., Shadick, N.A. and Reich, D. (2006). Principal components analysis corrects for stratification in genome-wide association studies. *Nature Genetics* **38**, 904–909.

Prineas, R.J. and Gillum, R. (1985). *US Epidemiology of Hypertension in Blacks*, chapter 2. Year Book, Chicago, IL, pp. 17–36.

Pritchard, J.K. and Donnelly, P. (2001). Case-control studies of association in structured or admixed populations. *Theoretical Population Biology* **60**, 227–237.

Qualheim, R.E., Rostand, S.G., Kirk, K.A., Rutsky, E.A. and Luke, R.G. (1991). Changing patterns of end-stage renal disease due to hypertension. *American Journal of Kidney Diseases* **18**, 336–343.

Reich, D.E. and Lander, E.S. (2001). On the allelic spectrum of human disease. *Trends in Genetics* **17**, 502–510.

Reid, D.D. (1971). The future of migrant studies. *Israel Journal of Medical Sciences* **7**, 1592–1596.

Rubin, D.B. (1984). Bayesianly justifiable and relevant frequency calculations for the applied statistician. *Annals of Statistics* **12**, 1151–1172.

Rudofsky, U.H. Lawrence, D.A. (1999). New Zealand mixed mice: a genetic systemic lupus erythematosus model for assessing environmental effects. *Environmental Health Perspectives* **107**, Suppl. 5, 713–721.

Smith, M.W., Patterson, N., Lautenberger, J.A., Truelove, A.L., McDonald, G.J., Waliszewska, A., Kessing, B.D., Malasky, M.J., Scafe, C., Le, E., de Jager, P.L., Mignault, A.A., Yi, Z., de The, G., Essex, M., Sankale, J.L., Moore, J.H., Poku, K., Phair, J.P., Goedert, J.J., Vlahov, D., Williams, S.M., Tishkoff, S.A., Winkler, C.A., De La Vega, F.M., Woodage, T., Sninsky, J.J., Hafler, D.A., Altshuler, D., Gilbert, D.A., O'Brien, S.J. and Reich, D. (2004). A high-density admixture map for disease gene discovery in African-Americans. *American Journal of Human Genetics* **74**, 1001–1013.

Stephens, J.C., Briscoe, D. and O'Brien, S.J. (1994). Mapping by admixture linkage disequilibrium in human populations: limits and guidelines. *American Journal of Human Genetics* **55**, 809–824.

Terwilliger, J.D. and Weiss, K.M. (1998). Linkage disequilibrium mapping of complex disease: fantasy or reality? *Current Opinion in Biotechnology* **9**, 578–594.

Thompson, E.A. and Guo, S.W. (1991). Evaluation of likelihood ratios for complex genetic models. *IMA Journal of Mathematics Applied in Medicine and Biology* **8**, 149–169.

Zimmet, P.Z. (1992). Kelly west lecture 1991. Challenges in diabetes epidemiology–from West to the rest. *Diabetes Care* **15**, 232–252.

36

Population Association

D. Clayton

Cambridge Institute for Medical Research, University of Cambridge, Cambridge, UK

This chapter relates the analysis of population-based studies of associations between genetic polymorphism and disease to the established literature on the analysis of epidemiological case-control studies. Dealing with confounding by use of stratified analyses and regression methods is discussed. The possible influence of hidden population structure is compared with the more generic problem of unmeasured confounding in epidemiology.

36.1   INTRODUCTION

This chapter concerns the detection and analysis of statistical association, *at the population level*, between a binary trait and genotype. Although the main interest is in disease traits, many of the methods are relevant to any binary trait. The only methods that are not more generally applicable are those that depend on the 'rare disease' assumption (that the relative frequency of disease in the population is small).

Population association between genotype at a particular locus and disease can arise in three ways:

1. the locus may be causally related to the disease, different alleles carrying different risks (*direct* association);

2. the locus may not itself be causal, but may be sufficiently close to a causal locus so as to be in linkage disequilibrium with it (*indirect* association); and

3. the association may be due to *confounding* by population stratification or admixture.

If the association has arisen for the last mentioned reason, it is of little scientific interest. It is therefore important to attempt to exclude such spurious association by appropriate design and/or analysis of studies.

The problem of confounding is discussed in every textbook of epidemiology. In the present context, confounding would arise if the population contained several ethnic groups,

Handbook of Statistical Genetics, Third Edition. Edited by D.J. Balding, M. Bishop and C. Cannings.
© 2007 John Wiley & Sons, Ltd.   ISBN: 978-0-470-05830-5.

if allele frequencies at the locus of interest differed between groups, and if disease frequency also differed between groups for reasons quite unrelated to the locus of interest. Such reasons could include genetic differences at a locus in another part of the genome or differences in exposure to environmental causes due to differing customs. It should not be forgotten that such confounding may act to create population association in the absence of a causal link (or linkage disequilibrium with a causal locus), but it may also act in the reverse manner to obscure a causal relationship.

A brief plan of the chapter is as follows. Section 36.2 introduces some notation and discusses statistical measures of association between disease risk and genotype at a diallelic locus. In Section 36.3 the epidemiological case-control study is briefly described and its application to genetic association studies discussed. Section 36.4 discusses simple statistical tests for association in such studies, and Section 36.5 shows how likelihood-based analyses can be carried out using standard software for logistic regression and the log-linear model. When stratification of the population is manifest, control for confounding in case-control studies may be achieved either by post-stratification during analysis or, at the design stage, by matching; these options are discussed in Section 36.6. But neither option is available if population stratification is hidden; Section 36.7 analyses the seriousness of this problem in practice and reviews approaches to dealing with it. The next two sections discuss problems arising as a result of increased polymorphism; Section 36.8 extends the discussion to the case of loci with more than two alleles, and Section 36.9 considers association with haplotypes formed by several closely linked loci. Finally, Section 36.10 briefly discusses some outstanding problems, notably the extension of these ideas to quantitative traits.

Sections 36.2–36.6 draw heavily on the epidemiological literature. The present discussion is of necessity brief, concentrating on matters particularly relevant to studies in genetic epidemiology; for more detailed treatments of this material, see Clayton and Hills (1993) or Breslow and Day (1980).

36.2 MEASURES OF ASSOCIATION

Table 36.1 introduces some notation for the case of a diallelic locus with alleles A and a. On the left, Table 36.1(a) shows the probability distributions of disease conditional upon genotype. Thus, π_{AA}, π_{Aa} and π_{aa} are the *penetrances* of the genotypes AA, Aa and aa respectively. On the right, Table 36.1(b) shows the distributions of genotype given presence or absence of disease in the population. These are denoted by $\gamma^{(1)}, \gamma^{(0)}$ and $\gamma^{(\cdot)}$, respectively.

Association is defined by differences between the penetrances π_{AA}, π_{Aa} and π_{aa} or, equivalently, between the distributions $\gamma^{(1)}$ and $\gamma^{(0)}$. However, the *strength* of association is most naturally expressed in terms of contrasts in the prevalences. A popular measure in epidemiology is the *relative risk* in which each prevalence is expressed relative to the risk in some 'reference' category. If allele a is the more common form, it would be natural to take genotype aa as reference and express the strength of association by the two relative risks:

$$\theta_{AA} = \frac{\pi_{AA}}{\pi_{aa}}, \quad \theta_{Aa} = \frac{\pi_{Aa}}{\pi_{aa}}.$$

Table 36.1 Population distributions of (a) disease given genotype, and (b) genotype given disease.

	Disease				Disease		
Genotype	Yes	No		Genotype	Yes	No	All
AA	π_{AA}	$1 - \pi_{AA}$		AA	$\gamma_{AA}^{(1)}$	$\gamma_{AA}^{(0)}$	$\gamma_{AA}^{(\cdot)}$
Aa	π_{Aa}	$1 - \pi_{Aa}$		Aa	$\gamma_{Aa}^{(1)}$	$\gamma_{Aa}^{(0)}$	$\gamma_{Aa}^{(\cdot)}$
aa	π_{aa}	$1 - \pi_{aa}$		aa	$\gamma_{aa}^{(1)}$	$\gamma_{aa}^{(0)}$	$\gamma_{aa}^{(\cdot)}$
	(a)				(b)		

However, we shall see that it is more convenient to measure strength of association in terms of *odds ratios*. The odds of disease contrasts the probability that disease is present with the probability that it is absent. Thus, for penetrance π, the odds of disease is $\pi/(1 - \pi)$ and the two odds ratios that describe association between disease and genotype are

$$\theta_{AA}^* = \frac{\pi_{AA}}{1 - \pi_{AA}} \bigg/ \frac{\pi_{aa}}{1 - \pi_{aa}}, \quad \theta_{Aa}^* = \frac{\pi_{Aa}}{1 - \pi_{Aa}} \bigg/ \frac{\pi_{aa}}{1 - \pi_{aa}}.$$

The convenience of the odds ratios stems from the relationship between the two parameterisations set out in Table 36.1(a) and (b). Denoting genotype by i, the connection between these two is given by

$$\gamma_i^{(1)} = \frac{\pi_i \gamma_i^{(\cdot)}}{\sum_i \pi_i \gamma_i^{(\cdot)}}, \quad \gamma_i^{(0)} = \frac{(1 - \pi_i)\gamma_i^{(\cdot)}}{\sum_i (1 - \pi_i)\gamma_i^{(\cdot)}},$$

so that

$$\frac{\gamma_i^{(1)}}{\gamma_i^{(0)}} = K \frac{\pi_i}{1 - \pi_i},$$

where K is constant across the three genotypes. Thus, ratios of genotype relative frequencies between people with and without disease are proportional to the corresponding odds of disease, and the odds ratio measures of association can also be written as

$$\theta_{AA}^* = \frac{\gamma_{AA}^{(1)}}{\gamma_{AA}^{(0)}} \bigg/ \frac{\gamma_{aa}^{(1)}}{\gamma_{aa}^{(0)}} = \frac{\gamma_{AA}^{(1)}}{\gamma_{aa}^{(1)}} \bigg/ \frac{\gamma_{AA}^{(0)}}{\gamma_{aa}^{(0)}}, \quad \theta_{Aa}^* = \frac{\gamma_{Aa}^{(1)}}{\gamma_{Aa}^{(0)}} \bigg/ \frac{\gamma_{aa}^{(1)}}{\gamma_{aa}^{(0)}} = \frac{\gamma_{Aa}^{(1)}}{\gamma_{aa}^{(1)}} \bigg/ \frac{\gamma_{Aa}^{(0)}}{\gamma_{aa}^{(0)}}.$$

The practical advantage of odds ratios is that the distributions $\gamma^{(1)}$ and $\gamma^{(0)}$ can be estimated from samples of persons with disease (cases) and of persons free of disease (controls). Particularly when the disease is rare, such studies are much more efficient than total population surveys.

When all penetrances are small, there is very little difference between relative risks and odds ratios. Further, the distribution of genotypes in disease-free subjects (controls) differs little from the population distribution (i.e. $\gamma^{(0)} \approx \gamma^{(\cdot)}$). This will normally be the case, although it may not be so for traits other than diseases.

Before moving on to discuss the design and analysis of population-based case-control studies, we should consider the case where we can assume that the population distribution of genotypes is in Hardy–Weinberg equilibrium (HWE). Then, denoting the relative frequencies of alleles A and a by $\alpha_A^{(\cdot)}$ and $\alpha_a^{(\cdot)}(= 1 - \alpha_A^{(\cdot)})$, respectively,

$$\gamma_{AA}^{(\cdot)} = (\alpha_A^{(\cdot)})^2, \quad \gamma_{Aa}^{(\cdot)} = 2\alpha_A^{(\cdot)}\alpha_a^{(\cdot)}, \quad \gamma_{aa}^{(\cdot)} = (\alpha_a^{(\cdot)})^2.$$

In general, if there is association between disease and genotype, this law will not hold within cases of disease or in controls (although in the latter case the discrepancy may be imperceptible for the reasons discussed above). For this reason, deviation from HWE in cases of disease is often taken as preliminary evidence for association. However, in one special case, association does not lead to deviation from HWE in cases. This occurs under the *multiplicative penetrance* model in which each copy of allele A multiplies risk by a factor ψ. Then $\theta_{Aa} = \psi$ and $\theta_{AA} = (\psi)^2$ and the distribution of genotypes in cases can be shown to follow the Hardy–Weinberg law, with modified allele frequencies:

$$\alpha_A^{(1)} = \frac{\psi\alpha_A^{(\cdot)}}{\alpha_a^{(\cdot)} + \psi\alpha_A^{(\cdot)}}, \quad \alpha_a^{(1)} = \frac{\alpha_a^{(\cdot)}}{\alpha_a^{(\cdot)} + \psi\alpha_A^{(\cdot)}}.$$

When disease is rare, $\alpha^{(\cdot)} \approx \alpha^{(0)}$ and ψ is very closely approximated by the odds ratio

$$\psi^* = \frac{\alpha_A^{(1)}}{\alpha_A^{(0)}} \bigg/ \frac{\alpha_a^{(1)}}{\alpha_a^{(0)}} = \frac{\alpha_A^{(1)}}{\alpha_a^{(1)}} \bigg/ \frac{\alpha_A^{(0)}}{\alpha_a^{(0)}}.$$

It seems most logical to refer to the parameters θ_{AA} and θ_{Aa} as *genotype relative risks* and the parameter, ψ, of the multiplicative model as a *haplotype relative risk*, although these terms have been used somewhat confusingly in the literature.

36.3 CASE-CONTROL STUDIES

In case-control studies of factors that might be associated with disease, the distributions of such factors are compared in a series of cases of the disease and in a control series. The fundamental assumption of such studies is that these two series of subjects may be used to provide unbiased estimates of the corresponding distributions amongst affected and unaffected members of some underlying population. This underlying population of interest is often called the *study base*. When this assumption is met, odds ratio measures of the strength of associations may be estimated, making use of the fact, demonstrated in Section 36.2, that odds ratios are stable whether defined in terms of the distribution of disease conditional upon risk factor or in terms of the distribution of risk factor conditional upon disease.

However, for many reasons, this fundamental assumption of the case-control study may not be met in practice, leading to biased findings. In the late 1960s and early 1970s there emerged an extensive literature cataloguing reasons for such bias. These fall into two broad classes:

1. *selection bias* caused by inappropriate sampling of cases and controls, and

2. *information bias* caused by differential measurement errors in cases and controls arising because measurements of risk factors are usually made when disease status is known to both subject and interviewer.

Studies of genetic association are not immune to either of these problems. For example, cases may be obtained by advertising nationally among clinicians treating the disease of interest, while controls may be recruited in the locality of the investigating centre. Such a design may easily introduce selection bias. A better procedure is to attempt to draw cases from a disease register that seeks to capture all cases of disease in a defined geographical area. The control group should then be a representative sample of the (disease-free) population of the same area. In most countries, however, there are no easy ways to draw such a sample. An alternative approach is to use matched designs; these will be discussed in Section 36.6.

Since genotype is invariant throughout life, genetic association studies are intrinsically less prone to information bias than are studies of environmental influences on disease. The only real danger is posed by poor laboratory procedures; if disease status of subjects is known to the technician who is scoring alleles, or if all cases and all controls are genotyped at different times, bias may result. Such bias is easily excluded by genotyping in random order, remaining blind to disease status.

Another problem in the interpretation of case-control studies has been termed *incidence/prevalence bias*. This is not really a bias at all but a failure to carefully define the nature of the disease variable. Thus, a table such as Table 36.1 could describe two rather different types of studies:

1. a cross-sectional study in which a population is surveyed at a fixed point in time and *current disease status* recorded, and

2. a longitudinal study in which a population of subjects, initially free of disease, are followed for a defined period and *new disease occurrence* recorded.

In the former case, π_{AA}, π_{Aa} and π_{aa} are disease *prevalences*, while in the latter case they are *incidences*. Incidence and prevalence are different quantities and may be influenced in different ways. Most importantly, a factor that is related to the duration for which a subject with disease remains in the population before death or migration will be reflected in prevalence even if it has no influence on incidence. Case-control studies based upon currently prevalent cases in a population measure effects of risk factors upon disease prevalence, while studies based around newly occurring cases in a defined period measure effects on disease incidence.

An example of the data arising from a case-control (incidence) study is shown in Table 36.2. The study was reported by Dunning *et al.* (1997) and concerns possible association of *common* polymorphisms in the *BRCA1* gene with breast cancer. Table 36.2(a) shows the frequency distribution of the Pro871Leu genotype in 800 cases and 572 controls. These frequencies can be used to estimate the corresponding population distributions and hence the odds ratios. Taking the Pro/Pro genotype as reference, the odds ratio estimates are

$$\hat{\theta}^*_{LL} = \frac{89}{342} \Big/ \frac{56}{266} = 1.236, \quad \hat{\theta}^*_{LP} = \frac{369}{342} \Big/ \frac{250}{266} = 1.148.$$

Since occurrence of breast cancer is a rare event, the odds ratios θ^*_{LL} and θ^*_{LP} closely approximate the corresponding genotype relative risks θ_{LL} and θ_{LP} in the population.

Table 36.2(b) shows allele frequencies in the 1600 chromosomes of cases and the 1142 chromosomes of controls. These can be used to estimate the allele distributions $\alpha^{(1)}$ and $\alpha^{(0)}$. Again, since breast cancer is rare, the allele frequencies in healthy controls are very

Table 36.2 Distributions of Pro871Leu polymorphism in the *BRCA1* gene yin breast cancer cases and in population controls.

Pro871Leu genotype	Subjects			Pro871Leu allele	Chromosomes	
	Case	Control			Case	Control
Leu/Leu	89	56		Leu	547	362
Leu/Pro	369	250		Pro	1053	782
Pro/Pro	342	266				
Total	800	572		Total	1600	1142
(a)				(b)		

nearly the same as in the population at large so that $\alpha^{(0)} \approx \alpha^{(\cdot)}$ and the haplotype relative risk may be estimated by the odds ratio

$$\hat{\psi}^* = \frac{547}{1053} \bigg/ \frac{362}{782} = 1.122.$$

The corresponding genotype relative risks, θ_{LL} and θ_{LP}, predicted by the multiplicative model are 1.254 and 1.122, respectively. The next section deals more formally with statistical inference concerning these parameters.

36.4   TESTS FOR ASSOCIATION

The statistical model underlying data such as those of Table 36.2(a) is that of two samples drawn from multinomial distributions. Denoting observed frequencies by f and genotype by the subscript i, the corresponding log likelihood is

$$\sum_i \left(f_i^{(1)} \log \gamma_i^{(1)} \right) + \sum_i \left(f_i^{(0)} \log \gamma_i^{(0)} \right).$$

By re-expressing the case genotype probabilities, $\boldsymbol{\gamma}^{(1)}$, in terms of $\boldsymbol{\gamma}^{(0)}$ and $\boldsymbol{\theta}^*$, this may be written as a function $\ell(\boldsymbol{\theta}^*, \boldsymbol{\gamma}^{(0)})$ of the control probabilities and the odds ratios. Since, by definition, $\theta_{aa}^* = 1$ and since $\sum_i \alpha_i^{(0)} = 1$, there are effectively only two free scalar parameters in each of the vectors $\boldsymbol{\theta}^*$ and $\boldsymbol{\gamma}^{(0)}$, the former being the parameter of interest and the latter being a 'nuisance parameter'.

There are two standard asymptotic tests of the hypothesis of no association, $H_0 : \boldsymbol{\theta}^* = \mathbf{1}$:

1. *The log-likelihood ratio (LLR) test.* Denoting the maximum- likelihood (ML) estimates of $\boldsymbol{\gamma}^{(0)}$ and $\boldsymbol{\theta}^*$ by $\hat{\boldsymbol{\gamma}}^{(0)}$ and $\hat{\boldsymbol{\theta}}^*$, and the ML estimate of $\boldsymbol{\gamma}^{(0)}$ under H_0 by $\hat{\hat{\boldsymbol{\gamma}}}^{(0)}$, this test statistic is defined as twice the difference between the corresponding log likelihoods:

$$2 \left[\ell(\hat{\boldsymbol{\theta}}^*, \hat{\boldsymbol{\gamma}}^{(0)}) - \ell(\mathbf{1}, \hat{\hat{\boldsymbol{\gamma}}}^{(0)}) \right].$$

2. *The score test.* If **u** represents the value of the vector of first derivatives of the log likelihood with respect to $\boldsymbol{\theta}^*$ evaluated at $\boldsymbol{\theta}^* = \mathbf{1}$ and $\boldsymbol{\gamma}^{(0)} = \hat{\boldsymbol{\gamma}}^{(0)}$, and **V** represents an estimate of its variance under H_0 (obtained by standard arguments from the matrix of second derivatives of the log-likelihood function), this test is defined by the quadratic form:

$$\mathbf{u}^t \mathbf{V}^\ominus \mathbf{u},$$

where $\ominus$ denotes a generalised inverse matrix.

Both of these statistics are asymptotically distributed as χ^2 on 2 degrees of freedom (df), and both can be expressed as simple functions of the observed frequencies, $f_i^{(1)}$ and $f_i^{(0)}$, and the corresponding 'expected' frequencies, $e_i^{(1)}$ and $e_i^{(0)}$, fitted under H_0:

$$\text{LLR test} = 2 \sum_i \left(f^{(1)} \log \frac{f^{(1)}}{e^{(1)}} + f^{(0)} \log \frac{f^{(0)}}{e^{(0)}} \right),$$

$$\text{Score test} = \sum_i \left[\frac{(f^{(1)} - e^{(1)})^2}{e^{(1)}} + \frac{(f^{(0)} - e^{(0)})^2}{e^{(0)}} \right].$$

The expected frequencies are calculated in the usual manner for tests of independence in contingency tables:

$$e_i^{(1)} = \frac{f_\cdot^{(1)} f_i^{(\cdot)}}{f^{(\cdot)}}, \quad e_i^{(0)} = \frac{f_\cdot^{(0)} f_i^{(\cdot)}}{f^{(\cdot)}}$$

(where $\cdot$ in subscript or superscript denotes summation). In our example of Table 36.2, the LLR and score tests are 2.056 and 2.055 respectively, corresponding to a P-value of approximately 0.36, so these tests do not suggest statistically significant association between breast cancer incidence and this polymorphism.

Asymptotic confidence intervals for estimates of odds ratios are provided by standard likelihood theory. It is normal to assume that the logarithm of an odds ratio estimate is asymptotically normally distributed with standard error estimated by the square root of the sum of reciprocals of the four frequencies used in the estimate. For example, for our estimate of θ_{LL} obtained in Section 36.3, the standard error of the log odds ratio is

$$\sqrt{\frac{1}{89} + \frac{1}{342} + \frac{1}{56} + \frac{1}{266}} = 0.189.$$

The approximate 95% confidence limits for the log odds ratio are $\log 1.236 \pm 1.96 \times 0.189$. These correspond to limits on the odds ratio of 0.85 and 1.79. Thus there is no suggestion that the Leu/Leu genotype is associated with increased risk of breast cancer.

Analysis of the chromosome count table, Table 36.2(b), follows very similar lines. However, for us to be able to legitimately ignore subject and treat chromosomes as independent observations, the assumption of HWE is essential. The test of the null hypothesis only requires that we assume HWE in the population since, under H_0, this ensures that both case and control distributions of genotype obey the Hardy–Weinberg law. However, for validity of the standard method of calculating a confidence interval for the odds ratio, we require the additional assumptions of (1) a rare disease (so that controls will be in HWE), and (2) the multiplicative model for penetrances (to ensure that cases are also in HWE).

For the data of Table 36.2(b), the LLR and score test statistics are 1.954 and 1.949 respectively. However, since these are tests of only a single parameter, they should be compared with the χ^2 distribution on 1 df, yielding a P value of approximately 0.16. The standard error of the log odds ratio estimate is

$$\sqrt{\frac{1}{547} + \frac{1}{1053} + \frac{1}{362} + \frac{1}{782}} = 0.0826.$$

Similar calculations as before lead to 95 % confidence limits for the haplotype relative risk of 0.95 and 1.32. As before, there is little evidence for association between breast cancer and this polymorphism.

Since the tests based on chromosomes has only 1 df, it must be expected to be more powerful that the 2 df tests – at least against alternative hypotheses, which are close to the multiplicative model. However, the need to assume HWE might be regarded as undesirable. If so, this assumption can be avoided by analysing the full genotype data of Table 36.2(a) using the multiplicative model

$$\theta_{LP}^* = \psi^*, \quad \theta_{LL}^* = (\psi^*)^2.$$

ML estimation of the parameter of this model can be carried out using logistic regression (see Section 36.5) and, by comparing maximised log likelihoods under null and alternative hypotheses, LLR tests are readily computed using this method. The score test of $H_0 : \psi^* = 1$ is also relatively well known, being equivalent to the Cochran–Armitage χ^2 test (1 df) for trend in the proportion of cases across rows of Table 36.2(a) (Armitage, 1955). In our example, the ML estimate of ψ^* is 1.125 and the standard error of its logarithm is 0.834, leading to confidence limits of 0.95 and 1.32 for ψ^*. The LLR test statistic is 1.991, while the score test takes the value 1.984. These results agree quite closely with those obtained by analysis of the chromosome counts (Table 36.2b).

Although the resistance of this approach to deviations from HWE is to be welcomed, such deviations cannot be entirely ignored. Deviation from HWE constitutes evidence of population stratification and, when this is present, there is a danger that such stratification could confound the association between disease and genotype. We will return to this problem in Section 36.7.

Just as there is a case for relaxing the HWE assumption while carrying out haplotype relative risk analyses based on the multiplicative model, there can be a case for making an HWE assumption when carrying out an analysis of genotype relative risks. The first three columns of Table 36.3 shows some data (also drawn from Dunning *et al.*, 1997) concerning another polymorphism in BRCA1.

The 2 df LLR test statistic is 11.718 and the (perhaps implausible) suggestion is that the homozygous Arg/Arg genotype is protective against breast cancer. However, because the Arg allele is uncommon, the frequency of the Arg/Arg genotype is very low in controls and, therefore, poorly estimated. Lathrop (1983) pointed out that the power of the 2 df test could be improved by assuming HWE in the population. With this assumption, and assuming a rare disease, both cases and controls will be in HWE at the null hypothesis, while under the alternative hypothesis only controls will be in HWE. The remaining columns show the expected frequencies under these assumptions. The LLR

Table 36.3 Distributions of Gln356Arg polymorphism in the *BRCA1* gene in breast cancer cases and in population controls, with expected frequencies assuming HWE.

Gln356Arg genotype	Observed		Expected (H_0)		Expected (H_1)	
	Case	Control	Case	Control	Case	Control
Arg/Arg	0	7	2.80	2.31	0	3.07
Arg/Gln	81	74	87.00	71.77	81	81.86
Gln/Gln	684	550	675.19	556.92	684	546.07

test for association compares the log likelihoods for these two fitted models and has 2 df. In this example this test statistic is 7.94 – a rather less extreme value than obtained with the simple 2 df test, which is inflated by the (probably chance) excess of Arg/Arg genotypes in controls.

The tests described so far in this section rely on asymptotic approximations which may be rather poor when there are small cell frequencies such as in Table 36.3. However, exact tests may be easily computed. The natural probability model is that the distribution of genotypes (or alleles) for case and control subjects (chromosomes) follow two multinomial distributions, which are identical under H_0. However this model is unsatisfactory for computing the null distribution of test statistics, owing to unknown 'nuisance' parameters – the common genotype (allele) population relative frequencies. This problem can be avoided by arguing conditionally upon the marginal genotype (allele) frequencies. The tables then follow hypergeometric distributions under H_0. This is the argument used in the construction of Fisher's exact test for 2×2 tables such as Table 36.2(b), but the same argument may be used to calculate exact P values for any of the statistics discussed above. In practice, complete evaluation of hypergeometric distributions is often too demanding and a simulation approach must be used. This proceeds by assigning the observed genotypes to cases and controls at random, recalculating the test statistic each time and counting the proportion of the time the statistic exceeds its value for the observed data. Random assignment of genotypes amongst subjects can be carried out in two ways. For a total of N subjects, we might

1. assign the N observed genotypes to cases and controls at random, or

2. assign the $2N$ observed alleles to cases and controls with no consideration of the pairing of alleles in the original genotypes.

The second of these approaches yields the P value under the HWE assumption, while the first yields a P value that makes no such assumption; this is more generally appropriate, and has the additional advantage of being rather easier to compute.

One further test should be discussed. This is a 1 df test in which the alternative model assumes dominance of one allele. This is a conventional test for association in the 2×2 contingency table in which cases and controls are classified as carriers of the dominant allele or non-carriers (i.e. homozygous for the recessive allele). In the absence of prior knowledge, this strategy will require two tests to be carried out since either allele could be dominant.

The question is, which test should be used? The 1 df tests based on the dominance model is frequently used in the study of single gene, or 'Mendelian' disorders while the

1 df multiplicative Cochran–Armitage test is more often preferred in studies of 'complex' disorders. This largely follows empirical experience and, perhaps, reflects the different mechanisms that apply in these settings. An additional consideration is that, when the polymorphism studied is not itself causal but related to disease via linkage disequilibrium with a nearby causal variant, dominance tends to be masked and the observed risk relationship becomes closer to the multiplicative penetrance prediction. Many authors have compared the power of these various tests under different model assumptions. Except in the case where the minor allele frequency is low, so that there are few subjects homozygous for the minor allele, the 2 df test against unrestricted alternatives behaves reasonably well against all alternatives.

36.5 LOGISTIC REGRESSION AND LOG-LINEAR MODELS

In Section 36.4 it was mentioned that maximisation of likelihoods for some models could be achieved by use of a computer program for logistic regression. However, since the 'response' variable measured in cases and controls is a genotype, which has *three* possible values, the relevance of logistic regression is not immediately apparent. This section explains this and also shows how another standard statistical technique, log-linear modelling, may be used to reproduce all the analyses described in Section 36.4. This will be useful for the extensions of this methodology to be discussed in later sections.

Although the natural way to model case-control data is in terms of probability distributions of genotype conditional upon disease status, reflecting the manner in which the data are generated, it has been shown that identical results are obtained from a likelihood-based analysis in which case-control status is regarded as the random outcome (Prentice and Pyke, 1979). The response variable is taken as the proportions of subjects with each genotype who are cases, and this is related via a logistic regression model to a design matrix expressing the precise model to be fitted. If π_i^* represents the probability that a subject, drawn at random from those subjects in the case-control study with genotype i, is a case. The logistic regression model is

$$\log \frac{\pi_i^*}{1 - \pi_i^*} = \mathbf{x}_i^{\mathrm{T}} \beta,$$

where $\mathbf{x}_i^{\mathrm{T}}$ is the ith row of the design matrix. With suitable choice of design matrix, the regression coefficients, β, are the logarithms of the odds ratio parameters discussed in Section 36.2.

Table 36.4 illustrates how two analyses of the data of Table 36.2 could be carried out using logistic regression. The first design matrix fits a fully saturated model and corresponds to the conventional 2 df test for association in the 3×2 contingency table. The coefficients of g_1 and g_2 in the model are the logarithms of the odds ratios θ_{LP}^* and θ_{LL}^* respectively, and the 2 df LLR test can be carried out by dropping g_1 and g_2 from the model. The second model is the multiplicative model in which $\theta_{\mathrm{LL}}^* = (\theta_{\mathrm{LP}}^*)^2$. The coefficient of g in this model corresponds to $\log \psi^*$ in our earlier notation, and the 1 df LLR test for trend can be carried out by dropping g from the model. The dominance

Table 36.4 Logistic regression of the Pro871Leu data.

Observed proportion	Design matrices				
	2 df model			1 df model	
	Constant	g_1	g_2	Constant	g
89/145	1	0	1	1	2
369/619	1	1	0	1	1
342/606	1	0	0	1	0

model can also be fitted in logistic regression, simply by coding the indicator g as 1 for carriers of the dominant allele and 0 otherwise (not shown).

This approach argues conditionally upon genotype and hence it is impossible to incorporate the HWE assumption, which is a model for the distribution of genotypes. A more flexible, but rather less convenient, alternative – log-linear modelling – allows a wider class of models to be fitted. In this approach, frequencies in a contingency table are assumed to be distributed as Poisson variates, the logarithms of their expected values obeying a linear model. Again it can be shown that this approach leads to likelihood inferences for odds ratios that are identical to those obtained under the assumption of multinomial distributions of genotypes conditional upon disease status.

Table 36.5 demonstrates how log-linear modelling can be used to reproduce two analyses discussed in Section 36.4. The cell frequencies of Table 36.2 are the response variable, and the first design matrix represents a saturated model in which disease status, genotype and their interaction are included. The parameters representing interaction in this model are the coefficients of $d.g_1$ and $d.g_2$ and these are the logarithms of the odds ratios θ_{LP}^* and θ_{LL}^* respectively. Dropping both these variables from the model provides the 2 df test for association between disease and genotype.

The second design matrix listed in Table 36.5 is a more restrictive model which assumes (1) the multiplicative mode for odds ratios, and (2) HWE equilibrium in both cases and controls. Note that this model must include an 'offset' (equivalent to the inclusion of a variable whose coefficient is constrained to take the value 1). Use of this design matrix reproduces the simple analysis of the 2×2 table of chromosome counts; the coefficient of the variable $d.g$ (the disease–genotype interaction term) is the odds ratio in this 2×2

Table 36.5 Log-linear modelling of the Pro871Leu data.

Observed frequency	Design matrices										
	2 df model						1 df + HWE model				
	Constant	d	g_1	g_2	$d.g_1$	$d.g_2$	Offset	Constant	d	g	$d.g$
89	1	1	0	1	0	1	0	1	1	2	2
369	1	1	1	0	1	0	$\log 2$	1	1	1	1
342	1	1	0	0	0	0	0	1	1	0	0
56	1	0	0	1	0	0	0	1	0	2	0
250	1	0	1	0	0	0	$\log 2$	1	0	1	0
266	1	0	0	0	0	0	0	1	0	0	0

table. An advantage of fitting the model to the full table of genotype counts rather than to the collapsed table of chromosome counts is that this provides tests of fit of modelling assumptions that are implicit in the latter analysis. An additional advantage is that the alternative analyses discussed in Section 36.4 can also be carried out quite easily. Thus, if the columns $d.g_1$ and $d.g_2$ in the first matrix are replaced by the single column $d.g$, the model assumes the multiplicative (1 df) model but does not assume HWE. Conversely, if the single column $d.g$ in the second matrix is replaced by the two columns $d.g_1$ and $d.g_2$, we have Lathrop's model which assumes HWE in controls but allows 2 df for disease–genotype association. As before, dominance models may also be considered with appropriate coding of the indicator g.

36.6   STRATIFICATION AND MATCHING

In the introduction to this chapter, the problem of confounding was briefly discussed; association between disease and genotype might be attributable to a third variable, not on a causal path between genotype and disease, which is independently related to both variables. In genetic epidemiology, most concern is usually related to the possibility of confounding by admixture or ethnic stratification of populations, but there are other possibilities. For example, a gene that is related to the tendency to smoke cigarettes (and such genes have been found) may be associated with lung cancer. In one sense such a gene is indeed a cause of lung cancer but, in the context of the question as to whether the gene is directly involved in the biology of cancer, cigarette smoking would be more likely to be regarded as a confounder.[1]

When the confounding variable has been measured in the study, it is relatively straightforward to deal with the problem during analysis. The classical method in epidemiology is by *stratification* of the analysis by the potentially confounding variable and testing for association between factor of interest (here genotype) and disease *within strata*. Figure 36.1 illustrates this idea.

Each stratum (ethnic group, for example) contributes a contingency table of the form discussed in Section 36.4 – either a 3×2 table of counts of subject by genotype or a 2×2 table of counts of chromosomes by allele.

Since each stratum, taken alone, may contain insufficient data to test for association, it is necessary to pool the evidence for association over strata. An obvious way of doing this

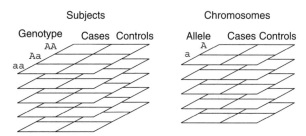

Figure 36.1   Stratified analysis of case–control studies.

---

[1] This example also illustrates why the concept of confounding is rather difficult to define precisely, since this depends crucially on the nature of causality.

would be to simply add the χ^2 values over strata, remembering to also add their degrees of freedom. However, the proliferation of degrees of freedom with this approach indicates that it is based on a very flexible model for association, with either two or one different parameter for each stratum. Since all these parameters will be imprecisely estimated, this approach must be expected to lack power.

An alternative approach is to adopt the model in which the parameters measuring association between genotype and disease are assumed to be constant across strata, although the distributions of both disease and genotype are allowed to vary among strata. With this model as the alternative hypothesis, tests of association have either 1 or 2 df. In the case of the 1 df tests, the score tests are well known and can be carried out using hand calculator. The test for association in a stack of 2×2 tables of allele counts is the Mantel–Haenszel test and an easily calculated estimate of their common odds ratio, ψ^*, is also available (Mantel and Haenszel, 1959). If we wish to avoid the HWE assumption, the 1 df score test is the stratified test for trend in proportions in the stack of 3×2 tables of genotype counts (Mantel, 1963); this is often known as the *Mantel extension test*. The 2 df score test for the stack of 3×2 tables is rather more difficult to calculate and is not explicitly described in the literature. The more general case of the $k - 1$ df test in a stack of $k \times 2$ tables has been described, but calculation of the variance of the score vector requires matrix inversion. An approximate procedure is to (1) calculate expected frequencies under the null hypothesis in each stratum separately, (2) add observed and expected frequencies over strata and (3) calculate a chi-squared test in this margin using the usual formula $\sum (O - E)^2 / E$. This approximation can be shown to be conservative – the value of χ^2 obtained this way is always slightly smaller than that obtained if the variance–covariance matrix of the score vector is calculated correctly.

LLR tests are more difficult to calculate than score tests, since the ML estimates of common odds ratios require iterative computation. However, the logistic regression and log-linear modelling approaches of Section 36.5 are easily extended to encompass stratified analyses.

Since logistic regression argues conditionally on genotype, this method explicitly allows association between stratum and genotype and to control for confounding it is necessary only to allow disease status to depend on stratum by including stratum in the regression model as a categorical variable (or 'factor'). If there are S strata, this introduces a further $S - 1$ parameters in the regression model. Thus the model of no association between disease and genotype within strata has S parameters, and the alternative model introduces a further 1 or 2 parameters.

Log-linear modelling is slightly more complicated since the relationship between genotype and stratum must be explicitly modelled; we require $S - 1$ parameters for the distribution of subjects between strata and, if HWE is assumed within strata, we require a further $S - 1$ parameters to model variation of allele relative frequencies amongst strata. As in the logistic regression approach, $S - 1$ parameters are needed to model association between disease status and stratum. Thus the model of no association between disease and genotype within stratum has a total of $3S$ parameters. Again, the alternative model introduces a further 1 or 2 parameters for association between disease and genotype. The design matrix for a two-stratum analysis is illustrated in Table 36.6.

Dropping the last column from the model produces a 1 df LLR test for association between genotype and disease within strata.

Table 36.6 Stratified analysis by log-linear modelling.

				Design matrix						
Stratum	Disease	Genotype	Offset	Constant	s	d	g	$s.d$	$s.g$	$(d.g)$
1	Case	AA	0	1	1	1	2	1	2	2
1	Case	Aa	log 2	1	1	1	1	1	1	1
1	Case	aa	0	1	1	1	0	1	0	0
1	Control	AA	0	1	1	0	2	0	2	0
1	Control	Aa	log 2	1	1	0	1	0	1	0
1	Control	aa	0	1	1	0	0	0	0	0
2	Case	AA	0	1	0	1	2	0	0	2
2	Case	Aa	log 2	1	0	1	1	0	0	1
2	Case	aa	0	1	0	1	0	0	0	0
2	Control	AA	0	1	0	0	2	0	0	0
2	Control	Aa	log 2	1	0	0	1	0	0	0
2	Control	aa	0	1	0	0	0	0	0	0

Computation of exact P values based on any of these test statistics may be obtained by simulation, randomly assigning genotypes to subjects within strata while ignoring disease status.

The use of regression models to control for confounding has much to recommend it, since such approaches extend without difficulty to deal with several potential confounders. For this reason, their use has become widespread in epidemiology. But whether analysis is by regression or by the more classical Mantel–Haenszel methods, it should be noted that the need to control confounding may result in some loss in precision of the study if the stratifying variable is strongly related to disease. In this case, some strata may have many fewer controls than cases, while others may have many fewer cases than controls. An optimal design would maintain the ratio of cases to controls across strata, and a study that is carried out so that this is so is called a *group matched* study.

Within a group matched study, there is no relationship between stratum and disease status and for some time it was believed that the matched design eliminates confounding so that the analyses discussed in this section are no longer necessary. Unfortunately this is not the case, owing to the non-collapsibility of the odds ratio as a measure of association; even when the distribution of disease status is constant over strata, and when the odds ratios are constant across strata, the odds ratios in the marginal table are closer to one than are the stratum-specific odds ratios. Likewise, the test for association in the marginal table is incorrect. Thus it is still necessary to include stratum effects when modelling data from a matched study. The only exception to this rule occurs in the degenerate case in which the distribution of genotype also does not vary among strata.

At the limit, matching can be so fine that each single case has its own control(s). This is an *individually matched* case-control study. Although, in principle, the analysis of such studies is the same as above, we encounter a new technical difficulty; since we must include stratum effects in the model and since we introduce a new stratum with each new case, the number of parameters in the model increases just as fast as the total study size. Unfortunately, the asymptotic properties of likelihood inference in logistic regression break down in these circumstances. The solution is to use *conditional* logistic regression – a variant of logistic regression analysis in which the parameters expressing stratum effects are eliminated from the likelihood by use of a conditional argument rather than by attempting to estimate them. Before concluding the topic of individually matched studies, it should be noted that an important special case occurs when the matching is by nuclear family so that cases are compared with unaffected sibling controls. This type of study is the 'sib TDT' study described by Spielman and Ewens (1998); such studies and their analysis are discussed in **Chapter 38** of this handbook.

We have seen that matching controls to cases at the design stage does not simplify analysis. It has also been shown elsewhere that the gains in efficiency achieved by matching are modest except when there is very strong confounding. These two facts might lead us to question its usefulness. However, there is another important motivation for matched designs. In Section 36.3 attention was drawn to the problem of selection bias in case-control studies and in particular to the lack of appropriate sampling frames for controls in most countries. However, matching can simplify the sampling problem. If controls are matched to cases, e.g. by general practitioner or by neighbourhood, it can be much easier to ensure that (1) all eligible cases within a given stratum are indeed captured in the study, and (2) that controls representative of the stratum are selected. It is these considerations rather than the desire for efficient control of confounding that accounts for the popularity of matching in case-control studies.

36.7 UNMEASURED CONFOUNDING

A criticism frequently levelled at case-control studies in genetic epidemiology is the possibility of confounding by *unmeasured* population stratification or admixture. But the possibility of unmeasured confounding is not unique to genetic epidemiology. Indeed, one of the most celebrated controversies of epidemiology surrounded Fisher's 'constitutional hypothesis', which suggested that unmeasured genetic factors could confound the observed association between smoking and lung cancer. Ultimately, such explanations can always be offered for observed associations. In general epidemiology, the only counter arguments are in terms of plausibility. An explanation in terms of unmeasured confounding must be plausible, both *biologically* and *quantitatively*. The quantitative plausibility of explanations based on unmeasured confounding has, not surprisingly, been the subject of much attention in the epidemiological literature, but has been largely ignored in the debate concerning population-based vs family-based association studies in genetic epidemiology. An early finding of such work was that putative confounders must be very strongly related to both disease and risk factor for their confounding effect to be appreciable (Bross, 1967).

To carry these arguments over into genetic epidemiology, it is most convenient to assume HWE and the multiplicative model for association, so that the theory can be expressed in terms of properties of 2×2 tables of chromosome counts. Let there be two hidden strata and assume that the odds ratios between disease status and allele is ϕ^* in each stratum. Then it can easily be shown that the *marginal* odds ratio is given by

$$\phi^* \left[\frac{\zeta p_A + (1 - p_A)}{\zeta p_a + (1 - p_a)} \right],$$

where p_A represents the proportion of control A alleles drawn from the high-risk stratum, p_a is the proportion of a alleles drawn from this stratum and ζ is the odds ratio between stratum and disease. The term in square brackets represents the factor by which the true odds ratio is inflated or deflated by confounding and has been termed the *confounding risk ratio* (Miettinen, 1972). Breslow and Day (1980) give examples of such calculations.

The problem of unmeasured confounding is no more serious when the factor of interest is a gene than when it is an environment exposure. Indeed it could be argued that it is considerably less serious, since it is possible to collect *evidence* for or against the existence of substantial admixture within a population. This can be checked by typing a number of genetic markers that are sufficiently distant from the locus of interest to be assumed to be in linkage disequilibrium with it. The presence of admixture is then indicated by deviation from HWE at each locus, by population associations amongst pairs of loci and, if disease risk differs among strata, by widespread association between genotype and disease – attributable to confounding. Although this possibility has been recognised for some time, formal methods of analysis have not been proposed until recently. Pritchard and Rosenberg (1999) propose testing for confounding by unobserved stratification by summation of χ^2 tests for association over markers. They then seek to adjust the P value for the candidate locus. A full description of approaches to modelling stratification and admixture is to be found in **Chapter 35** of this handbook.

A further approach, 'genomic control', is due to Devlin and Roeder (1999), who suggested that the effect of admixture and stratification is to inflate the 1 df Cochran–Armitage marker/disease association test statistics by a constant factor, which they designated λ. They showed that λ is expected to increase with the study size, essentially reflecting the increased sensitivity of large studies to hidden confounding, but argued that, given a large number of tests for markers widely spread on the genome, λ can be estimated empirically and the distribution of test statistic corrected. A simple estimate of λ is the ratio of the mean of the tests falling in the lower part of the distribution with its expectation under the χ^2 assumption. Devlin and Roeder suggested use of the smallest 50 % of test values, but others have suggested that 90 % can be used.

There are a number of difficulties with the idea of genomic control. First, although the type 1 error is corrected, this correction is achieved at the price of a loss of power to detect true associations. Secondly, there is an assumption of 'exchangeability' of markers; the estimate of λ will reflect inflation of tests due to small differences in allele frequencies at a large number of loci, due to 'genetic drift' at the point of separation of ancestral populations, However, some loci could have been under selection pressures, resulting in substantially larger differences. A final difficulty, although less serious given recent advances in high-throughput genotyping methods, is that accurate estimation of λ requires very large numbers of loci to be studied.

Devlin and Roeder (1999) discussed the case of 'cryptic Relatedness' between study subjects, leading to violation of the assumption of independent data points. They showed that this is serious when the relatedness is stronger between pairs of cases (and possibly pairs of controls) than it is between a case-control pair. Again, they suggested that the effect of this is to inflate the test statistics by a factor λ that increases with the size of the study. In principle, cryptic relatedness is no different from the stratification and admixture effects discussed by other authors. Cryptic relatedness is concerned with the effect of more recent coalescence of the genealogy of study subjects, while stratification represents ancient coalescence into a few ancestral populations. The difference is that, whereas the latter can be approached by modelling structure by using ancestry informative markers, cryptic relatedness will not be corrected by such an approach and genomic control remains the only remedy.

36.8 MULTIPLE ALLELES

The discussion to this point has been limited to the case of a diallelic genetic locus. This section discusses the case of a locus with $K > 2$ alleles. The most serious consequence that must be considered is the proliferation of possible genotypes that follows. A test based on genotype relative risks has $K(K + 1)/2 - 1$ df and lacks power even for quite modest values of K. However, the multiplicative model generalises naturally and allows tests based on haplotype relative risks. Denoting the genotype with alleles j and k by (j, k), the multiplicative model for genotype relative risks is

$$\theta_{(j,k)} = \psi_j \psi_k,$$

where $\psi_j, \psi_k (j, k = 1 \ldots K)$ are haplotype relative risks. One allele, usually the most common, is taken as reference so that the corresponding ψ_j takes the value 1.0 by definition. The global test of no association, $H_0 : \psi_j = 1, j = 1 \ldots K$ has $K - 1$ df. In the context of case-control studies, genotype and haplotype relative risks are closely approximated by the corresponding odds ratio parameters. In particular, taking allele 1 as reference, ψ_j is closely approximated by

$$\psi^* = \frac{\alpha_j^{(1)}}{\alpha_j^{(0)}} \Big/ \frac{\alpha_1^{(1)}}{\alpha_1^{(0)}} = \frac{\alpha_j^{(1)}}{\alpha_1^{(1)}} \Big/ \frac{\alpha_j^{(0)}}{\alpha_1^{(0)}}.$$

All of the methods discussed above in the case of a diallelic locus extend naturally. For example, the model that combines HWE and multiplicative assumptions can be fitted using the $K \times 2$ table of chromosome counts. Relaxation of the HWE assumption is most conveniently achieved by logistic regression. Denoting the probability that a subject, drawn at random from those subjects in a case-control study with genotype (j, k), is a case by $\pi_{(j,k)}^*$, the multiplicative model corresponds with the logistic regression model

$$\log \frac{\pi_{(j,k)}^*}{1 - \pi_{(j,k)}^*} = \beta_0 + \beta_j + \beta_k.$$

The design matrix has K columns, the first column being the unit vector, and the elements of remaining columns $j = 2 \ldots K$ taking values 0, 1 or 2 reflecting the number of times allele j occurs in each genotype. The dominance model is also readily extended to the multiple allele case, although there are more versions of the model to consider (since various combinations of alleles could be dominant). For tests for indirect association, the model of multiplicative effects of alleles is the one most frequently used.

Tests for association can be calculated by using logistic regression to calculate likelihood ratio tests. The score test, which generalises the Cochran–Armitage test, is essentially Hotelling's T^2 test, comparing the vector of means for the allele indicator variables between cases and controls (Xiong *et al.*, 2002; Chapman *et al.*, 2003; Fan and Knapp, 2003). To deal with observed potential confounders, the regression model can be extended by entering them into the regression model. The equivalent generalisation of the score test is a stratified version of the Hotelling's T^2 statistic.

The multiplicative model may also be defined by the property that the genotype relative risk for a heterozygous genotype is the geometric mean of the two homozygous genotypes defined by its alleles. In terms of odds ratios,

$$\theta^*_{(j,k)} = \sqrt{\theta^*_{(j,j)}\theta^*_{(k,k)}}.$$

A less restrictive approach, which nevertheless avoids the large number of degrees of freedom of the test between genotype frequencies, is to assume, in the case of a diallelic locus, only that the heterozygous genotype relative risk falls in the interval bounded by the two homozygous relative risks. The model could be fitted under these order constraints, but the asymptotic distribution of the LLR test statistic would be complex. An alternative approach is suggested by noting that, under the multiplicative model, the effects of the alleles on the two chromosomes are additive on the log odds scale. This can be extended by assuming additivity of effects *on some other scale*. The problem of the scale on which two effects are additive has received some attention in the epidemiological literature, since it is central to the discussion of *synergism* of risk factors.

A natural approach is suggested by generalised linear models (Nelder and Wedderburn, 1972); we may replace the logit transformation of the probabilities π^* by a more general 'link function', $g\left(\dfrac{\pi^*}{1-\pi^*}\right)$:

$$g\left(\frac{\pi^*_{(j,k)}}{1-\pi^*_{(j,k)}}\right) = \beta_0 + \beta_j + \beta_k,$$

a convenient choice of link function being the Box–Cox transformation of the odds (Box and Cox, 1964)

$$g(x) = \frac{x^\rho - 1}{\rho}, \quad \rho \neq 0,$$

$$= \log x, \quad \rho = 0.$$

This extension to the logistic regression model was proposed by Guerrero and Johnson (1982). In the present context, as $\rho \to \infty$, $\pi^*_{(j,k)}$ tends to the larger of $\pi^*_{(j,j)}$ and $\pi^*_{(k,k)}$, while as $\rho \to -\infty$ it tends to the smaller of these values. These extreme cases generalise the ideas of dominant and recessive inheritance respectively. When $\rho = 0$ the model reverts to the simple multiplicative model. For given ρ, this link function is either available or easily implemented in most generalised linear modelling programs. An important property of the Box–Cox transformation is that it is continuous in ρ through $\rho = 0$ so that, in principle, ρ can be treated as an extra model parameter and estimated by ML. In practice, it is more usual to repeat the analysis on a grid of values for ρ and to plot the resultant profile log likelihood. Note, however, that the usual asymptotic theory does not hold for LLR tests for association based upon maximisation with respect to ρ, since the likelihood is flat with respect to ρ under H_0.

A disadvantage of the above approach is that the effect parameters β_j, β_k are not invariant under case-control sampling and, following Breslow and Storer (1985), other authors have concentrated on generalised *relative* risk models, which, in the present context, take the form

$$\log \frac{\pi^*_{(j,k)}}{1 - \pi^*_{(j,k)}} = \beta_0 + \log \theta^*_{(j,k)},$$

$$h(\theta^*_{(j,k)}) = \beta_j + \beta_k.$$

With this model, a heterozygous genotype relative risk is a generalised mean of the two homozygous genotype relative risks:

$$\theta^*_{(j,k)} = h^{-1} \left[\frac{h\left(\theta^*_{(j,j)}\right) + h\left(\theta^*_{(k,k)}\right)}{2} \right].$$

Moolgavkar and Venzon (1987) have reviewed such approaches, criticising some earlier proposals on the grounds that the functional form of the model is not maintained under change of reference category. Their preferred choice of relative risk function is equivalent to the approach of Guerrero and Johnson (1982).

A widely used alternative to the $K - 1$ df test is to carry out K separate 1 df tests, each one focusing on a specific allele and combining all other alleles. This is an efficient approach when it is reasonable to assume that association is limited to a single allele, but it is necessary to correct the P value for multiple testing. However, this is not a straightforward matter since the tests are not independent. In the context of TDT tests this problem was discussed by Morris *et al.* (1997), who pointed out that the P value for the largest of the K correlated 1 df test statistics may be calculated by simulation from the randomisation distribution. The same approach can be applied in population case-control studies.

36.9   MULTIPLE LOCI

When the polymorphism of interest is not by itself functional but may be in linkage disequilibrium with a causal locus, it is sometimes easier to demonstrate association between disease and this region of the genome by using a more polymorphic marker. For example, a polymorphic marker provides a more detailed discrimination between haplotypes present in the population, with a correspondingly better chance of identifying the haplotype(s) carrying a causal variant. But it may happen that there are no highly polymorphic markers in the region of interest. However, single nucleotide polymorphisms (SNPs), which are diallelic, occur very frequently throughout the human genome, and an alternative to the use of a single marker is to relate disease to a *haplotype* formed by several closely linked SNPs. Even when a polymorphic marker is available, recent advances in genotyping technology have been such that, usually, typing several SNPs will be preferred. In this section, for simplicity, diallelic loci are assumed.

In general, the analysis of association involving extended haplotypes brings new possibilities and difficulties. However, in one case, the analytical problem becomes identical to that explored in the previous section. This is the case where K loci to be tested are in 'complete' linkage disequilibrium, defined by all pair-wise values of Lewontin's D' measure of linkage disequilibrium equal to 1. In this case, the markers represent K mutations, with no mutation occurring more than once and no recombination occurring in the region in the population history. In this case, only $K + 1$ different haplotypes are observed and the collection of markers behave in exactly the same way as a single polymorphic marker with $K + 1$ alleles. In this case, there is no phase uncertainty when resolving haplotypes (*i.e.* when assigning alleles along each chromosome). As an illustration, the logistic regression analysis of the model of multiplicative effects is achieved by regressing the disease status binary variable on indicator variables for each diallelic marker genotype, coded 0, 1 or 2. Similarly, the Hotelling's T^2 test with the same alternative model compares the vectors of means of these genotype scores for cases and controls.

Chapman *et al.* (2003) proposed that, when the pair-wise D' values are high but fall somewhat short of 1, this testing strategy can still be preferred to more detailed reconstruction of haplotypes. This is because there is little power to detect association with rare recombinant haplotypes, so that the expenditure in the test of the extra degrees of freedom that they require is not justified. With the availability of very large numbers of SNP markers and falling typing costs, it is likely that this situation will become commonplace. In the presence of more common recombinant haplotypes, however, full haplotype-based approach may be justified.

The main difficulty with haplotype-based analyses is that haplotypes are not directly observed owing to phase uncertainty. Family-based studies of association are not immune to this problem (Clayton, 1999), but studies based upon single individuals may be more seriously affected. If, in such a study, a subject is heterozygous at H of the loci considered, the observed genotype data are consistent with 2^H possible haplotype assignments. Methods based upon likelihood can be extended to deal with this, most conveniently by use of the EM algorithm. Each observed ambiguous genotype is expanded into all possible phases and, at the E step, its total observed frequency is divided between the possible phases according to their posterior probability, assuming HWE and current estimates of haplotype probabilities. A high-dimensional contingency table of imputed haplotypes by disease status can then be calculated, and the log-linear model fitted to this table (M step). However, some words of caution are necessary. Firstly, it is well known that the likelihood does not always have a unique maximum. When there are multiple maxima, the EM algorithm will converge to a local maximum and this may not be the global maximum. It is usually wise to repeat the model fit starting from different initial estimates. A second consequence of irregular likelihood surfaces is that LLR tests may not conform well to standard asymptotic theory. Nevertheless, exact P-values may be obtained by simulation by randomly permuting case-control status between subjects in the study (although this may be rather demanding in terms of computer time).

Score tests are generalised rather more easily to the case of unknown phase. The value of the 'score' vector, which would have been tested if the phase had been observed, is replaced by its posterior distribution of phase assignments calculated under the null hypothesis (Schaid *et al.*, 2002).

There is an extensive literature on further haplotype-based approaches to analysis that attempt to use estimated ancestral relationships between haplotypes to increase the power to detect and localise causal variants. These may be informal, and based on 'cladistic' classifications of haplotypes (see, e.g. Seltman *et al.*, 2001) or on more formal approximate models for ancestral recombination graphs. For detecting rare variants, several authors have proposed tests based upon extended haplotype sharing in cases (e.g. Tzeng *et al.*, 2003).

36.10   DISCUSSION

The discussion of previous sections has demonstrated that the analysis of population-based association studies can involve many of the statistical methods that have been developed for the analysis of multivariate categorical data. The main factors that give such analyses a distinctive flavour are

1. the fact that chromosomes are paired in subjects, and the implications of the HWE assumption (or its avoidance), and

2. the problem of unknown haplotype phase.

The latter problem, in particular, presents some challenging technical difficulties.

Most of this chapter has been concerned with inference from case-control studies under the 'rare disease' assumption. Although many of the methods described are applicable for general binary traits, those methods requiring the HWE assumption (in order, e.g. to resolve unknown haplotype phase) are more problematic. The log-linear models discussed here assume HWE conditional upon disease status, and this can only be expected to hold under the rare disease assumption or under the hypothesis of no association. In other cases, it would only be legitimate to assume HWE *marginally*. Since HWE holds under the null hypothesis, the size of tests constructed under conditional HWE models would be expected to be correct, but LLR and score tests may not have optimal properties and parameter estimation will be incorrect. For data from representative population samples, models for the joint distribution of genotype and disease phenotype must be specified in terms of the factorisation

$$\text{Prob (disease|genotype)} \times \text{Prob (genotype)},$$

where the HWE assumption is introduced in the second term. Because case-control sampling distorts the marginal distribution of genotypes, these studies require the rare disease assumption, which predicts the *conditional* assumption of HWE.

Lack of space prevents discussion of population association between quantitative traits and genotype. However, such a discussion would largely parallel that for discrete traits. Regression methods based on distribution of trait conditional upon genotype would usually be approached by assuming (or transforming to) normality of conditional distributions and using the classical (Gaussian) linear model. Proliferation of degrees of freedom may be avoided by assuming additivity of haplotype effects (i.e. zero dominance variance), and there is scope for extending such models by the incorporation of flexible 'link' functions. For methods that require us to assume HWE, the approach outlined in the previous paragraph will be satisfactory for the analysis of data drawn from representative population samples. However, it will be more difficult to incorporate such assumptions if sampling has been 'response based', e.g. if extreme trait values have been deliberately oversampled. There is scope for further work in this area.

Acknowledgments

The author is supported by a Welcome Trust Principal Research Fellowship.

REFERENCES

Armitage, P. (1955). Test for linear trend in proportions and frequencies. *Biometrics* **11**, 375–386.

Box, G. and Cox, D. (1964). An analysis of transformations. *Journal of the Royal Statistical Society Series B* **26**, 211–252.

Breslow, N. and Day, N. (1980). *Statistical methods in cancer research. volume I – the analysis of case-control studies*. IARC scientific publications. IARC, Lyon.

Breslow, N. and Storer, B. (1985). General relative risk functions for case-control studies. *American Journal of Epidemiology* **122**, 149–162.

Bross, I. (1967). Pertinency of an extraneous variable. *Journal of Chronic Diseases* **20**, 487–497.

Chapman, J.M., Cooper, J.D., Todd, J.A. and Clayton, D.G. (2003). Detecting disease associations due to linkage disequilibrium using haplotype tags: A class of tests and the determinants of statistical power. *Human Heredity* **56**, 18–31.

Clayton, D. (1999). A generalization of the transmission/disequilibrium test for uncertain haplotype transmission. *American Journal of Human Genetics* **65**, 1170–1177.

Clayton, D. and Hills, M. (1993). *Statistical models in epidemiology*. Oxford University Press, Oxford.

Devlin, B. and Roeder, K. (1999). Genomic control for association studies. *Biometrics* **55**, 997–1004.

Dunning, A., Chiano, M., Neil, R., Dearden, J., Gore, M., Oakes, S., Wilson, C., Stratton, M., Peto, J., Easton, D., Clayton, D. and Ponder, B. (1997). Common *BRCA1* variants and susceptibility to breast and ovarian cancer in the general population. *Human Molecular Genetics* **6**, 285–289.

Fan, R. and Knapp, M. (2003). Genome association studies of complex diseases by case-control designs. *American Journal of Human Genetics* **72**, 850–868.

Guerrero, V. and Johnson, R. (1982). Use of the Box-Cox transformation with binary response models. *Biometrika* **69**, 309–14.

Lathrop, G. (1983). Estimating genotype relative risks. *Tissue Antigens* **22**, 160–166.

Mantel, N. (1963). Chi-square tests with one degree of freedom: extension of the Mantel-Haenszel procedure. *Journal of the American Statistical Association* **58**, 690–700.

Mantel, N. and Haenszel, W. (1959). Statistical aspects of the analysis of data from retrospective studies of disease. *Journal of the National Cancer Institute* **22**, 719–48.

Miettinen, O. (1972). Components of the crude risk ratio. *American Journal of Epidemiology* **96**, 168–172.

Moolgavkar, S. and Venzon, D. (1987). General relative risk regression models for epidemiological studies. *American Journal of Epidemiology* **126**, 949–961.

Morris, A., Curnow, R. and Whittaker, J. (1997). Randomization tests of disease-marker associations. *Annals of Human Genetics* **61**, 49–60.

Nelder, J. and Wedderburn, R. (1972). Generalized linear models. *Journal of the Royal Statistical Society Series A* **135**, 370–384.

Prentice, R. and Pyke, R. (1979). Logistic disease incidence models and case-control studies. *Biometrika* **66**, 403–411.

Pritchard, J. and Rosenberg, N. (1999). Use of unlinked genetic markers to detect population stratification in association studies. *American Journal of Human Genetics* **65**, 220–228.

Schaid, D., Rowland, C., Tines, D., Jacobson, R. and Poland, G. (2002). Score tests for association between traits and haplotypes when linkage phase is ambiguous. *American Journal of Human Genetics* **70**, 425–434.

Seltman, H., Roeder, K. and Devlin, B. (2001). Transmission/disequilibrium test meets measured haplotype analysis: family-based association analysis guided by evolution of haplotypes. *American Journal of Human Genetics* **68**, 1250–1263.

Spielman, R. and Ewens, W. (1998). A sibship test for linkage in the presence of association: the sib transmission disequilibrium test. *American Journal of Human Genetics* **62**, 450–458.

Tzeng, J.-Y., Devlin, B., Roeder, K. and Wasserman, L. (2003). On the identification of disease mutations by the analysis of haplotype matching and goodness-of-fit. *American Journal of Human Genetics* **72**, 891–902.

Xiong, M., Zhao, J. and Boerwinkle, E. (2002). Generalized t^2 test for genome association studies. *American Journal of Human Genetics* **70**, 1257–1268.

37

Whole Genome Association

A.P. Morris and L.R. Cardon

Wellcome Trust Centre for Human Genetics, University of Oxford, Oxford, UK

Whole genome association (WGA) studies have been widely recognised as having great potential to identify genetic polymorphisms contributing to complex human diseases. With recent advances in single nucleotide polymorphism (SNP) genotyping technology, WGA studies using $> 10^5$ markers are being undertaken by many research groups worldwide with samples large enough to detect the modest genetic effects we expect for complex diseases. In this chapter, we review the key issues for the analysis of data from population-based WGA studies, building on the concepts introduced by Clayton (**Chapter 36**). We briefly discuss design issues and describe how to assess genotype quality from WGA genotyping technology. We discuss techniques for single-locus analysis and appropriate corrections that can be made to allow for multiple testing of the thousands of SNPs used in WGA studies. We describe how these methods could be extended to allow for environmental and other non-genetic risk factors, multiple SNPs, haplotypes, and epistasis. Finally, we emphasise the importance of replication of the results from WGA studies and discuss the prospects of this approach for complex disease gene mapping.

37.1 INTRODUCTION

The traditional approach to mapping disease genes has been linkage analysis, which studies the co-segregation of marker alleles with disease within large pedigrees or smaller family units such as affected sib pairs (see **Chapters 33** and **34**). This approach has proved to be successful for locating genes contributing to simple Mendelian disorders such as cystic fibrosis and Huntington's disease, where there is a clear relationship between phenotype and genotypes at the underlying functional polymorphism(s). However, linkage studies have proved less reliable for complex diseases, e.g. type 2 diabetes, where disease status may be difficult to define from multiple intermediate phenotypes, there may be multiple interacting genes underlying these phenotypes, and the effects of these genes may differ according to exposure to environmental and other non-genetic risk factors such as diet and smoking. As a result, individuals affected by complex diseases are less concentrated within families and affected family members are less likely to

Handbook of Statistical Genetics, Third Edition. Edited by D.J. Balding, M. Bishop and C. Cannings.
© 2007 John Wiley & Sons, Ltd.   ISBN: 978-0-470-05830-5.

share the same variants at the underlying functional polymorphisms than for Mendelian disorders.

Population-based association studies are more powerful than linkage studies for identifying genetic polymorphisms contributing to complex diseases, provided that the underlying causative variants are not very rare (Risch and Merikangas, 1996; Zondervan and Cardon, 2004). Association studies focus on identifying genetic markers that occur with different frequencies in samples of unrelated affected cases and unaffected controls, exploiting the fact that it is easier to ascertain large groups of affected individuals sharing a genetic risk factor for a complex disease across the whole population than within individual families, which would be required for linkage.

The success of association studies for disease gene mapping relies, in part, on genotyping the functional polymorphism(s) themselves (so-called *direct* association), or flanking genetic markers that are highly correlated with the functional polymorphism(s) (*indirect* association). Direct association studies focus on genotyping candidate loci with a relatively high prior probability of functional relevance, including non-synonymous polymorphisms, splice-site variants, and copy number polymorphisms. Conversely, indirect association studies incorporate panels of single nucleotide polymorphisms (SNPs), each of which is unlikely to be directly of functional relevance, but at sufficiently high density one or more is likely to be correlated with the underlying causative variants. This correlation is referred to as *linkage disequilibrium* (LD), generated as a result of the shared ancestry of a population of chromosomes at proximal loci. As a result, alleles at flanking loci tend to occur together, in cis, on the same chromosome, with each specific combination of alleles known as a *haplotype*.

37.1.1 Linkage Disequilibrium and Tagging

There are numerous measures of LD between a pair of SNPs, most based around the statistic $D = h - p_1 p_2$. In this expression, h denotes the population frequency of the MM haplotype, and p_1 and p_2 denote the population frequencies of the M allele at each SNP, where at each locus M and m denote the major and minor alleles, respectively. Under gametic phase equilibrium, $h = p_1 p_2$ so that $D = 0$. More generally, D takes values in the range $[-1, 1]$, but is highly dependent on population allele frequencies. To reduce this dependence, two commonly used measures of LD have been proposed:

$$r^2 = \frac{D}{p_1(1 - p_1)p_2(1 - p_2)},$$

and

$$D' = \begin{cases} \dfrac{D}{\max[-p_1 p_2 - (1 - p_1)(1 - p_2)]} & \text{if } D < 0 \\[3mm] \dfrac{D}{\min[p_1(1 - p_2), (1 - p_1)p_2]} & \text{if } D \geq 0 \end{cases}.$$

The statistic D' can take values in the range $[0, 1]$, where $D' = 1$ is consistent with no recombination between a pair of SNPs in the time since the mutations generating the polymorphisms occurred. This is referred to as *complete* LD, and implies that at least one of the four possible haplotypes has frequency 0. The squared correlation coefficient, r^2, between a pair of SNPs also takes values in $[0, 1]$. However, $r^2 = 1$ corresponds to *perfect* LD, where genotypes at one SNP can be used as proxies for genotypes at a second SNP, referred to as *genetically identical* by Lawrence *et al.* (2005). The value of r^2 is directly related to the power to detect association of a disease with a genetic marker in LD with a flanking functional polymorphism. For a more detailed discussion of LD measures, see **Chapter 27**.

With the increasing availability of high-density SNP genotyping technology, many empirical studies have been undertaken to characterise the extent and distribution of LD throughout the genome in different populations. Initial studies focused on specific genes and small genomic regions (Clark *et al.*, 1998; Johnson *et al.*, 2001; Reich *et al.*, 2001; Gabriel *et al.*, 2002), but later large genomic regions (Taillon-Miller *et al.*, 2000; Dunning *et al.*, 2000; Abecasis *et al.*, 2001) and whole chromosomes were screened (Patil *et al.*, 2001; Dawson *et al.*, 2002; Phillips *et al.*, 2003). The clear conclusion from these studies is that the extent of LD is extremely variable throughout the genome, and across different populations. Further, much of common human genetic variation can be arranged in blocks of SNPs in strong LD with each other, maintained by low levels of recombination, bounded by hotspots of meiotic crossover activity.

Knowledge of the patterns of LD throughout the genome aids study design, since markers can be selected so as to guarantee coverage of all common SNPs with some predetermined threshold of r^2. The advantage of this approach is that we need not genotype all SNPs in a candidate gene or region, or even the whole genome, but can focus on a smaller number of so-called tag SNPs from which we can recover much of the information about common human genetic variation. Tag SNPs can be selected within blocks of strong LD by selecting combinations of SNPs that jointly define all common haplotype variation. Alternatively, SNPs can be allocated to *bins* of strong LD, based on pair-wise r^2 measurements, and a single tag selected from each bin (Carlson *et al.*, 2004).

One of the key advances in the design of population-based association studies has been the publication of data from the International HapMap project (The International HapMap Consortium, 2005). The initial phase of the project genotyped more than 1 million evenly spaced SNPs genome-wide, in samples from four populations: (1) 30 Yoruba trios (two parents plus offspring) from Ibadan, Nigeria; (2) 30 trios from the CEPH collection from Utah, USA, all with north and west European ancestry; (3) 45 unrelated individuals from Beijing, China; and (4) 44 unrelated individuals from Tokyo, Japan. The second phase of the project genotyped a further 4.6 million SNPs in the same samples, so that the average inter-SNP spacing was less than 1 Kb. Genotype data from the project are publicly available, and can be downloaded for detailed analysis of LD to aid marker selection in association study design, and interpretation of the results of analysis.

37.1.2 Current WGA Studies

A key determinant of the success of population-based association studies to map complex disease genes is sample size. We expect the alleles underlying complex disease phenotypes to each have small individual marginal effects, requiring samples of the order of thousands of cases and controls. Improvements in high-throughput SNP genotyping technology have revolutionised the field, making such sample sizes feasible for large candidate regions, hundreds or thousands of candidate genes, or most recently, whole genome association (WGA) studies. The latest generation of mapping arrays consist of 10^5–10^6 genome-wide SNPs. The Affymetrix 500K GeneChip is based on randomly selected SNPs, and tags approximately 65 % of HapMap phase II polymorphisms with $r^2 > 0.8$ in the CEPH samples, although coverage is lower for the Yoruba samples, reflecting less extensive

LD in African populations (Barrett and Cardon, 2006). The Illumina HumanHap300K BeadArray includes more than 300K tag SNPs, selected to capture common variation among north and west European populations, covering about 75 % of HapMap phase II SNPs in the CEPH samples, but only 28 % among the Yoruba samples.

Despite the potential of association studies to identify polymorphisms contributing susceptibility to complex diseases, the success of initial screens of candidate genes or larger candidate regions was limited to a handful of major gene effects including *APOE* for Alzheimer's disease (Rubinsztein *et al.*, 1999) and *NOD2* for Crohn's disease (Hugot *et al.*, 2001). With more recent appreciation of the importance of sample size, study design and genotype calling the list of reported associations is rapidly expanding. Several candidate genes have been associated with type 2 diabetes in multiple studies, including *PPARG*, *KCNJ11*, and *TCF7L2* (Parikh and Groop, 2004; Grant *et al.*, 2006). However, several reported associations have been difficult to replicate, including the *G972R* polymorphism in the *IRS1* gene (Almind *et al.*, 1993), which has more recently been demonstrated to have no effect on the risk of type 2 diabetes in population-based studies (Florez *et al.*, 2004; van Dam *et al.*, 2004; Zeggini *et al.*, 2004). This demonstrates the need to develop statistical methods for WGA studies, with increased power to detect disease association, while minimising the occurrence of false positives.

One success story for WGA studies was the identification of a causal variant for age-related macular degeneration (AMD) among Europeans in the complement factor H (*CFH*) gene (Klein *et al.*, 2005). An initial genome scan of more than 100 K SNPs on the Affymetrix 111K GeneChip in just 96 cases and 50 controls revealed a strong association of a common intronic polymorphism in *CFH* with AMD (nominal p value $< 10^{-7}$). Investigation of the patterns of LD flanking this signal using HapMap phase I data revealed this SNP to be part of a block contained wholly within *CFH*. Resequencing of the block identified a polymorphism in strong LD with the associated SNP that represents a tyrosine–histidine change at amino acid 402. This effect size is much larger than we expect for complex diseases (relative risk of 7.4 for individuals homozygous for the mutant allele compared to those homozygous for the wild-type allele), and mapping for complex diseases, in general, is unlikely to be so straightforward. Nevertheless, with such a small sample size, and the poor coverage of the Affymetrix 111K GeneChip (31 % in HapMap phase II CEPH samples), this result is encouraging for WGA studies.

A recent WGA study for type 2 diabetes has identified four novel susceptibility loci in a Fench case–control cohort of more than 1300 individuals (Sladek *et al.*, 2007). The samples were initially genotyped using the Illumina Infinium Human1 BeadArray, which assays 110K gene-centric SNPs, in addition to the Illumina HumanHap300K BeadArray. Markers demonstrating significance of the order of $p < 10^{-4}$ for the gene-centric array and $p < 5 \times 10^{-5}$ for the 300K array were carried forward for genotyping in a second cohort of more than 2500 cases and almost 2900 controls. These SNPs included known associations in the *TCF7L2* gene (Grant *et al.*, 2006), but did not include polymorphisms in other previously identified genes such as *PPARG*. This demonstrates the difficulties in replicating results from WGA studies, where effect sizes may be small, studies may be underpowered, and there may be variable coverage of the genome by the genotyping technologies used.

The Wellcome Trust Case Control Consortium (WTCCC) represents one of the largest WGA studies to date. The main arm of the study consists of an indirect WGA screen of 2000 unrelated cases each of seven diseases (type 1 diabetes, type 2 diabetes, coronary heart disease, hypertension, bipolar disorder, rheumatoid arthritis, and inflammatory bowel disease) from across the United Kingdom, together with 1500 unrelated controls each from the 1958 British birth cohort and recruits from the UK national blood service. All samples have been genotyped using the Affymetrix 500K GeneChip Mapping Array. An additional direct association study of more than 15K non-synonymous coding SNPs for 1000 cases each of four additional diseases (breast cancer, autoimmune thyroid disease, multiple sclerosis, and ankylosing spondylitus) 1500 controls from the 1958 British birth cohort has also been undertaken as part of the WTCCC using a custom Infinium chip from Illumina. Analysis of the data generated by the WTCCC will point to the design of further studies for each disease, focusing on the most interesting positive signals for subsequent investigation.

In this chapter, we describe statistical methods for the analysis and interpretation of results from WGA studies. We begin by describing procedures to assess the quality of genotypes obtained from WGA genotyping technology. Then, we discuss techniques for single-locus analyses to detect association with disease, including appropriate multiple-testing corrections. Next, we describe extensions to allow for environmental effects and to include tightly linked multiple SNPs or haplotypes, and epistasic interactions across more widely spaced SNPs. Finally, we emphasise the importance of replication of the results from WGA studies, and discuss their prospects for complex disease gene mapping. Balding (2006) provides an additional review of the design of, and methods for the analysis of, WGA studies.

37.2   GENOTYPE QUALITY CONTROL

Data filtering to identify genotype errors is a critical aspect of WGA analyses, which can determine whether real discoveries are made or false positives plague interpretation. No experimental system involving biological material is without error, and with large numbers of both SNPs and study subjects in a single study, even the modest error rates of $<0.25\,\%$ expected with current technology can be important. If such errors were distributed randomly across both genotypes and phenotypes, their effect would be limited to a small loss of statistical power. However, because of the nature of the experimental technologies available and factors such as DNA quality and preparation, specific experimental conditions, skill of experimenter, incorrect automated assignment (or 'calling') of experimental intensity values into discrete genotype classes, and stochastic variation, errors are not always distributed randomly. Non-random distribution of errors can inflate type I error rates and/or reduce power. The difficulty often lies with the designation of a heterozygous genotype, which is incorrectly classified as homozygous or the genotype is labelled as missing because of assignment ambiguity.

Animal models of continuous characters, human data in complex traits and long-standing theory in biometrical genetics (Wright and Hastie, 2001; Valdar et al., 2006) all suggest that genetic influences on most multifactorial phenotypes follow an L-shaped distribution of effects, with a small number of alleles with large effects and a larger number of small effect sizes. For discrete human diseases, an odds ratio (OR) above about 1.25 might now be considered large and many more effects are expected to be smaller than this. These effect sizes mean that most WGA studies aim to identify very small differences in allele frequencies, often only a few percent, among phenotype groups. Accordingly, even small amounts of experimental error can have profound effects on the outcomes (Clayton et al., 2005; Barrett and Cardon, 2006), particularly in the presence of rare alleles. It may seem that an experimental ideal would be to identify individual genotype errors and correct them one at a time, but this is often difficult or impossible in the WGA setting. Therefore, many filtering procedures aim to identify specific SNPs yielding errors in multiple individuals (a problem with the marker) or individuals in the sample with errors across multiple SNPs (a problem with the DNA sample), and simply exclude them from the study.

Neutral genetic variants in a large random-mating population are expected to display Hardy–Weinberg equilibrium (HWE), under which expected genotype frequencies satisfy $\mathbf{E}(f_{MM}) = p^2$, $\mathbf{E}(f_{Mm}) = 2p(1 - p)$, and $\mathbf{E}(f_{mm}) = (1 - p)^2$, where p is the population frequency of allele M. Genotyping error can alter the observed frequencies from the expected proportions, and thus tests of deviations from HWE comprise a traditional approach to detecting genotyping errors and excluding markers with significant deviations (Weir, 1996). Such a test can be constructed using a Pearson goodness-of-fit statistic,

$$\text{HWE}X^2 = \sum_{g=mm,Mm,MM} \frac{(f_g - \mathbf{E}(f_g))^2}{\mathbf{E}(f_g)},$$

having an approximate χ^2 distribution under the null hypothesis of HWE, with 1 degree of freedom (df) because p is derived from the observed data.

There are three serious problems with the use of HWE tests in this manner: (1) natural selection and copy-number variants can also lead to significant deviations, and these may underlay true causal associations, which will be missed if the SNP is excluded because it 'fails' a HWE test; (2) the test is insensitive to the modest deviations that are most often observed in WGA studies; and (3) in the WGA setting of $> 10^5$ markers, an appropriate threshold of 'significance' is difficult to determine. As a result of these considerations, the most prudent use of HWE tests for genotyping error may be only to exclude the most egregious deviations by setting an extreme significance threshold such as 10^{-7} or less, and using exact tests for rare alleles (Weir *et al.*, 2005).

Missing data at individual genotypes is not uncommon, but when missing rates exceed about 5 % at a SNP, or about 10 % for an individual, there is reason to be suspicious of the assay or DNA sample and the most prudent course is to repeat the experiment for that SNP or sample (WTCCC, 2007). For case–control studies, in particular, markers with differences in missing data rates between cases and controls often yield false positives, which are sometimes striking in magnitude (Clayton *et al.*, 2005). Several statistics are available for testing the difference in missing data rates between cases and controls, e.g. one based on the normal approximation to the binomial distribution:

$$z = \frac{m_c - m_t}{\sqrt{m(1 - m)(1/n_c + 1/n_t)}},$$

where m_c and m_t are the proportions of missing genotypes among cases (c) and controls (t), samples sizes (missing + non-missing data) are labelled as n_c and n_t and m is the overall missing genotype rate at the marker.

Graphical displays are useful for identifying suspicious markers or samples. According to the L-shaped distribution of effect sizes, a study of 10^5–10^6 genetic markers genotyped on 1–2K cases and equal numbers of controls should reveal few genuine loci with single-locus p values below 10^{-6}, with many more associated loci lying in the range 10^{-3}–10^{-5}

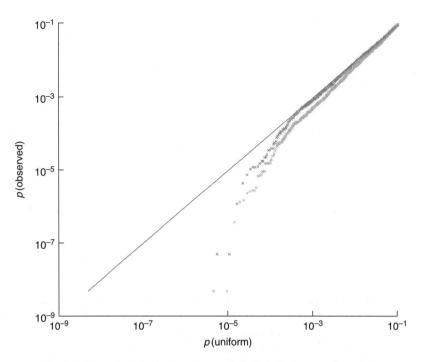

Figure 37.1 Probability–probability plot for association statistics from a WGA study of type 2 diabetes (T2D) genotyped for the Illumina Human1 BeadArray SNPs, from Sladek *et al.* (2007, supplementary Figure 4). Unadjusted *p* values for the maximum statistic over three tests of association are plotted against the expected uniform distribution of *p* values under the null hypothesis of no association, genome-wide. Systematic deviations from the $y = x$ line are indicative of the effects of population stratification, and some extreme deviations could reflect genotyping error. Corresponding values after genomic-control adjustment are also shown, and adhere to the $y = x$ line over most of the distribution; some of the very small *p* values that deviate from this line were subsequently confirmed to be associated with T2D. [Reprinted by permission from Macmillan Publishers Ltd: Sladek R *et al.*, A genome-wide association study identifies novel risk loci for type 2 diabetes. *Nature*, 2007; 445: 881–885.]

(Zondervan and Cardon, 2004). Accordingly, if one observes many highly significant loci in a particular study, they may be less likely to reflect real discoveries and more likely to reflect systematic genotype error in some of those markers. This logic has led to the widespread use of quantile–quantile (QQ) plots to examine the overall distribution of test statistics and assess whether there are too many data points in the tail. This approach, used to assess HWE deviation by Weir *et al.* (2005), was shown to be extremely effective in the case–control context by Clayton *et al.* (2005). To construct such plots, the ordered test statistics for association are plotted against the corresponding expected order statistics. For example, Clayton *et al.* (2005) plotted ascending values of the Cochran–Armitage test statistic (see Section 37.3) against $F^{-1}[i/(N + 1)]$, where $F[]$ is the χ_1^2 distribution function. Sladek *et al.* (2007) use a similar approach of plotting the unadjusted *p* values for the maximum statistic over three specific tests of association for each SNP against the expected uniform distribution (Figure 37.1). In either case, large values deviating from the $y = x$ null can be a symptom of genotyping error, as demonstrated by Clayton *et al.* (2005). This subjective screening procedure remains extremely useful to determine whether the data filters (missing data rates, HWE, allele frequency thresholds) in a particular study are sufficient to eliminate most of the problematic markers.

Another useful graphical procedure involves plotting, for each individual, the fraction of all markers that are heterozygous against the proportion of missing data for that individual (WTCCC, 2007). This plotting procedure can identify DNA samples that perform poorly in WGA genotyping, resulting in a high proportion of missing data and/or too few heterozygotes.

Genotyping so many markers in WGA studies enables identification of closely related individuals or duplicate DNA samples. Abecasis *et al.* (2001) showed that comparing the average number of alleles that are identical by state for two individuals with the variance of IBS sharing can identify duplicate samples or MZ twins, full-siblings, parent–offspring and half-siblings. In other words, for two individuals j and k having genotypes g_j and g_k at marker i of N, plot

$$\bar{x}_{\text{ibs}}^{(j,k)} = \frac{1}{2N} \sum_{i=1}^{N} x_i^{(j,k)},$$

against

$$s_{x_{\text{ibs}}^{(j,k)}}^2 = \frac{1}{N-1} \sum_{i=1}^{N} \left(x_i^{(j,k)} - \bar{x}_{\text{ibs}}^{(j,k)} \right)^2,$$

where

$$x_i^{(j,k)} = \begin{cases} 0 & \text{if} & g_j, g_k = mm, MM \\ 1 & \text{if} & g_j, g_k = mm, Mm \text{ or } Mm, MM \\ 2 & \text{if} & g_j, g_k = mm, mm \text{ or } Mm, Mm \text{ or } MM, MM. \end{cases}$$

More distant relatives are difficult to distinguish from unrelated individuals using this approach. See also Devlin and Roeder (1999) and Voight and Pritchard (2005).

37.3   SINGLE-LOCUS ANALYSIS

The current standard practice involves an individual test of each SNP typed in the WGA study, to identify any promising associations. We now review the main tests employed.

Consider a sample of unrelated cases, affected by the disease of interest, and unaffected controls, typed at a SNP with alleles denoted by M and m. We can represent the sample genotype data in a 2×3 contingency array (Table 37.1). Under the null hypothesis of no association with the disease, we expect the relative genotype frequencies to be the same in cases and controls. Thus, as described by Clayton (**Chapter 36**), we can construct a score test for association by calculating the standard Pearson's χ^2 statistic for independence of the rows and columns, given by

$$X_{\text{Gen}}^2 = \sum_{i=0,1,2} \sum_{j=A,U} \frac{(n_{ij} - \mathbf{E}[n_{ij}])^2}{\mathbf{E}[n_{ij}]}, \tag{37.1}$$

Table 37.1 Representation of SNP genotype data for a population-based association study in a 2×3 contingency array. The counts, n_{ij}, denote the observed sample frequency of individuals carrying i copies of allele m, and phenotype j, where $j = A$ corresponds to cases and $j = U$ corresponds to controls.

Genotype	Cases	Controls	Total
MM	n_{0A}	n_{0U}	$n_{0.}$
Mm	n_{1A}	n_{1U}	$n_{1.}$
Mm	n_{2A}	n_{2U}	$n_{2.}$
Total	$n_{.A}$	$n_{.U}$	$n_{..}$

where

$$\mathbf{E}[n_{ij}] = \frac{n_{i.}n_{.j}}{n_{..}}.$$

The test statistic, $X^2{}_{\text{Gen}}$, has an approximate χ^2 distribution with 2 df under the null hypothesis of independence (no association).

It may be of interest to investigate the association further by estimating the OR of disease for each genotype at the SNP. It is customary to calculate the sample OR relative to the most common genotype in controls as a baseline. For example, the OR for genotype mm relative to the baseline genotype MM is estimated by

$$\psi_{mm|MM} = \frac{n_{2A}n_{0U}}{n_{2U}n_{0A}}.$$

The variance of the log OR is approximately

$$\mathbf{V}[\ln \psi_{mm|MM}] = \frac{1}{n_{2A}} + \frac{1}{n_{2U}} + \frac{1}{n_{0A}} + \frac{1}{n_{0U}},$$

with corresponding 95 % confidence interval

$$\ln \psi_{mm|MM} \pm 1.96 \times \sqrt{\mathbf{V}[\ln \psi_{mm|MM}]}.$$

Case–control studies are retrospective in the sense that subjects are ascertained on the basis of their disease status, and are then genotyped. Since cases are over-sampled, the disease risk cannot be directly estimated. However, assuming the disease under investigation to be rare, the OR gives an approximation to the relative risk: an individual carrying genotype mm is approximately $\psi_{mm|MM}$ times more likely to develop the disease than an individual of genotype MM.

To reduce the df of the association test, we can focus on *allelic* effects by assuming that alleles at the SNP act independently in terms of disease risk. In other words, we assume a multiplicative model for disease penetrances so that $\psi_{mm|MM} = \psi^2_{Mm|MM}$, and hence a linear (additive) trend in the log-odds of disease with each copy of allele m. Under this assumption, we can test for association between the SNP and disease by means of the Cochran–Armitage trend test, given by

$$X^2_{\text{C–A}} = \frac{n_{..}[n_{..}(n_{1A} + 2n_{2A}) - n_{.A}(n_{1.} + 2n_{2.})]^2}{n_{.A}(n_{..} - n_{.A})[n_{..}(n_{1.} + 4n_{2.}) - (n_{1.} + 2n_{2.})^2]}. \tag{37.2}$$

Under the null hypothesis of no association between the SNP and disease, X_{C-A}^2, has an approximate χ_1^2 null distribution. We can also calculate the OR for allele m, relative to allele M, by

$$\psi_{m|M} = \frac{n_{1A}n_{0U}/(n_0. + n_1.) + n_{2A}n_{1U}/(n_1. + n_2.) + 4n_{2A}n_{0U}/(n_0. + n_2.)}{n_{0A}n_{1U}/(n_0. + n_1.) + n_{1A}n_{2U}/(n_1. + n_2.) + 4\sqrt{(n_{2A}n_{2U}n_{0A}n_{0U})}/(n_0. + n_2.)}.$$

The OR can be interpreted as follows: an affected individual is $\psi_{m|M}$ times more likely to carry genotype Mm than genotype MM at the SNP, and $\psi_{m|M}^2$ times more likely to carry genotype mm.

As discussed by Clayton (**Chapter 36**), the tests described above rely on asymptotic assumptions that are unlikely to hold when there are small genotype counts or the minor allele frequency is low. By conditioning on the marginal frequencies of the contingency array, exact tests can be constructed for the genotype-based or Cochran–Armitage statistics.

37.3.1 Logistic Regression Modelling Framework

For complex traits, we expect that disease risk might be modified by environmental effects that cannot be easily accommodated by the tests described above. A more flexible framework for modelling the relationship between disease phenotype and SNP genotype makes use of logistic regression techniques, as described by Clayton (**Chapter 36**). Consider, as before, a sample of unrelated cases and controls, yielding genotype data **G**. The logistic regression model is parameterised in terms of the log-odds of disease for each SNP genotype, denoted by $\boldsymbol{\beta}$. The log-likelihood of observed phenotype data, **y**, is given by

$$\ln f(\mathbf{y}|\mathbf{G}, \boldsymbol{\beta}) = \sum_i [y_i \ln \pi_i + (1 - y_i) \ln(1 - \pi_i)],$$

where y_i denotes the disease phenotype of the ith individual. The probability, π_i, that the ith individual is affected, given their genotype G_i at the SNP, is given by the *logit* link function,

$$\pi_i = \frac{\exp \eta_i}{1 + \exp \eta_i},$$

where the linear predictor is $\eta_i = \beta_{G_i}$.

Under the null hypothesis of no association of the SNP with the disease, we expect each genotype to have equal odds of disease, so that the linear component $\eta_i = \beta_0$. Evidence of association corresponds to deviation from this null model. Assuming that the alleles at the SNP act independently in terms of disease risk, in the same way as for the Cochran–Armitage trend test, there is an additive effect of the two alleles in the log-odds of disease. Under this additive model, treating allele M as baseline, the linear predictor for the ith individual is given by

$$\eta_i = \beta_0 + \beta_A z_{(A)i}, \tag{37.3}$$

where β_A denotes the additive effect of allele m, and $z_{(A)i}$ is an indicator variable representing the additive component of the ith genotype, summarised in Table 37.2.

Table 37.2 Coding of additive and dominance components of SNP genotypes.

Genotype	Additive component $z_{(A)i}$	Dominance component $z_{(D)i}$
MM	-1	0
Mm	0	1
Mm	1	0

Deviation from the additive model is referred to as *dominance*, and can be thought of as an interaction between the pair of alleles within each genotype at the SNP. Under this non-additive model, the linear predictor for the ith individual extends to

$$\eta_i = \beta_0 + \beta_A z_{(A)i} + \beta_D z_{(D)i}, \qquad (37.4)$$

where β_D denotes the dominance effect of allele m over allele M, and $z_{(D)i}$ is an indicator variable representing the dominance component of the ith genotype, summarised in Table 37.2.

We can test for association between disease and the SNP by comparing nested submodels of (37.4) with $\beta_A = 0$ and/or $\beta_D = 0$ using analysis of deviance. For example, the log-likelihood ratio statistic,

$$\Lambda_{\text{Gen}} = 2 \ln f(\mathbf{y}|\mathbf{G}, \hat{\beta}_0, \hat{\beta}_A, \hat{\beta}_D) - 2 \ln f(\mathbf{y}|\mathbf{G}, \hat{\beta}_0, \beta_A = 0, \beta_D = 0),$$

provides a genotype-based test of association, having an approximate χ_2^2 null distribution, and the maximum likelihood estimates, $\hat{\boldsymbol{\beta}}$, under each model are obtained numerically. The model in (37.4) is parameterised in a different way compared to the model used in the 2 df test described by Clayton (**Chapter 36**). However, although model parameter estimates are interpreted in different ways, the statistical tests are equivalent, and both are approximately equal to the Pearson 2 df test defined in (37.1).

Assuming that alleles at the SNP have independent effects on disease risk, we can form an additive test of association,

$$\Lambda_{\text{Add}} = 2 \ln f(\mathbf{y}|\mathbf{G}, \hat{\beta}_0, \hat{\beta}_A, \beta_D = 0) - 2 \ln f(\mathbf{y}|\mathbf{G}, \hat{\beta}_0, \beta_A = 0, \beta_D = 0),$$

which has an approximate χ_1^2 distribution and is equivalent to the Cochran–Armitage test (37.2). Within this framework, we can also test for deviations from the additive model of alleles at the SNP on the log odds of disease, given by the statistic

$$\Lambda_{\text{Dom}} = 2 \ln f(\mathbf{y}|\mathbf{G}, \hat{\beta}_0, \hat{\beta}_A, \hat{\beta}_D) - 2 \ln f(\mathbf{y}|\mathbf{G}, \hat{\beta}_0, \hat{\beta}_A, \beta_D = 0),$$

having an approximate χ_1^2 distribution under the null hypothesis of no dominance. We can test other disease models within this framework by imposing constraints in the model parameters. For example, to allow for a pure recessive effect of allele m, we constrain $\beta_D = -\beta_A$, leading to a test of association having an approximate χ_1^2 null.

Within the logistic regression framework, it is straightforward to incorporate covariates, $\mathbf{x}$, in the linear component, to allow for environmental effects. Specifically, the linear predictor of the ith individual extends to

$$\eta_i = \beta_0 + \sum_j \gamma_j x_{ij} + \beta_A z_{(A)i} + \beta_D z_{(D)i}, \qquad (37.5)$$

where x_{ij} is the response of the ith individual to the jth covariate, and γ_j is the corresponding regression coefficient. Under the null hypothesis of no association of the SNP with disease, each individual is equally likely to be affected in terms of their genotype, with risk modified only by the effects of the covariates. Thus, we can consider the same likelihood ratio tests of association by comparing sub-models of (37.5) subject to the constraints $\beta_A = 0$ and/or $\beta_D = 0$ by analysis of deviance, in the same way as above.

37.3.2 Interpretation of Results and Correction for Multiple Testing

A significant result in a test of association may suggest that the SNP itself is causative, directly influencing disease risk. However, such an assertion would need to be established via further functional studies. In fact, there are a number of possible alternatives.

1. Alleles at the SNP are correlated with alleles at the functional polymorphism, but do not directly influence disease risk. Such an indirect association occurs as a result of background LD between the two loci.

2. Alleles at the SNP and the functional polymorphism are confounded with underlying population structure that is not accounted for in the analysis. In the presence of structure, cases may be ascertained preferentially from one stratum of the population, because of higher disease prevalence and/or biased ascertainment. If SNP genotype frequencies vary between strata, we may detect an apparent association with disease in the population overall, even if there is no association within each of the individual population strata. This problem is addressed in more detail below.

3. For tests with significance level α, we expect $100\alpha\%$ of non-associated SNPs tested to show false positive signals of association.

The choice of significance level α should take account of multiple testing in WGA studies. For example, if we choose $\alpha = 5\%$ and test 20 independent SNPs for association with disease, we expect one of them to show significant evidence of association, even if none is truly associated with disease. It is crucial, therefore, to correct for multiple testing to maintain the type I error rate for the experiment overall (i.e. for all the SNPs tested in the association study).

The simplest approach to allow for multiple testing is to make use of a Bonferroni correction. Under this approach, each test is treated as independent, and the SNP-wise significance level is adjusted to achieve an overall experiment-wise type I error rate of $100\alpha\%$. Specifically, when testing N SNPs, we use a significance level of $100\alpha/N\%$ at each SNP. The disadvantage of this approach is that each test is assumed to be independent, whereas for WGA studies we expect adjacent SNPs to be correlated owing to background patterns of LD throughout the genome, and thus the Bonferroni correction will be conservative. An alternative correction that overcomes the problem of correlated tests makes use of the false discovery rate (FDR), by fixing the expected number of false positives among significant associations (Benjamini and Hochberg, 1995). Specifically, if we select an uncorrected SNP-wise significance level of α, the FDR is given by $N\alpha/k$, where k is the number of SNPs with a p value of less than α. Therefore, we can fix the SNP-wise significance level so as to obtain an overall FDR of 5%.

The most appropriate methods for correcting for multiple testing make use of permutation procedures. The null distribution of experiment-wise association statistics is generated by calculating the maximum association statistic over the genome for a large number of permutations of the original phenotype and genotype data. In the simplest procedures, phenotype labels are permuted while keeping genotypes fixed, thus maintaining the LD structure throughout the genome. Corrected p values for each SNP can then be calculated by comparing the observed test statistic with the distribution of the maximum test statistic from each permutation. For WGA studies, this process will be extremely computationally intensive. However, for an empirical experiment-wise significance level of the order of 5%, as few as 100 permutations of the data may be adequate.

A final approach to take account of multiple testing makes use of Bayesian statistical theory. Under this approach, we assign a prior probability of association to each SNP, reflecting our beliefs about the number of disease associated loci before we look at the observed genotype data. The advantage of the Bayesian framework is that we can allow our prior probability of association to vary across SNPs, reflecting their functional relevance or the results of previous linkage and/or association studies.

37.4   POPULATION STRUCTURE

One of the potential problems of population-based association studies is population structure, which, if not accounted for in the analysis, can inflate the false positive error rate for detecting associations. Consider a population consisting of two underlying strata, where the disease is common in stratum one, but rare in stratum two. If cases and controls are selected at random from the population overall, without regard to the underlying structure, cases will be preferentially selected from stratum one. As a result, SNPs with allele frequency differences between the strata will appear to be associated with disease, even if there is no association within each stratum.

One obvious approach to deal with the problem of structure is to match cases and controls for stratum of the population, e.g. on the basis of self-described ethnicity or geographical location. However, with migration and admixture between ethnic groups, these indicators may not reflect the complex nature of fine-scale structure within populations. Family-based association studies (see **Chapter 38**) provide a design-based solution to the problem, in which, for example, the two alleles of each parent of an affected child are matched, but this solution is expensive in terms of additional genotyping and limited to studies for which trios or other suitable family structures are available. For WGA studies, the problem of population structure can be overcome by statistical methods that account for the underlying structure in analyses. The most important of such methods are briefly reviewed here (see **Chapters 35** and **36** for more detail and further discussion).

One of the simplest methods to identify, and adjust for, structure in population-based WGA studies is genomic control (Devlin and Roeder, 1999). Under the null hypothesis of no disease association, the distribution of Cochran–Armitage test statistics is χ^2_1. However, in a stratified population, we expect that there would be allele frequency differences at many SNPs, genome-wide, and hence an excess of false positive signals of association. As a result, the observed distribution of association statistics will be inflated, with the magnitude of the inflation reflecting the extent of structure. The genomic control method takes account of structure by a linear rescaling of observed test statistics to approximately restore the χ_1^2 null distribution. Figure 37.1 shows a plot of p values for the WGA of Sladek *et al.* (2007) before and after genomic-control adjustment. This approach is appealing in its simplicity, but is limited to a simple test of association which, for example, do not incorporate environmental effects.

A more complex solution to the problem of population structure has been proposed by Pritchard *et al.* (2000a) and has come to be known as *structured association*. Under an admixture model, the proportion of an individual's genome that descends from each of K specific ancestral strata is treated as unknown. The posterior distribution of ancestry for each sampled individual is then approximated using Bayesian Markov chain Monte Carlo (MCMC) methods, using the STRUCTURE algorithm, based on genotype information from several hundred genome-wide SNPs (see **Chapter 30** for further details). Tests for association then compare allele frequencies between cases and controls within strata, implemented in the companion STRAT software (Pritchard *et al.*, 2000b). Alternatively, population ancestry could be included as covariates in a logistic regression framework to allow for more flexible modelling of genetic and non-genetic risk factors for complex disease. Potential disadvantages of the structured association approach include the following: (1) the number of ancestral subpopulations is unknown, and must be inferred using an *ad hoc* estimation procedure and (2) the MCMC algorithm is computationally intensive, and cannot in practice accommodate the numbers of markers used in WGA studies.

Setakis *et al.* (2006) have proposed accounting for population structure within the logistic regression model by treating the genotypes at genome-wide SNPs directly as covariates. Backward elimination or Bayesian shrinkage model selection techniques are employed to reduce the over-fitting problem arising from including many covariates in the model. Reich *et al.* (2006) have suggested using principal components analysis (PCA) to infer population structure and provide formal significance tests for between-strata differences. The eigenvectors and the corresponding loadings for each individual from the PCA can be used as covariates within a logistic regression framework to account for the underlying structure. A key advantage of this approach is computational efficiency, since PCA can be applied to data sets with $>10^5$ SNPs.

37.5 MULTI-LOCUS ANALYSIS

One of the most attractive features of SNPs for complex disease gene mapping is their abundance throughout the genome. However, each individual SNP provides little information about disease association unless it is highly correlated with the underlying (unobserved) causal polymorphism. Single-SNP analyses may thus be inefficient for mapping. For high-density panels of SNPs, such as those utilised in WGA studies, we expect that there would be correlations between several SNPs flanking a causal polymorphism, and so simultaneous analyses of multiple SNPs may jointly provide evidence of association for modest gene effects, even when the individual SNPs do not.

The logistic regression model described above provides a natural framework for multi-locus association analysis. Additive and dominance effects of all SNPs in the same gene or small genomic region can be fitted simultaneously in the logistic regression model. To allow for correlations between SNPs, and to reduce the problem of over-parameterisation, standard statistical model building techniques, such as forward selection and/or backward elimination, can be utilised to identify good combinations of SNPs to describe the association with disease. Such model building techniques tend to over-fit the observed phenotype and genotype data, and so appropriate correction for the selection process should be taken into account to avoid inflation of type 1 error, e.g. by permutation testing.

37.5.1 Haplotype-based Analyses

An alternative approach to multi-locus analysis is to focus on haplotype effects. Haplotypes are particularly attractive because much of diversity within blocks of LD is driven by mutation, rather than recombination. As a result, much of common genetic variation can be structured into haplotypes within blocks that are rarely disturbed by meiosis. Furthermore, Clark (2004) emphasises that the functional properties of a protein are determined by the linear sequence of amino acids, corresponding to DNA variation on a haplotype. For example, there is evidence that a combination of causal variants in cis in the *HPC2/ELAC2* gene increases the risk of prostate cancer (Tavtigian *et al.*, 2001). Finally, a rare causal allele may reside on a specific haplotype background that would not otherwise be identified through single-locus methods.

It is common to assume that each of the pair of haplotypes, H_{i1} and H_{i2}, forming the diplotype, H_i, of the ith individual, contributes independent effects to disease risk, where haplotypes are labelled according to their relative frequency in the population. Under this assumption, we can parameterise the logistic regression model in terms of the log odds of disease for each haplotype. Thus, the linear predictor of the ith individual is given by

$$\eta_i = \beta_0 + \sum_j \gamma_j x_{ij} + \beta_{H_{i1}} + \beta_{H_{i2}}.$$

In this expression, β_k denotes the log-OR of the kth most frequent haplotype, relative to the baseline haplotype, usually taken to be the most common, so that $\beta_1 = 0$. Furthermore, β_0 is the baseline log odds of disease and γ_j denotes the effect of the jth covariate. An affected individual is approximately $\exp[\beta_{H_{i1}} + \beta_{H_{i2}}]$ times more likely to carry diplotype H_i than to carry two copies of the baseline haplotype.

One of the obvious problems of haplotype-based analyses is that we do not observe the diplotypes, **H**, directly from the unphased genotype data. One solution to the problem would be to take a point estimate of the diplotype for each individual, using statistical methodology, such as PHASE (Stephens *et al.*, 2001; Stephens and Donnelly, 2003) or by maximum likelihood via implementation of the expectation–maximisation (E–M) algorithm (Excoffier and Slatkin, 1995), and to treat this estimate as if it were known. However, this does not take account of uncertainty in the haplotype reconstruction process, and as a result, the variances of model parameters are under-estimated, and the false positive error rate is inflated.

A better approach to allow for unknown phase is to consider the distribution of diplotypes consistent with each multi-locus genotype, denoted $f(H_i|G_i, \mathbf{h})$ for the ith individual, given unknown population haplotype frequencies **h**. The likelihood of observed phenotype data, $\mathbf{y}$, given the unphased genotype data, can then be expressed as

$$f(\mathbf{y}|\mathbf{G}, \mathbf{x}, \mu, \boldsymbol{\beta}, \boldsymbol{\gamma}, h) = \prod_i \sum_{H_i \in G_i} f(y_i | H_i x_i, \mu, \beta, \gamma) f(H_i | G_i h), \qquad (37.6)$$

where $H_i \in G_i$ denotes the set of diplotypes consistent with the observed multi-locus genotype of the ith individual. Under the null hypothesis of no association between SNP haplotypes and disease, the log odds of disease for each haplotype will be the same, and $\boldsymbol{\beta} = 0$. Thus, we can construct a likelihood ratio test of association by considering the deviance,

$$\Lambda_{\mathrm{Hap}} = 2 \ln f(\mathbf{y}|\mathbf{G}, \mathbf{x}, \hat{\mu}, \hat{\boldsymbol{\beta}}, \hat{\boldsymbol{\gamma}}, \hat{\mathbf{h}}) - 2 \ln f(\mathbf{y}|\mathbf{G}, \mathbf{x}, \hat{\mu}, \boldsymbol{\beta} = 0, \hat{\boldsymbol{\gamma}}, \hat{\mathbf{h}}),$$

where $f(\mathbf{y}|\mathbf{G}, \mathbf{x}, \hat{\mu}, \hat{\boldsymbol{\beta}}, \hat{\boldsymbol{\gamma}}, \hat{\mathbf{h}})$ and $f(\mathbf{y}|\mathbf{G}, \mathbf{x}, \hat{\mu}, \boldsymbol{\beta} = 0, \hat{\boldsymbol{\gamma}}, \hat{\mathbf{h}})$ are obtained by maximising the likelihood (37.6), respectively, over the set of parameters $\{\mu, \boldsymbol{\beta}, \boldsymbol{\gamma}, \mathbf{h}\}$, and over $\{\mu, \boldsymbol{\gamma}, \mathbf{h}\}$, subject to the constraint $\boldsymbol{\beta} = 0$. Under the null hypothesis of no association, the deviance Λ_{Hap} has an approximate χ^2 distribution with $d - 1$ df, where d is the number of distinct haplotypes consistent with the observed sample genotypes.

Zaykin *et al.* (2002) propose a two-stage strategy of first obtaining the distribution of diplotypes consistent with the unphased genotype of each individual by application of the E–M algorithm to the complete sample of cases and controls, but other algorithms for haplotype reconstruction, such as PHASE, could be used to obtain estimates of $f(H_i|G_i, \mathbf{h})$. In the subsequent test of association, the likelihood (37.6) is maximised only over the parameters μ, $\boldsymbol{\beta}$, and $\boldsymbol{\gamma}$. Schaid *et al.* (2002) use the same two-stage approach, but perform a score test of association between SNP haplotypes and disease, which is asymptotically equivalent to the likelihood ratio test of Zaykin *et al.* (2002).

37.5.2 Haplotype Clustering Techniques

One potential problem with haplotype-based analyses is lack of parsimony, since many haplotypes may be consistent with the observed unphased genotype data, some of which may be very rare. In the logistic regression model, a parameter is required for each haplotype, except the baseline, leading to a test of association with disease that, potentially, has many degrees of freedom. The effects of rare haplotypes will be difficult to estimate, and there may be a lack of power to detect association, particularly if only one or two haplotypes are at high or low risk of disease. We could resolve this problem by combining rare haplotypes into a single pooled category. However, a more satisfactory approach to reduce the dimensionality of the problem is to take advantage of the expectation that similar SNP haplotypes in a small genomic region tend to share recent common ancestry, and hence are likely also to share the same alleles at the underlying functional polymorphism(s). Therefore, similar SNP haplotypes are expected to have comparable disease risks, and thus could naturally be combined in the analysis.

A number of methods that group haplotypes according to some similarity measure and then assign the same risk to all haplotypes within the same cluster have been proposed (Templeton *et al.*, 1987; 1988; 1992; Templeton and Sing, 1993; Molitor *et al.*, 2003a; 2003b; Durrant *et al.*, 2004; Morris, 2005; 2006). In this way, we parameterise the logistic regression model in terms of the log ORs for each cluster, which will be less than the number of haplotypes, reducing the degree of freedom of the resulting test of association. Morris (2005) clusters haplotypes according to a Bayesian partition model

(Knorr-Held and Rasser, 2000; Denison and Holmes, 2001). The model is defined by specifying the number of clusters of haplotypes and the centre of each cluster, taken from the set of distinct haplotypes consistent with the observed multi-locus genotype data, without replacement. All remaining haplotypes are then assigned to the nearest cluster centre, where distance between a pair of haplotypes is measured in terms of unweighted SNP allele mismatches. Morris (2005) has developed a Bayesian MCMC algorithm, GENEBPM, to sample from the posterior distribution of haplotype clusters, and the corresponding cluster log ORs, allowing for the inclusion of covariates such as environmental risk factors. Output from the algorithm can be used to (1) estimate the log OR of disease for each distinct haplotype; (2) identify clusters of haplotypes with similar disease risks; and (3) estimate the posterior probability of haplotype association with disease. More recently, Morris (2006) has extended the GENEBPM algorithm to allow for deviations from the assumption of independent haplotype effects on disease risk.

The evolution of a sample of haplotypes within a population can be represented by means of a genealogical tree, most readily modelled by the coalescent process with recombination. A number of methods have been proposed for fine-scale association mapping with samples of unrelated cases and controls that take account of the shared ancestry of their chromosomes explicitly (Morris et al., 2002; 2004; Zollner and Pritchard, 2005; Minichiello and Durbin, 2006). These approaches are computationally intensive, making them impractical for the analysis of large WGA studies. Haplotype similarity can be represented by means of a dendogram, and can be thought of as representing their evolution in the population. Thus, haplotype clustering can be thought of as an approximation to the more complex population genetics processes underlying their evolution, providing computationally efficient algorithms that can be applied on the scale of WGA studies, while maintaining the principal features of the shared ancestry of a sample of chromosomes.

Multi-locus analysis of SNP haplotypes is appropriate within candidate genes or small genomic regions that have been subject to limited ancestral recombination. In the context of WGA studies, haplotype-based analyses are appropriate within blocks of strong LD. Haplotype-based analyses may provide additional information with higher-density genotyping in a follow-up study of associated regions from an initial genome scan. Employing clustering techniques may reveal patterns of haplotype similarity that help to refine the likely location of the underlying causal polymorphisms, and may identify individuals with high probability of carrying high-risk alleles who could be sequenced in functional studies.

37.6 EPISTASIS

The traditional definition of epistasis is the masking, or modification, of the effects of genotypes at one polymorphism by genotypes at a second one. One classical example of epistasis is the coat colour of Labrador dogs. One gene controls hair colour, differentiating between black and brown coats. However, a second gene determines the deposition of hair colour, where dogs carrying one genotype will be golden, irrespective of their black/brown genotype, masking the effects of the first gene. The existence of epistasis is not surprising since we expect the biological mechanisms underlying complex diseases to be extremely intricate, incorporating the effects of multiple genetic risk factors, acting together in some way. Furthermore, there is increasing evidence from model organisms, including *Drosophila melanogaster* and *Saccharomyces cerevisiae*, that epistasis occurs frequently, involves multiple polymorphisms, and may contribute large effects to the genetic component of phenotypic variation (Mackay, 2001; Brem and Kruglyak, 2005; Brem *et al.*, 2005; Storey *et al.*, 2005).

From a statistical viewpoint, epistasis corresponds to an interaction between genotypes at two or more loci. Thus, the logistic regression modelling framework described above for multi-locus analysis generalises naturally to allow for epistasis. For example, to model interaction between a pair of SNPs, j and k, the linear predictor for the ith individual can be expressed as

$$\eta_i = \beta_0 + \beta_{Aj} z_{(Aj)i} + \beta_{Dj} z_{(Dj)i} + \beta_{Ak} z_{(Ak)i} + \beta_{Dk} z_{(Dk)i}$$
$$+ \beta_{AAjk} z_{(Aj)i} z_{(Ak)i} + \beta_{ADjk} z_{(Aj)i} z_{(Dk)i} + \beta_{DAjk} z_{(Dj)i} z_{(Dk)i}$$
$$+ \beta_{DDjk} z_{(Dj)i} z_{(Dk)i}, \tag{37.7}$$

where the SNP genotype indicator variables are defined in Table 37.2. The parameters β_{Aj} and β_{Dj}, correspond to the additive and dominance *main effects*, respectively, of SNP j, with β_{Ak} and β_{Dk} interpreted in the same way for SNP k. The four interaction terms, $\beta_{AA}, \beta_{AD}, \beta_{DA}$, and β_{DD}, correspond to additive–additive, additive–dominance, dominance–additive, and dominance–dominance contributions to epistasis between SNPs j and k.

We can test for joint association of SNPs j and k with disease, allowing for epistasis between the two loci, by comparing the interaction model (37.7) and the null model $\eta_i = \beta_0$. Under the null hypothesis of no association of either SNP with disease, the difference in deviances between the two models has an approximate χ^2 distribution with 8 df. Furthermore, we can test for epistasis between SNPs j and k in their association with disease by comparing the interaction model (37.7) with a constrained model $\beta_{AA} = \beta_{AD} = \beta_{DA} = \beta_{DD} = 0$, where the difference in deviances now has 4 df. Alternatively, standard statistical model building techniques can be utilised to infer the best combination of main effects and interactions to describe the association with disease. With more than two SNPs, higher-order interactions could also be incorporated, although these effects are difficult to estimate without very large sample sizes, and are difficult to interpret.

In the presence of epistasis between SNPs, we would expect modelling interaction effects to lead to increased power over tests that include only the corresponding main effects. However, researchers are often reluctant to consider epistasis because of the fear that a less parsimonious model will decrease the power to detect association unless the interaction effects are large. Marchini *et al.* (2005) have shown that this fear is misplaced by demonstrating that testing for association allowing for additive main effects and additive–additive epistasis for each of $n(n-1)/2$ pairs of SNPs has greater power to detect association with disease than n single-locus tests with additive-only effects. This result holds for a range of interaction models, despite the additional burden of multiple testing via Bonferroni correction.

Evans *et al.* (2006) have investigated the use of two-stage approaches for testing for association between pairs of interacting SNPs and disease to reduce the multiple-testing burden of a complete two-dimensional scan of the genome. In the first stage, a single-locus test of association is performed at each SNP. Only those SNPs passing some predetermined threshold of significance are carried forward to the second stage of testing where various strategies could be employed to test for association of pairs of SNPs with disease, allowing for epistasis. For example, we could test for association between SNPs carried forward to the second stage, or we could consider all possible pairs of SNPs that contain at least one carried forward to the second stage of testing. The choice of a stringent level of significance for the first stage of testing reduces the number of tests performed overall, but also reduces the probability of detecting association of disease with a pair of SNPs made up of a strong epistatic component with minimal main effects. In fact, the results of detailed simulations over a wide range of models of epistasis suggest that two-stage strategies are, in general, less powerful than a two-dimensional scan of the genome, irrespective of the significance threshold at stage one, even while taking account of the additional correction for multiple testing.

An alternative approach to allow for epistasis in association studies is to make use of Bayesian model averaging techniques (Hoeting *et al.*, 1999). This approach can be used to obtain the joint posterior distribution of the main effects of all SNPs and interactions between all pairs of SNPs by considering the space of all possible models of association, conditional on the observed genotypes and disease phenotypes. Conti *et al.* (2003) describe a Bayesian MCMC algorithm to sample from the posterior distribution of models incorporating pair-wise and higher-order epistasis between multiple SNPs in a candidate gene, together with interaction with non-genetic risk factors. They demonstrate the utility of this approach to a population-based association study of colorectal polyps with candidate metabolic genes, allowing for epistasis between polymorphisms and interaction with non-genetic risk factors including smoking and consumption of well-done red meat. The main advantage of Bayesian model averaging approaches is that all SNPs can be considered simultaneously, rather than focusing on each pair of SNPs. However, these methods are likely to be extremely computationally intensive on the scale of WGA studies.

37.7   REPLICATION

Because of the small effect sizes of complex traits, the multiple-testing problem, and limited sample sizes and genotyping resources, most WGA studies have only marginal power to identify most real effects. As a result, many true results may be difficult to distinguish from chance results in the initial study. For example, one of the first WGA studies of Crohn's disease (Duerr *et al.*, 2006) clearly identified a single locus of large effect, but several other loci, which were subsequently shown to be genuinely associated, had significance levels of only $10^{-3}-10^{-5}$ and were initially missed (Cardon, 2006). These challenges highlight the fact that WGA scans are best considered as hypothesis-generating mechanisms, and replication studies are needed to validate and refine initial evidence from WGA studies.

In principle, replication studies are straightforward to design and implement: given assumptions of effect size and frequency, one can calculate sample sizes needed for a specified power, and then conduct careful validation experiments. In practice, however, replication has proved to be one of the most difficult areas in human complex trait genetics, one with a rich history of confusion and false claims. Surprisingly, this standard of failure has been observed in the simplest of studies, involving only single candidate genes with but a few genetic markers and statistical tests. In WGA studies, the challenge will be even greater, as large numbers of SNPs may need follow up.

For replication to serve as a useful measure of validation, association studies need to follow the standard principles of experimental design; i.e. replication studies should match the conditions of the initial study as closely as possible and test explicitly defined, refutable hypotheses. In the context of association studies, this means that the same SNPs, same statistical tests and same phenotypes should be tested in replication as those used in the initial WGA scans. Historically, however, investigators have often used nearby SNPs and conducted multiple statistical tests (which may or may not be reported) on samples with related but non-identical phenotypes, all under the guise of replication. Unsurprisingly, the results have been highly varied, with 'replication' sometimes claimed in error, sometimes missed, and occasionally, claimed with the opposite allele to that originally identified, a 'flip-flop' phenomenon that stretches biological credibility in the absence of unmeasured confounders (Patterson and Cardon, 2005; Lin *et al.*, 2007).

One of the main problems is that investigators have attempted to combine replication studies (using the same SNPs) with fine-mapping studies (using additional SNPs to try to find the strongest associated variants). This is an attractive strategy economically but creates challenges to the inference process since the former is fundamentally a hypothesis-testing exercise, while the latter assumes that the null hypothesis has already been rejected. In the association context, these two strategies are extremely difficult to disentangle. At present, the most robust strategy is to separate the two and conduct them in their natural sequential order (Clarke *et al.*, 2007).

Estimating the power and required sample size for a study to replicate WGA findings is complicated by inflation of effect sizes: for all but the largest genetic effects, the loci taken forward for replication will have an estimated effect size drawn from the upper tail of their effect-size distribution over replicate studies, a phenomenon known as the *Winner's Curse* (Bazerman and Samuelson, 1983). The process of selecting only the most significant loci for further scrutiny is known to invoke the Winner's Curse in both linkage (Goring *et al.*, 2001) and association studies (Lohmueller *et al.*, 2003). Accordingly, using these biased estimates to design replication studies results in overestimates of statistical power and underestimates of sample size requirements. This poses a power conundrum: replication studies will be underpowered if they use the biased WGA estimates of effect size, but if successful, they can provide unbiased effect-size estimates of the loci initially identified by WGA. Because many factors can influence the degree of bias, it is not easy to correct the effect-size estimates from WGA studies. Empirical evidence from the large number of studies currently under way will be extremely valuable in helping to provide rough brackets on the range of bias that is recoverable in replication studies of practical size and scope.

37.8 PROSPECTS FOR WHOLE-GENOME ASSOCIATION STUDIES

Initial reports of WGA studies began as early as 2002 (Ozaki *et al.*, 2002), but studies involving more comprehensive coverage of common variants began in 2005 (Klein *et al.*, 2005; Duerr *et al.*, 2006; Hampe *et al.*, 2007). The initial reports are promising, with each study identifying and validating several novel loci for different diseases. These studies demonstrate that WGA can be successful in identifying common variants for complex traits in humans. Given the chequered history of human genetic association studies (Cardon and Bell, 2001; Ioannidis *et al.*, 2001), this is a major advance in the field.

A large number of WGA studies, each involving thousands of individuals and $> 10^5$ genetic markers, are under way and likely to be reported over the next few years. If the initial trends continue, a number of new genes will soon be identified for complex traits, thereby uncovering new biological pathways and stimulating a cascade of focused experimental studies. For statistical genetics research, these studies raise a series of new opportunities and challenges. Knowledge of new genes will enable detailed studies of gene–gene and gene–environment interactions, determination of population-specific effects, and identification of further loci by conditioning on the initial findings. Moreover, owing to ongoing Biobank initiatives in a number of countries, in which $> 10^5$ individuals are being studied for a wide range of human conditions, it will be possible to conduct population-based assessments of specific gene or haplotype effects, evaluate longitudinal genetic effects, and study the impact of particular genes or gene combinations across traits (pleiotropy).

Ongoing technology advances suggest that WGA studies will not end with the current paradigm of 300–500K SNPs, or even with the immediate target of million-SNP panels. There is intense ongoing commercial research into resequencing the DNA of entire genomes of individuals, which, if achieved, would comprise the first true WGO design. If this technology eventually is able to be conducted with high accuracy and low cost so that it can be deployed in large samples, it will fill a void in the current generation of genetic studies: detection and annotation of rare genetic variants in humans. Present technologies focus on common genetic variants (typically $> 1\%$ frequency in particular populations), leaving the majority of human genetic variation unexplored. Rare variants are known to be of critical importance in human diseases (e.g. BRCA1 and BRCA2), so the upcoming resequencing information will be of immense biomedical importance. The statistical challenges facing such studies will be of a difficulty level matching the potential importance, requiring new designs, inferential frameworks, and theoretical models to exploit the emerging resource.

REFERENCES

Abecasis, G.R., Noguchi, E., Heinzmann, A., Traherne, J.A., Bhattacharya, S., Leaves, N.I., Anderson, G.G., Zhang, Y., Lench, N.J., Carey, A., Cardon, L.R., Moffatt, M.F. and Cookson, W.O.C. (2001). Extent and distribution of linkage disequilibrium in three genomic regions. *American Journal of Human Genetics* **68**, 191–197.

Abecasis, G.R., Cherny, S.S., Cookson, W.O. and Cardon, L.R. (2001). GRR: graphical representation of relationship errors. *Bioinformatics* **17**, 742–743.

Almind, K., Bjorbaek, C., Vestergaard, H., Hansen, T., Echwald, S. and Pedersen, O. (1993). Amino acid polymorphisms of insulin receptor substrate-1 in non-insulin dependent diabetes mellitus. *Lancet* **342**, 828–832.

Balding, D.J. (2006). A tutorial on statistical methods for population association studies. *Nature Reviews Genetics* **7**, 781–791.

Barrett, J.C. and Cardon, L.R. (2006). Evaluating coverage of genome-wide association studies. *Nature Genetics* **38**, 659–662.

Bazerman, M.H. and Samuelson, W.F. (1983). I won the auction but don't want the prize. *Journal of Conflict Resolution* **27**, 618–634.

Benjamini, Y. and Hochberg, Y. (1995). Controlling the false discovery rate: a practical and powerful approach to multiple testing. *Journal of the Royal Statistical Society* **B57**, 289–300.

Brem, R.B. and Kruglyak, L. (2005). The landscape of genetic complexity across 5,700 gene expression traits in yeast. *Proceedings of the National Academy of Sciences of the United States of America* **102**, 1572–1577.

Brem, R.B., Storey, J.D., Whittle, J. and Kruglyak, L. (2005). Genetic interaction between polymorphisms that affect gene expression in yeast. *Nature* **436**, 701–703.

Cardon, L.R., Bell, J.I. (2001). Association study designs for complex diseases. *Nature Reviews Genetics* **2**, 91–99.

Cardon, L.R. (2006). Genetics. Delivering new disease genes. *Science* **314**, 1403–1405.

Carlson, C.S., Eberle, M.A., Rieder, M.J., Yi, Q., Kruglyak, L. and Nickerson, D.A. (2004). Selecting a maximally informative set of single-nucleotide polymorphisms for association analyses using linkage disequilibrium. *American Journal of Human Genetics* **74**, 106–120.

Clark, A.G., Weiss, K.M., Nickerson, D.A., Taylor, S.L., Buchman, A., Stengard, J., Salomaa, V., Vartiainen, E., Perola, M., Boerwinkle, E. and Sing, C.F. (1998). Haplotype structure and population genetic inferences from nucleotide-sequence variation in human lipoprotein lipase. *American Journal of Human Genetics* **63**, 595–612.

Clark, A.G. (2004). The role of haplotypes in candidate gene studies. *Genetic Epidemiology* **27**, 321–333.

Clarke, G.M., Carter, K.W., Palmer, L.J., Morris, A.P., Cardon, L.R. (2007). Fine-mapping vs replication in whole genome association studies. *American Journal of Human Genetics*. (The revised version of the paper is currently being reviewed, and we expect a response imminently).

Clayton, D.G., Walker, N.M., Smyth, D.J., Pask, R., Cooper, J.D., Maier, J.D., Smink, L.J., Lam, A.C., Ovington, N.R., Stevens, H.E., Nutland, S., Howson, J.M.M., Faharn, M., Moor-head, M., Jones, H.B., Falkowski, M., Hardenbol, P., Willis, T.D. and Todd, J.A. (2005). Population structure, differential bias and genomic control in a large case-control association study. *Nature Genetics* **37**, 1243–1246.

Conti, D.V., Cortessis, V., Molitor, J. and Thomas, D.C. (2003). Bayesian modelling of complex metabolic pathways. *Human Heredity* **56**, 83–93.

van Dam, R.M., Hoebee, B., Seidell, J.C., Schaap, M.M., Blaak, E.E. and Reskens, E.J. (2004). The insulin receptor substrate-1 Gly972Arg polymorphism is not associated with type 2 diabetes mellitus in two population-based studies. *Diabetic Medicine* **21**, 752–758.

Dawson, E., Abecasis, G.R., Bumpstead, S., Chen, Y., Hunt, S., Beare, D.M., Pabial, J., Dibling, T., Tinsley, E., Kirby, S., Carter, D., Papaspyridonos, M., Livingstone, S., Ganske, R., Lohmussaar, E., Zernant, J., Tonisson, N., Remm, M., Magi, R., Puurand, T., Vilo, J., Kurg, A., Rice, K., Deloukas, P., Mott, R., Metspalu, A., Bentley, D.R., Cardon, L.R. and Dunham, I. (2002). A first generation linkage disequilibrium map of chromosome 22. *Nature* **418**, 544–548.

Denison, D.G.T. and Holmes, C.C. (2001). Bayesian partitioning for estimating disease risk. *Biometrics* **57**, 143–149.

Devlin, B., Roeder, K. (1999). Genomic control for association studies. *Biometrics* **55**, 997–1004.

Duerr, R.H., Taylor, K.D., Brant, S.R., Rioux, J.D., Silverberg, M.S., Daly, M.J., Seinhard, A.H., Abraham, C., Reueiro, M., Griffiths, A., Dassopoulos, T., Bitton, A., Yong, H., Targan, S., Wu Datta, L., Kistner, E.O., Schumm, L.P., Lee, A.T., Gregersen, P.K., Barmada, M.M., Rotter, J.I., Nicolae, D.L. and Cho J.H. (2006). A genome-wide association study identifies IL23R as an inflammatory bowel disease gene. *Science* **314**, 1461–1463.

Dunning, A.M., Durocher, F., Healey, C.S., Teare, M.D., McBride, S.E., Carlomango, F., Xu, C.F., Dawson, E., Rhodes, S., Ueda, S., Lai, E., Luben, R.N., van Rensburg, E.J., Mannermaa, A., Kataja, V., Rennart, G., Dunham, I., Purvis, I., Easton, D. and Ponder, B.A. (2000). The extent of linkage disequilibrium in four populations with distinct demographic histories. *American Journal of Human Genetics* **67**, 1544–1554.

Durrant, C., Zondervan, K.T., Cardon, L.R., Hunt, S., Deloukas, P. and Morris, A.P. (2004). Linkage disequilibrium mapping via cladistic analysis of SNP haplotypes. *American Journal of Human Genetics* **75**, 35–43.

Evans, D., Marchini, J., Morris, A.P. and Cardon, L.R. (2006). Two stage two locus models in genome wide association. *PLoS Genetics* **2**, e157.

Excoffier, L. and Slatkin, M. (1995). Maximum likelihood estimation of molecular haplotype frequencies in a diploid population. *Molecular Biology and Evolution* **12**, 921–927.

Florez, J.C., Burtt, N., de Bakker, P.I., Almgren, P., Tuomi, T., Holmkvist, J., Gaudet, D., Hudson, T.J., Schaffner, S.F., Daly, M.J., Hirschhorn, J.N., Groop, L., Altshuler, D. (2004). Association testing in 9,000 people fails to confirm the association of the insulin receptor substrate-1 G972R polymorphism with type 2 diabetes. *Diabetes* **53**, 3313–3318.

Gabriel, S.B., Schaffner, S.F., Nguyen, H., Moore, J.M., Roy, J., Blumenstiel, B., Higgins, J., de Felice, M., Lochner, A., Faggart, M., Lui-Cordero, S.N., Rotimi, C., Adeyemo, A., Cooper, R., Ward, R., Lander, E.S., Daly, M.J. and Althshuler, D. (2002). The structure of haplotype blocks in the human genome. *Science* **296**, 2225–2229.

Goring, H.H., Terwilliger, J.D. and Blangero, J. (2001). Large upward bias in estimation of locus-specific effects from genomewide scans. *American Journal of Human Genetics* **69**, 1357–1369.

Grant, S.F.A., Thorleifsson, G., Reynisdottir, I., Benediktsson, R., Manolescu, A., Sainz, J., Helgason, A., Stefansson, H., Emilsson, V., Helgadottir, A., Styrkarsdottir, U., Magnusson, K.P., Walters, G.B., Palsdottir, E., Jonsdottir, T., Gudmundsdottir, T., Gylfason, A., Saemundsdottir, J., Wilensky, R.L., Reilly, M.P., Rader, D.J., Bagger, Y., Christiansen, C., Gudnason, V., Sigurdsson, G., Thorsteinsdottir, U., Gulcher, J.R., Kong, A. and Stefansson, K. (2006). Variant of transcription factor 7-like 2 (TCF7L2) gene confers risk of type 2 diabetes. *Nature Genetics* **38**, 320–323.

Hampe, J., Franke, A., Rosenstiel, P., Till, A., Teuber, M., Huse, K., Albrecht, M., Mayr, G., De La Vega, F.M., Briggs, J., Gunther, S., Prescott, N.J., Onnie, C.M., Hasler, R., Sipos, B., Folsch, U.R., Lengauer, T., Platzer, M., Matthew, C.G., Krawczak, M. and Schreiber, S. (2007). A genome-wide association scan of nonsynonymous SNPs identifies a susceptibility variant for Crohn disease in ATG16L1. *Nature Genetics* **39**, 207–211.

Hoeting, J.A., Madigan, D., Raftery, A.E. and Volinsky, C.T. (1999). Bayesian model averaging: a tutorial. *Statistical Science* **14**, 382–417.

Hugot, J.P., Chamaillard, M., Zouali, H., Lesage, S., Cezard, J.P., Belaiche, J., Almer, S., Tysk, C., O'Moiran, C.A., Gassull, M., Binder, V., Finkel, Y., Cortot, A., Modigliani, R., Laurent-Puig, P., Gower-Rousseau, C., Macry, J., Colombel, J.F., Sahbatou, M. and Thomas, G. (2001). Association of NOD2 leucine-rich repeat variants with susceptibility to Crohn's disease. *Nature* **411**, 599–603.

Ioannidis, J.P., Ntzani, E.E., Trikalinos, T.A. and Contopoulos-Ioannidis, D.G. (2001). Replication validity of genetic association studies. *Nature Genetics* **29**, 306–309.

The International HapMap Consortium (2005). A haplotype map of the human genome. *Nature* **437**, 1299–1320.

Johnson, G.C.L., Esposito, L., Barratt, B.J., Smith, A.N., Heward, J., DiGenova, G., Veda, H., Cordell, H.J., Eaves, I.A., Dudbridge, F., Twells, R.C.J., Payne, F., Hughes, W., Nutland, S., Stevens, H., Carr, P., Tuomilehto-Wolf, E., Tunmilehto, J., Gough, S.C.L., Clayton, D.G. and Todd, J.A. (2001). Haplotype tagging for the identification of common disease genes. *Nature Genetics* **29**, 233–237.

Klein, R.J., Zeiss, C., Chew, E.Y., Tsai, J.Y., Sackler, R.S., Haynes, C., Henning, A.K., SanGiovanni, J.P., Mane, S.M., Mayne, S.T., Bracken, M.B., Ferrix, F.L., Ott, J., Barnstable, C. and Hoh, J. (2005). Complement factor H polymorphism in age-related macular degeneration. *Science* **308**, 385–389.

Knorr-Held, L. and Rasser, G. (2000). Bayesian detection of clusters and discontinuities in disease maps. *Biometrics* **46**, 13–21.

Lawrence, R., Evans, D.E., Morris, A.P., Ke, X., Hunt, S., Paolucci, M., Ragoussis, J., Deloukas, P., Bentley, D. and Cardon, L.R. (2005). Genetically indistinguishable SNPs and their influence on inferring the location of disease-associated variants. *Genome Research* **15**, 1503–1510.

Lin, P.I., Vance, J.M., Pericak-Vance, M.A. and Martin, E.R. (2007). No gene is an island: the flip-flop phenomenon. *American Journal of Human Genetics* **80**, 531–538.

Lohmueller, K.E., Pearce, C.L., Pike, M., Lander, E.S. and Hirschhorn, J.N. (2003). Meta-analysis of genetic association studies supports a contribution of common variants to susceptibility to common disease. *Nature Genetics* **33**, 177–182.

Mackay, T.F. (2001). The genetic architecture of quantitative traits. *Annual Review of Genetics* **35**, 303–339.

Marchini, J., Donnelly, P. and Cardon, L.R. (2005). Genome-wide strategies for detecting multiple loci that influence complex diseases. *Nature Genetics* **37**, 413–417.

Minichiello, M.J. and Durbin, R. (2006). Mapping trait loci by use of inferred ancestral recombination graphs. *American Journal of Human Genetics* **79**, 910–922.

Molitor, J., Marjoram, P. and Thomas, D. (2003a). Application of Bayesian spatial statistical methods to the analysis of haplotype effects and gene mapping. *Genetic Epidemiology* **25**, 95–105.

Molitor, J., Marjoram, P. and Thomas, D. (2003b). Fine scale mapping of disease genes with multiple mutations via spatial clustering techniques. *American Journal of Human Genetics* **73**, 1368–1384.

Morris, A.P. (2005). Direct analysis of unphased SNP genotype data in population-based association studies via Bayesian partition modelling of haplotypes. *Genetic Epidemiology* **29**, 91–107.

Morris, A.P. (2006). A flexible Bayesian framework for modelling haplotype association with disease allowing for dominance effects of the underlying causative variants. *American Journal of Human Genetics.* **79**, 679–694.

Morris, A.P., Whittaker, J.C. and Balding, D.J. (2002). Fine-scale mapping of disease loci via coalescent modelling of genealogies. *American Journal of Human Genetics* **70**, 686–707.

Morris, A.P., Whittaker, J.C. and Balding, D.J. (2004). Little loss of information due to unknown phase for fine-scale linkage disequilibrium mapping with single nucleotide polymorphism genotype data. *American Journal of Human Genetics* **74**, 945–953.

Ozaki, K., Ohnishi, Y., Iida, A., Sekine, A., Yamada, R., Tsunoda, T., Sato, H., Sato, H., Hori, M., Nakamura, Y. and Tanaka, T. (2002). Functional SNPs in the lymphotoxin-alpha gene that are associated with susceptibility to myocardial infarction. *Nature Genetics* **32**, 650–654.

Parikh, H. and Groop, L. (2004). Candidate genes for type 2 diabetes. *Reviews in Endocrine and Metabolic Disorders* **5**, 151–176.

Patil, N., Bem, A.J., Hinds, D.A., Barrett, W.A., Doshi, J.M., Hacker, C.R., Kautzer, C.R., Lee, D.H., Majoribanks, C., McDonough, D.P., Nguyen, B.T.N., Norris, M.C., Sheehan, J.B., Sten, N., Stern, D., Stokowski, R.P., Thomas, D.J., Trulson, M.O., Vyas, K.R., Frazer, K.A., Fodor, S.P.A. and Cox, D.R. (2001). Blocks of limited haplotype diversity revealed by high resolution scanning of human chromosome 21. *Science* **294**, 1719–1722.

Patterson, M. and Cardon, L. (2005). Replication publication. *PLoS Biology* **3**, e327.

Phillips, M.S., Lawrence, R., Sachidanandam, R., Morris, A.P., Balding, D.J., Donaldson, M.A., Studebaker, J.F., Ankener, W.M., Alfisi, S.V., Kuo, F.S., Camisa, A.L., Pazorov, V., Scott, K.E., Carey, B.J., Faith, J., Katari, G., Bhatti, H.A., Cyr, J.M., Derohannessian, V., Elousa, C., Forman, A.M., Grecco, N.M., Hoch, C.R., Kuebler, J.M., Lathrop, J.A., Mockler, M.A., Nachtman, E.P., Restine, S.L., Varde, S.A., Hozza, M.J., Gelfand, C.A., Broxholme, J., Abecasis, G.R., Boyce Jacino, M.T. and Cardon, L.R. (2003). Chromosome-wide distribution of haplotype blocks and the role of recombination hot spots. *Nature Genetics* **33**, 382–387.

Pritchard, J.K., Stephens, M. and Donnelly, P.J. (2000a). Inference of population structure using multilocus genotype data. *Genetics* **155**, 945–959.

Pritchard, J.K., Stephens, M., Rosenberg, N.A. and Donnelly, P. (2000b). Association mapping in structured populations. *American Journal of Human Genetics* **67**, 170–181.

Reich, D.E., Cargill, M., Bolk, S., Ireland, J., Sabeti, P.C., Richter, D.J., Lavery, T., Kouyoumjian, R., Farhadian, S.F., Ward, R. and Lander, E.S. (2001). Linkage disequilibrium in the human genome. *Nature* **411**, 199–204.

Risch, N. and Merikangas, K. (1996). The future of genetic studies of complex human diseases. *Science* **273**, 1516–1517.

Rubinsztein, D.C. and Easton, D.F. (1999). Apolipoprotien E genetic variation and Alzheimer's disease: a meta analysis. *Dementia and Geriatric Cognitive Disorders* **10**, 199–209.

Schaid, D.J., Rowland, C.M., Tines, D.E., Jacobson, R.M. and Poland, G.A. (2002). Score tests for association between traits and haplotypes when linkage phase is ambiguous. *American Journal of Human Genetics* **70**, 425–434.

Setakis, E., Stirnadel, H. and Balding, D.J. (2006). Logistic regression protects against population structure in genetic association studies. *Genome Research* **16**, 290–296.

Sladek, R., Rocheleau, G., Rung, J., Dina, C., Shen, L., Serre, D., Boutin, P., Vincent, D., Belisle, A., Hadjadj, S., Balkau, B., Heude, B., Charpentier, G., Hudson, T.J., Montpetit, A., Pshezhetsky, A.V., Prentki, M., Posner, B.I., Balding, D.J., Meyre, D., Polychronakos, C. and Froguel, P. (2007). A genome-wide association study identifies novel risk loci for type 2 diabetes. *Nature* **445**, 881–885.

Stephens, M. and Donnelly, P. (2003). A comparison of Bayesian methods for haplotype reconstruction from population genetic data. *American Journal of Human Genetics* **73**, 1162–1169.

Stephens, M., Smith, N.J. and Donnelly, P. (2001). A new statistical method for haplotype reconstruction from population data. *American Journal of Human Genetics* **68**, 978–989.

Storey, J.D., Akey, J.M. and Kruglyak, L. (2005). Multiple locus linkage analysis of genomewide expression in yeast. *PLoS Biology* **3**, e267.

Taillon-Miller, P., Bauer-Sardina, I.B., Saccone, N.L., Pulzel, J., Laitinen, T., Cao, A., Kere, J., Pilia, G., Rice, J.P. and Kwok, P.Y. (2000). Juxtaposed regions of extensive and minimal linkage disequilibrium in human Xq25 and Xq28. *Nature Genetics* **25**, 324–328.

Tavtigian, S., Simard, J., Teng, D., Abtin, V., Baumgard, M., Beck, A., Camp, N., Carillo AR, Chen Y, Dayananth P, Desrochers M, Dumont M, Farnham JM, Frank D, Frye C, Ghaffari S, Gupte JS, Hu R, Iliev D, Janecki T, Kort EN, Laity KE, Leavitt A, Leblanc G, McArthur-Morrison J, Pederson A, Penn B, Peterson KT, Reid JE, Richards S, Schroeder M, Smith R, Snyder SC, Swedlund B, Swensen J, Thomas A, Tranchant M, Woodland AM, Labrie F, Skolnick MH, Neuhausen S, Rommens J, Cannon-Albright LA. (2001). A candidate prostate cancer susceptibility gene at chromosome 17p. *Nature Genetics* **27**, 172–180.

Templeton, A.R., Boerwinkle, E. and Sing, C.F. (1987). A cladistic analysis of phenotypic associations with haplotypes inferred from restriction endonuclease mapping. I. Basic theory and an analysis of alcohol dehydrogenase activity in *Drosophila*. *Genetics* **117**, 343–351.

Templeton, A.R., Crandall, K.A. and Sing, C.F. (1992). A cladistic analysis of phenotypic associations with haplotypes inferred from restriction endonuclease mapping and DNA sequence data. III. Cladogram estimation. *Genetics* **132**, 619–633.

Templeton, A.R. and Sing, C.F. (1993). A cladistic analysis of phenotypic associations with haplotypes inferred from restriction endonuclease mapping. IV. Nested analyses with cladogram uncertainty and recombination. *Genetics* **134**, 659–669.

Templeton, A.R., Sing, C.F., Kessling, A. and Humphries, S. (1988). A cladistic analysis of phenotypic associations with haplotypes inferred from restriction endonuclease mapping. II. The analysis of natural populations. *Genetics* **120**, 1145–1154.

Valdar, W., Solberg, L.C., Gauguier, D., Burnett, S., Klenerman, P., Cookson, W.O., Taylor, M.S., Rawlins, J.N.P., Mott, R. and Flint, J. (2006). Genome-wide genetic association of complex traits in heterogeneous stock mice. *Nature Genetics* **38**, 879–887.

Voight, B.F. and Pritchard, J.K. (2005). Confounding from cryptic relatedness in case-control association studies. *PLoS Genetics* **1**, e32.

Weir, B.S. (1996). *Genetic Data Analysis II*. Sinauer Associates Inc: Sunderland, Massachusetts, USA, pp 112–133.

Weir, B.S., Cardon, L.R., Anderson, A.D., Neilsen, D.M. and Hill, W.G. (2005). Measures of human population structure show heterogeneity among genomic regions. *Genome Research* **15**, 1468–1476.

The Wellcome Trust Case Control Consortium (2007). A genome wide association study of 14,000 cases of seven common diseases and 3,000 shared controls. *Nature* (in press).

Wright, A.F. and Hastie, N.D. (2001). Complex genetic diseases: controversy over the Croesus code. *Genome Biology* **2**, COMMENT2007.

Zaykin, D.V., Westfall, P.H., Young, S.S., Karnoub, M.A., Wagner, M.J. and Ehm, M.G. (2002). Testing association of statistically inferred haplotypes with discrete and continuous traits in samples of unrelated individuals. *Human Heredity* **53**, 79–91.

Zeggini, E., Parkinson, J., Halford, S., Owen, K.R., Frayling, T.M., Walker, M., Hitman, G.A., Levy, J.C., Sampson, M.J., Feskens, E.J.M., Hattersley, A.T. and McCarthy, M.I. (2004). Association studies of insulin receptor substrate 1 gene (IRS1) variants in type 2 diabetes samples enriched for family history and early age of onset. *Diabetes* **53**, 3319–3322.

Zollner, S. and Pritchard, J.K. (2005). Coalescent-based association mapping and fine mapping of complex trait loci. *Genetics* **169**, 1071–1092.

Zondervan, K.T. and Cardon, L.R. (2004). The complex interplay among factors that influence allelic association. *Nature Reviews Genetics* **5**, 89–100.

38

Family-based Association

F. Dudbridge

MRC Biostatistics Unit, Institute for Public Health, Cambridge, UK

Family-based methods are a useful means to protect against the confounding effects of population stratification and other factors. The transmission/disequilibrium test is a test of linkage and association that is at once the most popular such method and the starting point for numerous extensions and generalizations, the most prominent of which are reviewed in this chapter. A logistic regression formulation leads to natural extensions for multiallelic markers, haplotypes and quantitative traits. When haplotypes are uncertain, special methods are described for testing for association within families. An approach suitable for general pedigree structures, which derives the conditional distribution of a covariance score function given the pedigree structure, is described. For quantitative traits, ordinary linear regression models are adapted for the family-based design and allow the modelling of linkage in the error variance. When testing association in the presence of linkage, some care is required to avoid bias due to correlated transmissions to siblings, and a likelihood-based solution to this problem that allows for uncertain haplotypes and missing parental genotypes is presented. Some ongoing work in the area is summarized.

38.1 INTRODUCTION

Family-based association studies include a broad range of methods that aim to detect association between genetic markers and disease or quantitative phenotypes, using families as the primary sampling unit. Most often, nuclear families consisting of the two parents and a number of full siblings are used, but extended pedigrees may also be used for testing association, as may subsets of nuclear families such as sib pairs or single parent families. As might be expected given the diversity of family structures, there is a large range of methods for family-based association; here only the most prominent ones will be discussed, and these are able to analyse all of the most common family structures.

The appeal of family-based association is due to several reasons. First, in common with all epidemiological studies, there is the need to match cases to controls at the population and, ideally, the individual level. As discussed in **Chapter 36**, inappropriate matching of cases to controls can lead to selection bias and confounding effects if gene frequencies differ between case and control populations. A situation of particular concern is population

Handbook of Statistical Genetics, Third Edition. Edited by D.J. Balding, M. Bishop and C. Cannings.
© 2007 John Wiley & Sons, Ltd.   ISBN: 978-0-470-05830-5.

stratification (see **Chapter 35**) in which the study population actually consists of a mixture of discrete sub-populations that differ both in gene frequency and disease prevalence. Case-control sampling might then recruit relatively more cases from the sub-populations with higher prevalence, leading to differences in marginal gene frequency between cases and controls even if a gene has no influence on disease.

Secondly, family-based association tests are confounded with linkage (Ott, 1989), in the sense that association can only be detected in the presence of linkage. While such confounding effects can present technical difficulties in interpreting results (Whittaker et al., 2000), they have the positive effect of confirming that truly associated markers are physically close to the causal genetic variant. This property also reinforces the intuitive view that genetic phenotypes are inherited and should therefore be detectable in families.

Another important advantage of family-based designs is that they allow parent of origin studies that allow for imprinting effects and interactions between maternal and foetal genotypes (Weinberg, 1999). Such analyses are simply not possible in population studies.

By choosing controls from within the same families as the cases, the problem of population stratification is largely eliminated, and the confounding effects of other factors can be substantially reduced. For example, members of the same family are likely to share a high proportion of environmental and dietary exposures, so that differences in phenotype are more likely to be due to genetic differences than to unmeasured confounders. Family-based designs are thus attractive from the point of view of identifying true causal effects. Their principal disadvantages arise from practical matters of recruitment and cost. Suitable family members may be unavailable for genotyping, for example in late-onset diseases in which both parents may be deceased. Family controls are over-matched, since they share, in expectation, up to one-half of their genome with the cases, leading to higher genotyping costs per unit of information as compared to unrelated controls (Cardon and Palmer, 2003). For these reasons, unrelated controls are currently favoured for large-scale genome association scans (see **Chapter 37**), but family-based controls remain useful, particularly for replication studies that can confirm previously suggested associations to a higher standard of evidence.

The current use of family-based controls originated with the haplotype relative risk method (Falk and Rubinstein, 1987). This uses a sample of affected cases together with each of their two parents, and for each case, the alleles present in the parents but not transmitted to the case are combined into a control genotype that can be analysed using standard methods for unrelated cases and controls (see **Chapter 36**). A variation, termed *haplotype-based haplotype relative risk*, treats the allele rather than the genotype as the unit of observation, again following standard methods for unrelated subjects (Terwilliger and Ott, 1992). The problem of estimating genotype relative risks in family-based designs was first addressed by Self et al. (1991), from which much of the regression methodology described below is derived.

The transmission/disequilibrium test (TDT) (Spielman et al., 1993) is a landmark in the development of family-based association, both as a point of departure for many extensions and generalizations, and as the baseline standard to which new methods are often compared. Its appeal lies in its simplicity, close relation to standard statistical tests and absolute protection from population stratification. In its simplest form, it is similar to the haplotype-based haplotype relative risk, except that it performs a matched analysis of untransmitted alleles versus transmitted alleles, rather than an unmatched analysis. This basic change allows the TDT to be viewed both as a test of association and as a test of linkage, and also protects against deviations from Hardy–Weinberg equilibrium that could be induced by non-random mating. Because of these properties, the TDT has become a very popular method and has spawned many extensions. Perhaps more than in other areas of statistical genetics, these methods have become identified with the software that implements them, owing to the variety of possible family structures and difficulties in fitting them to standard statistical models. Here, some of the more established methods will be reviewed, with references to the relevant software, and some comparisons will be drawn between these methods.

The structure of the chapter is as follows. Section 38.2 reviews the TDT and some of its properties. Section 38.3 sets out a logistic regression formulation of the TDT and describes how it can be used to include multiple alleles, genotypes and environmental covariates. Section 38.4 discusses the problem of uncertain haplotypes and missing parental genotypes, and describes the popular TRANSMIT program. In Section 38.5 some approaches for analysing general pedigrees are described, starting with the sib-TDT for sibships and progressing to a general approach implemented in the FBAT software. Methods for quantitative traits are discussed in Section 38.6, with emphasis on the popular QTDT software that allows joint modelling of linkage and association. Section 38.7 focuses on testing association in the presence of linkage; some solutions are given, including a likelihood-based program, UNPHASED, that relates closely to the original TDT. Finally, Section 38.8 gives a summary of ongoing work in this area.

38.2 TRANSMISSION/DISEQUILIBRIUM TEST

In its original form, the TDT considers the transmission of the variant allele of a biallelic marker from heterozygous parents to affected children. Table 38.1 shows the four cell counts relating to the transmissions from a parent to an affected child, arranged as a contingency table. The TDT treats the untransmitted allele as a matched control to the transmitted allele, in which case only the heterozygous parents are informative. The null hypothesis is that heterozygous parents transmit the two alleles with equal probability. Let $T = 1$ when the parent transmits the variant allele, and $T = 0$ otherwise; then, under equally likely transmission $E(T) = \frac{1}{2}$, $\mathrm{var}(T) = \frac{1}{4}$, and by applying the central limit theorem over the $n_{12} + n_{21}$ heterozygous parents, we have the TDT statistic

$$TDT = \frac{(n_{12} - n_{21})^2}{(n_{12} + n_{21})},\qquad(38.1)$$

which is asymptotically distributed as χ^2 with 1 degree of freedom.

Table 38.1 Counts of transmissions from n parents of affected children, used in calculating the transmission/disequilibrium test.

		Non-transmitted allele		
		Variant	Common	Total
Transmitted	Variant	n_{11}	n_{12}	$n_{1.}$
allele	Common	n_{21}	n_{22}	$n_{2.}$
	Total	$n_{.1}$	$n_{.2}$	n

It can be formally shown that equal transmission probability occurs either when there is no linkage between marker and disease, or when there is no association (Ewens and Spielman, 1995). This can be intuitively seen by considering the parental chromosomes that carry disease alleles that are penetrant in the child (assume that Mendelian segregation holds on chromosomes carrying non-penetrant alleles, leading to equal transmission probability). Figure 38.1 shows the four situations that are possible when a parent carries a penetrant allele and is heterozygous at the marker. When there is no linkage, either marker allele is transmitted with the disease allele with probability 1/2 regardless of the haplotype distribution in parental chromosomes: scenarios (a) and (b) are equally likely, as are (c) and (d). But when there is no association, there is no information on which marker allele occurs on the same haplotype as the disease allele, given that the parent is heterozygous: scenarios (a) and (c) are equally likely, as are (b) and (d). Therefore, either marker allele occurs on the disease haplotype with probability 1/2 and is then transmitted with probability 1/2 regardless of the recombination fraction. This means that the null hypothesis of the TDT may be taken to be either no linkage or no association.

Although the TDT is often used as a test of association, it was first proposed as a test of linkage (Spielman *et al.*, 1993). This was motivated by studies of the insulin gene in

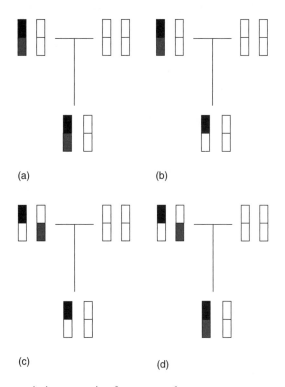

(a)   (b)

(c)   (d)

Figure 38.1 Four transmission scenarios for a parent heterozygous at a test marker. The disease locus is shown above the marker locus. Disease risk allele is shown in black, variant marker allele in grey: (a) marker allele on disease chromosome, no recombination; (b) marker allele on disease chromosome, with recombination; (c) marker allele on normal chromosome, no recombination; (d) marker allele on normal chromosome, with recombination.

Table 38.2 Counts of transmissions in n case–parent trios, used in calculating dominant and recessive versions of the TDT. Allele 1 denotes the common allele and allele 2 denotes the variant alleles.

		Offspring genotype			
		1/1	1/2	2/2	Total
	1/1 × 1/2	n_{11}	n_{12}	–	$n_{1.}$
Mating type	1/2 × 1/2	n_{21}	n_{22}	n_{23}	$n_{2.}$
	1/2 × 2/2	–	n_{32}	n_{33}	$n_{3.}$
	Total	$n_{.1}$	$n_{.2}$	$n_{.3}$	n

type-1 diabetes, which had shown population association in candidate-gene studies but little evidence for linkage in sib pairs. Thus, the presence of association was established and the TDT was able to show that this association coincided with linkage, as opposed to arising, say, from population stratification. This application has remained useful for candidate-gene studies, but in positional cloning studies that start with genome scans for linkage, it has been more appropriate to treat the TDT as a test of association for fine mapping. Now, with the move towards genome-wide association scans (see **Chapter 37**), the situation will again arise in which population association is demonstrated without evidence for linkage, and the interpretation of the TDT will revert to its original sense as a test of linkage. These considerations have an important bearing on how data are analysed in extended sibships or pedigrees, as discussed in Section 38.7.

The sampling units for the TDT are single parents of affected children, though both parents must be available for each child. This treats the alleles in the affected children as independent, which is true under the null hypothesis. However, when there is linkage and association, the alleles in the children are independent only under the multiplicative model of relative risk (see **Chapter 36**), so the TDT may not be the most powerful test in other models of risk. TDT statistics have been derived for dominant and recessive models that may be preferred in those instances (Schaid and Sommer, 1994). These treat the case–parent trio as the sampling unit, with the relevant cell counts given in Table 38.2. These give the χ^2 statistics

$$TDT_{\text{DOM}} = \frac{\left[n_{23} + n_{22} - \frac{3}{4}n_{2.} + n_{12} - \frac{1}{2}n_{1.}\right]^2}{\frac{3}{16}n_{2.} + \frac{1}{4}n_{1.}} \qquad (38.2)$$

and

$$TDT_{\text{REC}} = \frac{\left[n_{33} - \frac{1}{2}n_{3.} + n_{23} - \frac{1}{4}n_{2.}\right]^2}{\frac{1}{4}n_{3.} + \frac{3}{16}n_{2.}} \qquad (38.3)$$

38.3 LOGISTIC REGRESSION MODELS

The TDT can be derived from a logistic regression model that offers scope for extending the test in many ways. Recalling that the TDT is equivalent to a matched analysis of transmitted versus untransmitted alleles, the standard approach from epidemiology is to define a conditional logistic regression model with transmission as the random outcome and alleles as predictors. For a biallelic marker, the log-odds ratio β for transmission is given by

$$\text{logit}(p) = \log\left(\frac{p}{1-p}\right) = \beta, \tag{38.4}$$

where p is the probability that the variant allele is transmitted by a heterozygous parent. The likelihood for n heterozygous parents is

$$L = \prod_{i=1}^{n} \frac{\exp(\delta_i \beta)}{1 + \exp(\beta)}, \tag{38.5}$$

where $\delta_i = 1$ when parent i transmits the variant allele, and is 0 otherwise. Standard likelihood theory yields a score test for $\beta = 0$ that is identical to that obtained using the TDT.

For multiallelic markers, the model can be extended to fit a regression parameter to each pair of transmitted/non-transmitted alleles. However, this can result in a large number of parameters, and a more economical approach is to define marginal odds ratios for the transmission of each allele, and to fit these parameters in a model that conditions on the parental genotype. Thus, for alleles $i = 1, \ldots, k$, we define parameters $\beta_1, \ldots, \beta_k$ and for an i/j parent transmitting allele i we have

$$\text{logit}(p_{ij}) = \beta_i - \beta_j. \tag{38.6}$$

The likelihood contribution for such a parent is $\dfrac{\exp(\beta_i)}{\exp(\beta_i) + \exp(\beta_j)}$. The null hypothesis is that all $\beta_i = 0$, which may be tested using a joint score test, or by using a likelihood ratio test that compares the likelihood under the null to that obtained when it is maximized over all values of $\beta_1, \ldots, \beta_k$ (one parameter must be set to 0 for identifiability). This then yields maximum likelihood estimates of the odds ratios for transmission, which may be regarded as relative risks for disease (Cordell and Clayton, 2002). The model can be fitted using standard software for conditional logistic regression, and the null hypothesis has been shown to correspond to no linkage between marker and disease, or no association between any allele and disease (Sham and Curtis, 1995).

The regression model can be extended in many ways. A useful approach is to regard the analysis not in terms of transmissions from parents, but rather in terms of comparing cases, the affected children, to controls formed from the other combinations of parental alleles. The likelihood contribution from a full case–parent trio is equivalent to that obtained by matching the case to three controls corresponding to each of the other three child genotypes that could be formed from the parental genotypes. The genotypes may now be viewed as predictors and the genotype relative risks estimated, and covariate effects may also be added to the model (Self *et al.*, 1991). Additional markers can be included as covariates, to distinguish the effects of multiple loci in linkage disequilibrium (Cordell and Clayton, 2002).

The regression approach also permits a natural model for parent of origin effects. By defining separate relative risk parameters for the maternal and paternal alleles, tests can be constructed to compare their effects. For example, if a test of equality between maternal and paternal relative risks showed a significant difference, this would provide evidence of imprinting. If imprinting is present, say in the paternally inherited copy, then a test of the maternal relative risk alone would be more powerful than one of the marginal relative risk. Furthermore, the maternal genotype could be included as an additional covariate, allowing for maternal–foetal genotype interactions at the prenatal stage. Indeed, such an effect could confound a comparison of maternal and paternal risks, if not properly accounted for (Weinberg, 1999).

In parent of origin analyses, it is possible to increase the number of family-based controls by assuming symmetry of parental mating-type probabilities. In other words, the probability of the mother having genotype g_1 and the father having genotype g_2 is assumed to equal that of the mother having g_2 and the father having g_1. Under this assumption, further controls can be constructed by exchanging the genotypes of the parents. For example, suppose the mother has genotype 1/2 and the father 3/4, and the case has genotype 1/3. Then the family-based controls are 1/4, 2/3 and 2/4, and with the symmetry assumption we may also use 3/1, 4/1, 3/2 and 4/2 as controls, where the genotypes are ordered with the maternal allele first. This approach, termed *conditioning on exchangeable parental genotypes* (Cordell *et al.*, 2004) can lead to a substantial increase in effective sample size, compensating for the increased number of parameters that must be estimated in parent of origin models.

Provided the analysis is conditional on the parental genotypes, the TDT is robust to population stratification and will also be a test of linkage. However, it is useful to place the conditional model within an unconditional model, as this will be relevant for the discussion of missing data approaches. Assuming that case–parent trios are ascertained through the disease status of the child, the full likelihood of a trio is

$$L^{(f)} = \Pr(F, M, C | D) = \frac{\Pr(D | F, M, C) \cdot \Pr(F, M, C)}{\Pr(D)}, \tag{38.7}$$

where F, M, C are the genotypes of the father, mother and child, respectively, and D denotes that the child has disease. Assume that (1) conditional on the child genotype, the disease status of the child is independent of the parental genotypes; (2) there is Mendelian transmission, so all children are equally likely from the parents; then this simplifies to

$$L^{(f)} = \frac{\Pr(D|C) \cdot \Pr(F, M)}{\sum\limits_{f, m \in G} \Pr(f, m) \sum\limits_{c \in S(f, m)} \Pr(D|c)}, \tag{38.8}$$

where G is the set of all possible genotypes and $S(f, m)$ is the set of possible child genotypes for parents f and m. To relate this to the conditional likelihood used for the TDT, this can be written as

$$L^{(f)} = \frac{\Pr(D|C)}{\sum\limits_{c \in S(F, M)} \Pr(D|c)} \cdot \frac{\Pr(F, M) \sum\limits_{c \in S(F, M)} \Pr(D|c)}{\sum\limits_{f, m \in G} \Pr(f, m) \sum\limits_{c \in S(f, m)} \Pr(D|c)}$$

$$= L^{(c)} \cdot L^{(p)}. \tag{38.9}$$

In other words, the full likelihood is the product of a conditional likelihood and a parental contribution that gives the probability of observing the parents of an affected child. The parental likelihood includes a mating-type distribution $\Pr(F, M)$ that could be mis-specified under population stratification. Inference on the conditional likelihood, using a log-odds model for $\Pr(D|C)$, gives the conditional logistic regression TDT (38.6).

38.4 HAPLOTYPE ANALYSIS

The use of haplotypes to identify disease loci has received much attention in recent times (Schaid, 2004) because disease mutations are assumed to have arisen on a single founder chromosome whose haplotype should be more strongly associated to disease than any single marker. Furthermore, haplotype analysis can be used to detect and characterize multiple interacting loci, when haplotypes are constructed from disease alleles themselves.

In principle, haplotypes can be analysed in the same way as any multiallelic marker. There is however a practical difficulty in that haplotypes are usually not directly observed, but must be deduced from genotype data; and then the haplotypes may be ambiguous. For example, a subject that is heterozygous at two loci, that is, with genotypes 1/2 and 1/2, can have two possible pairs of haplotypes: 1–1 and 2–2, or 1–2 and 2–1. When haplotypes are ambiguous, statistical methods can be used to model the possible solutions, for example by maximum likelihood (see **Chapter 36**) or by coalescent modelling (see **Chapter 25**; Stephens *et al.*, 2001). In family data, there are additional complications, including the fact that many haplotypes are deducible after all, and that introducing a model of the haplotype distribution may not retain robustness to population stratification.

When there are complete genotype data available in a family, the haplotypes can often be deduced with certainty. Ambiguity only occurs when there are markers at which both parents are heterozygous for the same alleles (a situation known as *intercross* in experimental genetics) and the child is also heterozygous. Here, we cannot say which parent transmitted which allele, although the transmitted and untransmitted alleles themselves are identifiable. Haplotype ambiguity then occurs if another marker exists at which more than one allele is present in the parents. Because ambiguous haplotypes may be relatively uncommon, particularly for multiallelic markers, it may be possible to perform a standard TDT using just the deduced haplotypes. However, such an approach is biased unless we allow for the fact that not all families could be used (Dudbridge

Figure 38.2 Family in which haplotypes can be deduced, but would not be deducible from other children of the same parents.

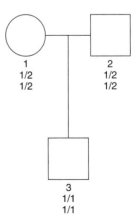

et al., 2000). Figure 38.2 shows a family in which the haplotypes can be deduced, so that we can score two transmissions of the 1–1 haplotype. However, for these parents, the haplotypes can only be deduced in this case and in the case where the 1–1 haplotype is not transmitted twice, so the expected transmission count is 1, with variance 1. We have seen that the usual TDT assumes an expected transmission count of 1 for two parents, with variance 1/2, so scoring just the certain haplotype transmissions would lead to an underestimate of the true variance and an inflated test statistic. Programs are available to adjust the TDT for this situation (Dudbridge *et al.*, 2000; Markianos *et al.*, 2001).

An alternative for analysing only the certain haplotypes is to fit a statistical model to the missing data, using maximum likelihood. A difficulty is that if we assume Hardy–Weinberg equilibrium, which is convenient for limiting the number of parameters, we may mis-specify the distribution of the uncertain haplotypes and lose robustness to population stratification. Clayton (1999) proposed a compromise solution that is implemented in the software TRANSMIT. The method uses a model for haplotype frequencies in a way that reduces its impact on inference on the relative risks.

Recall the factorization of the likelihood contribution from a case–parent trio

$$
L^{(f)} = \frac{\Pr(D|C)}{\sum\limits_{c \in S(F,M)} \Pr(D|c)} \cdot \frac{\Pr(F, M) \sum\limits_{c \in S(F,M)} \Pr(D|c)}{\sum\limits_{f,m \in G} \Pr(f, m) \sum\limits_{c \in S(f,m)} \Pr(D|c)}
$$

$$
= L^{(c)} \cdot L^{(p)}. \tag{38.10}
$$

Let $\boldsymbol{\beta}$ be the vector of haplotype log relative risks. Let $\boldsymbol{\gamma}$ parameterize the haplotype frequencies such that the frequency vector is $\dfrac{\exp(\boldsymbol{\gamma})}{\sum \exp(\gamma_i)}$. Thus, $\boldsymbol{\beta}$ specifies the model for $\Pr(D|C)$ and, assuming Hardy–Weinberg equilibrium, $\boldsymbol{\gamma}$ specifies the model for $\Pr(F, M)$. The parental likelihood depends on both $\boldsymbol{\gamma}$ and $\boldsymbol{\beta}$, whereas the conditional likelihood depends only on $\boldsymbol{\beta}$. The score function for the full likelihood is

$$
\mathbf{u}^{(f)} = \begin{pmatrix} \dfrac{\partial \log L^{(f)}}{\partial \boldsymbol{\beta}} \\ \dfrac{\partial \log L^{(f)}}{\partial \boldsymbol{\gamma}} \end{pmatrix} = \begin{pmatrix} \dfrac{\partial \log L^{(c)} L^{(p)}}{\partial \boldsymbol{\beta}} \\ \dfrac{\partial \log L^{(p)}}{\partial \boldsymbol{\gamma}} \end{pmatrix}, \tag{38.11}
$$

which could be used to construct score tests for $\boldsymbol{\beta}$, but this would depend strongly on the model for the parental likelihood. Clayton (1999) proposed instead a *partial score function*

$$
\mathbf{u}^{(*)} = \begin{pmatrix} \dfrac{\partial \log L^{(c)}}{\partial \boldsymbol{\beta}} \\ \dfrac{\partial \log L^{(p)}}{\partial \boldsymbol{\gamma}} \end{pmatrix}, \tag{38.12}
$$

in which inference about $\boldsymbol{\beta}$ depends only on the conditional likelihood. When there are ambiguous haplotypes, the partial score function becomes a weighted mean over the possible solutions, where the weights are the corresponding full likelihoods. Denoting the set of possible haplotype solutions by P, the partial score function can be written as

$$\mathbf{u}_P^{(*)} = \frac{\sum\limits_{j \in P} L_{(j)}^{(f)} \mathbf{u}_{(j)}^{(*)}}{\sum\limits_{j \in P} L_{(j)}^{(f)}}. \qquad (38.13)$$

This can be used to construct a score test for β, which depends on the haplotype frequency model only through the weights of the possible solutions. Indeed, when there are no uncertain data, the score test for β is the same as in the original TDT. All the extensions available in the logistic regression model can be cast in this framework, although the TRANSMIT program only implements tests of individual haplotypes and a global test of the whole set of haplotypes.

38.5 GENERAL PEDIGREE STRUCTURES

It is often the case that the available family data do not consist of only case–parent trios but include a variety of different pedigree structures. This can be a particular problem for late-onset traits, since parents of affecteds may be deceased at the time of diagnosis, and also for the testing of association in the presence of linkage, because the linkage structure needs to be accounted for in the analysis. A well-known problem affecting the TDT in an ostensibly simple situation arises in the case where one parent is missing (Curtis and Sham, 1995). Suppose the available parent is heterozygous and the child is homozygous. Then we might score the transmitted allele, but this overlooks the fact that, had the child been heterozygous, we would have been unable to identify the transmitted allele and so score the family. But, if we only score homozygous children, they are likely to be homozygous for the more common allele, which would lead to a seeming over-transmission of the common allele even if there were no linkage or association. This is one of many biases that can occur when data are available for some members of a family but not for others (Knapp, 2000).

Two general approaches have emerged to deal with the problems of missing family members. The first is to fit a statistical model to the missing data and conduct an analysis that takes all of the possible completions into account. Here, the approach used by TRANSMIT applies equally well to nuclear families with missing parents. The second approach is to construct test statistics that are unbiased under the null hypothesis while using only the available data. This approach retains complete robustness to population stratification and can be readily applied to arbitrary pedigree structures. Some methods of this type are described in this section.

As noted above, the case of missing parents in late-onset disease is of particular interest. In this case, unaffected siblings may be used as controls, as long as their relationship to affecteds is taken into account. One of the first such methods was the sib-TDT, which uses sibships consisting of at least one affected and one unaffected sib (Spielman and Ewens, 1998). It counts the number of variant alleles in affected sibs, calculating its mean and variance for each family under the assumption that the proportion of variant alleles is the same in affected sibs as it is in unaffecteds. These counts are summed over the set of families to form a z score. Several other sibling-based methods have been suggested (Curtis, 1997; Boehnke and Langefeld, 1998; Horvath and Laird, 1998), though the sib-TDT remains most closely related to the more recent approaches.

The pedigree disequilibrium test (PDT) combines the principles of the TDT and sib-TDT into a test for general pedigrees (Martin *et al.*, 2000). It splits a pedigree into a list of all case–parent trios and discordant sib pairs with genotype data. For trio j, X_{T_j} is defined as the number of transmissions of the variant allele minus the number of its non-transmissions. For sib pair j, X_{S_j} is defined as the number of copies of the variant allele in the affected sib minus the number in the unaffected sib. A measure of association for the pedigree is then

$$D = \frac{1}{N_T + N_S} \left[\sum_{j=1}^{N_T} X_{T_j} + \sum_{j=1}^{N_S} X_{S_j} \right], \tag{38.14}$$

where N_T and N_S are the total number of trios and discordant sib pairs, respectively. D has expectation 0 in any pedigree. After computing this measure for each of $i = 1, \ldots, N$ pedigrees, the PDT statistic

$$T = \frac{\sum_{i=1}^{N} D_i}{\sum_{i=1}^{N} D_i^2} \tag{38.15}$$

is asymptotically χ^2 with 1 degree of freedom. This gives a valid test of linkage or association in any pedigree structure, although some pedigrees are uninformative, notably the affected sib pair. The PDT has been adapted to test quantitative traits (Monks and Kaplan, 2000), haplotypes (Dudbridge, 2003) and genotypes (Martin *et al.*, 2003a).

A very flexible approach for constructing unbiased tests is implemented in the software FBAT (Lake *et al.*, 2000). The general principle is that under no linkage or no association, the covariance between genotypes and traits is 0. When T_{ij} denotes the trait value of person j in pedigree i and X_{ij} denotes some coding of its genotype, a score is constructed as

$$S = \sum_i S_i = \sum_i \sum_j T_{ij} X_{ij}. \tag{38.16}$$

The mean and variance of each S_i are computed with respect to its conditional distribution given the minimal sufficient statistic for the null hypothesis (Rabinowitz and Laird, 2000). In practice, this means enumerating all possible founder genotypes and considering the transmission patterns that result in the same information structure as in the observed data: an example is given below. After summation over pedigrees, the standardised score statistic

$$T = (S - E(S))[\text{var}(S)]^{-1}(S - E(S)) \tag{38.17}$$

may be referred to the χ^2 distribution with degrees of freedom equal to the rank of var(S), which is 1 for a biallelic marker.

The conditional distribution of S_i is derived separately for each nuclear family configuration, which is tedious but not intractable. A recursive algorithm is available for constructing distributions for general pedigrees. Table 38.3, following Rabinowitz and

Table 38.3 Conditional distributions used by FBAT when testing for linkage with one heterozygous 1/2 parent's genotype available.

Set of genotypes present in children	Conditional distribution
{1/1} or {1/2}	Observed data have conditional probability 1
{1/1, 1/2}	Random assignment of 1/1 and 1/2 that keeps the number of each invariant
{1/1, 2/2} or {1/1, 1/2, 2/2}	Randomly assign 1/1, 1/2 and 2/2 with probabilities 1/4, 1/2, 1/4, independently to each sib, discarding outcome without at least one assignment of 1/1 and one assignment of 2/2

Laird (2000), shows the derivation when testing for linkage of a biallelic marker when one parent's genotype is available. As noted above, when there is only one child, there is no way to score the family without incurring bias. However, when there are several children, their genotypes define a set of configurations whose probabilities can be computed exactly, assuming the null hypothesis. Each configuration gives rise to a value of S_i that can be used to obtain the mean and variance of S_i for that family type.

For example, suppose there is one heterozygous parent with three children with genotypes AA, AB and AA, respectively. Then, row 2 of Table 38.3 is appropriate, and denoting the value of S_i by $S_i(G_1, G_2, G_3)$ when the ordered genotypes of the children are G_1, G_2 and G_3, the mean of S_i is

$$\frac{1}{3}[S_i(AA, AB, AA) + S_i(AB, AA, AA) + S_i(AA, AA, AB)] \tag{38.18}$$

and its variance is

$$\frac{2}{9}[(S_i(AA, AB, AA))^2 + (S_i(AB, AA, AA))^2 + (S_i(AA, AA, AB))^2$$
$$- S_i(AA, AB, AA)S_i(AB, AA, AA) - S_i(AA, AB, AA)S_i(AA, AA, AB)$$
$$- S_i(AB, AA, AA)S_i(AA, AA, AB)]. \tag{38.19}$$

As described, the FBAT method is appropriate for any trait T_{ij}. For a disease, it is appropriate to code T_{ij} as 1 for affected and 0 for unaffected. Similarly X_{ij} may be any coding of genotypes: for biallelic markers, it is convenient to code X_{ij} as the number of variant alleles carried by the child, which corresponds to the multiplicative model favoured by the TDT. Dominant or recessive models may be coded using 1 for risk genotypes and 0 otherwise. Multivariate calculations are possible for multiallelic markers.

The FBAT approach is more flexible than the PDT, as it can use information from concordant sibships and also distinguish tests of linkage from tests of association. It has been subject to numerous extensions, including analysis of haplotypes (Horvath *et al.*, 2004) and multivariate traits (Lange *et al.*, 2003). Its strengths are the guaranteed protection from population stratification and applicability to any pedigree structure. Nevertheless, the approach has some limitations, including difficulty in estimating relative risk and sub-optimal handling of additional covariates. When these issues are important, the likelihood-based approaches described in Sections 38.3 and 38.7 may be more appropriate.

38.6   QUANTITATIVE TRAITS

So far, the emphasis has been on association to binary traits, but the attractions of the family-based design apply equally to quantitative traits. In the quantitative setting, population stratification is taken to mean a difference in the trait mean across subpopulations, together with a difference in gene frequency. It is usually assumed that the parametric form of the trait distribution is otherwise unchanged, such as a normal, and also that the variance and higher moments are also unchanged: therefore the effect of stratification can be thought of as a shift in location of the trait distribution.

Two types of regression models have been developed for family-based association of quantitative traits. The first uses logistic regression in a similar way as for discrete traits, treating transmission as the random outcome. The second uses linear regression with the genotype as the independent variable and the quantitative trait as the dependent variable, with adjustment to respect the family-based design. Generally, the methods based on linear regression are more sensitive to the assumption of normality, and may require more nuisance parameters than the methods using logistic regression; but they can be more flexible and powerful when their assumptions are met.

The logistic regression approach treats transmission as the random outcome, as for discrete traits, and treats the quantitative trait as an effect modifier for alleles acting as predictors (Waldman $et\ al.$, 1999). For a biallelic marker, the transmission probability of the variant allele from a heterozygous parent is given by

$$\text{logit}(p) = \alpha + Y_i\beta, \tag{38.20}$$

where Y_i is the trait value of child i, regression coefficient β is the transmission parameter of the variant allele, and the intercept α is included to account for association to a phenotype for which all the children have been selected. In an unselected sample, the intercept could be omitted.

For multiallelic markers, a conditional logistic regression model can be defined in a manner similar to (38.6). For a j/k parent transmitting allele j to child i,

$$\text{logit}(p_{jk}) = \alpha_j - \alpha_k + Y_i(\beta_j - \beta_k). \tag{38.21}$$

Models of this form are most sensitive to log-linear effects of the trait value on the transmission probability, and this is appropriate for small effects on normally distributed traits (Clayton and Jones, 1999). In this case, $Y\beta$ is the relative trait mean for a subject with trait Y carrying the variant allele, compared to a reference allele. If the traits are not zero-centred, it is prudent to include the intercept term α to implicitly subtract the mean from each trait value. This model can be extended to genotype association by using a polytomous logistic regression for multinomial outcomes (Kistner and Weinberg, 2004). These models can be fitted by standard software.

Linear regression is a natural framework for modelling association to quantitative traits. Here, the genotypes are treated as categorical variables that predict the expected trait value, and we test whether the mean is different for subjects having different genotypes. This is reminiscent of the one-way analysis of variance, but the main issue is to allow for different trait means in different population strata. Potentially, each family is in its own stratum, so the analysis must allow for the mean to differ between families.

A reasonable approach is to conduct ordinary linear regression with the genotypes normalised within families (Lunetta *et al.*, 2000),

$$Y_i = \alpha + \beta(G_i - \psi_i) + e_i, \tag{38.22}$$

where now α is the overall mean of Y, G_i is a code for the genotype of child i, ψ_i is the expected value of G_i given the mating type of the parents and an assumed genetic model, and e_i is the error. For example, for a biallelic marker, we might let G denote the number of copies of the variant allele. If one parent is homozygous for the variant allele ($G = 2$) and the other is heterozygous ($G = 1$), then we may define $\psi_i = \frac{3}{2}$ for that family. If both parents are heterozygous, we may have $\psi_i = \frac{1}{4} \cdot 0 + \frac{1}{2} \cdot 1 + \frac{1}{4} \cdot 2 = 1$. This coding is equivalent to a model proposed by Rabinowitz (1997), but the framework allows for other codings that are more sensitive to dominant or recessive effects (Lunetta *et al.*, 2000).

A more general model, implemented in the QTDT software, decomposes the total population association into two orthogonal components: the between-family association, and the within-family association (Abecasis *et al.*, 2000). The advantage of this model is that the within-family association is robust to population stratification, but in the absence of stratification, the same model can give a more powerful test of the total association. Furthermore, a difference between the between- and within-family association parameters can be taken as evidence of stratification. The method also allows for covariance between the traits of multiple siblings, in terms of variance components, which is not considered by the other models. Since the covariance between siblings depends upon linkage, the QTDT model allows linkage and association to be modelled and tested separately, in contrast to most other family-based designs (Cardon and Abecasis, 2000).

The model for the trait mean of child j in family i is

$$Y_{ij} = \alpha + \beta_b B_i + \beta_w(G_{ij} - B_i) + e_{ij}, \tag{38.23}$$

where β_b and β_w are between- and within-family coefficients for association, G_{ij} codes for the genotype of child j in family i, and B_i is an expected value of B_{ij} for family i. To allow for multiple children, B_i is defined as

$$B_i = \tfrac{1}{2}(G_{iF} + G_{iM}), \tag{38.24}$$

where G_{iF} and G_{iM} code for the father's and mother's genotypes, respectively, or when parents are not available

$$B_i = \frac{1}{n_i} \sum_j G_{ij}, \tag{38.25}$$

where n_i is the number of children in family i. The likelihood for a nuclear family is the multivariate normal density with the mean vector obtained by this model and the variance–covariance matrix constructed from variance components (see **Chapter 19**).

Likelihood ratios are used to test for within-family association ($\beta_w = 0$), total association ($\beta_w = 0$ and $\beta_b = 0$, with 2 degrees of freedom) and population stratification ($\beta_w = \beta_b$). Furthermore, β_w is a valid estimate of the additive genetic value of the test marker (Fulker et al., 1999; Abecasis et al., 2000).

This model can be regarded as a regression on the within-family component, with the intercept treated as a random effect that is modelled by the between-family component. An alternative approach would be to treat the intercept as a fixed effect that depends upon the parental mating type (Gauderman, 2003), so that the expected trait value within a family is specified directly:

$$Y_{ij} = \alpha_i + \beta_w G_{ij} + e_{ij}, \qquad (38.26)$$

where α_i are set to be equal for families having the same mating type. This approach has been shown to have improved power for detecting gene–gene and gene–environment interactions, although it requires more nuisance parameters than the model (38.23): this can be a serious problem for multiallelic markers or haplotypes.

A potential problem with the linear regression models is that they rely on normality of the residual distribution and, in the case of the full likelihood implemented in the QTDT software, on multivariate normality of the trait distribution. These assumptions may be violated if the underlying distribution is non-normal, or if the sample has been selected, for example, to have extreme trait values. A solution to this problem is to adopt a retrospective formulation in which the linear model is used to predict the probability of genotype given the trait, via Bayes' theorem (Liu et al., 2002). With some care, this approach can be implemented in standard software (Gauderman, 2003) but it is worth noting that, for small effects on normal traits, the logistic regression models (38.21) are approximately equivalent to the retrospective likelihood approach.

38.7   ASSOCIATION IN THE PRESENCE OF LINKAGE

It was earlier noted that the TDT and related methods can be regarded either as tests of linkage in the presence of association or tests of association in the presence of linkage. Often, the two interpretations are interchangeable: we are seeking evidence of a genetic effect, realised through linkage and association. In some contexts, however, the distinction is important. A situation that has received much attention occurs in positional cloning, when a genome scan for linkage identifies a broad genomic region and association methods are then used to fine map a gene by exploiting the short-range extent of linkage disequilibrium (see **Chapter 27**). In this case, family-based association methods must allow for the fact that linkage has been established for the tested markers, so that the test is strictly one of association.

To illustrate why this is necessary, consider a nuclear family with two children, typed for a marker that is completely linked to the disease gene (Figure 38.3). If there is no association, we have no information on which marker allele occurs together with the disease allele, given that the parent is heterozygous. But if there is complete linkage, the same marker allele will be transmitted to both children: in other words, the transmission to the second child is not independent of the first. Therefore, a standard TDT treating

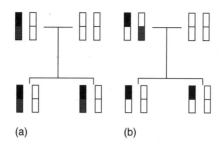

(a)   (b)

Figure 38.3 Two families that are equally likely under no association, given that the parent is heterozygous for the marker. Under complete linkage, only 2 or 0 transmissions of the variant allele can occur.

all transmissions as independent will inflate the evidence for association and increase the false-positive rate.

This problem only arises when there are multiple offspring in a family and the tested marker is linked. When all families have just one offspring, there is no need to allow for linkage. Most of the previously described methods are susceptible to the problem, except for the QTDT program, which explicitly models the correlation between children's traits in terms of the evidence for linkage (Abecasis *et al.*, 2000).

A common approach to allow for dependent observations is to use a variance estimate that treats the whole family as one sampling unit. When using the logistic regression models, this can be achieved via score or Wald tests of the odds ratios, counting all children independently but using a 'cluster' variance estimate instead of the analytic form (Gould *et al.*, 2006, chapter 1). The same approach is available in the TRANSMIT, PDT and FBAT programs. In general terms, the method involves calculating a score S_i and its expectation for each family, as if the children were independent. The variance is then estimated from the family-wise scores rather than from individual contributions, using the *empirical variance* estimator

$$\text{var}(S) = \sum_{i=1}^{N} [S_i - E(S_i)]^{\text{T}}[S_i - E(S_i)]. \tag{38.27}$$

Because this method is based on treating families as sampling units, the resulting score tests are unbiased.

A limitation of the empirical variance approach is that it does not easily extend to maximum likelihood estimation of missing data and uncertain haplotypes. Essentially, this is because the underlying likelihood still assumes that transmissions are independent, and the adjustment for correlated transmissions only occurs at the testing stage. While the TRANSMIT software implements an empirical variance estimate to allow for correlated transmission, it has been shown to be biased when parents are missing (Martin *et al.*, 2003b). These authors suggested a method to estimate the degree of linkage and thence adjust a score test; however, this method is difficult to implement in nuclear families with more than two children.

A likelihood-based solution to the problem of correlated transmission is based on the factorization of the nuclear family likelihood into conditional and parental contributions (38.9). Allowing for k affected children, the full likelihood for a nuclear family assuming independent transmissions is

$$L^{(f)} = \frac{\Pr(D_1, \ldots, D_k | C_1 \ldots, C_k)}{\sum\limits_{c_1,\ldots,c_k \in S(F,M)} \Pr(D_1, \ldots, D_k | c_1, \ldots, c_k)}$$

$$\cdot \frac{\Pr(F, M) \sum\limits_{c_1,\ldots,c_k \in S(F,M)} \Pr(D_1, \ldots, D_k | c_1, \ldots, c_k)}{\sum\limits_{f,m \in G} \Pr(f, m) \sum\limits_{c_1,\ldots,c_k \in S(f,m)} \Pr(D_1, \ldots, D_k | c_1, \ldots, c_k)}$$

$$= L^{(c)} \cdot L^{(p)}. \tag{38.28}$$

Inferences should be based on the conditional likelihood in order to retain the properties of the TDT, but the parental likelihood is needed to model the possible completions of missing data. In order to distinguish the two components, separate sets of relative risk parameters are specified in each component, with inference being drawn only on those in the conditional component. Let β be the vector of haplotype log relative risks in the conditional component, and φ those in the parental component. Let γ parameterize the haplotype frequencies as in (38.11), and let (g_1, g_2) be the two haplotypes comprising a genotype g. Assuming a multiplicative model for the joint risk of the children, we have

$$L^{(c)} = \frac{\prod\limits_{i=1}^{k} \exp(\beta_{C_{i1}} + \beta_{C_{i2}})}{\left[\sum\limits_{c \in S(F,M)} \exp(\beta_{c_1} + \beta_{c_2}) \right]^k} \tag{38.29}$$

and

$$L^{(p)} = \frac{\exp(\gamma_{F_1} + \gamma_{F_2} + \gamma_{M_1} + \gamma_{M_2}) \left[\sum\limits_{c \in S(F,M)} \exp(\varphi_{c_1} + \varphi_{c_2}) \right]^k}{\sum\limits_{f,m \in G} \exp(\gamma_{f_1} + \gamma_{f_2} + \gamma_{m_1} + \gamma_{m_2}) \left[\sum\limits_{c \in S(f,m)} \exp(\varphi_{c_1} + \varphi_{c_2}) \right]^k} \tag{38.30}$$

performing inference only on β. At this point, the children are still assumed to be independent. In order to relax this assumption, we then introduce the additional step of conditioning on an identity-by-descent (IBD) vector. This vector contains one entry for each transmission to each child, indicating whether the phase of that transmission is the same as that of the paternal transmission to the first child. The conditional likelihood can be written down without actually computing the IBD vector, and has the form

$$L^{(c|\text{IBD})} = \frac{\exp\left(\sum\limits_{i=1}^{k} \beta_{C_{i1}} + \beta_{C_{i2}} \right)}{\exp\left(\sum\limits_{i=1}^{k} \beta_{C_{i1}} + \beta_{C_{i2}} \right) + \exp\left(\sum\limits_{i=1}^{k} \beta_{C_{i1}} + \beta_{U_{i2}} \right) + \exp\left(\sum\limits_{i=1}^{k} \beta_{U_{i1}} + \beta_{C_{i2}} \right) + \exp\left(\sum\limits_{i=1}^{k} \beta_{U_{i1}} + \beta_{U_{i2}} \right)}, \tag{38.31}$$

where U_i denotes the genotype formed from the two alleles not transmitted to child i. The parental contribution becomes

$$L^{(p|\text{IBD})} = \frac{\exp(\gamma_{F_1} + \gamma_{F_2} + \gamma_{M_1} + \gamma_{M_2})}{\sum\limits_{f,m \in G} \exp(\gamma_{f_1} + \gamma_{f_2} + \gamma_{m_1} + \gamma_{m_2}) \left[\sum\limits_{c \in S(f,m)} \exp(\varphi_{c_1} + \varphi_{c_2}) \right]^k}$$

$$\cdot \left[\exp\left(\sum_{i=1}^{k} \varphi_{C_{i1}} + \varphi_{C_{i2}} \right) + \exp\left(\sum_{i=1}^{k} \varphi_{C_{i1}} + \varphi_{U_{i2}} \right) \right.$$

$$\left. + \exp\left(\sum_{i=1}^{k} \varphi_{U_{i1}} + \varphi_{C_{i2}} \right) + \exp\left(\sum_{i=1}^{k} \varphi_{U_{i1}} + \varphi_{U_{i2}} \right) \right] \qquad (38.32)$$

and the full likelihood $L^{(f|\text{IBD})} = L^{(c|\text{IBD})} \cdot L^{(p|\text{IBD})}$ is then maximised over the full set of parameters (β, φ, γ). This formulation has the effect of treating the whole family as one sampling unit, resulting in unbiased inference about β.

When the data are complete, the total likelihood factorises completely into conditional and parental components, yielding inference on β that is equivalent to the conditional logistic regression formulation of the TDT. When there are incomplete data, the likelihood contribution of a family is the sum of the likelihoods for the possible completions. Then the parental likelihood has the effect of weighting the conditional analysis, in an analogous manner to the TRANSMIT method. In fact, when φ is set to $\mathbf{0}$, the score function for the full likelihood is very similar to the partial score function used by TRANSMIT, with a slight difference in the weights used when the data are incomplete. The advantage of freely estimating φ is that the IBD vector can be identified in the parental contribution, which is not possible when $\varphi = \mathbf{0}$, thus eliminating the bias seen in TRANSMIT.

The model in (38.32) has been implemented in the software UNPHASED (Dudbridge, 2006). The likelihood formulation also allows estimation of odds ratios and testing of covariate effects, and extends readily to quantitative traits by treating the traits as effect modifiers as in (38.21).

It might be argued that, despite considerable efforts to devise valid tests of association in the presence of linkage, the problem is not particularly important. The reason is that, in many studies of complex disease, the evidence for linkage is fairly weak, resulting in only moderate dependence between transmissions to multiple children. Furthermore, given the weak evidence from traditional linkage analysis, it may be more worthwhile to seek stronger evidence for linkage, via association methods. Nevertheless, in situations in which the linkage evidence is particularly strong, for example, in some HLA-linked diseases, it is useful to have methods that are specifically designed to detect association.

38.8 CONCLUSIONS

Family-based methods are a useful and important means to protect against confounding effects in association studies. All the methods presented here compare, by various designs, the alleles transmitted to the study subjects to the alleles not transmitted by their parents. This genetic matching eliminates the possibility of population stratification, which is a concern in genetic studies because allele frequencies vary considerably between populations, owing to several factors including random drift (see **Chapter 31**). Furthermore, the shared environment within families ensures that other confounders, possibly not identified, are automatically controlled for. Together with the fact that family-based association can only be detected in the presence of linkage, these methods provide a benchmark that continues to be widely used in candidate-gene studies, and will serve as an important method for replication as genome-wide scans for population association become more widespread.

From the original TDT method (Spielman et al., 1993), the field has grown to encompass a wide range of applications, of which only the most prominent have been covered here. To a large extent, the field relies on customised software because many of the methods do not employ standard statistical models and data structures. The TRANSMIT (Clayton, 1999) and FBAT (Lake et al., 2000) programs are widely used for hypothesis testing, and the QTDT program (Abecasis et al., 2000) is widely used for testing and estimation of both linkage and association to quantitative traits. UNPHASED is a serious alternative for analysis of general nuclear families, providing estimation as well as hypothesis testing and allowing gene–gene and gene–environment interactions (Dudbridge, 2006). A further alternative is FAMHAP, which is faster but slightly less accurate (Becker and Knapp, 2004). Despite the proliferation of custom software, family-based analysis can be cast into standard models for conditional logistic regression, by regarding each nuclear family as a stratum with control subjects constructed from the other possible combinations of parental alleles. Provided these controls are constructed appropriately, a wide variety of analyses are possible using standard software (Cordell et al., 2004).

A number of topics have not been covered here, and are the subject of ongoing research. These include the combination of family-based and population-based studies (Epstein et al., 2005), and multiple imputation approaches for dealing with missing data (Kistner and Weinberg, 2005). Recent methods for extracting maximal information from incomplete data have attracted much interest (Rabinowitz, 2002; Allen et al., 2005) because they are robust to violations of simplifying assumptions such as Hardy–Weinberg equilibrium. Currently, their computational complexity is the greatest obstacle to more widespread use. Bayesian approaches have not been discussed here, but there is some work in this area (George and Laud, 2002; Bernardinelli et al., 2004; Denham and Whittaker, 2003). These methods may assume greater importance as investigators aim to incorporate external information on gene function and interaction into epidemiological studies.

REFERENCES

Abecasis, G.R., Cardon, L.R. and Cookson, W.O. (2000). A general test of association for quantitative traits in nuclear families. *American Journal of Human Genetics* **66**, 279–292.

Allen, A.S., Satten, G.A. and Tsiatis, A.A. (2005). Locally-efficient robust estimation of haplotype-disease association in family-based studies. *Biometrika* **92**, 559–571.

Becker, T. and Knapp, M. (2004). Maximum-likelihood estimation of haplotype frequencies in nuclear families. *Genetic Epidemiology* **27**, 21–32.

Bernardinelli, L., Berzuini, C., Seaman, S. and Holmans, P. (2004). Bayesian trio models for association in the presence of genotyping errors. *Genetic Epidemiology* **26**, 70–80.

Boehnke, M. and Langefeld, C.D. (1998). Genetic association mapping based on discordant sib pairs: the discordant-alleles test. *American Journal of Human Genetics* **62**, 950–961.

Cardon, L.R. and Abecasis, G.R. (2000). Some properties of a variance components model for fine-mapping quantitative trait loci. *Behavior Genetics* **30**, 235–243.

Cardon, L.R. and Palmer, L.J. (2003). Population stratification and spurious allelic association. *Lancet* **15**, 598–604.

Clayton, D. (1999). A generalization of the transmission/disequilibrium test for uncertain-haplotype transmission. *American Journal of Human Genetics* **65**, 1170–1177.

Clayton, D. and Jones, H. (1999). Transmission/disequilibrium tests for extended marker haplotypes. *American Journal of Human Genetics* **65**, 1161–1169.

Cordell, H.J., Barratt, B.J. and Clayton, D.G. (2004). Case/pseudocontrol analysis in genetic association studies: a unified framework for detection of genotype and haplotype associations, gene-gene and gene-environment interactions, and parent-of-origin effects. *Genetic Epidemiology* **26**, 167–185.

Cordell, H.J. and Clayton, D.G. (2002). A unified stepwise regression procedure for evaluating the relative effects of polymorphisms within a gene using case/control or family data: application to HLA in type 1 diabetes. *American Journal of Human Genetics* **70**, 124–141.

Curtis, D. (1997). Use of siblings as controls in case-control association studies. *Annals of Human Genetics* **61**, 319–333.

Curtis, D. and Sham, P.C. (1995). A note on the application of the transmission disequilibrium test when a parent is missing. *American Journal of Human Genetics* **56**, 811–812.

Denham, M.C. and Whittaker, J.C. (2003). A Bayesian approach to disease gene location using allelic association. *Biostatistics* **4**, 399–409.

Dudbridge, F. (2003). Pedigree disequilibrium tests for multilocus haplotypes. *Genetic Epidemiology* **25**, 115–121.

Dudbridge, F. (2006). UNPHASED user manual. Technical report 2006/05, MRC Biostatistics Unit, Cambridge.

Dudbridge, F., Koeleman, B.P., Todd, J.A. and Clayton, D.G. (2000). Unbiased application of the transmission/disequilibrium test to multilocus haplotypes. *American Journal of Human Genetics* **66**, 2009–2012.

Epstein, M.P., Veal, C.D., Trembath, R.C., Barker, J.N., Li, C. and Satten, G.A. (2005). Genetic association analysis using data from triads and unrelated subjects. *American Journal of Human Genetics* **76**, 592–608.

Ewens, W.J. and Spielman, R.S. (1995). The transmission/disequilibrium test: history, subdivision, and admixture. *American Journal of Human Genetics* **57**, 455–464.

Falk, C.T. and Rubinstein, P. (1987). Haplotype relative risks: an easy reliable way to construct a proper control sample for risk calculations. *Annals of Human Genetics* **51**, 227–233.

Fulker, D.W., Cherny, S.S., Sham, P.C. and Hewitt, J.K. (1999). Combined linkage and association sib-pair analysis for quantitative traits. *American Journal of Human Genetics* **64**, 259–267.

Gauderman, W.J. (2003). Candidate gene association analysis for a quantitative trait, using parent-offspring trios. *Genetic Epidemiology* **25**, 327–338.

George, V. and Laud, P.W. (2002). A Bayesian approach to the transmission/disequilibrium test for binary traits. *Genetic Epidemiology* **22**, 41–51.

Gould, W., Pitblado, J. and Sribney, W. (2006). *Maximum Likelihood Estimation with Stata*, 3rd edition. Stata Press, College Station, TX.

Horvath, S., Xu, X., Lake, S.L., Silverman, E.K., Weiss, S.T. and Laird, N.M. (2004). Family-based tests for associating haplotypes with general phenotype data: application to asthma genetics. *Genetic Epidemiology* **26**, 61–69.

Horvath, S. and Laird, N.M. (1998). A discordant-sibship test for disequilibrium and linkage: no need for parental data. *American Journal of Human Genetics* **63**, 1886–1897.

Kistner, E.O. and Weinberg, C.R. (2004). Method for using complete and incomplete trios to identify genes related to a quantitative trait. *Genetic Epidemiology* **27**, 33–42.

Kistner, E.O. and Weinberg, C.R. (2005). A method for identifying genes related to a quantitative trait, incorporating multiple siblings and missing parents. *Genetic Epidemiology* **29**, 155–165.

Knapp, M. (2000). The transmission/disequilibrium test and parental-genotype reconstruction: the reconstruction-combined transmission/disequilibrium test. *American Journal of Human Genetics* **64**, 861–870.

Lake, S.L., Blacker, D. and Laird, N.M. (2000). Family-based tests of association in the presence of linkage. *American Journal of Human Genetics* **67**, 1515–1525.

Lange, C., Silverman, E.K., Xu, X., Weiss, S.T. and Laird, N.M. (2003). A multivariate family-based association test using generalized estimating equations: FBAT-GEE. *Biostatistics* **4**, 195–206.

Liu, Y., Tritchler, D. and Bull, S.B. (2002). A unified framework for transmission-disequilibrium test analysis of discrete and continuous traits. *Genetic Epidemiology* **22**, 26–40.

Lunetta, K.L., Faraone, S.V., Biederman, J. and Laird, N.M. (2000). Family-based tests of association and linkage that use unaffected sibs, covariates, and interactions. *American Journal of Human Genetics* **66**, 605–614.

Markianos, K., Daly, M.J. and Kruglyak, L. (2001). Efficient multipoint linkage analysis through reduction of inheritance space. *American Journal of Human Genetics* **68**, 963–977.

Martin, E.R., Bass, M.P., Gilbert, J.R., Pericak-Vance, M.A. and Hauser, E.R. (2003a). Genotype-based association test for general pedigrees: the genotype-PDT. *Genetic Epidemiology* **25**, 203–213.

Martin, E.R., Bass, M.P., Hauser, E.R. and Kaplan, N.L. (2003b). Accounting for linkage in family-based tests of association with missing parental genotypes. *American Journal of Human Genetics* **73**, 1016–1026.

Martin, E.R., Monks, S.A., Warren, L.L. and Kaplan, N.L. (2000). A test for linkage and association in general pedigrees: the pedigree disequilibrium test. *American Journal of Human Genetics* **67**, 146–154.

Monks, S.A. and Kaplan, N.L. (2000). Removing the sampling restrictions from family-based tests of association for a quantitative-trait locus. *American Journal of Human Genetics* **66**, 576–592.

Ott, J. (1989). Statistical properties of the haplotype relative risk. *Genetic Epidemiology* **6**, 127–130.

Rabinowitz, D. (1997). A transmission disequilibrium test for quantitative trait loci. *Human Heredity* **47**, 342–350.

Rabinowitz, D. (2002). Adjusting for population heterogeneity and misspecified haplotype frequencies when testing nonparametric null hypotheses in statistical genetics. *Journal of the American Statistical Association* **97**, 742–751.

Rabinowitz, D. and Laird, N. (2000). A unified approach to adjusting association tests for population admixture with arbitrary pedigree structure and arbitrary missing marker information. *Human Heredity* **50**, 211–223.

Schaid, D.J. (2004). Evaluating associations of haplotypes with traits. *Genetic Epidemiology* **27**, 348–364.

Schaid, D.J. and Sommer, S.S. (1994). Comparison of statistics for candidate-gene association studies using cases and parents. *American Journal of Human Genetics* **55**, 402–409.

Self, S.G., Longton, G., Kopecky, K.J. and Liang, K.Y. (1991). On estimating HLA/disease association with application to a study of aplastic anemia. *Biometrics* **47**, 53–61.

Sham, P.C. and Curtis, D. (1995). An extended transmission/disequilibrium test (TDT) for multi-allele marker loci. *Annals of Human Genetics* **59**, 323–336.

Spielman, R.S. and Ewens, W.J. (1998). A sibship test for linkage in the presence of association: the sib transmission/disequilibrium test. *American Journal of Human Genetics* **62**, 450–458.

Spielman, R.S., McGinnis, R.E. and Ewens, W.J. (1993). Transmission test for linkage disequilibrium: the insulin gene region and insulin-dependent diabetes mellitus (IDDM). *American Journal of Human Genetics* **52**, 506–516.

Stephens, M., Smith, N.J. and Donnelly, P. (2001). A new statistical method for haplotype reconstruction from population data. *American Journal of Human Genetics* **68**, 978–989.

Terwilliger, J.D. and Ott, J. (1992). A haplotype-based 'haplotype relative risk' approach to detecting allelic associations. *Human Heredity* **42**, 337–346.

Waldman, I.D., Robinson, B.F. and Rowe, D.C. (1999). A logistic regression based extension of the TDT for continuous and categorical traits. *Annals of Human Genetics* **63**, 329–340.

Weinberg, C.R. (1999). Methods for detection of parent-of-origin effects in genetic studies of case-parents triads. *American Journal of Human Genetics* **65**, 229–235.

Whittaker, J.C., Denham, M.C. and Morris, A.P. (2000). The problems of using the transmission/disequilibrium test to infer tight linkage. *American Journal of Human Genetics* **67**, 523–526.

39

Cancer Genetics

M.D. Teare

Mathematical Modelling and Genetic Epidemiology, University of Sheffield Medical School, Sheffield, UK

Cancers result from an accumulation of inherited and somatic mutations yielding cells that have acquired the necessary characteristics for unregulated growth. The development of the tumour can be viewed as an evolutionary process, involving several classes of genes in tumour initiation and progression.

This chapter provides an overview of the development of cancer genetic models. The earlier mathematical and statistical models based on population and cancer family observations, are followed by evolutionary models applied to molecular genetic sequence data.

39.1 INTRODUCTION

Cancer is a genetic disease, in that a normal cell must undergo mutations to lose or gain functions allowing a tumour to develop. The earliest evidence indicating a genetic origin for cancer was reported in 1890. Abnormal configurations of chromosomes were observed in dividing cancer cells (von Hansemann, 1890), though at that time the link between chromosomes and inheritance was not known. This was followed by Theodore Bovari's experiments on the fertilisation of sea urchin eggs, strongly suggesting that individual chromosomes contained different information (Bovari, 1904). Bovari had developed procedures by which he could induce aberrant chromosome segregation during mitosis. He was then able to study the fate of these cells after mitosis. In most cases, the unequal distribution of chromosomes would lead to a detrimental effect on the cell. However, he noted that on rare occasions, some particular configurations of chromosomes would generate a cell with the ability of unlimited growth, and this effect could be passed on to its progeny. This suggested that tumours might arise through abnormal segregation of chromosomes to daughter cells (Bovari, 1914). He then went on to explore and develop this hypothesis, which led him to predict many aspects of genetic mechanisms in cancer development.

Handbook of Statistical Genetics, Third Edition. Edited by D.J. Balding, M. Bishop and C. Cannings.
© 2007 John Wiley & Sons, Ltd.   ISBN: 978-0-470-05830-5.

39.2 ARMITAGE–DOLL MODELS OF CARCINOGENESIS

39.2.1 The Multistage Model

In the early 1950s two hypotheses, regarding mechanisms for carcinogenesis, were put forward based on the studies of cancer mortality. Fisher and Holloman (1951) and Nordling (1953) studied the mortality rates for several tumour types and found that the logarithm of the death rate due to cancer increased in direct proportion to the logarithm of the age at death. More specifically they found that the death rate increased proportionately to the sixth power of the age at death. This relationship seemed to be limited to mortality rates between the ages of 25 and 75. It could be argued that data above age 75 were unreliable and cancer in children and young adults may be affected by other factors.

Fisher and Holloman proposed that this observed relationship between age and mortality could result if a colony of six or seven cancer cells was the critical mass necessary for independent growth to be sustained. Experimental data on the relationship between the dose of carcinogen and cancer incidence already exists and in general, the relationship between the two was more consistent with arithmetic rather than geometric data. (Berenblum and Shubik, 1949). Based on this apparent contradiction, Armitage and Doll (1954) felt that this made the Fisher and Holloman hypothesis 'untenable'. Nordling proposed an alternative hypothesis that the observed relationship could be explained if a single cancer cell was the result of seven successive and accumulated mutations. This model would explain the relationship if the probability of each mutation remains constant throughout the specified age range.

The development of the tumour can be represented as the following changes occurring in the cell (or its lineal descendants).

$$E_0 \xrightarrow{p_1} E_1 \xrightarrow{p_2} E_2 \cdots \xrightarrow{p_n} E_n.$$

E_0 represents the normal cell and E_n the malignant tumour cell. The p_i represent the rate of change or transition to the next stage. The reciprocal of these rates represents the average sojourn time, spent in years, in state or stage E_i. We denote as t, the time that the cell enters state E_n.

$$p_1 p_2 p_3 p_4 p_5 p_6 p_7 t^7. \tag{39.1}$$

The probability that a mutation of type i has occurred by time t is approximately given by $p_i t$, when each p_i is small. Thus if seven distinct mutations were required, the probability that all mutations have occurred by time t is as given above in (39.1). However, the model requires that the mutations occur in a specific order, and there are $n!$ of these. Therefore, the probability that a cell is in state E_n, at time t, is given by $p_1 p_2 \cdots p_n t^n / n!$

The probability of malignancy by time t can be thought of as equivalent to prevalence. The incidence is given by the derivative of the prevalence with respect to t to yield,

$$\frac{p_1 p_2 \cdots p_n t^{n-1}}{(n-1)!}. \tag{39.2}$$

Thus the double logarithmic plot should show a straight line with slope $(n-1)$. The incidence rate for an individual (as opposed to a cell) is given by (39.2) multiplied by

N_s, which is the average number of cells in the susceptible target tissue. The above result will be approximately true when the mutation probabilities (i.e. $p_i t$) are small.

This simple model has assumed that the specific mutations can be modelled by constant rates. Armitage and Doll (1954) went on to derive expressions allowing for variable mutation rates to illustrate what relationships might be expected in situations where variable exposure to mutagen influenced the mutation rate. Now assume that one, p_c say, varies with time. The probability that this cth change occurs in the time interval $(t_0, t_0 + dt_0)$, and the nth in interval $(t, t + \delta t)$ is given by the product of the following three probabilities:

1. The probability that $(c - 1)$ changes have occurred in the interval $(0, t_0)$ is given as derived above by, $p_1 \cdots p_{c-1} t_0^{c-1}/(c - 1)!$

2. The probability (conditional that exactly $c - 1$ changes occurred before t_0) that the cth change occurs in the small interval $(t_0, t_0 + dt_0)$, is given by $p_c(t_0)dt_0$, where $p_c(t_0)$ represents the mutation rate at time t_0.

3. The probability (conditional on the cth change before t_0) that the $(c + 1)$th, $(c + 2)$th, to nth changes occur in time (t_0, t) is given by

$$\frac{p_{c+1} p_{c+2} \cdots p_{n-1} (t - t_0)^{n-c-1} p_n}{(n - c - 1)!} dt.$$

This results in the expression for the total probability for the nth event occurring in the time interval $(t, t + dt)$ as follows:

$$\frac{p_1 \cdots p_{c-1} p_c(t) p_{c+1} \cdots p_{n-1} p_n t^{n-1}}{(n - 1)!} dt.$$

The $p_c(t)$ is the weighted mean of $p_c(t_0)$ over the whole time period, with the weight at time t_0 being proportional to $t_0^{c-1}(t - t_0)^{n-c-1}$, (which takes its maximum at $t_0 = t(c - 1)/(n - 2)$). Armitage and Doll argue that this slight modification to their constant rate model is enough to explain the deviations from the sixth power law seen in some tumours. Since the weighted mean of the varying mutation rate will depend upon the age, t, the overall incidence will not show a linear relationship with the sixth power of t.

The Nordling model would also explain that the experimentally observed data, i.e. rate of tumour incidence is directly proportional to the concentration of an effective carcinogen, if the effect of the carcinogen is to increase the mutation rate. Armitage and Doll (1954) attempted to test the hypothesis of Nordling by examining cancer mortality for cancer at a variety of different sites and for each sex. In particular they were interested to see if the hypothesis could explain data where the significance of carcinogenic factors was suspected to be variable. They examined the relationship between mortality rate and age for cancers at the commonest sites. These relationships are shown for a selection of common cancers but for more recent time periods (CancerStats, Cancer Research UK) in Figure 39.1. As trends in mortality can be influenced by improvements through treatment, incidence data only is presented. The log (incidence) for cancer of the colon and lung shows an approximately similar linear relationship with log (age). However, the log–log plot for female breast cancer deviates quite strongly from a linear relationship.

By inspection of the log–log plots they found that cancers fell into two main groups. Those that appeared consistent with proportional relationships, (in their study these were

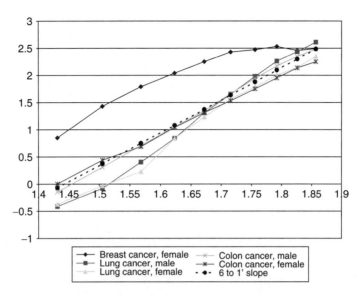

Figure 39.1 log(age) – log(incidence) plots for a selection of common site specific cancers.

cancers in the oesophagus, stomach, pancreas, colon and rectum) and those that did not. In the group that appeared to show linear relationships their regression coefficient for the slope of the line ranged from 5.0 to 6.5. These cancers show a similar relation to those reported by Fisher and Holloman and by Nordling. However, the other group sometimes showed a strong deviation from a linear relationship. These were cancers of the lung, bladder, female breast, ovary, cervix and uterus. Armitage and Doll pointed out that this latter group consists of cancers where the effects of causal factors are known to be variable. Tumours believed to be influenced by hormonal or endocrine secretions, might be expected to show evidence of age-specific effects. Incidence of bladder cancer was strongly associated with occupational hazards which may have both cohort and age-specific effects. A significant proportion of lung cancer was suspected to be due to the increasing rates of cigarette smoking and therefore exposure to this hazard was also not a constant with respect to time. Looking at the observed data in this way gave more support to the original hypothesis of Nordling (1953).

This multistage model appeared to fit the data well and could account for deviations from a constant mutation rate by allowing a variable rate affected by variations in exposure to carcinogens or mutagens. However, at the time, experimental data did not demonstrate evidence for more than two stages in tumour initiation, i.e. an exposure making an impact at an early versus late stage. As a consequence of this argument and suggestions made by Platt (1955), Armitage and Doll went on to offer a two-stage model for carcinogenesis.

39.2.2 The Two-stage Model

In this reduced model (Armitage and Doll, 1957), the first or early stage results in a change that gives one cell (and its descendants) a growth advantage over the adjacent normal cells. The clinical cancer then results from the second or late event occurring in this rapidly dividing 'clone' of cells.

This model assumes that the rates of both the first and second stage are influenced by carcinogenic exposures or agents. Assuming that the exposure is summarised by a 'dose' effect parameter d, the number of mutated cells will be proportional to the initiating dose and will also grow at an exponential rate depending upon the relative growth advantage of the mutated cells over the normal state. So the number of cells at time t will be n_t, which is proportional to $d_1 e^{kt}$, where k is a constant, and d_1 is the dose of the agent influencing the first stage. The resulting cancer incidence will be proportional to the dose effect influencing the second event (d_2) and the number of susceptible cells, i.e. $d_1 d_2 e^{kt}$.

Although individuals are likely to be exposed to the inducing agents throughout life, provided the dose remains constant the resulting incidence will be proportional to both concentrations. This model gives rise to a similar relationship between age and incidence as the multistage model.

This parsimonious two-stage model is able to account for all the features observed in cancer incidence plots. It explains the long latent period often observed after initial exposure to a carcinogen and, when incidence is low, a linear relationship can be seen between the concentration of initiating carcinogen and incidence by age. This two-stage model was also appealing as it reflected the biological process of a cell accumulating two mutated copies of the same gene or the reduction to homozygosity (which is discussed below).

39.2.2.1   The Philadelphia (Ph) Chromosome

Developments in cytogenetic techniques meant that by 1956 it was possible to unambiguously count the number of chromosomes in a normal cell (Tijo and Levan, 1956; Ford and Hamerton, 1956). With further technical improvements it became possible to distinguish and classify chromosome groups facilitating comparisons of karyotype in different cell types. This led to a major milestone in cancer genetics, the discovery of the *Ph* chromosome seen in cells from patients with chronic myelogenous leukaemia (Nowell and Hungerford, 1960). This was first reported as a shortened chromosome in the G chromosome group (this consisted of chromosomes 21 and 22; at the time there was no way to distinguish them). For the first time a specific chromosomal change was associated with a specific tumour type. Many years later it was found that the *Ph* chromosome was in fact a translocation between the long arm of chromosome 9 and a large part of chromosome 22 (Rowley, 1973). Work continued studying this translocation and it could be demonstrated that the translocation brought two genes together resulting in the production of a chimeric protein (Heisterkamp et al., 1983) which stimulates a growth signal pathway. This work confirmed that the genetic translocation was a critical event in the origin and cause of this cancer.

Over the 1950s and 1960s evidence appeared to accumulate that cancer could arise as a result of few mutations, though this argument was blurred by the fact that the early and late stage events could themselves be the result of several 'changes'. Work on childhood tumours seemed to confirm that a combination of inherited and somatic mutations played an essential role in tumour development (Burch, 1962; Falls and Neel, 1951; Crowe et al., 1956).

This research culminated in the key publication by Knudson (1971), where he performed a statistical analysis of retinoblastoma cases and proposed the 'two-mutation' hypothesis for cancer development. Knudson observed that familial cases and sporadic (non-hereditary) cases could arise from the same genetic mechanism. Experimental evidence was accumulating for genes that could suppress tumours (Harris *et al.*, 1969) and Knudson's hypothesis specifically relating to retinoblastoma would lead to the discovery of the first tumour suppressor gene (TSG).

Retinoblastoma is a tumour of the retina and can occur in one (unilateral) or both (bilateral) eyes. Knudson studied the reports of all the cases of retinoblastoma admitted to the MD Anderson Hospital between 1944 and 1969, consisting of 23 bilateral and 25 unilateral cases. In 14 of the bilateral cases, it was possible to estimate the number of distinct tumours in one eye. Knudson used other sources of data to derive population based estimates of the distribution of tumour types among gene carriers (unaffected, unilateral, and bilateral) of cases. The bilateral cases were assumed to be inherited, though the inherited lesion was not assumed to be fully penetrant as bilateral cases do occur in unaffected parents. Assuming that tumours develop according to a Poisson Process, Knudson found that the observed distribution of tumours, including the numbers in each eye, was consistent with a Poisson with rate $(m =)3$.

If n is the total number of target cells (i.e. those cells susceptible to mutation) contained by the two eyes (or retinae), then m/n is the probability that an (inherited) mutant cell develops into a primary tumour. Using prior estimates of n, yields an estimated tumour development rate (in the inherited form) of 0.75×10^{-6}. Assuming that this rate is the same in both the sporadic and hereditary form, this represents the probability of the second event.

Knudson then illustrated that the probability of the first mutation in sporadic tumours is approximately the same as the probability of a new mutation occurring in the germ line, which then passed on to future generations. The germinal mutation rate, μ_g, is approximately equal to 5×10^{-6} per generation. Assuming a generation time of 25 years, this means the sporadic mutation rate (expressed per year) is approximately 2×10^{-7}.

Conditional on the assumption that bilateral cases have all inherited the first mutation, if only one second step is involved in retinoblastoma development, the distribution of bilateral cases should follow an exponential. Therefore, the fraction of total bilateral cases that occurs in a given interval, the hazard rate, should be constant and the proportion of survivors (in the cohort of germinal or hereditary mutation carriers) is given by $S_h = \exp(-k_h t)$, where k_h is a constant. For the non-hereditary form where a single cell must acquire two mutations the proportion of survivors $S_n = \exp(-k_n t^2)$, k_n is a constant representing the non-hereditary mutation rate. Knudson found that the observed distribution of the ages at onset was consistent with his derived models, though the mutation rates looked slightly different in the bilateral and unilateral forms, which he argued could be accounted for by a small proportion of the unilateral cases being hereditary. Though other models could have accounted for such differences, this model suggested that the age distribution for familial and nonfamilial cases was resulting from the accumulation of the **same** genetic changes. Knudson went on to study the two-mutation model with application to other childhood cancers, but his theory was not confirmed through molecular studies until 1983, (see section on mutations resulting in loss of function).

39.2.2.2 Mutations Resulting in Gain of Function

In the 1960s it was known that cells in culture could be transformed by a number of viruses and retroviruses. Cellular genes were found to have similar sequences to those in the transforming retroviruses. These 'normal' genes were involved in regulating growth and differentiation, the inappropriate activation of which could lead to carcinogenesis. This class of genes was termed *proto-oncogenes*. When inappropriately activated they were termed *oncogenes*. Many oncogenes were identified through transfection experiments.

The *Ph* chromosome is an example of a specific balanced translocation giving rise to an activated oncogene (ABL–BCR fusion gene). Cytogenetic studies confirmed that chromosomal breakpoints seen in common translocations in other cancers were near to known proto-oncogenes (Rabbitts, 1994). Oncogenes are not only activated through translocation, some gain function through chromosomal amplification.

39.2.2.3 Mutations Resulting in Loss of Function

Gene-transfer studies seemed to present strong evidence for a single mutational step in carcinogenesis, but this is an artifact of the experimental design. However, evidence from observational studies such as Knudson's was more consistent with a two-stage process. In order for the inherited mutated allele to not have deleterious consequences on the organism, it was likely that this mutation was recessive at the cellular level. Comings (1973) further postulated that the two mutations might affect the same locus, resulting in loss of that gene function.

In 1983, through biochemical and molecular studies Cavenee *et al.* (1983) found that retinoblastoma development required loss of both copies of a specific region of chromosome 13. Thus the Rb1 locus was the first cloned example of a TSG, where, in contrast to the oncogene, loss of function is required for carcinogenesis. This result confirmed Knudson's two hit hypothesis, in that the familial and sporadic form of retinoblastoma were in fact due to the same genetic mechanism.

The success of the Rb story led to two major strategies for identifying further cancer predisposing genes, familial cancer linkage studies, and 'loss of heterozygosity' (LOH) studies. LOH studies stemmed from comparing genotypes observed in constitutional and tumour DNA in cancer patients. It was hypothesised that sections of chromosomes that are lost in tumours are likely to contain TSGs. If a genetic marker was located in a chromosomal region where one copy of a chromosome had been lost, the genotype would appear as homozygous as only the 'non-lost' allele would be detected. This observation would only be informative if the constitutional genotype was heterozygous and hence 'LOH'. There was great enthusiasm for studying familial forms of cancer as these could be expected to lead more quickly to the discovery of the target/predisposing genes. There were also many studies concentrating on patterns of allelic loss in tumours from panels of unrelated individuals. LOH profiles were found to be generally consistent within tumour types, but this approach alone led to few TSG discoveries (Presneau *et al.*, 2003).

However, genetic linkage studies (see **Chapter 33**) did lead to the discovery of many TSGs, though it became clear that single tumour suppressors may account for rare syndromes or a small fraction of the familial effect (Garber and Offit, 2005). In adult onset of disease, it can be very difficult to collect sufficient samples for informative linkage analysis. Often the pedigree comes to the attention of the investigators when many of the affected individuals are dead. Techniques were developed so that archived tumours could be used to reconstruct the constitutional genotypes. Through this strategy it became apparent that the observations of somatic change in the tumours could be used in a similar way to further informative meioses. This approach was built into a formal framework (Teare *et al.*, 1994; Rohde *et al.*, 1995; 1997), and derives from the linkage approach.

The tumour suppressor model assumes the high cancer risk in selected families is due to a dominantly inherited defective or predisposing allele. However, at the cellular level the defective gene acts recessively, as it requires the wild-type allele to be inactivated or lost. If a genetic marker is in linkage with the disease locus then it is expected to see over-segregation of one marker allele in affected individuals. By a simple extension to this, if the predisposing locus behaves as a tumour suppressor and the wild-type function is lost through loss of chromosomal material during mitosis, then the marker allele in phase with the wild type would also have a tendency to be lost in the tumour, and the marker allele in phase with the disease allele would have a tendency to be retained in the tumour. Besides the parameters used in classic linkage analysis this model requires parameters representing the probability of 'loss of genetic material' and somatic recombination. Features of the model are illustrated in Figure 39.2.

In the most general form we require three probabilities ($\lambda_{DD}, \lambda_{Dd}, \lambda_{dd}$) representing the probability of losing one marker allele in tumour, conditional on the genotype at the disease locus, and three similar probabilities ($\varepsilon_{DD}, \varepsilon_{Dd}, \varepsilon_{dd}$) representing the probability of losing both alleles at the marker locus conditional on the disease locus genotype. In the case where only one allele is lost the parameter ρ represents the probability that the marker allele lost in the tumour is in phase with the predisposing disease allele in the germ line. This parameter can be thought of as similar to θ, the probability of meiotic recombination (Table 39.1).

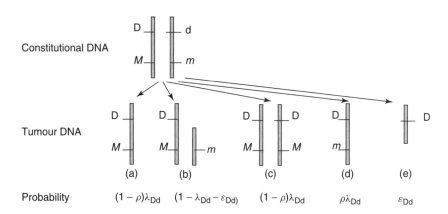

Figure 39.2 Examples of genotype configurations in tumour DNA, when marker is located on the same chromosome as TSG locus. Dd represents genotype at TSG locus, D = defective, d = wildtype; Mm represents heterozygous marker genotype. (a) Hemizygous, (b) retention of both alleles, (c) loss of wildtype chromosome followed by duplication, (d) somatic recombination and loss of chromosome strand bearing wildtype allele, (e) loss of both marker alleles. The model assumes that the wildtype TSG allele is lost.

Table 39.1 Probabilities of loss or retention of marker alleles in the tumour conditional on marker and TSG genotype.

Marker genotype	Constitutional TSG genotype					
	DD		Dd		dd	
MM	MM	$M-$	MM	$M-$	MM	$M-$
	$1 - \lambda_{DD} - \varepsilon_{DD}$	$1/2\lambda_{DD}$	$1 - \lambda_{Dd} - \varepsilon_{Dd}$	$(1 - \rho)\lambda_{Dd}$	$1 - \lambda_{dd} - \varepsilon_{dd}$	$1/2\lambda_{dd}$
	$-M$	$-$	$-M$	$-$	$-M$	$-$
	$1/2\lambda_{DD}$	ε_{DD}	$\rho\lambda_{Dd}$	ε_{Dd}	$1/2\lambda_{dd}$	ε_{dd}
Mm	Mm	$M-$	Mm	$M-$	MM	$M-$
	$1 - \lambda_{DD} - \varepsilon_{DD}$	$1/2\lambda_{DD}$	$1 - \lambda_{Dd} - \varepsilon_{Dd}$	$(1 - \rho)\lambda_{Dd}$	$1 - \lambda_{dd} - \varepsilon_{dd}$	$1/2\lambda_{dd}$
	$-m$	$-$	$-m$	$-$	$-M$	$-$
	$1/2\lambda_{DD}$	ε_{DD}	$\rho\lambda_{Dd}$	ε_{Dd}	$1/2\lambda_{dd}$	ε_{dd}
mm	mm	$m-$	mm	$m-$	mm	$m-$
	$1 - \lambda_{DD} - \varepsilon_{DD}$	$1/2\lambda_{DD}$	$1 - \lambda_{Dd} - \varepsilon_{Dd}$	$(1 - \rho)\lambda_{Dd}$	$1 - \lambda_{dd} - \varepsilon_{dd}$	$1/2\lambda_{dd}$
	$-m$	$-$	$-m$	$-$	$-m$	$-$
	$1/2\lambda_{DD}$	ε_{DD}	$\rho\lambda_{Dd}$	ε_{Dd}	$1/2\lambda_{dd}$	ε_{dd}

The cells of the table display the four possible genotypes in the tumour DNA, and the conditional probability of each genotype.

The original studies of LOH were so-called because it was commonly not feasible to detect loss of one out of two identical alleles; only loss of one out of two different alleles (i.e. reduction to homozygosity) could be detected. Loss of both marker alleles has been reported but we would expect it to be rare when the marker is close to a gene where at least one functional copy is essential. The probabilities in the table relate to phase-known genotypes but as in any linkage study, these can only be inferred from the segregation seen in the family. It is, therefore, apparent that the only informative situation for studying loss of alleles is when the constitutional genotype at the marker is heterozygous.

The benefits of including such observations in classic linkage analysis are mainly to increase power. If the cancer syndrome leads to individuals suffering multiple primary tumours, each tumour can be included as an independent observation. Affected parent–offspring pairs also became informative provided one tumour of the pair was observed (Rohde *et al.*, 1997). The tumour observations are considered in an extension to the penetrance function and the evidence for linkage is summarised in the same manner as classic linkage analysis with the lod score function (see **Chapter 33**) maximised over both θ and ρ. This method has so far had limited application as it is difficult to gain access to sufficient quantities of good quality tumour DNA. With current technological developments, especially with respect to whole genome amplification this approach may still prove to have a wide application. The method can also be extended to allow for other forms of loss of function such as epigenetic silencing (see **Chapter 40**).

39.2.2.4 Models Attempting to Identify the order of Acquired Mutations and hence Tumour Origin

As implied above, it is possible to develop cancer at more than one site (this is common in those predisposed to cancer development). However, when tumours are synchronously (clinically) detected these can be two independently arising tumours, or one primary and an associated metastatic tumour. If the tumours are of independent origin you would expect to see different patterns of mutations in them (even if the mutation targets are the same). However if one tumour is a 'clonal outcrop' related to an original primary, these tumours would be expected to share more mutation patterns. Establishing the true primary tumour may have important treatment implications.

An extension of the LOH linkage model was developed to consider the evidence that two concurrently diagnosed tumours (a and b) shared the same origin. In this setting the objective is not to map the predisposing locus but to assess how similar the LOH pattern is between two tumours from the same individual, and to infer which of the two cancers is the original. The model is formulated as shown in Table 39.2, a contribution to the likelihood is made for each observed (usually microsatellite) marker. The loss parameter index refers to c = common origin, a = occurs in a during metastasis, and b = occurs in b during metastasis. This model was applied to a data set consisting of 62 patients diagnosed with concurrent endometrial and ovarian cancer (Brinkman et al., 2004).

In the familial cancer setting statistical model development has focused on extending the mixed model (Lalouel et al., 1983; Antoniou and Easton, 2006). Such models can allow for effects of known risk genes through mutation screening and linkage data, whilst fitting models to the residual genetic component.

39.2.2.5 A General Theory of Carcinogenesis

Comings (1973) brought together many of the varied theories of carcinogenesis into a single model proposing that cancer arises when a cell has accumulated sufficient mutations to bypass the normal constraints on growth. By studying disease progression in Chronic Myelogenous Leukaemia patients Nowell (1976) observed that additional chromosomal alterations (besides the *Ph* Chromosome) accumulated with disease progression. These observations led him to propose the clonal evolution hypothesis for carcinogenesis. Under this model a cell suffers a first mutation, conferring a growth advantage. This generates a colony of genetically unstable daughter cells, with further mutations accumulating in successive generations. Thus more malignant subclones would evolve driven by selection through growth advantage.

Major support for this hypothesis came from molecular genetic studies on colorectal cancer by Vogelstein and co-workers (reviewed by Vogelstein and Kinzler, 1993). This form of cancer is recognised to develop through several clear stages of malignancy, from pre-malignant lesion, through to metastasis, and offers a means to study the accumulation of genetic changes directly by taking samples of early and late stage lesions. A number of key somatic changes are now well established, namely, the early event of LOH of chromosome 5p (the location of the APC gene), and the late event of TP53 mutation.

Table 39.2   Components of the likelihood for each compound allelic state.

Probability	At each examined locus, for a pair of tumours a and b
$(1 - \alpha)$	Not informative
$\alpha(1 - \lambda c)(1 - \lambda a)(1 - \lambda b)$	Informative, no allele loss detected
$\alpha(1 - \lambda c)\lambda a(1 - \lambda b)$	Informative, allele loss in a, not in b
$\alpha(1 - \lambda c)(1 - \lambda a)\lambda b$	Informative, allele loss in b, not in a
$\alpha\{\lambda c + (1 - \lambda c)\lambda a\lambda b[(1 - \rho)2 + \rho\ 2]\}$	Informative, loss of same allele
$\alpha[2(1 - \lambda c)\lambda a\lambda b(1 - \rho)\rho]$	Informative loss of different alleles

Recently, Luebeck and Moolgavkar (2002) used a multistage epidemiological model to fit the age-specific incidence of colorectal cancer in the US SEER registry (Surveillance, Epidemiology and End Results). Assuming a model of two-stage clonal expansion (TSCE) they used a maximum likelihood approach to estimate the number of mutations necessary in the pre-initiated state before the cell acquires the capacity for increased clonal expansion. They found evidence that the US age-specific incidence data was consistent with two rare events for the initiation stage, followed by a high frequency event resulting in rapid clonal expansion. One further rare event was required for the adenoma to progress to carcinoma.

As can be seen from the example plots in Figure 39.1, the risk of most common cancers increases with age. However, Peto and Mack (2000) reported evidence of high constant risk of breast cancer in twins and relatives of breast cancer patients. By plotting the incidence of breast cancer in relatives from the time that the family index case was diagnosed, their data appeared to suggest that the risk was constant, with respect to time and age of the relative at diagnosis. They showed a similar effect in twins and in studies of cancer incidence in the contra lateral breast. They proposed a 'molecular clock' model, such that breast cancers arise in a subset of susceptible women who, from a predetermined inherited age-point, are at high constant risk of breast cancer. This pattern was not observed in other familial cancers, and a later examination of a larger data set did not confirm the hypothesis, presenting stronger evidence that familial breast cancer risk did vary with age (Hemminki and Granström, 2002).

39.2.2.6 *Mathematical Modelling of Tumour Initiation and Progression*

It is now generally accepted that cancer arises as the result of mutations in three distinct classes of cancer susceptibility genes – gatekeepers, caretakers and landscapers (Vogelstein and Kinzler, 2002). Oncogenes and tumour suppressors belong to the gatekeeper class. Caretakers include DNA repair genes whose function is to ensure genome integrity. Mutations in caretaker genes lead to genetic instability allowing the cell to accumulate further mutations more rapidly. Landscapers do not themselves lead to abnormal cellular growth, but may provide the means for the tumour to resist the immune system and develop further stromal support systems. Mathematical models examining the abstract population dynamics of the evolving tumour have been developed by Nowak and colleagues (reviewed in Michor *et al.*, 2004).

Multicellular organisms are made up of a variety of cell types. Populations of specific cell types form '*compartments*' and these compartments of cells develop to perform or maintain organ specific functions. Homeostatic mechanisms exist to ensure that the total cell number within a compartment remains constant over time. Therefore, a balance between cell birth and cell death must be maintained. Tumourigenesis will follow if the cell birth rate exceeds the death rate.

Michor *et al.* (2004) have explored the population dynamics of cancer progression by using a Moran Process. This stochastic process imposes a constant population size. At time zero, all cells in the compartment are unmutated. The oncogene model requires the cell to acquire only one mutation to alter its fitness. At each time step a cell is randomly selected (proportional to its fitness) for duplication, the daughter cell replaces another randomly selected cell in the compartment. They demonstrate that large compartments (where number of cells is large) accelerate the accumulation of advantageous (for the cell) mutations, but slow down the accumulation of deleterious mutations. The converse is true for small compartments. This model thus predicts that the size of the compartment may be influential in determining the type of mutations that are observed.

When modelling the TSG element two mutations (each with distinct mutation rate μ_1 and μ_2) are necessary for the cell's fitness to be increased. In addition they assume that $\mu_1 > \mu_2$. They show that the probability that a cell arises with two mutated copies by time t, is again influenced by compartment size, denoted as N. When $N < 1/\sqrt{\mu_2}$, a cell with one mutated copy reaches fixation before a cell with two copies arises. Hence it takes two rate-limiting steps to inactivate a TSG in a small compartment. In moderately sized compartments where, $1/\sqrt{\mu_2} < N < 1/\sqrt{\mu_1}$, a cell with two mutated copies will emerge before the first mutation has reached fixation. In large compartments, where $N > 1/\sqrt{\mu_1}$, cells with one mutated copy will arise immediately and the dynamics is dominated by the waiting time for the second mutational step. Thus as the compartment size increases the TSG function is inactivated by two, one or zero rate-limiting steps.

Michor *et al.* (2004) went on to explore the dynamical effects of chromosomal instability (CIN) and TSG inactivation. The CIN event has an associated cost or fitness and the cell with CIN has an increased rate of LOH. In a small compartment, it takes two rate-limiting steps for a cell to accumulate two mutations in the TSG with or without CIN present. For a wide range of parameter values one or very few neutral CIN genes in the genome suffice to ensure that CIN initiates tumour formation in a pathway where a single TSG must lose function. One or several 'costly' CIN genes may initiate tumour formation in situations where functional loss of two TSGs is necessary. These simple mathematical models offer insightful explanations for some of the similarities and differences observed between cancers.

39.2.2.7 *Mutational Analyses of Tumour DNA*

The development of a cancer involves somatic mutation and selection, and population genetic approaches introduced in **Chapter 22** can be applied. Phylogenetic relationships (see **Chapter 16**) can be used in studies of many tumours for the accumulation of specific mutations within a single gene. Yang *et al.* (2003) formulated a likelihood based model to analyse the full range of mutations found in a TP53 database. Their approach specifically allows the mutation or codon substitution rate to vary depending upon both tumour type and functional domain. TP53 consists of six distinct functional domains Roemer (1999). Mutations are generally classified into missense (resulting in amino acid change), nonsense (non translation or truncation of protein) or silent (no amino acid change).

Parameters exist to represent relative substitution rates; (1) nonsense vs silent (2) missense vs silent and (3) transition vs transversion. These substitution rates can be decomposed into domain specific and dinucleotide specific rates, yielding likelihood ratio tests comparing nested hypotheses. Through applying this method to a TP53 database (consisting of over 15 000 tumour samples), they found strong evidence that the mutation rates were domain dependent. They also found evidence that the transition vs transversion bias varied across different tumour types.

The work by Yang *et al.* (2003), considered only the mutations detected in TP53 in many cancer types. However, the cancer cell is the direct result of the accumulation of mutations at many loci. In terms of the cancer modelling, it is important to be able to distinguish between target genes that are to be mutated (i.e. the mutation is pathogenic) and passenger mutations. Greenman *et al.* (2006), extend the likelihood approach of Yang *et al.* (2003) also deriving test statistics and a means to estimate the parameters incorporated into their model. The main emphasis of this parameterisation is to distinguish between those mutations which have driven the cell towards the tumour state and mutations which can be classed as hitch-hikers or passengers. They do this by explicitly describing the selection process separately from the mutation process. The set of silent mutations is used to estimate the mutation rates under the null (i.e. no association between cancer development and mutation). This method is in essence similar to epidemiological studies constructed to evaluate risk factors, whilst controlling for confounding factors. The strata in this setting takes account of mutation types (essentially six different types), as these may be dependent upon the cell and its environment.

As they are considering many genes, they classify non-silent mutations into three classes, missense and nonsense, as in Yang *et al.*, but have a further class for non-silent mutations at splice sites. Their approach assumes that a sample of cancers have been examined or screened for mutations within a specified and common set of cancer genes (or a screened genome). They are then able to test for evidence of a drive towards increased selection for specific non-silent mutation types. They apply this model to a series of 518 genes sequenced in 25 breast tumours (Stephens *et al.*, 2005). This set contained a total of 91 base substitutions in 71 genes. They found strong evidence of selection on rates for nonsense and splice-site mutations, and were able to estimate that 29.8 of these base substitutions were pathogenic.

39.2.2.8 *Future Directions*

The statistical methods presented here incorporate only a subset of cancer mutation observations. Methods need to be developed that can handle the full range of genetic alterations. In tumours these include epigenetic effects and DNA sequence level mutations ranging from a single base change to loss or duplication of whole chromosomes. Mathematical models suggest that accounting for the evolutionary dynamics and the size and structure of the tumour target tissue may assist in successfully distinguishing between hitch-hiker and functional change mutations.

ELECTRONIC RESOURCES

CancerStats, Cancer Research UK, http://info.cancerresearchuk.org/cancerstats/

REFERENCES

Antoniou, A.C. and Easton, D.F. (2006). Models of genetic susceptibility to breast cancer. *Oncogene* **25**, 5898–5905.

Armitage, P. and Doll, R. (1954). The age distribution of cancer and a multi-stage theory of carcinogenesis. *British Journal of Cancer* **8**, 1–12.

Armitage, P. and Doll, R. (1957). A two-stage theory of carcinogenesis in relation to the age distribution of human cancer. *British Journal of Cancer* **11**, 161–169.

Berenblum, I. and Shubik, P. (1949). An experimental study of the initiating stage of carcinogenesis, and a re-examination of the somatic cell mutation theory of cancer. *British Journal of Cancer* **3**, 109.

Bovari, T. (1904). *Ergebnisse Über die Konstitution der Chromatischen Substanz des Zelkerns.* Gustav Fischer, Jena.

Bovari, T. (1914). *Zur Frage der Entstehung Maligner Tumoren.* Gustav Fischer, Jena, pp. 1–64.

Brinkman, D., Ryan, A., Ayhan, A., McCluggage, W.G., Feakins, R., Santibanez-Koref, M.F., Mein, C.A., Gayther, S.A. and Jacobs, I.J. (2004). A molecular genetic and statistical approach for the diagnosis of dual-site cancers. *Jounal of National Cancer Institute* **96**, 1441–1446.

Burch, P.R.J. (1962). A biological principle and its converse: some implications for carcinogenesis. *Nature* **195**, 241–243.

Cavenee, W.K., Dryja, T.P., Phillips, R.A., Benedict, W.F., Godbout, R., Gallie, B.L., Murphree, A.L., Strong, L.C. and White, R.L. (1983). Expression of recessive alleles by chromosomal mechanisms in retinoblastoma. *Nature* **305**, 779–781.

Comings, D. (1973). A general theory of carcinogenesis. *Proceedings of the National Cancer Institute* **70**, 3324–3328.

Crowe, F.W., Schull, W.J. and Neel, J.V. (1956). *A Clinical, Pathological and Genetic Study of Multiple Neurofibromatosis.* Charles C. Thomas, Springfield, Ill.

Falls, H.F. and Neel, J.V. (1951). Genetics of retinoblastoma. *AMA Archives Of Ophthalmology* **46**, 367–389.

Fisher, J.C. and Holloman, J.H. (1951). A hypothesis for the origin of cancer foci. *Cancer* **4**, 916–918.

Ford, C.E. and Hamerton, J.L. (1956). The chromosomes of man. *Nature* **178**, 1020–1023.

Garber, J.E. and Offit, K. (2005). Hereditary cancer predisposition syndromes. *Journal of Clinical Oncology* **23**, 276–292.

Greenman, C., Wooster, R., Futreal, P.A., Stratton, M.R. and Easton, D.F. (2006). Statistical analysis of pathogenocity of somatic mutations in cancer. *Genetics* **173**, 2187–2198.

von Hansemann, D. (1890). Ueber asymmetrische Zelltheilung in epithel Krebsen und deren biologische Bedeutung. *Virchows Archiv fur Pathologische Anatomie* **119**, 299.

Harris, H., Miller, O.J., Klein, G., Worst, P. and Tachibana, T. (1969). Suppression of malignancy by cell fusion. *Nature* **223**, 363–368.

Heisterkamp, N., Stephenson, J.R., Groffen, J., Hansen, P.F., de Klein, A., Bartram, C.R. and Grosveld, G. (1983). Localisation of the c-abl oncogene adjacent to a translocation break point in chronic myelocytic leukemia. *Nature* **306**, 239–242.

Hemminki, K. and Granström, C. (2002). Risk for familial breast cancer increases with age. *Nature Genetics* **32**, 233.

Knudson, A.G. (1971). Mutation and cancer: a statistical study of retinoblastoma. *Proceedings of the National Academy of Sciences of the United States of America* **68**, 820–823.

Lalouel, J.M., Rao, D.C., Morton, N.E. and Elston, R.C. (1983). A unified model for complex segregation analysis. *American Journal of Human Genetics* **35**, 816–826.

Luebeck, E.G. and Moolgavkar, S.H. (2002). Multistage carcinogenesis and the incidence of colorectal cancer. *Proceedings of the National Academy of Sciences of the United States of America* **99**, 15095–15100.

Michor, F., Iwasa, Y. and Nowak, M.A. (2004). Dynamics of cancer progression. *Nature Reviews Cancer* **4**, 197–205.

Nordling, C.O. (1953). A new theory of the cancer inducing mechanism. *British Journal of Cancer* **7**, 68–72.

Nowell, P.C. (1976). The clonal evolution of tumour cell populations. *Science* **194**, 23–28.

Nowell, P.C. and Hungerford, D. (1960). A minute chromosome in human granulocytic leukemia. *Science* **132**, 1497.

Peto, J. and Mack, T.M. (2000). High constant incidence in twins and other relatives of women with breast cancer. *Nature Genetics* **26**, 411–414.

Platt, R. (1955). Clonal aging and cancer. *Lancet* **265**, 867.

Presneau, N., Manderson, E.N. and Tonin, P.N. (2003). The quest for a tumor suppressor gene phenotype. *Current Molecular Medicine* **3**, 605–629.

Rabbitts, T.H. (1994). Chromosomal translocations in human cancer. *Nature* **372**, 143–149.

Roemer, K. (1999). Mutant p53: gain-of – function oncoproteins and wild-type p53 activators. *Biological Chemistry* **380**, 879–887.

Rohde, K., Teare, M.D. and Santibanez Koref, M.S. (1997). Analysis of genetic linkage and somatic loss of heterozygosity in affected pairs of first-degree relatives. *American Journal of Human Genetics* **61**, 418–422.

Rohde, K., Teare, M.D., Scherneck, S. and Santibanez Koref, M.S. (1995). A program using constitutional loss of heterozygosity data to ascertain the location of predisposing genes in cancer families. *Human Heredity* **45**, 337–345.

Rowley, J.D. (1973). A new consistent chromosomal abnormality in chronic myelogenous leukemia. *Nature* **243**, 290–293.

Stephens, P., Edkins, S., Davies, H., Greenman, C., Cox, C., Hunter, C., Bignell, G., Teague, J., Smith, R., Stevens, C., O'Meara, S., Parker, A., Tarpey, P., Avis, T., Barthorpe, A., Brackenbury, L., Buck, G., Butler, A., Clements, J., Cole, J., Dicks, E., Edwards, K., Forbes, S., Gorton, M., Gray, K., Halliday, K., Harrison, R., Hills, K., Hinton, J., Jones, D., Kosmidou, V., Laman, R., Lugg, R., Menzies, A., Perry, J., Petty, R., Raine, K., Shepherd, R., Small, A., Solomon, H., Stephens, Y., Tofts, C., Varian, J., Webb, A., West, S., Widaa, S., Yates, A., Brasseur, F., Cooper, C.S., Flanagan, A.M., Green, A., Knowles, M., Leung, S.Y., Looijenga, L.H., Malkowicz, B., Pierotti, M.A., Teh, B., Yuen, S.T., Nicholson, A.G., Lakhani, S., Easton, D.F., Weber, B.L., Stratton, M.R., Futreal, P.A. and Wooster, R. (2005). A screen of the complete protein kinase gene family identifies diverse patterns of somatic mutations in human breast cancer. *Nature Genetics* **37**, 590–592.

Teare, M.D., Rohde, K. and Santibanez Koref, M.S. (1994). The use of loss of constitutional heterozygosity data to ascertain the location of cancer predisposing genes in cancer families. *Journal of Medical Genetics* **31**, 449–452.

Tijo, H.J. and Levan, A. (1956). The chromosome numbers of man. *Hereditas* **42**, 1–6.

Vogelstein, B. and Kinzler, K.W. (1993). The multi-step nature of cancer. *Trends in Genetics* **9**, 138–141.

Vogelstein, B. and Kinzler, K.W. (2002). *The Genetic Basis of Human Cancer*. 2nd edition. McGraw-Hill.

Yang, Z., Ro, S. and Rannala, B. (2003). Likelihood models of somatic mutation and codon substitution in cancer genes. *Genetics* **165**, 695–705.

40

Epigenetics

K.D. Siegmund

Department of Preventive Medicine, Keck School of Medicine, University of Southern California, Los Angeles, CA, USA

and

S. Lin

Department of Statistics, Ohio State University, Columbus, OH, USA

In recent years, epigenetic changes have been implicated to be associated with a number of complex human diseases. In particular, evidence is mounting that aberrant DNA methylation in the CpG islands of gene promoters is linked to cancer. These discoveries are in part propelled by high-throughput technologies, allowing one to interrogate specific CpG sites or profiling methylation patterns of the entire genome. This chapter provides a review of statistical treatments to several problems in epigenetics based on DNA-methylation data. The detail of methylation characterization varies from modeling DNA-methylation patterns in individual cells to high-throughput methylation profiling in human tissue. The challenge on how to integrate '-omics'-scale data, both genetic and epigenetic, is discussed.

40.1 A BRIEF INTRODUCTION

In almost every cell of a human's body the DNA content and nucleotide sequence are nearly identical. However, in order to specialize in function, each cell type expresses a characteristic subset of genes. For example, an epithelial cell in the colon expresses a different subset of genes from a ductal epithelial cell in the breast. Epigenetics, derived from Greek to mean 'upon' genetics, refers to the transmission of information regarding expression of genes to daughter cells at cell division. By contrast, genetic information is transmitted by the nucleotide sequence of DNA. Mechanisms conveying epigenetic information in humans are not fully understood, but are known to involve the interrelated processes of DNA methylation, histone modification and chromatin structure. Although it is the combination of these factors and others that results in gene expression or

Handbook of Statistical Genetics, Third Edition. Edited by D.J. Balding, M. Bishop and C. Cannings.
© 2007 John Wiley & Sons, Ltd. ISBN: 978-0-470-05830-5.

silencing, DNA methylation, a hallmark of epigenetic information, is the focus of this chapter.

Mammalian DNA methylation occurs when a methyl group is added to the cytosine residue of a CpG dinucleotide (Jaenisch and Bird, 2003). Indeed, normally, a large portion of the genomic DNA (except CpG islands (CGIs) within gene promoters) is methylated at CpG dinucleotides, which plays an important role in normal X-chromosome inactivation and genomic imprinting. X-chromosome inactivation is the phenomenon in which one of the X chromosomes in a female (either the maternally or paternally derived one) is randomly inactivated in an early embryonic cell, and that the same X in all cells descended from that cell are inactivated (Ross *et al.*, 2005). This clearly reflects its epigenetic nature as it is a heritable change in gene function (inactivation) without a change in the sequences on the X chromosome involved. It is noted that, although normal X inactivation is female related, X inactivation is not restricted to females. For example, it occurs in males with Klinefelter syndrome who have more than one X chromosome (Iitsuka *et al.*, 2001).

Genomic imprinting, also known as *parent-of-origin effect*, is another epigenetic factor. More than 1 % of all mammalian genes are known to be imprinted and catalogued (Morison *et al.*, 2001). In addition to DNA methylation, histone modification and differential packing density of DNA by histone proteins are other mechanisms known to be involved in the process of imprinting (Bartolomei and Tilghman, 1997; Strauch and Baur, 2005).

Before birth, the basic pattern of DNA methylation is established (Bird, 2002). These patterns are replicated at cell division along with DNA sequence. However, with age, methylated areas may lose methylation and unmethylated areas may gain methylation. For example, early studies of DNA methylation reported a global decrease in 5-methylcytosine content with increasing age (Wilson and Jones, 1983). More recent studies of selected CpG sequences have found age-related increases in DNA methylation (Ahuja *et al.*, 1998; Toyota *et al.*, 1999). Age-related changes are believed to be influenced by a combination of local DNA structure (methylation centers), gene expression level and environmental exposures (Ahuja and Issa, 2000). A recent study of monozygotic twins found that variation in DNA-methylation levels of twins increased with age (Fraga *et al.*, 2005) emphasizing the importance of nongenetic factors.

Many human diseases have an epigenetic component (e.g. cancer (Laird, 2005), Klinefelter syndrome (Iitsuka *et al.*, 2001), systemic lupus (Januchowski *et al.*, 2004; Patole *et al.*, 2005), and Rett syndrome (Kriaucionis and Bird, 2003)). Similar to changes seen in cells with aging, cancer is marked by a global loss of methylation (Ehrlich, 2002) that is seen in conjunction with increased (hyper)methylation of CGIs in the promoter regions of tumor suppressor genes (Jones and Laird, 1999). Although the study of cancer genetics has traditionally focused on identifying heritable DNA variants that increase cancer susceptibility, epigenetic studies have now shown that DNA methylation can be one 'hit' in the pathway to cancer (Jones and Laird, 1999) (see **Chapter 39**). A large number of epigenetic changes are observed in cancer cells, suggesting either a defect in the machinery for maintaining epigenetic signatures, or the clonal expansion of a single cell that has undergone a number of stochastic changes that have accumulated with age. The most well-known epigenetic signature defect in cancer is the CpG island methylator phenotype (CIMP) (Toyota *et al.*, 1999; Weisenberger *et al.*, 2006). CIMP is described by the hypermethylation of a number of CGIs in a subset of cancers. It was first identified for colorectal cancer but now has been reported for numerous other cancer sites (Issa, 2004).

In general, the epigenetic code is considered to be erased during meiosis (Chong and Whitelaw, 2004; Rakyan *et al.*, 2003). Researchers have argued that the erasing and reprogramming of the epigenetic state provides a 'clean slate' for the fertilized egg to restore its totipotency. However, for certain loci in plants the transmission of the epigenetic state through the germline has been demonstrated (Chandler and Stam, 2004). Now transgenerational epigenetic inheritance has also been observed in mice (Morgan *et al.*, 1999), rats (Anway *et al.*, 2005) and most recently humans (Chan *et al.*, 2006). These studies generate great interest since transgenerational epigenetic inheritance has been suggested as a mechanism for adaptive evolution (Belyaev *et al.*, 1981; Jablonka and Lamb, 1989; Monk, 1995). Although these processes are extremely important, in this chapter, we limit our attention to questions involving mitotic inheritance and aberrant DNA methylation in CGIs that occurs with ageing and cancer.

Implementation of state-of-the-art microarray and other high-throughput technologies has made possible the measurements of methylation signatures of multiple gene promoters simultaneously. Recently, the idea of a Human Epigenome Project has been conceptualized (Jones and Martienssen, 2005). As information from the Human Genome Project has already had a profound impact in both basic and translational sciences (Collins, 2004; Ponder, 2001), it is anticipated that the Human Epigenome Project, combined with high-throughput methylation assays and other types of epigenomic and genomic data, will facilitate our fundamental understanding of aberrant epigenetic mechanisms and propel research in areas such as cancer genetics (Jones and Martienssen, 2005; Rakyan *et al.*, 2004). Mathematical modeling and statistical methods have started to be developed to mine such massive amount of data, which could help contribute to the successful realization of the epigenome project.

In this chapter, we will focus on introducing, describing and discussing several problems and statistical treatments, one related to modeling methylation drift in human cell populations and others involving cancer genomics and high-throughput methylation data, either targeted interrogation or whole genome profiling. One of the overarching themes in many of the statistical treatments is finite mixture modeling, which, to some extent, reflects the heterogeneity nature of the data, the sample, and the underlying tumor progression mechanism.

40.2 TECHNOLOGIES FOR CGI METHYLATION INTERROGATION

There are a variety of platforms for analyzing DNA methylation including Bisulfite genomic sequencing (Frommer *et al.*, 1992), methylation-specific polymerase chain reaction (MSP) (Herman *et al.*, 1996), MethyLight (Eads *et al.*, 2000a), and chip-based technologies (Schumacher *et al.*, 2006; Khalili *et al.*, 2007). Some approaches sequence single DNA clones while others quantify methylation frequency in a DNA sample. Some technologies measure total genomic methylation content while others investigate a single CpG site or a region of linked CpGs. Several technologies are high throughput, making possible the measurements of methylation signatures of multiple genes simultaneously. In what follows, we briefly introduce two high-throughput platforms, namely, MethyLight and methylation microarrays.

40.2.1 MethyLight

MethyLight is high throughput in the number of samples (1000), and can analyze large sets of reactions (10–100). It has been widely used in the analysis of clinical samples (Eads *et al.*, 2000b; Virmani *et al.*, 2002; Weisenberger *et al.*, 2006; Widschwendter *et al.*, 2004) but is also amenable to the analysis of formalin-fixed samples such as those collected in large epidemiological studies (unpublished data). At the same time it is highly sensitive and has been proposed for the study of early detection of cancer (Laird, 2003).

MethyLight utilizes a fluorescence-based real time quantitative polymerase chain reaction (PCR) for measuring DNA methylation. The technology uses three different oligonucleotides, one forward and one reverse PCR primer and one hybridization probe. Quantitative values are determined from a standard curve of defined dilutions of a reference sample plotted as log quantity versus $C(t)$ value, the cycle number at which the fluorescent signal surpasses a detection threshold. The quantitative value for each experimental sample is derived from a linear regression on this standard curve (Eads *et al.*, 2000a). Variation in the DNA quantity and integrity and differences in efficiencies of reactions are controlled by normalizing against methylation-independent and methylated-reference reactions. Briefly, the quantitative measure from the experimental sample is normalized to that using a methylation-independent control reaction by computing the ratio. A reference reaction using an enzymatically methylated sample of SssI-treated sperm DNA is similarly normalized. The ratio of the normalized value for the experimental sample to that of the methylated-reference sample gives the percent of methylated reference (PMR).

MethyLight reactions are designed to detect molecules in which all CpGs (usually $\sim$8) are methylated. Because of this stringent criterion for detection, methylation is not found in some samples. This results in a distribution of the data that is quantitative, but having an 'excess' of zeros. This has motivated the development of novel statistical methods for cluster analysis described in Section 40.4.1.

40.2.2 Methylation Microarrays

One of the genomic strategies for efficient scanning of tumor genome for methylation alterations is CGI microarrays; the first of which is known as *differential methylation hybridization (DMH)* (Huang *et al.*, 1999). This first generation DMH arrayed GC-rich tags derived from a human CGI genomic library onto solid supports (e.g. nylon membranes). Then CpG DNA amplicons derived from paired tumor and normal samples (probes) are cohybridized to the microarray. Differentially methylated probes can then be identified, which signify methylation alterations of corresponding sequences in the tumor sample. There are a number of successful applications using this first generation of DMH arrays. For example, DMH was used to analyze DNA from 17 paired tumor and normal breast tissues, and it was observed that approximately 1 % of the CGI loci screened display tumor-specific hypermethylation (Yan *et al.*, 2001). More sophisticated DMH arrays that are CGI library based are now commercially available, including the Toronto human CpG 12K arrays (HCGI12K, University Health Network Microarray Center).

More recently, as new platforms are becoming increasingly available for building custom microarrays, other technologies for CGI (or large genomic regions) methylation profiling have mushroomed. Such works include that using the Affymetrix tiling microarrays on chromosomes 21 and 22 (Schumacher *et al.*, 2006) and the Agilent's custom

CpG promoter Methylation (CpGpM) arrays (Khalili *et al.*, 2007). Another technology that does not rely on the use of methylation sensitive restriction enzymes is that based on DNA immunoprecipitation (Keshet *et al.*, 2006; Weber *et al.*, 2005; Mukhopadhyay *et al.*, 2004). Briefly, sonicated DNA is immunoprecipitated using an anti-5-methylcytosine monoclonal antibody. Precipitated DNA from tumor samples and sonicated DNA from normal controls are then fluorescent labeled and cohybridized to the microarrays. There is a great deal of similarity between methylation microarray experiments and those performed using other types of microarrays, such as gene expression arrays discussed in **Chapter 6**–**Chapter 9**, but key differences exist, including sample preparation and experimental conditions.

40.3 MODELING HUMAN CELL POPULATIONS

Variable DNA methylation patterns observed in populations of morphologically identical cells can carry information about cell dynamics. Since DNA methylation patterns are copied fairly faithfully from one generation of cells to the next, random errors in methylation accumulated during cell division may record the histories of the cells. Molecular clock approaches, used to compare genomes between viruses, are now being used to make inference on human stem cells (Kim *et al.*, 2005a; 2005b; 2006; Yatabe *et al.*, 2001). The first study of this kind, set out to determine if stem cells in the colon are immortal (Yatabe *et al.*, 2001), is described below. Similar methods are being applied in studies of cancer to determine whether all cancer cells are immortal or whether cancer stem cells exist.

40.3.1 Background

The epithelial tissue from the human colon is sustained by millions of crypts, each containing approximately 2000 cells. Each crypt contains a mixture of stem cells and differentiated cells. The stem cells reside near the bottom of the crypt while their differentiated offspring migrate toward the colon surface. Mature cells are short lived and essentially all differentiated cells are replaced in about a week. Copy errors from cell division are maintained in the only long-lived cells, the stem cells.

The exact number of stem cells and their characteristics are unknown (Kim and Shibata, 2002). Specifically, it is unknown whether stem cells are immortal, each division resulting in two new cells, one stem cell and one cell that will differentiate, or whether they are defined by niches. Niches are regions containing cells that are externally directed to function as stem cells. One approach for inferring which of these two models may be at work in the human colon is to analyze DNA methylation patterns among crypt cell populations.

40.3.2 Methylation Patterns

DNA methylation patterns are determined using bisulfite genomic sequencing of cloned PCR products. The clones are sequenced at a series of 5–8 CpG sites. The data are coded such that '1' denotes methylated and '0' unmethylated CpGs. The code for a string of CpG sites is called a *tag*. An unmethylated tag of eight CpGs is represented by eight 0's

('00 000 000'). When studying N CpGs, there are a total of 2^N possible tags. In Yatabe *et al.* (2001), multiple clones are measured for each crypt (range 5–8) and multiple crypts per subject (range 7–9). The CpG regions studied are CpG-rich and are selected as they should normally be unmethylated in the colon.

Tags are summarized using three statistics: (1) the proportion of methylated sites (percent methylation), (2) the number of unique tags per crypt, and (3) the average Hamming distance, the average number of site differences between any two tags from the same crypt. In general, percent methylation captures information on the numbers of cell divisions since birth. Number of unique tags and average Hamming distance are measures of crypt diversity. The more stem cells or longer-lived stem cell lineages, the greater the number of unique tags and average Hamming distance.

For diploid genomes, a sensible use of average Hamming distance should only evaluate pairs of alleles sharing a common ancestor (within lineage). In Yatabe *et al.* (2001), the average Hamming distance is from a pool of alleles and represents a mixture of within- and between-lineage distance for two of the three loci studied (the third being on the X chromosome and measured only in males). The pure within-lineage average Hamming distance might be estimable from a statistical model or, in future studies, computed directly by using a nearby single nucleotide polymorphism to distinguish the different chromosomal lineages in the cell population.

Measures of cell diversity allow us to asses the histories of cells in the human colon. Under an immortal stem cell model, sequences will become increasingly diverse over time and we should observe similarity in the distances (or number of unique tags) among cells within crypts to those between crypts. For a niche, random loss and replacement of stem cells within a crypt eventually leads to a series of bottlenecks where all cells are related to a new most recent common ancestor. This leads to more closely related cells and would result in smaller distances (or fewer unique tags) among cells within a crypt than among cells between crypts.

40.3.3 Modeling Human Colon Crypts

Populations of cells in a human colon crypt are characterized using mathematical models from population genetics. More on such models can be found in **Chapter 22**. The parameter of primary interest is the probability that a stem cell divides asymmetrically, leaving one stem cell and one nonstem cell at each generation. For a niche, each stem cell division can result in 0, 1, or 2 stem cells as offspring. These occur with probabilities p_0, p_1, and p_2 respectively, where $p_0 + p_1 + p_2 = 1$. For an immortal lineage where the stem cell will never go extinct, $p_1 = 1$ and $p_0 = p_2 = 0$. Yatabe *et al.* (2001) consider a niche with $p_1 < 1$ and $p_0 = p_2 = (1 - p_1)/2$. The number of stem cells is constrained to be constant across all generations, but allows for a dynamic population where some stem cells go extinct and others spread throughout the crypt until their descendents populate the entire crypt. The models used for constraining the population size from one generation to the next arose from (Cannings, 1974; Karlin and McGregor, 1964). For details, the reader is referred to the supplementary material from (Yatabe *et al.*, 2001).

Other parameters in the model include the number of stem cells in the crypt and the frequency of methylation errors. Methylation errors can allow the addition of a novel methylation event or loss of a methyl group. The addition or loss of a methyl group can occur at different frequencies and are assumed to happen independently at each CpG site.

Future work may determine if these sites are indeed methylated independently or whether they depend on the methylation status of neighboring sites.

Quantities fixed in the analysis are the number of generations of cell division and the total number of cells in a crypt. The number of generations of cell division is a function of age of the individual (e.g. one division each day) and possibly exposures. For instance, in addition to age, the number of cell divisions in endometrial glands may depend on the number of live births of the woman and whether she is obese (Kim *et al.*, 2005b). Under the molecular clock hypothesis, a single choice of parameter values should yield the data observed for individuals of all ages and exposures.

As direct likelihood computation is difficult, rejection algorithms, a simulation-based approach, have been used for parameter estimation. In rejection algorithms the goal is to estimate the posterior distribution of the data given the parameters, $P(D|\theta)$. Methylation data are simulated under a proposed phylogenetic model (D') and compared to real methylation data (D). Parameters are selected such that the simulated data reflects the real data. As the chance of the real data being duplicated is extremely rare, typically summary statistics from the simulated and real data are compared (S' and S, respectively). The parameters are estimated by values for which the distance between the summary statistics for the simulated and real data is below some tolerance ($|S' - S| < \varepsilon$). This approach has been named *Approximate Bayesian Computation* (Beaumont *et al.*, 2002).

40.3.4   Summary

Methylation data in the human colon are consistent with the existence of stem cell niches, each niche containing multiple stem cells (Yatabe *et al.*, 2001). The existence of niches is also supported by studies of small intestine (Kim *et al.*, 2005a) and endometrial glands (Kim *et al.*, 2005b). In endometrial glands, an age-related association between mitotic age and methylation was reported; methylation was increasing with age prior to menopause and level thereafter. In addition, methylation was higher in obese women or those with lower parity, important epidemiologic risk factors in cancer. The analysis of random replication errors may provide new methods for studying cell proliferation with age and cancer. However, not all tissues show age-related changes in methylation. Methylation in the brain and heart do not increase with age (unpublished data), as would be expected in organs whose tissues do not divide in adults.

40.4   MIXTURE MODELING

Statistical modeling using finite mixtures of distributions has become one of the staples in analyzing large scale biological data, thanks to their flexibility. It provides a mathematically based approach for modeling a wide variety of random phenomena. Clustering or classification, either of features or of samples, is a class of problems often treated by such an approach. For instance, mixture modeling has been used extensively in gene expression analysis (see McLachlan *et al.*, 2006 and references therein). Analysis of methylation data has also made use of this versatile modeling framework. We briefly introduce the unified framework of finite mixture modeling next, which will be followed by three subsections; each describes a mixture modeling solution to an epigenetic problem.

We let $Y_i, i = 1, \ldots, n$, denote a random sample of size n of variable Y, which can be a scalar or a vector. Without loss of generality, we use $f(y)$ to denote the probability density function of Y, where $f(y)$ will be viewed as a probability distribution in the case that Y is a discrete random variable. Suppose Y_i is from a heterogeneous population with K subpopulations having component densities denoted by $f_k(y)$, $k = 1, \ldots, K$. Then the density $f(y)$ can be written as

$$f(y|\theta) = \sum_{k=1}^{K} \pi_k f_k(y|\theta_k), \tag{40.1}$$

where the π_k's are referred to as the *mixing proportions* (or weights) that are nonnegative and sum to 1; that is, $0 \le \pi_k \le 1$, $k = 1, \ldots, K$, and $\sum_{k=1}^{K} \pi_k = 1$. The parameter vector $\theta = (\theta_k, k = 1, \ldots, K)$, where θ_k is the parameter of the kth component density.

In the above formulation of the mixture model, the number of mixture components K is fixed. However, in many applications, such as in the case of learning the number of cancer subtypes classified by methylation data as in Section 40.4.1, K is unknown and has to be estimated along with the weights and the parameters of the component densities. Typically, in such situations the number of subgroups is determined using the Bayesian information criterion (BIC = $-2 \times$ log likelihood $+$ number of parameters $\times$ ln(number of observations)). The BIC allows the comparison of nonnested models having different numbers of clusters; the model with the lowest BIC is selected as best. By convention, differences in BIC greater than 2 are considered positive evidence for model differences, differences between 6 and 10 strong evidence and differences greater than 10 very strong evidence (Fraley and Raftery, 1998; Kass and Raftery, 1995).

40.4.1 Cluster Analysis

40.4.1.1 Background

Cancer patients with identical diagnoses show variable response to therapy. This has spawned the use of molecular analysis for the classification of disease subtypes. The use of cluster analysis, an approach to finding novel subgroups in data, has exploded. DNA methylation is amenable to disease classification as DNA-based signatures are more stable than alternate molecular features such as protein or RNA expression. Not only do DNA-methylation profiles differ across tissues in the body, but also among different cancer histologies from the same organ (Model *et al.*, 2001; Virmani *et al.*, 2002). This has motivated the use of DNA methylation for discovering novel disease subtypes.

One approach to clustering samples is to use finite mixture models. The number of subgroups is determined using the BIC and the mixing weights give the probability of belonging to each subgroup. Using these weights, samples can be classified by assigning them to the subgroup to which they have the greatest probability of belonging. Software exists for fitting mixtures of binary data and mixtures of normals. In order to accommodate data showing an excess of zeros as is commonly observed using the MethyLight technology, two novel approaches have been proposed.

40.4.1.2 The Bernoulli–Lognormal Mixture

DNA methylation is the outcome and we use the term *feature* to refer to the CpG regions studied. In the Bernoulli–lognormal mixture model, the outcome is modeled using a mixture of discrete and continuous components. A stringent conditional independence assumption is made such that within subgroup k, the methylation levels are independent across features. The likelihood for a single sample is the product of the density for each feature across all CpG regions. Suppressing the notation for feature, the distribution for a single measurement is

$$f_k(y|\theta) = (1 - p_k)^{I_{\{y=0\}}} \left(p_k \frac{1}{\sqrt{2\pi}\sigma} \exp\left(\frac{-(z - \mu_k)^2}{2\sigma^2} \right) \right)^{I_{\{y>0\}}}, \qquad (40.2)$$

where p_k is the probability of detecting methylation in class k, $I_{\{.\}}$ is an indicator function, which is equal to 1 if the condition in the brackets is satisfied, otherwise it is equal to 0, z is the log-transformed value of y if $y > 0$, μ_k is the mean of the (log-transformed) methylation level among the observations having positive measurements in group k, and σ^2 is the variance of the positive measurements. The variance could be allowed to vary by group or feature, but is assumed constant. For each feature $2 \times K$ parameters are estimated, a cluster-specific mean and probability of positive methylation. In total, the model estimates $F \times 2 \times K + 1$ parameters, where F is the number of features. The model has been fit using the expectation maximization (EM) algorithm and multiple starting values (Siegmund *et al.*, 2004).

40.4.1.3 Truncated Normal

MethyLight data have also been clustered using a truncated normal distribution. This model assumes that measurements of zero are due to the true methylation value falling below a threshold of detection; this threshold depends primarily on the biochemical properties of the reaction. The truncated normal density is

$$f_k(y|\theta) = \left(\frac{1}{\sqrt{2\pi}\sigma} \exp\left(\frac{-(z - \mu_k)^2}{2\sigma^2} \right) \right)^{I_{\{y>\tau\}}}, \qquad (40.3)$$

where z, μ_k, σ^2 and $I_{\{.\}}$ are as defined above and τ denotes the detection threshold that is unique to each MethyLight reaction. For each feature, $K + 1$ parameters are estimated, the cluster-specific means and the (common) truncation value. In total $F \times (K + 1) + 1$ parameters are estimated; fewer parameters than the Bernoulli–lognormal. A version of this model has been fit using Markov chain Monte Carlo (Marjoram *et al.*, 2006).

40.4.1.4 Summary

The Bernoulli–lognormal model was applied to a study of DNA methylation in 87 lung cancer cell lines (41 small cells and 46 nonsmall cells). The purpose was to demonstrate whether subtypes of cancer could be identified using DNA-methylation profiles alone. Cluster analysis was performed using a subset of seven CpG regions. Applying the BIC criterion provided very strong evidence for a two-cluster model; however, the classification of the samples to the known disease subtypes yielded high error rates ($\sim$20 %) (Siegmund *et al.*, 2004). This suggested that DNA-methylation profiles could distinguish cancer subtype but that larger numbers of features would be needed to reduce the classification error rate.

  The Bernoulli–lognormal and truncated normal mixture models have been evaluated via simulation studies. Not surprisingly, those studies demonstrated that the performance of both approaches relied on the key properties of the data being analyzed. Standard lessons were reported such as using the (correct) model that required the fewest number of parameters tended to result in the lowest classification error rates. This was true of the truncated normal model performing better than the Bernoulli–lognormal model when the proportion of zeros was due to a detection threshold (Marjoram *et al.*, 2006). However, this superiority was lost when the data included (unmodeled) correlation among genes within a subgroup, perhaps because the more flexible Bernoulli–lognormal model could absorb model misspecification with its added parameters.

40.4.2 Modeling Exposures for Latent Disease Subtypes

Typically novel disease subtypes identified by epigenetic profiles are characterized using external information. Sometimes the external information is a clinical endpoint such as survival or response to treatment. Other times, for instance when studying the etiology of disease, the external information might be an environmental exposure such as smoking history or dietary folate intake. Often the analysis proceeds by a two-step procedure of first, identifying the disease subtypes and second, associating the disease subtypes with outcomes or exposures. Such an analysis is simple and can be accomplished using standard statistical software. However, this two-step approach does not take measurement error of the first step into account. Simple solutions exist for getting unbiased estimates of association when the latent disease subtypes are the exposure for a clinical endpoint. In such situations the errors in variables problem relates to measurement error in the exposure and a quick fix is to substitute the membership probabilities π_k instead of a hard assignment into the most likely category as the exposure (Stram *et al.*, 2003). This method falls within the general class of single-imputation approaches utilized in epidemiological studies. However, this shortcut does not work when the latent disease subtype is the outcome. In that situation one solution is an extension of the finite mixture model.

Let **x** be vector of q exposures. In the extended finite mixture model

$$f_k(\mathbf{y}, \mathbf{x}|\boldsymbol{\theta}) = \sum_{k=1}^{K} \pi_k(\mathbf{x}) f_k(\mathbf{y}|\mathbf{x}; \theta), \qquad (40.4)$$

where $\pi_k(\mathbf{x})$ is the probability that a sample belongs in disease subtype k given exposures **x**. The probabilities $\pi(\mathbf{x}) = [\pi_1(\mathbf{x}), \ldots, \pi_k(\mathbf{x})]$ are modeled using polytomous logistic regression, logit $\pi(\mathbf{x}) = \alpha + \beta\mathbf{x}$, where α is a $(K-1)$-dimensional vector and β is a $(K-1) \times q$-dimensional matrix. A simplification of this model assumes that the exposure acts directly on the disease subtype and not on any individual feature such that conditional on disease subtype, the distribution of the measurements is independent of the exposure **x**, $(f_k(\mathbf{y}|\theta))$.

The extended finite mixture model is fit by Siegmund *et al.* (2006) using the EM algorithm. The parameter of interest is β, the log–odds ratio measuring the association between exposure and disease subtype. Its standard error is computed using the observed information matrix as described by Louis (1982). Siegmund *et al.* (2006) found that estimates from the extended finite-mixture model were unbiased and had the correct standard error estimates; however, adequate sample sizes were needed in order for the algorithm to converge. These results were compared to the naïve two-step analysis. When there is error in the identification of disease subtype, the estimate of the log–odds ratio was biased toward the null and its standard error underestimated by the two-step analysis.

When disease subtypes were distinct and could be determined from the methylation data with certainty, the two-step analysis was adequate.

The two approaches were compared using a real data set of colorectal adenomas. Researchers hypothesized that low folate availability would be associated with abnormal methylation across a number of CGIs in colorectal adenomas. An analysis of red blood cell folate level and DNA-methylation subtype in colorectal adenomas estimated an odds ratio (OR) of 0.31 (95 % confidence interval (CI) 0.08–1.26) from the extended finite mixture model ($n = 58$ case subjects). The OR estimates using the two-phase approach was 0.44 (95 % CI 0.15–1.28).

40.4.3 Differential Methylation with Single-slide Data

40.4.3.1 Background

Recent discoveries that *de novo* methylation of CGIs can be associated with multiple types of cancer have led to tremendous interests in whole-genome methylation profiling (Gebhard *et al.*, 2006; Ordway *et al.*, 2006; Weber *et al.*, 2005). One of the tasks of making sense of such methylation profiling is to uncover gene promoters that are hypermethylated in tumor samples. Such outcomes, combined with other biological findings and evidence, may lead to the capturing of 'tumor signatures'.

In some circumstances, especially in pilot studies, a single microarray may be used to probe the methylation pattern of a sample without either biological or experimental replicates. For gene expression profiling, there are a number of methods proposed for single-slide analysis. The earliest method in dealing with single-slide data was based on some fold–change criterion (Schena *et al.*, 1996). Several methods have been proposed to provide more formal statistical treatments to the problem, including the gene-pooling method of Chen *et al.* (1997), the Bayesian hierarchical $\gamma - \gamma$ model of Newton *et al.* (2001), and more recently, the normal–uniform (NU) mixture modeling approach of Dean and Raftery (2005) (see **Chapter 6** for more details on single-slide analysis tools.)

Since epigenomic methylation profiling using microarrays is a relatively young field compared to its mature sister of gene expression profiling (see **Chapters 6–8**), there are few methods developed specifically for methylation data. Although methods in the literature for gene expression analysis may be adapted for methylation data, systematic evaluations of their performances in the latter context are currently lacking. Furthermore, features of methylation data may be different from those of gene expression data, and, as such, it should not be taken for granted that relative performances of statistical methods for gene expression data would hold for methylation data. Due to these considerations, a novel mixture modeling approach was proposed (Khalili *et al.*, 2007).

40.4.3.2 A γ-normal-γ Model

CGI loci are to be classified into hypermethylated, hypomethylated, or nondifferentiated according to normalized logarithms of methylation intensity ratios of the tumor sample to the normal sample (y). Normalization methods for cDNA-gene expression arrays (see **Chapters 6–9**) are applicable to data from methylation microarrays.

Loci that are nondifferentiated are those that have equal methylation intensities in the tumor and the normal. On the other hand, loci that are hypermethylated will have positive log ratios ($y > 0$), whereas those with negative log ratios ($y < 0$) are hypomethylated. Each of the hypomethylated, nondifferentiated, and hypermethylated components can be modeled by a γ, a normal, and a γ distribution, respectively, leading to a gamma–normal–gamma (GNG) mixture as follows:

$$f(y|\theta) = \pi_1 G(-y|\alpha_1, \beta_1) I_{\{-y>0\}} + \pi_2 N(y|\mu, \sigma^2) + \pi_3 G(y|\alpha_2, \beta_2) I_{\{y>0\}}, \qquad (40.5)$$

where the weight parameters π_k $\left(\pi_k \geq 0, k = 1, 2, 3; \sum_{k=1}^{3} \pi_k = 1\right)$ represent the relative proportions of the probes belonging to these three components. Again, I is the indicator function, which is equal to 1 if the condition in the curly brackets is satisfied; otherwise it is equal to 0. The parameter vector $\theta = (\alpha_1, \beta_1, \mu, \sigma^2, \alpha_2, \beta_2, \pi_1, \pi_2, \pi_3; \sum \pi_k = 1)$ can be estimated using the EM algorithm. Once the parameters are estimated, the probability of each loci belonging to each of the hypomethylated, nondifferentiated, and hypermethylated components can be computed, leading to a classification rule based on these probabilities (Khalili *et al.*, 2007).

40.4.3.3 *Normal–uniform Model*

A two-component NU mixture model can be adapted from Dean and Raftery (2005) to classify CGI loci into differentially methylated or not.

The NU modeling philosophy differs from that of the GNG model in that the NU model lumps both the hypermethylated and the hypomethylated loci into a single component, whereas in GNG, these two types of loci are modeled separately to reflect their opposite–extremes feature. The normal component is intended to capture those loci that are nondifferentiated, as in the GNG model, and the uniform component is for those that are differentially methylated, either hyper- or hypomethylated. Thus, the distribution for the normalized log methylation intensity ratio is

$$f(y|\theta) = (1 - \pi)N(y|\mu, \sigma^2) + \pi U(y|a, b), \qquad (40.6)$$

where π represents the proportion of differentially methylated loci, μ and σ are the mean and standard deviation of the normal distribution, and $[a, b]$ is the interval on which the uniform distribution is defined. The parameter vector $\theta = (\mu, \sigma^2, a, b, \pi)$ can also be estimated using the EM algorithm. Bayes rule with 0–1 loss is used to classify loci into one of the two components (Dean and Raftery, 2005).

40.4.3.4 *Further Comments*

Applications of the GNG and NU models to three breast cancer cell lines to profile their individual methylation signature based on whole-genome methylation intensity data generated from the CpGpM arrays demonstrated better goodness of fit of the GNG model over the NU model. High reliability of the predicted hypermethylated loci based on the GNG model was also ascertained through a bootstrap resampling technique (Khalili *et al.*, 2007). The reason for the need of analyzing these three breast cancer cell lines separately as opposed to analyzing them together is that they are inherently heterogeneous biologically, and thus their methylation signatures are expected to be different as well. As discussed earlier, the uniform component of the NU model is designed to capture both the hyper- and hypomethylated loci, which are the two extremes in the data distribution. However, the uniform distribution inevitably also captures some of the loci that are nondifferentiated, which may affect the performance of the overall model. Nevertheless, it is premature to conclude that the GNG is a better model than the NU model for identifying CpG loci that are differentially methylated based on single-slide data, as the results are only from limited experiments and the comparisons are on a relative basis rather than relying on a 'gold standard'. For that matter, it is worth noting that other single-slide analytic methods in the gene expression literature may prove to be better alternatives for solving this epigenetic problem.

40.5 RECAPITULATION OF TUMOR PROGRESSION PATHWAYS

40.5.1 Background

Recapitulating pathways of tumor progression has significant implications in understanding the disease, in influencing the treatment decision, and in developing novel drug targets. Traditionally, tumor progression pathway studies have largely focused on discerning relationships among various stages, from precancerous to preinvasive/invasive, and to metastasis, using morphological data. For example, in breast cancer studies, several pathways have been proposed to describe the relationship between grades of ductal carcinoma in situ (DCIS) and grades of invasive ductal carcinoma (IDC) (Buerger et al., 1999; Gupta et al., 1997; Roylance et al., 2002). Interpreting pathways as compartmental models, (Sontag and Axelrod, 2005; Subramanian and Axelrod, 2001) used a set of differential equations to describe the flows of contents between compartments. In particular, they studied the relative plausibility of four pathways for describing the observed counts of DCIS and IDC concurrence. It was concluded that breast tumors are heterogeneous in nature, and pathways that depict simple progression patterns, such as linearity, do not provide adequate description of such tumor samples.

The heterogeneous nature of tumors suggests that clinical phenotypic data need to be coupled with molecular signatures in order to lead to a more satisfactory recapitulation of progression pathways. To this end, it is critical to note that recent advances highlight an important role of epigenetically mediated gene silencing in tumorigenesis (Baylin, 2005). The number of hypermethylated genes tends to increase in more malignant cells, and different tumor types may be marked by their unique epigenetic signatures. Thus, the notion of utilizing DNA-methylation profiles together with clinically observable phenotypes to recapitulate tumor progression was conceptualized (Wang et al., 2007). More importantly, the idea is practically feasible as these epigenetic marks are stable and heritable in tumor genomes (Baylin, 2005), and thus methylation data derived from tumors of different patients can be used as surrogates for examining progression patterns of tumors.

40.5.2 Heritable Clustering

A tumor progression pathway can be described by a directed graph with nodes corresponding to tumor stages and each directed edge denoting possible progression from one stage to another. In the current context, a node is depicted by the epigenetic (methylation) signature of the selected genes and the clinical variables (phenotypic data); see Figure 40.1

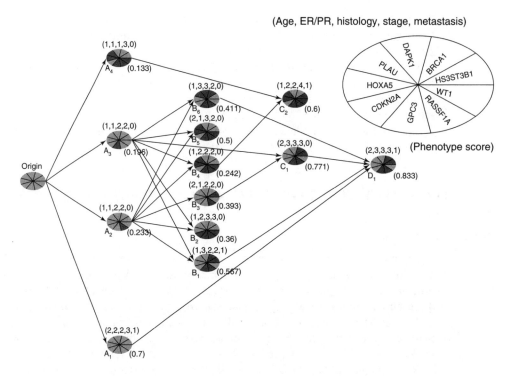

Figure 40.1   Example progression pathway network.

for an example. The heritable clustering algorithm of Wang *et al.* (2007) unfolds in three stages. First, the number of clusters (nodes of the pathway) is determined. Then, the samples are assigned to clusters and the cluster characteristics, both epigenetically and phenotypically, are determined. Finally, the clusters are organized into a pathway network to capture tumor progression to adhere to the notion that the hypermethylated loci acquired at each node are passed on to subsequent nodes and a progeny node is more aggressive phenotypically than its ancestor nodes. For determining the number of clusters, two parameters play a central role. One is the weight parameter (w) that is needed to balance the relative contributions from the methylation signature and the clinical phenotypes. The other is the within cluster similarity parameter (ε) to guarantee a certain degree of homogeneity. For each pair of (w, ε), the number of clusters $K(w, \varepsilon)$ from a clustering algorithm and the resulting total similarity $T_S(w, \varepsilon)$ can be determined. The number of clusters associated with the pair of parameters that maximizes the following objective function

$$f(w, \varepsilon) = \log(T_S(w, \varepsilon)) - \frac{K(w, \varepsilon)}{P + G}, \tag{40.7}$$

is the number of nodes of the pathway to be constructed, where P and G are the number of phenotypes and gene promoters, respectively.

Tumor samples are then classified into these nodes to define their characteristics, both epigenetically and phenotypically. Wang *et al.* (2007) considered four distance-based algorithms, H-clust, K-means, partitioning around medoids (PAM), and simulation-based (SIM). In addition, they also proposed an iterative likelihood based algorithm, which can account for dependencies among variables as well as incomplete data. Briefly, the idea is to group tumors with similar epigenetic and phenotypic characteristics according to their parameter vectors. That is, one assumes that tumors within a cluster (C_k) share the common distributional parameter vector $\theta^k = \{\theta_G^k, \theta_P^k\}$, which represents the cluster profile and will be updated iteratively. At each iteration, let $I_{k(t)} = 1$ if tumor t is in cluster C_k, otherwise it is 0. Thus, the joint likelihood is

$$
L\left(\theta_G^k, \theta_P^k, k = 1, \ldots, K | \mathbf{X}_t, \mathbf{Y}_t, t = 1, \ldots, T\right) = \prod_{k=1}^{K} \left\{ \prod_{t=1}^{T} \left[P\left(\mathbf{X}_t, \mathbf{Y}_t | \theta_G^k, \theta_P^k\right) \right]^{I_k(t)} \right\},
$$

(40.8)

where $\mathbf{X}_t, \mathbf{Y}_t, t = 1, \ldots, T$ are the observed methylation signature and phenotypes, respectively, T is the number of tumor samples, and K is the number of clusters. A tumor sample that has a larger likelihood in a different cluster than the one it is currently assigned to is a potential candidate for switching class membership.

Nodes with distinctive characteristics are then assembled into a directed graph to represent the pathway of tumorogenesis. The characteristics of a node are described by a center and a score, both epigenetically and phenotypically, signifying the 'severity' of the tumor stage it represents. Directed edges are used to connect any two nodes that have an ancestor–progeny relationship, where a node is an ancestor of another node (progeny node) if the progeny node has more severe characteristics, reflecting a temporal order.

Although strict epigenetic heritability is observed in Wang *et al.* (2007) in building the pathway, one may entertain the idea that methylation is reversible by lifting this stringent requirement.

40.5.3   Further Comments

A key utility of the heritable clustering algorithm is that the resulting pathway offers an opportunity for scientists to visualize the relationships between methylated gene promoters, observed clinical phenotypes, and their progression patterns. By utilizing both epigenetic signature and phenotypic information, pathway of tumor stages can be refined to better reflect its temporal ordering. For a panel of genes that are tumor-associated or tumor suppressor genes, hypermethylation in the promoters may lead to tumorigenesis or tumor progression, and thus their interplay with observable clinical characteristics can lead to better capturing of progression pathways. For instance, tumors with more aggressive phenotypes tend to exhibit higher level of methylation in such genes. In the same vein, other biological data types, such as histone modification, promoter sequence data and protein–DNA-binding data, which interact with methylation in a complex fashion, may also help to further define the pathways and offer a more system approach to recapitulate tumor progression. As the application of the algorithm to a set of primary breast cancer data demonstrates, the likelihood approach offers a more flexible and competitive algorithm for obtaining tight clusters than distance-based clustering algorithms (Wang *et al.*, 2007). Furthermore, the resulting pathway network portrays complex and nonlinear interplays between the methylation signature of the selected genes and the phenotype data, echoing the conclusion drawn by Sontag and Axelrod (2005) regarding nonlinearity and heterogeneity of tumor progression.

40.6 FUTURE CHALLENGES

A major challenge facing researchers studying complex traits is the integration of data from the various '-omics' platforms. Numerous biological and biomedical areas are propelled by recent advances of high-throughput technologies. In addition to whole-genome methylation profiling, a relatively new comer, whole genome profiling of DNA variations (single-nucleotide polymorphism (SNP) arrays), genetic changes in terms of copy number alterations (comparative genomic hybridization (CGH) arrays), transcriptom profiling (gene expression arrays), protein–DNA binding (chromatin immunoprecipitations (ChIP)–chip), and RA-omics (micro-RNA arrays), are platforms that provide large amount and diverse types of genomic and epigenomic data. Each of these contains valuable information on different aspects of the whole biological system, but more importantly, if all information is considered jointly in a truly integrated fashion, then the whole is greater than the sum of its parts. Such data integration is a goal of the new NIH-directed initiative, the Cancer Genome Atlas.

One of the earliest integrated approaches is that of combining DNA variation, gene expression, and phenotypic data to dissect complex traits (Schadt et al., 2005). A popular way of studying the interplay of the diverse data types is through treating the gene expression intensity as a quantitative trait, the expression quantitative trait loci (eQTL) approach (Schadt et al., 2003; Wayne and McIntyre, 2002). Similar eQTL approach was also used to study natural variation in human gene expression and potential regulatory role of SNPs (Morley et al., 2004). Transcription regulation modeling is another area in which diverse data types have been jointly considered. Microarray gene expression data, ChIP–chip protein-DNA-binding data, and promoter sequence data are all believed to play a role in transcription regulation, but their combined information has proved to be much more powerful in inferring regulatory elements and complex regulatory networks (Bar-Joseph et al., 2003; Liu et al., 2002; Sun et al., 2006). Integrating methylation profiling data with other types of epigenomic and genomic data is the natural next step to increase the power for various predictive investigations, including disease diagnostics, prognostics and therapeutics. However, cross-platform integration is by no means a trivial exercise, as different data types are interrelated in a complex manner through a highly sophisticated, yet unknown, biological system. As such, modeling of such a complex system is highly challenging. Such interdisciplinary problem requires knowledge in biology, technology, statistical modeling, and computational skills. From the statistical perspective, hierarchical modeling holds promises to tackling such problems. Large scale capable computational methods such as Markov chain Monte Carlo and approximate Bayesian computation are key to successful implementations of cross-platform models.

In conclusion, the field of epigenetics is in its infancy compared to the field of genetics. Although the impact epigenetic states will have on dissecting complex traits is still unknown, it is clear that genetic and environmental factors are not its only determinants and DNA sequence is not its only heritable unit.

Acknowledgments

The authors would like to thank Drs. Pearlly Yan, Victoria Cortessis, Darryl Shibata, and Peter W. Laird for their comments on an earlier draft. This work was supported by NIH grants CA097346 and U54CA113001, NSF grants DMS-0306800 and DMS-0112050 and NIEHS grant 5P30 ES07048.

REFERENCES

Ahuja, N. and Issa, J.P. (2000). Aging, methylation and cancer. *Histology and Histopathology* **15**, 835–842.

Ahuja, N., Li, Q., Mohan, A.L., Baylin, S.B. and Issa, J.P. (1998). Aging and DNA methylation in colorectal mucosa and cancer. *Cancer Research* **58**, 5489–5494.

Anway, M.D., Cupp, A.S., Uzumcu, M. and Skinner, M.K. (2005). Epigenetic transgenerational actions of endocrine disruptors and male fertility. *Science* **308**, 1466–1469.

Bar-Joseph, Z., Gerber, G.K., Lee, T.I., Rinaldi, N.J., Yoo, J.Y., Robert, F., Gordon, D.B., Fraenkel, E., Jaakkola, T.S., Young, R.A. and Gifford, D.K. (2003). Computational discovery of gene modules and regulatory networks. *Nature Biotechnology* **21**, 1337–1342.

Bartolomei, M.S. and Tilghman, S.M. (1997). Genomic imprinting in mammals. *Annual Review of Genetics* **31**, 493–525.

Baylin, S.B. (2005). DNA methylation and gene silencing in cancer. *Nature Clinical Practice Oncology* **2**(Suppl. 1), S4–S11.

Beaumont, M.A., Zhang, W. and Balding, D.J. (2002). Approximate Bayesian computation in population genetics. *Genetics* **162**, 2025–2035.

Belyaev, D.K., Ruvinsky, A.O. and Borodin, P.M. (1981). Inheritance of alternative states of the fused gene in mice. *The Journal of Heredity* **72**, 107–112.

Bird, A. (2002). DNA methylation patterns and epigenetic memory. *Genes and Development* **16**, 6–21.

Buerger, H., Otterbach, F., Simon, R., Poremba, C., Diallo, R., Decker, T., Riethdorf, L., Brinkschmidt, C., Dockhorn-Dworniczak, B. and Boecker, W. (1999). Comparative genomic hybridization of ductal carcinoma *in situ* of the breast-evidence of multiple genetic pathways. *The Journal of Pathology* **187**, 396–402.

Cannings, C. (1974). The latent roots of certain Markov chains arising in genetics: a new approach, I. Haploid models. *Advances in Applied Probability* **6**, 260–290.

Chan, T.L., Yuen, S.T., Kong, C.K., Chan, Y.W., Chan, A.S.Y., Ng, W.F., Tsui, W.Y., Lo, M.W.S., Tam, W.Y., Li, V.S.W. and Leung, S.Y. (2006). Heritable germline epimutation of MSH2 in a family with hereditary nonpolyposis colorectal cancer. *Nature Genetics* **38**(Suppl. 10), 1178–1183, (advanced online publication).

Chandler, V.L. and Stam, M. (2004). Chromatin conversations: mechanisms and implications of paramutation. *Nature Reviews Genetics* **5**, 532–544.

Chen, Y., Dougherty, E.R. and Bittner, M.L. (1997). Ratio-based decisions and the quantitative analysis of cDNA microarray images. *Journal of Biomedical Optics* **2**, 364–374.

Chong, S. and Whitelaw, E. (2004). Epigenetic germline inheritance. *Current Opinion in Genetics and Development* **14**, 692–696.

Collins, F.S. (2004). The case for a US prospective cohort study of genes and environment. *Nature* **429**, 475–477.

Dean, N. and Raftery, A.E. (2005). Normal uniform mixture differential gene expression detection for cDNA microarrays. *BMC Bioinformatics* **6**, 173.

Eads, C.A., Danenberg, K.D., Kawakami, K., Saltz, L.B., Blake, C., Shibata, D., Danenberg, P.V. and Laird, P.W. (2000a). MethyLight: a high-throughput assay to measure DNA methylation. *Nucleic Acids Research* **28**, E32.

Eads, C.A., Lord, R.V., Kurumboor, S.K., Wickramasinghe, K., Skinner, M.L., Long, T.I., Peters, J.H., Demeester, T.R., Danenberg, K.D., Danenberg, P.V., Laird, P.W. and Skinner, K.A.

(2000b). Fields of aberrant CpG island hypermethylation in Barrett's esophagus and associated adenocarcinoma. *Cancer Research* **60**, 5021–5026.

Ehrlich, M. (2002). DNA methylation in cancer: too much, but also too little. *Oncogene* **21**, 5400–5413.

Fraga, M.F., Ballestar, E., Paz, M.F., Ropero, S., Setien, F., Ballestar, M.L., Heine-Suner, D., Cigudosa, J.C., Urioste, M., Benitez, J., Boix-Chornet, M., Sanchez-Aguilera, A., Ling, C., Carlsson, E., Poulsen, P., Vaag, A., Stephan, Z., Spector, T.D., Wu, Y.Z., Plass, C. and Esteller, M. (2005). From the cover: epigenetic differences arise during the lifetime of monozygotic twins. *Proceedings of the National Academy of Sciences of the United States of America* **102**, 10604–10609.

Fraley, C. and Raftery, A.E. (1998). How many clusters? Which clustering method? Answers via model-based cluster analysis. *Computer Journal* **41**, 578–588.

Frommer, M., Mcdonald, L.E., Millar, D.S., Collis, C.M., Watt, F., Grigg, G.W., Molloy, P.L. and Paul, C.L. (1992). A genomic sequencing protocol that yields a positive display of 5-methylcytosine residues in individual DNA strands. *Proceedings of the National Academy of Sciences of the United States of America* **89**, 1827–1831.

Gebhard, C., Schwarzfischer, L., Pham, T.H., Schilling, E., Klug, M., Andreesen, R. and Rehli, M. (2006). Genome-wide profiling of CpG methylation identifies novel targets of aberrant hypermethylation in myeloid leukemia. *Cancer Research* **66**, 6118–6128.

Gupta, S.K., Douglas-Jones, A.G., Fenn, N., Morgan, J.M. and Mansel, R.E. (1997). The clinical behavior of breast carcinoma is probably determined at the preinvasive stage (ductal carcinoma *in situ*). *Cancer* **80**, 1740–1745.

Herman, J.G., Graff, J.R., Myohanen, S., Nelkin, B.D. and Baylin, S.B. (1996). Methylation-specific PCR: a novel PCR assay for methylation status of CpG islands. *Proceedings of the National Academy of Sciences of the United States of America* **93**, 9821–9826.

Huang, T.H., Perry, M.R. and Laux, D.E. (1999). Methylation profiling of CpG islands in human breast cancer cells. *Human Molecular Genetics* **8**, 459–470.

Iitsuka, Y., Bock, A., Nguyen, D.D., Samango-Sprouse, C.A., Simpson, J.L. and Bischoff, F.Z. (2001). Evidence of skewed X-chromosome inactivation in 47,XXY and 48,XXYY Klinefelter patients. *American Journal of Medical Genetics* **98**, 25–31.

Issa, J.P. (2004). CpG island methylator phenotype in cancer. *Nature Reviews Cancer* **4**, 988–993.

Jablonka, E. and Lamb, M.J. (1989). The inheritance of acquired epigenetic variations. *Journal of Theoretical Biology* **139**, 69–83.

Jaenisch, R. and Bird, A. (2003). Epigenetic regulation of gene expression: how the genome integrates intrinsic and environmental signals. *Nature Genetics* **33**, 245–254.

Januchowski, R., Prokop, J. and Jagodzinski, P.P. (2004). Role of epigenetic DNA alterations in the pathogenesis of systemic lupus erythematosus. *Journal of Applied Genetics* **45**, 237–248.

Jones, P.A. and Laird, P.W. (1999). Cancer epigenetics comes of age. *Nature Genetics* **21**, 163–167.

Jones, P.A. and Martienssen, R. (2005). A blueprint for a human epigenome project: the AACR human epigenome workshop. *Cancer Research* **65**, 11241–11246.

Karlin, S. and McGregor, J. (1964). Direct product branching processes and related Markov chains. *Proceedings of the National Academy of Sciences of the United States of America* **51**, 598–602.

Kass, R.E. and Raftery, A.E. (1995). Bayes factors. *Journal of the American Statistical Association* **90**, 773–795.

Keshet, I., Schlesinger, Y., Farkash, S., Rand, E., Hecht, M., Segal, E., Pikarski, E., Young, R.A., Niveleau, A., Cedar, H. and Simon, I. (2006). Evidence for an instructive mechanism of de novo methylation in cancer cells. *Nature Genetics* **38**, 149–153.

Khalili, A., Potter, D., Yan, P., Li, L., Gray, J., Huang, T.H. and Lin, S. (2007). Gamma-normal-gamma mixture model for detecting differentially methylated loci in three breast cancer cell lines. *Cancer Informatics* **2**, 43–54.

Kim, J.Y., Siegmund, K.D., Tavare, S. and Shibata, D. (2005a). Age-related human small intestine methylation: evidence for stem cell niches. *BMC Medicine* **3**, 10.

Kim, J.Y., Tavare, S. and Shibata, D. (2005b). Counting human somatic cell replications: methylation mirrors endometrial stem cell divisions. *Proceedings of the National Academy of Sciences of the United States of America* **102**, 17739–17744.

Kim, K.M. and Shibata, D. (2002). Methylation reveals a niche: stem cell succession in human colon crypts. *Oncogene* **21**, 5441–5449.

Kim, J.Y., Tavare, S. and Shibata, D. (2006). Human hair genealogies and stem cell latency. *BMC Biology* **4**, 2.

Kriaucionis, S. and Bird, A. (2003). DNA methylation and Rett syndrome. *Human Molecular Genetics* **12**, Spec No. 2, R221–R227.

Laird, P.W. (2003). The power and the promise of DNA methylation markers. *Nature Reviews Cancer* **3**, 253–266.

Laird, P.W. (2005). Cancer epigenetics. *Human Molecular Genetics* **14**, Spec No. 1, R65–R76.

Liu, X.S., Brutlag, D.L. and Liu, J.S. (2002). An algorithm for finding protein-DNA binding sites with applications to chromatin-immunoprecipitation microarray experiments. *Nature Biotechnology* **20**, 835–839.

Louis, T.A. (1982). Finding the observed infromation matrix when using the EM algorithm. *Journal of the Royal Statistical Society, Series B* **44**, 226–233.

Marjoram, P., Chang, J., Laird, P.W. and Siegmund, K.D. (2006). Cluster analysis for DNA methylation profiles having a detection threshold. *BMC Bioinformatics* **7**, 361.

McLachlan, G.J., Bean, R.W. and Jones, L.B. (2006). A simple implementation of a normal mixture approach to differential gene expression in multiclass microarrays. *Bioinformatics* **22**, 1608–1615.

Model, F., Adorjan, P., Olek, A. and Piepenbrock, C. (2001). Feature selection for DNA methylation based cancer classification. *Bioinformatics* **17**(Suppl. 1), S157–S164.

Monk, M. (1995). Epigenetic programming of differential gene expression in development and evolution. *Developmental Genetics* **17**, 188–197.

Morgan, H.D., Sutherland, H.G., Martin, D.I. and Whitelaw, E. (1999). Epigenetic inheritance at the agouti locus in the mouse. *Nature Genetics* **23**, 314–318.

Morison, I.M., Paton, C.J. and Cleverley, S.D. (2001). The imprinted gene and parent-of-origin effect database. *Nucleic Acids Research* **29**, 275–276.

Morley, M., Molony, C.M., Weber, T.M., Devlin, J.L., Ewens, K.G., Spielman, R.S. and Cheung, V.G. (2004). Genetic analysis of genome-wide variation in human gene expression. *Nature* **430**, 743–747.

Mukhopadhyay, R., Yu, W., Whitehead, J., Xu, J., Lezcano, M., Pack, S., Kanduri, C., Kanduri, M., Ginjala, V., Vostrov, A., Quitschke, W., Chernukhin, I., Klenova, E., Lobanenkov, V. and Ohlsson, R. (2004). The binding sites for the chromatin insulator protein CTCF map to DNA methylation-free domains genome-wide. *Genome Research* **14**, 1594–1602.

Newton, M.A., Kendziorski, C.M., Richmond, C.S., Blattner, F.R. and Tsui, K.W. (2001). On differential variability of expression ratios: improving statistical inference about gene expression changes from microarray data. *Journal of Computational Biology* **8**, 37–52.

Ordway, J.M., Bedell, J.A., Citek, R.W., Nunberg, A., Garrido, A., Kendall, R., Stevens, J.R., Cao, D., Doerge, R.W., Korshunova, Y., Holemon, H., McPherson, J.D., Lakey, N., Leon, J., Martienssen, R.A. and Jeddeloh, J.A. (2006). Comprehensive DNA methylation profiling in a human cancer genome identifies novel epigenetic targets. *Carcinogenesis* **27**, 2409–2423.

Patole, P.S., Zecher, D., Pawar, R.D., Grone, H.J., Schlondorff, D. and Anders, H.J. (2005). G-rich DNA suppresses systemic lupus. *Journal of the American Society of Nephrology* **16**, 3273–3280.

Ponder, B.A. (2001). Cancer genetics. *Nature* **411**, 336–341.

Rakyan, V.K., Chong, S., Champ, M.E., Cuthbert, P.C., Morgan, H.D., Luu, K.V. and Whitelaw, E. (2003). Transgenerational inheritance of epigenetic states at the murine Axin(Fu) allele occurs after maternal and paternal transmission. *Proceedings of the National Academy of Sciences of the United States of America* **100**, 2538–2543.

Rakyan, V.K., Hildmann, T., Novik, K.L., Lewin, J., Tost, J., Cox, A.V., Andrews, T.D., Howe, K.L., Otto, T., Olek, A., Fischer, J., Gut, I.G., Berlin, K. and Beck, S. (2004). DNA methylation profiling of the human major histocompatibility complex: a pilot study for the human epigenome project. *PLoS Biology* **2**, e405.

Ross, M.T., *et al.* (2005). The DNA sequence of the human X chromosome. *Nature* **434**, 325–337.

Roylance, R., Gorman, P., Hanby, A. and Tomlinson, I. (2002). Allelic imbalance analysis of chromosome 16q shows that grade I and grade III invasive ductal breast cancers follow different genetic pathways. *The Journal of Pathology* **196**, 32–36.

Schadt, E.E., Lamb, J., Yang, X., Zhu, J., Edwards, S., Guhathakurta, D., Sieberts, S.K., Monks, S., Reitman, M., Zhang, C., Lum, P.Y., Leonardson, A., Thieringer, R., Metzger, J.M., Yang, L., Castle, J., Zhu, H., Kash, S.F., Drake, T.A., Sachs, A. and Lusis, A.J. (2005). An integrative genomics approach to infer causal associations between gene expression and disease. *Nature Genetics* **37**, 710–717.

Schadt, E.E., Monks, S.A., Drake, T.A., Lusis, A.J., Che, N., Colinayo, V., Ruff, T.G., Milligan, S.B., Lamb, J.R., Cavet, G., Linsley, P.S., Mao, M., Stoughton, R.B. and Friend, S.H. (2003). Genetics of gene expression surveyed in maize, mouse and man. *Nature* **422**, 297–302.

Schena, M., Shalon, D., Heller, R., Chai, A., Brown, P.O. and Davis, R.W. (1996). Parallel human genome analysis: microarray-based expression monitoring of 1000 genes. *Proceedings of the National Academy of Sciences of the United States of America* **93**, 10614–10619.

Schumacher, A., Kapranov, P., Kaminsky, Z., Flanagan, J., Assadzadeh, A., Yau, P., Virtanen, C., Winegarden, N., Cheng, J., Gingeras, T. and Petronis, A. (2006). Microarray-based DNA methylation profiling: technology and applications. *Nucleic Acids Research* **34**, 528–542.

Siegmund, K.D., Laird, P.W. and Laird-Offringa, I.A. (2004). A comparison of cluster analysis methods using DNA methylation data. *Bioinformatics* **20**, 1896–1904.

Siegmund, K.D., Levine, A.J., Chang, J. and Laird, P.W. (2006). Modeling exposures for DNA methylation profiles. *Cancer Epidemiology, Biomarkers and Prevention* **15**, 567–572.

Sontag, L. and Axelrod, D.E. (2005). Evaluation of pathways for progression of heterogeneous breast tumors. *Journal of Theoretical Biology* **232**, 179–189.

Stram, D.O., Leigh Pearce, C., Bretsky, P., Freedman, M., Hirschhorn, J.N., Altshuler, D., Kolonel, L.N., Henderson, B.E. and Thomas, D.C. (2003). Modeling and E-M estimation of haplotype-specific relative risks from genotype data for a case-control study of unrelated individuals. *Human Heredity* **55**, 179–190.

Strauch, K. and Baur, M.P. (2005). Parent-of-origin, imprinting, mitochondrial, and X-linked effects in traits related to alcohol dependence: presentation group 18 of genetic analysis workshop 14. *Genetic Epidemiology* **29**(Suppl. 1), S125–S132.

Subramanian, B. and Axelrod, D.E. (2001). Progression of heterogeneous breast tumors. *Journal of Theoretical Biology* **210**, 107–119.

Sun, N., Carroll, R.J. and Zhao, H. (2006). Bayesian error analysis model for reconstructing transcriptional regulatory networks. *Proceedings of the National Academy of Sciences of the United States of America* **103**, 7988–7993.

Toyota, M., Ahuja, N., Ohe-Toyota, M., Herman, J.G., Baylin, S.B. and Issa, J.P. (1999). CpG island methylator phenotype in colorectal cancer. *Proceedings of the National Academy of Sciences of the United States of America* **96**, 8681–8686.

Virmani, A.K., Tsou, J.A., Siegmund, K.D., Shen, L.Y., Long, T.I., Laird, P.W., Gazdar, A.F. and Laird-Offringa, I.A. (2002). Hierarchical clustering of lung cancer cell lines using DNA methylation markers. *Cancer Epidemiology, Biomarkers and Prevention* **11**, 291–297.

Wang, Z., Yan, P., Potter, D., Eng, C., Huang, T.H. and Lin, S. (2007). Heritable clustering and pathway discovery in breast cancer integrating epigenetic and phenotypic data. *BMC Bioinformatics* **8**, 38.

Wayne, M.L. and McIntyre, L.M. (2002). Combining mapping and arraying: an approach to candidate gene identification. *Proceedings of the National Academy of Sciences of the United States of America* **99**, 14903–14906.

Weber, M., Davies, J.J., Wittig, D., Oakeley, E.J., Haase, M., Lam, W.L. and Schubeler, D. (2005). Chromosome-wide and promoter-specific analyses identify sites of differential DNA methylation in normal and transformed human cells. *Nature Genetics* **37**, 853–862.

Weisenberger, D.J., Siegmund, K.D., Campan, M., Young, J., Long, T.I., Faasse, M.A., Kang, G.H., Widschwendter, M., Weener, D., Buchanan, D., Koh, H., Simms, L., Barker, M., Leggett, B., Levine, J., Kim, M., French, A.J., Thibodeau, S.N., Jass, J., Haile, R. and Laird, P.W. (2006). CpG island methylator phenotype underlies sporadic microsatellite instability and is tightly associated with BRAF mutation in colorectal cancer. *Nature Genetics* **38**, 787–793.

Widschwendter, M., Siegmund, K.D., Mullcr, H.M., Fiegl, H., Marth, C., Muller-Holzner, E., Jones, P.A. and Laird, P.W. (2004). Association of breast cancer DNA methylation profiles with hormone receptor status and response to tamoxifen. *Cancer Research* **64**, 3807–3813.

Wilson, V.L. and Jones, P.A. (1983). DNA methylation decreases in aging but not in immortal cells. *Science* **220**, 1055–1057.

Yan, P.S., Chen, C.M., Shi, H., Rahmatpanah, F., Wei, S.H., Caldwell, C.W. and Huang, T.H. (2001). Dissecting complex epigenetic alterations in breast cancer using CpG island microarrays. *Cancer Research* **61**, 8375–8380.

Yatabe, Y., Tavare, S. and Shibata, D. (2001). Investigating stem cells in human colon by using methylation patterns. *Proceedings of the National Academy of Sciences of the United States of America* **98**, 10839–10844.

Part 7

Social and Ethical Aspects

41

Ethics Issues in Statistical Genetics

R.E. Ashcroft

Queen Mary's School of Medicine and Dentistry, University of London, London, UK

The ethical, legal and social issues concerning genetic research (genethics) are extensive and complex. This chapter reviews some of the central issues in current debates. The first part of the chapter considers the scope of genethics and the relationship between ethics, morality, professional conduct and genetics research. It then considers the relationship between risk-control and benefit-maximising models of governance in genetics research. The second part of a chapter, using a case study of the UK Biobank, looks at the main issues in governance of genetic databases, including scientific value of the research, recruitment, consent and mental capacity, voluntariness and incentives to participate, feedback of research results, confidentiality and security. The third part of the chapter looks at issues in scientific conduct of research, concentrating on stewardship and benefit sharing with host communities. The fourth part looks briefly at societal issues, concentrating on two case studies: geneticisation and reductionism, and the issue of race in genetic research.

41.1 INTRODUCTION: SCOPE OF THIS CHAPTER

The ethical, social and legal issues arising in genetic research and its applications are so extensive that they have generated a whole field of research and scholarship, often referred to as ethical, legal and social implications (ELSI) of genetics or 'genethics' (Clarke and Ticehurst, 2006; Sherlock and Morrey, 2002). A few examples of topics discussed in this literature include ethical debates about bio-safety of genetically modified crops, the morality of patenting genetically engineered living creatures, the obligations researchers in human biodiversity may have to share the benefits of any commercialisable discoveries with the donors of samples and the communities of which they are members, the moral limits on modification of the human genome, and the question of whether prenatal or pre-implantation genetic testing amounts to discrimination against the disabled or a new variant eugenics.

These issues are so diverse and complex that to cover them all adequately would require book-length treatment. In the present chapter I concentrates on issues of primary concern to statistical geneticists. This is a necessarily selective survey. I concentrates on an issue that is of considerable practical concern in genetic research – the ethical governance on

Handbook of Statistical Genetics, Third Edition. Edited by D.J. Balding, M. Bishop and C. Cannings.
© 2007 John Wiley & Sons, Ltd. ISBN: 978-0-470-05830-5.

research using biobanks. This topic cuts across many other topics in the ethics of human genetic research as well as touching on some aspects of research on non-human organisms. The second part of this chapter discusses the ethics of research using biobanks so far as this relates to the interests of sample donors. The third part of this chapter discusses the ethics of such research so far as this relates to the relationships between researchers. The final part of the chapter discusses the ethics of research so far as this relates to the interests of society.

Before beginning an examination of these issues, it will be useful to review, in this first part of the chapter, what the main sources of ethical argument are in the literature.

41.1.1   What is Ethics?

Ethics can be defined as philosophical inquiry into the values, rules of conduct and character traits, which are involved in right action, doing good and living well. It is often contrasted with morality, which is the commonly shared set of rules and principles shared within a community and taken for granted in assessing one's own behaviour and that of others. As defined, ethics can be thought of as systematic inquiry into the foundations of morality and – where necessary – correction of the principles of morality in the light of reason and evidence (Benn, 1998).

Ethics in this philosophical sense should not be confused with 'professional ethics'. Professionals such as doctors or lawyers may refer to conduct as 'unethical', by which they mean that it violates the formal or informal norms of expected behaviour by members of that profession, as laid out in codes of conduct or as inculcated through professional training. This usage of the term ethics is analogous to the term morality as defined above. Professional ethics (or, as I would prefer to say, professional morality) is referred to as such to distinguish it from common morality, which is the morality shared by most members of a community whether or not they are members of a profession. Philosophical ethics may address questions of professional ethics, but professional ethics need not be philosophical. In recent years there has been a considerable growth in teaching of and research into professional ethics – for example, medical ethics, which is now taught formally in all medical schools in the United Kingdom. Professional ethics is developed in close liaison with the legal and regulatory requirements on professionals.

One important part of philosophical ethics is bioethics. Bioethics can be defined as the application of principles and methods of analysis of philosophical ethics to the analysis of moral and social problems arising in the life sciences and medicine. Genethics is therefore part of bioethics.

The methods of bioethics are generally analytic and philosophical. Nevertheless, in recent years considerable attention has been paid to the need for ethical arguments to make use of the best quality social-scientific evidence: as has been said, good ethics needs good facts. This is particularly important in the area of population and public health genetics, since studies involve large numbers of participants, and are occasions of potential controversy. Empirical research has some potential both to clarify what issues are at stake in the participant community and also to build trust and confidence in the aims and processes of research. Examples of such research include the large scale social survey research using the Euro barometer survey into public attitudes and values concerning biotechnology (Bauer and Gaskell, 2002) and qualitative interview studies into research participants reasons for agreeing or refusing to take part in biobank research (Haimes and Whong-Barr, 2004). The Wellcome Trust has funded a major programme of social research on genethics aspects of biobank research, summarised in Haimes and Williams (2006) and on specific issues in pharmacogenomics research, summarised in UK Pharmacogenetics Study Group (2006). Recent collections presenting empirical research from the United Kingdom and other European countries are, Tutton and Corrigan (2004) and Ashcroft and Hedgecoe (2006).

In addition to analytical and empirical methods, research in genethics necessarily overlaps with research in law and public policy. There has been a considerable growth in research in medical and biotechnology law over the past twenty years, using both standard case and statute law resources and also, increasingly, the jurisprudence of human rights. In the last ten years there have been a number of significant international human rights declarations concerning genetics, and these are directly relevant to population genetic research.

41.1.2 Models for Analysing the Ethics of Population Genetic Research

There are a number of different ways of conceiving what ethically is at stake in population genetics research, each of which goes some way to shaping the regulatory and ethical framework in which contemporary genetic research is done. We can start by categorising these into two broad groups: benefit-maximisation models and risk control models. Benefit maximisation-models of the ethics of population genetic research focus on the benefits of such research, and seek ways to maximise these. Risk-control models identify specific risks which may be involved in genetic research and seek to control and in many cases minimise these. While these might be seen as complementary in that benefit-maximisation models recognise risk-based constraints, and risk control models try to avoid strong risk aversion, which might dilute any benefits accruing to such research to the point of futility, nevertheless they differ in emphasis and orientation.

41.1.2.1 Risk Control Models

For historical reasons, I thinks it is arguable that risk-control models of genethics have dominated discussion. We can identify a number of factors explaining this. The first is the history of eugenics, especially in the way eugenic ideas were used to justify forced sterilisations, barriers to 'mixed' or 'inappropriate' marriages, racial discrimination, abuse of the mentally ill and learning disabled, up to mass killings of the genetically 'unfit' under the German Third Reich (Paul, 2002). Any genetic research in humans since the Second World War has needed to establish the distance between its aims and those of eugenics. Genetic research and genetic medicine is widely seen as an area in which serious risks of personal harm and social injustice need to be forestalled or overcome. A second factor is the history of the ethics of research on human subjects. Again because of the Nazi experience (and parallel episodes in the Japanese empire, and subsequent morally problematic experiments carried out during the Cold War by NATO and Warsaw Pact countries and elsewhere), research on human beings has been considered intrinsically risky (Rhodes, 2005). To the extent that population genetics for medical or non-medical purposes involves engaging with human subjects, it is seen as a form of research on human subjects and is governed by the research ethics risk control paradigm. Next, because of the scale of modern population genetic research two further types of risk have come to the fore. The first is the role of the state in funding, facilitating and regulating genetics research. Concerns of modern citizens about state interference in their lives are read across into the field of genetics research. The second is the role of the commercial private sector in developing technologies using genetic research and its applications. Concerns about the self-interested behaviour of corporate actors are similarly read across into the field of genetics research. These types of risk are exemplified by concerns about data protection, confidentiality and privacy, intellectual and real property rights in samples, data and discoveries, benefit sharing with donors of samples or data, exploitation of poor individuals or groups (especially transnationally), state coercion and surveillance, sharing of information with agencies or individuals for non-medical purposes and so on (Bauer and Gaskell, 2002). Finally, there is a general concern with any kind of research into the biological characteristics of human beings, that such research undermines human identity or human dignity. This kind of risk is often invoked in religious contexts, but it has also been influential in the framing of certain kinds of regulation, most notably international patent law, and law and regulations prohibiting germline gene therapy and reproductive cloning (Beyleveld and Brownsword, 2001). Unlike the first four kinds of risk, this is not clearly linked to any specific historical or political experience, but may be linked to a more general historical epoch often referred to as 'modernity', which many intellectual historians associate with secularisation or with the 'scientific revolution' of the sixteenth century. Further, this sort of risk is generally considered qualitatively rather than quantitatively. Notionally it cannot be traded off against other benefits or risks, because undermining human dignity (for instance by undermining the grounds for considering human beings fundamentally equal and members of the same human family) would be absolutely wrong. This sort of risk is held in mind especially in international human rights declarations.

At this stage, my purpose in describing these five kinds of risk is not to argue about their scope, significance and centrality to the ethics of genetics research, but rather to argue that concentrating on these kinds of risk makes sense to many analysts of the ethics of genetics for historical and political reasons. Because they are seen to be salient, they frame and shape much of the discussion of the ethics of genetic research, and much of the regulatory framework for genetic research is fixed by the concern for these risks and solutions developed in other contexts for managing them.

As well as identifying these five kinds of risk by the context in which they arise and the sorts of harm or moral danger involved, they can also be categorised by the level at which they operate: individual, family, social group, society as a whole, humanity as a whole. There is a pronounced tendency in the literature to concentrate in risks of harm to identifiable individuals, and to regulate that risk of harm through a dependency on the informed consent mechanism. As seen, it is harder to regulate risks which operate at group level, and the troubles attending the extension of informed consent models to group protection are well known and difficult.

41.1.2.2 *Benefit Maximisation Models*

Perhaps surprisingly, benefit-maximisation models are rarely articulated formally, but are instead the 'common sense of science'. Benefit maximisation arguments can appear in various forms, from appeals to the future health benefits that will accrue if particular research lines are (successfully) pursued, to appeals to a right to freedom of scientific inquiry, to more bluntly economic arguments about the inefficiency of research ethics regulations. In the current state of ethical debate it is arguable that benefit-maximisation models function in two ways: first, as a corrective to (excessive) caution in risk-control models, and second, as substantive arguments about the best way to get maximum value out of particular research resources. An example of the latter type of argument is the argument about whether privately funded entrepreneurship and 'pay-per-view' style models of access to genetic databases by researchers, or alternatively open source, publicly funded databases, are more efficient and effective in producing high quality genetic science and technologies (Sulston and Ferry, 2002; Rabinow, 2003).

41.2 A CASE STUDY IN ETHICAL REGULATION OF POPULATION GENETICS RESEARCH: UK BIOBANK'S ETHICS AND GOVERNANCE FRAMEWORK

In order to understand the ways in which the different kinds of research risk frame the governance of population genetic research, while retaining an intention to do research that is maximally beneficial, it is helpful to consider a case study. The UK Biobank is a major initiative in studying the interaction between genes and environment in a health context, using a large cohort study drawn from the UK population. It is described on the project's website as follows:

> UK Biobank is a long-term national project to build the world's largest information resource for medical researchers. It will follow the health of 500 000 volunteers aged 40–69 in the UK for up to 30 years.

> The UK Biobank is funded by the Department of Health, the Medical Research Council, the Scottish Executive, and the Wellcome Trust medical research charity. The project will help approved researchers to develop new and better ways of preventing, diagnosing and treating common illnesses such as cancer, heart disease, diabetes and Alzheimer's disease.[1]

On the website, the need for the UK Biobank is explained as follows:

> UK Biobank is a medical research study of the impact on health of lifestyle, environment and genes in 500 000 people currently aged 40–69 from all around the UK. This age group is being studied because it involves people at risk of developing serious diseases – including cancer, heart disease, stroke, diabetes, dementia – over the next few decades.

> The UK National Health Service treats the single largest group of people anywhere in the world, and keeps detailed records on all of them from birth to death. Consequently, follow-up of UK Biobank participants through routine medical and other records will allow identification of those who develop a wide range of disabling and life-threatening conditions. This will make UK Biobank a uniquely valuable resource.

> Scientists have known for many years that our risks of developing different diseases are due to the complex combination of different factors: our lifestyle and environment; our personal susceptibility (genes); and the play of chance (luck). Because UK Biobank will involve thousands of people who develop any particular disease, it will be able to show more reliably than ever before why some people develop that disease while others do not. This should help to find new ways to prevent death and disability from many different conditions.[2]

Recruitment and participation are described thus:

> People to invite into UK Biobank are identified from central registries. The only information provided, in confidence, to UK Biobank is name, address, sex, date of birth, and general practice. These details are processed centrally (in accordance with the Data Protection Act) and an invitation letter sent directly by UK Biobank. General practitioners are advised that people registered with their practices may be invited to take part.

Taking part in UK Biobank involves participants in:

- Attending a local study assessment centre for about 90 minutes to answer some simple questions, to have some standard measurements, and to give small samples of blood (about 2 tablespoons) and urine;
- Agreeing to allow their health to be followed for many years by UK Biobank directly through routine medical and other records;
- Being re-contacted by us to answer some additional questions and/or attend a repeat assessment visit (which would be entirely optional).

> A very wide range of tests will be done on the blood and urine samples for approved medical research, and it is impossible to predict all of the uses to which the samples might be put during the next few decades. But, since none of these individual test results will be fed back to participants, their doctors or anyone else, taking part in UK Biobank should not have any adverse effects on participants (including their employment status or ability to get insurance).

> By analysing the answers, measurements and samples collected from participants, medical researchers will be able to work out why some people develop particular diseases while others do not. Although taking part in UK Biobank may not help participants directly, it should give future generations a much better chance of living their lives free of diseases that disable and kill.[3]

The UK Biobank has a detailed governance framework, which I will discuss and describe in detail below. For the moment, let us examine the issues this initial presentation of the UK Biobank highlights as being ethically important or of interest to the general reader of its website.

41.2.1   The Scientific and Clinical Value of the Research

The first thing emphasised in the public presentation of the Biobank is the *expected value* of the research. Participants are invited to take part, and the public (and scientific community) are invited to support the project, on the basis that the Biobank will produce important new knowledge, which will make a significant contribution to the understanding, prevention, diagnosis and treatment of major diseases.

In this regard, UK Biobank is typical of most human subjects research involving large numbers of people (Tutton and Corrigan, 2004). To succeed, it must recruit a large number of people, and secure their consent to a variety of investigations, some of which may be painful or inconvenient, and some of which may require long-term contact. Biobank is presented as a moral enterprise. It is thought not only to be scientifically interesting, but also to be potentially useful and helpful in advancing a vital interest that all share, their interest in being healthy and in receiving good medical care. A potential participant might ask him or herself, 'why should I volunteer?' One answer might be: to help scientists do something they think is really interesting. For some research participants, this would be a sufficient reason to take part. Many people do think that science is intrinsically valuable, and that it is exciting and honourable to play a part in its advance. Yet in fact this appeal to shared scientific curiosity is relatively rare in modern biomedical research. A risk of selling science to the public on the grounds of utility is that scientific projects may fail, or may fail to produce the expected or hoped for results. In clinical trials research, researchers are aware of the 'therapeutic misconception' held by many patients, who believe that they are receiving the treatment judged to be best for them, when in fact they are participating in a trial which may or may not benefit them personally. This issue is less pertinent in epidemiological research, but something similar may apply at the level of society at large: social participation in research like Biobank is not guaranteed to benefit anyone individually or as a collective. Yet researchers have an incentive to emphasise the chances of success, and to play down the chances of failure.

Participants in research like Biobank who take part on the grounds of its expected usefulness may perceive this utility in one of three ways. First, they may perceive participation as useful to them personally, here and now. Second, they may perceive participation as generating benefits, which will be useful to people like them (possibly including themselves personally) in future. Third, they may perceive participation as generating benefits, which may benefit people in future, without consideration of their own personal interests. It is only rarely the case that there are direct personal benefits to participating in epidemiological research, but see the discussion of feedback of results to participants below. On the other hand, researchers and research participants alike frequently emphasise the second and third kind of reason to take part as good reasons.

Participation on the basis of benefits to people like myself (possibly including me) in future could be called *solidary participation*: I take part out of solidarity with people I recognise as my peers and whose suffering I want to ease, on the basis that they would do the same for me, or that I hope that they would. Participation on the basis of solidarity is a theme that has long been emphasised in epidemiology, at least in the post-war period in which epidemiology in the United Kingdom and other welfare states was linked institutionally to social medicine, to socialised healthcare, and to social movements such as the Trades Union movement (Ashcroft *et al.*, 2000). Solidarity may be presented in a more or less moralised form. On the one hand, many commentators would see solidarity as the basis of any moral relationship with others, since what motivates us to help others is an understanding of the plight of others based on empathy. On the other hand, solidarity could be a much more pragmatic motive, where people reason on a quasi-contractual basis that they are obliged to put into a social relationship more or less what they get out of it. Some recent work argues that people have a duty to participate in research, because they have benefited to date from research in which others have participated. Solidary participation on this understanding is no more than enlightened self-interest.

Participation on the basis of benefits to others without consideration of the relationship they may have to people can be described as altruistic participation. Appeals to pure altruism are relatively rare, just as it was noted that appeals to shared curiosity are rare. In practice, most appeals for research participation appeal either to direct self-interest (through payment to participants in healthy volunteer research, or through an improved chance of receiving the treatment that is most effective or safest in a clinical trial) or to solidarity. The reasons for this are complex and are suspected to be not very well understood, but part of the explanation may be the dominance of a 'risk control' approach to research ethics. The influential Declaration of Helsinki and most other statements of research ethics emphasise that the welfare of the individual subject shall take precedence over the interests of science and society. A consequence of this emphasis is that any approach to research recruitment emphasises the interests of the participant rather than the benefits to science or society, even where the interests of the participant are only minimally affected, or where those interests are entirely consistent with maximising the scientific and social benefits.

So far i have considered the ways in which appeal to scientific value figure in arguments persuading participants to take part in Biobank, and in persuading the public to accept Biobank as a national scientific project using public resources. It is important also to emphasise the other side to this argument, which is that for these reasons to hold, it must actually be the case that Biobank represents good value for money and a worthwhile investment of scientific resources of time, talent and facilities. In general terms, it is accepted that scientific projects must be well founded in terms of posing a well-defined question, to which the answer is not already known, and which, if answered, would generate new knowledge of scientific importance. The twin mechanisms of systematic review (to establish what is already known, and to what degree of methodological confidence) and peer review (to establish the credentials of the research team, the likely scientific value of the research and the value for money the project offers) are intended to ensure that scientific projects put to the public to invite participation are well founded scientifically, and thus that the claims made about the utility of the research are as well supported as is possible.

An important issue in genetic research, however, is that in many cases genetic sample collections are built up not in the context of a specific research project, with tightly designed objectives and research questions, but as resources for use in future unspecified research (Gibbons et al., 2007). Sample collections may be made of those collected for clinical or other purposes, and are turned into research resources retrospectively; or they may be collected for use in one research project, and then reused in other future research projects for other purposes; or they may directly be collected as a sample bank, with the research uses to be specified later. For example, a researcher may build a collection of samples taken from the brains of people with Parkinson's disease, for use by Parkinson's disease researchers, without any specific research project in mind at the point of collection. In a commercial clinical trial, the pharmaceutical company sponsoring the research may build a collection of samples from participants in the trial for pharmacogenetics research either prospectively or retrospectively, to help interpret their trial results and assist in licensing applications. Here, the questions of utility may be more diffuse, and peer review, if practised at all, would focus more on the governance of the collections, and on the research projects, which are subsequently proposed that intend to use the collections.

We have seen how Biobank and other research projects start their approach to ethical governance and recruitment by emphasising the scientific and public value of their projects. The next topic addressed is the methodology of recruitment.

41.2.2 Recruitment of Participants

Biobank are at pains to emphasise how individuals are identified as possible participants. This has two elements: selection and identification. Potential participants are selected on the basis of the inclusion and exclusion criteria, which define the study population in light of the research objectives of the study. Biobank intends to enrol 500 000 individuals, between 40 and 69 years of age, from all over the United Kingdom. Those excluded from the study are those who are unable to give informed consent (for example, because of diminished mental capacity), those too ill to take part in data collection and those who study recruiters deem uncomfortable with any of the conditions of participation. The latter exclusion essentially is intended to exclude people whom it is felt would prefer not to consent, or seem not to understand what is required of them, but who nonetheless seem for one reason or another to be giving their consent. These are very broad inclusion criteria, as is appropriate for a study, which intends a rather comprehensive analysis of the relationship between physical and mental health, lifestyle and environmental factors and genotype. Other studies might have more restrictive inclusion criteria. Case-control studies, for instance, where 'cases' must meet defined clinical or other criteria for inclusion and where 'controls' must be comparable to selected cases, will have much more specific inclusion and exclusion criteria. In the context of controlled clinical trials there are important issues concerning the fairness of inclusion and exclusion criteria in terms of the distribution of the risks and benefits of research to (non-)participants. This is less of a significant issue in epidemiological research, but does arise in the context of the interpretation of findings, which will be described below.

A more important question concerns how individuals eligible for participation are identified as suitable for recruitment, and how they are then approached. Some studies simply ask people to volunteer, through placing advertisements. It is up to each individual to decide whether they are eligible and interested, and then to contact researchers. For most purposes this is not adequate in terms of constructing an unbiased sample for research, so the practical issue is how to identify potential research participants, and then to contact them, without breaching norms of confidentiality and law relating to data protection. If potential participants are identified through public information (such as electoral rolls or telephone directories), this is rarely a significant issue. However, where participants may be identified through information not in the public domain, such as personal medical records, the situation may be complicated. For a researcher who would not have access to this information routinely (for instance, in the case of medical records, because he or she does not have clinical care of these patients), individual consent might be required for researcher access to these records; but the researcher can only know which records (or which patients) he or she would like to see, and thus whose consent is needed, when he or she has had the chance to look at the records. In practice, pragmatic solutions to this Catch-22 need to be found: either someone who has right of access to the set of records preselects the individuals for the researcher and makes the first approach to them to obtain consent to the researcher approaching them, or actually to enrol them into the study, or the researcher is given a contract with the record-holding organisation that allows them to see records and places them under a contractual duty not to misuse the records outside the terms of the contract. Under some circumstances, it might be that a research project is of such public importance that enrolment of participants may take place without their consent (if this involves data extraction from records only). Many countries have regulations in place to permit this. However, in the case of genetics research, while this method might be used to allow data extraction to identify individuals who are potential research participant, actual recruitment into the research would almost always involve a direct approach to the individuals and the seeking of their consent.

The principal exception to these rules about recruitment of participants are where the samples to be used are de-identified, in such a way that re-identification of individuals is impossible. Normally this would occur when a sample bank is used, rather than samples being collected for a specific research project.

41.2.3 Consent

The central element to the ethics of genetic sample based research is – inevitably – consent. Consent in epidemiological and genetic research has been much debated in recent years (O'Neill, 2002; McHale, 2004; Gibbons *et al.*, 2005). This is in part because genetic research poses new issues, and in part because of the social and economic trends providing the context against which genetic research takes place now.

The standard requirements for a valid informed consent concern (a) the mental capacity of the individual to give consent to the research participation; (b) the voluntariness of the decision; and (c) the information necessary to make a decision (Jackson, 2006).

Mental capacity is a complex topic, and could be treated at length, but for present a few simple points are sufficient. Mental capacity is the ability to understand and retain the information one is given to make a decision, to believe it, and to weigh it and make a choice. A consequence of this definition is that capacity is relative to the nature of the decision being made: it is much easier to show that someone has the capacity to buy a newspaper than it is to show that that they have the capacity to consent to a heart transplant. So one does not simply have 'mental capacity', but rather, under a particular set of circumstances, one can be said to have the capacity to make this sort of decision. In a medical context, adults are presumed to have mental capacity to make any decision that they may be asked to make, and considerable evidence and inquiry may be needed to show that they lack mental capacity to make that decision. This is true even of people with a psychiatric disorder. In the case of children, for the most part, the opposite is true: a child under 18 (or, for many purposes, under 16) has to be shown to have the ability to make a decision, and is presumed to lack that capacity. For medical research purposes, these assumptions about adults, children and mental capacity are controversial. Should it, for instance, be assumed that research is necessarily harder to understand than ordinary clinical treatment? Surely not. But some research is highly complex. Moreover, what may be difficult to understand is that enrolment in research is voluntary, that according to all research guidance it is clear that patients are not to be compelled to take part in research, or threatened with poor treatment if they refuse, or forbidden to leave research once it is started. Patients (in particular) may believe that participation is either obligatory, or necessarily in their best interests, or that they 'owe' their doctor something, when none of these need be true. (This, again, is the 'therapeutic misconception' mentioned above.) So what may be the stumbling block in determining capacity is not the complexity of the technical information, but the changed relationship between researcher and participant (where it was doctor–patient, it is now a rather different relationship between researcher and participant).

We can debate whether this is an issue of capacity, or rather one of voluntariness. In practice, since most epidemiological research is non-therapeutic, people who lack capacity are simply excluded from research (unless the research links specifically to the reasons for their incapacity – as in some psychiatric genetic research), since participation in research can rarely be shown to be in the individual's best interest. This would be the test of whether someone can be enrolled in research when they lack mental capacity. A finer grained interpretation of this condition can be developed, as in guidelines from the UK Medical Research Council and other similar bodies (Medical Research Council, 1991). On this interpretation, a person lacking capacity to consent to a research project can be included in the research if it is in their best interests, or if the risks are minimal and this research cannot be conducted in people with capacity and this research would benefit others in future with a similar condition and participation is not *against* the individual's interests. Under some circumstances research, which is not minimal risk can be permitted, but there would have to be a compelling case made about why this research was essential for the welfare of people suffering from this condition under investigation and there was no other way to carry it out or to resolve the problem under study.

A related difficult issue concerns the achievement at a later date of capacity by someone enrolled in a study while lacking capacity. For example, a child might be enrolled in a study at birth. When the child reaches majority, it is sometimes argued (as in the Icelandic DNA database, for example) that the now adult individual should have the power to withdraw the consent given on their behalf, or to ratify it. This raises complex issues concerning what it actually means to leave a study in this context, as it will discussed below, but also the wider question of how far 'proxy consent' can really be considered valid, and how far parents (or carers) can choose authoritatively for their children (or incapacitated relative) (Ross, 1998; Archard, 2004).

The voluntariness condition is less controversial. In practice, people may be under a variety of pressures to participate, but these would not normally amount to coercion. It is well known that some people are more successful at recruiting research participants than others, other things being equal, and how far this is due to personal charm or to conveying the importance of the research or a low tolerance to taking no for an answer is often difficult to discern! Ethical concern tends to be directed at formal obstacles to voluntariness. Since the Nuremberg Code of 1947, all research ethics guidelines have insisted that people who take part in research should be free to leave it at any time, in the interests of allowing people to change their minds or go back on decisions they regret or – in the worst case – were suborned into making (Marks, 2006). This poses a difficult challenge in epidemiological research. Since such research depends on the collection of large data-sets and the analysis of such sets as collections of data, often longitudinally, it is not always clear what 'freedom to leave' means. Clearly it can mean that once a person leaves the study, no new data on them can be collected, but existing data can continue to be used. Or it might mean, more permissively, that once they leave, further data on them may be collected from routine data without further contact with the person. Or it might mean, more restrictively, that once they leave existing data on that person should be removed from the data-set (UK Biobank Ethics and Governance Council, 2006). The standard view is that the existing data can be used once the person has decided to leave the study, and that it cannot be removed. To remove the data would compromise the integrity of the data-set, and may actually be impossible for technical reasons, if data once added are de-identified. In more principled terms, it is not clear that the person has a right to request that their data should be removed: firstly, that they agreed to that data being used and cannot retrospectively revoke that agreement, and secondly, that although the data may be derived from them, they are not owned by them, but by the researcher in whom intellectual property in that data resides. What is being negotiated here is the meaning of 'leaving the study'. The normal approach here is to be reasonably explicit about what agreeing to take part in the study amounts to, and about what leaving the study amounts to. Nonetheless, it is possible to imagine circumstances under which someone loses their trust in the researcher or the research project to the extent that they feel that they have been misled. They might then reasonably say that the data was collected under false pretences, and that it should be deleted. This might arise in a study of race/ethnic differences, for instance, where a person feels that the study undermined the dignity or reputation of their race/ethnic group in a way they could not foresee and were not warned about. Similarly, it might arise where a child was recruited into a study by his or her parents in early childhood, and may at adulthood wish to remove his or her data from the study, which were collected without his or her consent.

A second issue of voluntariness concerns incentives to participate. These may be both formal and informal. Formal incentives, in the form of payments or offers of services in exchange for agreement to participate, are controversial for two main reasons. Firstly, many epidemiologists would feel that people should participate in research for solidary or altruistic reasons, and offering payments, gifts or in-kind exchanges devalues such reasons for participation. They would argue that this discourages people from participating in research unless there is something in it for them. Secondly, ethicists tend to worry that people who take part when there is an incentive scheme are doing so in order to get the incentive, rather than with full understanding of the risks and benefits of participation. The greater the incentive, the greater the risk that this may occur, up to the point where people put themselves knowingly in danger merely because they want or need the incentive. Of course, this is the situation with most phase I drug research and many kinds of employment. But this is felt to be regrettable, if necessary, and not something to be expanded. One hard issue here is that it is almost, if not actually, impossible to draw a line separating reasonable from coercive levels of inducement.

Informal incentives to participate comprise reasons to participate in the research, which are formally part of the research protocol itself, but which are attractive to participants for reasons unconnected with the research. For example, in many epidemiological studies, such as Biobank, participation in the research can involve the collection of vital statistics and medical examinations for the purposes of collecting baseline and study time-point data, but which are also (potentially) useful health screening data for the research participants. Many studies are attractive to participants because they are getting free, or more than usually convenient, 'health checks'. Sometimes, participants may also believe that they are getting access to tests or checks on their health that are not normally available. This is particularly the case in some genetic studies investigating risk factors for common diseases. A patient in a study of the genetic risks for colon cancer, for example, may believe that they will be given individual feedback on their genotype, which would allow them to know whether they were at risk of developing colon cancer.

These informal incentives are highly problematic, and much discussed in the literature (Haimes and Williams, 2006). It is generally accepted that data that can be interpreted 'at the bedside' and which are collected on an identifiable basis should be returned to the participant on request, although if the participant does not want them, there is no need to force these data on them. In most epidemiological research, the researcher is not seeing the participant as a patient, and the researcher has no specific duty of care to the participant, such that if the participant has a high blood glucose level they should be counselled to see their general practitioner about possible diabetes mellitus. Indeed, for many of the tests and measurements that may be done in a research project, the participant would need to be counselled and consent obtained *to the medical test* rather than to research data collection only if this were done. Nonetheless, if data is collected, which incidentally suggests that the participant would be wise to see a doctor, it is generally good practice to pass this on. More controversy surrounds the feedback of research findings to patients, which only arise out of the analysis of the research data itself. So should it appear that a particular genotype is at greater risk of colon cancer than normal, there is not thought to be any obligation to tell individual research participants with this genotype that they need to see their doctor for further counselling and possible investigations. In part, this is because the quality of such genetic testing in research is not at clinical grade. Participants should normally be kept up to date with the progress of the research, in terms of key findings for instance. But this level of feedback about the group as a whole would not have implications for specific individuals. Or rather, it does not unless the individual were to know that the genotype demonstrated to be predictive of colon cancer happened to be the one he or she carried (Johnstone and Kaye, 2004).

One way to resolve all of this is to be quite explicit up front about what individuals will be fed back and what they will not be fed back, and what participants should do if they have further (medical) questions. However, this does not entirely address the point of principle: do individuals have a right to the data that concerns their health and genetic constitution? Do the researchers have a duty to give it to them? Within a medical relationship, doctors are entitled to withhold information from patients if it may be misinterpreted or if it would be psychologically damaging to the patient to receive a piece of information, which they are ill-prepared to cope with. This entitlement is somewhat controversial even within medicine (it is the so-called 'therapeutic privilege'). Where the researcher has no medical relationship with the participant it is even less clear what the answer is. Current practice is not to feed back individual data directly to the patient, but this may change. In the present context, what is clear is that participants who think they have an informal incentive to participate in a research project, which is that they will get early access to innovative tests, need to be clearly told what they can and cannot expect to receive and why.

This takes us to the most complex issue of all, what is the nature of the information participants should receive in order that they can give a valid consent. The importance of information is twofold. First, the quality of the information and style in which the information is presented has a major influence on the potential participant's ability to understand what is being proposed, and consequently on the likelihood that they agree to take part. Second, the information given defines what it is the participant is consenting to, and hence what that consent authorises the researcher to do in terms of investigations, and how data and samples may be used in research. The latter issue has given rise to heated debate.

One important feature of much epidemiological research is that data and samples hold much of their value because they can be reused and reanalysed, either directly or in combination with data collected for other purposes. Thus, a tissue sample collection created for research on the consumption of salt and cardiovascular disease could be useful to a researcher interested in the vascular aspects of Alzheimer's disease. Yet the consent taken for building the collection for the first purpose would not necessarily authorise the use of the samples for the second purpose. A sample collection obtained and managed by one research group could be useful to a different research group, possibly in another country. A sample collection built up with public money could be transferred to a spin-out company created by publicly funded researchers, and the samples become a valuable commercial resource; yet the samples were collected on the basis that these would be a public resource rather than the capital for a commercial venture. In each case, the consent volunteered by research participants may not authorise these changes of use, yet each change might be seen by the researchers (and indeed by ethics committees and research funders) as useful ways of getting the best value out of their investment in building up the sample resource.

The problem here is that consent, to be valid in law, needs to be quite specific, whereas what researchers typically need is a consent that is quite broad and durable in terms of what it authorises in the short and long term. Different approaches have been tried to get around this problem. One approach is to try to create 'broad' consents, which allow participants to consent to a class of uses of their data and sample, rather than to only a narrow and specific use within a tightly defined protocol. The broad consent is supported by independent oversight by a research ethics committee and a governance process, which protects the interests of sample donors or data subjects in place of the protection, which a series of narrow and specific consents would give. This is largely the approach taken by UK Biobank. Another approach is to require a new consent each time a new use of the samples is proposed. In practice this would normally require each individual to be recontacted by the research team. This is advantageous in terms of ensuring that each new investigation is formally authorised by each participant, and it provides a mechanism for the participant to leave, by reminding them that their continued participation is optional. On the other hand, it is cumbersome and expensive, may cause attrition in the study population, and may even be burdensome on participants who are willing to continue as long as they are not bothered too often. This approach is taken in longitudinal studies where recontact would take place in the normal course of events, for new questionnaires or invasive investigations to take place. But in most epidemiological research, where direct contact with participants is unusual after initial recruitment, this approach is unpopular for pragmatic reasons.

The broad consent approach seems popular with the research community, and in many ways is supported by empirical evidence about what research participants want to know. What it does not address is another issue, viz. the ways consent addresses the issues of most concern to research participants. Consent to participate in research, being developed on the basis of the risk-control model appropriate to clinical trials, tends to focus on the risks and benefits of research participation, the nature of the investigations a participant will undergo, and other issues concerned with the personal safety and integrity of the individual. However, for many research participants these are not the only important issues. At least some research participants are interested in issues such as the possible commercialisation of research findings, the possible uses of their data or samples in research of which they may not approve. For instance, while a participant might be entirely happy for their samples to be used in genetic research in cardiovascular disease, they might disapprove of research in the genetic basis of intelligence, and thus the reuse of their samples donated for the former by researchers working on the latter. Now, it is reasonable to say that although participants are the donors of the samples, their donation of the samples involves ceding control of those samples (ownership, if you like, although notions of property in biological samples are controversial too), and beyond certain limits they have no further say in what is done with those samples. Nevertheless, protecting and promoting trust in the research enterprise and in specific researchers may involve giving assurances to sample donors about the kinds of use that are foreseen in the long term, beyond the terms of narrow and specific consent in the short term and what kinds of use are excluded. UK Biobank, for example, supplements specific informed consent with a commitment to long-term engagement with participants through public announcements, websites, newsletters and other methods.

41.2.4 Confidentiality and Security

One of the more practically important issues of concern to participants, which can have major consequences for the governance of research sample collections, is confidentiality (Laurie, 2002). A general principle of the management of confidential information, such as medical records and genetic information, is that those who have access to it should have access to it only on a need to know basis. This protects the subject of the information from disclosure of information to parties who should not have access to particular items of information. In the context of research, this means that personal information collected or extracted for research purposes should be recorded in a way which minimises the extent to which the information allows identification of the individual data subjects. There is a tension in epidemiological research between the protection of individual privacy and confidentiality by deidentification measures up to full anonymisation, and retention of sufficient information that would allow informative analysis of data-sets, linkage of different items of information, and (re)contact of individuals (for research purposes, or to disclose clinically relevant information to the individual).

Two approaches are popular. One is to protect privacy through recording information in minimally identifying form, relative to the kind of information required for the research project (or sample bank). The advantage of this approach is that it builds in privacy protection in a once-for-all way. The disadvantage is that it is relatively inflexible to changes of protocol or for reuse or reanalysis. The other approach, often combined to some extent with the first, is to protect privacy through access controls and through coding so that linkage between records can be managed on a limited basis but database queries are not permitted, which would allow identification of individuals. This is a technology-based approach, although it is usual for senior management of the project, or, if it exists, the external governance board, to 'hold' the key, which allows deanonymised relinkage, and that this is done only in accordance with a defined protocol. One concern with this approach is that technology-based solutions may be subject to attack, and that in some ways they provide false reassurance about the security of a system which can always be subject to human intervention or human error.[4]

41.3 STEWARDSHIP

Moving away from research participant related ethical issues, we turn to the value of the research resource created in research, which could be a data-set, a sample collection or indeed a research protocol, which partners can sign up to in whole or in part. Much debate has centred on the question of whether and how intellectual property rights should be vested in such resources, and on how the value of these resources can be maintained in the long term. There is no straightforward answer to these questions, but there is consensus on one point, which is that resources created with public or charitable money should normally be regarded as open access resources, and that licensing fees (if any) should only be levied to cover the costs of processing the requests and upkeep of the research resource, on a non-profit basis. Open resources such as these typically also stipulate that exclusive access licenses will not be granted, and often some requirement is made that research findings and sometimes any additional data or samples collected in the course of licensed research using the resource should be deposited with the resource for future users to access. Some of the challenges here then turn not on use of and access to the research resource, but in finding ways to guarantee its long-term viability. Samples need to be stored, which costs money; and access systems also need maintenance and oversight. At the end of a project lifetime, or when key personnel leave or retire, sample or data collections can fall into disuse or disrepair. Most public research funders (such as the UK Medical Research Council) now require detailed plans to be made for the stewardship of research resources beyond the lifetime of their established funding or identified management.

This issue of the stewardship of a collection relates to wider issues of scientific research integrity: scientists are generally expected to share their data and teach their techniques. This is for three reasons. Firstly, this provides external researchers with a way of checking that research results are valid and non-fraudulent. Second, it allows for the sharing of best practice, so as best to allow the rapid development of science. Third, it encourages the sense of there being a 'community of science' engaged in a common endeavour to advance humanity. Many critics of the commercialisation of science, especially the use of restrictive contracts and intellectual property rights, are generally concerned that this process undermines the three objectives stated here. On the other hand, commercialisation provides its own incentives for entrepreneurship and ingenuity, and can provide wider social benefits in terms of stimulating the technological application of scientific knowledge.

41.3.1 Benefit Sharing

As well as the value of the research resource to researchers, and in many cases to commercial companies, a database or sample collection represents an investment of time and effort on the part of the participants. In many settings, particularly in the developing world, there is a strong sense that researchers owe a duty of reciprocity to their participants to share any financial or clinical benefits of research with their host communities. This may be negotiated as part of 'community consent'. In developed world settings, researchers normally argue that they are dealing with individuals, that the contribution of any given individual on an identifiable basis to a particular project is minimal, and that it is the collected research resource, which has value as a composite. In addition, the research may have been hosted using public infrastructure, and any commercial benefits deriving from the infrastructure are taxed in such a way as to ensure the reinvestment of part of the proceeds in the State. However, these arguments carry less weight where researchers are doing research in a resource poor setting in another country. In such situations it is arguable that the ratio between benefit accruing to the researcher and sponsor and that accruing to research participants is far out of balance, and to expect participants to donate samples altruistically when this may be the only exploitable resource they have is unreasonable and unfair. Benefit sharing agreements may be quite complex and difficult to enforce, especially when they are made between groups and researchers rather than between individuals and researchers. Sometimes benefit sharing may involve money payments, but more often involve benefits in kind, such as the provision of hospital facilities to a community, or other medical services (Parry, 2004).

41.3.2 Community Involvement

Community involvement has been advocated increasingly in recent years (Hansson, 2005; Haimes and Whong-Barr, 2004). In some developing world settings, community consultation (sometimes involving 'community consent') has been seen as essential to the 'license to do research' in a setting, partly in view of the history of colonial exploitation of poor or vulnerable communities. In the developed world, community consultation has been seen as a useful method for building support for a project, encouraging recruitment, and allowing feedback to participants. On occasion, researchers allow community consultation to play a part in the governance of a project, although a more common approach is to have one or two community members as members of the project steering committee or ethics governance board.

41.4 WIDER SOCIAL ISSUES

Aside from the impact of the research on individual participants or on host communities, there are wider social issues raised by genetics research. As noted at the beginning of this chapter, genetics has many social and ethical issues associated with it. For present purposes, I will concentrate on just two: geneticisation and race/ethnicity in genetics research.

41.4.1 Geneticisation

Geneticisation is the process of transforming diseases or other physiological or psychological traits into traits defined by their genetic causes or risk factors. This is a complex social process, rarely involving straightforward genetic reductionism, but often involving a focussing of attention on the genetic factors influencing a trait and away from other social or biological factors. Genetic reductionism is, roughly, the attempt to explain any physiological or psychological trait exclusively or principally on the basis of genetic factors (Moss, 2003; Sarkar, 1998). For example, starting from the observation that most crimes of violence are committed by men, one could seek to show that the explanation of this is genetic (possibly invoking evolutionary mechanisms) and then to identify specific genes (or gene variants), which account for this, and then proceed to identify individuals with these genes (or gene variants) as being potentially violent criminals, and managing them accordingly. Geneticisation of violent crime need not take as straightforward an approach as this (Wasserman and Wachbroit, 2001). It might well recognise that many factors other than genes are relevant in the causal pathway to violent criminal acts, but nonetheless emphasise that genes are possibly the factor most mutable and controllable in a societal response to violent crime. To take a more medical example, there are many risk factors for cardiovascular disease, from diet, to stress, to genetic constitution. A geneticised approach to cardioprevention would focus on identifying alleles creating higher than average risk of cardiovascular disease, and screen for the presence of these alleles in a population, and tailor prevention strategies accordingly. This would contrast with a more classical public health approach of changing the behaviour of an entire population in order to reduce the number of cases of heart disease (Khoury *et al.*, 2000).

Geneticisation is clearly a process of some importance, and to the extent that it is more sophisticated than simple genetic reductionism, and allows for non-deterministic causality between gene and phenotype, a language of risk rather than certainty, and the role of other social and biological mediating factors, it causes less concern than genetic reductionism. Genetic reductionism can be criticised on many grounds, but socially the principle risk is that by focussing on heritable characters, it causes increased attention to be paid to reproductive policy and less attention to be paid to social policy. Geneticisation need not have this consequence. Nevertheless, attention to genetic factors over social factors may have the consequences of increased emphasis on individual health and behaviour over social conditions, increased emphasis on personal responsibility for health rather than on collective responsibility for health services and environmental conditions, emphasis on high technology interventions over low-cost social interventions and so on (Novas and Rose, 2000; Hedgecoe, 2001; ten Have, 2001).

While these criticisms have merit in many cases, they risk downplaying the utility that genetic research can have even in social public health. In some respects it is better to think of the geneticisation thesis as a descriptive claim about an ongoing social process rather than as a direct normative critique of such a process. That said, it does capture a worry that is frequently expressed about an individualistic trend in modern healthcare. Interestingly, this contrasts with the arguments frequently raised by genetics researchers about the social basis of their research in trust, community cooperation, benefit sharing and open source data collection and so on.

41.4.2   Race, Ethnicity and Genetics

Any scientific research on racial or ethnic difference is inherently controversial, but genetic research, because of the history of eugenics and biological racism, is especially so (Macbeth and Shetty, 2001; Ellison and Goodman, 2006; see also **Chapter 31**). Central elements of the controversy include the following. Firstly, there is controversy over whether any biological sense can be attributed to socially prevalent conceptions of race, and hence whether scientific inquiry into purported biological concepts of race can have any rational justification (Marks, 1995; Nature Genetics, 2004). Second, even allowing that some biological concept of human variation along race-like lines can be defended, there is a vexed question over whether the biological concepts map onto the concepts used in ordinary social life (Smart *et al.*, 2006; Royal, 2006). If they do, can science be seen as supporting social attitudes to racial difference which may be morally and politically problematic, and if they do not, does using a language so open to misinterpretation not confuse issues in a dangerous way (Ashcroft, 2006). Third, assuming some rigorous and value-neutral concept of racial or ethnic difference can be established, which is biologically useful, do the ways in which these findings are then applied in medicine and applied science make sense? And so on. One of the lessons of this complex debate is quite general across human genetics research, which is that genetic research into the biological bases of traits which are complex in their structure and in their meanings in society is very difficult to carry out with 'clean hands'. This raises large issues about whether certain questions should not be investigated at all, or only with great care, and what 'approaching an issue with great care' means in terms of the social responsibilities of scientists.

41.5 CONCLUSIONS

This chapter has necessarily been highly selective and rather discursive. It intends to give a summary overview of some of the more important practical and social issues in the conduct of statistical genetic and genetic sample based research. Although some of the issues discussed are complex and confusing, three things remain clear. First, that public trust in medical and biological research remains relatively high, in part because of the care taken to engage with them about science and to maintain high ethical standards. Although many of the issues in this chapter may seem frustratingly unresolved at the level of theory, in practice there is considerable consensus about many of them. For example, the UK Biobank Ethics and Governance Framework has been published for about a year at the time of writing, following extensive consultation with the academic community, commissioned surveys and focus groups, and direct public consultation, and seems to have general support. Second, there is now an extensive and growing literature of philosophical, ethical, legal and empirical research which can help frame and illuminate the issues and help policy-makers, scientists and the public resolve them. Third, that it remains crucial that scientists remain engaged with these debates, both in informing them and in steering them in directions which will both control risks to participants and society, and maximise the benefits that genetic research will generate.

Acknowledgments

The author would like to thank Erica Haimes, Michael Parker and Susan Gibbons for valuable help with references.
1. http://www.ukbiobank.ac.uk/about.php Accessed 22-1-2007
2. http://www.ukbiobank.ac.uk/about/why.php Accessed 22-1-2007
3. http://www.ukbiobank.ac.uk/about/why.php Accessed 23-1-2007
4. For a treatment of this issue in a crime novel, see Indridason (2004)

REFERENCES

Archard, D. (2004). *Children: Rights and Childhood*. Routledge, London.

Ashcroft, R.E. (2006). In *Race in Medicine: From Probability to Categorical Practice*, T.H. Ellison and A.H. Goodman, eds, *The Nature of Difference: Science, Society and Human Biology* CRC Press/Taylor & Francis, Boca Raton, FL, 135–153.

Ashcroft, R.E., Jones, S. and Campbell, A.V. (2000). Solidarity in the UK welfare state reforms. *Health Care Analysis* **8**, 377–394.

Ashcroft, R.E. and Hedgecoe, A. (eds) (2006). Genetic databases and pharmacogenetics: social, policy and ethical issues. Symposium. *Studies in History and Philosophy of Biological and Biomedical Sciences* **37**, 499–601.

Bauer, M.W. and Gaskell, G. (eds) (2002). *Biotechnology: The Making of a Global Controversy* Cambridge University Press, Cambridge.

Benn, P. (1998). *Ethics*. UCL Press, London.

Beyleveld, D. and Brownsword, R. (2001). *Human Dignity in Bioethics and Biolaw*. Oxford University Press, Oxford.

Clarke, A. and Ticehurst, F. (eds) (2006). *Living with the Genome: Ethical and Social Aspects of Human Genetics*, Palgrave MacMillan, Basingstoke and New York.

Ellison, G.T.H. and Goodman, A.H. (eds) (2006). *The Nature of Difference: Science, Society and Human Biology*, CRC Press/Taylor & Francis, Boca Raton, FL.

ten Have, H.A. (2001). Genetics and culture: the geneticization thesis. *Medicine, Health Care and Philosophy* **4**, 295–304.

Gibbons, S.M.C., Helgason, H.H., Kaye, J., Nõmper, A. and Wendel, L. (2005). Lessons from European population genetic databases: comparing the law in Estonia, Iceland, Sweden and the United Kingdom. *European Journal of Health Law* **12**, 103–133.

Gibbons, S.M.C., Kaye, J., Smart, A., Heeney, C. and Parker, M. (2007). Governing genetic databases: challenges facing research regulation and practice. *Journal of Law and Society* **34**(2), 163–189.

Haimes, E. and Whong-Barr, M.T. (2004). Key issues in genetic epidemiology: lessons from a UK based empirical study. *TRAMES* **8**, 150–163.

Haimes, E. and Williams, R. (2006). *Review of Research on Human Biological Sample Collections*. Wellcome Trust, London.

Hansson, M.G. (2005). Building on relationships of trust in biobank research. *Journal of Medical Ethics* **31**, 415–418.

Hedgecoe, A. (2001). Ethical boundary work: geneticization, philosophy and the social sciences. *Medicine, Health Care and Philosophy* **4**, 305–309.

Indridason, A. (2004). *Tainted Blood*. Harvill, London.

Jackson, E. (2006). *Medical Law: Text, Cases and Materials*. Oxford University Press, Oxford.

Johnstone, C. and Kaye, J. (2004). Does the UK Biobank have a legal obligation to feedback individual findings to participants? *Medical Law Review* **12**, 239–267.

Khoury, M.J., Burke, W. and Thomson, E.J. (eds) (2000). *Genetics and Public Health in the 21st Century: Using Genetic Information to Improve Health and Prevent Disease*, Oxford University Press, Oxford.

Laurie, G. (2002). *Genetic Privacy: A Challenge to Medico-Legal Norms*. Cambridge University Press, Cambridge.

Macbeth, H. and Shetty, P. (eds) (2001). *Health and Ethnicity*, Routledge, London.

Marks, J. (1995). *Human Biodiversity: Genes, Race, and History*. Aldine de Gruyter, New York.

Marks, S.P. (ed) (2006). *Health and Human Rights: Basic International Documents*, Harvard School of Public Health and Harvard University Press, Cambridge, Mass.

McHale, J.V. (2004). Regulating genetic databases: some legal and ethical issues. *Medical Law Review* **12**, 70–96.

Medical Research Council (1991). *The Ethical Conduct of Research on the Mentally in Capacitated*. Medical Research Council, London.

Moss, L. (2003). *What Genes Can't Do?* MIT Press, Boston.

Nature Genetics (2004). *Genetics for the Human Race* **36**(Suppl. 1), S1–S60.

Novas, C. and Rose, N. (2000). Genetic risk and the birth of the somatic individual. *Economy and Society* **29**, 485–513.

O'Neill, O. (2002). *Autonomy and Trust in Bioethics*. Cambridge University Press, Cambridge.

Parry, B. (2004). *Trading the Genome: Investigating the Commodification of Bioinformation*. Columbia University Press, New York.

Paul, D.B. (2002). Is human genetics disguised eugenics? In *The Philosophy of Biology*, D. Hull and M. Ruse, eds, Oxford University Press, Oxford. pp. 536–551.

Rabinow, P. (2003). *French DNA: Trouble in Purgatory*. University of Chicago Press, Chicago.

Rhodes, R. (2005). Rethinking research ethics. *American Journal of Bioethics* **5**(1), 7–28.

Ross, L.F. (1998). *Children, Families and Health Care Decision-Making*. Oxford University Press, Oxford.

Royal, C.D.M. (2006). 'Race' and ethnicity in science, measurement and society. *Biosocieties* **1**, 325–328.

Sarkar, S. (1998). *Genetics and Reductionism*. Cambridge University Press, Cambridge.

Sherlock, R. and Morrey, J.D. (eds) (2002). *Ethical Issues in Biotechnology*, Rowman and Littlefield, Lanham, MD.

Smart, A., Tutton, R., Ashcroft, R.E., Martin, P.A. and Ellison, G.T.H. (2006). Can science alone improve the measurement and communication of race and ethnicity in genetic research? Exploring the strategies proposed by *Nature Genetics*. *Biosocieties* **1**, 313–324.

Sulston, J. and Ferry, G. (2002). *The Common Thread: Science, Politics, Ethics and the Human Genome*. Corgi, London.

Tutton, R. and Corrigan, O. (eds) (2004). *Genetic Databases: Socio-Ethical Issues in the Collection and Use of DNA*, Routledge, London.

UK Biobank Ethics and Governance Council (2006). UK biobank ethics and governance framework, version 2.0, `http://www.ukbiobank.ac.uk/ethics/efg.php` (accessed 22-01-2007).

UK Pharmacogenetics Study Group (2006). Policy issues in pharmacogenetics `http://www.york.ac.uk/res/pgx/publications/` (accessed 22-01-2007).

Wasserman, D. and Wachbroit, R. (eds) (2001). *Genetics and Criminal Behavior*, Cambridge University Press, Cambridge.

42

Insurance

A.S. Macdonald

Department of Actuarial Mathematics and Statistics, Heriot-Watt University, Edinburgh, UK

As soon as DNA-based genetic testing became available, questions were asked about its use by insurers to discriminate against carriers of deleterious alleles. The statistical interest in these questions lies mainly in the application of genetic epidemiology to the actuarial models used to price life and health insurance. This chapter surveys the relevant research, including early work mostly focussed on single-gene disorders, and more recent attempts to predict the relevance of multifactorial genetics to insurance practice.

42.1 PRINCIPLES OF INSURANCE

42.1.1 Long-term Insurance Pricing

Insurance is an unusual branch of commerce because the basis of its sound operation is mathematical in nature. Just as airlines (say) must take the mathematics of aerodynamics as they find it, insurance companies may be in peril if they ignore the actuarial mathematics and statistics that govern their businesses. But actuaries do not deal with precise and impersonal qualities like airflow over a wing; they deal with people and their personal attributes, that lead to an assessment of the risk of making a claim under an insurance contract. Some personal attributes (sex, race, disability) give rise to stronger sensitivities than others (age, weight, smoking habits) and the experience of the last 10 years suggests that any kind of personal genetic information falls in the first group.

The mathematical nature of insurance, and life insurance in particular, is straightforward. The simplest contract, called *whole-of-life insurance*, pays an agreed sum (the sum assured) immediately on the death of the insured person. Suppose a person wishes to buy whole-of-life insurance with sum assured S. The problem is to determine the premium P to be paid at outset (in practice, premiums are usually payable monthly but we ignore this complication). Death being certain, so the insurer will pay out S at some future time with certainty. Intuitively the 'fair' premium before allowing for expenses, profit, etc. appears

Handbook of Statistical Genetics, Third Edition. Edited by D.J. Balding, M. Bishop and C. Cannings.
© 2007 John Wiley & Sons, Ltd.   ISBN: 978-0-470-05830-5.

to be S. However, this takes no notice of the insurer's ability to invest the premium P at interest. Suppose the insurer can earn interest at rate i per year on invested funds. If the insured person dies after exactly T years, the insurer will then have assets of $P(1 + i)^T$, and will have to pay out S. The 'fair' premium that balances the books is clearly:

$$P = S(1 + i)^{-T}. \tag{42.1}$$

Future cashflows discounted to allow for interest, such as $S(1 + i)^{-T}$, are called *present values*. Equation (42.1) is an example of the actuarial 'principle of equivalence', in which the present values of future income and future outgo are equated, giving an equation to be solved for the unknown P. The problem here is that, except in very unusual circumstances, T is not known in advance.

The problem is resolved by regarding T as a random variable, taking non-negative values, possibly with an upper limit to represent a feasible lifespan. Then $(1 + i)^{-T}$ is also a random variable, and the insurer's books will balance 'on average' if the premium charged is the expected value $E[S(1 + i)^{-T}]$. In other words, if the timing of any future cashflows is uncertain, we equate the expected present values (EPVs) of income and outgo.

This simple principle is widely applied in practice to insurances extending over very long terms, typically covering the event of death (life insurance), onset of a serious disease (critical illness insurance), inability to work (income protection or disability insurance) and the need for long-term care (long-term care insurance). In all cases, the times or ages at which the insured events may occur is a crucial element of the pricing problem. This is what distinguishes life and related insurances from other familiar contracts such as home or motor insurance, which usually do not extend cover beyond a year or two. In the following, we will use the generic term *life insurance* to include all long-term insurance covering illness, disability or death.

The opportunity to invest funds at interest is only one reason why the times of insured events are important. Most life insurance contracts do not run for life, but expire at some advanced age, often an anticipated retirement age. This introduces the possibility that no payment will be made at all. For example, Table 42.1 shows the probabilities that a healthy man of selected ages will die within 35 years, according to the AM92 life table (based on the observation of men with certain types of life insurance contracts in the United Kingdom in 1991–1994). This shows that many insurances will cover quite rare events, whose occurrence or non-occurrence depends on the random future lifetime T, or its analogues. It is evident that good estimates of the distribution of T (denoted $F_T(t)$) and its analogues are needed. Table 42.1 also shows that the financial consequences of inaccurate estimation of $F_T(t)$ could be severe; e.g. attributing, to a person of any given

Table 42.1 Probability that a healthy man will die within 35 years, based on the AM92 life table.

Age	Probability
20	0.042
25	0.067
30	0.111
35	0.186

age, the mortality of a person 5 years younger, would erroneously reduce the probability of their death within 35 years by about one-third, with a corresponding effect on premiums.

42.1.2 Life Insurance Underwriting

Estimating $F_T(t)$ is the basic problem of survival analysis; $F_T(t)$ itself is described in a life table. Although life tables are often based on very large data sets (such as national registries or insurance companies' records) it is evident at once that heterogeneity is an issue. The following are some of the factors known to influence mortality and morbidity (some may be mutually confounding): age, sex, socio-economic class, nationality, education, occupation, diet, smoking and drinking habits, exercise, obesity, medical history, relatives' medical histories, marital status, whether insured or not. For clarity, we suppose individuals are labelled $i = 1, 2, \ldots$, and with the ith person we associate a vector $\mathbf{z}_i = (z_i^1, z_i^2, \ldots, z_i^r)$ representing the values of these or other covariates. The distribution of T is now $F_{T|\mathbf{z}}(t|\mathbf{z})$, depending on the covariates.

Underwriting is the process of obtaining information about risk factors (covariates) and adjusting premiums accordingly. Faced with a multitude of possible risk factors, the underwriter will usually focus on a small number, chosen because of known major effect, correlation with other risk factors (hence suitability as a proxy), and availability of data. Using more risk factors than necessary leads to extra expense. Suppose the first u risk factors are chosen as the basis for underwriting, and define $\mathbf{z}_i' = (z_i^1, \ldots, z_i^u)$ and $\mathbf{z}_i'' = (z_i^{u+1}, \ldots, z_i^r)$. Let the distribution of $\mathbf{z}''$ given $\mathbf{z}'$ be $F_{\mathbf{z}''|\mathbf{z}'}(\mathbf{z}''|\mathbf{z}')$, among those assumed to buy insurance. Suppose the present value of some insurance cashflow depending on T is $\pi(T)$. Then the premium that the ith person ought to be charged under the equivalence principle is the EPV:

$$P(\mathbf{z}_i') = \int_{\mathbf{z}''} \int_t \pi(t) \; \mathrm{d}F_{T|\mathbf{z}}(t|(\mathbf{z}_i', \mathbf{z}'')) \; \mathrm{d}F_{\mathbf{z}''|\mathbf{z}'}(\mathbf{z}''|\mathbf{z}_i'). \tag{42.2}$$

The actual calculation does not go quite like this, since $\mathbf{z}''$ is generally unobserved. Define:

$$G_{T|\mathbf{z}'}(t|\mathbf{z}') = \int_{\mathbf{z}''} F_{T|\mathbf{z}}(t|\mathbf{z}) \; \mathrm{d}F_{\mathbf{z}''|\mathbf{z}'}(\mathbf{z}''|\mathbf{z}'). \tag{42.3}$$

Then the calculated premium ought to be:

$$P(\mathbf{z}_i') = \int_t \pi(t) \; \mathrm{d}G_{T|\mathbf{z}'}(t|\mathbf{z}_i'). \tag{42.4}$$

It is possible that more than u risk factors were modelled, given past data, and as a result of the usual model selection process $\mathbf{z}'$ was chosen as having adequate explanatory power. Then the reduction step in (42.3) is, in principle, carried out explicitly and could be revisited if needed. It is equally possible that only the risk factors $\mathbf{z}'$ were ever investigated, so $G_T(t; \mathbf{z}')$ was estimated directly and (42.3) represents completely unobserved averaging.

Viewed as a prediction problem, (42.2) and (42.4) give correct answers only if the distributions that appear in them are appropriate. Specifically: (1) the distribution that *has* been estimated, $G_{T|\mathbf{z}'}(t|\mathbf{z}')$, is the same in the future as it was in the past, or changes in a predictable way; and (2) the density that averages out the unobserved heterogeneity, $F_{\mathbf{z}''|\mathbf{z}'}(\mathbf{z}''|\mathbf{z}')$, is unchanging over time. We know that (1) is far from true, because there have been great changes in mortality rates over calendar time. Fortunately these have mostly been decreases (longer lifetimes than predicted) so errors have been profitable for sellers of life insurance. (Not so for sellers of pensions and annuities.) The truth, or otherwise, of (2), is impossible to establish. It may be summed up as follows: premiums will be charged correctly, if the underlying characteristics of the population who actually buy insurance in future, are the same as those of the population used to estimate $G_{T|\mathbf{z}'}(t|\mathbf{z}')$. This explains why insurers often prefer to estimate life tables based on their own experience, rather than external data, even if the latter is much more extensive. The insurer may not know (for example) that heavy smokers are overrepresented among their customers. But as long as they use life tables based on their own data (or adjust industry standard tables appropriately) and as long as this behaviour does not change, they do not need to know. However, ignorance of this kind can be uncomfortable.

If, for some reason, the distribution of the unobserved risk factors among those who actually do buy insurance is $F^*_{\mathbf{z}''|\mathbf{z}'}(\mathbf{z}''|\mathbf{z}')$ rather than $F_{\mathbf{z}''|\mathbf{z}'}(\mathbf{z}''|\mathbf{z}')$, leading to a different theoretical premium $P^*(\mathbf{z}'_i)$, a profit or loss will arise as a result of actually charging $P(\mathbf{z}'_i)$.

42.1.3 Familial and Genetic Risk Factors

Some of the most important risk factors used in life and health underwriting arise from an applicant's personal history, which we interpret broadly to include lifestyle and habits as well as medical history. The most common approach is to charge different premiums based on a small number of well-understood risk factors such as age, sex and smoking habits, but otherwise to limit closer investigation of cases to a small proportion indicated by significantly adverse responses to questions about health and medical history. However, in more recent times 'preferred lives' underwriting has spread quite widely, in which insurers more actively seek out persons with an advantageous risk profile. Leigh (1990) gives a review of underwriting practice, while Brackenridge and Elder (1998) is a standard reference on life insurance underwriting.

Since nearly all life and health insurance is bought by economically active adults to protect dependents or to secure financial transactions, the genetic disorders that are relevant are those with late onset, such that a person may be completely healthy at ages when insurance is purchased. Thus most interest is focused on a small set of relatively rare single-gene dominantly inherited disorders. In 1996, the genetics advisor to the Association of British Insurers listed eight disorders as relevant to insurance, namely Huntington's disease, familial adenomatous polyposis, hereditary breast cancer, early-onset Alzheimer's disease, myotonic dystrophy, multiple endocrine neoplasia, hereditary motor and sensory neuropathy and adult polycystic kidney disease. The last of these was soon removed, since it is normally diagnosed by ultrasonography rather than by a DNA test.

42.1.4 Adverse Selection

Losses arising through unobserved heterogeneity are called *adverse selection*, a term in widespread use in economics. It can arise accidentally, or as a result of one party to a contract deliberately withholding information from the other, or through the actions of competitors in the market. For example, a motor dealer may conceal a car's poor service record, or their customer may conceal their own poor credit record. Similarly, a life insurer may conceal high charges behind complicated terms and conditions, or the buyer of life insurance may fail to mention a poor health record. Sellers try to protect themselves by demanding information – e.g. credit checks and insurance underwriting – with sanctions if it is not given honestly and accurately. Buyers, who are often relatively powerless individuals, look to laws, regulations and professional advisers for their protection.

A classic example of the insurance industry's reaction to the mere possibility of adverse selection arose in the United Kingdom, when in 1981 one life insurer introduced separate premium rates for smokers and non-smokers. Those for non-smokers were lower than any competitors' premium rates (which was the point of the exercise) and those for smokers were correspondingly higher. Other companies faced the prospect of losing all their non-smoking customers to the innovator, and attracting an all-smoking clientele, while charging premiums that assumed a 'normal' mix of smoking habits. Within a few years, almost all companies charged different rates to smokers and non-smokers. There is no evidence that adverse selection actually did appear, but the threat was enough.

The question of why someone wants to purchase insurance lies behind insurers' concerns about adverse selection. By its nature, insurance covers reasonably rare events, so a modest premium can secure a large amount of cover. Someone who knows they are at high risk – either because they possess information and realise its implications, or because they intend to commit fraud – undermines the sound business model. Not surprisingly, insurers tend to resist suggestions that they should eschew the use of any risk factor they think might be relevant, or that they be instructed to move a covariate from $\mathbf{z}'$ over to $\mathbf{z}''$. In reality, this only matters if the risk factor in question is likely to influence the desire for insurance and motivate the informed individual to purchase it. Sometimes this possibility is very plausible: intimations of mortality as we grow old mean that life insurance without age discrimination would hardly be a feasible proposition. Sometimes it is rather implausible: most smokers are aware that, as a group, they die earlier than non-smokers but: (1) does that knowledge impel them to buy life insurance? and (2) if not, would the sound basis of the business be threatened if we disallowed discrimination against smokers?

The question that has fuelled the genetics and insurance debate is the following: where does personal medical information in general, and genetic information in particular, lie on this spectrum?

42.1.5 Family Medical Histories

For a long time before DNA-based genetic tests were developed, insurers had routinely used family medical history in underwriting. The chief source of information was (and is) a question on the proposal form which typically will ask about: (1) the ages at death and causes of death of the applicant's parents; and (2) whether a parent or sibling has suffered from a disorder that might be inherited. Inquiries rarely if ever extend beyond first-degree relatives, partly because of the difficulty of verifying and interpreting extended pedigrees.

Such family medical histories may disclose a risk of a Mendelian disorder, but they are also strongly predictive of the risk of common diseases, such as coronary heart disease (CHD). In the latter case, the familial link need not be genetic at all, but could lie in shared environment, habits or socio-economic class.

42.1.6 Legislation and Regulation

Sex and disability are qualities that have, in the past, been associated with discrimination that is now deprecated in most modern societies, to the extent of being outlawed in many jurisdictions. However, it is common for insurance companies to be allowed to use sex and disability as risk factors, provided they can show evidence of a statistical or actuarial nature that confirms their relevance. In some countries with such provisions, a commission or tribunal may assess the evidence; in others the courts.

The question of whether various types of genetic information should be given similar protection has been widely considered. Nys *et al.* (2002) give a recent survey of practice in the European Union. The cases of Australia, Sweden and the United Kingdom are perhaps of unusual interest, because in each country the task has fallen to a governmental or legal commission, with some important differences of approach.

1. In the United Kingdom, the Human Genetics Commission (HGC) advises the government on all aspects of genetics, including insurance questions. The insurance industry has, since 1996, agreed to a moratorium on the use of DNA-based test results in underwriting. Its current form, running from 2005 to 2011, has the following key features: (1) insurers will not ask anyone to take a genetic test; (2) insurers will not ask about test results acquired in the course of research trials; and (3) insurers will not ask about existing test results unless a life insurance policy exceeds £500 000 sum assured, or other forms of insurance exceed £300 000 sum assured or equivalent. However, the use of family medical history is unrestricted.

   The UK government has set up the Genetics and Insurance Committee (GAIC) which is responsible for deciding what genetic test results may be used for underwriting, when sums assured exceed the limits mentioned above. It will consider applications submitted to it by the insurance industry, against three criteria: (1) *technical relevance*: does the test accurately measure the genetic information? (2) *clinical relevance*: does a positive result in the test have likely future adverse implications for the health of the individual? (3) *actuarial relevance*: does a positive result justify increased premiums? Thus for the first time in the United Kingdom, discriminatory pricing has to be justified in advance, rather than defended in court if challenged. To date GAIC has approved only one test for use, that is for Huntington's disease in the case of life insurance.

2. In Sweden, similar arrangements are in place but family medical history may not be used. In addition, long-term health insurance for children is sold in Sweden, and genetic information may not be used at all in respect of these contracts. This is a rare example of insurance possibly covering early-onset and recessive disorders.

3. The Australian Law Reform Commission (ALRC) reviewed all aspects of human genetics and their reports (ALRC, 2001; 2002; 2003) are particularly useful references. In respect of insurance, they recommended a system quite similar to that in the United Kingdom.

Many of the social policy aspects of genetics and insurance are discussed in Daykin *et al.* (2003) and Doble (2001) and official reports such as ALRC (2001; 2002; 2003), HCSTC (1995; 2001), HGAC (1997) and HGC (2002).

42.1.7 Quantitative Questions

By its nature, insurance practice seeks to quantify risk, and for that purpose to build models of individual life histories. In this respect, actuaries have much the same interests as epidemiologists. Two quantitative questions address the debate that has grown up around genetics and insurance:

1. If insurers were allowed to use genetic information (including family history) to underwrite insurance premiums, by how much would those premiums increase?

2. If insurers were not allowed to use genetic information to underwrite insurance premiums, what additional costs might they face as a result of adverse selection?

42.2 ACTUARIAL MODELLING

42.2.1 Actuarial Models for Life and Health Insurance

Life and health insurances pay benefits either on the occurrence of an event, or during the continuation of some status. The simplest example is death: life insurance will pay a lump sum upon death, while an annuity or pension will continue to be paid until death. Other events or states that can lead to benefits being paid are suffering one of a specified list of severe illnesses (critical illness insurance); being unable to work (disability or income protection insurance) and inability to care for oneself (long-term care insurance). In each case, the actuary needs a model of an insured person's life history, including all relevant events.

The simplest such model was introduced in Section 42.1.1 to illustrate the statistical principle of insurance, namely that a person's remaining lifetime is represented by a positive random variable T. A traditional life table is simply a tabulation of the survival function $S_T(t)$ associated with T, defined by $S_T(t) = P[T > t]$. Clearly this is the complement of the distribution function $F_T(t) = P[T \leq t]$. Assuming T to have a density $f_T(t)$ on a suitable age range, the hazard rate $\lambda(t)$ is defined by $\lambda(t) = f_T(t)/S_T(t)$. The hazard rate has the following intuitive interpretation: conditional on being alive at time (or age) t, the probability of dying by time $t + dt$ is approximately $\lambda(t)dt$, for small dt. Figure 42.1 illustrates this model, representing death as a transition between an 'alive' state and a 'dead' state, governed by the hazard rate $\lambda(t)$.

Faced with more complicated types of insurance, other events have to be modelled and a useful approach is to represent events in a life history as transitions between suitably defined states. For example, in Figure 42.2 the 'able' state represents fitness to work, and the 'ill' state represents inability to work. Under the simplest disability insurance policy, the insured person would pay premiums while 'able' and receive regular benefits to replace lost earnings while 'ill'. It quickly becomes impractical to specify such extended models using random times between transitions – analogues of T – as the basic quantities. For

Figure 42.1 A two state model of mortality.

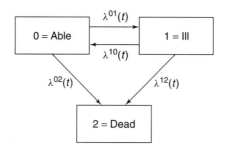

Figure 42.2   A model of illness and death.

example, the number of events (transitions) that may occur in one person's lifetime in the model of Figure 42.2 is not even bounded. Instead it is much more convenient to specify the transition intensities between each pair of states, here denoted $\lambda^{ij}(t)$ between states i and j. These are natural quantities to estimate using occurrence-exposure rates, and if they depend only on the insured person's age the resulting model is Markov and most quantities of actuarial interest can be computed numerically as the solutions of linear ordinary differential equations. We mention the two most important examples (see Hoem (1988) and Norberg (1995) for details).

1. Let $p^{ij}(s, t)$ be the probability of being in state j at time t, conditional on being in state i at time s ($s \leq t$). Then Kolmogorov's forward equations are:

$$\frac{\partial}{\partial t} p^{ij}(s, t) = \sum_{k \neq j} p^{ik}(s, t)\lambda^{kj}(t) - \sum_{k \neq j} p^{ij}(s, t)\lambda^{jk}(t). \tag{42.5}$$

2. Suppose a lump sum $b^{ij}(t)$ is payable on transition from state i to state j at time t, and a continuous payment at rate $b^{i}(t)$ per year is made if the insured person is in state i at time t (whether these cashflows are payable to or from the insured person just depends on their signs; we assume positive cashflows are paid to the insured person). Let $V^{i}(t)$ be the amount the insurance company must hold in reserve at time t, to honour its future obligations, if the insured person is then in state i. (This is called the *prospective policy value*, and may be thought of as the expected value of the insurer's future outgo, net of premiums received, allowing for interest.) Suppose interest can be earned at the instantaneous rate δ *per annum*. Then Thiele's equations are:

$$\frac{\mathrm{d}}{\mathrm{d}t} V^{i}(t) = \delta V^{i}(t) - b^{i}(t) - \sum_{j \neq i} \lambda^{ij}(t)\left(b^{ij}(t) + V^{j}(t) - V^{i}(t)\right). \tag{42.6}$$

In words, the reserve that must be held earns interest, but is depleted by the continuous payments made as long as the insured person remains in state i, and also by the lump sum payable on transition into any other state j. However, when such a transition occurs the reserve that was held, $V^{i}(t)$, is released, but the reserve $V^{j}(t)$ required by presence in the new state must now be held instead.

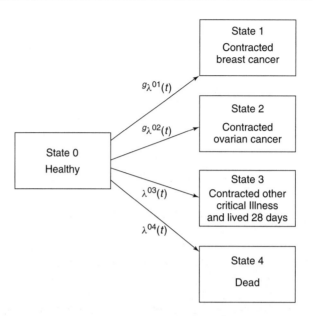

Figure 42.3 A model suitable for pricing critical illness insurance, with breast and ovarian cancer selected as particular causes of claims. The superscript 'g' represents genotype.

For completeness, to illustrate the other insurance contract of practical importance, Figure 42.3 shows a model suitable for pricing and reserving for critical illness insurance. Here, breast and ovarian cancers are picked out as particular causes of claiming; all other causes are combined into a transition into a single state. By letting the transition intensities that govern onset of breast and ovarian cancer depend on genotype g, we can measure the impact of known deleterious genotypes on insurance prices. Similar models can be set up in respect of any genetic disorder, in respect of which the epidemiology is advanced enough to furnish reasonable estimates of the onset rates. What might be regarded as 'reasonable' is a question that could have a bearing on the decisions made by GAIC in the United Kingdom.

42.2.2 Parameterising Actuarial Models

Actuarial researchers almost never have access to pedigree data. Rare exceptions occur when epidemiologists publish details of the pedigrees they have used (Gui and Macdonald, 2002a; Espinosa, 2006). Therefore, they rely on the epidemiological literature. Given the nature of the models described previously, the parameters needed are: (1) age-dependent rates of onset (equivalently, penetrance estimates); (2) duration-dependent rates of mortality post-onset (age dependence may have to be considered too); and (3) mutation frequencies.

Questions concerning known mutation carriers do not require knowledge of the mutation frequencies, but questions concerning incomplete information do. This arises in two circumstances:

1. The cost of adverse selection depends on the size of the group who may gain access to insurance below cost.

2. When a genetic disorder occurs as a rare subset of a common disorder (such as breast cancer) then a family history does not identify the presence of a mutation, it just alters the probability that a mutation is present.

Onset rates and mortality rates may be extracted from the literature by whatever means present themselves, including when necessary being read from graphs (Macdonald (2003b) discusses many of these issues). Occasionally point estimates of onset rates are given (e.g. Ford *et al.*, 1998) which can be used directly, but relative risks, Kaplan-Meier estimates and odds ratios are also common. In many cases ascertainment bias is likely to be present, that even the authors of the articles in question were unable to control, in which case the premium rates obtained from the parameterised model are overstated, and some form of sensitivity analysis is advisable, e.g. by assuming onset rates to be 50 % or less of those estimated.

42.2.3 Market Models and Missing Information

Models such as those illustrated in Figures 42.1 to 42.3 may be used to represent the life history of a person who possesses an insurance contract of the appropriate type. They are adequate for determining premiums and reserves, provided the model has been fully parameterised. However, they have to be extended to address questions of adverse selection or underwriting based on family history.

Adverse selection depends on whether a person decides to buy insurance at a given price, on the basis of the information they possess. Both the information and the decision, therefore, have to be represented in the model. Figure 42.4 illustrates such a model, for the simplest case of critical illness insurance. It represents a person's life history in an insurance market. They are assumed to have a known genotype g_i, and the label i indexes the states and the transition intensities. They start in state $i0$ at (say) age 20. As time passes, they may decide to buy insurance in the normal way (state $i1$) or they may decide to have a genetic test (state $i2$) and, once they know that their genotype is g_i, then decide to buy insurance (state $i3$). Once insured, they remain insured until some suitable age, say 60 or 65, during which time they will receive the benefit if they avoid death (state $i5$) and suffer a critical illness (state $i4$).

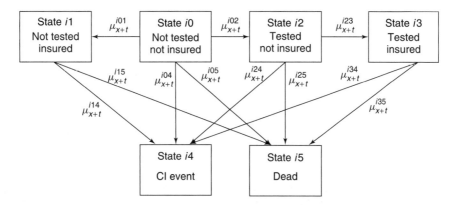

Figure 42.4 A Markov model of insurance purchase and critical illness (CI) insurance events for a person with genotype g_i.

This model assumes the genotype g_i to be fixed, even if not known to the individual at outset. Each possible genotype defines a subset of the population of people who might be active in the insurance market, and we set up such a model for each of them, with transition intensities adjusted to represent each genotype. The probabilities of being in any of the subgroups are just the population frequencies of the genotypes.

In fact we need to refine the partition of the population beyond that defined by genotype, for two reasons: (1) we wish to allow for underwriting based on family history; and (2) genetic testing is most likely to be taken up by people who have a family history of the relevant disorder. An example of such a refinement is illustrated in Figure 42.5, in the case of a dominantly inherited disorder of purely genetic origin (such as Huntington's disease). In such a case, we assume that 'family history' means an affected parent or sibling, and that a healthy person with a family history is 'at risk' but inherited the mutation with probability 1/2. The proportions assumed to be in each subgroup depend on the population mutation frequency, bearing in mind that members of both 'at risk' subgroups have an affected parent.

This model is flexible enough to answer many questions relating to single-gene disorders.

1. The rate at which insurance is purchased in the 'not at risk' subgroup determines the size of the insurance market.

2. The rate at which insurance is purchased by persons who have a family history, or an adverse test result, and the average amount purchased, determines the extent of adverse selection.

3. The rate of genetic testing, given a family history, can be adjusted to suit the clinical approach to the disorder.

4. Underwriting classes are defined by collections of 'insured' states, within each of which the same rate of premium is payable. For example, if insurers underwrite on the basis of family history and there is no genetic testing, states 21, 23, 31 and 33

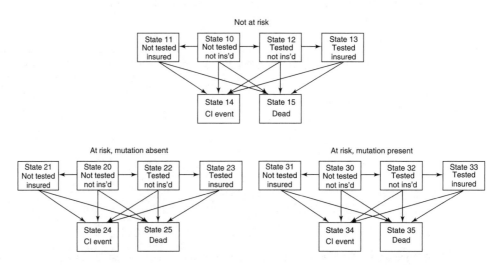

Figure 42.5 A Markov model allowing for a family history of a Mendelian disorder.

would form the 'family history' underwriting class. The effect of introducing genetic testing, and moratoria on the use of genetic information, can then be assessed by changing the composition of the underwriting classes. See Macdonald (2003a) for details.

The algorithm applied to the model is straightforward: the frequency p determines the occupancy probabilities in states 10, 20 and 30 at birth. Kolmogorov's equations solved forward yield the occupancy probabilities in all states at all relevant ages, then Thiele's equations solved backwards from the age at which insurance expires yield EPVs of future cashflows at all ages back to the age at which insurance purchasing commences. In fact, since Thiele's equations give the EPVs in all states, including the insured states, this model subsumes that of Figure 42.3.

One slightly technical point arises in connection with premium rates. If level premiums were charged depending on the age at which insurance was purchased, then at any age x, the rate of premium cashflow in each insured state would be an average over the rates of premium charged at all earlier ages, involving the age-dependent rate of purchase and the probabilities of remaining in the insured state since purchase. This is feasible, but can be avoided simply by charging a rate of premium equal to the weighted average intensity from the insured states comprising an underwriting class, into the corresponding 'critical illness event' states, the weights being the occupancy probabilities and the amounts assured. The resulting cashflows are adapted to the Markov framework.

The model can be extended in several ways to meet more demanding problems. Some examples are as follows (Gui, 2003; Gui *et al.*, 2006; Gutiérrez and Macdonald, 2004; 2007; Lu, 2006).

1. Heterogeneous genetic disorders, where different mutations have different penetrances, can be handled by partitioning the population into more subgroups, adding an extra pair of 'at risk' subgroups for each distinguishable mutation.

2. Genetic disorders that account for a small proportion of a common disease, such as breast cancer, cause the problem that family history does not identify mutation-carrying families. However, if the definition of 'family history' is precise enough, e.g. 'two or more first-degree relatives with breast or ovarian cancer before age 50', the onset of a family history can be treated as an event in the insured person's life history, and represented by transition into a 'family history' state.

3. By adding a transition from onset of the disorder to the 'dead' state, life insurance can be modelled. This usually introduces duration dependence, therefore a semi-Markov model.

4. In the case of progressive disorders, such as Huntington's disease, a critical illness insurance claim will usually come much later than onset of the first symptoms, since it will depend on some advanced level of disability. If post-onset survival is modelled, as in (3) above, then an accelerated lifetime model derived from this may be used to model the timing of the insurance claim.

42.2.4 Modelling Strategies

The rarity of single-gene disorders of high relevance to insurance makes it implausible that adverse selection in a well-developed insurance market will be so great as to disrupt it.

This can be tested quite simply by choosing parameters for the model that are guaranteed to be extreme, e.g. that 2 % of the population carry high-penetrance mutations and that anyone with an adverse genetic test result will be quite likely to purchase insurance quickly. If the cost of adverse selection, represented for example by premium rate increases, is small, we can be fairly confident that adverse selection is manageable. The advantage of this 'top-down' strategy is that it does not require detailed studies of individual disorders to be undertaken, and for that reason it was the approach used first. Macdonald (1997; 1999) suggested that 10 % was a very conservative upper limit for premium rate increases in life insurance. Since improving longevity has resulted in life insurance premiums falling by considerably more than this over two decades, that particular market seems reasonably safe.

Some caveats must be entered. A similar outcome may be unlikely in a small, less mature insurance market. But if extreme assumptions suggest that adverse selection could be expensive, we learn nothing from the 'top-down' approach and must turn to the more laborious 'bottom-up' strategy, in which the major relevant disorders are modelled individually, to try to obtain a realistic estimate of likely costs. In the long run this is a better basis for discussing genetics and insurance anyway, so it is a program worth undertaking.

42.2.5   Statistical Issues

As mentioned above, the actuarial researcher rarely has access to the data underlying published estimates of penetrance or onset rates. This makes it difficult to estimate sample variances or confidence intervals for any of the actuarial quantities derived from an epidemiological study, which might have been based on a relatively small number of subjects. Perhaps because they spring from different traditions in actuarial science, premium rates for non-life insurance (house, motor, etc.) have long been treated as estimates in a fully statistical framework, but premiums for life insurance have not. It seems appropriate to do so when considering medical underwriting on the basis of epidemiological data.

Premium rates either are or are based upon EPVs, which are the solutions of Thiele's equations, so it is clear that they are very complicated functions of whatever data underlie the transition intensities of the model. Two approaches, both based on bootstrapping, have been suggested.

1. If a fully parametric model is used with pedigree data (Elston, 1973) the information matrix will be estimated and can be used to bootstrap values of the parameters, assumed to be multivariate normal.

2. If a Kaplan-Meier survival curve is obtained, and the underlying event times, numbers of events and numbers at risk are available, various methods of simulation may be employed. Lu (2006) and Lu et al. (2007) found that simple resampling and the so-called weird bootstrap (Andersen et al., 1993) gave similar results, based on studies of polycystic kidney disease.

It is clear that the absence of information about the sampling distribution of premium rates is one of the least satisfactory aspects of actuarial modelling of genetic disorders, but it is equally hard to overcome this disadvantage.

42.2.6   Economics Issues

An important parameter in the models shown in Figures 42.4 and 42.5 is the rate at which insurance is purchased normally, in the absence of any genetic or familial risk. In particular, it influences the severity of any adverse selection. Very few studies exist from which this quantity could be estimated, so in most applications of the model a plausible but hypothetical value has been used. This is flawed from the point of view of economic theory, because if adverse selection increases the price of insurance for everybody, the demand for insurance on the part of low-risk individuals should decrease, hence the 'normal' rate of insurance purchase in the model. Since this increases the proportion of high-risk individuals purchasing insurance, the premium should increase yet again, and there is a risk of entering an 'adverse selection spiral', whose worst outcome would be that the price of insurance should rise to the level of the highest-risk individuals.

Several authors have formulated economic market models to explore the nature of any equilibria that emerge if genetic information is withheld from insurers, see Doherty and Thistle (1996), Hoy and Polborn (2000) and Hoy and Witt (2005). The last of these considered a specific disorder, namely breast cancer and the *BRCA1/2* genes. They modelled a life insurance market in which family background, disclosed to the insurer, produced 13 categories of risk, but other genetic information (test results) was known only by the applicant. In the presence of a high-risk group, the equilibrium insurance premium was up to 297 % of the population weighted probability of death. De Jong and Ferris (2006) suggested a very simple statistical adverse selection model based on the correlation between a risk factor and the random amount of insurance sought, and used it to model adverse selection. In the absence of data relating demand for insurance to genetic risk their main example was the imposition of unisex pricing in the market for annuities. Pauly *et al.* (2003) did obtain estimates of how demand for life insurance might vary with BRCA1/2 genotype, applied in Viswanathan *et al.* (2007). Macdonald and Tapadar (2006) used a simple utility model to map out risk thresholds below which adverse selection was unlikely to appear, given a uniform insurance contract driven by fixed need. They suggested that multifactorial disorders would not lead to an adverse selection spiral.

42.3   EXAMPLES AND CONCLUSIONS

42.3.1   Single-gene Disorders

Tables 42.2 and 42.3 show some examples of critical illness insurance premium rates given by models for selected single-gene disorders. They are expressed as percentages of the standard premiums a healthy person would pay for the same contract. In a sense these are worst-case examples, because premium increases in respect of life insurance are often ameliorated by the significant chances of survival post-onset, but modelling of life insurance is less complete.

In Table 42.2 we assume an adverse genotype to be known; this therefore represents the hypothetical consequences of information that insurers in many countries may not use. In each case a range is shown because of uncertainties within the models, of a non-statistical nature. For example, different epidemiological studies may be used to parameterise the

Table 42.2 Examples of premium ratings for CI insurance for female mutation carriers, as percentages of standard rates..

Disorder	Gene	Age 30 Term 30 years (%)	Age 40 Term 20 years (%)	Age 50 Term 10 years (%)
APKD	APKD1/2	335–435	332–468	297–422
BC/OC	BRCA1	381–1110	355–1112	259–740
BC/OC	BRCA2	252–578	296–768	352–1056
EOAD	PSEN1	866–2040	1032–3022	1076–3714
HD	IT15 (40 CAG)	165–268	169–298	133–247
HD	IT15 (45 CAG)	635–1181	455–1018	224–630
HD	IT15 (50 CAG)	1002–2007	591–1372	276–840

Sources: APKD, Gutiérrez and Macdonald (2003); BC/OC, Macdonald *et al.* (2003); EOAD, Gui and Macdonald (2002b); HD, Gutiérrez and Macdonald (2004).

Table 42.3 Examples of premium ratings for CI insurance for female applicants with a family history, as percentages of standard rates.

Disorder	Gene	Age 30 Term 30 years (%)	Age 40 Term 20 years (%)	Age 50 Term 10 years (%)
BC/OC	BRCA1/2	103–184	–	102–158
EOAD	PSEN1	432–769	363–605	153–198
HD	IT15	203–296	142–202	107–128

Sources: BC/OC, Macdonald *et al.* (2003); EOAD, Gui and Macdonald (2002b); HD, Gutiérrez and Macdonald (2004).

model (APKD), onset rates may be reduced if ascertainment bias is likely to be present (breast/ovarian cancer, EOAD) or it may be uncertain at what stage in the progression of a disease an insurance claim would succeed (Huntington's disease). Since insurers would decline to offer cover once the risk indicates a premium greater than about 350 % of the standard, it is clear that large numbers of mutation carriers would be denied cover. (But note that life insurance cover would generally be offered up to about 500 % of standard premiums, so access to life insurance would be less restricted, albeit at a price.)

Table 42.3 assumes that only the presence of a family history (usually an affected parent) is known; this therefore represents the consequences of information that insurers in many countries may use. Access to critical illness cover is greatly eased, especially at older ages. This is because premium rates are averaged over carriers and non-carriers of mutations, and as persons at risk get older and remain healthy, the chance that they are a mutation carrier falls to well below 1/2. There may also be an element of averaging over more and less severe variants of a disease, since the precise causative mutations are assumed to be unknown.

Table 42.4 shows examples of percentage premium increases that might fall upon all policyholders, if severe adverse selection resulted from limiting access to certain genetic test results. The cost is in two parts if family history as well as genetic test results may not be used, because just adding high-risk individuals to the 'standard' insurance pool will raise premiums, even if nobody changes their behaviour. The meaning of 'severe'

Table 42.4 Examples of percentage premium increases ratings for females in a large CI insurance market, caused by severe adverse selection.

Disorder	Gene	'Lenient' Moratorium on genetic tests (%)	Moratorium on family history	
			Pooling cost (%)	Adverse selection cost(%)
APKD	APKD1/2	0.051–0.072	0.203–0.273	0.111–0.126
EOAD	PSEN1	0.014–0.021	0.077–0.118	0.039–0.084
HD	IT15	0.014–0.025	0.038–0.069	0.041–0.066

Sources: APKD, Gutiérrez and Macdonald (2003); EOAD, Gui and Macdonald (2002b); HD, Gutiérrez and Macdonald (2004).

adverse selection is described in the studies cited, but it perhaps implies a level of rational economic behaviour, not to mention ability to afford insurance, that a sceptic might doubt would happen in practice. Hence these very small percentages argue against adverse selection being a serious threat to the critical illness insurance market, let alone the life insurance market. However, this does depend on: (1) the rarity of relevant disorders; and (2) the significant size of these insurance markets: it might not hold in other circumstances. The basic statistical laws that govern insurance, mentioned in Section 42.1.1, have not altered; it just happens that the numbers are small.

42.3.2 Multifactorial Disorders

The models described above represent single-gene disorders well, because genotype and family history partition the population into a small number of subgroups. They are in principle capable of representing multifactorial disorders, but practical limitations are imposed by increasing complexity, and the fact that current epidemiology does not so often lead directly to usable onset rates. Therefore, such actuarial research as has been done has always incorporated some hypothetical features.

1. Macdonald *et al.* (2005a; 2005b) modelled CHD using three risk factors assumed to be static (sex, smoking status and body mass index) and three assumed to be dynamic (i.e. changing through life: diabetes, two levels of hypercholesterolaemia and three levels of hypertension). The model was parameterised using the Framingham data, which does not include any genetic covariates. The hypothetical genetic component of the model was to assume that some genotype would increase the intensities of transition through the worsening risk factors by a factor of 5 or 50. Table 42.5 shows some examples of the results using a factor of 5 in the form of the percentage *extra* premium that would be charged, above the standard rate (therefore 110 % in Table 42.2 or 42.3 would be +10 % here). The first column shows risk factors already present in the person applying for insurance. What is most remarkable is the negligible increase in premiums for someone presenting no risk factors. Other extra premiums are reasonably consistent with many underwriting guidelines. Table 42.6, by contrast, shows the effect of increasing by five times the direct risks of stroke or heart attack (represented by the transition intensities into those states). The extra premiums are markedly increased. The conclusion is that multifactorial disorders that

Table 42.5 Premium ratings for males, non-smokers, normal body mass index (BMI) aged 35 at entry with policy term 10 years, under hypothetical assumptions of genetic influence increasing the incidence of risk factors five times.

Risk factors	Premium rating factors with five times the incidence rate of				
	None (%)	H'chol (%)	H'tension (%)	Type 1 diabetes (%)	Type 2 diabetes (%)
No risk factors	+0	+5	+13	+2	+6
H'chol Cat 1	+3	+12	+17	+5	+9
Type 1 diabetes	+298	+304	+314	–	–
Type 2 diabetes	+67	+73	+84	–	–
H'tension Cat 1	+6	+12	+29	+8	+12
H'chol Cat 2	+25	–	+45	+28	+32
H'chol Cat 1 and type 1 diabetes	+302	+313	+320	–	–
H'chol Cat 1 and type 2 diabetes	+71	+82	+89	–	–
H'chol Cat 1 and H'tension Cat 1	+9	+19	+34	+11	+15
H'tension Cat 1 and type 1 diabetes	+305	+313	+336	–	–
H'tension Cat 1 and type 2 diabetes	+74	+82	+105	–	–
H'tension Cat 2	+34	+44	+53	+37	+41
H'chol Cat 2 and type 1 diabetes	+330	–	+356	–	–
H'chol Cat 2 and type 2 diabetes	+99	–	+125	–	–
H'chol Cat 2 and H'tension Cat 1	+34	–	+70	+37	+41
H'tension Cat 1, H'chol Cat 1, type 1 diabetes	+309	+322	+342	–	–
H'tension Cat 1, H'chol Cat 1, type 2 diabetes	+78	+91	+112	–	–
H'tension Cat 2 and H'chol Cat 1	+40	+56	+60	+42	+47
H'tension Cat 2 and type 1 diabetes	+342	+354	+368	–	–
H'tension Cat 2 and type 2 diabetes	+111	+124	+137	–	–
H'tension Cat 3	+81	+95	–	+84	+90
H'chol Cat 2, H'tension Cat 1, type 1 diabetes	+342	–	+388	–	–
H'chol Cat 2, H'tension Cat 1, type 2 diabetes	+111	–	+158	–	–
H'chol Cat 2 and H'tension Cat 2	+81	–	+108	+84	+89
H'tension Cat 2, H'chol Cat 1 and type 1 diabetes	+349	+370	+376	–	–
H'tension Cat 2, H'chol Cat 1 and type 2 diabetes	+118	+139	+146	–	–
H'tension Cat 3 and H'chol Cat 1	+89	+111	–	+91	+97
H'tension Cat 3 and type 1 diabetes	+407	+423	–	–	–
H'tension Cat 3 and type 2 diabetes	+176	+193	–	–	–
H'tension Cat 2, and H'chol Cat 2 and type 1 diabetes	+402	–	+438	–	–
H'tension Cat 2, and H'chol Cat 2 and type 2 diabetes	+172	–	+207	–	–
H'tension Cat 3 and H'chol Cat 2	+147	–	–	+150	+157
H'tension Cat 3, H'chol Cat 1, type 1 diabetes	+416	+445	–	–	–
H'tension Cat 3, H'chol Cat 1, type 2 diabetes	+185	+214	–	–	–

Source: Macdonald *et al.* (2005b).

Table 42.6 Premium ratings for males, non-smokers, normal BMI aged 35 at entry with policy term 10 years, under hypothetical assumptions of genetic influence increasing the incidence of CHD and stroke five times.

	Premium rating factors with five times the incidence rate of						
				CHD modified by the presence of			
						Type 1	Type 2
Risk factors	None (%)	CHD (%)	Stroke (%)	H'chol (%)	H'tension (%)	diab (%)	diab (%)
No risk factors	+0	+142	+37	+26	+29	+1	+5
H'chol Cat 1	+3	+156	+40	+157	+35	+4	+8
Type 1 diabetes	+298	+481	+351	+331	+334	+481	–
Type 2 diabetes	+67	+250	+121	+100	+104	–	+250
H'tension Cat 1	+6	+168	+44	+36	+168	–	+11
H'chol Cat 2	+25	+267	+62	+267	+74	+26	+33
H'chol Cat 1 and type 1 diabetes	+302	+499	+355	+499	+342	+499	–
H'chol Cat 1 and type 2 diabetes	+71	+268	+124	+268	+111	–	+268
H'chol Cat 1 and H'tension Cat 1	+9	+184	+47	+184	+184	+10	+15
H'tension Cat 1 and type 1 diabetes	+305	+513	+361	+343	+513	+513	–
H'tension Cat 1 and type 2 diabetes	+74	+283	+130	+113	+283	–	+283
H'tension Cat 2	+34	+298	+83	+82	+298	+35	+43
H'chol Cat 2 and type 1 diabetes	+330	+641	+384	+641	+392	+641	–
H'chol Cat 2 and type 2 diabetes	+99	+411	+153	+411	+161	–	+411
H'chol Cat 2 and H'tension Cat 1	+34	+309	+73	+309	+309	+35	+44
H'tension Cat 1, H'chol Cat 1, type 1 diabetes	+309	+534	+365	+534	+534	+534	–
H'tension Cat 1, H'chol Cat 1, type 2 diabetes	+78	+304	+134	+304	+304	–	+304
H'tension Cat 2 and H'chol Cat 1	+40	+324	+89	+324	+324	–	+49
H'tension Cat 2 and type 1 diabetes	+342	+681	+414	+403	+681	+681	–
H'tension Cat 2 and type 2 diabetes	+111	+451	+183	+173	+451	–	+451
H'tension Cat 3	+81	+450	+213	+147	+450	+82	+93
H'chol Cat 2, H'tension Cat 1, type 1 diabetes	+342	+694	+398	+694	+694	+694	–

(*continued overleaf*)

Table 42.6 (*continued*).

H'chol Cat 2, H'tension Cat 1, type 2 diabetes	+111	+465	+167	+465	+465	–	+465
H'chol Cat 2 and H'tension Cat 2	+81	+528	+130	+528	+528	+82	+96
H'tension Cat 2, H'chol Cat 1 and type 1 diabetes	+349	+714	+421	+714	+714	+714	–
H'tension Cat 2, H'chol Cat 1 and type 2 diabetes	+118	+485	+190	+485	+485	–	+485
H'tension Cat 3 and H'chol Cat 1	+89	+485	+220	+485	+485	+90	+102
H'tension Cat 3 and type 1 diabetes	+407	+881	+598	+489	+881	+881	
H'tension Cat 3 and type 2 diabetes	+176	+651	+368	+259	+651	–	+651
H'tension Cat 2, and H'chol Cat 2 and type 1 diabetes	+402	+976	+474	+976	+976	+976	–
H'tension Cat 2, and H'chol Cat 2 and type 2 diabetes	+172	+747	+243	+747	+747	–	+747
H'tension Cat 3 and H'chol Cat 2	+147	+771	+278	+771	+771	+148	+166
H'tension Cat 3, H'chol Cat 1, type 1 diabetes	+416	+925	+608	+925	+925	+925	–
H'tension Cat 3, H'chol Cat 1, type 2 diabetes	+185	+696	+377	+696	+696	–	+696
H'tension Cat 3, H'chol Cat 2 and type 1 diabetes	+491	+1293	+682	+1293	+1293	+1293	–
H'tension Cat 3, H'chol Cat 2 and type 2 diabetes	+260	+1064	+452	+1064	+1064	–	+1064

Source: Macdonald *et al.* (2005b).

act by modifying risk factors for a disease, rather than the disease outcome itself, present significantly more manageable insurance risks. In particular, the kinds of simple models of disease onset applicable in the study of single-gene disorders may be misleading if they are naively applied to the study of multifactorial disorders.

2. Great interest currently centres on large-scale prospective cohort studies in several countries. In the United Kingdom, the Biobank project will recruit 500 000 people aged 40–70, and track them for 10 years, with linkage to national health registers. The resulting data will be made available for analysis in later studies, presumed to be of nested case-control type. Macdonald *et al.* (2006) introduced a hypothetical 2×2 gene–environment model of heart attack risk, and simulated outcomes of the UK Biobank study. The question of interest was as follows: will the kind of case-control

studies likely to follow UK Biobank identify different levels of risk that both insurers and GAIC (in the United Kingdom) might regard as relevant and reliable? The answer was 'doubtful'. Only very large studies had sufficient power to pick out the modestly elevated risk, and that was in the context of a simplified genetic model more akin to a single-gene disorder than to a genuinely multifactorial disorder. This and other studies have found no evidence to suggest that increasing knowledge of multifactorial disorders will lead to changes in insurance practice.

REFERENCES

ALRC (2001). *Protection of Human Genetic Information.* Issues Paper No. 26. Australian Law Reform Commission. www.alrc.gov.au.

ALRC (2002). *Protection of Human Genetic Information.* Discussion Paper No. 66. Australian Law Reform Commission. www.alrc.gov.au.

ALRC (2003). Essentially yours: the protection of human genetic information in Australia. Report No. 96, Australian Law Reform Commission. www.alrc.gov.au.

Andersen, P.K., Borgan, O., Gill, R.D. and Keiding, N. (1993). *Statistical Models Based on Counting Processes.* Springer-Verlag, New York.

Brackenridge, R. and Elder, J. (1998). *Medical Selection of Life Risks.* Macmillan.

Daykin, C.D., Akers, D.A., Macdonald, A.S., McGleenan, T., Paul, D. and Turvey, P.J. (2003). Genetics and insurance – some social policy issues (with discussions). *British Actuarial Journal* **9**, 787–874.

De Jong, P. and Ferris, S. (2006). Adverse selection spirals. *ASTIN Bulletin* **36**, 589–628.

Doble, A. (2001). *Genetics in Society.* Institute of Actuaries in Australia, Sydney.

Doherty, N.A. and Thistle, P.D. (1996). Adverse selection with endogeneous information in insurance markets. *Journal of Public Economics* **63**, 83–102.

Elston, R.C. (1973). Ascertainment and age at onset in pedigree analysis. *Human Heredity* **23**, 105–112.

Espinosa, C. (2006). Ascertainment bias in estimating rates of onset of early-onset Alzheimer's disease: a critical illness and life insurance application Ph.D. dissertation, Heriot-Watt University, Edinburgh.

Ford, D., Easton, D.F., Stratton, M., Narod, S., Goldgar, D., Devilee, P., Bishop, D.T., Weber, B., Lenoir, G., Chang-Claude, J., Sobol, H., Teare, M.D., Struewing, J., Arason, A., Scherneck, S., Peto, J., Rebbeck, T.R., Tonin, P., Neuhausen, S., Barkardottir, R., Eyfjord, J., Lynch, H., Ponder, B.A.J., Gayther, S.A., Birch, J.M., Lindblom, A., Stoppa-Lyonnet, D., Bignon, Y., Borg, A., Hamann, U., Haites, N., Scott, R.J., Maugard, C.M., Vasen, H., Seitz, S., Cannon-Albright, L.A., Schofield, A., Zelada-Hedman, M. and The Breast Cancer Linkage Consortium (1998). Genetic heterogeneity and penetrance analysis of the BRCA1 and BRCA2 genes in breast cancer families. *American Journal of Human Genetics* **62**, 676–689.

Gui, E.H. (2003). Modelling the impact of genetic testing on insurance – early-onset Alzheimer's disease and other single-gene disorders Ph.D. dissertation, Heriot-Watt University, Edinburgh.

Gui, E.H. and Macdonald, A.S. (2002a). A Nelson-Aalen estimate of the incidence rates of early-onset Alzheimer's disease associated with the Presenilin-1 gene. *ASTIN Bulletin* **32**, 1–42.

Gui, E.H. and Macdonald, A.S. (2002b). Early-onset Alzheimer's disease, critical illness insurance and life insurance. Genetics and Insurance Research Centre Research Report 02/2, Heriot-Watt University, Edinburgh.

Gui, E.H., Lu, B., Macdonald, A.S., Waters, H.R. and Wekwete, C.T. (2006). The genetics of breast and ovarian cancer III: a new model of family history with applications. *Scandinavian Actuarial Journal* 338–367.

Gutiérrez, M.C. and Macdonald, A.S. (2003). Adult polycystic kidney disease and critical illness insurance. *North American Actuarial Journal* **7**(2), 93–115.

Gutiérrez, M.C. and Macdonald, A.S. (2004). Huntington's disease, critical illness insurance and life insurance. *Scandinavian Actuarial Journal* 279–313.

Gutiérrez, M.C. and Macdonald, A.S. (2007). Adult polycystic kidney disease and insurance: a case study in genetic heterogeneity. *North American Actuarial Journal* **11**(1), 90–118.

Hoem, J.M. (1988). The versatility of the Markov chain as a tool in the mathematics of life insurance. In *Transactions of the 23rd International Congress of Actuaries*, Helsinki, 171–202.

HCSTC (1995). *House of Commons Science and Technology Committee, Third Report: Human genetics: The Science and its Consequences.* H.M.S.O., London.

HCSTC (2001). House of Commons Science and Technology Committee, Fifth Report: Genetics and insurance. www.publications.parliament.uk/pa/cm200001/cmselect/cmsctech/174/17402.htm.

Hoy, M. and Polborn, M. (2000). The value of genetic information in the life insurance market. *Journal of Public Economics* **78**, 235–252.

Hoy, M. and Witt, J. (2005). *Welfare effects of banning genetic information in the life insurance market: the case of the BRCA1/2 genes.* University of Guelph, Discussion Paper 2005-5.

HGAC (1997). *The Implications of Genetic Testing for Insurance.* Human Genetics Advisory Commission, London.

HGC (2002). *Inside information: Balancing interests in the Use of Personal Genetic Data.* The Human Genetics Commission, London.

Leigh, T.S. (1990). Underwriting – a dying art? (with discussion). *Journal of the Institute of Actuaries* **117**, 443–531.

Lu, L. (2006). Some actuarial and statistical investigations into topics on genetics and insurance. Ph.D. dissertation, Heriot-Watt University, Edinburgh.

Lu, L., Macdonald, A.S. and Wekwete, C.T. (2007). Premium rates based on genetic studies: how reliable are they? *Insurance: Mathematics and Economics.*

Macdonald, A.S. (1997). How will improved forecasts of individual lifetimes affect underwriting? *Philosophical Transactions of the Royal Society Series B* **352**, 1067–1075 and (with discussion) *British Actuarial Journal* **3**, 1009–1025 and 1044–1058.

Macdonald, A.S. (1999). Modeling the impact of genetics on insurance. *North American Actuarial Journal* **3**(1), 83–101.

Macdonald, A.S. (2003a). Moratoria on the use of genetic tests and family history for mortgage-related life insurance. *British Actuarial Journal* **9**, 217–237.

Macdonald, A.S. (2003b). Genetics and insurance: What have we learned so far? *Scandinavian Actuarial Journal* 324–348.

Macdonald, A.S., Pritchard, D.J. and Tapadar, P. (2006). The impact of multifactorial genetic disorders on critical illness insurance: a simulation study based on UK Biobank. *ASTIN Bulletin* **36**, 311–346.

Macdonald, A.S. and Tapadar, P. (2006). Multifactorial genetic disorders and adverse selection: Epidemiology meets economics. Submitted.

Macdonald, A.S., Waters, H.R. and Wekwete, C.T. (2003). The genetics of breast and ovarian cancer II: a model of critical illness insurance. *Scandinavian Actuarial Journal* 28–50.

Macdonald, A.S., Waters, H.R. and Wekwete, C.T. (2005a). A model for coronary heart disease and stroke, with applications to critical illness insurance underwriting I: the model. *North American Actuarial Journal* **9**(1), 13–40.

Macdonald, A.S., Waters, H.R. and Wekwete, C.T. (2005b). A model for coronary heart disease and stroke, with applications to critical illness insurance underwriting II: applications. *North American Actuarial Journal* **9**(1), 41–56.

Norberg, R. (1995). Differential equations for moments of present values in life insurance. *Insurance: Mathematics and Economics* **17**, 171–180.

Nys, H., Dreezen, I., Vinck, I., Dierickx, K., Dequeker, E. and Cassiman, J.-J. (2002). *Genetic Testing*. European Commission, Brussels.

Pauly, M., Withers, K., Viswanathan, K.S., Lemaire, J., Hershey, J., Armstrong, K. and Asch, D.A. (2003). *Price Elasticity of Demand for Term Life Insurance and Adverse Selection*. National Bureau of Economic Research, Working Paper 9925.

Viswanathan, K.S., Lemaire, J., Withers, K., Armstrong, K., Baumritter, A., Hershey, J., Pauly, M. and Asch, D.A. (2007). Adverse selection in life insurance purchasing, due to the BRCA 1/2 genetic test and elastic demand. *Journal of Risk and Insurance*.

43

Forensics

B.S. Weir

Department of Biostatistics, University of Washington, Seattle, WA, USA

The use of DNA profiles for human identification often requires statistical genetic calculations. The probabilities for a matching DNA profile can be evaluated under alternative hypotheses about the contributor(s) to the profile, and presented as likelihood ratios. It is conditional probabilities that are needed: the probabilities of profiles given that they have already been seen, and these depend on the relationships between known and unknown people. The algebraic treatment is greatly simplified when it can be assumed that allelic frequencies have Dirichlet distributions over populations. The growing size of DNA profile databases has led to empirical verification of the probabilities of finding pairs of people with the same profile.

43.1 INTRODUCTION

Human individualisation based on the genome exploits the fact that everyone except for identical twins is genetically distinguishable. Moreover, human genetic material is found in every nucleated cell in the body and can be recovered from samples as diverse as bone, blood stains, saliva residues, nasal secretions and even fingerprints. DNA may be recovered from very old samples that have been well preserved, and DNA signatures may even be preserved over successive generations.

Genetic markers have been used for human individualisation since the discovery of blood groups, and statistical genetic arguments have long played a large part in parentage dispute cases. The role of statistical genetics in forensic science increased sharply in the late 1980s when DNA markers began to be used and an emphasis shifted from excluding specific people as being the sources of evidentiary stains to making probability statements about genetic profiles if these people were not the sources. As more markers have been developed, it has become less likely that two people would share the same DNA profile and therefore more likely that people will be convicted on the basis of DNA evidence. Given the serious nature of the charges of many crimes where genetic markers are used, and the serious consequences of conviction for these crimes, there has been considerable scrutiny paid to the statistical genetic arguments

Handbook of Statistical Genetics, Third Edition. Edited by D.J. Balding, M. Bishop and C. Cannings.
© 2007 John Wiley & Sons, Ltd.   ISBN: 978-0-470-05830-5.

upon which forensic probabilities are based. These arguments are reviewed in this chapter.

A common situation is where DNA is recovered from a biological sample left at the scene of a crime, and there is reason to believe the sample is from the perpetrator of the crime. DNA is also extracted from a blood or saliva sample from a suspect in the crime and is found to have the same profile as the crime sample. An immediate question is how much evidence against the suspect is provided by this matching, and a naive answer might be that the probability of the suspect having the evidentiary profile, if he was not the perpetrator, is the population proportion of that profile. High values for this proportion would favour the suspect and low values would not. How is the proportion to be estimated?

For genetic profiles based on single loci, it is quite feasible to estimate the population proportion of any genotype on the basis of a moderate-sized sample of profiles from that population. The size of the sample would need to be greater for highly variable loci where there are many different genotypes. There are issues of how the population is to be defined, and then how it is to be sampled. As the number of loci used for identification increases, the numbers of genotypes increases substantially and no sample can hope to capture all genotypes. Current practice in the United States is to use a set of 13 short tandem repeat (STR) loci, each with at least 9 alleles and 45 genotypes. The number of 13-locus profiles is therefore considerably more than 10^{21}, so that less than one profile in a trillion of all possible profiles exists anywhere in the world. Although there may not, therefore, be much value in declaring that a particular profile is rare, a first attempt to attach a frequency to a 13-locus profile might be to multiply together the frequencies of the 26 constituent alleles, along with a factor of 2 for every heterozygote. The implied assumption of allelic independence, within and between loci, is a statistical genetic issue.

A more satisfactory approach is to ask the question: What is the probability of an unknown person chosen from the population having a particular genetic profile given that the profile has been seen already for the suspect? Calculating this conditional probability needs to take into account the relationship between the known suspect and the unknown person. This relationship may be due to close family membership or to shared evolutionary history. Once again, these are statistical genetic issues.

43.2 PRINCIPLES OF INTERPRETATION

Evett and Weir (1998) suggested that genetic evidence be interpreted according to three principles:

- *First Principle*: To evaluate the uncertainty of any given proposition it is necessary to consider at least one alternative proposition.

- *Second Principle*: Interpretation is based on questions of the kind 'What is the probability of the evidence given the proposition?'

- *Third Principle*: Interpretation is conditioned not only by the competing propositions, but also by the framework of circumstances within which they are to be evaluated.

The first of these principles leads to the use of likelihood ratios LRs, as will soon be shown. The second is meant to draw a distinction from the very common 'prosecutor's fallacy' (Thompson and Schumann, 1987) of quoting probabilities of a proposition given the evidence. The third principle recognises the difference in the strength of evidence of a blood stain at the scene of a crime having a profile matching that of a suspect, and the evidence of bloodstain in the clothing of a suspect away from the crime scene with a profile matching that of a victim.

The propositions mentioned in the principles refer, in this context, to the source of the genetic profile in the evidentiary stain. For the immediate discussion these will be taken to be

H_p: The profile is from the suspect.
H_d: The profile is from some other person.

Suppose G_s and G_c are the profile types from the suspect and the crime stain, and they are found to match. Then the evidence E is these two profiles: $E = (G_s, G_c)$. Consideration of alternative propositions is carried out by comparing the probabilities of E under these propositions by means of the likelihood ratio

$$LR = \frac{Pr(E|H_p)}{Pr(E|H_d)} = \frac{Pr(G_s, G_c|H_p)}{Pr(G_s, G_c|H_d)}$$

$$= \frac{Pr(G_c|G_s, H_p)}{Pr(G_c|G_s, H_d)} \times \frac{Pr(G_s|H_p)}{Pr(G_s|H_d)}$$

$$= \frac{1}{Pr(G_c|G_s, H_d)},$$

with the last step depending on the assumption that the profiles must be found to match when they have a common source, and on recognition that $Pr(G_s)$ does not depend on H_p or H_d.

The forensic question of attaching weight to matching genetic profiles has therefore reduced to the statistical genetic question of determining the probability of a profile given that (the same) profile has been seen already. This conditional probability will be referred to as the *match probability*. It would be a substantial simplification to assume the two profiles were independent and work only with the probability of the crime stain profile:

$$LR = \frac{1}{Pr(G_c|H_d)} = \frac{1}{Pr(G_c)}.$$

The use of LRs for DNA evidence has been described in several textbooks: Aitken (1995), Robertson and Vignaux (1995), Royall (1997), Schum (1994), Evett and Weir (1998), Balding (2005), Buckleton *et al.* (2005) and Lucy (2005).

The odds form of Bayes' theorem relates the probabilities of the propositions after the evidence to the probabilities prior to the evidence:

$$\frac{Pr(H_p|E)}{Pr(H_d|E)} = \frac{Pr(E|H_p)}{Pr(E|H_d)} \times \frac{Pr(H_p)}{Pr(H_d)},$$

which can be stated as

$$\text{Posterior odds on}\,H_p = \text{LR} \times \text{prior odds on}\,H_p.$$

It needs to be stressed that the results in this section apply only to the situation where the DNA evidence refers to material left at the crime scene by the perpetrator, and there is no DNA evidence at the scene that does not provide a match to the profile of the suspect. If the evidence refers to a bloodstain found on the clothing of the suspect, e.g. and the stain has a DNA profile matching that of the victim then additional factors need to be considered: What is the probability that the victim's blood would be transferred during the crime? What is the probability that the suspect would have non-self blood on his or her clothing? What is the probability that non-self blood on the suspect's clothing would match that of the victim?

43.3   PROFILE PROBABILITIES

Although it is conditional profile probabilities (match probabilities) that are needed, these are most directly calculated as joint divided by marginal probabilities. Dropping the H_d symbol for convenience:

$$\Pr(G_c|G_s) = \frac{\Pr(G_c, G_s)}{\Pr(G_s)}.$$

Profile probabilities are therefore needed to calculate match probabilities.

At a single locus A with alleles A_u in proportion p_u, the probabilities P_{uv} for genotypes $A_u A_v$ within a single population may be parameterised as

$$P_{uv} = \begin{cases} p_u^2 + f p_u(1 - p_u), & u = v, \\ 2 p_u p_v - 2 f p_u p_v, & u \neq v. \end{cases} \tag{43.1}$$

As the loci used for individualisation are unlikely to be under the influence of selection, the use of a single inbreeding coefficient f for all genotypes may be thought reasonable, although genotype-specific coefficients should strictly be used for loci with allele-specific mutation processes (Graham et al., 2000). Expressing genotype proportions as functions of allele proportions is necessary for highly variable loci, when many genotypes are not seen in a sample and their population frequencies are difficult to estimate.

43.3.1   Allelic Independence

Considerable attention has been paid to the issue of whether f can be assumed small enough to ignore. Standard testing procedures rarely find significant departures from the hypothesis that f is zero (Weir, 1992; Maiste and Weir, 1995; Zaykin et al., 1995), although this may not be very surprising. The goodness-of-fit test has the statistic

$$X^2 = \sum_u \frac{n(\tilde{P}_{uu} - \tilde{p}_u^2)^2}{\tilde{p}_u^2} + \sum_{u \neq v} \frac{n(\tilde{P}_{uv} - 2\tilde{p}_u\tilde{p}_v)^2}{2\tilde{p}_u\tilde{p}_v},$$

where $\tilde{P}_{uv}$ and $\tilde{p}_u$ are the sample genotype and allele proportions in a sample of n individuals. For a locus with m alleles it is distributed as χ^2 with $m(m-1)/2$ df when

f is zero. When f is not zero, X^2 has non-centrality parameter

$$\lambda = \sum_u [nf^2(1-p_u)^2] + \sum_{u \neq v} [n2f^2 p_u p_v]$$

$$= (m-1)nf^2.$$

and this allows the power of the test to be calculated. For a locus with $m = 2$ alleles, the test has one degree of freedom and λ needs to be at least 10.5 for the power to be at least 90 % when the significance level is 0.05. In other words, the sample size must be at least $10.5/f^2$. A sample of 105 000 would therefore be needed to have 90 % chance of detecting an f as small as 0.01. Even an f value of 0.03 would require a sample of over 11 000. In Table 43.1, the sample sizes needed for 90 % power to detect $f = 0.05$ are shown for a range of values of m, along with the powers for that f value when $n = 1000$. It is not likely that Hardy–Weinberg disequilibrium, at the level usually thought to exist in human populations, will be detected with samples of 1000 or less.

What should be done when a test does cause rejection of the null hypothesis? Forensic agencies routinely test for allelic independence at six or more loci in three or more samples. A conventional 5 % significance-level test would be expected to give at least one significant result even if there was independence at all loci in all populations. However, to ignore single rejections on that basis calls into question the logic of performing the test in the first place. Requiring each of t tests to meet a $0.05/t$ significance level in order to declare rejection, the Bonferroni correction, seems unduly conservative. Zaykin *et al.* (2002) examined multiple-testing issues, and showed that Fisher's procedure for combining p values may be more informative. When a hypothesis is true, the p value (the probability of the data or of data with at least as much departure from the hypothesis) has a uniform distribution and minus twice its logarithm has a χ^2 distribution with 2 df. A set of t independent tests can be said to indicate that at least one of the t hypotheses is false if the sum of the $-2\ln(p)$ values exceeds the critical value of the χ^2 distribution with $2t$ df. As an example, LR tests for allelic independence at seven loci were performed on data in three samples (Scholl *et al.*, 1996) and the p values are shown in Table 43.2. For a 5 % significance level, ignoring the multiple-testing issue would lead to rejection of independence at GYPA in the Navajo sample and maybe at D7S8 in that sample. Applying

Table 43.1 Hardy–Weinberg test properties when $f = 0.05$.

m	df	For 90 % power		Power if $n = 1000$
		λ	n	
2	1	10.5	4200	0.35
3	3	14.2	2840	0.44
4	6	17.4	2320	0.50
5	10	20.5	2050	0.54
6	15	23.6	1888	0.57
7	21	26.6	1773	0.60
8	28	29.6	1691	0.63
9	36	32.6	1630	0.64
10	45	35.6	1582	0.65

the Bonferroni correction for the seven tests conducted for the Navajo data would lead to no rejections in that sample; applying the correction to the three tests conducted for GYPA would lead to rejection at that locus; applying the correction to all 21 test would cause no rejections in the whole set. Fisher's procedure, however, suggests dependence in the Navajo sample, but not at the GYPA locus.

An alternative to testing hypotheses about f is to make statements about its value in the relevant population, by means of either a point estimate or a posterior probability distribution (Ayres and Balding, 1998; Shoemaker *et al.*, 1998). Given that human populations do have non-zero values of f, there is some appeal to making probability statements about f lying in a certain range rather than simply failing to reject the false hypothesis that it is zero. Other comments about conventional hypothesis tests were made by Evett and Weir (1998). Although there have been concerns in the past about values of f in forensic calculations, the concerns may have been overstated. As is discussed below, the quantity of interest is the match probability or the probability of a profile given that it as already been seen, and this is a statement about four alleles per locus, whereas f is used in profile probabilities which are statements about two alleles per locus. The distinction was made by Ayres and Overall (1999).

43.3.2   Allele Frequencies

Apart from some uncertainty about f values, the difficulty in using 43.1 is that the allele frequencies are also unknown. This problem is due to uncertainty about the population to which these frequencies apply. The population is sometimes referred to as being that of 'possible perpetrators' as (under H_d) the unknown perpetrator belongs to the population. Under H_p, the suspect must also belong to this population. Usually, however, the population is not defined with sufficient detail to suggest a sampling strategy for estimating allele frequencies from that group. Even an eyewitness description of the appearance of the perpetrator may not delineate the population very precisely. There is also the practical issue of collecting a new sample from which to estimate allele frequencies for every new crime.

Instead, it is customary to base calculations on samples collected from some population that is larger than the relevant population, and which may well have substructure. Typically forensic agencies work with samples of people described by broad racial labels such as 'Caucasian' even when these labels refer to admixed groups, as is the case with 'African

Table 43.2 p values for tests of allelic independence.

Locus	Sample			$-2\sum \ln(p)$
	Navajo	Peublo	Sioux	
LDLR	0.377	0.397	0.599	0.566
GYPA	0.014	0.470	1.000	0.122
HBGG	0.136	0.168	0.790	0.235
D7S8	0.052	1.000	0.804	0.385
GC	0.259	0.124	0.213	0.125
HLA-DQA1	0.438	0.368	0.562	0.569
D1S80	0.750	0.559	0.211	0.563
$-2\sum \ln(p)$	0.031	0.430	0.812	0.216

American'. Assignment of a person to such groups is generally by self reporting, and the samples tend to be drawn from the geographic area served by the agency. Equation 43.1 is now to be understood to apply to a subpopulation, i, of some larger sampled population:

$$P_{uvi} = \begin{cases} p_{ui}^2 + f_i p_{ui}(1 - p_{ui}), & u = v, \\ 2p_{ui}p_{vi} - 2f_i p_{ui}p_{vi}, & u \neq v. \end{cases} \qquad (43.2)$$

A simplifying assumption is that all f_i values are equal (and maybe equal to zero), and that all p_{ui} values have the same expected value p_u. Taking expectations over subpopulations:

$$E(p_{ui}^2) = p_u^2 + p_u(1 - p_u)\theta,$$

where θ or F_{ST} is a measure of population structure. The genotype proportions in the whole population are the expected values of the subpopulation values:

$$P_{uv} = E(P_{uvi}) = \begin{cases} p_u^2 + Fp_u(1 - p_u), & u = v, \\ 2p_u p_v - 2Fp_u p_v, & u \neq v, \end{cases} \qquad (43.3)$$

where F or F_{IT} is the total inbreeding coefficient, in contrast to the within-population inbreeding coefficient $f = (F - \theta)/(1 - \theta)$. If the subpopulations are not inbred, then $f = 0$ and θ could be used in place of F. The population-wide genotype frequencies in 43.1 may be taken to apply, on average, to any subpopulation.

Although 43.3 appears to circumvent the need to know subpopulation allele frequencies, it does raise other issues. It is not possible to estimate F or θ from data only at the whole-population level. Estimation would require observations at the subpopulation level – in which case this overall formulation would not be needed. In practice, a numerical value is assigned to F or θ.

Secondly, it is not possible to estimate p_u^2 as the square of an estimate of p_u as that would ignore the variation in p_u that is being described by θ. Suppose the ith subpopulation has allele A_u with frequency p_{ui}, and it forms a proportion w_i of the whole population ($\sum_i w_i = 1$). A sample of size n individuals from the whole population has n_i individuals from the ith subpopulation. With random sampling, it might be supposed that $n_i = nw_i$. The sampling properties of $\tilde{p}_{ui}$, the proportion of A_u alleles among the $2n_i$ alleles from the ith subpopulation, are (Weir, 1996),

$$E(\tilde{p}_{ui}) = p_u$$

$$Var(\tilde{p}_{ui}) = p_u(1 - p_u)\left(\theta + \frac{1 - \theta}{2n_i}\right)$$

$$Cov(\tilde{p}_{ui}, \tilde{p}_{vi}) = -p_u p_v\left(\theta + \frac{1 - \theta}{2n_i}\right).$$

Assuming the subpopulations to be independent, therefore,

$$E(\tilde{p}_u) = p_u$$

$$E(\tilde{p}_u^2) = p_u^2 + p_u(1 - p_u)\left(\theta \sum_i w_i^2 + \frac{1 - \theta}{2n}\right)$$

$$E(2\tilde{p}_u \tilde{p}_v) = 2p_u p_v - 2p_u p_v \left(\theta \sum_i w_i^2 + \frac{1-\theta}{2n} \right), \quad u \neq v,$$

so that estimating squares or products of allele frequencies by squares or products of estimated allele frequencies causes overestimation for homozygotes and is therefore conservative, but causes underestimation for heterozygotes and is therefore prejudicial. The effects will be small for small θ or for highly structured populations (small w_i).

43.3.3 Joint Profile Probabilities

The joint genotype probabilities needed to calculate match probabilities require account to be taken of the relationships among sets of four alleles, two per individual, in the same way that single genotype probabilities require information about the relationship of pairs of alleles. In the random-mating situation, where allelic relationships do not depend on the arrangements of alleles within genotypes, there are four measures of allelic relationship: θ, γ, δ and Δ for pairs, triples, quadruples and two pairs of alleles, respectively (Cockerham, 1971).

The probability that four alleles are all of type A_u is just the allele frequency p_u if they are all identical by descent, and this identity situation has probability δ. At the other extreme, four alleles are all A_u with probability p_u^4 if there is no identity among them, and this situation has probability $(1 - 6\theta + 8\gamma + 3\Delta - 6\delta)$. Such arguments lead to the joint genotype probabilities

$$\begin{aligned}
\Pr(A_u A_u, A_u A_u) &= (1 - 6\theta + 8\gamma + 3\Delta - 6\delta)p_u^4 \\
&\quad + 6(\theta - 2\gamma - \Delta + 2\delta)p_u^3 \\
&\quad + (4\gamma + 3\Delta - 7\delta)p_u^2 + \delta p_u, \quad\quad\quad (43.4) \\
\Pr(A_u A_v, A_u A_v) &= 4(1 - 6\theta + 8\gamma + 3\Delta - 6\delta)p_u^2 p_v^2 \\
&\quad + 4(\theta - 2\gamma - \Delta + 2\delta)p_u p_v (p_u + p_v) \\
&\quad + 4(\gamma - \delta)p_u p_v, \quad u \neq v. \quad\quad\quad (43.5)
\end{aligned}$$

These expressions are greatly simplified under the assumption, made in coalescent approaches, of evolutionary stationarity. Under that assumption, the set of allele frequencies has a Dirichlet distribution over populations with the consequence that the probability of drawing allele A_u from a population given that n previously drawn alleles contained n_u of that type is

$$\Pr(A_u | n_u \text{ among } n) = \frac{n_u \theta + (1 - \theta)p_u}{1 + (n-1)\theta}. \quad\quad\quad (43.6)$$

The Dirichlet assumption also implies that $\gamma = 2\theta^2/(1+\theta)$, $\delta = 6\theta^3/(1+\theta)(1+2\theta)$, and $\Delta = \theta^2(1+5\theta)/(1+\theta)(1+2\theta)$ and it leads to the match probabilities given by Balding and Nichols (1994):

$$\Pr(A_u A_v | A_u A_v) = \begin{cases} \dfrac{[2\theta + (1-\theta)p_u][3\theta + (1-\theta)p_u]}{(1+\theta)(1+2\theta)}, & u = v, \\[3mm] \dfrac{2[\theta + (1-\theta)p_u][\theta + (1-\theta)p_v]}{(1+\theta)(1+2\theta)}, & u \neq v. \end{cases} \quad\quad\quad (43.7)$$

Table 43.3 Effects of population structure on match probabilities.

		Reciprocal of match probability			
		$\theta = 0$	$\theta = 0.001$	$\theta = 0.01$	$\theta = 0.03$
$p = 0.01$	Heterozygote	5000	4152	1295	346
	Homozygote	10 000	6 439	863	157
$p = 0.05$	Heterozygote	200	193	145	89
	Homozygote	400	364	186	73
$p = 0.10$	Heterozygote	50	49	43	34
	Homozygote	100	96	67	37

A feature of the Dirichlet distribution is that the frequencies of all pairs of different allelic types have negative correlations. This is trivially true for biallelic single nucleotide polymorphism (SNP) loci but it cannot be true for STR loci affected by stepwise mutation (Graham *et al.*, 2000) although the consequences on match probabilities may not be large for rare genotypes.

The match probabilities in 43.7 are greater than the profile probabilities in 43.3 for allele frequencies less than 0.5. Although a profile is rare, as soon as it has been seen there is an increased probability that there is another copy of it in the population. It needs to be stressed that the equations apply on average for any subpopulation within the population. Some numerical consequences of allowing for population structure are shown in Table 43.3, in the case where all alleles at a locus have the same frequency. The table shows that even small values of θ can have an appreciable effect when allele frequencies are small.

A larger dependency between DNA profiles occurs when two people are in the same family. Brothers, e.g. have at least a 25 % probability of sharing the same genotype at any locus. If only family relatedness is considered and neither relative is inbred, then 43.4 and 43.5 are replaced by

$$\Pr(A_u A_u, A_u A_u) = k_0 p_u^4 + k_1 p_u^3 + k_2 p_u^2 \tag{43.8}$$

$$\Pr(A_u A_v, A_u A_v) = 4k_0 p_u^2 p_v^2 + k_1 p_u p_v (p_u + p_v) + 2k_2 p_u p_v, \quad u \neq v. \tag{43.9}$$

Here k_0, k_1, k_2 are the probabilities that the relatives share 0, 1, 2 pairs of alleles identical by descent. Values for these coefficients for some common types of relationship are given in Table 43.4 and the match probabilities are

$$\Pr(A_u A_v | A_u A_v) = \begin{cases} k_0 p_u^2 + k_1 p_u + k_2, & u = v, \\ 2k_0 p_u p_v + \frac{1}{2} k_1 (p_u + p_v) + k_2, & u \neq v. \end{cases}$$

To complete this section, consider the situation of non-inbred relatives in a population with population structure parameter θ where allele frequencies follow the Dirichlet distribution. Equations 43.4 and 43.5 are modified to

$$\Pr(A_u A_u, A_u A_u) = k_0 \Pr(A_u A_u A_u A_u) + k_1 \Pr(A_u A_u A_u) + k_2 \Pr(A_u A_u)$$

$$\Pr(A_u A_v, A_u A_v) = 4k_0 \Pr(A_u A_u A_v A_v) + k_1 [\Pr(A_u A_u A_v) + \Pr(A_u A_v A_v)]$$

$$+ 2k_2 \Pr(A_u A_v), \quad u \neq v.$$

Table 43.4 Identity by descent probabilities for common non-inbred relatives.

Relationship	k_0	k_1	k_2
Identical twins	0	0	1
Full-sibs	$\frac{1}{4}$	$\frac{1}{2}$	$\frac{1}{4}$
Parent–child	0	1	0
Double first cousins	$\frac{9}{16}$	$\frac{3}{8}$	$\frac{1}{16}$
Half-sibs*	$\frac{1}{2}$	$\frac{1}{2}$	0
First cousins	$\frac{3}{4}$	$\frac{1}{4}$	0
Unrelated	1	0	0

*Also grandparent–grandchild and uncle–nephew.

as given by Fung *et al.* (2003). It needs to be stressed that these results hold only for non-inbred relatives –siblings whose parents are first cousins, e.g. can both be homozygous because their four alleles all descend from a single allele carried by their parents' grandparents. The allelic-set probabilities in these equations refer to the generation to which the relatives' most recent common ancestors belong. The match probabilities become

$$
\Pr(A_u A_v | A_u A_v) = \begin{cases} k_0 \dfrac{[2\theta + (1-\theta)p_u][3\theta + (1-\theta)p_u]}{(1+\theta)(1+2\theta)} \\ \quad + k_1 \dfrac{2\theta + (1-\theta)p_u}{1+\theta} + k_2, \qquad u = v, \\[2em] k_0 \dfrac{2[\theta + (1-\theta)p_u][\theta + (1-\theta)p_v]}{(1+\theta)(1+2\theta)} \\ \quad + k_1 \dfrac{2\theta + (1-\theta)(p_u + p_v)}{2(1+\theta)} + k_2, \qquad u \neq v. \end{cases}
$$

Parameters p_u and θ are assumed to have the same value in successive generations, so that the same level of approximation holds as in 43.7.

43.4 PARENTAGE ISSUES

The previous section considered relationships among pairs of people caused by evolutionary processes and/or by family membership. Family relatedness lies at the heart of issues concerning parentage, whether these arise in civil paternity suits or in forensic situations such as incest. The genetic evidence often consists of three genetic profiles: those of the mother (G_m), her child (G_c) and the alleged father (G_a) although the child's genotype may be replaced by its paternal allele (A_p). The child's maternal allele is either known from a comparison of G_m and G_c or is assumed to be either of the mother's alleles with equal probability when mother and child are both heterozygotes for the same allele – in either case the paternal allele is deduced by subtraction from G_c. The LR, or paternity index (PI), can be written in several ways, such as

$$
\mathrm{PI} = \frac{\Pr(A_p | G_m, G_a, \text{ A is the father of C})}{\Pr(A_p | G_m, G_a, \text{ A is not the father of C})}.
$$

The numerator of this expression follows immediately from Mendelian laws, but the denominator depends on the relationship between the three people: mother, alleged father and actual father.

If there is no immediate family relatedness among mother, alleged father and actual father, but they can all be considered to be random members of a population for which the Dirichlet distribution of the last section holds, then the various values of the PI are shown in Table 43.5. The first scenario in that table, e.g. has PI of $1/\Pr(A_u|A_u A_u A_u A_u)$, and the case where mother, child and alleged father are all $A_u A_v$ has a PI of $1/[\Pr(A_u|A_u A_v A_u A_v) + \Pr(A_v|A_u A_v A_u A_v)]$. These can be found from 43.6. Setting $\theta = 0$ leads to the classic results of $1/p_{A_p}$ or $1/2p_{A_p}$, according to whether the alleged father is homozygous or heterozygous for the parental allele A_p. When $\theta \neq 0$, however, it is necessary to consider the genotype of the mother as well as that of the alleged father.

Another extension to classic theory is to allow the alleged father to be related to the actual father for the PI denominator calculations. Provided the alleged father is not inbred and population structure can be ignored the relatedness between these two men can be described by the identity by descent coefficients shown in Table 43.4. If the alleged father is homozygous for the paternal allele A_p, the PI is $2/[(2k_2 + k_1) + (k_1 + 2k_0)p_{A_p}]$, and if he is heterozygous for the paternal allele the PI is half that value. The classic PI values result when $k_2 = k_1 = 0$. The quantity $(k_2/2 + k_1/4)$ is called the *coancestry* or *kinship coefficient of the two relatives*.

Table 43.5 Paternity index for population described by θ.

G_m	G_c	G_a	PI
$A_u A_u$	$A_u A_u$	$A_u A_u$	$\dfrac{1 + 3\theta}{4\theta + (1 - \theta)p_u}$
		$A_u A_w, w \neq u$	$\dfrac{1 + 3\theta}{2[3\theta + (1 - \theta)p_u]}$
	$A_u A_v$	$A_v A_v$	$\dfrac{1 + 3\theta}{2\theta + (1 - \theta)p_v}$
		$A_v A_w, w \neq v$	$\dfrac{1 + 3\theta}{2[\theta + (1 - \theta)p_v]}$
$A_u A_v$	$A_u A_u$	$A_u A_u$	$\dfrac{1 + 3\theta}{3\theta + (1 - \theta)p_u}$
		$A_u A_w, w \neq u$	$\dfrac{1 + 3\theta}{2[2\theta + (1 - \theta)p_u]}$
	$A_u A_v$	$A_u A_u$	$\dfrac{1 + 3\theta}{4\theta + (1 - \theta)(p_u + p_v)}$
		$A_u A_v$	$\dfrac{1 + 3\theta}{4\theta + (1 - \theta)(p_u + p_v)}$
		$A_u A_w, w \neq u, v$	$\dfrac{1 + 3\theta}{2[3\theta + (1 - \theta)(p_u + p_v)]}$

43.5 IDENTIFICATION OF REMAINS

Family relatedness may also be involved for identification of remains. A simple forensic situation is where a bloodstain is found in the car of a person suspected of having murdered a victim, but there is no body. The DNA profile of the bloodstain could be compared to that from material known to be from the victim or it could be compared to profiles determined for the victim's relatives. In that case, calculations similar to those for parentage issues are made. The need to identify bones has received attention in recent years with efforts to identify repatriated Korean and Vietnam War remains. STR profiles were used to identify 26 of 61 bodies following the Waco, Texas fire disaster (Clayton *et al.*, 1995) and 139 of 141 individuals in the 1996 Spitsbergen air accident (Olaisen *et al.*, 1997). There was complete success in identifying tissue recovered from the sites of the 1996 TWA 800 (Ballantyne, 1997) and the 1998 Swissair 111 (Robb, 1999) airplane disasters. A very large task confronted forensic scientists associated with the 2001 World Trade Center disaster, and a personal account is given by C.J. Brenner at http://www.dna-view.com. By the end of April 2005, remains of 1592 of the 2749 missing people had been identified and DNA profiling was the sole means of identification in 86 % of those identifications.

As an example of the type of LR calculation that might be undertaken in such cases, suppose that a DNA profile G_x is produced from remains thought to belong to a man for whom profiles are available from his wife G_w, his child G_c and his two parents G_m and G_f. The competing hypotheses, H_p and H_d, are that the remains either are or are not from the missing man. The LR is

$$\text{LR} = \frac{\Pr(G_c, G_x, G_w, G_m, G_f | H_p)}{\Pr(G_c, G_x, G_w, G_m, G_f | H_d)}$$

$$= \frac{\Pr(G_c | G_x, G_m, H_p)\Pr(G_x | G_m, G_f, H_p)}{\Pr(G_c | G_w, G_m, G_f, H_d)\Pr(G_x)}.$$

The probabilities of profiles G_w, G_m, G_f do not depend on the two hypotheses and so cancel out of the ratio. The probability $\Pr(G_x)$ depends on allele frequencies and the remaining three probabilities follow from simple Mendelian calculations. Algebraic expressions in more complex cases are tedious to derive, but Brenner (1997) has developed software that will perform the appropriate symbolic manipulations. There is a growing use of Bayesian networks, or Probabilistic Expert Systems, to provide numerical solutions following the construction of a graph linking all profiles (Dawid and Evett, 1997; Evett *et al.*, 2002; Mortera, 2003).

43.6 MIXTURES

Forensic samples may contain material from more than one contributor. A common situation is for evidence collected in rape cases where material from the victim, possible consensual partners, and the perpetrator(s) may all be present. Even if some of these people contributed only a small proportion of the DNA in the sample, improved technology has made it easier to detect their alleles in the mixed profile. The genetic evidence E for mixed-stain cases is the set of alleles found among all the people who have either been typed directly or whose type is inferred because they are considered to have contributed to the stain. Consideration of the alleles from people who may have been typed even though they are excluded and/or are hypothesised not to have contributed to the stain is necessary to allow for the effects of population structure.

In line with the principles of evidence interpretation, there needs to be alternative propositions that specify the numbers and profiles of contributors to the evidentiary sample. Some of these contributors will be known and typed people, and some will be unknown people. Those contributors, together with any typed people who are known (under the proposition) not to be contributors, contain among them a set of alleles whose probability depends on the separate allele proportions and the population structure parameter θ according to results such as those in 43.6. There is also a factor of 2 for each known heterozygote, and a term for the number of ways of arranging all $2x$ alleles from x unknown people into pairs. There may be different sets of alleles from unknown people under some propositions, and the probabilities for these sets must be added together. The LR is the ratio of probabilities under alternative propositions.

Much of the complexity in dealing with mixtures can be removed by a mnemonic notation, as laid out in Table 43.6 (Curran *et al.*, 1999; Evett and Weir, 1998). There are sets of alleles (not necessarily distinct) that occur in the crime sample (C). For a particular proposition there may be alleles (T) carried by typed people declared to be contributors, alleles (W) carried by unknown contributors to the sample, as well as alleles (V) carried by people declared not to have contributed to the sample. There are corresponding sets of distinct alleles, and these sets are indicated by a g subscript. Note that the same person may be declared to be a contributor to the sample under one proposition, and declared not to be contributor under another proposition. Note also that the word 'known' in Table 43.6 refers to a value specified by the proposition under consideration.

The alleles in the evidence profile are carried either by typed people declared to be contributors or by unknown people, so that C is the combination (union) of sets T and W. For a given proposition, the probability of the evidence profile depends also on the alleles carried by people who have been typed but are declared by that proposition not to have contributed to the profile. For a proposition in which there are x unknown contributors, the probability is $P_x(T, W, V|C_g)$. This probability is for all $2n_C + 2n_V = 2n_T + 2n_W + 2n_V$ alleles in the sets T, W, V, among which allele A_u occurs $c_u + v_u = t_u + w_u + v_u$ times. The probabilities are added over all possible $n_x = (c + r - 1)!/[(c - 1)!r!]$ distinct sets of w_u. As listed in Table 43.7, c is the number of distinct alleles in C_g and r is the number of alleles carried by unknown people that can be any one of these c alleles.

Generating the n_x sets W is a two-stage process. Some of the alleles in each set must be present: these are the alleles in the set C_g that are not in set T_g. Other alleles are not under this constraint because they already occur in T_g, and there are r_u copies of A_u alleles in this unconstrained set. It is a straightforward computing task to let r_1 range over the integers $0, 1, \ldots, r$, then let r_2 range over the integers $0, 1, \ldots, r - r_1$, then let r_3 range over the integers $0, 1, \ldots, r - r_1 - r_2$, etc. The final count r_c is obtained by subtracting the sum of $r_1, r_2, \ldots, r_{c-1}$ from r. The total number of A_u alleles in set W is $\sum_{u=1}^{c} w_u = 2x$ where $w_u = r_u$ for those alleles in both C_g and T_g, and $w_u = r_u + 1$ for alleles in C_g but not in T_g.

For any ordering of the $2x = \sum_u w_u$ alleles in W, successive pairs of alleles can be taken to represent genotypes and there are $(2x)!/(\prod_{u=1}^{c} w_u!)$ possible orderings. This is

Table 43.6 Notation for mixture calculations.

Alleles in the profile of the evidence sample

C	The set of alleles in the evidence profile
C_g	The set of distinct alleles in the evidence profile
n_C	The known number of contributors to C
h_C	The unknown number of heterozygous contributors
c	The known number of distinct alleles in C_g
c_u	The unknown number of copies of allele A_u in C
	$1 \le c_u \le 2n_C$, $\sum_{u=1}^{c} c_u = 2n_C$

Alleles from typed people that H declares to be contributors

T	The set of alleles carried by the declared contributors to C
T_g	The set of distinct alleles carried by the declared contributors
n_T	The known number of declared contributors to C
h_T	The known number of heterozygous declared contributors
t	The known number of distinct alleles in T_g carried by n_T declared contributors
t_u	The known number of copies of allele A_u in T
	$0 \le t_u \le 2n_T$, $\sum_{u=1}^{c} t_u = 2n_T$

Alleles from unknown people that H declares to be contributors

W	The sets of alleles carried by the unknown contributors to C
x	The specified number of unknown contributors to C: $n_C = n_T + x$
$c - t$	The known number of alleles that are required to be in W
r	The known number of alleles in W that can be any allele in C_g, $r = 2x - (c - t)$
n_x	The number of different sets of alleles W, $n_x = (c + r - 1)!/[(c - 1)!r!]$
r_u	The unknown number of copies of A_u among the r unconstrained alleles in W
	$0 \le r_i \le r$, $\sum_{i=1}^{c} r_i = r$
w_u	The unknown number of copies of A_u in W: $c_u = t_i + u_u$, $\sum_{u=1}^{c} u_u = 2x$
	If A_u is in C_g but not in T_g: $u_u = r_u + 1$. If A_u is in C_g and also in T_g: $w_u = r_u$

Alleles from typed people that H declares to be non-contributors

V	The set of alleles carried by typed people declared not to be contributors to C
n_V	The known number of people declared not to be contributors to C
h_V	The known number of heterozygous declared non-contributors
v_i	The known number of copies of A_i in V: $\sum_i v_i = 2n_V$

the number of possible sets of unknown genotypes that have each allelic set W. Although it is the genotypes that correspond to the x unknown people, it is the set of $2x$ alleles that determine the probability, in combination with the $2n_T + 2n_V$ alleles among the known people. Because the n_T typed people all have specified genotypes, there is just a factor of 2 for each heterozygote, and there is a factor of 2 for each heterozygote among the set of n_V non-contributors.

Using 43.7, the probabilities of the set of alleles in the evidence is

$$P_x(T, W, V | C_g) = \sum_{r_1=0}^{r} \sum_{r_2=0}^{r-r_1} \cdots \sum_{r_{c-1}=0}^{r-r_1-\ldots-r_{c-2}}$$

$$\times \frac{(2x)! 2^{H_T + H_V} \prod_{u=1}^{c} \prod_{j=0}^{t_u+w_u+v_u-1} [(1-\theta)p_u + j\theta]}{\prod_{u=1}^{c} w_u! \prod_{j=0}^{2x+2n_T+2n_V-1} [(1-\theta) + j\theta]}. \qquad (43.10)$$

Table 43.7 Effects of family relatedness on match probabilities.

Relationship	Match Probability
Homozygotes $A_u A_u$	
Full-sibs	$\dfrac{(1+p_u)^2 + (7+7p_u - 2p_u^2)\theta + (16 - 9p_u + p_u^2)\theta^2}{4(1+\theta)(1+2\theta)}$
Parent and child	$\dfrac{2\theta + (1-\theta)p_u}{(1+\theta)}$
Half-sibs	$\dfrac{[2\theta + (1-\theta)p_u][2 + 4\theta + (1-\theta)p_u]}{2(1+\theta)(1+2\theta)}$
First cousins	$\dfrac{[2\theta + (1-\theta)p_u][1 + 11\theta + 3(1-\theta)p_u]}{4(1+\theta)(1+2\theta)}$
Unrelated	$\dfrac{[2\theta + (1-\theta)p_u][3\theta + (1-\theta)p_u]}{(1+\theta)(1+2\theta)}$
Heterozygotes $A_u A_v$	
Full-sibs	$\dfrac{\begin{array}{c}(1+p_u+p_v+2p_up_v)\\ +(5+3p_u+3p_v-4p_up_v)\theta + 2(4-2p_u-2p_v+p_up_v)\theta^2\end{array}}{4(1+\theta)(1+2\theta)}$
Parent and child	$\dfrac{2\theta + (1-\theta)(p_u+p_v)}{2(1+\theta)}$
Half-sibs	$\dfrac{\begin{array}{c}(p_u+p_v+4p_up_v)\\ +(2+5p_u+5p_v-8p_up_v)\theta + (8-6p_u-6p_v+4p_up_v)\theta^2\end{array}}{4(1+\theta)(1+2\theta)}$
First cousins	$\dfrac{\begin{array}{c}(p_u+p_v+12p_up_v)\\ +(2+13p_u+13p_v-24p_up_v)\theta + 2(8-7p_u-7p_v+6p_up_v)\theta^2\end{array}}{8(1+\theta)(1+2\theta)}$
Unrelated	$\dfrac{2[\theta + (1-\theta)p_u][\theta + (1-\theta)p_v]}{(1+\theta)(1+2\theta)}$

Likelihood ratios are formed as the ratios of two such probabilities.

Every person typed is declared to be either a contributor or a non-contributor. The number of people typed, and the alleles they carry among them, are the same for every proposition. For this reason, $n_T + n_V$, $H_T + H_V$ and $w_u + v_u$ will be the same in the probabilities for each proposition.

If population structure is ignored, and θ is set to zero, 43.10 reduces to

$$P_x(\mathcal{T}, \mathcal{W}, \mathcal{V}|C_g) = \sum_{r_1=0}^{r} \sum_{r_2=0}^{r-r_1} \cdots \sum_{r_{c-1}=0}^{r-r_1-\cdots-r_{c-2}} \frac{(2x)!2^{H_T+H_V}}{\prod_{u=1}^{c} w_u!} \prod_{u=1}^{c} p_u^{t_u+w_u+v_u},$$

as an alternative to the expression given by Weir *et al.* (1997). The value of LR now depends only on the numbers and frequencies of the alleles carried by unknown contributors. There is no need to consider the genotypes of typed people, whether or not they contribute to the evidence sample. This is different to the situation where population structure is taken into account – then the genotypes of all typed people are needed.

The arguments made for incorporating non-contributors can be extended. Several people may be typed during the course of an investigation. Even if they are excluded from being contributors, they provide information for the probability calculations when they can be considered to belong to the same subpopulation as (some of) people not excluded. They make their contribution to the calculation via allelic set $\mathcal{V}$.

Gill *et al.* (2006) discussed the complications for interpreting mixtures when some of the alleles in the mixed profile may be masked by typing artefacts such as stutter or may have dropped out completely and are not detected. A complete analysis needs to take into account the relative amounts of DNA inferred to be present at each of the alleles observed to be in the mixture. Having to allow for unseen alleles reduces the possibility of being able to exclude a potential contributor to the mixture simply because that person's alleles are not detected. Great care needs to be taken to avoid prejudicial conclusions if it is decided to ignore those loci in a profile for which interpretation is difficult or alleles are suspected of not being detected.

43.7 SAMPLING ISSUES

43.7.1 Allele Probabilities

In place of the unknown allele probabilities, it is usual to use the allele proportions obtained from a sample of individuals from the population (not from the specific relevant subpopulation). This leads to sampling variation in calculated match probabilities. Suppose the probability P_l at locus l is estimated as $\tilde{P}_l$ and that loci can be treated as being independent. Then the multilocus match probability $P = \prod_l P_l$ is estimated as $\tilde{P} = \prod_l \tilde{P}_l$ and the central limit theorem allows $\ln(\tilde{P})$ to be regarded as having a normal distribution. A 95 % confidence interval for P is, therefore, $(\tilde{P}/C, C\tilde{P})$ where $\ln(C) = 1.96\sqrt{\mathrm{Var}[\ln(\tilde{P})]}$. The task is to estimate the variance of $\ln(\tilde{P})$. From the assumed independence of loci,

$$\mathrm{Var}[\ln(\tilde{P})] = \mathrm{Var}[\sum_l \ln(\tilde{P}_l)] \approx \sum_l \mathrm{Var}(\tilde{P}_l)/P_l^2.$$

As θ is generally assigned a numerical value such as 0.03, rather than being estimated from sample data, it will be assumed constant. For a profile that is homozygous for allele u at locus l:

$$\mathrm{Var}(\tilde{P}_l) \approx \left(\frac{\partial \tilde{P}_l}{\partial \tilde{p}_{lu}}\right)^2 \mathrm{Var}(\tilde{p}_{lu})$$

and for a profile heterozygous for alleles u and v:

$$\mathrm{Var}(\tilde{P}_l) \approx \left(\frac{\partial \tilde{P}_l}{\partial \tilde{p}_{lu}}\right)^2 \mathrm{Var}(\tilde{p}_{lu}) + \left(\frac{\partial \tilde{P}_l}{\partial \tilde{p}_{lv}}\right)^2 \mathrm{Var}(\tilde{p}_{lv}) + 2\left(\frac{\partial \tilde{P}_l}{\partial \tilde{p}_{lu}}\right)\left(\frac{\partial \tilde{P}_l}{\partial \tilde{p}_{lv}}\right)\mathrm{Cov}(\tilde{p}_{lu}, \tilde{p}_{lv}).$$

The variances and covariances of for allele proportions are from Section 43.3.2

$$\mathrm{Var}(\tilde{p}_u) = p_{lu}(1 - p_{lu})\left(\theta \sum_i w_i^2 + \frac{1 - \theta}{2n_l}\right),$$

$$\mathrm{Cov}(\tilde{p}_{lu}, \tilde{p}_{lv}) = -p_{lu}p_{lv}\left(\theta \sum_i w_i^2 + \frac{1 - \theta}{2n_l}\right),$$

where n_l individuals are scored at locus l. Note that these include the binomial sampling variance within a population as well as the Dirichlet variance between populations. This second term seems necessary when the probabilities $\tilde{P}_l$ are estimated by substituting sample allele proportions into 43.7, since those equations are designed to incorporate evolutionary variation among populations.

The required variance for profile $A_{lu}A_{lv}$ is

$$
\text{Var}[\ln(\tilde{P}_l)] = \begin{cases}
\begin{aligned}
& p_{lu}(1 - p_{lu})(1 - \theta)^2 \left(\theta \sum_i w_i^2 + \frac{1-\theta}{2n_l} \right) \\
& \times \left(\frac{1}{3\theta + (1-\theta)p_{lu}} + \frac{1}{2\theta + (1-\theta)p_{lu}} \right)^2,
\end{aligned} & u = v, \\[2em]
\begin{aligned}
& (1-\theta)^2 \left(\theta \sum_i w_i^2 + \frac{1-\theta}{2n_l} \right) \left(\frac{p_{lu}(1 - p_{lu})}{[\theta + (1-\theta)p_{lu}]^2} \right. \\
& \left. - \frac{2p_{lu}p_{lv}}{[\theta + (1-\theta)p_{lu}][\theta + (1-\theta)p_{lv}]} + \frac{p_{lv}(1 - p_{lv})}{[\theta + (1-\theta)p_{lv}]^2} \right),
\end{aligned} & u \neq v.
\end{cases}
$$

In practice, the allele probabilities p_{lu} are replaced by sample values $\tilde{p}_{lu}$. When $\theta = 0$, the variances reduce to

$$
\text{Var}[\ln(\tilde{P}_l)] = \begin{cases}
\dfrac{2(1 - p_{lu})}{n_l p_{lu}}, & u = v, \\[1.5em]
\dfrac{(p_{lu} + p_{lv} - 4p_{lu}p_{lv})}{2n_l p_{lu}p_{lv}}, & u \neq v.
\end{cases}
$$

43.7.2   Coancestry

As mentioned earlier, the parameter θ that features so prominently in many of the genetic forensic calculations is generally assigned a numerical value rather than being estimated from forensic databases. Part of the reason for this is the difficulty in obtaining data from subpopulations to allow estimation of θ, the very quantity introduced to avoid having to collect such data in the first place. If observations were available at the crime-relevant subpopulation level, there would not be a need for θ in 43.7.

There are occasions, however, when multiple samples are available from the same major racial classification. Population geneticists have long estimated θ or F_{ST} from such data (e.g. Cavalli-Sforza *et al.*, (1994); **Chapter 29**). A classical procedure for estimation rests on the method of moments (Weir and Cockerham, 1984), generally within a hierarchical sampling framework such as alleles within individuals, individuals within subpopulations, and subpopulations within populations. Recent work by Weir and Hill (2002) has re-examined this classical approach, and has considered the effects of assuming that allele frequencies are normally distributed over populations. For large sample sizes, θ can be estimated for locus l as

$$
\hat{\theta} = \frac{1}{(r - 1)(m_l - 1)} \sum_{i=1}^{r} \sum_{u=1}^{m} \frac{(\tilde{p}_{liu} - \bar{p}_{lu})^2}{\bar{p}_{lu}}.
$$

In this equation, $\tilde{p}_{liu}$ is the sample frequency of allele A_{lu} at locus l in the ith of r samples, and u ranges from 1 to m_l. The quantity $\bar{p}_{lu}$ is the average frequency of A_{lu} over the whole dataset. Estimates can be averaged over loci under the assumption that θ

is the same for each locus. The average estimate is distributed as θ times a χ^2 distribution with $d = (r-1) \sum_l (m_l - 1)$ df, so that a 95 % confidence interval for θ is

$$\left(\frac{d\hat{\theta}}{X_{0.975}}, \frac{d\hat{\theta}}{X_{0.025}} \right),$$

where $X_{0.025}, X_{0.975}$ are the 2.5th and 97.5th percentiles of the χ^2 distribution with d df. This asymmetric confidence interval is analogous to the asymmetric posterior probability intervals found by Ayres and Balding (1998) from a Bayesian perspective. It would be possible to incorporate this variation into expressions for the sampling variation of estimated profile probabilities.

43.8   OTHER FORENSIC ISSUES

There are several issues in addition to those already covered that affect the interpretation of DNA evidence. The various problems described in this section could all be avoided by starting with the principles of evidence interpretation and proceeding to an evaluation of conditional profile probabilities. Future directions are indicated by efforts to assign individuals to phenotypic classes or ethnic groups, and to place DNA profile statements in a hierarchy of propositions.

43.8.1   Common Fallacies

By far the most common fallacy associated with the forensic use of DNA is the Prosecutor's fallacy mentioned earlier. Virtually every media account of courtroom testimony contains statements of the form 'Based on the DNA evidence, the forensic scientist testified that there was only one chance in a million that someone other than the defendant left the crime sample' whereas the scientist most probably stated the result correctly as 'If the defendant did not leave the crime sample, there is only one chance in a million that it would match his type.' Weir (2000) has described the frustration in trying to have journalists write carefully, and even writers for *Science* are not immune from the fallacy (*Science* 278:1407, 1997).

The defense attorney's fallacy (Thompson and Schumann, 1987) is less common but may appear compelling to a jury. If the profile in question has a probability of 1 in 1 million, and the defendant lives in a country of 100 million, the defense may claim that he is merely 1 person among the 100 who have that profile. They may even suggest there is a 99 % chance he is not the source of the crime sample. The fallacy arises from a confusion between the probability of the profile under H_d (and hence the LR) and the posterior probability of H_d. It implies the assignment of equal prior probabilities to everyone in the country, and it shows a lack of appreciation for the distribution of the number of matches about the expected value. A full discussion of the related 'island problem' has been given by Dawid (1994) and Balding and Donnelly (1995).

43.8.2   Relevant Population

Estimated profile or match probabilities refer to a population. They are not proportions of people in that population with the profile but instead refer to the probability with which a person in the population either has the profile or matches the profile from someone already typed. Which population is meant? As the calculations usually refer to the alternative proposition H_d, it is clear that the suspect's ethnicity, e.g. does not define the population. That person did not contribute the crime profile, so his characteristics do not define the population to which the actual contributor belongs (Lewontin, 1993; Weir and Evett, 1992; 1993). However, there is a tendency to concentrate on calculations based on allelic frequencies in a sample of people with the same racial designation as the suspect.

It is proper to define the population by the circumstances of the crime, and these may point to a geographic region or an ethnic group. Usually the conditional probability $\Pr(G_c|G_s, H_d)$ will be greater when the actual contributor and the suspect belong to the same (sub)population, so that focusing on the suspect's racial group has an element of conservativeness. Balding (1999) has presented a more satisfactory approach of allowing the unknown contributor under H_d to belong to one of very many different sets of people defined by the relationship to the suspect: the suspect's siblings, his other relatives, other members of his subpopulation, other members of his racial group, or anyone outside his racial group.

43.8.3 Database Searches

One of the few remaining statistical areas of debate in the forensic uses of DNA concerns the effects of database searches. The treatment so far in this chapter has considered the situation where a suspect has been identified by an investigation and is then typed and found to have a DNA profile matching that of a crime sample. For crimes with no suspect, it is now possible to search large databases of profiles from known individuals such as previously convicted offenders. Does the evidence have any less value against a person who becomes a suspect because his profile was identified after such a search, than when the person was a suspect before being found to match? The confusion has arisen because of the observation that the expected number of matches increases with the size of the database, and there has been a recommendation (National Research Council, 1996) that a profile probability of P should be modified to NP when a suspect is identified by searching a database of size N.

Balding and Donnelly (1996) and Donnelly and Friedman (1999) present careful analyses to show that the LR for the database search case cannot be greater than for the single suspect case. The evidence against the one person who goes to trial is essentially unaltered by the means by which he was identified. It is not the database that is on trial. In responding to Stockmarr (1999), Evett *et al.* (2000) show how the confusion can be avoided by adhering to the Principles of Evidence Interpretation and recognising the difference between LRs and posterior odds. There could well be an argument made that the prior odds against the suspect are less if he was identified by a database search, so that the posterior odds will also be less since the LR is unaltered.

43.8.4 Uniqueness of Profiles

As predicted by Stigler (1995), DNA profiles are now regarded as being as valuable and reliable as fingerprints. Part of this acceptance rests on the very successful use of DNA profiling to identify remains after mass disasters such as airplane crashes, and was not hindered by the anomalous verdict in the criminal trial of O.J. Simpson (Weir, 1995). With this growing acceptance has come less of a need to present statistical arguments.

Moreover, the increasing number of loci used for forensic DNA profiles has made any statistics appear to be beyond credibility. Maybe for these reasons, there has been a move by the Federal Bureau of Investigation (FBI) to dispense with numbers (reported in Science 278:1407, 1997).

Some relevant discussion was given by Kingston (1965) long before the advent of DNA profiling. If a particular item of evidence has a probability P, then he assumed that the unknown number x of occurrences of the profile in a large population of N people is Poisson with parameter $\lambda = NP$: $\Pr(x) = \lambda^x e^{-\lambda}/x!$. If λ is large the binomial distribution would be needed. Both Poisson and binomial require profiles from different people to be independent, so neither can be strictly true for DNA profiles. Suppose a person with the particular profile commits a crime, leaves evidence with that profile at the scene, and then rejoins the population. A person with the profile is subsequently found in the population. A simple model says that the probability that this person is the perpetrator is $1/x$. Although x is not known, it must be at least one, so the probability that the correct person has been identified is the expected value of $1/x$ given that $x \geq 1$:

$$\Pr(\text{correct}) = \frac{\sum_{x \geq 1} \frac{1}{x} \Pr(x)}{\sum_{x \geq 1} \Pr(x)} \approx 1 - \frac{\lambda}{4}.$$

For the United States, with a population of about 3×10^8, a profile with a probability of 10^{-10} would give $\lambda = 0.03$ and a probability that the correct person had been identified of 0.9925. Kingston (1965) went on to determine the probability that at least two people in the population have a particular profile given that at least one person is known to have it. This probability is

$$\Pr(\text{not unique}) = \frac{\Pr(x \geq 2)}{\Pr(x \geq 1)} \approx \frac{\lambda}{2}.$$

For the USA example, the probability that the profile is unique, $(1 - \lambda/2)$, is 0.985.

An alternative calculation was provided by Balding (1999). He supposed that a person (the perpetrator) is sampled anonymously and randomly from a population of $(N + 1)$ people and found to have a certain DNA profile. Each other person has the same probability P of having that profile. A second person (the suspect) is sampled from the population and (event E) is found to have the same profile as the first. Event U is that the second person matches the first and that there is no other person in the population with the same profile. It is possible (event G), with probability $1/(N + 1)$ that the second person is actually the first person. The conditional probability $\Pr(G|E)$, from Bayes' theorem, is $1/(1 + NP)$ since the probability of E given that G did not occur is just P. The probability of U given both G and E is the probability that none of N people have the profile: $(1 - P)^N$, so

$$\Pr(U|E) = \Pr(U|G, E)\Pr(G|E) = \frac{(1 - P)^N}{1 + NP} > 1 - 2\lambda.$$

For the USA example, this lower bound is 0.94.

It is not clear which of the three probabilities (the two of Kingston or that of Balding) the FBI wished to determine, but they did want a value of 0.01 which requires $P = 1/25N$ or $P = 1/50N$ for Kingston and $P = 1/200N$ for Balding. The FBI then reduced P by a factor of 10 to account for uncertainty in estimating its value. There are two problems with this approach (Weir, 1999; 2001). In the first place, all N profiles in a population are not independent and the Poisson/binomial calculation ignores all the issues of conditional probabilities, population structure and relatedness discussed above. The more serious problem may be that of perception – there is quite a difference between telling a jury that the suspect has been identified as the perpetrator by his DNA profile and telling them that there is a 1 % chance someone else has this profile. The absoluteness implied by statements of identity is not a statistical concept.

A related set of calculations has to do with multiple occurrences of *any* profile, not a particular profile, in a database. This is the so-called birthday problem. The probability that at least two of a sample of n people have the same unspecified birthday (or profile) for the case where every birthday (or profile) has the same probability P, is

$$\text{Pr(at least one match)} = 1 - \text{Pr(no matches)}$$

$$= 1 - \{1(1 - P)(1 - 2P) \cdots [1 - (n - 1)p]\}$$

$$\approx 1 - \prod_{i=0}^{n-1} e^{-iP} \approx 1 - e^{-n\lambda/2}.$$

In the classic birthday problem, $P = 1/365$ and the probability exceeds 50 % once n gets as large as 23. For the USA example of $P = 10^{-10}$ the chance of some profile being replicated in the population of $N = 3 \times 10^8$ is essentially 100 %. The Arizona Department of Public Safety reported a 'near match' in a database of 65 493 for a profile that had an estimated probability of 1 in 3.1 billion. Using that probability, the chance of finding two matching profiles in the database would be 50 % so it would not have been surprising if they had found a complete match. There are about two billion pairs of people in the Arizona database, so the occurrence of one matching pair is seen to still be a rare event. It is important to stress that the birthday problem requires two or more copies of some birthday (profile), not two or more copies of a particular profile.

There is an interesting twist to the notion expressed above that DNA profiling has become as well-accepted as has fingerprinting. Publications have begun to appear that apply LR approaches to fingerprint minutiæ(Neumann *et al.*, 2006). Such work is needed to assess the evidential contribution of fingerprints that may be partial, distorted or with poor signal to noise ratio.

43.8.5 Assigning Individuals to Phenotypes, Populations or Families

An individual's phenotype depends ultimately on his genetic makeup, and there have been hopes that DNA profiles could be used to identify the ethnicity or physical appearance for the unknown donor of an evidentiary stain. There is the trivial example of sex-linked markers being used to determine gender, or at least X,Y chromosomal composition. There has also been some progress in predicting phenotype when that is limited to red hair (Grimes *et al.*, 2001) because many cases of red hair can be associated with the genotype at a single locus, the melanocortin-1 receptor. Grimes *et al.* conclude 'Given that between 5.3 and 11 % of the population of the British Isles are reported to have red hair, a positive test result could have a major impact on the course of an investigation.'

Much less certainty can be attached to inferring ethnic origin. Lowe *et al.* (2001) used the UK Forensic Science Service panel of six STR loci to suggest a means of reducing the number of interviews needed to resolve a case. They first estimated the probability of the crime scene profile G for each ethnic group r for which they have allele frequencies. If I indicates an investigator's prior information and $\Pr(r|I)$ his or her prior probabilities for ethnic origin of the crime-scene stain donor, Bayes' theorem provides posterior probabilities of

$$\Pr(r|G, I) = \frac{\Pr(G|r, I)}{\sum_s \Pr(G|s, I)\Pr(s|I)}.$$

If an investigator faces the task of interviewing and/or DNA testing all members of a population until the true donor of the stain is found, then the expected number of interviews may be reduced if the posterior probabilities are used. For example, if the posterior probability is highest for ethnic group R, then preferentially interviewing members of that group will be efficient if the probability is greater than the proportion of people in the population who are of origin R.

Shriver *et al.* (1997) adopted an alternative procedure of looking for DNA loci that are particularly discriminating. They claimed evidence for a set of 10 loci, selected from a set of 1000, that would provide good discrimination between African Americans and European Americans. Their methods and conclusions were challenged by Brenner (1998), who points to the danger of selecting the most discriminating loci from a large number of loci based on the performances in small samples. There is the possibility that 'From among 1000 loci, one could similarly find a set of 10 loci that differentiate the 9-year olds from the 10-year olds in the local playground.' Nevertheless, Brenner supports the work of Lowe *et al.* (2001) that the STR loci used by forensic scientists have good potential for ethnic assignment.

Genetic markers can also be used to assign individuals to families and this is of particular use following mass disasters. An alternative use of markers for making inferences about relatedness is that of familial searching. Bieber *et al.* (2006) describe the situation where an evidentiary profile is compared to every profile in a database and the LR for the two profiles being from (specified) relatives or from unrelated people is calculated. High LRs are suggestive of relatedness, Bieber *et al.* show by simulation when the database did contain one true brother or father of the donor of the evidence sample that, about half the time, the highest LR identified the true relative. To achieve 90 % probability of identifying a subset of the database that included the true relative, it may be necessary to consider at least the 100 highest LRs. Other issues in the estimation of relatedness of relationship from genotypes were reviewed by Weir *et al.* (2006).

43.8.6 Hierarchy of Propositions

This chapter has been concerned mainly with the statistical procedures to quantify the strength of the evidence provided by sets of DNA profiles, especially in the case of matching profiles. These statistical methods are generally straightforward, but they do not capture all the forensic issues. Evett *et al.* (2002) have introduced the concept of a 'hierarchy of propositions' The highest-level propositions have to do with the guilt or innocence of a defendant, whereas the lowest-level propositions have to do with whether a DNA sample is from the defendant. The authors conclude with the substantial point that 'even if the inference with regard to whether or not the source of a sample of DNA is effectively indisputable, the inference with regard to whether the defendant is the offender may be subject to considerable uncertainty'.

43.9   CONCLUSIONS

Quantifying the evidentiary value of matching genetic profiles requires a comparison of the probabilities of the evidence under alternative propositions. These probabilities, in turn, require the probability that an unknown person has the profile given that a known person has the profile, and this is a statistical genetic quantity. The match probability depends on the relationship between the two people, whether this is due to joint family membership, or simply joint membership in some population.

Even though the propositions concerning the origins of a crime stain may list only one person, other people may have been typed as part of the investigation. Their profiles affect the match probabilities under a general model that allows for population structure. All expressions are considerably simplified when the population can be assumed to be have reached an equilibrium stage under the action of evolutionary forces, but this theory is presently in place only for single loci. An assumption is made that different loci are independent, so that match probabilities may be multiplied over loci.

There are many advantages to a general and coherent approach to the interpretation of genetic evidence. This approach, expounded upon by the authors already cited is finding acceptance by forensic scientists in many countries, but it requires a willingness to abandon simplistic approaches that appeal to expedience. The interests of justice, in the long run, will best be served by sound statistical theories.

REFERENCES

Aitken, C.G.G. (1995). *Statistics and the Evaluation of Evidence for Forensic Scientists*. John Wiley & sons, New York.

Ayres, K.L. and Balding, D.J. (1998). Measuring departures from Hardy-Weinberg: a Markov Chain Monte Carlo method for estimating the inbreeding coefficient. *Heredity* **80**, 769–770.

Ayres, K.L. and Overall, A.D.J. (1999). Allowing for within-subpopulation inbreeding in forensic match probabilities. *Forensic Science International* **103**, 207–216.

Balding, D.J. (1999). When can a DNA profile be regarded as unique? *Science and Justice* **39**, 257–260.

Balding, D.J. (2005). *Weight-of-evidence for DNA Profiles*. John Wiley & Sons, Chichester.

Balding, D.J. and Donnelly, P. (1995). Inference in forensic identification. *Journal of the Royal Statistical Society Series A* **158**, 21–53.

Balding, D.J. and Donnelly, P. (1996). Evaluating DNA profile evidence when the suspect is identified through a database search. *Journal of Forensic Sciences* **41**, 603–607.

Balding, D.J. and Nichols, R.A. (1994). DNA match probability calculation: how to allow for population stratification, relatedness, database selection and single bands. *Forensic Science International* **64**, 125–140.

Ballantyne, J. (1997). Mass disaster genetics. *Nature Genetics* **15**, 329–331.

Bieber, F.R., Brenner, C.H. and Lazer, D. (2006). Finding criminals through DNA of their relatives. *Science* **312**, 1315–1316.

Brenner, C.H. (1997). Symbolic kinship program. *Genetics* **145**, 535–542.

Brenner, C.H. (1998). Difficulties in the estimation of ethnic affiliation. *American Journal of Human Genetics* **62**, 1558–1560.

Buckleton, J.S., Triggs, C.M. and Walsh, S.J. (2005). *Forensic DNA Evidence Interpretation*. CRC Press, Boca Raton, FL.

Cavalli-Sforza, L.L., Menozzi, P. and Piazza, A. (1994). *The History and Geography of Human Genes*. Princeton University Press, Princeton, NJ.

Clayton, T.M., Whitaker, J.P. and Maguire, C.N. (1995). Identification of bodies from the scene of a mass disaster using DNA amplification of short tandem repeat (STR) loci. *Forensic Science International* **76**, 7–15.

Cockerham, C.C. (1971). Higher order probability functions of identity of alleles by descent. *Genetics* **69**, 235–246.

Curran, J., Triggs, C.M., Buckleton, J. and Weir, B.S. (1999). Interpreting DNA mixtures in structured populations. *Journal of Forensic Sciences* **44**, 987–995.

Dawid, A.P. (1994). *Aspects of uncertainty: a tribute to D.V. Lindley*, P.R. Freeman and A.F.M. Smith, eds. John Wiley & Sons, Chichester, pp. 159–170.

Dawid, A.P. and Evett, I.W. (1997). Using a graphical method to assist the evaluation of complicated patterns of evidence. *Journal of Forensic Sciences* **42**, 226–231.

Donnelly, P. and Friedman, D. (1999). DNA database searches and the legal consumption of scientific evidence. *Michigan Law Review* **97**, 931–984.

Evett, I.W., Foreman, L.A. and Weir, B.S. (2000). Letter to the Editor concerning a paper by A. Stockmarr. *Biometrics* **56**, 1274–1275. Response to Devlin, *Biometrics* **56**, 1277.

Evett, I.W., Gill, P.D., Jackson, G., Whitaker, J. and Champod, C. (2002). Interpreting small quantities of DNA: the hierarchy of propositions and the use of Bayesian networks. *Journal of Forensic Sciences* **47**, 520–530.

Evett, I.W. and Weir, B.S. (1998). *Interpreting DNA Evidence: Statistical Genetics for Forensic Science*. Sinauer, Sunderland, MA.

Fung, W.K., Carracedo, A. and Hu, Y.-Q. (2003). Testing for kinship in a subdivided population. *Forensic Science International* **135**, 105–109.

Gill, P., Brenner, C.H., Buckleton, J.S., Carracedo, A., Krawczak, M., Mayr, W.M., Morling, N., Prinz, M., Schneider, P.M. and Weir, B.S. (2006). DNA commission of the International Society of Forensic Genetics: Recommendations on the interpretation of mixtures. *Forensic Science International* **160**, 90–101.

Graham, J., Curran, J. and Weir, B.S. (2000). Conditional genotypic probabilities for microsatellite loci. *Genetics* **155**, 1973–1980.

Grimes, E.A., Noake, P.J., Dixon, L. and Urquhart, A. (2001). Sequence polymorphism in the human melanocortin 1 receptor gene as an indicator of the red hair phenotype. *Forensic Science International* **122**, 124–129.

Kingston, C.R. (1965). Applications of probability theory in criminalistics. *Journal of the American Statistical Association* **60**, 70–80.

Lewontin, R.C. (1993). Which population?. *American Journal of Human Genetics* **52**, 205.

Lowe, A.L., Urquhart, A., Foreman, L.A. and Evett, I.W. (2001). Inferring ethnic origin by means of an STR profile. *Forensic Science International* **119**, 17–22.

Lucy, D. (2005). *Introduction to Statistics for Forensic Scientists*. John Wiley & Sons, Chichester.

Maiste, P.J. and Weir, B.S. (1995). A comparison of tests for independence in the FBI RFLP databases. *Genetica* **96**, 125–138.

Mortera, J. (2003). *Highly Structured Stochastic Systems*, P.J. Green, N.L. Hjort and S. Richardson, eds. Oxford University Press, Oxford.

National Research Council (1996). *The Evaluation of Forensic DNA Evidence*. National Academy Press, Washington, DC.

Neumann, C., Champod, C., Puch-Solis, R., Egli, N., Anthonioz, A., Meuwly, D. and Bromage-Griffiths, A. (2006). Computation of likelihood ratios in fingerprint identification for configurations of three minutiæ. *Journal of Forensic Sciences* **51**, 1255–1266.

Olaisen, B., Sternersen, M. and Mevåg, B. (1997). Identification by DNA analysis of the victims of the August 1996 Spitsbergen civil aircraft disaster. *Nature Genetics* **15**, 404–405.

Robb, N. (1999). 229 people, 15,000 body parts: pathologists help solve Swissair 111's grisly puzzles. *Canadian Medical Association Journal* **160**, 241–243.

Robertson, B. and Vignaux, G.A. (1995). *Interpreting Evidence: Evaluating Forensic Science in the Courtroom*. John Wiley & Sons, Chichester.

Royall, R. (1997). *Statistical Evidence: A Likelihood Paradigm*. Chapman and Hall, London.

Scholl, S., Budowle, B., Radecki, K. and Salvo, M. (1996). Navajo, Pueblo, and Sioux population data on the loci HLA-DQA1, LDLR, GYPA, HBGG, Gc, and D1S80. *Journal of Forensic Sciences* **41**, 47–51.

Schum, D.A. (1994). *Evidential Foundations of Probabilistic Reasoning*. John Wiley & Sons, New York.

Shoemaker, J., Painter, I.S. and Weir, B.S. (1998). A Bayesian characterization of Hardy-Weinberg disequilibrium. *Biometrics* **25**, 235–254.

Shriver, M.D., Smith, M.W., Jin, L., Marcini, A., Akey, J.M., Deka, R. and Ferrell, R.E. (1997). Ethnic-affiliation estimation by use of population-specific DNA markers. *American Journal of Human Genetics* **60**, 957–964.

Stigler, S.M. (1995). Galton and identification by fingerprints. *Genetics* **140**, 857–860.

Stockmarr, A. (1999). Likelihood ratios for evaluating DNA evidence when the suspect is found through a database search. *Biometrics* **55**, 671–677.

Thompson, W.C. and Schumann, E.L. (1987). Interpretation of statistical evidence in criminal trials–the prosecutors fallacy and the defense attorneys fallacy. *Law and Human Behavior* **11**, 167–187.

Weir, B.S. (1992). Independence of VNTR alleles defined as fixed bins. *Genetics* **130**, 873–887.

Weir, B.S. (1995). DNA statistics in the Simpson matter. *Nature Genetics* **11**, 365–368.

Weir, B.S. (1996). *Genetic Data Analysis II*. Sinauer, Sunderland, MA.

Weir, B.S. (1999). Are DNA profiles unique? In Proceedings of the 9th International Symposium on Human Identification, Orlando, Florida, pp. 114–117. http://www.promega.com/geneticidentity/proceed.html.

Weir, B.S. (2000). In *Statistical Science in the Courtroom*, J. Gastwirth, ed. Springer-Verlag, New York.

Weir, B.S. (2001). DNA match and profile probabilities: Comment on Budowle (2000) and Fung and Hu (2000). Forensic Science Communications 3. http://www.fbi.gov/hq/lab/fsc/backissu/jan2001/weir.htm.

Weir, B.S., Anderson, A.B. and Hepler, A.M. (2006). Genetic relatedness analysis: modern data and new challenges. *Nature Reviews Genetics* **7**, 771–780.

Weir, B.S. and Cockerham, C.C. (1984). Estimating F-statistics for the analysis of population structure. *Evolution* **38**, 1358–1370.

Weir, B.S. and Evett, I.W. (1992). Whose DNA? *American Journal of Human Genetics* **50**, 869.

Weir, B.S. and Evett, I.W. (1993). Reply to Lewontin. *American Journal of Human Genetics* **52**, 206.

Weir, B.S. and Hill, W.G. (2002). Estimating F-statistics. *Annual Review of Genetics* **36**, 721–750.

Weir, B.S., Triggs, C.M., Starling, L., Stowell, L.I., Walsh, K.A.J. and Buckleton, J.S. (1997). Interpreting DNA mixtures. *Journal of Forensic Sciences* **42**, 113–122.

Zaykin, D., Zhivotovsky, L.A. and Weir, B.S. (1995). Exact tests for association between alleles at arbitrary numbers of loci. *Genetica* **96**, 169–178.

Zaykin, D., Zhivotovsky, L.A., Westfall, P.H. and Weir, B.S. (2002). Truncated product method for combining *p*-values. *Genetic Epidemiology* **22**, 170–185.

Reference Author Index

Subject Index